Evolution of the *Earth*

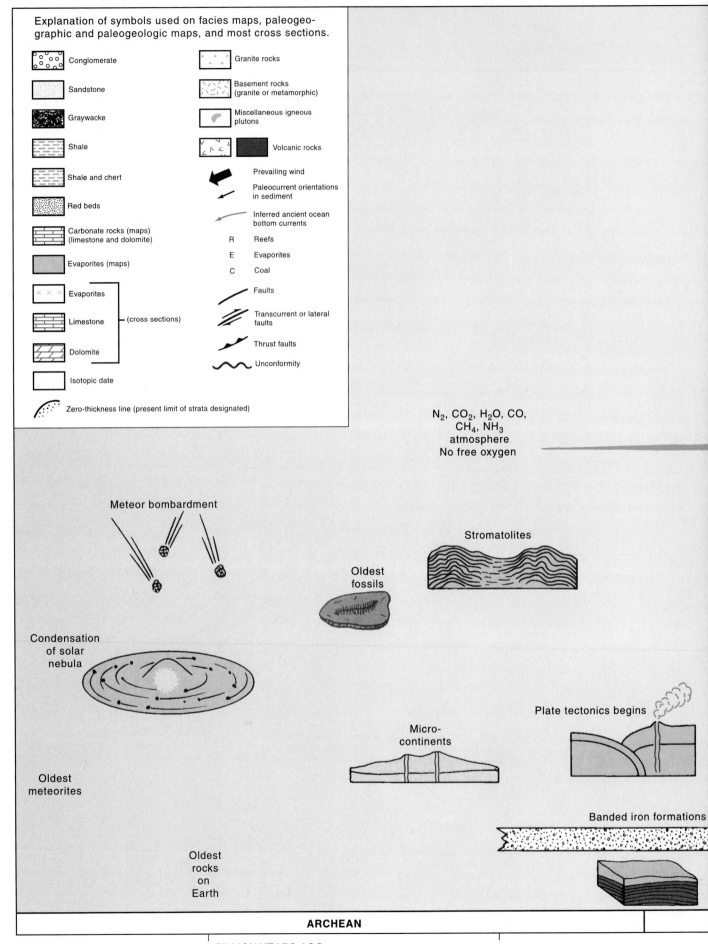

Explanation of symbols used on facies maps, paleogeographic and paleogeologic maps, and most cross sections.

Conglomerate

Sandstone

Graywacke

Shale

Shale and chert

Red beds

Carbonate rocks (maps)
(limestone and dolomite)

Evaporites (maps)

Evaporites

Limestone — (cross sections)

Dolomite

Isotopic date

Zero-thickness line (present limit of strata designated)

Granite rocks

Basement rocks
(granite or metamorphic)

Miscellaneous igneous
plutons

Volcanic rocks

Prevailing wind

Paleocurrent orientations
in sediment

Inferred ancient ocean
bottom currents

R   Reefs

E   Evaporites

C   Coal

Faults

Transcurrent or lateral
faults

Thrust faults

Unconformity

N_2, CO_2, H_2O, CO,
CH_4, NH_3
atmosphere
No free oxygen

Meteor bombardment

Oldest
fossils

Stromatolites

Condensation
of solar
nebula

Plate tectonics begins

Micro-
continents

Oldest
meteorites

Banded iron formations

Oldest
rocks
on
Earth

ARCHEAN

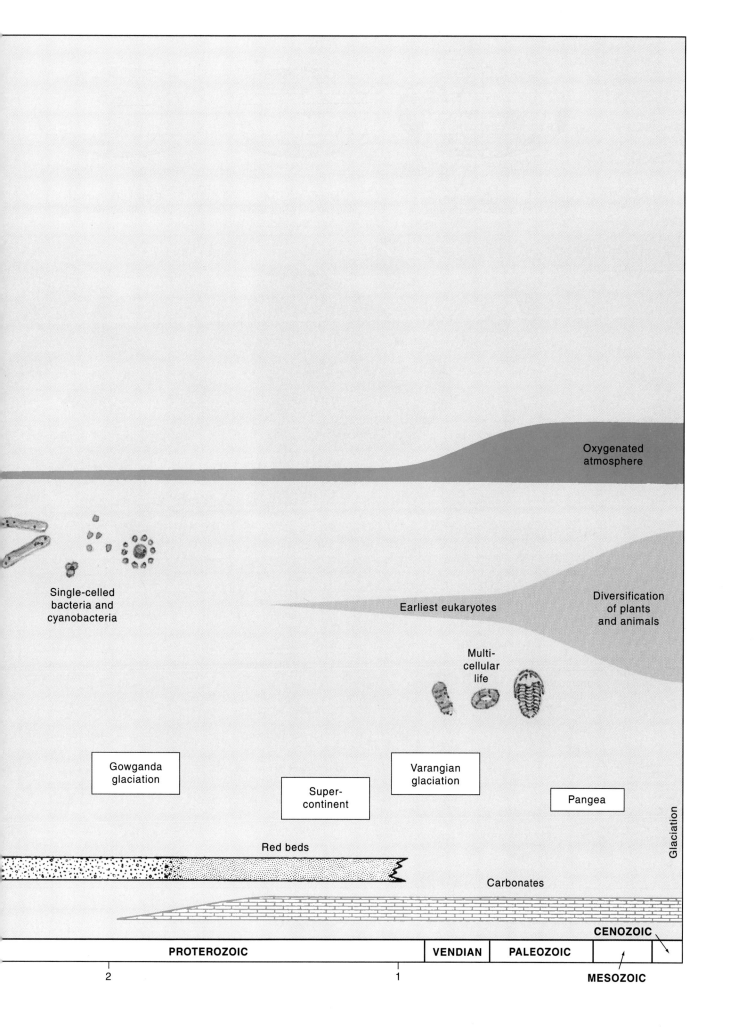

Oxygenated
atmosphere

Single-celled
bacteria and
cyanobacteria

Earliest eukaryotes

Diversification
of plants
and animals

Multi-
cellular
life

Gowganda
glaciation

Super-
continent

Varangian
glaciation

Pangea

Glaciation

Red beds

Carbonates

CENOZOIC

PROTEROZOIC | VENDIAN | PALEOZOIC

MESOZOIC

2

1

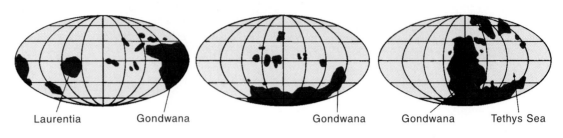

Laurentia   Gondwana

Late Cambrian
520 million years ago

Gondwana

Middle Silurian
430 million years ago

Gondwana   Tethys Sea

Early Late Permian
260 million years ago

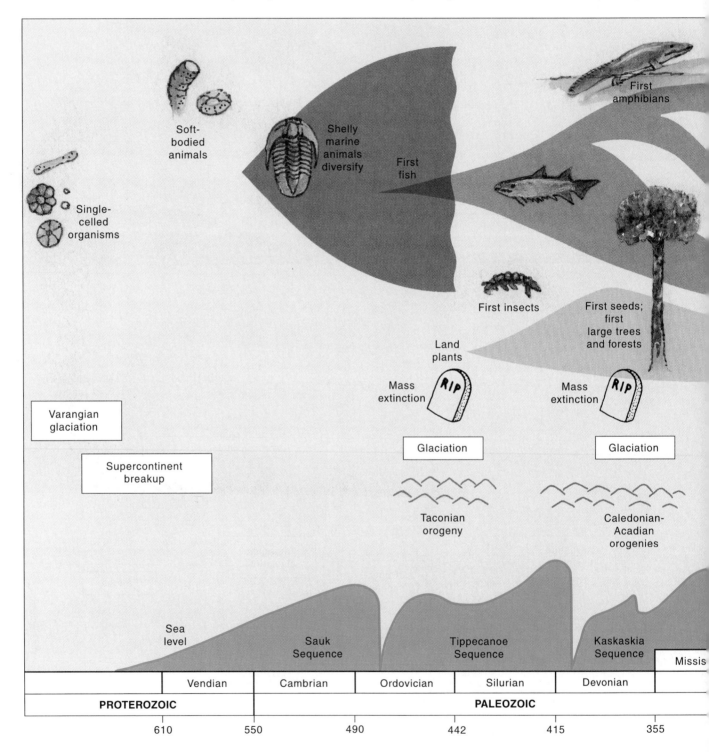

Single-celled organisms

Soft-bodied animals

Shelly marine animals diversify

First fish

First amphibians

First insects

Land plants

First seeds; first large trees and forests

Mass extinction **RIP**

Mass extinction **RIP**

Varangian glaciation

Supercontinent breakup

Glaciation

Glaciation

Taconian orogeny

Caledonian-Acadian orogenies

Sea level

Sauk Sequence

Tippecanoe Sequence

Kaskaskia Sequence

Missis

	Vendian	Cambrian	Ordovician	Silurian	Devonian	
PROTEROZOIC			**PALEOZOIC**			

610   550   490   442   415   355

Pangea

Triassic into Jurassic
240–195 million years ago

Mid-Tertiary
40–25 million years ago

Present

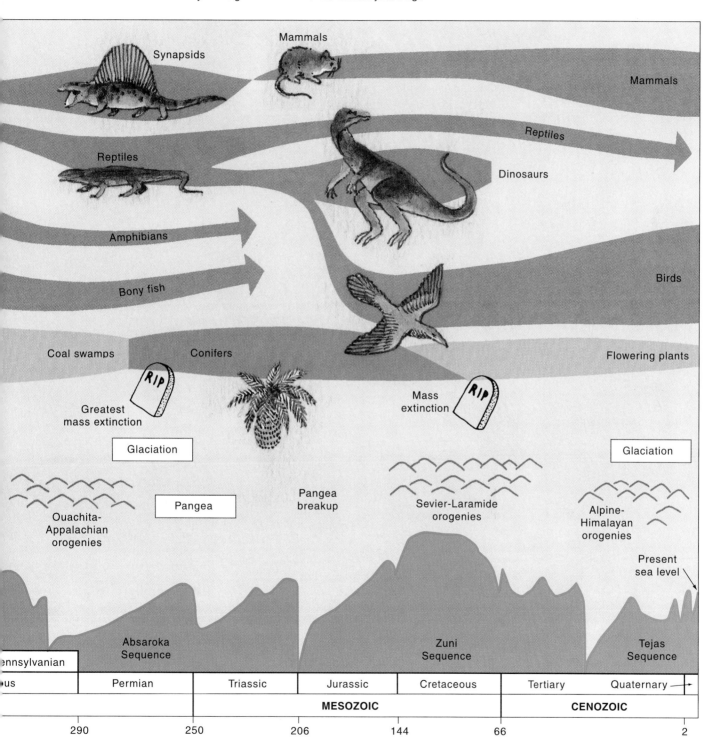

Synapsids

Mammals

Mammals

Reptiles

Reptiles

Dinosaurs

Amphibians

Bony fish

Birds

Coal swamps

Conifers

Flowering plants

RIP

Greatest
mass extinction

Mass
extinction

RIP

Glaciation

Glaciation

Pangea
breakup

Sevier-Laramide
orogenies

Ouachita-
Appalachian
orogenies

Pangea

Alpine-
Himalayan
orogenies

Present
sea level

Absaroka
Sequence

Zuni
Sequence

Tejas
Sequence

ennsylvanian							
us	Permian	Triassic	Jurassic	Cretaceous	Tertiary	Quaternary	→
			MESOZOIC		**CENOZOIC**		

290   250   206   144   66   2

Seventh Edition

Evolution of the Earth

Donald R. Prothero
Occidental College

Robert H. Dott, Jr.
University of Wisconsin, Madison

Mc Graw Hill **Higher Education**

Boston Burr Ridge, IL Dubuque, IA Madison, WI New York
San Francisco St. Louis Bangkok Bogotá Caracas Kuala Lumpur
Lisbon London Madrid Mexico City Milan Montreal New Delhi
Santiago Seoul Singapore Sydney Taipei Toronto

The McGraw·Hill Companies

Higher Education

EVOLUTION OF THE EARTH, SEVENTH EDITION

Published by McGraw-Hill, a business unit of The McGraw-Hill Companies, Inc., 1221 Avenue of the Americas, New York, NY 10020. Copyright © 2004, 2002, 1994, 1988, 1981, 1976, 1971 by The McGraw-Hill Companies, Inc. All rights reserved. No part of this publication may be reproduced or distributed in any form or by any means, or stored in a database or retrieval system, without the prior written consent of The McGraw-Hill Companies, Inc., including, but not limited to, in any network or other electronic storage or transmission, or broadcast for distance learning.

Some ancillaries, including electronic and print components, may not be available to customers outside the United States.

 This book is printed on recycled, acid-free paper containing 10% postconsumer waste.

1 2 3 4 5 6 7 8 9 0 QPD/QPD 0 9 8 7 6 5 4 3

ISBN 0–07–252808–7

Publisher: *Margaret J. Kemp*
Sponsoring editor: *Thomas C. Lyon*
Developmental editor: *Lisa A. Leibold*
Executive marketing manager: *Lisa L. Gottschalk*
Senior project manager: *Mary E. Powers*
Lead production supervisor: *Sandy Ludovissy*
Lead media project manager: *Judi David*
Media technology producer: *Renee Russian*
Senior coordinator of freelance design: *Michelle D. Whitaker*
Cover/interior designer: *Elise Lansdon*
Cover illustration: *Luis V. Rey*
Senior photo research coordinator: *Lori Hancock*
Photo research: *David Tietz*
Supplement producer: *Brenda A. Ernzen*
Compositor: *Carlisle Communications, Ltd.*
Typeface: *10/12 Times Roman*
Printer: *Quebecor World Dubuque, IA*

Library of Congress Cataloging-in-Publication Data

Prothero, Donald R.
    Evolution of the earth / Donald R. Prothero, Robert H. Dott, Jr. — 7th ed.
        p.   cm.
    Includes index.
    ISBN 0–07–252808–7
        1. Historical geology. I. Dott, Robert H., 1929– . II. Title.

QE283.P75   2004
551.7—dc21   2003051153
                                                  CIP

www.mhhe.com

Dedication

This book is affectionately dedicated to
Nancy Dott and Teresa LeVelle
and to our children
for tolerating us while we worked long
hours on this book.

Contents in Brief

Contents

Chapter 1

Chapter 2

Chapter 3

Chapter 4

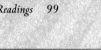

The Relative Geologic Time Scale and Modern Concepts of Stratigraphy   67

Chapter 5

The Numerical Dating of the Earth   87

Chapter 6

The Origin and Early Evolution of the Earth   101

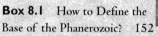

Chapter 7

Mountain Building and Drifting Continents   121

Chapter 8

Cryptozoic History: An Introduction to the Origin of Continental Crust   151

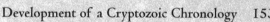

Chapter 9

Early Life and Its Patterns   181

Chapter 10

Earliest Paleozoic History: The Sauk Sequence—An Introduction to Cratons and Epeiric Seas   207

Chapter 11

The Later Ordovician: Further Studies of Plate Tectonics and the Paleogeography of Orogenic Belts   231

Chapter 15

Cenozoic History: Threshold of the Present   413

Preface

Tiem present and time past

Are both perhaps persent in time future,

And time future contained in time past.

T.S. Eliot, *Four Quartets*

About This Book

Evolution of the Earth was developed in the late 1960s from a second-semester course taught by Robert II. Dott, Jr., and Roger L. Batten at the University of Wisconsin, Madison, taken mostly by first- and second-year college students representing a broad spectrum of science and nonscience backgrounds. Other parts of the book reflect Don Prothero's 22 years of teaching in small liberal arts college.

When the first edition of *Evolution of the Earth* was published in 1971, there was a vacuum in historical geology instruction. The long-standing approach was to present a dry relation of lists of formation and fossil names for many areas, but with little or no unifying theme and little emphasis upon interpretation and hypotheses testing. That approach was aptly characterized by the late L.L. Sloss as "The Roll Call of the Ages." We selected, instead, the overall evolution of the earth—both physical and organic—as a central theme. It was fortunate that the unifying plate-tectonic theory appeared in time to provide more coherence to physical history.

The general response of our own and others' students has been most pleasing. *Evolution of the Earth* has not escaped the notice of combatants in the continuing controversy between evolutionists and creationists, either. We receive both praise and condemnation, for besides being criticized for endorsing evolution, we also have been scolded for being too soft-spoken on the issue. The worst fate is to be ignored.

Approach

The basic philosophy of *Evolution of the Earth* remains unchanged in this new edition. We are committed to the belief that introductions to the sciences should be primarily conceptual rather than informational. *Evolution of the Earth* reveals the logical framework of geology, shows relations of the science to the totality of human knowledge, and gives some idea of what it is like to be a participant in the discipline. Our experience indicates that students are stimulated by exposure to scientific controversies, occasional spicy personal feuds, or an amusing faux pas. In this edition, these items are mostly presented in boxes.

In keeping with our preference for a "How do we know?" rather than "What do we know?" approach, we stress the assumptions made by earth historians, the kinds of evidence gathered, the tools for gathering that evidence, and the processes of reasoning and limitations of hypotheses that are required to reconstruct and interpret the past. At the same time, we have tried to emphasize the uniqueness of geology as an intellectual discipline. Its historical nature sets geology apart from the more familiar nonhistorical sciences; therefore, we have deliberately tried to draw out and illustrate this uniqueness.

As readers use the book, we hope that they will keep in mind the following three maxims: (1) new concepts of time, (2) the universality of irreversible evolutionary changes, and (3) the importance throughout time of ecological interactions—feedbacks—between the evolution of living forms and the physical world.

Features

- The areas of hottest controversy, such as mass extinctions, dinosaur endothermy, the origin of life, and controversies over late Proterozoic tectonics and glaciation, are given separate sections, so that students can appreciate the different sides of the debates.
- The citations at the end of each chapter have been expanded to include many primary references for those who want to go straight to the source.

- The endpapers bring all of geologic history into a single time frame.
- The appendix on classification includes several up-to-date phylogenetic diagrams, which clearly show the hierarchical organization of life.

Ancillaries

Instructor's Manual

Free to adopters, the *Instructor's Manual* contains suggestions for supplementary aids, laboratory activities, and sample test questions and can be downloaded from our website at *http://www.mhhe.com/prothero7/*. You can obtain a password for this material from your McGraw-Hill sales representative.

Transparencies

One hundred acetate transparencies of key text illustrations are available to qualified adopters.

Using This Book

Initial encounters with the sciences generally involve major shifts of scales—both spatial and temporal. We have had success with students by beginning with discussions of familiar geologic processes that have affected humans directly in historic times. We then shift into truly geologic frames of reference. For the more science-oriented student, we recommend simple homework problems that deal with geologic time and the rates of a wide variety of geologic processes. Suggested examples are contained in the *Instructor's Manual* available at *http://www.mhhe.com/prothero7/*.

The changes in the seventh edition reflect our own experiences, together with the constructive criticisms of many other teachers at a variety of colleges having widely different requirements. We have also adapted to changing times and rapid advances in science. We believe that this new edition provides great flexibility for instructors and students. *Evolution of the Earth* assumes only a prior knowledge of general geology comparable to a high-school earth science or first-semester college physical geology course. We have retained the classical foundations of earth and life history, but certain material, especially of a chemical and physical nature, which seems necessary to us for an adequate understanding of modern geology, may prove difficult for some readers. We are confident that a skillful teacher can help students through such material by drawing out the essential highlights without intimidating the students. The glossary should aid all readers, but especially those with limited backgrounds.

Seventh Edition Changes

The seventh edition has been significantly updated and revised to reflect the changing nature of many areas of earth and life history.

- *Chap. 6:* updated section on the origin of the moon
- *Chaps. 6 and 8:* new content on the oldest rocks and minerals on Earth and their implication for liquid water and a cool crust at 4.4 billion years ago
- *Chap. 9:* new content about inertial polar wandering and the "Cambrian explosion"
- *Chaps. 8, 9, 10, 12, 13, 14, and 15:* the latest paleogeographic maps of Pannotia, Rodinia, and later supercontinents
- *Chaps. 14 and 15:* the latest on the Permo-Triassic and Cretaceous-Tertiary extinctions
- *Chap. 16:* the late Paleocene Thermal Maximum and gas hydrates
- *Chap. 17:* the "global conveyor belt" and the causes of the ice ages
- *Chap. 17:* the latest discoveries of human ancestors, including *Sahelanthropus* at 6 million years ago

Historical Emphasis

The historical elaboration of the development of basic principles for interpreting earth and life history invoked in the early chapters is continued in streamlined form in later ones. Students and instructors will find this feature valuable for several reasons. First, it provides a reference for the reader's own discoveries. Second, it helps to clarify that science is a human activity and that the quest for the understanding of nature is an ongoing, open-ended process in which readers themselves can participate. Finally, the historical emphasis reveals the cultural relationship of the science, which are unusually rich in the case of geology. Historical geology provides both a great opportunity and an obligation to present a broad integration of diverse materials and to clarify the relevance of science. In our experience, nonscience students especially appreciate this material. The first four chapters in which the historical approach is most prominently incorporated can be read rather quickly and understood by most students with a minimum of classroom elaboration.

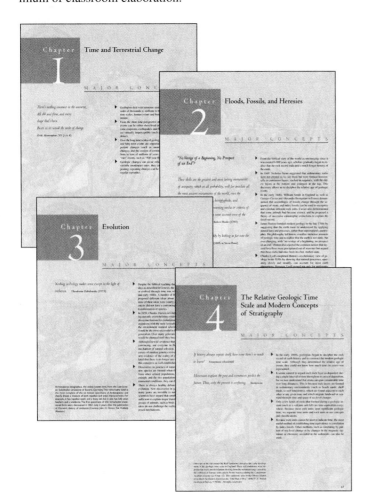

Relationships to Physical Geology

Chapters 5 through 7 present some of the more complex principles and concepts that are necessary to interpret earth and life history. These are very important to the remainder of the book, and experience shows that they require classroom supplementation.

We realize that these chapters overlap with books on physical geology. However, we feel that it is important to have them here as essential introductions to earth history. They also help to show the tie between physical and historical geology.

Topical Approach to Chronology

After the background chapters, which themselves contain abundant earth history, we prefer to bring up new concepts within the context of the historical record as each becomes important to the discussion. This is our rationale for reserving the principal discussion of mountain building until the first major orogeny of the Paleozoic is treated under Ordovician history; mountain building is further elaborated in several additional locations. Reefs are not explored in detail until discussion of the middle Paleozoic, when large fossil reefs built by animals first became widespread. Similarly, hypotheses for extinctions are elaborated upon under the late Paleozoic and the end of the Mesozoic.

In developing new emphasis in historical geology, we have minimized the encyclopedic "Roll Call of the Ages." Chapters 8 and 10 through 16 treat the historical record for the earth with North American data integrated more closely with those for other continents than in prior editions. But whether a given discussion concerns North America or some other part of the world, the purpose is always to illustrate how geologists unravel and interpret the historical record, rather than to burden the reader with a formidable mass of factual detail and terminology. Instead, although a broad chronological framework is provided throughout, chief emphasis is placed upon interpretation of physical and biological environments, reconstruction of paleogeograhy, and understanding of evolution. One unifying theme throughout is that of *overall chemical evolution of the earth.* Interaction or feedback between the living and the nonliving is emphasized, as are major stratigraphic and tectonic patterns that recur in both time and space. In this way, each of the North American chapters tends to be topical as well as chronological, with only one or a few unique circumstances developed fully in each, instead of including more mundane stratigraphic detail. Factual detail that is presented is for documentation only; students should not be expected to memorize much of it but, rather, to understand the principles and logic of the historical arguments for which the facts provide essential evidence.

The concluding chapter reminds the reader of the three fundamental maxims that follow from a study of earth history, which were stated at the beginning of this preface. Brief illustrations of these maxims are extrapolated into humanity's future in order to suggest some conclusions that have relevance to all members of the human race.

Acknowledgments

It is impossible to acknowledge all of the countless individuals who have contributed directly or indirectly to the completion of this book. We especially thank W. P. Gerould and the late L. L. Sloss for originally encouraging Dott and Batten to embark on the writing of the first edition. We thank many other colleagues, including those acknowledged in previous editions, for advice on many topics. Our colleagues have been most generous with their photographs, and their contributions are acknowledged in the captions.

We thank the reviewers of the seventh edition for their careful examination and useful suggestions for improvement:

Ajoy K. Baksi, *Louisiana State University*
Melinda Hutson, *Portland Community College*
David T. King, Jr., *Auburn University*
Arthur C. Lee, *Roane State Community College*
Robert Titus, *Hartwick College*

We also thank the reviewers of the previous editions:

Emily CoBabe, *The University of Massachusetts, Amherst*
John Ernisse, *Clarion University, Pennsylvania*
David Golz, *California State University, Sacramento*
William Hammer, *Augustana College*
David Hickey, *Lansing Community College*
Raymond Ingersoll, *University of California, Los Angeles*
Markes E. Johnson, *Williams College*
Peter Kresan, *The University of Arizona, Tucson*
Lawrence Krissek, *The Ohio State University*
Elana Leithold, *North Carolina State University*

Shannon O'Dunn, *Grossmont-Cuyamaca Community College*
William Orr, *University of Oregon, Eugene*
Vicki Pedone, *California State University, Northridge*
Greg Retallack, *University of Oregon, Eugene*
Bruce Runnegar, *University of California, Los Angeles*
David Schwimmer, *Columbus College, Georgia*
Gregory Wheeler, *California State University, Sacramento*
Sherwood Wise, Jr., *Florida State University*
Jim Woodhead, *Occidental College*
Tom Yancey, *Texas A&M University*

We thank the many people at McGraw-Hill who made extraordinary efforts to produce this book. Lisa Leibold, our developmental editor, saw this edition through from start to completion.

And I have written books on the soul,

Proving absurd all written hitherto,

And putting us to ignorance again.

(So said Robert Browning's skeptical philosopher, Cleon.)

Donald R. Prothero
Robert H. Dott, Jr.

About the Authors

Donald R. Prothero is chairman and professor of geology at Occidental College in Los Angeles, California, and lecturer in geobiology at the California Institute of Technology. He is on the editorial board of *Geology* magazine and serves on the Penrose Conference Committee of the Geological Society of America. In 2000, he served as vice president of the Pacific Section SEPM (Society for Sedimentary Geology), and he served as program chair of the Society of Vertebrate Paleontology from 1998 to 2003. He has been a Guggenheim and NSF Fellow and a Fellow of the Linnean Society, and in 1991 received the Schuchert Award of the Paleontological Society for outstanding paleontologist under the age of 40. In addition to more than 200 scientific papers on subjects ranging from mammalian evolution to Neogene radiolarians to magnetostratigraphy to chert petrography, he has authored or edited 17 other books, including *The Evolution of Perissodactyls* (Oxford University Press, 1989, co-edited with R.M. Schoch), *Interpreting the Stratigraphic Record* (W.H. Freeman, 1990), *Eocene-Oligocene Climatic and Biotic Evolution* (Princeton University Press, 1992, co-edited with W.A. Berggren), *The Eocene-Oligocene Transition: Paradise Lost* (Columbia University Press, 1994), *The Terrestrial Eocene-Oligocene Transition in North America* (Cambridge University Press, 1996, co-edited with R.J. Emry), *Sedimentary Geology* (W.H. Freeman, 2003, with F. Schwab), *Bringing Fossils to Life: An Introduction to Paleobiology* (McGraw-Hill, 2003), *Magnetic Stratigraphy of the Pacific Coast Cenozoic* (Pacific Section, SEPM, 2001), *Horns, Tusks and Flippers: The Evolution of Hoofed Mammals* (John Hopkins University Press, 2003, with R. Schoch), *and From Greenhouse to Icehouse* (Columbia University Press, 2003, with L.C. Ivany and E.A. Nesbitt).

Robert H. Dott, Jr., is the Stanley A. Tyler Emeritus Professor of Geology at the University of Wisconsin at Madison. He was president of the Society for Sedimentary Geology (SEPM) from 1981 to 1982 and president of the History of Earth Sciences Society (1990). He has received the Wisconsin Student Association's Outstanding Teacher Award (1964); the American Association of Petroleum Geologists (AAPG) President's Award (1956) and Distinguished Service Award (1984); Honorary Member (1987) and the Twenhofel Medal (1993) from SEPM; and the Parker Medal from the American Institute of Professional Geologists (1992). He received the History of Geology Award (1995) and the L.L. Sloss Award for Sedimentary Geology (2001) from the Geological Society of America. He has been an AAPG Distinguished Lecturer (1985), Visiting Lecturer in the People's Republic of China (1986), Cabot Distinguished Lecturer at the University of Houston (1986–1987), Visiting Erskine Fellow at Canterbury: University of New Zealand (1987), and Sigma Xi National Lecturer (1988–1989). He has published extensively on many aspects of sedimentary geology and in the history of geology. Besides *Evolution of the Earth*, Professor Dott has co-authored *Humphrey Davy on Geology* (University of Wisconsin Press, 1980, with R. Siegfried), has co-edited *Modern and Ancient Geosynclinal Sedimentation* (SEPM, 1974, with R.H. Shaver), and has edited *Eustasy: The Historical Ups and Downs of a Major Geological Concept* (Geological Society of America, 1992).

Evolution of the Earth

I

Time and Terrestrial Change

MAJOR CONCEPTS

There's nothing constant in the universe,

All ebb and flow, and every

shape that's born

Bears in its womb the seeds of change.

Ovid, *Metamorphoses,* XV (A.D. 8)

▶ Geologists deal with immense spans of time, typically of the order of thousands to millions to billions of years. On these time scales, human events and human history are but a brief instant.

▶ From the short time perspective of human history, geologic events can be either short-lived and dramatic (such as volcanic eruptions, earthquakes, and floods) or so slow that they are virtually imperceptible (such as crustal uplift or subsidence).

▶ Over the long time scales of geologic history, both short-term and long-term events are important, although the most important changes (such as mountain building, sea-level changes, and the erosion of continents) take place over millions to tens of millions of years. In this framework, even "rare" events, such as "500-year floods," occur frequently.

▶ Geologic changes can occur either at constant (linear) or variable (nonlinear) rates; they can be repeating or nonrepeating; repeating changes can be rhythmic (periodic) or irregular (episodic).

Benjamin Franklin advised that the only thing certain is death and taxes, but *change* is the greatest certainty of all! Even if you have not experienced personally anything so dramatic as an earthquake or a flood, from the news you know that volcanoes such as Mount St. Helens suddenly erupt after long periods of silence, and history lessons recall past dramatic earth changes. Life on earth also has changed, and life crises have inspired such literary accounts as the biblical story of Noah's Flood. Understandably, people always have been deeply impressed by so-called catastrophic geologic events, whereas subtler, slower changes, even if ultimately more profound, have gone unnoticed. Thus, the people who left the footprints shown in Fig. 1.1 undoubtedly were more concerned about the volcano erupting nearby than about the age of the landscape where they lived.

A natural preoccupation with cataclysmic happenings has shaped the development of geologic concepts, particularly as they bear upon the age and history of the earth. Therefore, we shall begin with a few examples of important geologic processes that have acted during the past several thousand years of recorded human history. First, we shall discuss examples of sudden and violent events and then of subtler, slowly acting ones. Consideration of such changes in the context of a familiar span of time seems to provide the best basis for approaching matters of greater antiquity. The goal of this chapter, then, is mind expansion—that is, stretching your concept of time beyond the ordinary human experience into the geologist's frame of reference. At first it may seem like a trip on a magic carpet, but after reading a few more chapters, you should feel comfortable with millions and billions of years of geologic time.

Dramatic Geologic Events in Human History

In August 1883, after 200 years of dormancy, a small volcanic island named Krakatoa, located between Java and Sumatra in Indonesia (Fig. 1.2), disappeared in what was long considered the most violent explosion ever witnessed by humanity. A 300-meter- (1,000-foot-) deep hole was formed in the bottom of the sea where the island had stood. The explosion was heard 5,000 kilometers (3,000 miles) away. Volcanic ash, which was blown 80 kilometers high and fell on ships 2,000 kilometers from the island, darkened the sky for 2 days (Fig. 1.3). Huge **tsunamis,** or sea waves, 40 meters (130 feet) high, crashed upon surrounding islands, drowning at least 35,000 people and destroying 300 towns. Fine ash blown high into the atmosphere caused red sunsets in Europe and a cooling of global climate by half a degree Celsius for a few years.

Another geologic catastrophe recorded in history books was the eruption of Vesuvius volcano in Italy in A.D. 79. Earthquakes warned of volcanic activity, and many people fled from the Roman city of Pompeii at the foot of the mountain near present-day Naples. Those who lingered were killed suddenly by poisonous volcanic gases and then buried with their houses, furniture, and even pets beneath volcanic ash and mudflows.

A volcanic event of still greater magnitude occurred in the eastern Mediterranean region about 3,600 years ago. Santorin, or Thera, is a crescent-shaped island between Greece and Crete (Fig. 1.4). It is a volcanic crater partially submerged beneath the sea—a puny remnant of a once lofty mountain. Many small eruptions have occurred at Thera within historic times, but old lava flows and thick volcanic ash deposits tell of a more violent earlier history. Sampling by oceanographers of deep-sea sediments in the eastern Mediterranean has revealed widespread buried ash layers that become thicker toward Thera, their obvious source. Dating

Figure 1.1 The oldest known human footprints from 3.6 million-year-old rocks at Laetoli in northern Tanzania. The footprints show a well-developed arch and no divergence of the big toe, clearly demonstrating that this ancient hominid, *Australopithecus afarensis,* had the upright, bipedal, free-striding gait of modern humans at this early date. The trackway covers 70 meters and shows two adults walking side-by-side across a plain covered by wet volcanic ash, which hardened to preserve the footprints. A third set of footprints appears to belong to a child who walked in the footsteps of one of the adults. The footprints to the far left belong to an extinct three-toed horse. (© *John Reader/Photo Researchers.*)

Figure 1.2 1979 eruption of Krakatoa, photographed by the Kraffts, a famous husband-and-wife team of volcanologists who died in the eruption of Mt. Unzen in Japan in 1991. *(Krafft-Explorer/Photo Researchers.)*

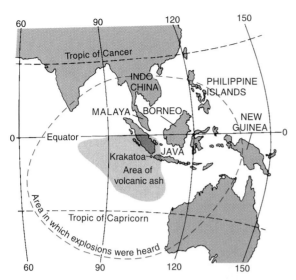

Figure 1.3   Krakatoa volcano and areas of audible noise and significant ash fall from its 1883 eruption. *(From Volcanoes of the Earth, Revised edition, by Fred M. Bullard, Copyright © 1976. By permission of the University of Texas Press.)*

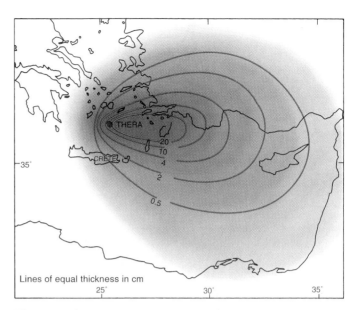

Figure 1.4   Location of the volcano Thera relative to historically prominent areas of the eastern Mediterranean. The shaded zone represents probable fallout pattern of volcanic ash, which erupted about 1500 B.C. and nearly wiped out the Minoan civilization of Crete. Lines of equal thickness of ashfall shown; thickness in cm. *(Location of the volcano Thera by Floyd McCoy.)*

indicates that the last major eruption occurred around 1650 B.C. The material erupted was *three or four times greater* than from Krakatoa.

Thera's eruptions are of special interest not only for their geologic magnitude but also for their possible historical implications. The earliest European civilization—called Minoan—began on Crete only 125 kilometers (75 miles) south of Thera, and a well-preserved Minoan settlement has been excavated on Thera itself. The Cretan civilization collapsed suddenly about 1500 B.C. for reasons that eluded archaeologists for generations. Greek archaeologist Spyridon Marinatos postulated in the 1930s that eruptions of Thera may have ended the Minoan culture. Severe earthquakes accompanying the eruptions would have destroyed many buildings, and tsunamis generated by the tremors could reach northern Crete without warning in a mere 20 minutes. As around Krakatoa, such waves could do further damage. Prevailing winds blew great volumes of ash southeastward, as shown in Fig. 1.4. Remnants of this ash have long been known on the islands of Crete and Rhodes and have been found also in northeastern Africa within Nile delta sediments. It is intriguing to speculate that Thera's ash may have caused the "days of darkness" of the Old Testament and that the flooding of Thera's crater may have inspired the myth of sinking Atlantis.

Volcanism is very common in the great rift valley region of East Africa, which stretches from Ethiopia to South Africa. As Africa began to rip apart, the rift valley filled with the lakes and flood plains inhabited by our earliest ancestors. Massive eruptions have blanketed large areas of the valley many times in the past 10 million years. About 3.5 million years ago, an eruption of a volcano known as Sadiman, near a site called Laetoli in northern Tanzania, preserved a remarkable story. Because the volcanic ash was wet and muddy, it contains the tracks of elephants, rhinoceroses, extinct three-toed horses and saber-toothed cats, gazelles, monkeys, hares, guinea fowl, and even insects and millipedes. The

most remarkable tracks, however, were made by some of our earliest ancestors (Fig. 1.1). Two individuals of *Australopithecus afarensis* (a famous specimen is commonly known as "Lucy") apparently walked side-by-side across the wet ash bed, leaving a trackway that can be traced at least 9 meters (30 feet). At one point, something must have caught their attention, because they stopped, turned, and then resumed their northward trek. After the tracks dried, they were blanketed by another ashfall, so they were preserved. In 1976, Mary Leakey and her colleagues discovered the tracks and spent years slowly unearthing them. From measurements of the footprints and the length of the stride, we can estimate that these individuals were about 1.2 meters (4 feet) tall. They had reduced big toes, like ours, so they could not grasp branches with their feet and climb as an ape can. Most important, they walked fully erect as we do today. Contrary to long-held expectations, the earliest fossils and tracks of our ancestors demonstrate that our upright posture was one of the first innovations to appear in human evolution, long before our enlarged brain, complex social behavior, or other hallmarks of humanity.

Volcanic eruptions continue to inconvenience and sometimes endanger people. Take, for example, the 1973 eruption of Heimaey volcano in Iceland, whose lava buried many houses (Fig. 1.5), and the 1980 eruption of Mount St. Helens in Washington state.

Equally dramatic have been devastating earthquakes such as occurred in San Francisco in 1906 (Fig. 1.6), near Yellowstone Park in 1959, and in Alaska in 1964. Though forgotten today, a violent earthquake felt all over western Europe and northwestern Africa in 1755 and a resulting tsunami that destroyed Lisbon made far greater impact on people's minds than any of these other

Figure I.5   Basaltic lava flow encroaching upon houses In Heimaey, Iceland, during a 1973 eruption. *(Courtesy of Campbell Craddock, University of Wisconsin.)*

Figure I.6   With curious irony, this statue of famous geologist Louis Agassiz (see Chapter 16) was toppled from its mounting at Stanford University during the 1906 San Francisco earthquake. It has since been remounted, little worse for its Humpty-Dumpty-like fall. *(From Lawson, et al., 1908, The California earthquake of 1906, Carnegie Institute of Washington.)*

Figure I.7   *Gemini* spacecraft photo looking southeast across northern Egypt with the vegetated, dark-colored Nile River valley in the lower right and the dark, straight-sided Red Sea in the upper left; Arabia lies on the horizon beyond. The diagonal white bank is a cloud streak. *(Courtesy NASA: photograph 66-63533.)*

Subtle Geologic Events in Human History

Climate

Human civilization appeared more or less simultaneously along the lower Nile River in Egypt (Fig. 1.7), in the Tigris and Euphrates valleys in present Iraq, on the Indus delta of northwest India, and in northeastern China. The eastern Mediterranean–Middle East region is today a harsh, arid country that seems an unlikely cradle for agricultural societies, yet when the great Ice Age glaciers covered northern Europe prior to 10,000 years ago, this region enjoyed a climate that was comfortably warm and more humid than now. For example, former river valleys and dried lakes lie buried beneath the sands of the Sahara. Climatologists believe that an atmospheric high-pressure cell then centered over the ice sheet pushed moist westerly winds south of the Mediterranean Sea. Knowing this, we should not be surprised that early humans spread across the area and there developed their culture rapidly. After the great glaciers retreated, the Middle East climate harshened drastically. By 5000 to 3000 B.C., major cultural centers had become restricted to the large, through-flowing river systems, such as the Nile, Tigris-Euphrates, and Indus, that were more or less independent of the increasing drought. It is from just such areas that our earliest written records have come. They tell of already highly developed agricultural systems with elaborate irrigated plantations and advanced urban centers, as in the Fertile Crescent of Mesopotamia (Fig. 1.8).

events. The timing of that catastrophe on All Saint's Day deeply impressed superstitious Europeans.

That earth tremors are the rule is indicated by the fact that annually about 800,000 earthquakes too mild to be felt by people are recorded worldwide. At least 10,000 people are killed each year by more severe earthquakes.

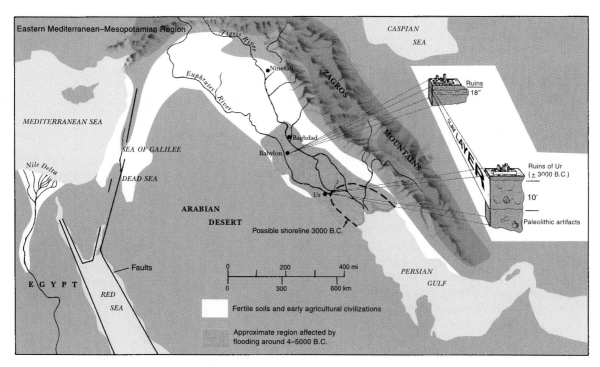

Figure 1.8   Eastern Mediterranean-Mesopotamian region showing the Fertile Crescent (white), centers of early agrarian civilizations, and the known extent of the post-Paleolithic flood deposit of the Tigris-Euphrates valley. Ur apparently was built near the shoreline of the Persian Gulf, and it was long thought that the "clay layer" was marine—a record of Noah's flood. The clay is now thought to be river flood deposit instead, but it still may have contributed to the flood myths found in many Middle Eastern cultures.

Glaciers and Sea Level

Other relatively slow geologic changes have also occurred during human history. Measurements at widely scattered seaports beginning around 1890 demonstrated that sea level was rising over the entire earth. Today sea level is rising at least 2 mm per year. Since 1890, that would mean a total rise of about 225 millimeters (9 inches), certainly not devastating. But if long continued, this increase would be of some concern in the future, especially on an earth rapidly becoming crowded by a staggering population boom. You might well ask why should sea level be changing at all. We do not have far to look for the explanation. Over approximately the same 60-year interval, the climate was warming. For example, winters had been much milder than in the "good old days" of great-grandfather's youth, when people skated across a completely frozen lower Hudson River from New York City to New Jersey. As the general climate warmed from 1890 to about 1950, by from 0.5 to 5°C (depending upon latitude), glaciers in high mountains over most of the earth began melting, in some cases retreating as much as 16 kilometers (10 miles). The world's largest ice masses, those of Greenland and Antarctica, apparently retreated slightly at their margins, too. The obvious effect of so much ice melting was to raise sea level.

If we assume that melting will continue indefinitely and then gaze into a crystal ball, we can make some startling predictions. From estimates of the total volume of water still locked up in glacial ice, it is calculated that sea level would be raised through complete melting by *at least 50 meters more above its present level.* This would, of course, flood most coastal areas where many of the world's largest concentrations of civilized activity are found. In North America, New York, Miami, New Orleans, Houston, and Seattle all would be submerged. Memphis, now 560 kilometers from the sea, would become a major port, and Washington, D.C., would suffer particularly interesting consequences. The Pentagon, only 11 meters above sea level at its base, would become useless, and the Washington Monument would show just its top above water as a useful navigational marker. If you live near a coast, consider how your hometown would fare with a 50-meter rise of sea level.

Now consider the opposite effect of glaciation—namely, the worldwide lowering of sea level during maximum glacial advances. The last glacial maximum occurred about 20,000 to 25,000 years ago, and from evidence of submerged ancient shoreline features, it is inferred that sea level must have been approximately 100 meters (300 feet) lower than at present.

Geologists have studied the changes of sea level that occurred during the past 20,000 years and have attempted to link these with historical records. An average rise of 100 centimeters per century occurred from about 17,000 up to 6,000 years ago. About 4000 B.C., rapid rise culminated, and all subsequent rise has averaged only 12 to 15 centimeters per century. Clearly, the last phase of rapid rise may have caused the flooding of some early Bronze Age coastal settlements, and it is interesting to find reference to a deluge in the legends of many separate ancient cultures including Greek, Babylonian, Hindu, and Hebrew. Since 3000 B.C., however, it is known that the Tigris-Euphrates delta has *advanced* southward perhaps as much as 175 kilometers (100 miles); today it advances 25 meters per year.

Could recent small oscillations of sea level have given rise to the many ancient flood legends? If the biblical story were correct, there should have been simultaneous flooding of all other great ancient coastal civilizations. But it has not been possible to establish any synchroneity of deluges reported in the many ancient traditions. Probably these events resulted from periodic local river floods, which caused devastation of the great cultural centers concentrated on low-lying deltas of large rivers. Even today the Tigris-Euphrates delta experiences frequent floods, and two-thirds of Bangladesh, on the immense Ganges delta, disappears annually beneath several meters of water during the monsoon season. Both regions are low and swampy for hundreds of miles inland.

For the past 6,000 years, maximum vertical fluctuations of only about 3 meters above or below present sea level have occurred. Obviously a rise of only 3 meters could not have inundated the whole then-known world; only coastal areas would have been inundated. On the other hand, if we look ahead, a slight warming of climate for the next 1,000 years or so might result in an additional rise of at least 50 meters. Perhaps we should look still further ahead, however, and prepare instead for cooler times, for our present "interglacial" stage should last only about 10,000 more years, at which time large ice sheets may begin forming and sea level falling once more.

Crustal Changes

Much of the low European coastline has been flooded slowly at least since the 1600s, when tidal records were first made. This is to be expected from what we have just seen about postglacial flooding. In central Scandinavia (Fig. 1.9), on the other hand, an awareness developed about the year 1700 that the coastline there had retreated seaward, suggesting an apparent fall of sea level of more than 1 meter in the previous century. It had been assumed that the earth's solid crust was fixed until, in 1765, a Finnish surveyor proved that the Baltic coastline was changing more in the north than in the south. He concluded that the Scandinavian crust must be rising rather than sea level simply falling. This is portrayed in the uplift area of the upper right part of Fig. 1.9, where we see that the northern Baltic shore is rising three times faster than the southern.

Amsterdam records indicate that sea level is apparently rising 20 centimeters (8 inches) per century there. But this is about twice the rate deduced for the worldwide effect of glacial retreat. How can we reconcile this discrepancy? Where the apparent rate of rise seems "too large," we conclude that there also has been some subsidence, or sinking, of the land, as in the brown area of Fig. 1.9. Conversely, where the apparent rate of rise of sea level seems "too small," the coastal land must have risen as sea level has risen. Therefore, the study of any coast must consider crustal changes as well as worldwide sea-level fluctuations. *Crustal*

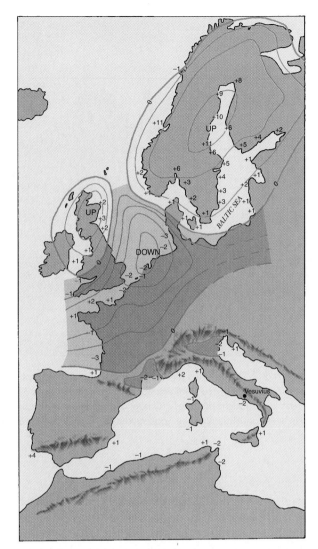

Figure 1.9   Relative coastal subsidence (negative, shaded) and uplift (positive, unshaded) in northwestern Europe in the 10,000 years since the last glacial advance. Numbers indicate known rate of present coastal movement in millimeters per year. *(Drawing by Tagawa from "The changing level of the sea" by R. Fairbridge in SA, 1960.)*

warping, or bending, may cause either local **subsidence** (sinking) or **uplift** (rising) (Box 1.1).

A long history of careful measurements shows that the North Sea region subsided, with the consequence that the low countries were flooded. Simultaneously, most of interior central Europe as well as the northern two-thirds of Great Britain and Scandinavia rose at a geologically rapid rate (Fig. 1.9). Borings in the Netherlands indicate that old shoreline sediments, known from their fossils to have accumulated roughly 1 million years ago, now lie 2,000 meters below present sea level (i.e., beneath 2,000 meters of younger deposits). The crust of the earth there subsided about 2,000 meters in 1 million years, or an average of about 2 millimeters per year. Measurements in central Scandinavia indicate

Box 1.1

Human Beings as Geologic Agents

Human beings sometimes have contributed to geologic change, usually unwittingly. It is well known that historic Venice is sinking at a present rate of 6 to 10 millimeters per year. At the most, 1.5 millimeters is due to worldwide sea-level rise, and only 2 millimeters or less can be attributed to regional crustal subsidence (Fig. 1.9). What, then, explains the alarming flooding of this historic city? It has been shown that withdrawal of ground water has lowered the water table 20 meters (65 feet) beneath a nearby industrial center during the past 30 years. Thus, about 80 percent of the sinking of Venice is caused by rapid removal of water by humans, which has caused sediments to compact beneath the area under the weight of overlying material. Pumping of water has been reduced, and the sinking has slowed.

Near Los Angeles, California, human activity encouraged encroachment of the sea for 30 years. The second largest oil field in California extends beneath the coast under Long Beach harbor. This area is the only satisfactory harbor in the greater Los Angeles region; therefore, a large naval base, an automobile assembly and shipping facility, and other industrial activities are concentrated along its waterfront. Over the years, extraction of petroleum and natural gas from poorly cemented sandstones more than 1,000 meters below the surface has lowered the pressure in the pores of the strata and has allowed the sediment grains to compact; subsidence and flooding has resulted. You can imagine the consternation of irate naval and industrial landowners! To stem further subsidence, petroleum companies today replace extracted petroleum with water pumped into the sandstones.

People have been geologic agents in this region in other ways. First, they persist in stripping vegetation from hillsides for new sub-urban housing tracts, which encourages landsliding during heavy rains. Both natural and human-caused fires denude the mountain slopes, further aggravating the erosion problem. Housing developments also continue to be erected across known active faults and old landslides. Human activities have had a more subtle effect upon southern California beaches. Prior to the great wartime building boom of the 1940s, rivers continually replenished the beach sands, which were slowly swept southward along the coast until they were flushed into deep water via a submarine canyon. Metropolitan expansion brought complete usage of normal runoff waters, so that rivers are dry except during rare winter downpours. Furthermore, the rivers are now confined by artificial cement channels, which deprive them of sand, which would otherwise be added from their banks. The beaches have been starved of sand, so winter storm waves are taking their toll on the shore and some choice homes. This calamity has necessitated elaborate sand-drift traps and expensive barging and pumping to refeed the beaches.

Alarming depletion of natural water supplies in the arid Southwest and the necessity for transporting water hundreds of miles at fantastic expense is another story of humans versus nature. Most ground water now being extracted in such areas fell as rain thousands of years ago and is being withdrawn more rapidly than it is being replenished. As pressure on finite resources increases, other regions must also take notice, for many people today are living precariously on borrowed water—borrowed from the past without any adequate reinvestment program for the future's mushrooming thirsty population.

---

that parts of the North Sea and Baltic coasts now rise as much as 10 to 11 millimeters per year. The maximum total rise has been 250 meters (825 feet) since the retreat of glacial ice from the region about 10,000 years ago. The average apparent rate of rise, then, has been about 2.5 meters per century, an astoundingly rapid rate, and much faster than the simultaneous rate of subsidence of the nearby low countries (Fig. 1.9).

Submarine topography and dredgings indicate that much of the North Sea floor and the English Channel formed a low, swampy land connecting Britain with the Continent at the end of the last glaciation. Glacial ice had then melted entirely from Britain and must have been shrinking rapidly from Scandinavia. Although sea level had begun rising rapidly, it did not flood this region until 6,000 or 7,000 years ago, when, you will recall, we estimate that the present general level was achieved. Easy migration of plants, animals, and early humans was possible across this area from the Continent to Britain for at least 3,000 years. As low areas were flooded, the former ice cap centers of Scandinavia and Britain rose in response to removal of the great load of ice that had weighed down the crust there. This cause-and-effect is duplicated for North America, where measurements show that most of east-central Canada is rising in response to deglaciation.

How Fast Is Fast? Time and Rates of Geologic Processes

Having reviewed examples of different geologic processes active within the brief span of human history, we shall now begin to investigate the matter of geologic time. Here we step into a new frame of reference for most people. For example, the time until the next school holiday seems an eternity for children, and a decade seems moderately long to most adults. The *Pioneer* space probe required a decade to cross most of the radius of the solar system. It was launched from earth in 1972 and took 1 year to reach Jupiter but 10 more years to reach Neptune; it passed Pluto and left the solar system in 1983. Halley's comet, however, comes by only once in a lifetime—that is, every 76 years. (Coincidentally, it was passing the earth when Mark Twain was born and again when he died.)

Time is to the geologist much as the immensity of space is to the astronomer. Five thousand years of recorded human history seems a long time, and if we tell you that humans first appeared between 1.5 and 2 million years ago, that will seem to you respectably antique. What do people think, however, when we tell

Table I.I	Comparison of Rates of Some Geologic Processes				
Slowest ————————————————————————————▶ Fastest					
Average Erosion of Continent	**Cutting of Grand Canyon**	**Postglacial Rise of Sea Level**	**Rise of Scandinavia**	**Advance of Tigris-Euphrates Delta**
0.03 mm per year	0.7 mm per year	5 mm per year	10 mm per year	25,000 mm per year

them that *the earth is almost 3,000 times older?* And our solar system has now revolved around the Milky Way galaxy to the same relative position that it occupied 200 million years ago?

Experience shows that the magnitude of geologic time is new to the uninitiated, so geologists find analogies with more familiar measures of time to be helpful. Our favorite is to compare geologic time with 1 calendar year. If we consider New Year's Day as representing 5 billion years ago (roughly the earth's birthday), we would find that the record of January through October—over three-fourths of earth history—is exceedingly obscure! The first recognizable fossil animals appeared a mere 600 to 700 million years ago, or in mid-November; extinction of the dinosaurs occurred on Christmas day; and *Homo sapiens* appeared at 11 P.M. on December 31. All of recorded human history occurred in the last few seconds of New Year's Eve. The oldest known living things, a California creosote bush 11,700 years old, a bristlecone pine tree 4,900 years old, and a Sequoia tree almost 4,000 years old, all sprouted and Christ was born within those last few seconds. Subsequent events—such as the entire industrial revolution, discovery of electricity, and invention of the automobile, airplane, and television—hardly seem worth mentioning on this scale.

Examples of Rates

To illustrate geologic time still another way, let us perform some simple arithmetic to discover approximate rates of various geologic changes. It is important to realize that many such rates can now be measured and that we find great variation in them. There is also great variation in the regularity or frequency of recurrence of different geologic changes.

Recall that sea level was approximately 100 meters lower about 20,000 years ago and has risen with minor fluctuations since. The average rate of rise would be 0.005 meter, or 5 millimeters, per year. In other words, it would take an average of 66 years to rise 1 foot. On the other hand, looking to the future and taking the view that sea level will rise another 50 meters if existing ice caps melt, and assuming the same average rate of melting and rise as mentioned previously, it should take but a mere 10,000 years to produce the final deluge. Table 1.1 summarizes for comparison the rates of four other common geologic processes.

Besides the question "How fast is fast?" we also must ask "How frequent is frequent?" Floods are among the most familiar geologic events. River flood plains are as much a part of a total watershed system as the main channels, and it should be obvious to anyone living near a large river that, as a shrewd Louisiana geologist once remarked, "Flood plains are for floods!" But people have long ignored this simple fact and have continued to build houses, towns, and factories on flood plains, which are inevitably destined to be doused.

On the carefully controlled Mississippi River, severe floods are infrequent enough that inhabitants become complacent. In 1965, a combination of large ice jams and sudden melting of heavy snow produced a terrible flood. No old-timers could remember such an event. Then, only 28 years later in the summer of 1993, even worse flooding occurred (Fig. 1.10). Floods illustrate clearly the limitation of human time perspectives. If events like these occur, on average, only once every few centuries, there could be no record of them in the folklore of a young nation.

Figure I.10   The upper Mississippi River flood of 1993 at Davenport, Iowa—classed as a "500-year flood." Circular structure just above the bridge in the foreground is a rained-out baseball stadium. *(©Michael Springer.)*

In 1966, the river Arno flooded Florence, Italy, causing enormous damage to priceless art treasures. Florence has a much longer historical memory than America; the danger of flooding there was recognized over 400 years ago by Michelangelo, who advised a nephew not to buy a house in a certain district because "every year the cellars flood." Since A.D. 1300, moderate floods have occurred at Florence about every 25 years and massive floods about every 100 years. Only three superfloods like that of 1966 have occurred during the past six centuries, however, or approximately one every 200 years. Similarly, "unprecedented" North Sea flooding of the Netherlands in 1953 was caused by meteorological circumstances estimated to occur only once every 400 years.

Estimates of frequencies of events like these are based upon the *probability* of a flood in a certain region. If you think as a gambler does, there is a virtual certainty of one superflood occurring at Florence *sometime* within any 200-year period; put another way, there is less than a 1 percent chance of such a flood occurring every year. It is like weather forecasting. There may be a 100 percent chance of a storm occurring *somewhere* within a region *sometime* during a given day, but an exact prediction of where and when is not possible.

Another seemingly unprecedented flood occurred in 1976 in Big Thompson Canyon, Colorado. Less than 3 years later, people had begun to reoccupy devastated canyon-bottom areas. Either they are suicidal or they are incurable gamblers, for they are playing the odds that it was a *300- or 500-year flood.* Though by human standards such floods are *rare events,* these "catastrophes" must be considered frequent and normal on a geologic time scale.

An important question might be asked here. Are the rates we measure today for the rising of some land areas adequate to produce great mountains over a reasonable span of geologic time, or do such features require abnormally faster rates? Said another way, are we today witnessing a period of active mountain making in areas such as California, or is this a relatively static time?

During his great round-the-world voyage in *H.M.S. Beagle,* Charles Darwin experienced several earthquakes. In 1835, following an especially severe one, he noted an uplift of the Chilean coast of about a meter, as well as nearby volcanic eruptions. He reasoned that a rise of a few feet per shock might produce, over geologic time, whole mountain ranges. Fig. 1.11 is a simple graph projecting some rates of measured uplift in several parts of California, obviously a relatively active region. This graph was prepared by an American geologist to show what changes of elevation might be expected 200,000 years in the future, a very short time geologically. Obviously we must conclude that present, known rates of uplift can indeed produce mountains in a geologically brief time.

Different Ways of Growing and Changing

In the section "Examples of Rates," we talked only of average rates of change; for simplicity, therefore, we extrapolated constant rates over long times. Fig. 1.11 illustrates a graph showing this type of constant change, which plots as a straight line and, so, is called *linear change.* However, we know that most natural processes do not

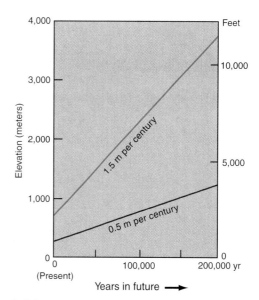

Figure 1.11 Present known average rates of uplift of two lowland areas in southwestern California extrapolated into the future at constant or linear rates. Obviously, mountains could result in a few hundred thousand years. *(Adapted from J. Gilluly; 1949: with permission of the Geological Society of America.)*

proceed at constant rates. In nature, linear change is quite the exception. For example, erosion of a newly uplifted mountain range tends to proceed fastest early and then slow down after uplift ceases, because stream gradients flatten and their ability to erode decreases. Growth in mammals slows as adulthood is approached. These are both examples of *nonlinear changes.*

We hear a great deal about the human population "explosion," and a glance at Fig. 1.12 reveals that the multiplication of *Homo sapiens* is another example of nonlinear change. We see that human population growth was very slow at first; it must have taken hundreds of thousands of years to break the first million mark. Records exist only for the past few centuries, but it is clear that the first documented doubling of population took two centuries (1650–1856); the second doubling a little less than one century; and today it doubles in only 30 years! The present growth rate is equivalent to adding the population of Britain annually. In the year 2009, people probably will number around 8 billion—almost double the present count.

Another point worth noting is that such *nonlinear growth* results even if the *rate* of addition of people remains constant. If the birthrate were to increase, then total population size would grow even faster.

All the changes discussed up to this point move in one general direction (increase of size, decrease of elevation, etc.); that is, they are *nonrepeating changes.* Many changes in nature, however, are *repeating* ones. Commonly these repeat with a more or less regular time period, such as the fluctuations of the tides and changes of seasons. Such repeating changes are called *periodic* or *rhythmic, events.* Events that do not recur with a regular time period are called **episodic events.**

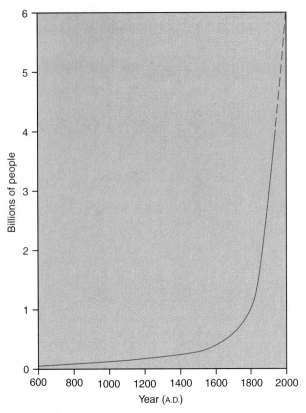

Figure 1.12   Growth curve for human population of the world, example of a nonconstant, or nonlinear, rate of change. *(Data from United Nations estimates.)*

Catastrophic Versus Uniform Views of Change and the Age of the Earth

Ancient peoples quite understandably were impressed only by rapid and violent geologic processes. This led to a catastrophic view of the earth (commonly called **catastrophism**), which dominated European thought until 1850. Changes were thought to occur suddenly, rapidly, and devastatingly by a series of catastrophes. The most famous example of such a supposed catastrophe was, of course, the great Deluge of the Old Testament. Others included volcanic eruptions at Pompeii and Thera as well as the Lisbon earthquake.

A parallel idea consistent with catastrophism was a universally accepted belief that the earth was but a few thousand years old. This conception is revealed in Shakespeare's *As You Like It*, written around 1600, where reference is made to the age of the earth being about 6,000 years. In 1654, one of several more definite pronouncements came from Anglican Archbishop Ussher, who announced that, based upon his analysis of genealogies in the Scriptures and of astronomical cycles, the world had been created in the year 4004 B.C. on October 23. Some years later, another authority fixed an equally precise date for the great Deluge, November 18, 2349 B.C. At that time, practically all leading European intellectuals believed implicitly in a literal interpretation of the Bible, so it is not surprising to find that a 6,000-year age for the earth was gospel for fully 200 years.

In the seventeenth century, the Scriptural account of Creation was a point of departure for European science. Humans were considered early comers, who lived for a time in paradise. As punishment for their sins, the Deluge was sent to throw the earth's surface into chaos. Countless creatures suddenly were buried in sediments, violent waves and currents scoured out valleys and piled up rocks to form mountains, and the earth was wracked by earthquakes and volcanic eruptions. After the Flood, the mountains began to be worn down and valleys to be filled.

When geology emerged as a vigorous intellectual pursuit around 1800, an alternative concept of earth change began to develop. It was then argued that the same common processes being observed were largely responsible for all past changes imprinted in the rock record. A noncatastrophic, or more *uniform,* view of change thus emerged, which provided no role for unknown, miraculous, or supernatural processes. It also regarded the sudden, violent phenomena as no more—and possibly less—important overall than more continuous and milder processes. The most uniform processes operate on a regularly recurring, continual, day-to-day basis, such as the constant pounding of surf on a shoreline and the daily rise and fall of the tides. But most of the less frequently occurring processes also must be considered geologically normal or uniform. For example, the world annually experiences about 150 major earthquakes, 20 volcanic eruptions, and countless floods, so none of these episodic events can be considered unusual or catastrophic.

Today we envision the earth as dynamic, ever changing, and evolving under the influences of many complex physical, chemical, and biological processes, which show great variations of rates and intensities. These processes compete with one another to produce a situation tending toward balance, or **equilibrium,** among them. But the equilibrium is always being disturbed by new changes. The resulting modifications are in part nonrepeating changes and in part repeating ones. Thus, American geologist C. R. Van Hise reflected this newer view of the earth in 1898 when he wrote that "the earth is not finished, but is now being, and will forevermore be re-made."

A corollary of the uniform change concept, which had great impact upon Western society in the 1800s, was its inescapable implication that the earth must, in fact, be much more than a few thousand years old in order to allow enough time for all observable ancient changes to have occurred in noncatastrophic ways. We begin to see here the seeds of a social controversy whose repercussions affected our entire civilization, with the age of the earth and the nature of geologic changes as principal points of disagreement. Galileo's telescope already had stretched human minds to consider an infinity of space and other possible worlds. Now geology suggested an infinity of time and history.

There is a subtler but even more fundamental issue here as well. Some cultures even today present stories of the earth and of human life that differ markedly from our modern scientific version. These stories have evolved and have been handed down through countless generations, primarily as oral accounts. Most have some basis in factual evidence and simply represent differing perceptions of history based upon a great variety of cultural and environmental influences. From the point of view of most scientists, such accounts seem heavily charged with mythical elements, tend to be rigidly dogmatic, and lack both disciplined reasoning from evidence and the critical testing of hypotheses. Through the application of what is called the scientific method of analysis, we have developed a quite different view of the development of the earth and of life, which we generally believe to be mostly free of mythical influences, yet, even within our own supposedly sophisticated and science-conscious culture there persist groups, such as the Flat Earth Society, which cling to earlier, simplistic views of the earth. Such groups either deny or reinterpret the same evidence that is incorporated into modern scientific interpretations of earth and life. It is easy to be smug about primitive ideas of origins, but the history of science itself is riddled with inappropriate dogmatism and discarded scientific hypotheses that once were in vogue but that ended up on the intellectual ash heap. Thus, we should not be arrogant about any of our favorite "correct" explanations of nature, even today. Science constantly reexamines or tests such explanations by a continual, open-ended process of refinement. Although science has an excellent track record in answering many questions about nature, there is still much that is not understood. Indeed, there is no reason to suppose that the study of nature will ever reach "final" conclusions.

In Chap. 2, we shall examine the philosophical crisis between the catastrophic and uniform views of the earth as we begin to develop the first principles upon which to base interpretations of the fascinating diary of our evolving planet. Throughout the book, we shall also look at a few examples of important discarded scientific hypotheses, both because such a historical approach conveys a truer picture of science and because it helps us to imagine ourselves in the role of the discoverers and interpreters who developed the fundamental concepts. What better way is there to learn those concepts?

Summary

Geologic Events in Human History

- **Dramatic events,** such as volcanic eruptions, earthquakes, river floods, and tsunamis, greatly impressed early peoples, leading naturally to a catastrophic view of earth history.
- **Subtle events** producing slow geologic changes, such as the rise or fall of sea level and the gentle warping up **(uplift)** and down **(subsidence)** of the earth's crust, were long overlooked.

Geologic Time and Rates of Change

- Most geologic phenomena must be measured on a scale unfamiliar to the layperson. For example, although a "500-year flood" is a **rare event** on the human scale of years and decades, it is commonplace on the geologic scale of millions of years. And continuous, subtle changes, such as the rise of mountains and the erosion of an entire continent, which are imperceptible to humans, produce important results on the geologic scale of time. Changes may show different patterns:
  1. **Linear change** occurs at a constant rate, whereas the more common **nonlinear change** does not.

  2. Some geologic changes are **nonrepeating,** but others are **repeating.**
  3. Of repeating changes, those that repeat in a regular fashion through time are **periodic,** or **rhythmic;** those that do not recur regularly are **episodic.**
- All these different kinds of geologic changes with their variable rates compete and result in a tendency toward **equilibrium,** but a balance is rarely maintained for long spans of geologic time.

Catastrophic Versus Uniform Views of Geologic Change

- The catastrophic view of the earth dominated for centuries until about 1850, when a more uniform view of geologic change developed. The catastrophic view backed by Scriptural authority estimated the earth to be only 6,000 years old, but the uniform view, with its greater emphasis upon slow rather than rapid changes, required that the earth be much older. The nineteenth-century controversy between these two was the beginning of a wider conflict between the scientific and more mythical approaches to understanding nature. This conflict has not disappeared entirely even today.

Readings

Alaska's Good Friday earthquake. 27 March 1964. Washington, D.C. U.S. Geological Survey, Circular 491.

Brice, W. R. 1982. Bishop Ussher, John Lightfoot and the age of Creation: *Journal of Geological Education* 30, 18–24.

Bullard, F. M. 1976. *Volcanoes of the earth.* Austin: University of Texas Press.

Dott, R. H., Jr. 1983. Episodic sedimentation—How normal is average? How rare is rare? Does it matter? *Journal of Sedimentary Petrology* 53:5–23.

Doumas, C., ed. 1978. *Thera and the Aegean world.* Warminster, England: Aris and Phillips.

Fairbridge, R. W. 1960. The changing level of the sea. *Scientific American,* May, 70–79.

Man against volcano: The eruption of Heimaey, Vestmann Islands, Iceland. Washington, D.C.: U.S. Geological Survey INF-75-22.

Morris, H. M., and Whitcomb, J. C. 1961. *The Genesis Flood.* Philadelphia: Presbyterian and Reformed Publishing.

Vitaliano, D. B. 1973. *Legends of the earth—Their geologic origin.* Bloomington: Indiana University Press.

Young, D. A. 1977. *Creation and the Flood, an alternative to flood geology and theistic evolution.* Grand Rapids, Michigan: Baker Book House.

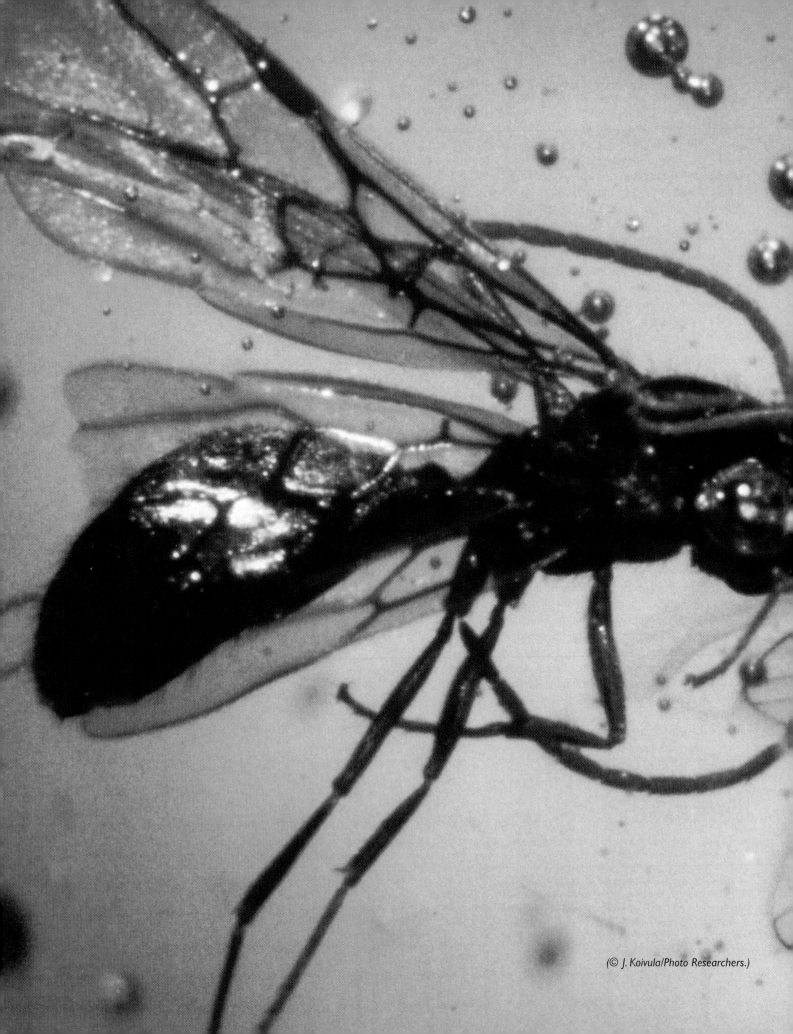

Chapter 2

Floods, Fossils, and Heresies

"No Vestige of a Beginning, No Prospect of an End"?

These shells are the greatest and most lasting monuments of antiquity, which in all probability, will far antedate all the most ancient monuments of the world, even the pyramids, obelisks, mummys, heiroglyphicks, and coins. . . . Nor will there be wanting media or criteria of chronology which may give us some account even of the time when [they formed]. Robert Hooke (1703)

The mind seemed to grow giddy by looking so far into the abyss of time. John Playfair (1805, at Siccar Point)

▶ From the biblical view of the world as unchanging since it was created 6,000 years ago, scholars gradually began to realize that the rock record indicated a much longer history of the earth.

▶ In 1669, Nicholas Steno suggested that sedimentary rocks were not created as we see them but were formed horizontally in continuous layers, stacked in sequence, with the oldest layers at the bottom and youngest at the top. This discovery allows us to decipher the relative age of geologic events.

▶ In the early 1800s, William Smith in England as well as Georges Cuvier and Alexandre Brongniart in France demonstrated that assemblages of fossils change through the sequence of strata, and index fossils can be used to recognize and correlate different rock units. Cuvier also demonstrated that some animals had become extinct, and he proposed a theory of successive catastrophic extinctions to explain the fossil record.

▶ James Hutton founded modern geology in the late 1700s by suggesting that the earth must be understood by applying natural laws and processes, rather than supernatural catastrophes. His philosophy led him to visualize immense amounts of geologic time and to realize that the earth is not static, but ever changing, with "no vestige of a beginning, no prospect of an end." Hutton also rejected the common notion that layered lava flows were precipitated out of seawater but argued that these rocks had once been in a hot, molten state.

▶ Charles Lyell completed Hutton's revolutionary view of geology in the 1830s by showing that natural processes, operating slowly and steadily, can account for most earth phenomena. However, Lyell argued not only for uniformity of natural processes but also for uniformity of the rates at which they operate. For over a century, many geologists were reluctant to accept the evidence of large-scale natural processes (such as gigantic floods or asteroid impacts) that operate at unusually high rates and intensities.

Until the late eighteenth century, the biblical Genesis was so perfectly rationalized with geologic facts that only a handful of people questioned the conventional geology. For example, it was the adding up of ages of successive generations recorded in the Bible that formed the basis for the generally accepted 6,000-year age for the earth. Even up to 1800, agreement of biblical revelation with reason seemed adequate to most people. The presence of abundant fossil marine creatures in strata strewn over the land offered proof of the Flood. The earth was thought to have been a paradise with a mild, uniform climate and a flat surface before the sinful fall of humanity. The Flood was believed to have piled up debris to make mountains, which in turn disrupted the flow of air to cause a harshening of climate. Moreover, erosion of the land implied a decay of the earth, which would culminate in the end of the world as foretold in certain biblical passages. Martin Luther even predicted pessimistically that the ultimate catastrophe would occur by 1560.

Until fossils were studied in detail, there was little reason to suspect that there had been distinct historical epochs very different from the present and that Genesis might not provide an adequate account of earth history. How fossils form and what they can tell us about the past are the first subjects of this chapter. Then we shall discuss some early theories to explain the overall development of the earth. We shall see that fossils were central to those theories as well as to the adoption of the uniform view of earth change in place of the catastrophic view, and we shall show why today we view the entire earth as evolutionary—the nonliving as well as the living realm. Finally, we shall explore briefly the important difference between a historical science, such as geology, and more familiar nonhistorical physics and chemistry. To reconstruct a past that no human was around to witness is one of the finest challenges we can imagine. How exciting it is to imagine the ground shaking as a dinosaur passes near us 100 million years ago or the lapping of a rising Cambrian tide around our ankles 500 million years ago!

What Is A Fossil?

Early Questions

Realization came slowly that explanations of nature must be based upon detailed evidence from the earth itself. Geology's awakening began with revelations about fossils and strata in the late 1600s. But it was a slow awakening. One of the most difficult issues was a search for some explanation of how the earth could sustain life in the face of obvious decay of the landscape. It was not until the late 1700s that a reasoned hypothesis of continual earth rejuvenation resolved this paradox.

The study of earth history received a boost from a great interest in the past generated by the discovery and excavation of Pompeii in the eighteenth century and the publicizing of Egyptian, Mayan, and other ancient civilizations in the early nineteenth century. It was natural that a fascination with earlier human civi-

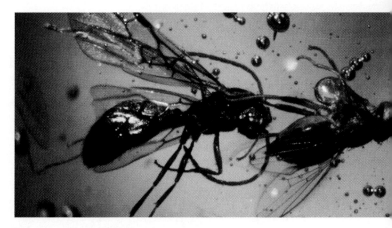

Figure 2.1 Well-preserved wasp and fly entombed in crystallized tree resin known as amber, formed about 30 million years ago. This highly unusual mode of fossilization preserves even the finest details of hairs and wing veins, so that these insects can be compared with living specimens. *(© J. Koivula/Photo Researchers.)*

lizations would stimulate interest in still older remains of life on earth. The use of fossils for dating strata and drawing geologic maps was also accelerated by the practical needs of the industrial revolution for more mineral resources.

Among early peoples, there were many controversies concerning geologic matters. Besides supposed catastrophic floods, probably the most important things that puzzled the ancients were fossils. What is a fossil? The word means simply "something dug up," but, as generally used, a **fossil** is any recognizable evidence of preexisting life. Fossils are preserved in many different ways, varying from perfect preservation of an entire organism (as in Fig. 2.1) to mere impressions. Shells of animals; bones; plant structures, such as petrified tree trunks; impressions of plant leaves, soft worms, or jellyfish; and even tracks, burrows, and fecal material formed by animals qualify as fossils.

Some of the first ideas about fossils came from Greece. Large bones were interpreted as relics of a former race of heroic human giants, but seashells hundreds of feet above sea level and miles inland posed weightier questions. Had the sea once covered the land and then receded, or had these objects "grown" in the rock much as mineral crystals grow? Perhaps the animals had crawled into cracks in the rock and died, then were converted into stone. The Greeks supposed that fossils were much younger than the rocks in which they occurred.

Centuries later, some people thought that fossils grew in the rocks from seeds, and it was suggested that they grew from fish spawn caught in cracks in the rocks during the great Flood. Such ideas, which seem quaint to us today, were firmly rooted in the conceptions of nature accepted in their times. Viewed in historic context, these ideas made perfect sense.

Four major questions about fossils were posed by early speculations: (1) Are fossils really organic remains? (2) How did they get into the rocks? (3) When did they get there—while the rock was being formed or long after? (4) How did they become petrified? Let us examine each of these.

Da Vinci's Insight

The most important early record of a careful interpretation of fossils was that left by Italian genius Leonardo da Vinci, one of the most original natural philosophers of the early Renaissance. About A.D. 1500, he recognized that fossil shells in north Italy represented ancient marine life, even though they were found in strata exposed many miles from the nearest seashore. In opposition to the popular view that fossils had been washed in by the biblical Deluge, he argued that clams could not travel hundreds of kilometers in 40 days. Moreover, many of the shells certainly could not have been washed inland great distances because they were too fragile. He also pointed out that the fossils found in these ancient strata were still intact and resembled the living communities of organisms that he observed at the coast. Da Vinci further noted that there were many distinct layers that were fossil-rich separated by completely barren, unfossiliferous ones. This suggested to him that many events, such as seasonal river floods, were represented, rather than a single, worldwide deluge.

Steno's Principles

A century later, another resident of Italy, Nicholas Steno, sought a natural explanation of fossils. His writings, unlike da Vinci's, however, were widely circulated and had immediate impact. Steno concluded that fossils formed together with the rocks in which they occur—mostly slowly, bit by bit, layer by layer. Thus, strata must contain a decipherable chronological record of earth history, and in 1669 Steno stated three of the most basic principles for the analysis of that history.

First, Steno recognized the importance of horizontal layering, or **stratification,** the most conspicuous single property of sedimentary rocks (Fig. 2.2). Because particles settle from a fluid in proportion to their relative weight, the heavier ones first and so on, any change in size of particles causes layers to develop. Second, Steno appreciated the importance of dissolution and precipitation of chemically soluble sedimentary materials. In short, he recognized the tendency for uniformity of texture and composition in individual strata, and he inferred that the characteristics of different strata reflected changes of such conditions as temperature, wind, currents, and storms.

Finally, and most significantly, all sedimentary strata consisting of particles and shells could not be of the same age and have existed from the beginning of time but must have been deposited layer by layer, one on top of another. Therefore, in a sequence with many layers of stratified rocks, a given layer must be older than any overlying ones. We now call this the principle of superposition of strata. Together with two other principles, it forms the crux of interpretation of geologic history from the sedimentary rock record. Briefly stated, Steno's three principles are

Figure 2.2   Conspicuous stratification, the most characteristic feature of sedimentary rocks, interpreted fully by Nicholas Steno in the seventeenth century (sandstone and thin shales of late Paleozoic Itarare Formation, Brazil). *(Courtesy J. C. Crowell.)*

1. *Principle of superposition:* In any succession of strata not later deformed, the oldest stratum lies at the bottom, with successively younger ones above. This is the ultimate basis of relative ages of all strata *and* the fossils they contain.
2. *Principle of original horizontality:* Because sedimentary particles settle from fluids under gravitational influence, most stratification originally must be horizontal; steeply inclined strata, therefore, have suffered subsequent disturbance.
3. *Principle of original lateral continuity:* Strata originally extended in all directions until they thinned to zero or terminated against the edges of their original area, or **basin of deposition.**

The Grand Canyon illustrates all three principles (Fig. 2.3). First, the upper three-fourths of the canyon walls expose horizontal strata undisturbed since their deposition long ago. At the bottom of the canyon, however, one can see tilted strata, which, by the second principle, must have been originally horizontal. They are separated from the horizontal sequence by a discontinuity, or break in the rock sequence, called an **unconformity,** which here represents an ancient surface of erosion—a buried landscape. From the first principle, we see that the oldest stratum in either sequence must lie at the bottom. Furthermore, we conclude that the tilted sequence as a whole is older than the flat sequence, for

Figure 2.3   View in the Grand Canyon, Arizona, near Shinumo Creek illustrating Steno's principles of superposition, original horizontality, and original lateral continuity. Two discordant boundaries within the strata are indicated by arrows; these are called unconformities. The upper one separates older, tilted (late Proterozoic) strata from younger, flat ones (Paleozoic). The lower unconformity separates deeply eroded metamorphic rocks below (older Proterozoic) from tilted strata above. (© *John S. Shelton.*)

the former underlies the latter. Finally, the third principle is illustrated by the fact that each stratum exposed on the far wall has its counterpart on the nearer walls, and it requires little imagination to see that each was continuous across the present canyon area prior to cutting of the gorge by the Colorado River.

Although the original lateral continuity of strata is obvious in the Grand Canyon, in most areas it is less so, yet everywhere

that we see the eroded or broken edges of strata exposed, we know that originally they were more extensive laterally than they appear to be today. This fact is of paramount importance in trying to understand ancient conditions on earth from a much eroded and fragmentary stratified record.

Utility of Fossils

Fossils and Geologic Mapping

Although predicted around 1700 (see Box 2.1), proof of the utility of fossils for identifying specific strata and thus allowing the **correlation** of strata between widely scattered localities was slow in coming. It was tied closely to the first geologic maps. By the mid-eighteenth century, geologists had collected many fossils and had begun mapping earth materials. As early as 1723, English naturalist John Woodward suggested an identity, or a correlation, of certain strata on the European mainland with those in Britain on the basis of "great numbers of shells and other productions of the sea." A crude geologic map published in 1746 (Fig. 2.5), although primarily intended to show minerals, recorded many fossil localities and showed the continuity of chalk strata beneath the English Channel, as suggested by Woodward (Fig. 2.6).

Just before 1800, a surveyor named William Smith was surveying land in England. From visits to mines, he recognized a widespread regularity in the rock succession, which already had been demonstrated locally in coal mines (Fig. 2.7). Smith finally proved the enormous utility of fossils. In 1796, he wrote of the "wonderful order and regularity with which nature has disposed of these singular productions [fossils] and assigned to each its class and its peculiar Stratum." Culmination of his effort was preparation of the first geologic map of high quality; some sheets were completed before 1800.

Box 2.1

Robert Hooke and the Meaning of Fossils

At the same time that Steno's writings appeared, famous British scientist Robert Hooke also argued that fossils were truly organic, and he joined other challengers of the conventional wisdom that all had been emplaced by the single biblical Flood. Hooke studied and meticulously illustrated fossil shells (Fig. 2.4), making extensive use of the microscope, which had recently been invented. Most significant was Hooke's suggestion, published in 1703, that fossils might be useful for making chronological comparison of rocks of similar age, much as old Roman coins were used to date successive human historical events in Europe (see quotation at the beginning of the chapter). Hooke speculated that species had a fixed "life span," for many of the fossils he studied had no known living counterparts. This was one of the earliest hints of extinction of species, since most people assumed that all life was created about 6,000 years ago and was still living. But extinctions would be evidence of imperfections in the Creation, a concept that was difficult to accept in Hooke's day.

It was nearly a century before Hooke's idea was tested and proven valid. In Hooke's day, no one paid much attention to detailed study of individual strata. Until the late eighteenth century, natural scientists were preoccupied with grander problems. For example, combat still continued with the Deluge enthusiasts, known as diluvialists, who held such interesting ideas as the suggestion that the Flood had first dissolved all antediluvian matter, except fossils, and then reprecipitated the sediments in which the fossils became encased. But where were the remains of people killed by the Flood? This question was of sufficient moment that a Swiss naturalist, J. Scheuchzer, in 1709 excitedly interpreted a giant salamander skeleton as the fossil remains of a man drowned by the Deluge.

First things first—clearly science was not yet ready to *use* fossils! Slowly, doubts arose over the geologic importance of the Flood. Rather than being rejected outright, however, it became regarded as a brief corrective redesign late in the Creation.

Figure 2.4 Seventeenth-century illustrations of fossil ammonoids (*Cornua ammonis* or "snake-stones") prepared by Robert Hooke. *(From posthumous works of Robert Hooke (1635–1703), 1703.)*

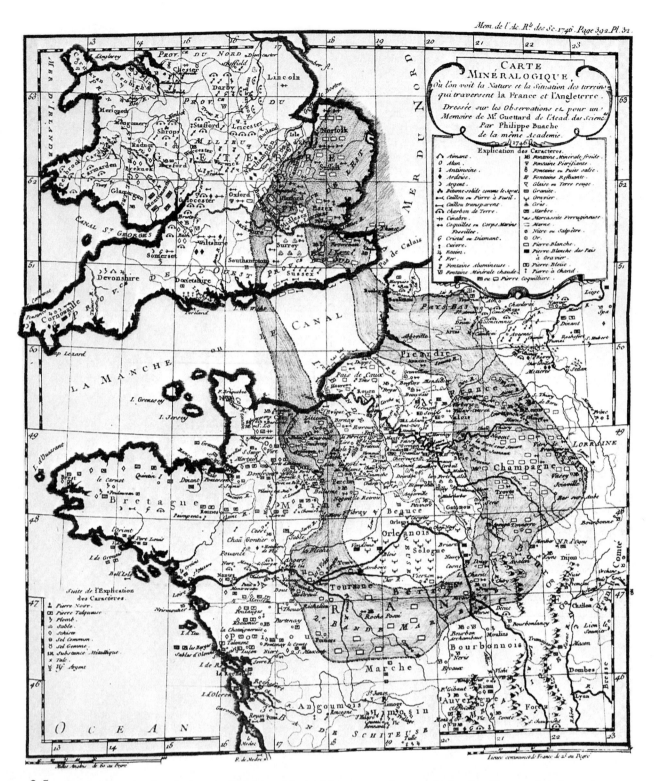

Figure 2.5   Guettard's early mineralogic map of France and England, forerunner of true geologic maps. Note symbols for rock types, ores, coal; shaded oval band represents the chalk-bearing strata. *(From Histoire de L'Academie Royal des Sciences, Paris, 1746; prepared with assistance from A. L. Lavoisier.)*

Because Smith was working in a region where natural rock exposures are rare, mines and excavations were of great importance in helping him form his ideas. He was engaged in canal excavations, which were being dug to facilitate the transport of manufactured goods during the early industrial revolution. Smith was able to trace and map different strata according to their color and mineral composition as well as their distinctive fossils. As Smith's mapping extended across England (Fig. 2.8), he could recognize most strata over long distances by their fossils, much as one might recognize bygone eras from the distinctive styles of old coins, bottles, or cans in the layers of different metropolitan rubbish heaps.

Smith's map and its accompanying table of strata, completed in 1815, were a cartographic masterpiece. His prowess as a careful observer is exemplified by his distinction of fossils still in their original place from badly abraded ones that would be less reliable for tracing individual strata because they had been transported and mixed after death. The map immediately proved useful, not only for planning canals, quarries, and mines but also as a guide to soils. Its author long since had recognized that soil develops from underlying bedrock and, therefore, reflects some characteristics of the rock. Smith gradually was able to establish a detailed sequence of strata, which then enabled him to predict distribution of different rock bodies in new, unmapped country.

Meanwhile, in the Paris region, the usefulness of fossils in geologic mapping was discovered by Georges Cuvier and Alexandre Brongniart, two scientists working together and apparently unaware of Smith's work (Fig. 2.8). Like Smith, they also developed an essentially modern geologic map (published 1811), showing the distribution of various rock divisions based largely upon the similarity of fossils found in particular strata. Cuvier wrote:

> These fossils are generally the same in corresponding beds, and present tolerably marked differences of species from one group of beds to another. It is a method of recognition which up to the present has never deceived us.

The simultaneous but independent discovery by Smith in England and Cuvier and Brongniart in France that strata could be distinguished by their fossils was a "scientific breakthrough." After this giant step was taken around 1810, the use of fossils

Figure 2.6   The famous chalk deposits of the White Cliffs of Dover, formed from the skeletons of trillions of microscopic algae called coccolithophorids (see Fig. 14.41) which bloomed in the shallow seas of the Cretaceous. These chalk deposits crop out all over southeastern England and northern France (see Fig. 2.5). *(D. R. Prothero.)*

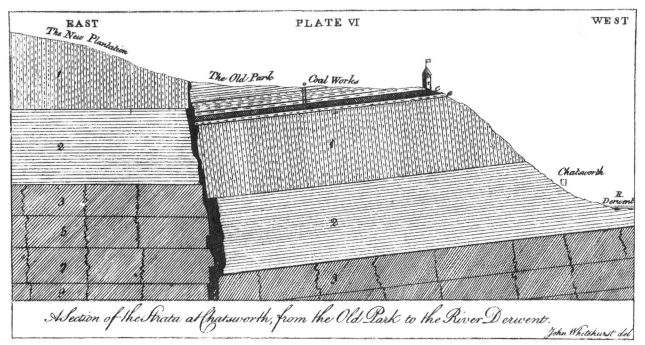

Figure 2.7   An early cross section showing the stratigraphic sequence and faulting in a coal field near Newcastle, England. The author had a quaint explanation of the tilting and faulting shown: subterranean heat caused expansion and cracking of the earth; seawater then leaked downward to be converted to steam, resulting in sudden convulsions of the crust, climaxed by the biblical Deluge. *(From John Whitehurst, 1778, An inquiry into the original state and formation of the earth deduced from facts and the laws of nature. Printed by J. Cooper, 1778.)*

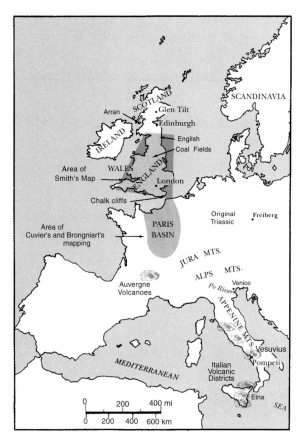

Figure 2.8 Index map of Europe showing areas of the Smith and Cuvier-Brongniart geologic maps, as well as some localities referred to in this chapter and Chap. 4.

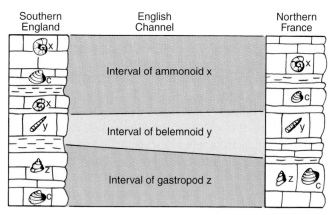

Figure 2.9 The use of fossils for correlation across the English Channel of strata studied by Smith in England and Cuvier and Brongniart in France. Fossils x, y, and z are useful index fossils, but clam c is not, because it ranges through the entire sequence.

was accepted quickly. The subdivision, tracing, and mapping of European strata went forth at a great pace with stimulation from the industrial revolution.

Principle of Fossil Correlation and Index Fossils

The first geologic maps were based largely upon the ability to compare and trace individual strata using their unique fossils to help identify them. In other words, mapmakers correlated strata containing similar fossils. Paleontologic, or fossil, correlation involves the most important new principle developed after those of Steno—the *principle of fossil correlation*—which states that like assemblages of fossils are of like age, and therefore the strata containing them are also of like age (Fig. 2.9). Implicit here is the extension of Steno's principle of superposition to fossils such that, in a succession of strata containing fossils, obviously the lowest fossils are older than those occurring in overlying strata.

Application of the principle of fossil correlation involves what we now call **index fossils.** An index fossil is one that is particularly useful for correlation of strata; that is, it is an index, or a label, for a particular stratum or group of strata. For a fossil to be

considered a stratigraphic index, it must be (1) easily recognized (i.e., unique), (2) widespread in occurrence, and (3) restricted to a very limited thickness of strata. To have utility, fossils must meet all these conditions, but few fossils meet all three criteria. Many are difficult to distinguish from others, some are not widespread, and many were too long-lived (that is, they occur through great thicknesses of strata without change, as does fossil c, in Fig. 2.9).

What makes correlation with index fossils work? Neither Smith nor Cuvier formally set down all the logic in Hooke's 1703 suggestion, which they proved, but today we can do this in retrospect. The utility of an index fossil requires these conditions: *first,* that once a species became extinct it never reappeared; and *second,* that no two species are identical. The method, as Cuvier observed, "up to the present has never deceived us." And this is still true after nearly 200 years of additional experience.

Explanations of Change Among Fossils

Catastrophism

Before mapping with Brongniart, Cuvier had compared fossil bones with the skeletons of modern elephants and showed that archaic elephants once roamed the Paris region. Apparently some past inhabitants of the earth had vanished completely—Cuvier had proven the extinction of former species! Thus, it was no surprise to him that the fossils in the older strata around Paris showed "differences of species from one group of strata to another." Moreover, several unconformities punctuate the rock sequence there, and these must reflect profound changes. As you can see in Fig. 2.10, there are alternations of sandstones, shales, and limestones. In addition, there are several zones of ancient soil with fragments of plants in them as well as several breaks in deposition marked by concentrations of wood, shell fragments, and small pebbles. The record of sedimentation is not a smooth, continuous one but tells a story of interruptions

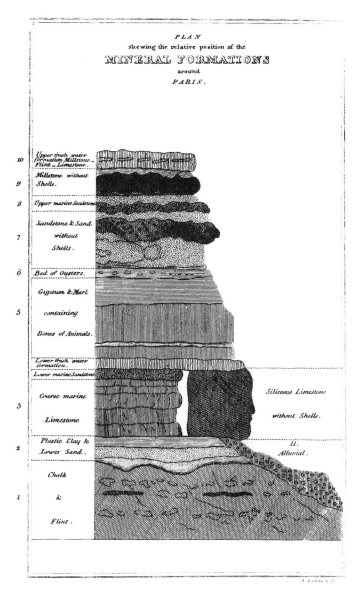

PLAN
Shewing the relative position of the
MINERAL FORMATIONS
around
PARIS.

10 Upper fresh water formation Millstone Flint .. Limestone.
9 Millstone without Shells.
8 Upper marine Sandstone
7 Sandstone & Sand without Shells.
6 Bed of Oysters.
5 Gypsum & Marl containing Bones of Animals.
   Lower fresh water formation.
   Lower marine Sandstone.
3 Coarse marine Limestone
2 Plastic Clay & Lower Sand.
1 Chalk & Flint.

Siliceous Limestone without Shells.

11.
Alluvial.

Figure 2.10   Stratigraphic columnar section of strata studied by Cuvier and Brongniart around Paris. Several unconformities are present—for example, between numbers 1 and 2, 6 and 7, and 8 and 9—representing breaks between marine and nonmarine conditions. It is interesting to note that, in the translation of this work from the French, some English editor added what is obviously Napoleon Bonaparte's profile to the cliff for unit 3. *(From Georges Cuvier (1769-1832), 1818,* Essay on the Theory of the earth; *English Translation.)*

of deposition accompanied by abrupt changes of fossil assemblages. Cuvier assumed that portions of assemblages in successive groups of strata had become extinct before succeeding strata were deposited. He thought there were many wholesale catastrophic extinctions of millions of organisms caused by violent oscillations of the sea, hence the name *catastrophism*. Cuvier noted that fossils in successively younger strata were more like modern organisms, and so successive extinctions must have eliminated many now unknown

species. With each catastrophe, life moved progressively toward the ultimate "perfection" of the modern scene.

Cuvier's hypothesis seemed a reasonable and complete explanation of fossils based upon both local evidence and then popular assumptions about earth change as discussed in Chap. 1. It is significant that he did not feel constrained by a single biblical Deluge, however, but invoked as many floods as the rock evidence seemed to require. Cuvier's scientific stature gave much impetus to the philosophy of catastrophism, but that view was soon to undergo important changes and to be challenged by alternatives. Whereas Cuvier had implied that all organisms that ever lived were created early and then subjected to selective extinctions, another Frenchman, Alcide D'Orbigny, postulated about 1859 that new species were created following each of twenty-seven catastrophes. It was the combined effects of successive partial extinctions followed by new creations that led ultimately to the life forms of today, according to this modified theory, known as *catastrophism and special creations*. It said, in effect, that there was little or no connection between the index fossils of successive groups of strata.

Descent by Evolution

The alternative to a catastrophic explanation of the fossil record is that there *was* a connection among the different fossils in successive strata. In other words, younger organisms were the descendants of older ones, and species had changed somehow through time. Some lines of descent ceased through extinction, but others survived the catastrophes described by Cuvier and D'Orbigny. Even before Cuvier, several observers had noted a general tendency for younger fossils to resemble living organisms more and more. For example, Frenchman Giraud-Soulavie as early as 1780 subdivided strata into five "ages," based upon the gross characteristics of their fossil assemblages. Rocks of the "first age" included no forms analogous to living ones, the "second" included both extinct and living marine forms, the "third epoch" had shells of modern types only, the "fourth" was plant-bearing, and the "fifth" was distinguished by mammal remains. This was one of the first attempts to subdivide the stratigraphic rock record into broad divisions based upon changes in the development of life. It is a precursor to a geologic timescale and implies an irreversible trend of development. Change through time via a linked, unidirectional series of events is called *evolutionary change*. What happened yesterday influenced what happens today, but not vice versa.

Organic evolution, the alternative to Cuvier's and D'Orbigny's catastrophism, was not new in their time. The general concept that species somehow changed or evolved to other species had been suggested at least 100 years earlier, and two of Cuvier's own colleagues were ardent evolutionists who continually feuded with him. Apparently no urgent need was felt for such a theory, however, until more factual evidence had accumulated and the intellectual climate had begun to change around 1800 (see Box 2.2). We shall discuss evolution extensively in Chap. 3.

Box 2.2

Mark Twain Winks at Paleontology

Speculation and feuding about organic evolution and extinctions began with the earliest suggestions nearly 200 years ago, and they continue today. The prominence of this controversy in the late nineteenth century is revealed by Mark Twain's use of fossils and evolution as a vehicle for editorializing with his characteristically acidic pen on the arrogance of the human species:

> It was foreseen that man would have to have the oyster. . . . Very well you cannot make an oyster out of whole cloth, you must make the oyster's ancestor first. This is not done in a day. You must make a vast variety of invertebrates to start with—belemnites, trilobites, Jebusites, Amalekites, and that sort of fry, and put them to soak in a primary sea, and wait and see what will happen. Some will be a disappointment—the belemnites; the Ammonites, and such; they will be failures, they will die out and become extinct, in the course of the nineteen million years covered by the experiment, but all is not lost, for the Amalekites will fetch the homestake; they will develop gradually into encrinites, and stalictites, and blatherskites, and one thing and another as the mighty ages creep on . . . the Archaean and the Cambrian Periods . . . and at last the first grand stage in the preparation of the world for man stands completed, the oyster is done. An oyster has hardly any more reasoning power than a scientist has; and so it is reasonably certain that this one jumped to the conclusion that the nineteen million years was a preparation for *him*; but that would be just like an oyster. (From the essay *Was the World Made for Man?*)

First Unified Hypotheses of the Earth

The Cosmogonists

Modern science began in the seventeenth century partly as an endeavor to understand God through His work, Nature, as well as through His Word, the Bible. The discoveries and explanations of the solar system by Galileo with his telescope and by Newton with his mathematics represent the roots of this new, systematic approach to the eternal human search for understanding. Edmund Halley's application of Newton's principles of gravitation to calculate the regularity of passage of comets helped to confirm the conviction that nature could be understood by rigorous analysis. In 1682, Halley predicted (correctly) the return in 1758 of the comet later named for him. Moreover, he offered a scientific explanation of a biblical story by suggesting that a collision of some other comet with the earth might have caused Noah's Flood.

Long before the value of index fossils for correlation had been demonstrated by Smith and Cuvier, a few people already had offered theories of the earth that were no longer strictly tied to Scriptures but drew, instead, upon evidence from nature and from the emerging principles of Newton and others. Several Renaissance thinkers, such as Descartes and Leibniz, attempted to formulate all-inclusive hypotheses, called *cosmogonies* to explain the origin and development of earth, life, and the entire universe. Though riddled with speculation, these early explanations had a long-lasting influence. The cosmogonists assumed: (1) that the earth originally was hot and glowed like the sun; (2) that, as the earth cooled, a primitive, hard crust formed; (3) that the water and atmosphere became segregated according to relative densities; and (4) that the interior had a fiery core surrounded concentrically by solid material. The concept of global cooling seemed to be supported by occasional volcanic eruptions from the interior, the increase of temperature in deep mines, and the discovery of tropical animal and plant fossils in the Arctic.

Buffon's Break with Genesis

One of the most creative eighteenth-century thinkers in natural science was Frenchman G. L. de Buffon, who published, beginning in 1749, a thirty-four-volume work entitled *Histoire Naturelle*. Accepting the idea of a molten origin, Buffon suggested that planets originated through detachment of hot portions of the sun by collision or near-collision with a large comet, an idea doubtless derived from the great interest in comets stimulated by Halley's studies of their periods see (Fig. 6.2). Subsequent earth history, he postulated, consisted of cooling through six distinct epochs totaling 75,000 years. This estimate was calculated from the results of simple experiments with heated steel spheres, which he thought could provide an analogy for the earth's cooling rate. Buffon was one of the first to question the literal significance of the Six Days of Creation and to postulate a great age for the earth. "A year is to God as a thousand years to man." Buffon represented a major turning point in the history of geologic thought by his rejection of Scripture as a literal source of geologic insight and by his extension of the supposed age of the earth.

The First True Geologic Chronology

Ideas commonly have arisen independently in different minds thousands of miles apart. Such apparently was the case with the first concepts of natural subdivisions within the rocks of the earth's crust. Whereas Buffon's epochs constituted a speculative *chronology* (succession of events), several contemporaries working in Germany, Scandinavia, Italy, Switzerland, and England were recognizing and naming natural local rock divisions. In the English coal fields as early as 1719, a succession of strata was recognized and illustrated in detailed cross sections. In the middle eighteenth century, German J. Lehmann distinguished stratified, gently tilted fossiliferous deposits from more primitive, unfossiliferous, steeply tilted ones cut by dikes and ore veins (material congealed from fluids that were injected along fractures). He believed that all had formed in the sea but that primitive rocks

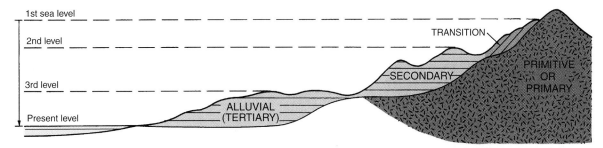

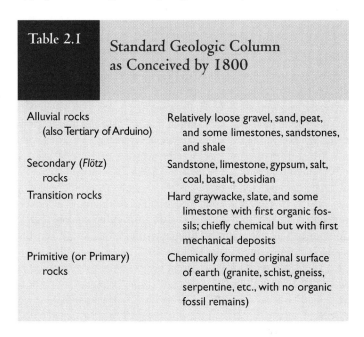

Figure 2.11   Early neptunian concept of deposition of strata by a receding, universal ocean. Note that, as drawn here from early descriptions of the neptunian theory, positions of the strata seem to defy superposition; that is, successively younger ones do not everywhere directly overlie older. Later neptunians, such as A. G. Werner, envisioned the older strata as dipping and flattening beneath younger ones away from the mountains.

were deposited in disorder during the early Creation, whereas the fossiliferous strata were formed during the Deluge. A third group of deposits postdated the Deluge and were formed by accidents of nature such as earthquakes, landslides, and volcanic eruptions. He assumed that these divisions matched the epochs of Genesis.

In 1759, an Italian authority on mining and mineralogy, Giovanni Arduino, formally distinguished the following rocks in northern Italy: Primitive Mountains with unfossiliferous schists—altered sedimentary rocks—and veins of the high Alpine mountain core; Secondary Mountains with limestone and shale containing abundant fossils; Tertiary Mountains with richly fossiliferous clay, sand, and limestone of the low hills; and finally the youngest Volcanic Rocks. This scheme, which clearly was influenced by Lehmann's, became the basis of the first formal stratigraphic standard for subdivision of the rocks of the earth's upper crust. The universality of historical development of the entire earth's surface long had been assumed, and it already was recognized that fossiliferous marine strata similar to those of Italy were widespread over much of Europe. Therefore, it was natural for Arduino's Primitive, Secondary, Tertiary, and Volcanic to be applied throughout Europe and even America.

Neptunism

A. G. Werner

In 1787, one of the most influential men in the history of geology, Abraham Gottlob Werner, published a general theory of the origin of the earth's crust. Werner's predecessors had postulated a universal ocean, in which most known rocks were formed. Since Viking times, it had been known that the water of the northern Baltic Sea was retreating, for new rocks kept appearing around its shores; it was natural to suppose this reflected a worldwide change. All mountains were assumed to be original irregularities of a chaotic early Creation, and the succession of strata was presumed to reflect subsequent gradual shrinking of the sea and emergence of land. Because of the great emphasis upon the sea for explaining all the crust, this theory earned the apt nickname **neptunism** (for Neptune, Roman god of the sea) (Fig. 2.11).

Werner adopted the early neptunian concept of the earth and amplified it through a more detailed accounting of the chronological succession of strata. But Werner was not a catastrophist. He regarded the earth as much older than humanity and did not bother

Table 2.1	Standard Geologic Column as Conceived by 1800
Alluvial rocks (also Tertiary of Arduino)	Relatively loose gravel, sand, peat, and some limestones, sandstones, and shale
Secondary (*Flötz*) rocks	Sandstone, limestone, gypsum, salt, coal, basalt, obsidian
Transition rocks	Hard graywacke, slate, and some limestone with first organic fossils; chiefly chemical but with first mechanical deposits
Primitive (or Primary) rocks	Chemically formed original surface of earth (granite, schist, gneiss, serpentine, etc., with no organic fossil remains)

correlating stratigraphic history with Scripture. Werner regarded Primitive rocks, such as granite and schist (altered or metamorphosed sediments of diverse kinds) as unfossiliferous and entirely of chemical origin (Table 2.1). Because they appear in the high cores of mountain ranges, they must be the earliest rocks precipitated from the sea. As the water subsided, fossiliferous Transition rocks were deposited. These Transition rocks included the first mechanically formed (i.e., fragmental) impure sandstones, called graywackes, as well as some chemical limestone. Werner attributed the inclination of these strata (Fig. 2.11) to initial deposition on the irregular Primitive crustal surface or to collapse into subterranean caverns. Lower topographically were flat-lying rocks, later classified as Secondary. These comprised sandstones, limestones, gypsum, rock salt, coal, and basalt. Finally, in the lowlands and valley bottoms, the youngest, or Alluvial, deposits were formed as the sea receded to its present position; they are nonmarine. (Arduino's Tertiary and Volcanic were not recognized by Werner as separate divisions.) Werner's grouping of rocks by supposed relative age—Primitive, Transition, Secondary, Alluvial—soon became almost universally accepted as a standard for worldwide comparison. He had succeeded in mobilizing, at last, a conscious scientific desire to study the history of the earth.

View of the Chaine des Puys near Clermont seen obliquely from the Summit of the Puy de la Rodde.

Figure 2.12   Auvergne volcanoes, south-central France. Note the perfect volcanic cones and humocky-surfaced lava flows extending out from two of the craters near the center. Fig. 2.8 shows the location of this area. *(From G. P. Scrope, 1825,* Considerations on Volcanoes. London, W. Phillips, 270 p.*)*

Heated Basalt Controversy

The neptunian scheme suffered from several obvious shortcomings, such as a gigantic water-disposal problem. Wernerians ignored many of these, even while proudly announcing that they rejected hypothesis and speculation and had built their entire case upon irrefutable facts, a pious claim pressed with equal vigor by the opposition.

Most difficult and controversial was the origin of basalt, a dark, fine-grained rock crystallized from hot, molten lava. To Werner, volcanoes all were recent and had no great importance in the history of the earth. This was in spite of dramatic evidence marshalled by the previous generation of antineptunian Italian geologists, who thought that all land was formed initially by volcanic eruptions such as they themselves had seen build new islands in the Mediterranean. Werner, instead, endorsed an old idea that volcanic eruptions, when they do occur, originate from combustion of buried coal seams. Therefore, volcanoes could have occurred only after deposition of the great Secondary coal layers. He contended stubbornly to his last day that basalt found interstratified with sediments had been precipitated from the universal ocean. Where it looked like lava, it had been fused by combustion of adjacent coals.

It may seem astounding that most geologists could seriously entertain the neptunian origin of basalt, but it is even more surprising to realize that two Frenchmen had published volcanic interpretations of basalts in central France *before* Werner first published his theory (Fig. 2.12)! Together with the Italians, those men established the volcanic origin of basalt, to which three of Werner's own students soon were converted. Subsequent discovery of volcanic rocks interstratified with Secondary and Primitive deposits proved the vulcanists' claim that volcanism had occurred throughout earth history.

Plutonism—the Beginning of Modern Geology

James Hutton

In the last quarter of the eighteenth century, geology finally was coming of age, and the name *geology* itself came into general use. A great Scottish innovator, James Hutton, was a gentleman farmer and geologist. Farming generated an interest in the earth, and Hutton soon became obsessed with discovering an explanation of how the environment for land plants and animals could be maintained in the face of obvious destructive forces acting to wear away the landscape. He announced his revolutionary theory in 1785, when Werner's influence was at its peak. Eventually Hutton's theory was to provide the foundation of modern geology.

An Old Dynamic Earth

Hutton early recognized that rocks exposed to the atmosphere tend to decay and produce gravel and soil. He noted by analogy that many strata contain debris derived from older rocks, which apparently had decayed in the past. Moreover, he appreciated that sediments just like ancient sedimentary rocks were still forming, chiefly in the sea. These observations led him to a cyclic view of earth change, in which construction of new products neatly balances destruction of old. He saw that the earth could renovate itself. Here, at last, was an optimistic theory of continual rejuvenation to replace pessimistic medieval dogmas of doom and decay. Hutton's balance between destruction and rejuvenation represents, in modern terms, a tendency for equilibrium between competing processes. Metaphorical notions of equilibrium were common in Hutton's day, too. For example, the industrial revolution stimulated a popular comparison of nature to a gigantic machine set in motion by some external intelligence. Hutton then saw the earth as a great heat engine, with volcanoes as its safety valves. He also envisioned the earth as a cyclical machine, like a clock whose pendulum swings to and fro; thus, mountains are raised and then eroded, only to be raised and eroded again and again.

Hutton's greatest contribution was the original full appreciation that the earth is internally dynamic and ever changing. This was in sharp opposition to the Wernerian view that rocks had formed on a static, solid earth foundation. This was really the central conflict with neptunism! To Hutton, it was absurd that steeply inclined Transition and Secondary strata in the mountains originally had been deposited in such positions. He reasoned instead that these strata had been formed horizontally and then tilted and crumpled later by internal earth forces.

Hutton also attacked catastrophism. He recognized that valleys were cut by rivers rather than by the sloshings of great oceanic floods. He objected to the belief that "everything must be deluged" and believed instead that, given enough time, modern earth processes were capable of having produced the record of the past.

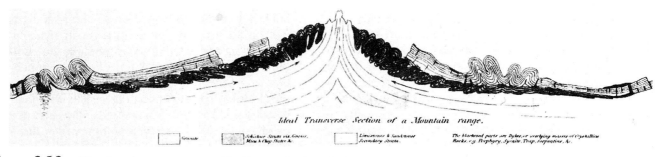

Ideal Transverse Section of a Mountain range.

Figure 2.13   The plutonist concept initiated by Hutton of upheaval of mountains and folding of strata as results of internal heat. Note the discordance between the upper (Secondary) strata and the more contorted schists (dark). *(From G. P. Scrope, 1825, Considerations on Volcanoes. London, W. Phillips, 270 p.)*

Hutton saw that, if we assume that the earth is only 6,000 years old, we *must* subscribe to catastrophic earth change to explain geologic facts, as we noted in Chap. 1. The fresh condition of a 2,000-year-old Roman wall in northern England, however, convinced him that ordinary geologic processes—by human standards—act slowly. Because rocks seemed to reveal only the results of ordinary processes such as those visible and acting at the present, a great deal more time must have been required for the almost imperceptible changes to have accomplished great geologic work. "What more can we require?" said Hutton. "Nothing but time."

"Nothing in the Strict Sense Primitive"

Hutton's first quarrel with neptunians was over the latter's different chronological rock the divisions, especially the Primitive. A uniqueness for any of divisions was counter to Hutton's own observations as well as to his concept of an earth in dynamic equilibrium and acted upon by the same—that is, uniform—processes through all time. The presence of pebbles and sand within Scottish schists otherwise regarded as Primitive attested to the existence of still older rocks somewhere from which those fragments had been eroded. So-called Primitive rocks could not be entirely of chemical origin, as the neptunians claimed; neither were they unique in lacking fossils. Seemingly there was "no vestige of a beginning, no prospect of an end," wrote Hutton, and he could recognize "nothing [that was] in the strict sense primitive."

Basalt

Hutton was impressed early with the importance of subterranean heat, whose existence he inferred from hot springs and volcanoes. He interpreted baked coal seams as having been heated by basaltic dikes that cut across them. In a well-exposed basalt sill interlayered with sediments, he found clear evidence both of the basalt's hot origin and of forcible intrusion after the sediments had formed. The finer-grained margins of the basalt suggested chilling by the enclosing strata, which, in turn, showed evidence of baking next to the basalt.

Hutton's igneous interpretations of basalt and natural rock glass were reinforced experimentally in 1792 by a friend, Sir James Hall, who melted basalt between 800° and 1200°C, only to produce glass with rapid cooling but basalt again with slow cooling. This represents one of the first applications in geology of experimentation, which is one of the most powerful techniques of scientific investigation.

Granite and Mineral Veins

Hutton knew that mineral veins occur in areas of Primitive granites and schists. (Granite is a coarsely crystalline, light-colored igneous rock that crystallized slowly below the earth's surface.) In 1764, he observed an unlayered mass of granite surrounded by layered schists in northern Scotland and thought that the granite, as well as the complex minerals in associated metallic veins, could have formed only through crystallization from hot fluid material. Sea water could not hold all the different compounds simultaneously in solution; therefore, veins and dikes were not simple oceanic precipitates in open cracks, as the neptunians preached. Influenced by the work of a prominent chemist friend, Joseph Black, Hutton shrewdly inferred that great pressure deep in the earth affects chemical reactions; compounds that are gaseous at high temperature and low pressure, such as those expelled to the atmosphere from volcanoes, would not be lost as gases from hot solutions under great pressure at depth. Moreover, there could be intense heat without fire at such depths due to the pressure. Hutton's interpretation of a hot origin for granite eventually displaced the neptunian view. Great appeal to subterranean heat to produce basalt, granite, and mineral veins earned for his theory the nickname **plutonism** (for Pluto, god of the lower, infernal world).

Evidence of Upheaval of Mountains

Hutton proposed that heat also consolidated sediments and then caused upheaval of rocks to form mountains by thermal expansion of the crust (Fig. 2.13). Granites, veins, and basalts were assumed to have formed at such times. He deduced that there must be dikes of granite cutting across, or intruding, older rocks. Having predicted their ultimate discovery, he scoured Scotland for examples and found many such dikes transecting schists in central and western Scotland. His elation could not have been greater had they been veins of solid gold.

Hutton also discovered what he had predicted and long sought: a great unconformity in the stratified rock record (Fig. 2.14). He suspected that the contact between Secondary and older rocks might reveal clear evidence of great upheaval of mountains from the ancient sea floor. In all three exposures that he discovered, pebbles worn from older rocks occurred just above the unconformity. In Hutton's examples, the dip of strata below the unconformity was steep, and from Steno's principle of original horizontality one concludes that severe upheaval had tilted the rocks and formed mountains, which

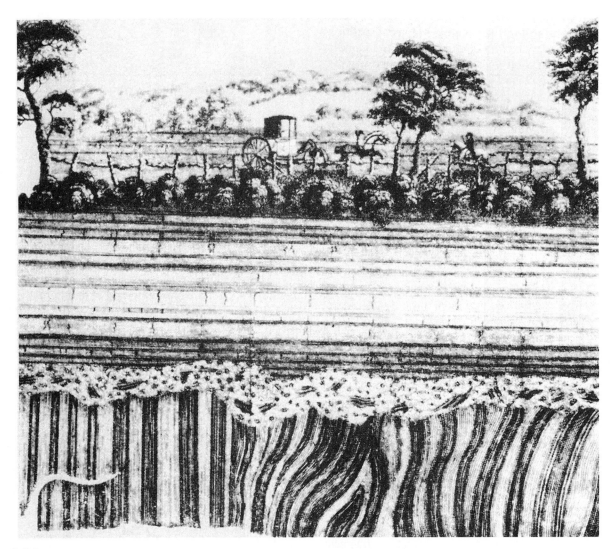

Figure 2.14   Hutton's unconformity with basal "puddingstone," or conglomerate, along the River Jed, south of Edinburgh, Scotland, the second such example discovered by James Hutton in 1787. *(From* Theory of the earth, *1795.)*

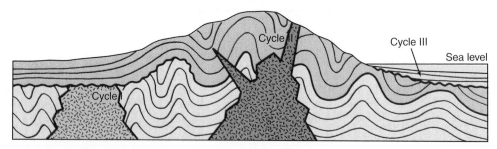

Figure 2.15   Diagrammatic representation of Hutton's concept of cyclic deposition, upheavals, intrusion of granites, and development of unconformities (wavy lines) to produce a continuous progression of new landscapes born from the wastes of old. *Cycle I* strata were folded, intruded by granite, and then eroded. *Cycle II* strata were then deposited and in turn upheaved, folded, and intruded by granite. Erosion is now reducing the *Cycle II* mountains, resulting in deposition of *Cycle III* sediments.

were then worn away; their remnants were later covered by younger, nearly flat strata.

The unconformity, together with intrusive granite dikes, proved conclusively that mountains are neither static bumps left from a primeval earth surface nor relics of the Deluge but were formed by repeated dynamic convulsions of the crust (Fig. 2.15). Thus, through observation, prediction, and verification, Hutton had marshalled compelling support for his theory of continual earth rejuvenation through repeated cycles of erosion and deposition followed by upheavals.

Charles Lyell's Uniformity of Nature

Despite the Huttonian revolution, most people still endorsed neptunism in the early 1800s. For example, a Cambridge University professor told his class, "There is enough water in the crystals in my museum to extinguish all the fires of the plutonists." Such water, which formed vapor bubbles as the minerals crystallized, was proof enough for the good professor that basalt had precipitated from seawater. Moreover, Hutton was a poor writer and received bad reviews because of his anticatastrophist preachings (Fig. 2.16). His contention of great antiquity of the earth, expressed in the subtitle to this chapter, was like a red flag to a bull. In the eyes of many educated people of the early 1800s, no science was valid if it did not support and clarify Scriptures, which Hutton had refrained from doing, even though he believed in grand design or wisdom in nature.

The eventual flowering of the new geology was due in large part to British geologist Charles Lyell's great book entitled *Principles of Geology,* which was revised twelve times between 1830 and 1875. Lyell, who was born the same year that Hutton died, painstakingly illustrated the concept of uniformity of natural processes through time and the corollary necessity for a great antiquity of the earth. He travelled extensively in Europe and North America and was able to show that geologic processes observed today can be assumed to have operated throughout the past (Fig. 2.17). On Sicily in 1828, Lyell found Mt. Etna to be especially instructive. People have been living with that huge, restless volcano for thousands of years, so its eruptive history is exceptionally well documented in human writings going back at least to 425 B.C. Moreover, Lyell found that ancient river gravels interstratified with some of Etna's prehistoric lavas contained bones of several extinct mammal species, including extinct mammoths, and that the oldest lavas are interspersed with marine limestones containing a fossil assemblage containing 10 percent extinct species. To Lyell, these facts, together with the enormous volume of volcanic rock, implied a considerable antiquity for Etna. Moreover, the limestones beneath the mountain are still relatively young geologically, for their index fossils, most of whose species are still living, represent only the later Tertiary division. Truly, the earth must be very old indeed!

Lyell adopted a more extreme view of uniformity than had Hutton. He probably did so as a strategy to provide the most powerful possible antidote for catastrophism because he felt a sense

Figure 2.16 Caricature drawn in 1787 of "... the eminent old geologist, Dr. James Hutton, rather astonished at the shapes which his favorite rocks have suddenly taken"—the faces of several antagonists. *(From John Kay, 1842, Edinburgh portraits.)*

Figure 2.17 One of many noncatastrophic, historic geologic changes documented by Charles Lyell. The ancient village of Eccles in Norfolk, England, was beginning to be attacked by the North Sea around 1600. By 1839, coastal dunes had buried all but the top of the tower (left), but by 1862 wave and wind attack had exposed the ruins again (right). *(From Charles Lyell, 1867, Principles of geology; 10th ed., v. 1, pp. 514–515.)*

Box 2.3

Uniformitarianism in the Public Eye

The Kelvin-Lyell controversy spanned nearly half a century and attracted much attention. Mark Twain presented a characteristic spoof of Lyell's uniformitarianism in *Life on the Mississippi*, but consider the following, written by prominent British geologist W. J. Sollas in 1877:

> One is told that a certain boy grows one-tenth of an inch in a year, and that he is four feet high. The geologist then argues ten years to grow one inch, 480 years for four feet, therefore the boy is 480 years old. A person who knows more about the boy in question comes forward to inform us that when he was younger, he grew much faster, and that, after all, his age is only ten years. The geologist replies, it may be so. Let me see, one-tenth of an inch in a year, on the doctrine of uniformity, one inch in 10 years. No, you are clearly mistaken; 10 years is far too short a time. The organic processes of growth could not possibly have occurred in this time. I grant you my own calculation is not rigorously exact; let us say then our boy is but 400 years of age. (On evolution in geology: *The Geologic Magazine,* v. 4, pp. 1–7.)

of mission for "freeing the science from Moses." What he was reacting to so strongly is exemplified by the following passage published by Cuvier as late as 1831:

> The thread of the operations is broken, the march of nature is changed; and not one of her agents now at work would have sufficed to have effected her ancient works. *(Discourse on the Revolution of the Globe)*

Of course, Lyell embraced the uniformity of known causes or processes throughout time, which Hutton had preached so vigorously. Actually that view was implied in many earlier writings, and some catastrophists held it as well. To those catastrophists, it was not the *kinds* of processes that had been so different at past times but only their *intensity*. Lyell's extreme brand of uniformity, which was nicknamed **uniformitarianism** after his book appeared, incorporated not only Hutton's limited *uniformity of causes* but also a *uniformity of intensity* of changes as well. This second, peculiarly Lyellian brand of ultrauniformity, or **gradualism,** amounted to a delicately balanced, or equilibrium, view of the entire earth! According to Lyell, some violent events, such as earthquakes, do occur, *but no more frequently in one epoch than in another.* If one continent was large, another simultaneously was small; if the climate here was cold, it was balanced there by being hot. By such rationalizing, he envisioned that conditions remained essentially constant through time *for the earth as a whole.*

Lyell worried specifically about Hutton's cycles of erosion followed by upheaval of mountains, which seemed too spasmodic, so he suggested that the "energy of subterranean movements has been almost uniform as regards the whole earth." The sites of mountains may shift, but an overall balance of mountain making is maintained. Apparently, for Lyell to acknowledge any irregular variation of intensity was to leave the door dangerously ajar for catastrophism; therefore, only uniformly repeating changes confined within a very narrow range of variation of rate and intensity were permissible. Although the earth had been constantly changing throughout geologic time, overall it had remained in beautiful equilibrium.

Evolution—Lyell's Ultimate Challenge

In addition to successfully selling the new geology, Lyell's writings also had great impact upon Victorian writers and theologians. John Ruskin, for example, wrote in 1851, "I could get along very well if it were not for those geologists. I hear the clink of their hammers at the end of every bible verse." Also of great significance was Lyell's influence upon his younger friend, Charles Darwin. Ironically, however, Lyell himself had great difficulty in abandoning a fixed-species concept in favor of Darwin's theory of organic evolution (first published in 1859). To accept change of one species into another in an irreversible line of descent seemed to him contradictory to his own unbending devotion to a uniformly repeating earth. Evolution, which has direction, and Lyell's strict steady state, which does not, were incompatible concepts.

It must have been a rude shock for Lyell that his most vigorous challenge came not from the catastrophists but from Britain's most distinguished physicist. Sir William Thomson, the Lord Kelvin, saw Lyell's strict uniformitarianism as perpetual motion—an earth machine that never ran down was a physical absurdity! Kelvin reasoned that the energy reservoir of the entire solar system must have been greater in the past and was gradually being dissipated. His position implied significant differences in past conditions and, thus, a nonrepeating, or evolutionary earth, rather than a perfectly cyclical one. Equilibrium may be achieved temporarily but cannot persist over geologically significant spans of time (Box 2.3).

As Lyell and Kelvin squabbled, geologists became increasingly aware that the intensity of geologic processes had, indeed, varied greatly through time. Moreover, in support of Kelvin, it became clear from the rock record that conditions on earth had changed irreversibly, so that in certain respects different epochs of geologic time were unique—history was not as strictly uniform as Lyell had contended. Examples now known of important nonrepeating events include the change from an early oxygen-poor

Figure 2.18 Many geologists were skeptical of Lyell's cyclic concept of geologic history. In one passage, Lyell (1830) suggested that the great extinct reptiles of the Mesozoic might someday return to rule the earth. British geologist Henry de la Beche satirized Lyell's extreme uniformitarianism in this cartoon of Professor Ichthyosaurus lecturing to his class about a strange creature from a previous creation. In the caption, the professor says, "You will at once perceive that the skull before us belonged to some of the lower order of animals. The teeth are very insignificant, the power of the jaws trifling, and altogether it seems wonderful how the creature could have procured food." The skull is, of course, that of a human.
(From Charles Lyell, 1867, Principles of geology; 10th ed., v. 1, pp. 514–515.)

Earth history can be viewed very broadly as a great competition among different processes of change, such as the uplift of mountains, the soil formation and growth of plants on those mountains, and the erosion of the same mountains. The sum total of all such dynamic competitions *tends toward* whatever ultimate equilibrium configuration will allow a minimum expenditure of energy. Thus, a river tries to erode mountains and adjust the gradient of its valley so that the discharge of water and its sediment burden can be transported with the maximum efficiency, that is, with the minimum expenditure of energy (Ol' Man River must be lazy). In nature, major changes—such as the upheaval of new mountains—are always occurring, however, and these changes tend to upset the full achievement of equilibrium. Tireless nature starts anew to establish a new equilibrium adjusted to the new conditions.

The concept of an evolutionary earth ever striving for equilibrium but never achieving it for long is the single most important idea presented in this book, and we shall return to it repeatedly. History never repeats itself exactly, and the net effect of cumulative, irreversible changes has resulted in the evolution of the nonliving realm as well as living organisms. Indeed, the evolution of each realm has affected that of the other. This "evolving together," or coevolution, has involved what is termed *feedback,* which is a convenient way of summarizing cause-and-effect relationships in nature. **Feedback** characterizes any system in which the result of a process directly modifies the system's further development. That is, the result "feeds back into" the system and causes either an amplification or a reduction of the process and changes the conditions for equilibrium in that process. An example is an increase of food supply that results in a population explosion of grazing animal species A, which is accompanied by an increase in predator species B.

atmosphere to an oxygen-rich one, the apparent origin of life under conditions that no longer exist, evidence of larger and more frequent meteorite impacts in the distant past than today, and the complete extinction of large groups of organisms, such as the dinosaurs and marine reptiles (Fig. 2.18).

Equilibrium and Feedback

The concept of equilibrium is of enormous importance in science. It derives from the rigorous laws of thermodynamics, which describe in mathematical terms the flow and conservation of energy in chemical and physical systems. Equilibrium is easy to grasp for a static system in which all material is at rest and nothing is happening, but in nature practically nothing is static. Development of the concept of equilibrium, or a **steady-state condition** in dynamic systems was a major advance for geology. In an open system, such as a beach, there is a constant flow of material and energy, but the system looks much the same from day to day. Energy is constantly being expended, and sand is constantly in motion, yet the beach does not change in form except during severe storms, which only temporarily upset the equilibrium. Most large rivers with steady discharges also illustrate dynamic equilibrium. Today Lyell's ultrauniform earth would be viewed as an extreme example of a steady state.

Doctrine of Uniformity Today

The idea of some sort of uniformity in nature through time is absolutely basic to the analysis of earth history. It is important, therefore, to understand clearly what this means to modern geology. Now that we have seen that there were several different shades of meaning for both catastrophism and uniformity during the nineteenth century, we can simplify the variations as follows:

1. Uniformity of natural causes or processes throughout time
   a. Only those causes still operating have operated in the past.
   b. Some causes not still operating were important in the past.
2. Uniformity of both intensity and rate throughout time.

Some extreme catastrophists required *both* presently unknowable or supernatural (i.e., scientifically inexplicable) causes *and* extreme variations of intensity. At the other extreme was Lyell's gradualism, which allowed only presently acting causes operating always with their present intensities as well. Between these extremes were many catastrophists who allowed only natural causes but believed in extreme variations of intensity. And there were uniformists who allowed some variation of intensity but *also* acknowledged the possibility in the past of some causes that are no longer operating; those still were "natural causes," however, in the sense of being scientifically understandable.

The names *catastrophism* and *uniformitarianism* should be used only in their nineteenth-century context, for today should envision neither extremely violent events with supernatural causes nor a rigidly uniform, steady-state earth. Rather, modern geology sees the earth as being evolutionary and having changed through an irreversible, or evolutionary, chain of cumulative historical events. The only assumption made today is that *the principles of nature have been uniform through time.* This uniformity is an assumption made about nature and, so, is a doctrine, rather than a logically proven law. As Hutton wrote in 1788, "the uniformity of nature, even if not strictly true, is necessary for our clear conception of the system of nature." Such an assumption is absolutely necessary to the scientific analysis of earth history and is merely a special geologic application of the *doctrine of simplicity* common to all sciences—namely, that the simplest possible explanation is always to be preferred unless (or until) there is compelling evidence that demands a more complex hypothesis. It should be noted that simplicity does not exclude the possibility that either past *intensities* or *kinds* of processes may have been different from those of the present. The great Scablands floods in Washington State 22,000 years ago (see Box 2.4) exemplify the necessity of invoking intensities unknown today.

The most appropriate name for our modern doctrine of uniformity is **actualism** (from the French, meaning "actual or present causes"), which both predates and is less ambiguous than uniformitarianism. The following, very modern statement of actualism made by Hutton's friend John Playfair in 1802 makes one wonder why the Lyellian controversy ever arose:

> Amid all the revolutions of the globe the economy of Nature has been uniform, and her laws are the only thing that have resisted the general movement. The rivers and the rocks, the seas, and the continents have been changed in all their parts; but the laws which describe those changes, and the rules to which they are subject, have remained invariably the same.

By assuming actualism, we can reason by analogy from the study of geologic processes acting today either in the field or in the laboratory to gain clues about the past effects of such processes. We do *not* assume that those processes always acted with the same rates and intensities, for Lyell's hypothesis of gradualism has failed the test of evidence.

Why so much fuss about what finally may seem simple and straightforward? Our extensive treatment was necessary because actualism provides the logical basis for understanding earth history and because of great confusion and publicity achieved of late by various *neocatastrophists* ("new catastrophists"), such as Immanuel Velikovsky (author of *Worlds in Collision,* 1950, and *Earth in Upheaval,* 1955) and the more extreme views of the antievolution Creationists. These people persist in representing modern geology as still assuming the extreme gradualism of Lyell, which is absurd. Geologists today routinely accept sudden,

violent, and even historically unique events as perfectly consistent with modern earth theory.

Scientific Methodology of Geology

Historical Science

Physics and chemistry are concerned almost exclusively with phenomena controlled by presumably universal natural systems that are largely nonhistorical—that is, *independent of the time at which they operate.* Geology, astronomy, and biology, on the other hand, are more historical. When a geologist focuses only upon *present* processes and characteristics of earth materials, he or she is an applied physicist or chemist. But when a geologist interprets a series of *past* events, he or she becomes a unique historical scientist. While assuming all physical, chemical, and biological theory, the geologist's chief goal is the reconstruction and explanation of history. Moreover, the prediction of results from known causes in the sense that is so important in nonhistorical science is replaced in geology by the interpretation of ancient causes from their historical results. Prediction does have a special role, too, however. From some set of clues, one may predict that such and such a situation should exist somewhere in the field even if never before seen. Hutton did this when he predicted that an unconformity should be visible in certain places in Scotland and then went on a successful search for it.

Actualism provides the connecting thread between the present and the past, for it allows us to reconstruct events never witnessed by humans. Geologists employ comparisons, or analogies, between modern and ancient situations (and among ancient ones as well). This process gave rise long ago to the cliché that the *present is the key to the past.* In fact, the reverse is equally true, for the geologic record provides a long time perspective for assessing the present earth, which we now know is unique in its climate, topography, and life. It is simple common sense to compare both ways, because human history provides such a short observational span of time, compared with geologic time.

Geologic Reasoning

The nonscientist needs to appreciate the difference between scientific proof and absolute proof in logic, which is the difference between *probability* and *certainty.* Science deals with probabilities; that is, one hypothesis seems for the moment more correct than another. Therefore, scientific explanation is never complete, for it produces only an approximate working model of nature as we understand it at a particular time and within the social framework of that time. Science is badly misrepresented when it is portrayed as giving final and dogmatic truths.

Contrary to a popular textbook myth, it is rare for a scientist to make very many observations without already having a tentative, or working, hypothesis in mind to test. There is a constant feedback among observation, hypothesis, and experiment. Moreover, some breakthroughs have resulted from intuitive flashes based upon skimpy evidence. For example, Hutton had

Box 2.4

Gradualism Meets the Scablands

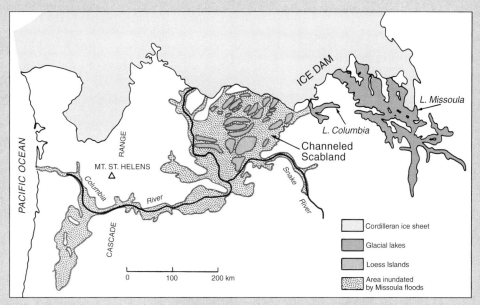

Figure 2.19   Map of the Pacific Northwest, showing areas of the Channeled Scablands and Glacial Lake Missoula, as well as the ice dam formed by the extension of the Canadian ice sheet.

Lyell's extreme version of uniformitarianism, which confused uniformity of processes (actualism) with uniformity of rates (gradualism), had a profound influence on generations of geologists. For a century after the publication of his *Principles of Geology,* the geologic community was reluctant to accept evidence of events of gigantic scale. The suggestion of a catastrophe seemed a throwback to the pre-Lyellian days of supernaturalism and a retreat from testable science.

In 1927, iconoclastic American geologist J. Harlen Bretz challenged the geologic establishment with a controversial idea. For years he had been studying a region in central Washington known as the "Channeled Scablands," (Fig. 2.19). Underlain by thick lava flows erupted about 15 million years ago, the Scablands had been deeply scoured, apparently by waters of enormous power, so that the cover of windblown glacial dust (loess) had been stripped away, exposing the basaltic bedrock. Deep, vertical-walled canyons known as *coulees* were gouged into the hard basalt; this gouging was difficult to explain by the gradual erosion of normal rivers. In places, side channels, which had been carved into the plateau, dropped abruptly into the coulees with no gradual gradient, much like the hanging valleys formed by glaciers (Fig. 2.20A). These abrupt drops made sense only if the coulees had been filled to the brim and the water had roared across the divides between canyons. Some plateaus still had a thin cover of loess remaining, as if they had been islands surrounded by the floodwaters (Fig. 2.20B). Other places had gravels from sources hundreds of miles away and boulders far too large to have been carried by any stream of normal strength.

To Bretz, a flood of enormous power clearly had once crossed the Scablands. When he presented his interpretation at scientific meetings, however, geologists could not conceive of anything that could have produced such catastrophic floods. Underlying their skepticism and rejection of Bretz was a deep belief in gradualism. Catastrophic floods violated their understanding of uniformitarianism, and they preferred a more gradualistic explanation of the Scablands, which required only normal stream erosion. One geologist wrote that "the erosion features of the region are so large and bizarre that they defy description." However, he was not willing to abandon gradualism. "I believe that the existing features can be explained by assuming normal stream work of the ancestral Columbia River. . . . Before a theory that requires a seemingly impossible quantity of water is fully accepted, every effort should be made to account for the existing features without employing so violent an assumption."

Geologists were also skeptical about the source of such an enormous volume of water. Bretz suggested that a volcanic eruption under the glaciers to the north might have caused sudden melting, but most geologists found this unconvincing. Unfortunately, Bretz did not pursue this important part of his argument, insisting that his "interpretation of the channeled scablands should stand or fall on the scabland phenomena themselves."

The Scablands remained controversial for decades until new evidence emerged. In the 1950s, aerial photography showed that there were ripple marks up to 7 meters (22 feet) in height and spaced 130 meters (425 feet) apart (Fig. 2.20C). These ripples were so huge that Bretz had walked over them for years and did not recognize them under the sagebrush. From this evidence, hydraulic engineers could estimate that the maximum discharge of water had been about 752,000 cubic feet per second, racing close to the speed

—Continued on page 34

Continued from page 33—

A.

Figure 2.20   Landscapes of the Channeled Scablands. *A:* The water once flowed over the steep cliff (now known as Dry Falls) and scoured the basin below (now filled with a small lake), forming a cataract many times the size of Niagara Falls. *B:* The contrast between the scoured basaltic bedrock channels (left and center) and the untouched uplands covered with glacial loess soils (right) is very striking. *C:* Giant ripples on the ancient gravel bars along the Columbia River. The road and farm buildings give a sense of scale. *(Courtesy Alan L. Mayo, GeoPhoto Publishing Company.)*

B.

C.

limit of our modern highways and capable of moving boulders 11 meters (36 feet) in diameter!

Although Bretz's ideas have been vindicated his opponents were right in some ways. Bretz had failed to find a source for the floodwaters; yet, since the 1880s, geologists had known that the canyons in western Montana had been filled with glacial meltwaters to form glacial Lake Missoula (Fig. 2.19). These waters were impounded behind an ice dam formed by the snout of a southbound glacier in the valley of Lake Coeur d'Alene, Idaho. Bretz originally postulated a single flood, but other investigators had shown there were multiple events. When the stairstep terraces of Lake Missoula were discovered, they proved that the ice dam had melted and released its waters several times.

Late in Bretz's life, geologists studied the Scablands as models of the gigantic erosional features discovered on the surface of Mars. Bretz had become a pioneer in planetary geology without even intending it. In 1969, at the age of ninety-six, he received the Penrose Medal, the highest award of the Geological Society of America. Unlike most scorned scientists throughout history, Bretz had lived to see his ideas accepted and acclaimed as the basis for the emerging science of planetary geology. Today, with the controversy over asteroid impacts and their possible relation to mass extinction, catastrophic phenomena are more popular than ever see (Chap. 14). Modern catastrophism however, still requires the uniformity of processes (actualism). Only the *scale* of such processes, such as extraterrestrial impacts, conflicts with the gradualistic bias that geologists inherited from Lyell.

formulated his entire theory and deduced from it the existence of intrusive dikes and unconformities before he saw either feature in the field.

The disciplined intellectual endeavor called science is a process of analysis of nature that is based upon the formulation of and testing of hypotheses derived from observations. Experiments simplify observation by imposing controlled conditions. Scale models of objects or processes are familiar and of obvious value—say, for experimenting with the transport of sand by wind or water. But some phenomena are too large or too complex for scale modeling, so we may choose to make detailed observations of a modern natural system, such as a large river. Mathematical modeling also is promising for the investigation of complex phenomena that cannot be scaled down in the laboratory, for it allows one to see how a large number of factors should interact with one another over long spans of geologic time. Geologists use mathematical modeling, for example, to investigate the effects of a simultaneous worldwide rise of sea level, local subsidence of the crust, and climatic change on erosion and sedimentation (as discussed in Chap. 1).

Geologic phenomena involve so many interacting factors and such a fragmentary historical record that it is commonly difficult to formulate a definitive test for a particular hypothesis. Because of this, most geologists consciously try to construct multiple working hypotheses so as to force themselves to examine a problem in several different ways. This increases the chance of discovering unsuspected new evidence that might be easily overlooked if only a single, pet explanation is considered. In the final analysis, the study of earth history from fragmentary clues has much in common with the solution of a good mystery. Sherlock Holmes captured this spirit perfectly in *The Five Orange Pips,* when he said, "The ideal reasoner would, when he had been shown a single fact in all its bearings, deduce from it not only all the chain of events which led up to it, but also all the results which would follow from it . . . as Cuvier could correctly describe a whole animal from the contemplation of a single bone."

Summary

Fossils and Stratigraphy

- Fossils include any record of preexisting life. They may be preserved as original or altered shells, bones, teeth, wood, impressions of soft tissue, or trace fossils (e.g., tracks, burrows). Besides providing evidence of the existence, evolution, and extinction of past life forms, fossils also give insight into past environmental conditions and provide tools for establishing like age or **correlation** of strata between localities.
- **Steno's principles** are
  1. **Superposition.** In any succession of strata, the oldest lies at the bottom, with successively younger ones above.
  2. **Original horizontality.** When formed, most stratification is horizontal.
  3. **Original lateral continuity.** Strata originally extended in all directions as far as the edges of their **basin of deposition.**
- **Fossil correlation,** first demonstrated by Smith and Cuvier, is the comparison of strata having similar fossil assemblages to establish their similar age. Physical properties of strata (e.g., color, grain size) are generally used in conjunction with fossils for correlation. **Index fossils** are those that are useful for correlation; they must have been short-lived, widespread, and easily recognizable.
- **Hypotheses to explain differences between fossil assemblages** are
  1. **Catastrophism and special creations.** Catastrophes caused selective extinctions of some species; subsequently special creations produced new species, which were different from and unrelated to those in older strata.
  2. **Organic evolution.** Extinctions of some species and changes (evolution) of other species from one time to the next explained the fossil record. A biological connection was maintained as each successive assemblage of survivors was descended from the prior assemblage.

Unified Hypotheses for the Solid Earth

- **Neptunism** considered the solid earth to be static, or passive. All rocks, including those now called igneous and metamorphic, were deposited from a universal ocean, which gradually receded during earth history. The first formal geologic **chronology,** consisting of **Primitive, Secondary,** and **Tertiary** divisions, seemed to be explained adequately by neptunism.
- **Plutonism** of Hutton considered the solid earth to be dynamic and invoked internal heat to explain igneous and metamorphic rocks as well as the upheaval of mountains. Erosion of landscapes and subsequent burial under new sediments resulted in breaks in the rock record called **unconformities.** By recurring countless times in a repetitive fashion, these rejuvenating processes guaranteed that habitable land would exist at all times, even though decay is the ultimate fate of any specific land area.

Catastrophism Versus Uniformity in Nature

- **Catastrophism** always invoked extreme intensity of violent processes and in some versions also called upon supernatural causes, which are unknown (and apparently unknowable) today.

- **Uniformitarianism** was the nineteenth-century antithesis of catastrophism. In the strict Lyellian form, which may be called **gradualism,** it insisted not only upon a uniformity through time of natural causes or processes but also upon a uniformity or very narrow variations of intensity and rates of change.
- **Actualism** assumes only a **uniformity of kinds of natural causes**—that the principles, or laws, of nature have been constant through time. *Modern geology assumes only this!* It recognizes that intensities and rates of change have varied greatly and that there were some natural processes active in the past that may no longer be important on earth.
- **Neocatastrophists** have revived nineteenth-century catastrophism and tend to allege incorrectly that modern geology still endorses the extreme Lyellian form of nineteenth-century uniformitarianism.

Steady-State Versus an Evolutionary Earth

- The **steady-state earth** of Lyell had energy being dissipated through processes such as erosion, volcanism, and mountain building, but the globe as a whole was considered to be at equilibrium and not changing through time in an overall sense. Its history had no direction. Kelvin saw the physical absurdity of such a perpetual motion machine. Because its energy reserves have to run down, the earth must have changed significantly through time in an irreversible fashion—the earth must be evolutionary.
- **Evolutionary change** is a profound concept. Contrary to the popular cliché, history never repeats itself exactly. Although individual phenomena have recurred many times, sequentially linked events are not repeatable *in exactly the same way.* Evolutionary change involves such linked events in an irreversible sequence with direction. Moreover, the evolution of the earth has involved **feedback** between living and nonliving realms, so that the evolution of the one has affected the evolution of the other.

Geology as a Historical Science

- Geology assumes the laws of physics and chemistry and tests hypotheses based upon observed evidence, but geology differs from the more familiar, nonhistorical sciences. Actualism provides the connection from present to past through analogical reasoning to allow the inference of ancient causes from their historical results. **Multiple working hypotheses** are tested constantly by a complex process with feedback among observation, hypothesis, experiment, and then new observations.

Readings

Adams, F. D. 1938. *The birth and development of the geological sciences.* New York: Dover. (Paperback, Dover, 1954.)

Albritton, C. C., Jr. 1967. *Uniformity and simplicity.* Geological Society of America Special Paper 89.

Baker, V. R. 1973. *Paleohydrology and sedimentology of Lake Missoula flooding in eastern Washington.* Geological Society of America, Special Paper 144.

——— and D. Nummedal. 1978. *The Channeled Scabland.* Washington, D.C.: NASA Planetary Geology Program.

Berggren, W. A., and J. A. Van Couvering, eds. 1984. *Catastrophes and earth history: The new uniformitarianism.* Princeton, N.J.: Princeton University Press.

Bretz, J. H. 1923. The Channeled Scabland of the Columbia Plateau. *Journal of Geology* 31:617–49.

———. 1927. Channeled Scabland and the Spokane flood. *Journal of the Washington Academy of Sciences* 17:200–211.

———. 1969. The Lake Missoula floods and the Channeled Scablands. *Journal of Geology* 77:505–43.

Cuvier, G. 1818. *Essay on the theory of the earth, with mineralogical notes by Professor Jameson and observations on the geology of North America by Samuel L. Mitchill.* New York: Kirk & Mercein.

Davies, G. L. 1969. *The earth in decay.* New York: American Elsevier.

Gould, S. J. 1978. *The panda's thumb.* New York: Norton. (Essays on Lyell and the great Scablands debate.)

———. 1983. *Hen's teeth and horses toes.* New York: Norton. (Essays on Steno, Hutton, and Cuvier.)

———. 1984. Toward a vindication of punctuational change. In *Catastrophes and earth history: The new uniformitarianism,* edited by W. A. Berggren and J. A. Van Couvering. Princeton, N.J.: Princeton University Press.

———. 1986. Evolution and the triumph of homology, or why history matters. *American Scientist* 74:60–70.

———. 1987. *Time's arrow, time's cycle.* Cambridge, Mass.: Harvard University Press.

Greene, J. C. 1959. *The death of Adam.* Ames: Iowa State University Press, chaps. 1–3 (Paperback, Mentor Books, 1961.)

Greene, M. T. 1982. *Geology in the nineteenth century: Changing views of a changing world.* Ithaca, N.Y.: Cornell University Press.

Hutton, J. 1795. *Theory of the earth with proof and illustrations.* London: Cadell & Davies.

Laudan, R. 1987. *From mineralogy to geology: The foundations of a science, 1650–1830.* Chicago: Chicago Univeristy Press.

Lyell, C. 1830–1875. *Principles of geology,* 12 eds., London: J. Murray.

Morris, H. M. 1974. *Scientific creationism.* San Diego: Creation-Life.

Playfair, J. 1802. *Illustrations of the Huttonian theory.* London: Cadell & Davies.

Porter, R. 1977. *The making of geology—Earth science in Britain 1660–1850.* Cambridge: Cambridge University Press.

Rudwick, M. J. S. 1972. *The meaning of fossils: Episodes in the history of palaeontology.* London and New York: MacDonald-Elsevier.

Rupke, N. A. 1983. *The great chain of history.* Clarendon N. J., Oxford University Press.

Secord, J. A. 1986. *Controversy in Victorian geology.* Princeton, N.J.: Princeton University Press.

Velikovsky, I. 1955. *Earth in upheaval.* New York: Doubleday.

Warshofsky, F. 1977. *Doomsday—The science of catastrophe.* New York: Readers Digest Press.

Winchester, S. 2001. *The map that changed the world: William Smith and the birth of modern geology.* New York: HarperCollins.

(Courtesy John H. Ostrom, Yale Peabody Museum.)

Chapter 3

Evolution

Nothing in biology makes sense except in the light of evolution.

Theodosius Dobzhansky (1973)

▶ Despite the biblical teaching that all life was created in six days as described in Genesis, the notion that life had changed or evolved through time was widespread in the late 1700s and early 1800s. A number of British and French scientists proposed different ideas about how life had evolved, but none of their ideas were widely accepted because these scientists did not have a convincing mechanism to explain the transformation of species.

▶ In 1859, Charles Darwin revolutionized biology by providing not only overwhelming evidence that life had evolved but also a mechanism for evolutionary change. He proposed that organisms with the most favorable features were selected by the environment (natural selection), and these individuals would be the most successful in leaving offspring in the next generation. Over many generations, successful descendants would be changed until they became a new species.

▶ Although Darwin's evidence that evolution had occurred was convincing, not everyone in the late 1800s accepted his mechanism of natural selection. In the early 1900s, the discovery of modern genetics revived Darwin's ideas. So much new evidence of the reality of evolution has now accumulated that there is no longer any doubt that life has evolved; this concept is as well established as the law of gravity.

▶ Discoveries in genetics of natural populations showed that new species are formed when they are genetically isolated from other related populations. If isolation persists long enough, and the two populations live under different environmental conditions, they can diverge to form new species.

▶ There is always healthy debate about the mechanisms of evolution. New discoveries in genetics have suggested that many genes are invisible to natural selection. Also, some scientists have argued that small changes in species are insufficient to explain major transformations among different groups of animals, such as birds arising from reptiles. These ideas do not challenge the reality of evolution, only its proposed mechanisms.

According to philosopher Thomas Kuhn, science progresses not by the slow accumulation of knowledge until final truth is reached but by major intellectual breakthroughs—what he called *scientific revolutions*. For example, in ancient times, astronomers thought the stars, planets, and sun circled the earth. As theories to explain the complicated motions of heavenly bodies became more and more intricate, it became increasingly difficult to make sense of the motions of the planets. Then in 1543 came Copernicus, with his radical proposal that the sun, not the earth, is the center of our solar system. Suddenly all the puzzling observations were explained by a much simpler theory.

Biology underwent its most significant scientific revolution in 1859 with the publication of Charles Darwin's *On the Origin of Species by Means of Natural Selection*. Before that time, few scientists questioned the idea that all species had been created by God thousands of years ago and had since remained unchanged. Evidence gradually accumulated that made this notion more and more difficult to defend, but the grip of convention and religious dogmatism was too strong to allow any change in the prevailing belief about nature. Darwin's book was like a lightning bolt that galvanized not only biology but all of Western society. Suddenly species were dynamic entities that changed through time and were constantly adapting to their environment. All life was interrelated, and—most important to humans—we were part of the animal kingdom. The debate raged for years, but by the time Darwin died in 1882, the scientific community and most educated people had become convinced of the reality of evolution.

The Evolution of Evolution

Today we use the word *evolution* to mean virtually any kind of continuous, unidirectional change, from the evolution of the universe to the evolution of ideas. The idea of *organic evolution* simply states that all life on earth is descended from other life, sometimes transformed from very different ancestors. In other words, all life is linked by a process of ancestry and descent; all life is interrelated. Glimmerings of this notion go back at least as far as the Greeks in the fifth century B.C., who saw the similarities in the organs of animals and realized that all animals were built on a common plan. To some Greek philosophers, the concept that "everything changes" was fundamental to their thought.

However, the dominance of the Church in Western thought stifled discussion of evolution for over a thousand years. According to Christian belief, all animals and plants were created as we see them today and there is no possibility that species could change. During the 1700s, however, it became more and more difficult for scientists to reconcile this static view of nature with their expanding knowledge of the biological world. When Linnaeus first standardized biological classification in 1758, about 8,000 species were named and recognized by Europeans. Soon voyages and expeditions of discovery were rapidly expanding the catalogue of known species of animals and plants, and puzzling patterns were beginning to emerge. How could all of these animals have fit onto Noah's ark? Even more

puzzling, why did faraway places such as Australia and South America have their own, unique types of animals and plants rather than the familiar animals of Europe? If every kind of animal on earth had dispersed from Mt. Ararat, the supposed landing site of Noah's ark (now in Turkey), why did no European animals end up in Australia instead of the pouched mammals that are found there?

A second line of evidence that species had changed emerged from the earth itself. Steno, Hooke, and others showed that the fossil record was a treasure trove of strange forms, which indicated that the earth had changed. How else did seashells get to the top of the Apennine Mountains? By the early 1800s, Cuvier had demonstrated that some animals, such as mammoths and mastodons, were unquestionably extinct. They were so big that no explorer could have missed them, even in the most remote and least explored corners of the globe. The religious community was appalled at the idea that a caring God would let any of His creations become extinct, but by the early nineteenth century the evidence was incontrovertible. More important, Cuvier, Smith, and other geologists provided increasing evidence for radical changes in the earth's past. During these great "revolutions," as Cuvier called them, life had changed repeatedly, and many extinct species had been replaced by later ones. How could this be reconciled with the notion that all species had been created only once, as described in Genesis? By 1859, most geologists had come to the conclusion that Genesis could not explain the fossil record, but they had no other explanation for it. Indeed, some fossils (for example, the half-reptile, half-bird *Archaeopteryx*) were discovered that challenged the notions that all forms of life were discrete creations (Fig. 3.1).

Buffon's Evolution

Georges de Buffon (see Chap. 2) was the first influential man to convince some of his fellow scientists and philosophers that evolution must have occurred. He defined the concept of species for the first time, noting that each species exists as a separate entity and demonstrating that no species could interbreed with another. This is a fundamental concept in biology and has taken on an even greater importance since the development of genetics. Buffon stated that the environment directly caused variation in offspring. Although we now know that variation is not caused by environment, Buffon was the first to understand the important role of environment in evolution.

Buffon's writings are full of contradictions, but gradually through his life he moved toward the conclusion that all forms of life must have evolved from fewer forms. His one, most important contribution is that all organisms are closely related to the environment in which they live and that the environment somehow molds and changes the organisms. He noted that there are also changes due to inheritance of characteristics from both parents. Thus, he established two of the fundamental attributes of evolution, adaptation and inheritance; he asked the right questions at the right time in history. He put these two concepts together, saying that modifications of organisms are due to inheritance of characters and changes caused by environment;

Figure 3.1 *Archaeopteryx lithographica,* the oldest known bird, from the Late Jurassic Solenhofen Limestone of Bavaria, Germany. This remarkable fossil is the most complete of the six known specimens of *Archaeopteryx* and clearly shows a mixture of both reptilian and avian characteristics. For example, it has reptilian teeth and a bony tail, but it also has fully avian feathers and a wishbone. The first specimens of this remarkable transitional form were discovered in 1861, only 2 years after the publication of Darwin's theory of evolution. *(Courtesy of John H. Ostrom, Yale Peabody Museum.)*

these in turn produced a natural descent, which resulted in life as we see it today. In reading his statement, we are not convinced that we should classify this as a theory of evolution, for he did not explain *how* new characters could arise or *how* the environment could cause modifications *that could be inherited.*

Thus, the conditions for a theory of evolution acceptable to the scientific community were established by Buffon by the middle of the eighteenth century as a reasonable, but vague, concept to explain life as we see it. The majority of the scientific community, including Charles Darwin, were strong dissenters and put their finger on the most serious problem for evolutionists, that of *time.* How could all life become so diversified in such a short period of time since the Creation? Experiments involving breeding of domesticated plants and animals failed to produce a new species. Cuvier firmly stated that there are no changes in the short run of time. He studied the anatomy of cats mummified with peo-ple in Egyptian tombs and found them to be identical to those living in his day. Thus, no change had occurred over a long period of time, and he concluded no evolution is possible. The geologists realized that they needed more time also, and Buffon, as we saw in Chap. 2, suggested that the earth was 75,000 years old. The problem of the span of time available for evolution was not settled until much later.

Erasmus Darwin: Did He Begin It All?

Buffon strongly influenced one of the important intellectuals of Great Britain, Erasmus Darwin, the grandfather of Charles. As an animal experimenter, he saw that there are many changes going on during **ontogeny** (the development of an individual): a caterpillar changes into a butterfly and a tadpole into a frog. He noted that, in breeding, changes can be recognized from generation to generation, although the changes are very small.

Erasmus Darwin went on to say that there are many forces controlling evolutionary change. He maintained that mating and the various weapons, ornaments, and protective devices (deer antlers, etc.) attendant on mating were important dynamic forces of evolution. Another point he emphasized was that species became adapted to their environment primarily by the method of gathering food. For example, the development of the elephant's trunk was to aid in tearing grass and moving it up to the mouth. The third element is security: protective coloration, wings, legs, shells, spines, and so on. The results of these forces were new characteristics *acquired to fit the organism to its environment and passed on to the organism's descendants.* Thus, Darwin realized that there is a strong relationship between environment and heredity.

But, again, his statement, like Buffon's, lacks one main criterion for a full-blown theory of evolution: how do the characteristics arise? He does tell us why they might have developed and that, if the characteristics make organisms more fit, they are passed on to the offspring. These ideas strongly influenced the early thinking of Charles.

Lamarckian Evolution

Jean Baptiste de Monet, the chevalier de Lamarck (1744–1829), was one of the greatest naturalists of the eighteenth century. Originally famed as a botanist, he was entrusted with the French Natural History Museum's collections of "insects and worms"—the invertebrates that no other natural historian was interested in. From this background, he laid the foundations of modern invertebrate zoology and was one of the first of the "natural philosophers" to see all life—both plant and animal—as an integrated whole, for which he coined the term *biology.* Although he contributed many other ideas to modern biological thought, his name is now associated with only one of them: his theory of evolution.

His thinking represents the culmination of a long series of philosophers who attempted to explain all the universe in a single concept. Lamarck's theory of evolution is one of the most misunderstood, partly because post-Darwinian scientists attached his name to a single erroneous concept that was originally only a

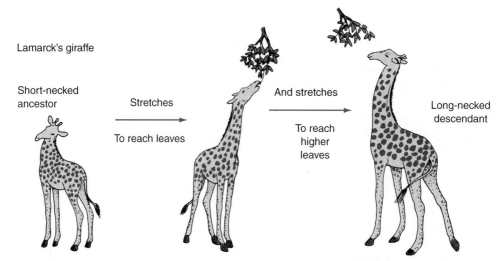

Lamarck's giraffe

Short-necked
ancestor

Stretches

To reach leaves

And stretches

To reach
higher
leaves

Long-necked
descendant

Figure 3.2   Lamarck's explanation of evolution as applied to giraffes. Ancestral short-necked giraffes must stretch their necks to reach leaves. After some generations, the giraffes have become a long-necked species, due to continual stretching for food. *(After E. O. Wilson et al., 1973,* Life on Earth, *Sinauer. Reprinted by permission of Edward O. Wilson.)*

small part of his full range of thought. He summarized his philosophy in a book published in 1809. The thread going through it is that the fundamental aspect of nature is change. For example, minerals do not have a fixed composition but are only stages in a continuous flux of disintegration and recombination. He believed that the species concept was invented by humans and that species are not natural groupings. Note how different this idea of species is from Buffon's fixed species. Life, Lamarck said, is a great stream of gradual complication. He believed that this system was replenished each day by spontaneous generation from inorganic material, which filled up the system from the bottom, the simplest forms today being the most recently formed. The most complicated animals are the oldest groups. Even more amazing was his belief that no animal group ever becomes extinct. If some group (or better, stage) should be destroyed locally, it will be quickly replaced by those forms coming up from below, just as a hole fills up with water.

Inheritance of Acquired Characteristics

The only facet of Lamarck's theory still remembered is called the **inheritance of acquired characteristics.** Lamarck perceived that the effects of the environment somehow could be inherited by offspring. Buffon and Erasmus Darwin saw this, but neither could provide a mechanism to explain it. Lamarck found such a mechanism, although few believed him.

Lamarck did not believe, as did Buffon, that the environment directly causes animals to be different; rather, he thought that changes in the environment lead to changes in the needs of organisms. Changes in the needs mean that the animals will be forced to change their habits. If these changes become constant—that is, if the environment continues to change, animals will acquire new habits, which will last as long as the needs that gave rise to them. New habits will give rise to new characteristics and thus a new species will arise. This was one of the few mechanisms suggested in the nineteenth century for the creation of new characteristics. It

is a subtle concept in a way, because it requires a mysterious inner force in the organism to develop a new feature, but it gets around arguments about the direct effect of the environment.

One classic example Lamarck used to explain his theory is how the giraffe got its long neck (Fig. 3.2). The giraffe, Lamarck theorized, started out as a grazing animal; for whatever cause, there came to be a shortage of ground cover. Giraffes turned to eating leaves on trees, and, as more and more turned to this source of food, the supply of leaves decreased. The giraffes were forced to stretch their necks continuously to reach higher leaves. Thus evolved the long neck, and voilà, a new species was born. This is more or less a complete example of a theory of evolution. The theory accounts for the effects of a changing environment, a mechanism for developing a new characteristic (hence a new species), and a way whereby these changes are passed on to the offspring—a far more satisfying concept than Buffon's rather ill-defined direct environment approach. Lamarck also threw in a few "laws" to help amplify his idea. The first is built into other theories as well—namely, that everything gained by the **phylogeny** (the evolutionary history of an organism) through the influence of nature is preserved by heredity, provided that the modifications are common to both parents. The second idea states that, the more an organ is used (a stretched neck, for example), the stronger it becomes and the larger it will be in proportion to the duration of its use.

Lamarck was opposed by the powerful Cuvier, who saw to it that, when Lamarck died in 1829, the best of Lamarck's ideas were either forgotten or buried under the weight of ridicule. By the time *On the Origin of Species* was published in 1859, most scientists had only a distorted notion of what Lamarck had really said. Ironically, Darwin himself, despairing of a mechanism to explain inheritance, began to believe in the inheritance of acquired characteristics toward the end of his life. So did most other late nineteenth-century biologists. In the 1890s, however, German scientist August Weissman (1834–1914) showed that chopping off the tails of twenty generations of mice did not result in

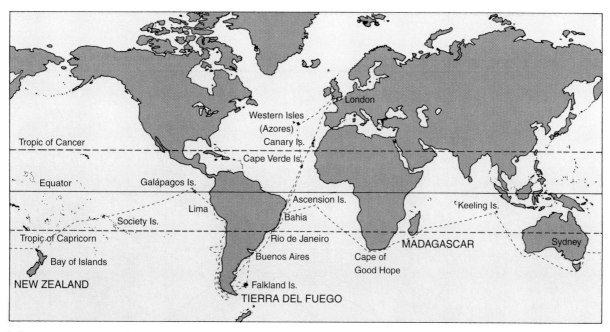

Figure 3.3   The route of *H. M. S. Beagle*, 1831–1836. *(After the endpapers of the Pallas facsimile of Darwin's* Journals of researchers: *London, 1839; New York, Hafner, 1952.)*

mice with shorter tails. Similarly generations of shepherds have cut off the tails of their sheep to keep flies from infesting the anal area, but sheep continue to grow tails. By 1900, it had become well established that changes in the germ line (reproductive cells, eggs and sperm) could result in changes in the body but that changes in the body could not work their way back into the genes in the germ line. For example, if you work out at the gym and become a body builder, your muscular strength cannot be passed on directly to your children. The notion of inheritance of acquired characteristics seemed dead forever. Unfortunately, we attach the name of Lamarck to this concept, although it was not original to him, and it was peripheral to his main ideas. Most of Lamarck's contemporaries, and even Charles Darwin, believed in it!

Charles Darwin and Natural Selection

Lamarck produced his theory in 1809. That year is certainly an important one because it was on February 12, 1809, that two men were born who had a profound influence on the nineteenth century: Abraham Lincoln, emancipator of slaves, and Charles Darwin, who we might call the emancipator of our then primitive ideas on the development of life. Darwin attended the University of Edinburgh as a medical student, at which time he studied geology and became thoroughly grounded in all of natural history. He later became dissatisfied with the course of his life and went to Christ College at Cambridge University as a divinity student, acquiring a great interest in botany and the collecting of insects. The single most important event that was responsible for his gaining a notion of evolution, according to him, was the fortunate trip that he made in the years 1831 to 1836 aboard the *H. M. S. Beagle,* one of the early worldwide scientific cruises. This trip repre-

sents a milestone of scientific investigation; it was the first to study the bewildering variety and strange adaptation of plants and animals of the tropics and the Southern Hemisphere (Fig. 3.3).

Darwin observed the fantastic variety of species found in the world, the tremendous numbers of individuals in each species, and the amount of competition among populations for food. From his observations, he deduced that living organisms are enormously fertile, yet curiously the number of individuals in a given population appears to be fairly constant. Populations are stable in size, he said, because there is a tremendous struggle for life, because of rivals or prey-predator relationships and the pressures of the inorganic environment. Further, populations show a great deal of variation among individual members. The result is that some organisms seemed to be better fitted for life than others. Notice that this is an idea that Erasmus Darwin had. Charles was profoundly influenced by this idea and from reading Thomas Malthus's *An Essay on the Principles of Population,* which was written in 1798. Malthus said that populations increase geometrically while food increases arithmetically, and there is a continuous struggle for food. Malthus, of course, applied this to human populations, saying that, when the population increases have reached the saturation point for food, then famine occurs and the population is reduced.

In the struggle for life, not all offspring survive; in fact, there is a high rate of mortality, and only those that are best fit for life survive. Darwin observed that the reason that some organisms are not fit for life is that they have inherited characteristics that prevent them from surviving in their environment. It appears that nature selects the animals best fit for the environment and thus "molds" the animals. Through time, populations become better fitted to their environments as less well adapted members fail to reproduce offspring as successfully. These ideas, which Darwin called **natural selection,** are his chief contributions to evolutionary thought.

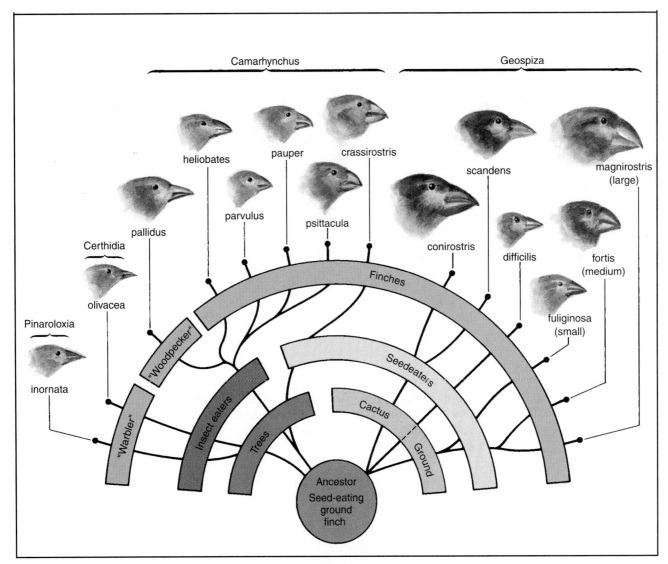

Figure 3.4   The adaptive radiation of Galápagos finches, first described by Charles Darwin after his *Beagle* voyage. From one common ancestor blown in from the South American mainland, at least four different genera of finches *(Camarhynchus, Geospiza, Certhidia, Pinaroloxia)* and more than fourteen species (lowercase names) have evolved in the past 5 million years. By developing a variety of types of beaks, different finches have become specialized for different diets. *(After G. G. Simpson and W. S. Beck, 1965, Life, Harcourt Brace World, Inc.)*

Darwin recognized, as did Lamarck, that environments change, necessitating continual character modifications in populations. Darwin saw that, as an environment changes, nature selects those individuals best fitted for the particular changes occurring; *only the fittest survive.* But this is an oversimplification of his statement, and Darwin recognized that there are many different possibilities. For example, there may be no observable environmental change, yet evolution seems to go on; populations are continually and *gradually* changing toward being better adapted or progressively adapted to fit their environment. Evolution is likely to move in the direction of greater adaptation; that is, as a result of natural selection, species will be dominated by individuals who have features equipping them to function more effectively in their environment.

How did Darwin come to these conclusions? One of the stops made by the *Beagle* was in the Galápagos Islands. While

there, Darwin collected a variety of birds including finches (Fig. 3.4). When he returned to London, he recognized that there were thirteen species of finches and further observed that each species appeared to be confined to one island. These finches were like none other in the world; they obviously were most closely related to, but distinct from, those in South America. The Galápagos finches are the only known terrestrial bird species found on the islands, and they are living in environments unknown elsewhere to the family of finches. Darwin was greatly puzzled by this but gradually came to the view that they had evolved from perhaps a single species that had accidentally arrived from South America. Although finches are unable to fly great distances, their ancestors somehow had managed to get to the islands, which are 500 miles away from the Ecuadorian coast. Because there was no competition from other birds; they became adapted to habitats not otherwise occupied. This example of how geographic isolation

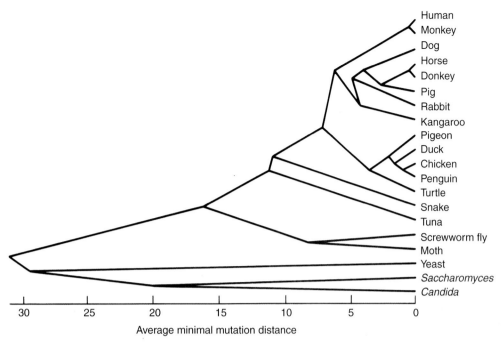

Figure 3.5   The branching pattern of life as revealed by the molecule known as cytochrome c. The molecular similarity between distantly related organisms is shown by sequence of branching and is remarkably similar to the evolutionary family tree of organisms deduced from anatomical evidence. Even molecules that Darwin could not have known about provide evidence of evolution. *(After W. M. Fitch and E. Margoliash, 1967, Science, v. 155, pp. 279–284. Copyright ©1967 American Association for the Advancement of Science. Reprinted with permission.)*

combined with natural selection result in new species of finches was one of Darwin's triumphs but one which he did not sense while in the Galápagos or on the *Beagle*.

Evidence of Evolution

Charles Darwin was not the first to argue that life had evolved or to propose the idea of natural selection. Why, then, does he get most of the credit? For one thing, Darwin's book had the good fortune of being published at the right time. Fifteen years before, amateur naturalist Robert Chambers anonymously published a book entitled *Vestiges of the Natural History of Creation*. Although it pointed to the evidence for the transformation of species, this book was so harshly ridiculed that it became infamous. The world was not yet ready for evolution, and Chambers's biology was so naive that natural historians had no trouble picking him apart.

Recent scholarship has revealed another reason that Darwin waited for 20 years to publish his ideas. In the 1830s and 1840s, evolution was a radical French notion favored by the working-class scientists in London. The conservative scholars and clergy at Oxford and Cambridge, protective of their social class, associated evolution with other dangerous, revolutionary ideas. As a Cambridge-trained, wealthy gentleman, Darwin was naturally reluctant to associate with the radicals until he was forced to publish in 1859.

More important, Darwin was an established naturalist with a good reputation, who had all the evidence marshalled in a convincing fashion in one place. From the time he first wrote down his

ideas in 1838, Darwin spent 20 years compiling and fortifying the case for evolution. His book had two main goals: *establishing the fact that life had evolved* and *proposing a mechanism for it*, natural selection. He did both with such overwhelming evidence that he convinced the world, where all his predecessors had failed. In the 1850s, the reality of evolution was becoming more and more obvious to many scientists, but Darwin mustered all the evidence available at the time. These arguments included the following:

1. *The branching organization of life.* From the time of Linnaeus's classification, it was clear that nature was organized into a great hierarchy of groups nested within groups. Humans are part of the group, the primates, that also includes apes, monkeys, and lemurs. The primates, in turn, are part of a larger group, the mammals. The mammals cluster with the birds, reptiles, amphibians, and fish to form a larger group, the vertebrates. The vertebrates, along with insects, molluscs, corals, sponges, and other invertebrates, form the animal kingdom. The very process of diagramming this structure suggested a branching family tree. Why would this be so unless life *did* have a family tree?

    This branching structure was deduced in Darwin's time from the obvious anatomical features of animals and plants. In our generation, striking proof of this branching structure has been added from evidence Darwin could not have anticipated: the molecules in our bodies. Comparisons of the molecular structure of a great number of organisms have repeatedly shown the same pattern of branching suggested by anatomy (Fig. 3.5). For example, our genetic code is 98 percent identical to that of the chimpanzee and

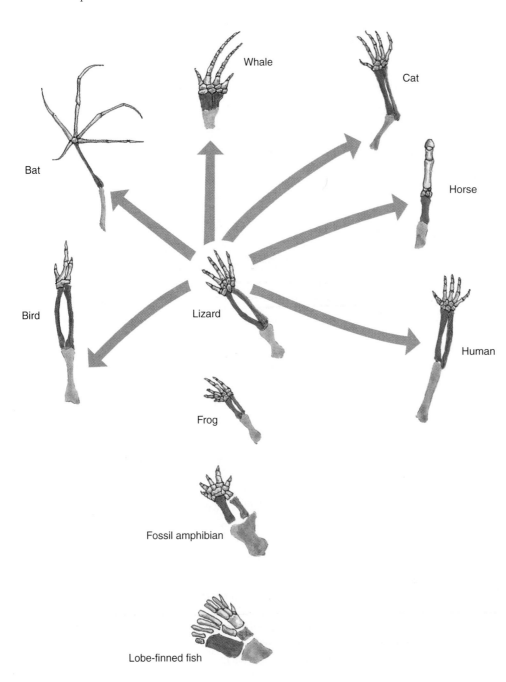

Figure 3.6 Homology of the forelimbs of various vertebrates. The corresponding bones of different animals are indicated by similar colors. The green limb bone is the humerus, or upper arm bone; the red bone is the radius and the blue the ulna of the lower arm; tan indicates the wrist bones and the phalanges (fingers). Although these limbs function in completely different ways (bat wings for flying, whale flippers for swimming, horse limbs for running, human hands for grasping), they are structurally similar and derived from the same embryonic tissues. The simplest explanation for this consistent structural plan is that all vertebrates are derived from a common ancestor, the lobe-finned fishes. (*After T. M. Berra, 1990,* Evolution and the myth of creationism, *Stanford University Press.*)

progressively less similar to the codes of other apes, Old World monkeys, New World monkeys, lemurs, and other mammals. Our immune systems react most strongly to antibodies from apes or monkeys, which are very similar to molecules found in humans, and weakly to rabbit and rat antibodies, which are very different from ours in molecular structure. This selectivity makes sense only if these molecules have been inherited from common ancestors. More closely related animals have had less opportunity for their molecules to evolve and change.

2. **Homology.** Organs with strikingly different functions are commonly built from the same basic parts. For example,

the forelimbs of vertebrates are used for many purposes—grasping in apes and humans, flying in birds and bats, flippers for swimming in seals and dolphins—but they all have the same underlying structure of bones, muscles, and nerves (Fig. 3.6). Why would this be true unless they all had a common ancestor? Certainly there are other ways to construct a wing. (Take insects, for example.) When flying evolved in vertebrates, however, the existence of certain fundamental structures in the vertebrate common ancestor dictated that only a limited number of types of wings could be built. Similar elements derived from a common ancestor are called *homologous* elements. Thus, our upper arm bone,

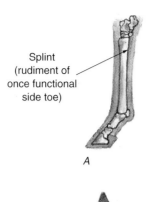

Splint
(rudiment of
once functional
side toe)

A

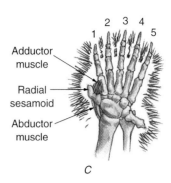

Adductor
muscle

Radial
sesamoid

Abductor
muscle

C

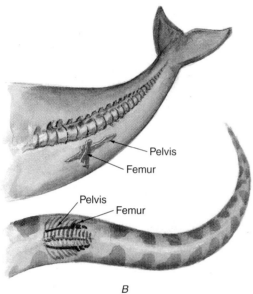

Pelvis

Femur

Pelvis

Femur

B

Figure 3.7   Examples of imperfection in animal design. *A–B:* Vestigial organs show that some organisms still retain useless remnants as reminders of their evolutionary past. For example, the splint bones (remnants of side toes) of horses show they once had three toes on each foot. The tiny pelvis and thigh bones buried inside the trunks of whales and in snakes show they both once had functional hind limbs. *C:* The famous example of the "panda's thumb." Pandas, like all other carnivorous mammals, have all five of their fingers (numbered digits) fused into a paw. But pandas use their paws to strip leaves off the bamboo shoots they eat. Since their true thumb (digit 1) cannot grasp branches, they have a modified wrist bone, the radial sesamoid, as a crude "sixth finger" or makeshift "thumb" to serve as a clumsy substitute. Many similar examples of jury-rigged adaptation clearly show that evolution is opportunistic, and any adaptation just good enough to allow survival will suffice. Such imperfections do not make sense if life was created by a divine designer. *(A–B after T. M. Berra, 1990,* Evolution and the myth of creationism, *Stanford University Press. C after S. J. Gould, 1980,* The panda's thumb, *Norton.)*

or humerus, is homologous with the humerus in all other vertebrate forelimbs. The supports of the bat's wing are homologous with our finger bones. By contrast, structures that perform the same function but are not derived from the same evolutionary origin are known as **analogous** structures. The wings of insects are analogous with those of birds or bats.

3. *Vestigial structures and other imperfections.* People often marvel at the apparent perfection of nature and at the amazing adaptation of each organism to its environment. Darwin pointed out, however, that nature's *imperfections* are more revealing than the examples of good design. Both a divine creator and natural selection could produce animals that are well adapted to their environment, but why would a wise creator give animals handicaps or useless structures? Darwin pointed out that many animals had tiny remnants, or *vestiges,* of features that they no longer used. In your body, you have stunted tail bones that no longer function and an appendix that is now largely useless. Horses have tiny splint bones on either side of their toes, which are remnants of ancient side toes (Fig. 3.7A). When these break, the horse is crippled for life. Whales and pythons have remnants of hips

and hind legs buried deep in their bodies, even though neither uses them anymore (Fig. 3.7B). Why would these features exist unless whales and snakes had evolved from animals that once had hind limbs, horses from three-toed ancestors, or humans from apes with tails?

Other naturalists besides Darwin have emphasized that nature is full of poor designs or of structures that work just well enough for the animal to survive. The panda, for example, has its five fingers fused into a paw, like all other carnivorous mammals. However, the panda is a carnivore that has returned to herbivory, spending more of its time eating bamboo. It cannot use its thumb, fused to the rest of the paw, to strip off bamboo leaves while feeding. Instead, it has a modified wrist bone, which serves as a makeshift "thumb," giving it a weak grasp suitable only for stripping leaves (Fig. 3.7C). Certainly this apparatus does not work as well as a true thumb, but given the panda's evolutionary background, that's all it has to work with. The improvised thumb works just well enough for the panda to survive.

4. *Embryonic history.* The great German embryologist Karl Ernst von Baer (1792–1876) showed that the development

Fish   Salamander   Tortoise   Chicken   Pig   Cow   Rabbit   Human

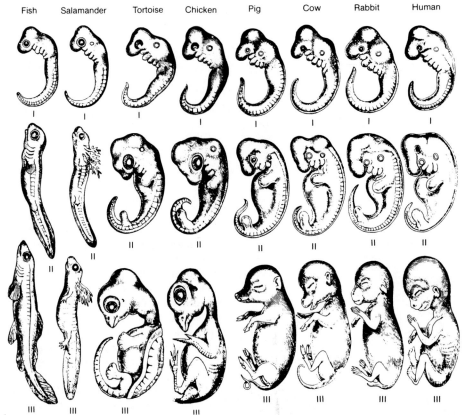

Figure 3.8   Embryos of different vertebrates at comparable stages of development. The earlier stages of development *(top row)* are strikingly similar in every group, regardless of their adult anatomy. Note that each embryo begins with a similar number of gill arches (pouches below the head) and a similar vertebral column. In later stages of development, these structures are modified to yield different adult forms. *(From G. J. Romanes, 1910, Darwin and after Darwin, Open Court, Chicago.)*

of animals is powerful evidence of their ancestry. Whether they develop into fish, amphibians, or humans, all vertebrate embryos start out very similar, with gill slits and a long tail (Fig. 3.8). In fish and amphibians, these gills and tail may persist into adults, but in the embryonic development of humans, these structures are lost. Why would humans have embryos with gills and a tail unless their ancestors also once had these features?

The embryonic history, or ontogeny, of vertebrates retains so much of their evolutionary history, or phylogeny, that German embryologist Ernst Haeckel (1834–1919) postulated his famous "biogenetic law" in 1866. In Haeckel's words, "ontogeny recapitulates phylogeny"— embryonic development repeats evolutionary history. Although this is true in a very general way, it is not true in detail. Many features of adult fish, for example, never appear in human embryos, and other features, such as the yolk sack, the umbilical cord, and the amniotic membrane that surrounds the fetus, are embryonic features that never appear in adults of any species. Haeckel postulated that organisms evolved by adding stages to the end of their embryonic development. We now know that new stages can

be added at the beginning or at any time in development. More important, phylogeny is the *result* of a series of ontogenies, not their cause. Although the extreme elements of Haeckel's ideas are now discredited, von Baer's clear demonstration of the evidence from embryology is still one of the most powerful arguments for the reality of evolution.

5. *Biogeography.* The explorations of nineteenth-century naturalists added many species to the growing catalogue of life. Many regions of the world have unique assemblages of plants and animals found nowhere else. How could these have all migrated from Mt. Ararat? Did creation take place separately on each continent? The finches of the Galápagos Islands, for example, are unique. To Darwin, it made no sense for God to have created new finches for just these islands, when it was simpler to explain them as descendants of ancestral finches blown in from the South American mainland.

Even more striking is the fact that, on many continents, unrelated animals have evolved to occupy similar ecological niches. For example, Australia was originally inhabited by pouched mammals, or **marsupials,** such as kangaroos or koalas. These marsupials developed forms that mimic those of wolves, cats, wolverines, woodchucks,

Placentals

Wolf
(*Canis*)

Ocelot
(*Felis*)

Flying squirrel
(*Glaucomys*)

Groundhog
(*Marmota*)

Anteater
(*Myrmecophaga*)

Mole
(*Talpa*)

Mouse
(*Mus*)

Marsupials

Tasmanian wolf
(*Thylacinus*)

Native cat
(*Dasyurus*)

Flying phalanger
(*Petaurus*)

Wombat
(*Phascolomys*)

Anteater
(*Myrmecobius*)

Mole
(*Notoryctes*)

Mouse
(*Dasycercus*)

Figure 3.9   Extraordinary examples of ecological parallelism between Australian pouched mammals, or marsupials, and their placental mammal counterparts on other continents. Even though marsupials and placentals diverged more than 100 million years ago, similar ecological niches on different continents caused these very different groups of mammals to converge on similar body shapes and ecology. (From Life, an Introduction to Biology, *2nd edition by George S. Simpson and William S. Beck, © 1965. Reprinted with permission of Brooks/Cole.*

anteaters, moles, monkeys, and even flying squirrels—yet all marsupials are unrelated to their non-Australian counterparts (Fig. 3.9). The only logical explanation for this was that Australia was originally colonized by marsupials, which then evolved to fill the niches filled in other parts of the world by true wolves, cats, and the rest. There are hundreds of other examples of this *ecological convergence,* both on earth today and in the geologic past. They are impossible to explain by migration from Noah's ark and make sense only if different kinds of organisms colonized regions unoccupied by competitors and evolved to fill vacant ecological niches.

These lines of evidence convinced Darwin and his contemporaries that evolution had occurred. In the past 140

years, the evidence has only gotten stronger. Even more convincing, scientists have actually seen evolution occur in many different organisms. Subtle evolutionary changes have been abundantly demonstrated in the laboratory, especially with the fruit fly *Drosophila.* A number of new species of plants have evolved in the past century. Insects are constantly evolving new species in response to predators and pesticides. Microorganisms evolve in a matter of hours in response to changes in their environment. Colds and flu come back every year because the viruses that cause them have evolved a new outer coat to make them unrecognizable to our immune system. There is no longer any doubt that life has evolved or that evolution has occurred—we see it around us every day.

Reality of Evolution

Unfortunately, most people blur the distinction between the *reality* that evolution has occurred and *theories or mechanisms* of *how* it has occurred. The biggest problem is the popular misunderstanding of science and how it operates. Hollywood has created a mythology of scientists in white lab coats, working with their bubbling beakers and lightning sparks, marching slowly toward truth. In fact, scientists rarely try to prove something true. More often, they are trying to prove something false!

Philosophers long ago pointed out that, if you want to know if something is true, no amount of confirming observations can ever prove it (in the strict logical sense). For example, thousands of white swans will never prove the statement that "all swans are white." However, a single contrary piece of evidence can prove something false. The famous black swans of Australia, for example, falsify the statement that all swans are white. Scientists conduct much of their search for knowledge by using the **hypothetico-deductive method.** They propose ideas—hypotheses—that might explain how nature operates; then they search for tests that might prove the ideas wrong, eliminating competing hypotheses by deduction. If they falsify one hypothesis, they move on to another. Scientists continually try to knock down their hypotheses until the only ones remaining are ideas that have many confirming tests—they are well *corroborated.* This does not mean that scientists have reached final truth—only that the current set of hypotheses has withstood many critical tests and is the best set of explanations scientists have. A set of hypotheses that withstands many tests and explains a wide range of data reaches the status of theory.

Unfortunately, the public's mistaken idea about science and truth leads to much confusion. Strictly speaking, everything in science must be open to testing, and no conclusion is final. In practical terms, some ideas in science are so well tested that they are the same as "fact" in everyday language. The theory that the earth revolves around the sun is still subject to being falsified, but it is as much a fact as anything else we accept, although we cannot yet stand outside our solar system and watch it happen. The same is true of the theory of evolution. There are so many confirming lines of evidence and so many failed attempts to falsify it that, in popular language, evolution is "fact." To the general public the word *theory* has come to mean something very sketchy or tenuous, such as the various "theories" of how President Kennedy was assassinated. When presidential candidate Ronald Reagan said in 1980 that evolution "is a theory, a scientific theory only, and it has in recent years been challenged in the world of science and is not yet believed in the scientific community to be as infallible as it once was believed," he reflected this popular misconception.

As we have seen, the *reality* of evolution has been demonstrated beyond doubt, but the *mechanisms* of evolution are still under some debate. Darwin proposed one mechanism, natural selection, but there are skeptics within the scientific community. This is healthy. Science stagnates when all the ideas have reached the status of "fact" and there is no dispute. Argument and testing of hypotheses are the only way that science moves forward. As the Reagan quote demonstrates, people who do not understand the scientific process regard the arguments about *mechanisms* among evolutionary biologists as evidence of doubt whether evolution has occurred. Nothing could be further from the truth! The existence of these disputes indicates that the field is dynamic and is not accepting any dogmas as proven. Even if we never reach a complete understanding of the true mechanism for evolution, however, we could not change the fact that it has occurred. We still have no effective mechanism for explaining what gravity is, yet this does not change the fact that objects fall toward the earth or that heavenly bodies travel in paths predicted by Newton's laws of gravitation.

Genetics and the Evolutionary Synthesis

Although most of Darwin's contemporaries were eventually convinced of the reality of evolution, many did not think natural selection was an adequate explanation for how life has changed. In particular, scientists knew little about how favorable variations were inherited. Darwin learned about inheritance and artificial selection in domestic animals by raising pigeons. Like most other breeders of domestic animals, he knew that unusual characteristics in an animal would disappear unless the animal was bred with another having the same characteristic. Random interbreeding would rapidly blend out any desired variant in domestic animals. Darwin worried that this kind of blending inheritance meant that evolutionary novelties would be rapidly diluted when a new type of organism interbred with the main population. He died in 1882 without knowing the solution to his problem, and late in life he even embraced the inheritance of acquired characteristics (misnamed "Lamarckism") as a possible mechanism.

Darwin's theory of evolution was incomplete. What it failed to explain was the first appearance of a new characteristic, such as a new color in pigeons or a longer neck in giraffes. Once such a characteristic appeared, his mechanism of natural selection could explain its success in later generations, eventually becoming a typical characteristic of a new species. Ironically one of Darwin's contemporaries had the solution. He was an obscure Austrian monk named Gregor Mendel (1822–1884), who lived in a monastery in Brunn, in what was once Czechoslovakia. Mendel had taken up gardening as a hobby and began a series of experiments in cross-breeding pea plants. By sheer luck, his pea plants did not show blending inheritance. Instead, it was possible for some features in the parental generation (such as white flowers) to disappear in the second generation of red flowers but then reappear in the third generation (see Box 3.1). Red and white flowers did not blend to form pink flowers. Instead, both characteristics remained discrete, and the less dominant color could reappear in later generations unaltered. Mendel discovered the basic laws of heredity and showed that they could work in simple, predictable fashion. More important for Darwinians, these laws showed that some evolutionary novelties were not "blended out" but could reappear as discrete characteristics after being suppressed for generations. If these novelties reappeared when natural selection favored them, then they could lead to evolutionary novelties.

Box 3.1

Mendelian Genetics and Molecular Biology

Before 1900, most breeders were unable to unlock the secret of heredity because they focused on organisms with complex patterns of inheritance. By accident, Gregor Mendel experimented on plants with simpler patterns of heredity. Pea plants vary only in flower color, seed color, smoothness of seed surface, and size. Most of these variants, however, are in one of only two states. The gene for flower color, for example, has two states (or *alleles*), red allele and white allele. When Mendel crossed a plant with red flowers and one with white, he got nothing but red flowers in the next generation. When he crossed plants of the second generation, however, he got a ratio of three red flowers to one white flower in the third generation. Red and white flowers had not blended to form pink flowers. Instead, the white flower demonstrated *discrete* inheritance, reemerging after disappearing for a generation.

Mendel conducted many such experiments with other characteristics of his pea plants. Eventually he realized that the 3:1 ratio consistently appeared whenever he backcrossed the offspring of the crosses between two different types (Fig. 3.10). He explained this as follows. If the purebred red flower had two red alleles (*CC*) in the first generation and the white flower has two white alleles (*cc*), then the red egg cells have nothing but *C* and the white pollen nothing but *c*. All their offspring would be *Cc*. If red is *dominant* over white, the red color is the only one expressed. In the third generation, crossing *Cc* with *Cc* plants produced four possible combinations: *CC, cC, Cc,* and *cc*. The

—Continued on page 52

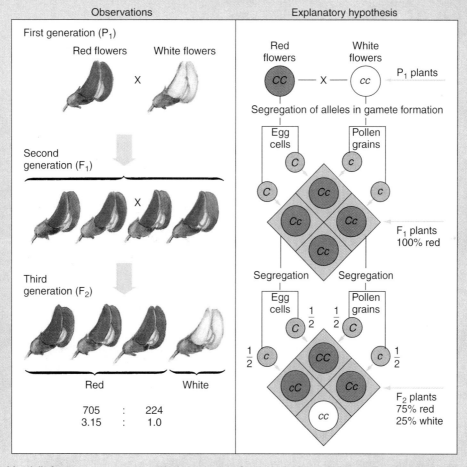

Figure 3.10   Mendel's famous experiments with the genetics of pea plants. According to prevailing notions of inheritance, red- and white-flowered peas should have blended their genes to form pink flowers. Instead, Mendel found that the first generation of peas grown from red and white parents were entirely red. The next generation typically yielded three red flowers for every white flower. Mendel suggested that each flower had two possible genetic *factors* (now called different alleles of each gene), a dominant allele (*C*) or a recessive allele (*c*). In the first generation, red flowers had two *CC* alleles, and white flowers had two *cc* alleles. Mixtures of *Cc* produced red flowers. When red and white flowers were crossbred, they produced all *Cc* descendants in the second generation. When those descendant plants were then cross-fertilized 75 percent of the third generation were either *CC* or *Cc* (producing red flowers) and 25 percent were *cc* (with white flowers). This experiment showed that inheritance was not blending but discrete, so that features could disappear for a few generations and then reappear at a time when natural selection might favor it. Discrete inheritance was important to Neo-Darwinism, since Darwin's critics had argued that blending inheritance would gradually dilute any favorable variations after a few generations for crossbreeding with individuals who did not have the new variation. *(After G. G. Simpson and W. S. Beck, 1965, Life, Harcourt Brace World, Inc.)*

Continued from page 51—

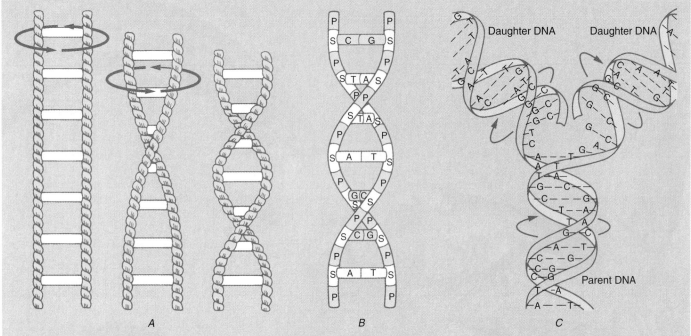

Figure 3.11 *A:* The DNA molecule is similar to a rope ladder, with rigid steps, which has been twisted. *B:* The two strands of the DNA "ladder" are made of sugars (S) and phosphates (P). The "rungs" of the "ladder" are made of nucleotides either cytosine (C) and guanine (G), or by adenine (A) and thymine (T). *C:* Copying of the DNA molecule occurs when the two original strands "unzip" from each other. Additional nucleotides then find their appropriate match and link up to form daughter DNA. *(After S. M. Stanley, 1986,* Earth and life through time, *Freeman.)*

first three have the dominant *C* allele, so they are red, but the fourth has nothing but the *recessive c* allele, so it is white.

Once Mendel proved that heredity was discrete and not blending, genetics had a simple foundation to build upon. More complex patterns of inheritance could now be deciphered. The genetics of many more organisms, from humans to rabbits to fruit flies to bacteria, became better known. Still the biochemical basis for heredity was a mystery. By 1944, biologists had shown that a substance known as DNA found in the nucleus of the cell must be the carrier of genetic information. Several labs raced to decipher the molecular structure of DNA and to explain how it could carry such complex coded information. In 1953, James Watson and Francis Crick argued that the DNA molecule must be arranged like a twisted rope ladder, with the "rope" coiled into two spirals called a double helix (Fig. 3.11). The rigid "rungs" of the ladder were the "alphabet" of the code. In the past 40 years, Watson and Crick's discovery has been confirmed and amplified. Presently the greatest effort in molecular biology is focused on reading the entire genetic code of humans the human genome project), which would lead to enormous breakthroughs in evolutionary anthropology, in medicine, and in the prevention and treatment of genetic defects. The most striking feature of the genetic code is that it is universal to life (Fig. 3.12). With a few possible exceptions, every organism on earth, from bacteria to fungi to plants to humans, use the same genetic code. This is one of the most powerful arguments for the fact that all life is interrelated and once had a common ancestor.

Second letter

	U	C	A	G	
U	UUU UUC } Phe UUA UUG } Leu	UCU UCC UCA CCG } Ser	UAU UAC } Tyr UAA UAG } Stop	UGU UGC } Cys UGA Stop UGG Trp	U C A G
C	CUU CUC CUA CUG } Leu	CCU CCC CCA CCG } Pro	CAU CAC } His CAA CAG } Gin	CGU CGC CGA CGG } Arg	U C A G
A	AUU AUC } Ile AUA AUG Met	ACU ACC ACA ACG } Thr	AAU AAC } Asn AAA AAG } Lys	AGU AGC } Ser AGA AGG } Arg	U C A G
G	GUU GUC GUA GUG } Val	GCU GCC GCA GCG } Ala	GAU GAC } Asp GAA GAG } Glu	GGU GGC GGA GGG } Gly	U C A G

(left axis: **First letter**; right axis: **Third letter**)

Figure 3.12 The genetic code. All of the twenty amino acids found in all life are coded by a sequence of three nucleotides (adenine, guanine, cytosine, or uracil). Although sixty-four possible combinations are possible with three positions and four letters, only twenty amino acids are commonly produced. Notice that, in many cases, the third letter of the sequence is irrelevant. For example, all combinations that begin with "UC" produce serine. Thus, mutations in the third letter are often invisible to natural selection and thus considered adaptively neutral. Three codes (UAA, UAG, UGA) stop the copying or translation of the amino acid sequence. The abbreviated amino acids are Alanine, Arginine, Asparagine (Asn), Aspartic acid, Cysteine, Glycine, Glutamic acid, Glutamine (Glu), Histidine, Isoleucine (Ile), Leucine, Lysine, Methionine, Phenylalanine, Proline, Serine, Threonine, Tyrosine, Tryptophan (Trp), and Valine.

Mendel was the first true geneticist, but he published only one paper on his discoveries in an obscure journal in 1865. This paper was ignored, not only because of its obscurity but also because it was full of mathematics, which was unfamiliar to naturalists of the time. Fifteen years after Mendel's death, his work was independently rediscovered by three scientists. Genetics had grown enormously by 1900, and the time was ripe for Mendel's ideas. Based on the foundation of Mendel's laws of heredity, genetics came to dominate evolutionary theory in the early twentieth century. By the late 1930s, scientists had a large body of knowledge on the genetics of animals such as fruit flies, but most geneticists saw little connection between their work and traditional Darwinian natural selection. Some were impressed with the fact that most changes in genes (**mutations**) were harmful. Others were impressed with abrupt, large-scale changes known as *macromutations* and felt these changes were the stuff of evolution. Most geneticists thought that natural selection performed only a minor role as a sorting mechanism, incapable of generating new forms.

From this eclipse of traditional Darwinism arose a new school of thought, which came to be called the **synthetic theory of evolution,** or **Neo-Darwinism.** The critical breakthrough was the development of mathematical population genetics in the 1930s and 1940s by Sir Ronald Fisher, Sewall Wright, and J. B. S. Haldane. Just like Lamarck and his neck-stretching giraffe, genetics had concentrated on big mutations within individuals. But Darwinian evolution is about small changes accumulating in populations. By constructing mathematical and statistical models of the genetics of populations, these people showed that natural selection could result in evolutionary change. Mutations, even though they are rare and mostly harmful, could still produce enough evolutionary novelty for natural selection to act upon.

In 1937, Russian geneticist Theodosius Dobzhansky integrated genetics with the rest of evolutionary biology in his book *Genetics and the Origin of Species.* Soon other scientists were seeing the connections between their fields and Neo-Darwinism. Ernst Mayr, a specialist in tropical birds, developed the connection between Darwinism and species in nature in his 1942 book *Systematics and the Origin of Species.* Paleontologist George Gaylord Simpson published *Tempo and Mode of Evolution* in 1944. Simpson showed that the fossil record was consistent with the new evolutionary synthesis. By the time of the centennial of *On the Origin of Species* in 1959, the modern evolutionary synthesis was in full flower.

The modern synthetic theory of evolution is very similar to Darwin's conception, with the addition of modern genetics. According to Neo-Darwinians, evolution begins with the variability of natural populations. Populations vary because each member has a slightly different genetic code, or **genotype.** The genetic code is contained in the nucleus of every cell of a plant or animal, in molecules known as **DNA (deoxyribonucleic acid).** The variations in genotype are expressed in the variations in the **phenotype,** or the physical characteristics of the individual organism. Populations are continually supplied with new variation by several mechanisms. One is mutation, which directly changes the details of the genetic code. Also, the strands of DNA can get twisted, reversed, deleted, or duplicated, resulting in a new genetic code. Finally, the fact that sexual organisms are always combining two different genetic codes (one from each parent) each time they mate successfully means that each offspring bears a new genetic code formed by **recombination** of the two codes of its parents. Whatever the source of variation, most natural populations have a tremendous number of different phenotypes. They are all struggling to survive against the limitations of their environment and competition from other species as well as members of their own species.

Lamarck tried to explain the origin of the giraffe's long neck by suggesting that each *individual* giraffe stretched until the neck became longer by use. By contrast, Darwinians would postulate a *population* of primitive giraffe ancestors, some with longer necks and some with shorter ones (Fig. 3.13). When climatic changes or competition from other herbivorous mammals made low-growing plants scarce, only giraffes with longer necks could reach the higher branches beyond the reach of most other herbivores. Those giraffes with the longer necks would be favored by natural selection to survive and breed, passing along their long-neck genes to their offspring. Over enough generations, sustained natural selection for longer necks produces a population of giraffes with generally long necks.

Natural selection has been observed many times in the wild and in the laboratory. The most famous example is the case of the peppered moths (Fig. 3.14). Found in the Midlands of England, these moths once had a speckled black and white pattern on a dull yellow background, a coloration that concealed them from predatory birds when the moths rested on the speckled, moss-covered bark of the plane tree (sycamore). During the industrial revolution, air pollution covered many of the tree trunks with black coal soot. On blackened trees downwind from smoking factories, the normal peppered moths became scarce and a black mutant became common. These dark peppered moths were well concealed against blackened tree trunks, but light-colored moths were now conspicuous and readily spotted by birds. In a classic series of experiments, H. B. D. Kettlewell placed different colors of moths on different backgrounds and watched which ones the birds ate. Naturally the conspicuous form was always the more vulnerable to predation. When the use of oil and natural gas reduced coal soot in the English Midlands in the 1970s, the frequency of dark moths decreased and that of light-colored moths increased as their background resumed its normal light color. The peppered moth vividly demonstrates how selection (in the form of bird predation) is constantly choosing survivors among the natural variation found in all populations.

The Origin of Species

Experiments with fruit flies and peppered moths prove that natural selection changes gene frequencies over time to produce populations with new phenotypes. To Darwin and many of his followers, the proof that species could change over time was adequate to explain the origin of species. This explanation would be sufficient if species were continually transformed into a single daughter species, but how does simple transformational change explain a species splitting into two or more daughter species?

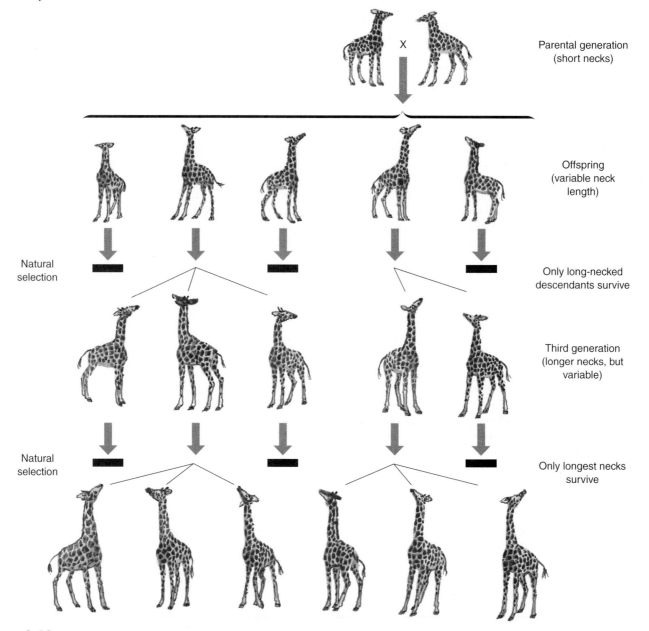

Figure 3.13   Darwin's explanation of evolution as applied to giraffes. Instead of stretching their necks and passing it on directly to their descendants (see Fig. 3.2), each generation of short-necked giraffes had slight variations in the lengths of their necks. Only those with the longer necks survived the process of natural selection and produced the next generation, which started out with slightly longer necks. The long-necked individuals in the second generation were again favored by natural selection, resulting in even longer necks in the third generation.

Figure 3.14   The dark and light varieties of peppered moth on a lichen-covered plane (sycamore) tree near Liverpool, England. *(© Stephen Dalton/Photo Researchers.)*

The discoveries of population genetics in the 1930s showed that large populations tend to be genetically stable. Any new variant is quickly swamped by interbreeding with all the normal individuals in the population. Small populations, by contrast, can evolve rapidly even if only a few members have an unusual characteristic and become the parents of most later generations. This is especially true of populations descended from the original settlers on small islands. Their gene frequencies can differ greatly from those of mainland populations because the founders of the island population may have had an unusual sample of genes relative to the mainland populations they left. This phenomenon is known as the **founder effect.** For example, the blood types of human populations of different Polynesian islands vary greatly between islands because each was settled by only a few boat loads of people who differed from their parent populations. In fact, physical isolation is not even necessary; any kind of genetic isolation of a small inbred population will do. For example, people who belong to small, inbred religious sects, such as the Amish or Mennonites, typically have very different gene frequencies from other Americans because they are all descended from a small number of founders and rarely marry outside their religion.

Clearly genetic isolation was essential if populations were to speciate into new forms different from the parent population. Ernst Mayr developed these arguments further. He suggested that most species arose during some form of genetic isolation from the main parent population. Typically this occurred when a small portion of the main population became separated by a geographic barrier, such as mountain, desert, or ocean. This *peripheral isolate* then continued to diverge genetically, so that, if the barrier later disappeared, the new species would have changed enough that it would no longer interbreed with the ancestral population. Mayr called this **allopatric** (Greek for "different homeland") **speciation**—new species form when a larger population is split up into subpopulations occupying different home ranges. By contrast, populations that live together (**sympatric**) are continually interbreeding and have little chance to differentiate genetically.

In the 60 years since Mayr's (1942) book, the allopatric model continues to be favored as the primary mechanism of speciation in nature. However, it has been shown that the geographic barrier need not be as big as a mountain range. Any barrier that prevents various members of a species from interbreeding will do. Such organisms as the wingless morabine grasshoppers of central Australia move so little distance in their lifetimes that different species are separated by a hybrid zone only a few hundred meters wide. In fact, a sudden mutation that causes organisms to change their mating behavior will stop interbreeding, so it is possible for genetic isolation to occur even in nearly sympatric populations.

Despite his book's title, Darwin never addressed the problem of the *origin* of species as we now understand the concept. The allopatric speciation model argues that large populations are relatively static and immune to change. They should not transform gradually into new species. Neo-Darwinian speciation theory predicts that evolutionary novelties should appear relatively quickly in small populations, typically on the periphery of the geographic range. Thus, species should arise suddenly (in the geologic sense) from static main populations.

The Fossil Record and Evolution

Although Mayr's book was published in 1942 and was read by generations of paleontologists, they were remarkably slow to see the implications of speciation theory for the fossil record. Expecting Darwinian gradual transformations from one species to another, they kept trying to find examples of **phyletic gradualism** in the fossil record long after most biologists had ceased to believe in it. In 1972, 30 years after Mayr's book, Niles Eldredge and Stephen Jay Gould pointed out that the allopatric speciation model made a very different prediction about the fossil record. If species really arose in peripheral isolates, then they should appear suddenly in our incomplete fossil record, migrating in from their places of isolation, where fossilization is less likely. More important, Eldredge and Gould predicted that most species would remain unchanged over long periods of time, punctuated only rarely by speciation events. Eldredge and Gould called their model **punctuated equilibrium,** and it caused a debate in evolutionary biology and paleontology that still rages 30 years later.

Since 1972, many examples of both phyletic gradualism and punctuated equilibrium have been documented and debated. Sometimes the same fossil record has been used to support both hypotheses. For example, the evolution of the horse was first documented in the 1870s as an example of gradualism (Fig. 3.15A). Horses were portrayed as having evolved in a straight line from tiny, four-toed *Eohippus* of the Eocene to living horses of the genus *Equus*. This evolutionary trend showed not only an increase in size through 50 million years but also other changes related to becoming better adapted for running and grazing on gritty grasses. As additional horse fossils were collected, however, this gradualistic, straight-line trend of one lineage was revised. More and more fossil horse specimens showed that many species of horse had lived side-by-side. In the middle Miocene in Nebraska, for example, one locality produced 12 species of horse! Recent work has shown that most of these species were static through most of their long histories. The classic, oversimplified model of horse evolution as a gradual transformation has been replaced with a "bushy" family tree, with many species coexisting and remaining unchanged for millions of years (Fig. 3.15B). This is exactly what the punctuated equilibrium model predicts.

Although this debate is not over, several conclusions have become clear. First of all, the fossil record yields examples of *both* phyletic gradualism and punctuated equilibria, although the relative dominance of one over the other is still controversial. More important, Eldredge and Gould pointed out something that paleontologists had known (but not emphasized) for over a century: most fossil species do indeed appear suddenly and then persist unchanged for millions of years. In fact, this long-term stability over millions of years, through well-documented climatic changes, greatly exceeds the stability of species conceded by traditional Neo-Darwinism. To some paleontologists, species are more than just populations and genes. They are real entities that seem to have some kind of internal stabilizing mechanism preventing much phenotypic change, even when selection forces change. Clearly the fossil record produces some unexpected results that are not yet

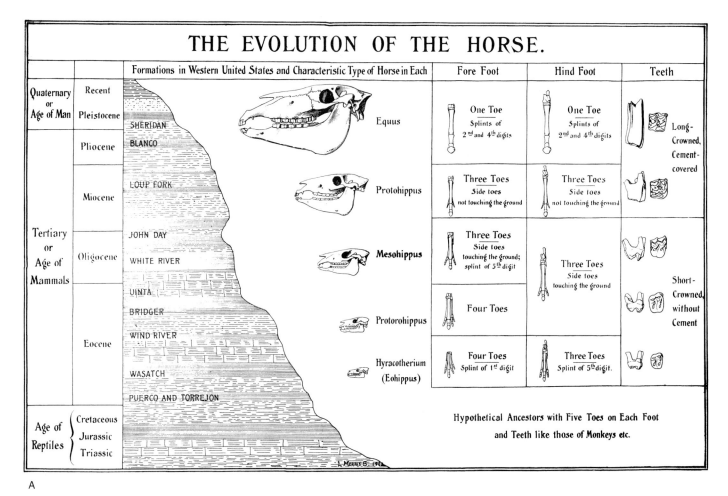

A

Figure 3.15A   Traditional representation of horse evolution, which showed a simple linear sequence through geologic time. Although the anatomical details were correct, this illustration gives the false impression that there was a simple pattern of a single lineage evolving to produce the modern horse. *(After Matthew, 1926. Image #: 35522 The American Museum of Natural History Library.)*

consistent with everything we know about living animals and laboratory experiments. This is good news. If the fossil record taught us nothing that we didn't know already from biology, there wouldn't be much point to evolutionary paleontology.

The fossil record demonstrates some phenomena extraordinarily well. Over the 3.5 billion years during which life has evolved, many patterns are repeated. When organisms speciate, they must develop traits that differentiate them from their ancestors. This is known as **divergence.** Normally, the difference between two closely related species is slight. However, at certain times in earth history, the planet has had many new ecological niches for exploitation. In this case, speciation can be very rapid and extreme divergence occurs. For example, after the extinction of the dinosaurs, the world was essentially empty of large land vertebrates. Surviving mammals in the Paleocene were all tiny, insectivorous beasts, mostly rat-sized or smaller. They found themselves in a world full of vacant ecological niches ready for a variety of forms, including herbivores of every size, tree-dwelling fruit and seed eaters, and meat eaters. Within a few million years,

mammals began to diverge into a great variety of dietary types of different sizes, resulting in an evolutionary "explosion" known as an **adaptive radiation** (Fig. 15.44). By the beginning of the Eocene, fewer than 10 million years later, mammals occupied nearly every available niche, from terrestrial herbivores and carnivores, to tree-dwelling primates and rodents, to flying bats and even whales!

The availability of ecological niches leads not only to divergence but also to its opposite, **convergence.** In this case, two or more unrelated animals have independently converged on the same body form because it represents an unexploited ecological opportunity. For example, we have seen how the Australian continent, settled only by marsupials, had no competition from placental mammals found in the rest of the world (placentals give birth after a long development inside the uterus, as humans do). Consequently the niche for wolves, cats, moles, and so on were open and eventually occupied by pouched mammals, which are extraordinary mimics of their non-Australian counterparts (Fig. 3.9).

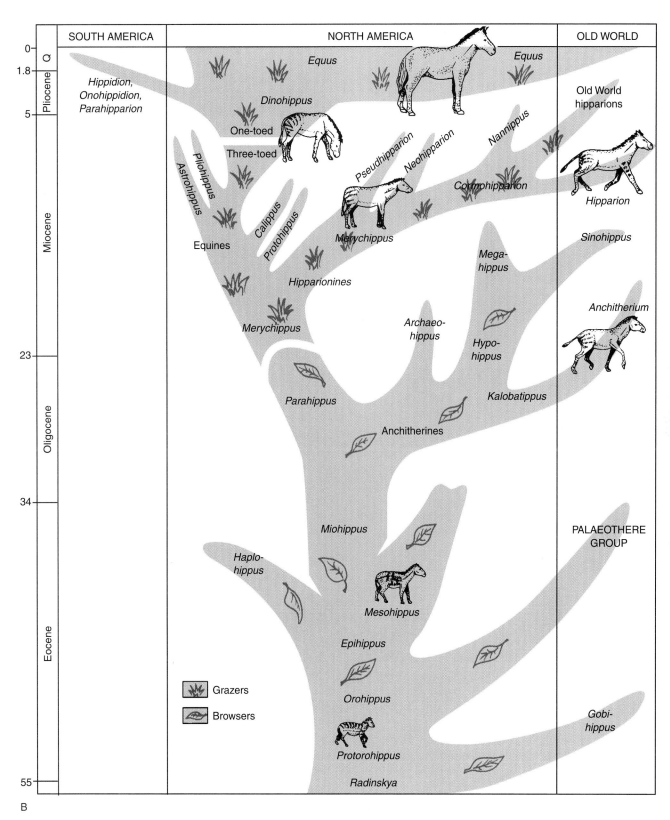

Figure 3.15B   *Current version of horse evolution. The evolution of the modern horse is the result of multiple lineages branching and then going extinct or splitting to form new lineages. Throughout horse evolution, there were several species of horses living at any given time in the geologic past.*

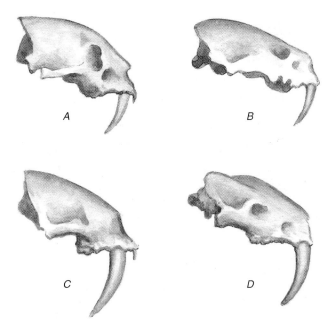

Figure 3.16   Parallel evolution of saber-toothed carnivorous mammals. The saber-toothed adaptation appeared independently at least four times during mammal history, so clearly it was a very successful lifestyle. A: The early Oligocene nimravid *Hoplophoneus*, which was catlike in body form but apparently related to dogs and bears. True cats are represented by B: the Pliocene cat *Machairodus* and C: the Pleistocene cat *Smilodon*, famous from the La Brea tar pits. D: The saber-toothed pouched mammal *Thylacosmilus* from the Pliocene of South America. Although it closely resembles the true cats, its closest relatives are other pouched mammals, such as opossums and kangaroos. Not shown is the saber-toothed creodont *Apataelurus*, which was a member of an extinct group of archaic carnivorous mammals distantly related to living carnivorans. (*Modified from G. G. Simpson, 1941*, American Museum Novitates, *no. 1130.*)

A single taxonomic group may have repeatedly evolved descendants that converge on the same body form. For example, in the microfossils and ammonites, certain elaborate shell types have reevolved from simpler forms after a mass extinction. This is known as **iterative** (repeated) **evolution.** In other cases, animals of different taxonomic origins have repeatedly converged on a body form that is not currently occupied by a living species. For example, there are no surviving saber-toothed cats; they became extinct at the end of the last ice age about 9,000 years ago. However, the niche for saber-toothed carnivorous mammals must have been an extraordinarily successful one over the geologic past because it has been occupied at least four times in the past 40 million years (Fig. 3.16). In addition to the famous saber-toothed cat of the ice ages, saber-toothed beasts developed in the nimravids, another extinct family of carnivorans unrelated to true cats, and in another group of extinct carnivorous mammals, the creodonts, only distantly related to living carnivorans. During the liocene in South America, there was even a pouched marsupial which developed saberlike canine teeth!

Challenges to the Neo-Darwinian Synthesis

As soon as a science seems to have all the answers, it becomes stagnant and dull. New unsolved problems and controversies are essential to scientific progress. Stale dogmas that have been accepted for too long must be challenged. To some evolutionary biologists, the Neo-Darwinian synthesis approached this level of stagnation about the time of the centennial of *On the Origin of Species* in 1959. All the major problems seemed to be solved, and all that remained was fitting little pieces into the grand picture. As one might expect of a dynamic science, this unanimity did not persist. The past 40 years have seen a number of challenging new views to revitalize evolutionary theory.

Neutralism

One of the first challenges came from genetics. Traditional Neo-Darwinism holds that selection is all-powerful, invisibly operating on organisms all the time. This view became known as **panselectionism** (*pan* means "all" or "everything" in Greek). But in 1966, experiments on genetic variability showed that organisms have far more genetic information than they need. Apparently most genetic variability does not produce phenotypic characteristics that selection can act upon. If selection cannot detect these genetic differences, then they must be invisible to selection, or adaptively neutral. This school of thought came to be known as **neutralism.**

At one level, there is no question that most genetic changes are adaptively neutral. For example, the sixty-four possible combinations in the genetic code (Fig. 3.12) specify only twenty amino acids, plus a few "stop" codes. In most cases, the first two nucleotides of the triplet specify the amino acid, and a change in the third nucleotide makes no difference. For example, if the first letter is cytosine (*C*) and the second guanine (*G*), the amino acid will be argenine no matter what the third letter is. It is easy to see that the great majority of genetic changes (especially if they happen in the third position) could happen without being detected by selection.

If neutralism is correct and most genetic changes occur without interference from natural selection, then random mutations can keep occurring without being weeded out. Random mutations might eventually accumulate until they are big enough to result in new characteristics. At this point, selection may or may not operate, but the central issue is that neutralists thought that some characteristics might arise through random mutations, with only minimal interference from selection. Richard Lewontin suggested examples of characteristics that, although they provide selective advantage, are neutral in that they provide the same selective advantage as other possible combinations. For example, African black and white rhinos have two horns, one on the nose and one on the forehead. Indian rhinos, on the other hand, have only one horn, on the nose. Lewontin did not deny that natural selection certainly works for a horn for protection in battle. Whether a rhino has one or two horns appears to be selectively neutral—both rhinos survive with their respective number of horns, and there is no apparent advantage or disadvantage to having one horn or two. The number of horns is an accident of history; African rhinos had two-horned ancestors, and Indian rhinos had one-horned ancestors. In the geologic past, both one-horned and two-horned rhinos have lived together, and there is no evidence that one was selected over the other on the basis of its horn.

Traditional Neo-Darwinians concede that a large portion of the genetic code must be adaptively neutral, but they argue that much of the supposedly neutral genetic information may provide subtle selective advantages that we cannot yet detect. Case studies have shown unexpected uses for variations previously thought to be useless. In addition, genes have complex systems of coding for phenotypes. Many genes code for more than one feature simultaneously. If selection maintains a gene because it codes for one very important feature, all the other features it determines may be "carried along," even if they are neutral or even slightly harmful.

The panselectionist-neutralist controversy continues to rage, although evolutionary biology clearly has retreated from the extreme panselectionist position it held 40 years ago.

Inheritance of Acquired Characteristics Revisited

Although the inheritance of acquired characteristics had been thoroughly discredited by 1959, a few evolutionary biologists were unhappy with the Neo-Darwinian alternative. Natural selection on variable populations does not seem to provide much opportunity for rapid response to environmental changes. Many generations must die for most innovations to become a dominant part of the genotype. If, however, some mechanism of "feedback" allowed changes in the environment to result directly in genetic changes, then organisms could adapt much more quickly.

Several experiments suggest that some characteristics *may* be incorporated directly into the genetic code. Take the immune system, for example. Whenever you are exposed to a disease, your immune system develops antibodies, which kill the foreign infection. However, you acquire this immunity during your lifetime—you are not born with it in your genetic code. Several recent experiments suggest that mice that have been exposed to a specific disease and then become immune to it pass their immunity on to their offspring. These experiments are still very controversial, but the flexibility and rapid response of the immune system have long been a puzzle to Neo-Darwinism.

Whatever the eventual conclusion of this inherited immunity controversy, molecular biology has produced evidence that some genetic information can get from the phenotype back into the genotype. Barbara McClintock won the Nobel Prize by documenting "jumping genes," which can move from one spot on a DNA strand to another spot on the same strand, changing the phenotypic results. Other experiments have clearly shown that external DNA can be incorporated into a cell and possibly into the host DNA. In one case, different bacteria appeared to exchange bits of genetic material, a switch that enabled them all to evolve a new mutation rapidly. A recently discovered group of viruses known as retroviruses (such as the HIV virus that causes AIDS) copy their genetic information into the genetic code of their host. Could this be a mechanism that allows environmental changes to be translated directly back into the genetic code? These discoveries are still too recent and controversial to evaluate, but clearly the genetic code holds more surprises in store for us.

Macroevolution

Neo-Darwinists have clearly demonstrated that natural selection can cause small-scale evolution within species, or **microevolution** (such as changing the eye color or wing shape of fruit flies), but does natural selection explain the origin of major evolutionary changes, or **macroevolution?** How did complex structures such as the eye or the bird's wing evolve? How do we make the jump from, say, a running reptile to a flying bird? What good is "half a wing"? Natural selection experiments have shown that it is easy to change small features (such as the veins on the wings of fruit flies) but impossible to change fruit flies into something else. In some experiments, fruit flies subjected to long-term selection adapted at first, then died off rather than evolve into something new. As punctuated equilibrium has suggested, species are not infinitely flexible. They seem to have limits to their ability to adapt, and then they either die off or change radically, so how does evolution make radical changes?

Neo-Darwinists have suggested that some changes are not as impossible as they first appear. For example, in 1871, one of Darwin's critics, St. George J. Mivart, argued that a complex structure such as the eye could not have evolved by gradual stages. What good is half an eye? The spectrum of life suggests that there are intermediate stages to complex organs, however. Within the molluscs, for instance, there is a full range of eye types, including simple, light-sensitive spots; dimples with light receptors; and well-developed eyeballs with lenses as good as our own Fig. 3.17). The complex eye of the octopus could easily have evolved from ancestors with simpler eyes.

Another important explanation is **preadaptation.** Organs adapted for one function may be "taken over" and used for a new function. For example, the earliest fish had no jaws. Embryonic evidence shows that jaws were derived from parts of the first and second gill arch, which originally supported the gill openings in a fish (Fig. 3.18). The earliest jawed fish took over one structure (the gill arch) and opportunistically transformed it into another, the jaws. To cite another example, the bones of the middle ear (the "hammer," "anvil," and "stirrup") are embryologically derived from the jaw joint (Fig. 3.19). This seems less bizarre when you realize that some reptiles hear with their lower jaw when they are in contact with the ground. (Snakes cannot hear the snake charmer—they are swaying in response to his body motions.) Fossil evidence and embryology show that these bones became smaller and smaller in the ancestors of mammals until they disappeared from the jaw and became part of the hearing mechanism.

Another example is the evolution of the bird's wing, which seems impossibly complex at first. However, feathers first evolved from scales, apparently for insulation (one of their main functions in birds even today). When ancestors of birds began to flap their forelimbs, they already had scales modified for creating aerial lift with minimum weight. Feathers apparently were preadapted for insulation and later became useful for flight.

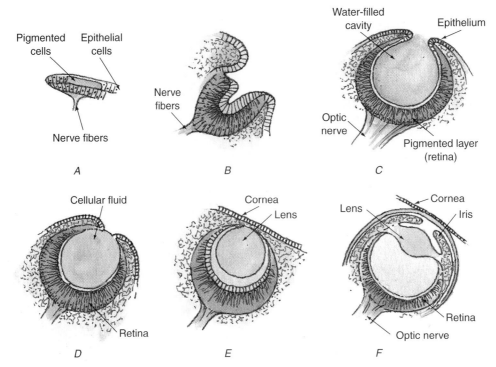

Figure 3.17   Stages in the evolution of the eye as demonstrated by the molluscs. Darwin's critic, St. George Mivart, argued that natural selection could not explain a structure as complex as the vertebrate eye. What good is half an eye, if some of the critical parts are missing? But the stages of complexity of molluscan eyes clearly show that each intermediate step could be fully functional and allow descendants to evolve with more complex eyes. *A:* Simple pigment spot for sensing changes in light intensity. *B:* Folding of pigment cells concentrates their activity, improving light detection. *C:* A partly closed, water-filled cavity of pigment cells, which forms images on the pigmented layer, as does a pinhole camera. *D:* Transparent cellular fluid is used instead of water, forming a barrier that protects the pigmented layer (retina) from injury. *E:* A thin film or transparent skin covers the eye for further protection. *F:* The complex eye found in squids, with an adjustable iris diaphragm and a focusing lens. *(Modified from H. W. Conn, 1900, The method of evolution, Putnam's.)*

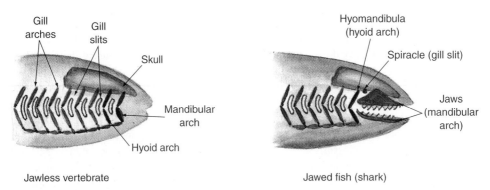

Figure 3.18   The transformation of the vertebrate gill arches into jaws. Ancestral jawless vertebrates had a series of gill arches below their braincase and spine, which supported the gill slits between them. When vertebrates needed a structure to improve their food-gathering capacity, they modified their first gill arch (the mandibular arch) to become the upper and lower jaws. The second gill arch (the hyoid arch) later became the support for the jaws. This transformation not only is consistent with the fossil evidence of early jawed fish but also occurs during embryonic development. Early embryos are jawless, but their jaws form from gill arches during their later developmental stages. *(Modified from S. L. Luria et al., 1981, A view of life, Benjamin Cummings.)*

It is difficult to transform *adults* of most organisms into new species, but the *embryonic stages* of life are not so stereotyped. Small changes in the developmental timing of maturation of the body with respect to sexual maturity can result in big changes in body form. In some cases, these changes allow species to escape

from the trap of their adult adaptation. For example, a Mexican salamander known as the axolotl never seemed to complete its metamorphosis into a land-dwelling salamander with lungs (Fig. 3.20). Instead, it became a sexually mature adult while retaining its juvenile gills. When French scientists experimentally fouled

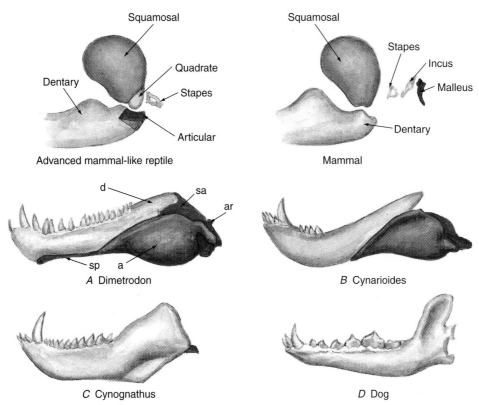

Figure 3.19   Transformation of the jawbones into ear bones during mammalian evolution. The ancestors of mammals, such as the synapsids *Dimetrodon, Cynarioides,* and *Cynognathus* had a number of accessory bones (colored) in addition to the tooth-bearing dentary bone. As synapsid jaws encountered increased stresses of a stronger bite and more complex chewing, the accessory bones were a source of mechanical weakness in the jaw. These bones gradually shrank during synapsid evolution, so that early Permian synapsids, such as *Dimetrodon,* had relatively large accessory jaw elements, but early Triassic synapsids, such as *Cynognathus,* had nearly lost them altogether. In true mammals, such as the dog, these additional jaw elements disappeared altogether, except for the articular bone, which had been the main jaw joint along with the quadrate bone of the skull (top). When the squamosal bone of the skull came into direct contact with the dentary and developed a new jaw joint, it made the old quadrate/articular joint unnecessary. However, many reptiles (and possibly synapsids) hear through their jaws and conduct the sound to the inner ear through the jaw joint and stapes (stirrup bone, a remnant of the hyoid arch in Fig. 3.18). When the quadrate and articular bones became unnecessary as jaw joints, they retained their role in sound conduction and became the incus (anvil) and malleus (hammer) of the middle ear. *(Modified from S. L. Luria et al., 1981, A view of life, Benjamin Cummings.)*

Figure 3.20   The Mexican salamander *Ambystoma,* known to the Aztecs as the axolotl. During normal conditions, the axolotl retains the larval gills into sexual maturity, allowing it to retain the aquatic existence of its larvae. However, when harsh conditions come along, it completes its metamorphosis into an adult with lungs, allowing it to find a new lake and survive. The evolutionary flexibility of changing the timing of sexual maturity with respect to development allows an organism to make radical evolutionary changes with minimal genetic change and is probably the major mechanism for the evolution of new body forms. *(From Dumeril, 1867, Ann. Sci. Nat. Zool., v. 7, pp. 229–234.)*

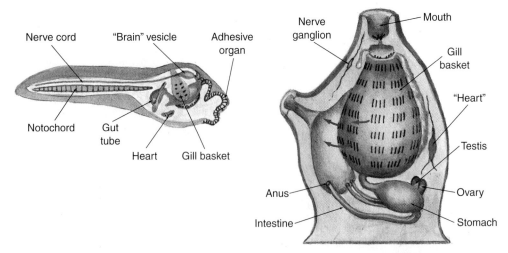

Figure 3.21   The ancestral vertebrate may have evolved by changing developmental timing, as in the axolotl. Our closest living invertebrate relatives are the tunicates, whose adult forms are sedentary colonial filter feeders *(right)*. However, tunicate larvae *(left)* are very "tadpole"-like, with a fishlike swimming tail and a long, cartilaginous notochord, the precursor of the bony vertebral column. If our tunicate ancestors had become sexually mature during their swimming larval stages, they would have produced descendants that are very similar to the most primitive, jawless vertebrates, such as the hagfish. *(Modified from S. L. Luria et al., 1981, A view of life, Benjamin Cummings.)*

the axolotl's ponds, the salamanders suddenly "grew up" and lost their gills for lungs. Apparently the axolotl has slowed down its developmental timing, retaining juvenile features into sexual maturity so that it can exploit an ecological niche in ponds that is unavailable to a land-dwelling salamander.

This mechanism may have even resulted in the origin of vertebrates. One of our closest invertebrate relatives is a tiny, saclike organism known as the tunicate (Fig. 3.21). The adults are attached colonial forms that filter seawater through their gill basket and feed on the detritus. How could this immobile organism with no eyes or tail be ancestral to fishes? Its larva, however, is very fishlike, with a long tail supported by a rod of cartilage, much like that of the earliest fishes. Vertebrates apparently arose when the juveniles of tunicates became sexually mature and never completed their metamorphosis.

We have seen that the apparently large gaps between organisms are not so large when the potential of preadaptation and flexibility in developmental timing are factored in. Obviously not all examples of macroevolution have been fully explained. The mechanisms suggested here, however, point to ways in which such changes could have occurred. Clearly many problems await breakthroughs in research. Evolutionary biology is not stagnant but, rather, full of conflict, controversy, and surprises, which make it one of the most interesting areas of modern science.

Evolution and Creationism

When Darwin's book was published in 1859, it caused a tremendous uproar among educated people all over the Western world. Most did not object to the scientific evidence for the evolution of animals and plants but were outraged that humans were not separate from the rest of the animal kingdom. However, Darwin's case was so strong that, by the time he died in 1882, much of the

educated world had accepted the reality of evolution, although not all had accepted natural selection as its mechanism. When Cornell University President Andrew Dickson White published *A History of the Warfare of Science with Theology in Christendom* in 1896, he assumed that religion would never again interfere in the domain of science.

Evolution remains the best scientific explanation for the diversity and distribution of the life that we see today and in the fossil record. Our scientific goal is to seek physical and biological causes for the changes that life has undergone and to integrate those causes with the physical history of the earth in order to arrive at a satisfactory understanding of diversity. Since its publication, *On the Origin of Species* has been remarkably successful in moving us toward that goal. Scientific explanations contain the seeds for their own refutation through hypothesis testing; therefore, the scientific process is open-ended and self-correcting. Darwin's theory of evolution qualifies as a good hypothesis because it generated predictions that led to fruitful new observations and experiments, which provide tests of the theory. Indeed, recent probing of Darwinism has led to the new ideas discussed in this chapter, which challenge the original hypothesis that natural selection is the only mechanism of evolutionary change. Even if the theory were to be proved wrong, it would still have to be regarded as a good hypothesis because it has provided such a fruitful stimulus for research for a century and a half.

The evolutionary view of life has been so long accepted by the scientific community that it would be unnecessary to reaffirm it were organic evolution not under attack since about 1970 by the advocates of so-called "scientific creationism". This book is not the place to attempt a full discussion of that movement, but it does seem necessary to state that the basis of "scientific creationism" is the belief that all life forms were divinely created simultaneously only a few thousand years ago and that they have undergone only trivial modifications since that time. Because the

Creationist position requires a supernatural event caused by supernatural forces no longer in operation, criticism of it lies beyond the reach of natural science, for such claims can be neither proved nor falsified. To support their position, Creationists have had to deny the validity of radioactive dating of rocks, to appeal to a single universal flood to explain the occurrence of all fossils as well as mountains and canyons, and even to suggest that light from distant stars was created on its way toward the earth. It is as if the entire universe was created, together with all your memories, only minutes before you read this paragraph and, for whatever reason, was made to appear very old. Whatever personal appeal such a view may have, there is no place for it in science because it cannot be subjected to the same kinds of testing that correct and strengthen scientific explanations.

In the United States, the battle has focused on the teaching of evolution in schools. In the famous 1925 "Scopes monkey trial," the state of Tennessee outlawed the teaching of evolution in public schools. The last of these "monkey laws" was finally overturned by the Supreme Court in 1967, for the simple reason that creationism is a religious belief and the Constitution forbids the government to favor one religion over another. Creationists then changed their tactics, omitting the direct references to God from their "public school editions" of textbooks. In 1982, an Arkansas Federal Court saw through this smokescreen and struck down their attempt to sneak religion into science classes. Since that time, Creationists have taken yet another tack. They are now appealing to our society's notion of fairness, arguing that their ideas should be given "equal time" with other ideas (such as evolution) that compete with their worldview.

After repeated losses in the courts, the Creationists took a new approach. For public consumption, they omit all reference to religion and call their ideas by a new label: "intelligent design." They point to example after example of complex natural systems and argue that they show intelligent planning that cannot be explained by evolution. In a nutshell, the "intelligent design" argument is a rehash of the old "natural theology" arguments of William Paley in the late 1700s. Paley argued that nature appeared to be highly complex and intelligently designed, and to his mind, this suggested that it had to have had a Designer. Natural theologians assumed that the Designer had to be the Judaeo-Christian God, although they never considered the beliefs of religions and cultures that do not have a divine creator of the universe.

Ironically the entire natural theology/intelligent design argument was discredited by Scottish philosopher David Hume in 1779, twenty years before Paley's work was published. Hume pointed out that, even if one accepts the idea that nature appears to be intelligently designed, it does not necessarily follow that the Designer was the Judaeo-Christian God—it could have been the deity of another culture or religion or even a committee of deities. Even more fatal to the intelligent design argument are the numerous examples of poor design or of jury-rigged features in nature that work just well enough for organisms to survive but are not well designed in any engineering sense. As Hume points out, these examples of poor design create problems for those who would argue that nature reflects a divine Designer. If the design is poor, does that mean the divine Designer was incompetent, not trying hard enough, or not paying attention?

Why should Creationism not be given equal time in our public schools? The debate over *whether* evolution has occurred has long since been settled in scientific circles, even though the mechanism is not fully understood. Ideas that have no scientific credibility should not be included in science classes. Otherwise, any notion held by any minority could demand "equal time" in science classes and hopelessly clutter the curriculum. For example, members of the Flat Earth Society believe that the earth is flat and lies at the center of the universe and that photos taken by the astronauts showing a spherical earth are a hoax! Should these beliefs also be given equal time in geology and astronomy textbooks, even though they have been rejected for almost 500 years? As for uncertainty about evolutionary mechanisms, we still believe in the validity of gravity, even though its mechanism is only dimly understood.

Instead of destroying one's appreciation of nature, the modern evolutionary perspective broadens it for many people. The panorama of the birth and death of stars and galaxies, of 4.6 billion years of earth history, and of the spectacle of the ascent of life is for them more inspiring and enriching than a narrow, stagnant world only 6,000 years old. Moreover, because creationism closes the door on further inquiry, it gives none of the satisfaction for the human spirit that is provided by the continuing challenge of the scientific exploration of nature. In the concluding sentence of *On the Origin of Species,* Darwin wrote: "There is grandeur in this view of life, . . . whilst this planet has gone cycling on according to the fixed laws of gravity, from so simple a beginning endless forms most beautiful and most wonderful have been and are being evolved."

Summary

Evolution
- **Organic evolution** is the cumulative change of organisms through time.

History
- Buffon's view of evolution was based on his observation that the environment directly caused **variations** in parents that were passed on to offspring.

- Erasmus Darwin introduced the idea that changes occurred during the early development of the individual and that new characters were acquired to make the individual better fit for its environment.
- Lamarck was the first evolutionist to address the problem of how a new characteristic arose. He said that, as the environment changed, the needs of organisms changed and thus their habits changed. These new habits induced the

formation of new features, such as the longer neck of the giraffe, and these new **acquired characteristics** were inherited by offspring.

Darwin

- Charles Darwin's chief contribution to evolutionary theory was the concept of natural selection. When new characteristics appear that provide even a slight advantage to an organism in its environment, that advantage will be selected for through reproduction.
- **Natural selection** is one of the few processes of evolution that can be demonstrated. For example, the protective coloration found in the peppered moths of the English Midlands enabled darker-colored moths to survive predation by birds on plane trees darkened by the soot of the industrial revolution.
- Darwin showed that evolution occurred by many lines of evidence, including the following:
  1. The **branching organization of life** suggests a branching family tree.
  2. The organs of animals may be used for different functions, but they have a common **homologous** anatomical pattern due to their common ancestry.
  3. Many animals retain **vestigial** structures, remnants of very different ancestral anatomy.
  4. The **embryonic history** of an organism retains features of the organism's evolutionary history.
  5. The **biogeographic distribution** of organisms reflects evolution in many different places, not dispersal from the supposed landing site of Noah's Ark.
- Evolution has been observed many times, so is as well established in science as the fact of gravity or the fact of a spherical earth revolving around the sun. Although scientists may argue about *how* evolution occurred, there is no longer any doubt in the scientific community that evolution *has* occurred.
- In 1865, Mendel discovered the hereditary unit—later named the **gene**—and showed how new characteristics are transmitted to offspring. Mendel's work was rediscovered in 1900, leading to the birth of modern **genetics.** By 1953, the basis of heredity in the molecule **DNA** had been discovered. From molecular biology, we have learned that random changes in genes, called **mutations,** produce new characteristics. If these provide better **adaptation,** natural selection preserves them.

Evolutionary Theory

- In modern evolutionary biology, species are thought to arise when they become **genetically isolated** in small populations, allowing new genes to become dominant. These small, peripheral **allopatric** populations may eventually return to their homeland, but by then they may have become new species, which can no longer interbreed with their ancestral species.
- Darwin envisioned evolution as small, gradual changes within a species over long periods, leading to new species by very small steps. This is known as **phyletic gradualism.**
- The allopatric speciation model suggests that species do not arise gradually in the main population but, rather, differentiate in peripheral populations and then suddenly reappear in the ancestral region. As a result, the fossil record is **punctuated** by the seemingly sudden appearance of new species, after which these species persist through long periods of **equilibrium.**
- During evolution, organisms can either **diverge** from a common body plan or **converge** from different body plans onto the same anatomy. Rapid divergence can lead to an **adaptive radiation** of many new species into an "empty" environment, eventually diversifying to fill smaller and smaller **ecological niches.**
- Although Darwinian natural selection is widely accepted as the primary mechanism of evolution by most evolutionary biologists, there are always new scientific challenges about *how* evolution has occurred. Some argue that many genes are **selectively neutral,** allowing random changes to accumulate without being selected for or against. Others have suggested that some acquired characters may be able to re-enter the genes, at least in single-celled organisms.
- The small-scale genetic and anatomical changes abundantly demonstrated in laboratory experiments are examples of **microevolution.** Some scientists suggest that large-scale evolutionary change, or **macroevolution,** is more than simple microevolution scaled larger. For example, the difficulty of inventing radically new anatomical structures (such as a wing or an eye) might be explained if they were **preadapted** for some other use and then opportunistically taken over for a new function. In other cases, changes in **developmental timing** allow an organism to escape the ecological trap of its adult anatomy by reproducing while still in its juvenile form.

Readings

Darwin, C. 1859. *On the origin of species by means of natural selection, or the preservation of favoured races in the struggle to survive.* London: John Murray.

Dawkins, R. 1982. *The extended phenotype: The gene as a unit of selection.* New York: Oxford University Press.

———. 1987. *The blind watchmaker.* New York: Norton.

Desmond, A. 1991. *The politics of evolution: Morphology, medicine, and reform in Radical London.* Chicago: Chicago University Press.

———, and J. Moore. 1991. *Darwin: The life of a tormented evolutionist.* New York: Warner.

Dobzhansky, T. 1937. *Genetics and the origin of species.* New York: Columbia University Press.

Eldredge, N. 1982. The monkey business. *A scientist looks at Creationism.* New York: Washington Square.

———. 1985. Time frames. New York: Simon and Schuster.

———. 2000. The triumph of evolution and the failure of Creationism: New York: Freeman.

———, and S. J. Gould. 1972. Punctuated equilibria: An alternative to phyletic gradualism. In *Models in paleobiology,* edited by S. J. Gould. San Francisco: Freeman Cooper.

Frazzetta, T. H. 1975. *Complex adaptations in evolving populations.* Sunderland, Mass.: Sinauer.

Futuyma, D. J. 1979. *Evolutionary biology.* Sunderland, Mass.: Sinauer.

———. 1983. *Science on trial, the case for evolution.* New York: Pantheon.

Gould, S. J. 1977. *Ontogeny and phylogeny.* Cambridge, Mass.: Harvard University Press.

———. 1978. *Ever since Darwin.* New York: Norton.

———. 1980. *The panda's thumb.* New York: Norton.

———. 1983. *Hen's teeth and horse's toes.* New York: Norton.

———. 1985. *The flamingo's smile.* New York: Norton.

———. 1991. *Bully for Brontosaurus.* New York: Norton.

———. 1993. *Eight little piggies.* New York: Norton.

———. 2002. The structure of evolutionary theory. Cambridge Mass.: Harvard University Press.

———, and N. Eldredge, 1977. Punctuated equilibria: The tempo and mode of evolution reconsidered. *Paleobiology,* 3:115–51.

Hoffman, A. 1989. *Arguments on evolution.* New York: Oxford University Press.

Hull, D. L. 1973. *Darwin and his critics.* Chicago: University of Chicago Press.

Kitcher, P. 1982. *Abusing science: The case against Creationism.* Cambridge, Mass.: MIT.

Mayr, E. 1942. *Systematics and the origin of species.* New York: Columbia University Press.

———. 1982. *The growth of biological thought.* Cambridge, Mass.: Harvard University Press.

Minkoff, E. C. 1983. *Evolutionary biology.* Reading, Mass.: Addison-Wesley.

Numbers, R. L. 1992. *The Creationists.* New York: Knopf.

Ruse, M. 1982. *Darwinism defended. A guide to the evolution controversies.* Reading, Mass.: Addison-Wesley.

Simpson, G. G. 1944. *Tempo and mode in evolution.* New York: Columbia University Press.

Stanley, S. M. 1979. *Macroevolution: Patterns and process.* San Francisco: Freeman.

———. 1981. *The new evolutionary timetable.* New York: Basic Books.

Wesson, R. 1991. *Beyond natural selection.* Cambridge, Mass.: MIT.

Wills, C. 1989. *The wisdom of the genes, new pathways in evolution.* New York: Basic Books.

The Relative Geologic Time Scale and Modern Concepts of Stratigraphy

MAJOR CONCEPTS

If history always repeats itself, how come there's so much to learn? **Anonymous schoolchild**

Historians explain the past and economists predict the future. Thus, only the present is confusing. **Anonymous**

▶ In the early 1800s, geologists began to decipher the rock record of earth history and to construct the modern geologic time scale. Although they determined the relative age of events, they could not know how much time (in years) was represented.

▶ It seems natural to regard each rock layer as deposited during a single interval of time throughout its area of deposition, but we now understand that strata can span considerable time over long distances. This is because rock layers are formed in sedimentary environments (such as beach sands, shelf muds, or reef limestones), which are found adjacent to each other at any given time and which migrate landward or seaward through time and space if sea level changes.

▶ Only a few kinds of rocks that formed during a geologic instant (such as a volcanic ash fall) are time-equivalent everywhere. Because most rock units span significant geologic time, we separate time units and rock units in our concepts and classifications.

▶ Because rock units cannot be used to indicate time, the most useful method of establishing time equivalence is correlation by index fossils. Other methods, such as correlating by pattern of sea-level change or by changes in the magnetic signature or chemistry recorded in the sediments, can also be used.

This chapter illustrates the application of the principles discussed in Chap. 2 to the development of the relative geologic time scale—the standard calendar for earth history—and then introduces some additional concepts necessary to a modern analysis of sedimentary rocks, within which most of the record of earth history is preserved. First we shall see how geologic mapping in northern Europe led to the modern relative time scale by the application of the principle of superposition and the principle of fossil correlation. Next we shall digress to consider several important corollaries to superposition, which also provide evidence of relative age. Then we shall consider the important distinction between the rock record riddled with gaps (unconformities) and the time represented by both the rocks and the gaps. Next we shall consider the influences of environment upon both sediments and organisms, along with the profound consequences of shifting environments, such as when the sea encroaches upon or retreats from large areas. We shall examine briefly several new types of relative time scales based upon criteria other than fossils, each of which will then be illustrated in later chapters. Finally, we shall explore the very spasmodic, or episodic, nature of the stratigraphic record, which reflects great variations of rates and intensities of deposition, evolution, structural changes, and sea-level fluctuations.

Figure 4.1 Outcrops of the Devonian Old Red Sandstone critical to the early development of the geologic time scale in England. These red sandstones were deposited in rivers and flood plains flowing from the mountain range caused by the collision of Europe with eastern North America during the Caledonian-Acadian orogeny (see Chap. 12). This sandstone spire in the Orkney Islands of northern Scotland is known as the "Old Man of Hoy." *(Courtesy of the Geological Survey of Great Britain.)*

Early Mapping and Correlation of Strata

After publication of Smith's and Cuvier's first geologic maps, studies proceeded at a rapid pace. Though the Wernerian chronology (review Table 2.1) continued to be used as a standard of reference for supposed relative age, local names for strata became more and more numerous. Naming of distinctive rock bodies was a natural by-product of mining and mapping. The names developed unconsciously as a shorthand and reflected geographic localities or peculiar rock types (Fig. 4.1). At first, such names were used informally and only locally, but as studies were extended, the lateral continuity of certain distinctive strata became apparent, and it was natural that some names were extended more widely. At the same time, fossils were being collected and studied more and more, and strata with similar assemblages were correlated from widely separated areas even where complete physical continuity between them could not be observed because of discontinuous out-crops. In this way, the geologic map of all western Europe gradually developed.

As the more important named divisions were extended, relationships among different major groups of strata gradually became clear through application of the principle of superposition and its corollaries. A group of Secondary strata named Juras for the Jura Mountains of France and Switzerland, when mapped northward, were found to overlie another group named Trias in central Germany and to underlie a third group in France named Cretaceous. By application of Steno's principle of original lateral continuity, the Cretaceous name was extended from France across to the chalk cliffs of Dover (Chap. 2) in England. Mean-

while, the coal-bearing strata of Britain were named Carboniferous, and correlation allowed extension of that name to the coal-bearing rocks on the Continent.

Thus, by tracing strata in exposed rock, in mine tunnels, and between drill holes, the superpositional relationships of all these groups of strata could be established over the vast region shown in Fig. 4.2. Contrary to disclaimers by neocatastrophists, the relative ages of all these rocks are known with certainty through superposition, so they provide a sound basis for a standard geologic time scale.

By trial and error, the old rock divisions became subdivided into clearly traceable, named groups of related strata. The names applied to each group, once arranged in their proper vertical succession according to superposition, provided a more detailed, practical, and more formal chronology, which gradually displaced the older Wernerian scheme. There was a period during the early nineteenth century when out-of-date concepts overlapped with the growing new chronology. For example, although the idea of a single major deluge had yielded to growing geologic evidence to the contrary, many people still clung to the idea of a nearly universal but brief, recent flood, and it became commonplace to classify strata as antediluvial and postdiluvial (before and after the flood). Such a flood was thought essential to account for large, "erratic" boulders scattered widely over northern Europe. The boulders and associated clay or drift, called Diluvium, were assumed to have been "drifted in" by icebergs during the flood. Many prominent geologists—the diluvialists—staunchly held this belief, which was

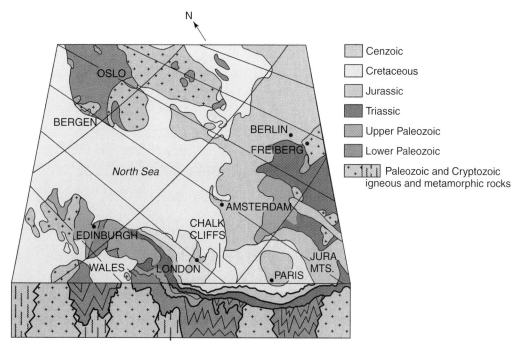

Figure 4.2   The geology of northwestern Europe where much of the geologic time scale was developed. Note the superpositional relations and lateral extent of major rock divisions as well as several major unconformities. Mapping of natural surface outcrops was supplemented by artificial excavations for canals and quarries as well as subsurface data from mines and drill holes to provide the basis for the time scale.

not displaced by the glacial theory for such deposits until after 1840. As late as 1829, the formal name Quaternary was proposed in good Wernerian tradition to include these young deposits. The occurrence of primitive Paleolithic human artifacts in such deposits led Lyell to conclude that humans had existed at least through most of Quaternary time.

Modern Relative Time Scale

At first, there was no conscious thought of building a new time scale, but there developed the need to classify and organize material into a manageable, orderly form. Such a formal study belongs to the subscience of stratigraphy.

   In 1835, two British geologists, Adam Sedgwick and Roderick Murchison, undertook to name formally the entire European succession. They began in Wales, where they named two divisions, the older the Cambrian (for Cambria, the ancient Roman name for Wales) and the younger the Silurian (for an ancient Welsh tribe, the Silures) (Table 4.1). Sedgwick and Murchison next extended their work southward to Devonshire. There they encountered unfamiliar strata with marine fossils that seemed intermediate between Silurian and Carboniferous and, so, perhaps were contemporaneous with the nonmarine Old Red Sandstone that lies in a similar stratigraphic position in Wales. They named these intermediate rocks Devonian and found deposits of nonmarine Old Red type interstratified with some marine Devonian layers in South Wales, confirming correlation of the two (Fig. 4.3).

   Sedgwick recognized the need for a formal stratigraphic classification with different levels of subdivisions based upon

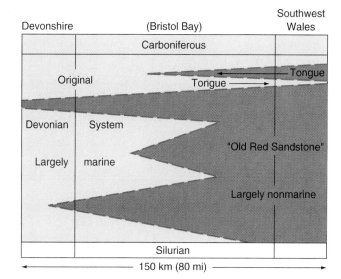

Figure 4.3   Relations of the nonmarine Old Red Sandstone facies of Wales to marine Devonian facies of Devonshire, as inferred by Sedgwick and Murchison. Intertonguing of marine and nonmarine deposits proved the Devonian age of the Old Red Sandstone. (Note the tongues in the upper right.) These facies also proved that different environments of deposition existed simultaneously.

more rational criteria than those of the archaic scheme. He proposed that large divisions be based upon the broad characteristics of fossils that were by now rather well known. Each of the large divisions would include a number of smaller subdivisions. The Paleozoic Era (meaning "early" or "old life") comprised the

Table 4.1	The Modern Relative Geologic Time Scale Compared with the Archaic Scale (See also End Papers)		
Archaic Scale as Applied in Britain	**MODERN SCALE**		
	Eras	**Periods**	**Epochs**
Alluvium Quaternary (1829)* Diluvium	Cenozoic	Neogene (1853)	Holocene (or Recent) (1885) (1883) Pleistocene (1839) Pliocene (1833) Miocene (1833)
Tertiary (1759)	(Recent Life) (1841)	Paleogene (1866)	Oligocene (1854) Eocene (1833) Paleocene (1874)
Secondary (1759)	Mesozoic (Middle Life) (1841)	Cretaceous (1822) Jurassic (1795) Triassic (1834)	
(Old Red Sandstone)	Paleozoic	Permian (1841) Carboniferous (1822) Devonian (1837)	North America: { Pennsylvanian (1891) Mississippian (1870)
Transition (of Werner) (1786)	(Ancient Life) (1838)	Silurian (1835) Ordovician (1879) Cambrian (1835) Vendian (1952) (see Chaps. 8 and 9)	
Primitive or Primary (1759)	Cryptozoic or Precambrian (local subdivisions are used, but their worldwide correlation is difficult; further discussion appears in Chap. 8.)		

The "Eon (1931)" label appears vertically in the Eras column beside Cenozoic through Mesozoic, and "Phanerozoic" appears vertically spanning Eras through the Paleozoic rows.

*Dates indicate when the divisions were named.

Cambrian, Silurian, and Devonian divisions. The Paleozoic eventually included all divisions dominated by invertebrate animals of related types (Table 4.1). We now know that invertebrate animal fossils first appear as impressions in strata called Vendian, which occur below Cambrian rocks. Today the entire fossiliferous span from Vendian to the present is conveniently called the Phanerozoic Eon (meaning "visible life"). The interval preceding the Phanerozoic (an interval that encompasses about 80 percent of total earth history), has been given different names over the years, including Cryptozoic ("obscure life") and Precambrian. (See Box 8.1 for further discussion of this problem.)

As Table 4.1 shows, major time scale divisions based solely upon fossil life are called eras of geologic time. Their subdivisions

are called periods. Younger periods are subdivided into still smaller divisions called epochs. Professional geologists require still finer subdivisions, but these need not concern us here (Box 4.1).

Also in the 1830s, several epochs of the old Tertiary division, which by then had been incorporated as a system, were defined and named (Table 4.1) on the basis of relative percentages of living species represented in their fossil assemblages. For example, the Eocene contained 3 to 4 percent, Miocene 17 percent, and the newer Pliocene 90 to 95 percent species that are still living today. Soon the Mesozoic Era ("middle life") and Cenozoic Era ("recent life") were named. The Permian Period was named next, and, with the illogical retention of archaic Tertiary and Quaternary as periods of the Cenozoic Era, the time scale was

Box 4.1

Nothing Quite Like a Good Fight . . .

Development of the geologic time scale was anything but an orderly affair. One of geology's many stormy feuds arose over the division and naming of strata. Besides its amusement value, this incident carries an important message about the difficulty of stratigraphic subdivision, especially in the early days. Sedgwick and Murchison, who were fast friends during the 1830s, later fell out bitterly over definition of the boundary between the Cambrian and Silurian Systems in Wales (Fig. 4.4). As they mapped closer to one another, it became

apparent that each had included one particular group of rocks in "his" system. Sedgwick's topmost Cambrian overlapped Murchison's lowermost Silurian (Fig. 4.4). What to do? These otherwise proper English gentlemen ceased speaking to each other, and the problem was unresolvable during their lifetimes. Much later, a compromise was struck by removing the rocks in question and erecting a new system, called Ordovician (Table 4.1).

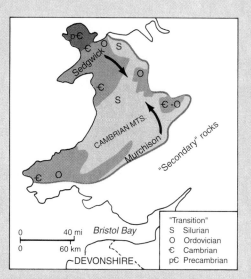

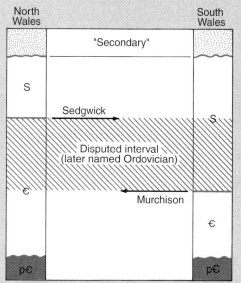

Figure 4.4 Geology of Wales *(left)*, where Sedgwick and Murchison worked and feuded over correlation of Cambrian and Silurian strata *(right)*. The Old Red Sandstone marks the base of the archaic Secondary division (see Table 4.1).

essentially complete. Attempts have been made to replace "Tertiary" and "Quaternary" (Table 4.1), but these terms still appear because geologists find them useful.

Even though the time scale grew haphazardly, it has evolved into an organized, workable scheme of classification. It illustrates the success of applying the principles of superposition, original horizontality, and original lateral continuity and using similar fossil assemblages. Corollaries to superposition developed by Hutton also allowed the establishment of relative ages of unconformities, igneous bodies, and faults (see Box 4.2).

Final proof of the inadequacy of Werner's theory of the earth also came with the new chronology. Werner's Primitive rocks were shown to include Paleozoic, Mesozoic, and even some Cenozoic ones. His Transition ended with the Silurian in Britain, Carboniferous in Germany, and Eocene in the Alps. Clearly rock type, structure, and metamorphism were not, after all, reliable indices of age, though this important fact was slow to be recognized.

Rocks Versus Time

It is necessary to make clear distinctions between abstract geologic time and the tangible rock record. Time is continuous, whereas the rock record is riddled with unconformities of varying magnitude, yet all that we are to know of history must be gleaned from the imperfect rock record. Today we distinguish formally the Cambrian System of rocks from the Cambrian Period of time. The rocks of this system in Wales, where they were named, provide a world standard for correlation of rocks anywhere else that, on the basis of similar fossils, are judged to have formed during the same period of the Paleozoic Era.

A pure time division, such as the period, must include the time represented by all gaps as well as the time represented by the preserved rocks in the corresponding universal rock division (the system, in this case). Fig. 4.6 illustrates this difference. It follows that, if elsewhere a more complete sequence of Cambrian rocks were found with fewer and smaller unconformities

Box 4.2

Corollaries to Superposition for Determining Relative Age

More Huttonian Insights

In developing his theory of the earth, Hutton also formulated additional principles for determining relative age, which can be considered *corollaries to the principle of superposition.* Fig. 4.5 is a cross section through Scotland illustrating the features studied by Hutton, and it will serve to clarify these corollaries. Hutton's great unconformity can be seen beneath the Devonian Old Red Sandstone. Rocks below are Silurian, so it is apparent that, sometime in late Silurian and/or early Devonian time, profound mountain building occurred across Scotland. The fact and age of the mountain-building event are clear on the southeastern (right) end of the cross section from the superpositional relations of the Old Red Sandstone overlying the unconformity. But to the north (left) of the Highland Boundary Fault, things are not so definite. Upheaval, metamorphism, and igneous activity occurred, but when? Did these changes happen at the same time that the unconformity developed farther south (right)? Hutton's corollaries to superposition provide the answer.

Relative Age from Cross-Cutting Relationships

The *principle of cross-cutting relationships* states that any igneous rock or any fault must be younger than all rocks through which it cuts. In Fig.4.5, the "Caledonian granites" must be younger than the metamorphic rocks they intruded. Similarly the dikes shown as heavy, black, steeply inclined lines must be the youngest rocks shown because they cut all others, including the "Caledonian granites." The five faults shown in Fig. 4.5 must be relatively young, too, because they displace all rocks except the dikes. Because faults tend to be active for long periods, their cross-cutting relationships date only the last fault movement; other evidence is required to date first movements. (Note that the relative ages of the faults and dikes cannot be determined because they are not in contact with each other.)

Relative Age from Included Fragments

The *principle of included fragments* states that any rock represented by fragments included in another rock must be older than the enclosing one. It applies equally to inclusions of foreign rocks contained in igneous bodies or to grains or pebbles in sediments. Thus, inclusions of "Primitive" metamorphic rocks in the "Caledonian granites" of Fig. 4.5 further prove that the granites are younger than the metamorphics. In turn, because the Old Red Sandstone at the right side of Fig 4.5 contains pebbles of Caledonian-type granite and schist, it must be younger than both of those rocks exposed at the left.

It was this kind of reasoning that allowed Hutton to conclude that, indeed, the formation of mountains accompanied by the intrusion of granites occurred all across Scotland prior to erosion of this unconformity and the deposition of the Old Red Sandstone. Exactly the same reasoning is routine today.

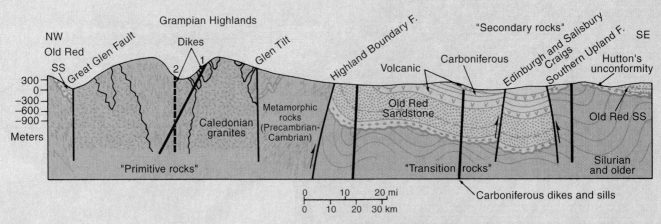

Figure 4.5   Cross section across Scotland showing superposition, cross-cutting, and included-fragment relationships. Hutton was able to show that the "Primitive" and "Transition," rocks were folded, intruded by granites, uplifted, and then deeply eroded before the Old Red Sandstone was deposited unconformably upon them. *(Adapted from Geological map of Great Britain.)*

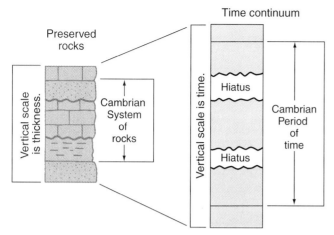

Figure 4.6 The relation of a preserved rock record with discontinuities and the abstract time continuum corresponding to that rock record plus its unconformities. A tangible rock record exists only for the colored portion at the right, whereas the blank areas (hiatuses, or gaps) represent unconformity intervals.

and more fossils, it might provide a better world standard of reference than does that of Wales. The original European standards are so firmly established by long usage, however, that such changes are rarely made.

Thus, what we have are time divisions and rock divisions. The most generally accepted stratigraphic classification scheme is as follows:

Relative Time Divisions	Equivalent Universal Rock Divisions
Eon	Eonothem
Era	Erathem
Period	System
Epoch	Series
Age	Stage

Rock divisions are termed universal because they comprise at least continentwide if not worldwide standards, as opposed to purely local rock units such as the Old Red Sandstone. Though the rocks of Wales that define the Cambrian System have certain characteristics of texture and color, all rocks considered of like age elsewhere regardless of *their* local characteristics are also classified as part of the Cambrian System because they have similar index fossils. Comparison among these systems is made only in terms of age equivalence.

The Formation

The most basic local unit of stratigraphy is the **formation,** first defined in Germany about 1770. Its original definition as a distinctive series of strata that originated through the "same formative processes" is still valid 200 years later. Formations must be distinctive in appearance in order to be easily recognizable. Ideally they are named for a type locality where they are normally displayed in a well-exposed type section (the most typical or defining locality or section).

Many practical problems arise in defining formations. Designation of upper and lower limits, or contacts, may be difficult, as shown in Fig. 4.7. Obviously the definition of a formation is somewhat arbitrary and is influenced by scale factors. A *stratum* is the thinnest rock layer observable; formations include more than one stratum, and groups contain many formations.

Characteristics chosen to define a formation include one or more of the following: (1) composition of mineral grains, (2) color, (3) textural properties (size of grains, etc.), (4) thickness and geometry of stratification, (5) character of any organic remains, and (6) outcrop character. The sum total of such characteristics is referred to as the **lithology** of a formation (from *lithos,* meaning "rock").

Lateral Variations

So far, nothing has been said about the time relationships of formations or of their lateral (horizontal) relationships. Even after neptunism had been laid to rest, there persisted a belief in simple,

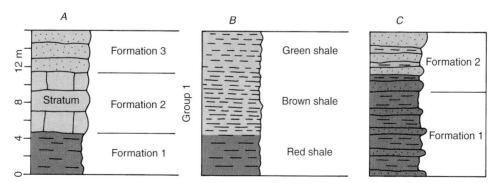

Figure 4.7 Designation of formations by rock type is clear in A. It is not so in B because of subtle gradations of color, or in C because of intimate interstratification of two lithologies. In the last two cases, some arbitrary division must be chosen. In C, scale is important, for a single, thin sandstone layer hardly has utility as a formation. (Dots represent sandstone, dashes shale, and the brick pattern limestone.)

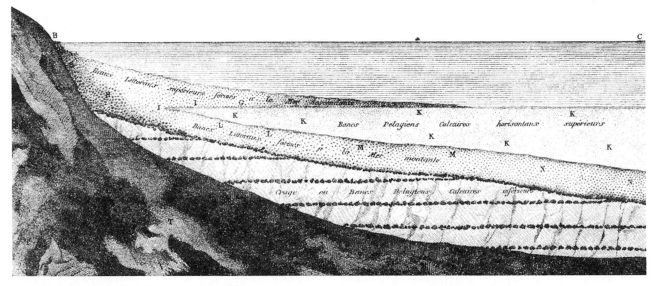

Figure 4.8   Lavoisier's diagram of the relationship of coarse littoral *(Bancs Littoraux)* and finer pelagic *(Bancs Pelagiens)* sediments to the northern French coastline. Lavoisier recognized that gravel can be moved only by waves near shore, whereas finer sediments are carried into deeper water. He also recognized that distinctive organisms inhabit each environment. But if sea level rose, flooding the land *(la Mer Montante)*, both littoral and pelagic sediments would migrate landward. Conversely, if sea level fell *(la Mer descendante)*, they would shift seaward. *(From A. Lavoisier, 1789, Memoires d'Academie Royal Sciences.)*

universal patterns to the history of the entire earth. It was natural to assume that the stratigraphic divisions being named in Europe—and destined to become the standard systems—were universally present and of the same age worldwide (i.e., synchronous). That geologists found similar fossils in any given system over much of Europe seemed ample evidence of universality. Many conceived of a layer-cake arrangement, with each formation assumed to extend indefinitely without change. This premise had been the cornerstone of Werner's time scale, and it was not easily given up because it had the seductive appeal of simplicity.

It is not clear when geologists first began to suspect that distinctive, locally named formations could not extend unchanged indefinitely. The earliest clear records occur in 1789 French writings, which show a recognition that similarity of fossils in similar sedimentary rocks might reflect environmental factors rather than strict age equivalence. This important inference was based upon comparison with modern environments, as portrayed in Fig. 4.8. Shallow, near-shore marine sediments tend to be coarser and contain organisms adapted to rough water, whereas contemporaneous offshore, deeper, quieter marine sediments are finer and contain delicate, bottom-dwelling organisms together with floating and swimming forms. The sedimentary products of each environment have unique characteristics, *even though they accumulate at the same time and grade imperceptibly into one another.* In other words, formations might vary in sediment type, or lithology, laterally as well as vertically.

In the 1830s, Sedgwick and Murchison determined in Britain that the dominantly nonmarine deposits of the Old Red Sandstone of Wales are synchronous with marine Devonian deposits farther south by finding the two types inter-stratified, as shown in Fig. 4.3. Plant and fish remains of the Old Red Sandstone reflect environments markedly different from the environment of the marine fossils of the Devonian. Obviously there must be some fundamental lateral difference related to contrasting environments that existed simultaneously side-by-side. Fig. 4.3 showed that two different lithologies can grade laterally into one another in a complex manner called **intertonguing.** We may designate tongues of one lithology penetrating into one another, as shown at the upper right corner of Fig. 4.3. Similarly, a tongue of *Bancs Littoraux* is illustrated near the top of Fig. 4.8 extending offshore from left to right.

Depositional Environments and Sedimentary Facies

Definition of Sedimentary Facies

Full realization of the fundamental importance of lateral variations of strata and the relationships between such variations and depositional environments did not come until a well-exposed example of lateral changes of both lithology and fossils was described in 1838 in Switzerland, where intertonguing of two lithologies containing different fossil assemblages could be seen clearly on a mountainside. The term **sedimentary facies,** which means merely "total aspect" of strata, was coined to characterize these differences of lithology, and the two facies were linked to different depositional environments.

A particular three-dimensional body of sediment designated as a facies grades by some change in its properties, such as a

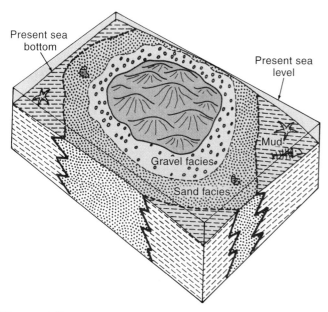

Figure 4.10   Modern ripple marks exposed at low tide, Sapelo Island, Georgia. The straight, sharp, symmetrical ripple crests are typical of those formed by waves in shallow water. Note also holes formed by burrowing organisms. *(R. H. Dott, Jr.)*

Figure 4.9   Sedimentary facies around a hypothetical island showing tendency for coarser sediments to occur in strongly agitated, near-shore environments. Bottom-dwelling organisms in the different environments differ considerably, also. (The upper surface of the diagram is a map of modern bottom-sediment types showing lateral variations only at a moment in time; sides are cross sections showing facies relationships through time.)

gradual increase in the ratio of mud to sand, into another, adjacent facies (Fig. 4.9). Facies are not named formally, as are formations, but are identified simply by their dominant lithology (e.g., the sand and mud facies in Fig. 4.9). Because the facies concept relates sediments to their depositional environments, the study of both sediments and organisms in modern environments obviously provides important clues for understanding ancient facies. For example, the comparison of features preserved in ancient strata, such as ripple marks, with their counterparts forming in modern environments (Fig. 4.10) have long provided analogies for such interpretations.

Regional Analysis of Facies

Fig. 4.9 shows facies patterns on a local scale, but the great importance of the facies concept for the analysis of earth history becomes clearer on a regional scale. To illustrate, let us return to Hutton's Scotland. Fig. 4.5 shows the unconformable overlap of the Devonian Old Red Sandstone across deformed Silurian and older rocks in Scotland. Fig. 4.11 shows a sedimentary facies map (A) and a restored cross section (B) for the Devonian of Europe. The map distinguishes present distribution of Devonian rocks and, by applying the principle of **original lateral continuity** to the facies patterns, suggests how much more extensive the Devonian facies probably were (diagonal patterns). Cross sections, which show vertical variations among rocks, are important complements to facies maps.

From Fig. 4.11, we see that mountain building profoundly disturbed the northwestern margin of Europe during Silurian and Devonian time, whereas the region to the southeast remained more stable. This contrast is reflected by (1) more and greater unconformities in the west; (2) coarse and thick nonmarine sediments ("Old Red") in the west; and (3) finer and more marine deposits in the east. Vigorous erosion of the "Caledonian Mountains" shed immense volumes of coarse debris eastward; greatest thicknesses are nearest the mountains (Fig. 4.11B), and the longest gap due to erosion also is nearest the mountains (Fig. 4.11C). Uplift of the mountains was spasmodic, for a major unconformity divides the Old Red Sandstone at the left end of the section. A lot of history is revealed here!

Transgression and Regression by the Sea

Preservation of large volumes of most strata requires subsidence of the earth's crust during deposition, although for marine sediments a rise of sea level might conceivably allow some preservation without subsidence. Examples from the human historical period in northern Europe show how regional facies patterns tell us about major changes of land and sea. As indicated in Chap. 1, the North Sea coast of Holland has been submerging since the last glacial advance. The submergence is due to both worldwide rise of sea level and local land subsidence, but the sedimentary result is the same in any case. Holes have been drilled and dug in Holland for centuries, producing an accurately dated Pleistocene and Holocene succession (Fig. 4.12). Advance of the sea over the land, termed *transgression,* caused a continuous shift of environments and their sedimentary and biological products landward. (Retreat of the sea from a land area, called *regression,* has the opposite effects.) The result of this transgression is a more or less continuous, nearly flat layer of sandy near-shore deposits and

Figure 4.11
Devonian facies of
Europe. *A:* Facies
map shows areas of
preserved strata as
well as inferred
restorations where
erosion has
removed them
(diagonal pattern).
B: Restored cross
section showing
facies, thickness,
and variations in
unconformities.
Effects of
Caledonian
mountain-building
event shows at left.
C: Time diagram
contrasting the
preserved strata
versus the large
gap due to
unconformities.
*(Adapted from R.
Brinkmann, 1960,
Geologic evolution
of Europe; by
permission of
Ferdinand Enke
Verlag; and
L. J. Wills, 1951,
Palaeogeographical
atlas; by permission
of Blackie and
Son Ltd.).*

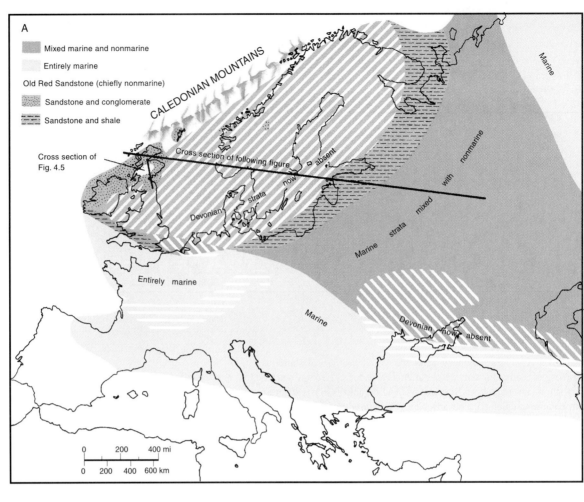

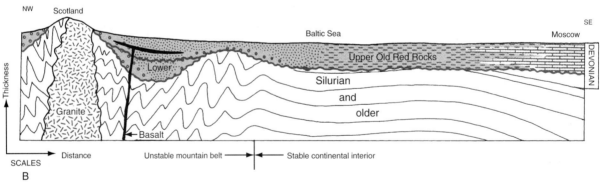

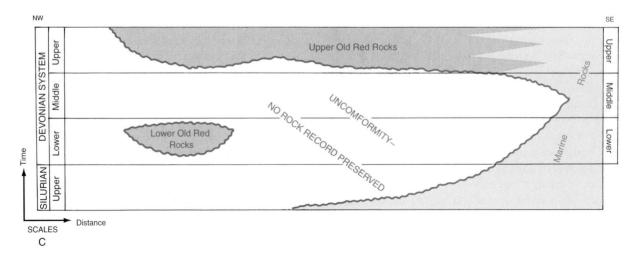

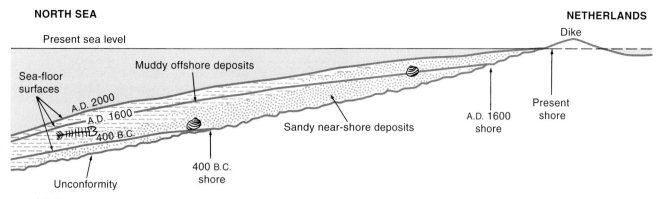

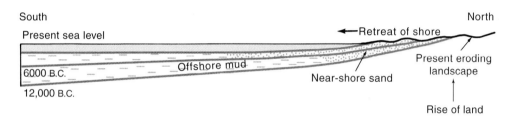

Figure 4.12   Historic example of transgression on the Netherlands coast, showing landward shift of facies dated by archaeological and carbon 14 evidence. Note migration of characteristic molluscan animals with the sandy facies; the fish, being a swimming form, was independent of the bottom environment and might be found in either facies. (Angle of slope is exaggerated.)

Figure 4.13   Historic example in the northern Baltic Sea of regressive facies—that is, the seaward shift of different sediment types as environments shifted southward *(left)* during the past 14,000 years due to the rise of Scandinavia since the melting of glaciers.

finer offshore ones. The relation between these two sandy and muddy facies is complex, but it involves a seaward gradation along any single time datum, such as the present sea floor, from sand to mud. This gradation is due to a seaward decrease of agitation of the water by waves and currents. A *time datum* is any line or surface of equal age that can be traced widely within a group of strata. The time lines labeled in Fig. 4.12 are examples, as is the Thera volcanic ash layer blown across the eastern Mediterranean region about 1500 B.C. (see Fig. 1.4). A geologically instantaneous time datum, such as a volcanic ash layer, is an especially valuable reference time line because it can be traced from one facies into another, as in Fig. 4.12.

If the Netherlands deposits are preserved and lithified, what should future geologists designate as formations? The obvious choices would be a sandstone formation and a shale formation, but note that these two would *in part grade laterally into one another.* Some tongues of sand penetrate into the generally muddy facies, and vice versa. The sand tongues may reflect periods of violent storms, which moved sand farther offshore than normal.

The time datum shows that neither formation is of exactly the same age everywhere along the cross section. Each is older at the seaward than at the landward end because of the progressive transgression by the sea. On the other hand, at any point some part of one is the same age as part of the other. Therefore, *formations neither extend indefinitely laterally nor are everywhere of exactly the same age.*

There are two basic types of facies patterns. A **transgressive facies pattern** is illustrated in both Figs. 4.8 and 4.12. Trans-

gressive facies generally (1) reflect shrinkage of land area, (2) are preceded by an erosional unconformity, (3) show a landward shift of adjacent facies through time, and (4) tend to become finer upward at any one geographic locality. A **regressive facies pattern** generally results from a relative fall of sea level or a rise of land level. Its characteristics are the opposite of those of the transgressive pattern; thus, regressive facies (1) reflect enlargement of land area, (2) are followed by an erosional unconformity, (3) show a seaward shift of facies through time, and (4) tend to become coarser upward at any one locality. Fig. 4.13 illustrates a regressive facies pattern due to the recent rise of northern Scandinavia as a result of deglaciation (compare Fig. 1.9).

About 100 years ago, German geologist Johannes Walther studied the relationships of modern facies to their environments. He noted that inevitably environments shift position through geologic time. As they do so, the respective sedimentary facies of adjacent environments succeed one another in vertical sequences. As an important consequence of this fact, *the vertical progression of facies will be the same as corresponding lateral facies changes.* Therefore, in a transgressive facies pattern, just as the sequence becomes finer upward, the finer facies also spreads laterally in the direction of transgression (Fig. 4.12). Conversely, just as a regressive sequence becomes coarser upward, the coarser facies also spreads laterally in the direction of regression (Fig. 4.13). This important linking of vertical and lateral facies changes is called **Walther's Law,** and it is of fundamental importance in interpreting ancient geography (paleogeography).

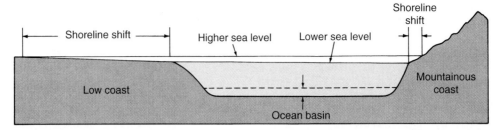

Figure 4.14   Contrasting effects of sea-level fluctuations on low versus steep coastlines due to upwarping of the ocean basin. Roughly half of the total continental surface today lies within a few hundred meters of sea level; thus, a modest sea-level change produces a profound change in land area. Many such modifications have occurred in the past.

Local Versus Worldwide Transgression and Regression

As was hinted at in Chap. 1, large-scale changes of sea and land levels have been extremely important in earth history. What causes could produce such colossal changes? Worldwide sea-level changes (sometimes called *eustatic* changes) may result from fluctuating continental glaciation or large-scale warping of deep ocean basins. The relief of the land will dictate enormous differences in response to sea-level changes as expressed by the magnitude of shoreline movements during transgression or regression. Low, flat land, such as the Netherlands, would be inundated or drained very widely by only a slight relative change of sea level, whereas a bold coastline such as that of California would be little affected (Fig. 4.14).

Local changes such as mountainous uplifts or crustal subsidence can result in local transgression or regression as well. The plot thickens, however, if more than one cause occurs simultaneously; the effects along a given coast will be more complex and may even give us some surprises. For example, transgression in the Netherlands has been accentuated by subsidence of the earth's crust at the same time that postglacial sea level has been rising worldwide (Fig. 4.12). Conversely, the Baltic Sea coast simultaneously has experienced regression in spite of the postglacial sea-level rise because the crust there has risen more rapidly than has sea level (Fig. 4.13).

There are other complications that may result from relative rates of sedimentation and erosion along a coastline. Very rapid sedimentation due to uplift of a distant inland area or a climatic change that accelerates erosion may cause local seaward retreat of a shoreline independent of structural warping of the crust or worldwide changes of sea level. For example, rapid advance of the Tigris-Euphrates delta has caused the northern shore of the Gulf of Arabia to retreat from the land toward the gulf perhaps as much as 175 kilometers (100 miles) in the past 3,000 years, even as sea level has risen about 4 meters worldwide (Fig. 4.15). The delta advance resulted from accelerated sediment transport due to climate change and to human agricultural practices inland (see also Fig. 1.8).

Ideally we would like to be able to distinguish worldwide causes of transgressions and regressions from local changes, but such a distinction often cannot be made. Obviously from the fragmentary evidence preserved locally in ancient rocks, we cannot

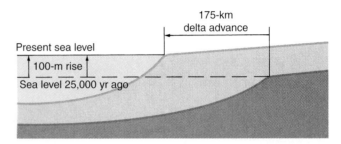

Figure 4.15   Advance of the Tigris-Euphrates delta 175 kilometers into the Gulf of Arabia during the past 3,000 years in spite of the worldwide rise of sea level of about 4 meters during that same period. Thus, in spite of the worldwide postglacial transgression, rapid sedimentation has caused local regression here. Not to true scale (compare Fig. 1.8).

determine the ultimate cause of every transgressive or regressive facies pattern; we see only an *apparent sea-level change* reflected. A great deal of continentwide or even worldwide information is required to test and reject possible alternative explanations.

Biostratigraphic Concepts

When an American geologic pioneer identified thick strata in southeastern California as Cambrian a century ago, he was performing a correlation with Wales based upon index fossil assemblages. An interval of strata characterized by a distinctive index fossil is termed a **fossil zone** (Fig. 4.16). Correlation using fossil zones assumes that the interval of time during which the distinctive species of a given zone lived was synchronous over the entire geographic range of that species. Experience suggests that, at the level of precision of correlation so far achieved, this assumption of simultaneous geographic dispersal is generally valid. We know, for example, that free-floating larval stages of many marine animals are dispersed rapidly by ocean currents and that the spores and seeds of many plants are rapidly dispersed by winds. The time represented by individual fossil zones varied considerably, but where we have an independent measure of numerical time (see Chap. 5), typical durations were of the order of 0.5 to 2 million years.

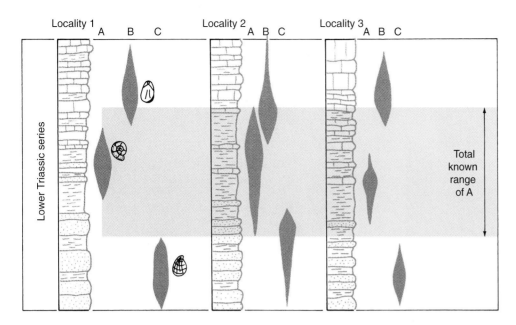

Figure 4.16 Correlation using three different index fossils. Brown patterns indicate ranges and relative abundances for each species. A single fossil zone is shown in blue. In practice, assemblages are more useful than a single species, but close attention must be paid to overlapping stratigraphic ranges of index fossils. Note also that both the range and the maximum development (widest brown pattern) of a single species vary slightly from place to place.

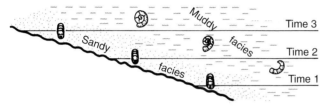

Figure 4.17 Significance of contrasting rates of evolution and rates of environmental shift due to transgression. The brachiopod *Lingula* (Fig. 9.18A) in the sandy facies evolved very slowly and is a poor index fossil; it has migrated with its shifting sandy environment and is unchanged biologically after millions of years. Cephalopods found in the muddy facies, however, were swimming forms free of the bottom environment. They also evolved rapidly; therefore, their species are admirable index fossils for times 1, 2, and 3.

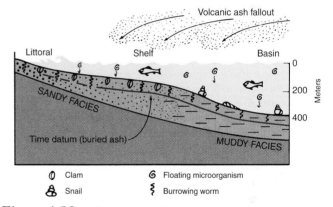

Figure 4.18 The problem of correlation between different sedimentary facies. Bottom-dwelling organisms tend to be restricted to but one facies, whereas floating and swimming organisms are independent of facies; therefore, the latter tend to be the best index fossils for correlation among different facies. Geologically instantaneous time datums such as a volcanic ash layer allow correlation between facies.

The facies concept makes it clear that evolution through time is not the only change operating, for environmental changes also must be considered as factors affecting distribution of fossils, as emphasized in the preceding discussion. Environmental changes are, geologically speaking, relatively short-term affairs and, so, are not very significant when considering the fossil assemblages of the large, universal rock divisions. For smaller time divisions, however, the vagaries of local environmental influences must be considered, because environmental changes may have been more rapid than evolutionary ones. Fig. 4.17 shows how different rates of evolution may compare with environmental change. Incidentally, fossil zones do not necessarily correspond to rock formations. Some formations may contain parts of several zones, but others may span only a fraction of one zone. Lithologic formation boundaries need not have any relationship to the biostratigraphic boundaries of fossil zones.

Because sedimentary environment influences bottom-dwelling organisms profoundly, such creatures may be of little value in correlation of strata of adjacent different facies. Fossil types notoriously restricted to one or a few lithologies are called **facies fossils.** For example, certain clam species today burrow only in beach sands, whereas others live only in muddy tidal flats. These would be less than perfect index fossils because of their narrow preference for a particular sediment type. It follows, then, that the best index fossil is one that lived independent of the bottom environment where sediments form. Obviously these would be floating or swimming organisms (Fig. 4.18). Fortunately

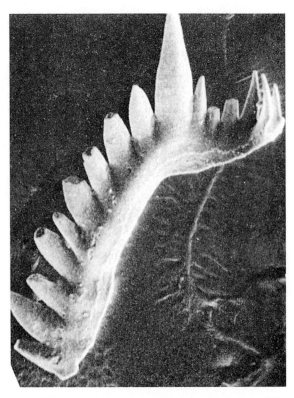

Figure 4.19   Lateral view of a conodont. Conodonts are small anatomical parts of an extinct eel-like swimming vertebrate, according to recent discoveries in Scotland. Evolutionary structural changes occurred rapidly, and conodonts are widespread in different rock types, so they are ideal index fossils. This conodont is from the Permian of Texas (magnified 110 times). *(Courtesy D. L. Clark.)*

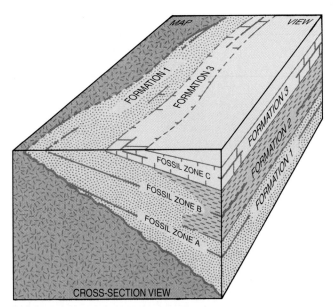

Figure 4.20   Three-dimensional relationships of formations and index fossil zones; top surface shows a geologic map constructed by establishing physical continuity of formations between isolated outcrops. Age correlation by fossil zones shows that formation 3 is synchronous everywhere, but formations 1 and 2 vary in age due to lateral facies changes. (Compare with Figs. 4.12 and 4.17.)

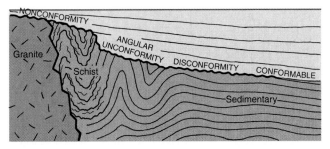

Figure 4.21   Variations in unconformities. Note that one type may change laterally to another. *Nonconformities* are underlain by igneous or metamorphic rocks; *disconformities* show little or no angular discordance.

there are several such groups of fossils that also evolved rapidly and therefore serve well for correlation even between different sedimentary facies (Fig. 4.19). Figure 4.20 shows three fossil zones and their relations to a transgressive facies sequence. Fossil zone B represents the ideal case of an index fossil that occurs in two different facies. Conversely many facies-dependent fossils are useful environmental indicators for the sediments in which they occur.

Unconformities

Some Refinements

Just as we have refined our concepts of the formation and of index fossils, we shall now refine our treatment of unconformities. Like formations, unconformities can be traced and mapped to establish their physical continuity; they also can be studied from the standpoint of age. Continuity and age are established by mapping the relationships of strata and fossils immediately above and below the unconformable contact. It generally turns out that, like some formations, an unconformity surface varies in age from place to place (see Fig. 4.11C). More important, the total time in-

terval represented by the discontinuity may vary greatly; at one place, there may be far more rock record missing than at another point along the same surface, as illustrated in Fig. 4.11C. An unconformity may even disappear laterally into a continuous, unbroken, conformable sequence of strata (Fig. 4.21). Unconformities, then, show lateral and vertical differences as important as those of rock units.

Today we distinguish among several types of unconformities (Fig. 4.21). The unconformity studied by Hutton in Scotland is called an *angular unconformity* with a conspicuous angle of discordance between the older and younger strata. Such an unconformity shows that severe deformation occurred before the

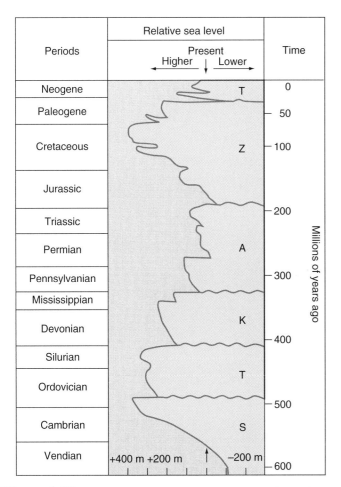

Periods	Relative sea level	Time

Figure 4.22 Six unconformity-bounded sequences (letters) from which a worldwide sea-level fluctuation curve for the past 600 million years has been inferred (commonly known as the "Vail" curve). Note two maxima (high levels) about 500 and 75 million years ago, and three minima (low levels) about 600 million years, 200 million years, and the present. Maximum estimated fluctuation, were about 350 meters above and about 150 meters below present level.

unconformity was buried. *Disconformity* refers to the opposite case, where there is no discordance between strata below and above the discontinuity surface. *Nonconformity* is sometimes used to designate examples wherein underlying rocks are igneous or metamorphic. What is more important than the names is that some unconformities represent profound upheaval and deep erosion, as in mountain making (e.g., Fig. 4.5), whereas others represent little or no structural disturbance, simply erosion or even nondeposition without erosion. This last type may form below sea level without exposure and may represent long time intervals.

Unconformity-Bounded Sequences

In the 1950s, American geologist L. L. Sloss suggested a new way of studying the stratigraphic record, which emphasizes unconformities. Sloss designated large-scale, laterally extensive rock units in central North America as *sequences,* which include

many formations as packages bounded by exceptionally profound regional unconformities. No restricted time connotation or synchroneity is attached to the unconformable sequence boundaries; in fact, they are known to be of varying age from place to place, as in Fig. 4.11C. Unconformity-bounded sequences constitute an additional type of stratigraphic division—a regional rock unit, but not a universal time division; conventional period or era boundaries are ignored in defining them. From hard-won experience, certain unconformities are found to be the most important among many. Six major unconformities clearly punctuate the Phanerozoic record in a very meaningful way (Fig. 4.22).

Two developments in the petroleum industry have given new importance to the sequence concept during the past 20 years. These were the perfection of high-resolution seismic techniques for probing the subsurface structure of the crust and the application of these techniques in large-scale exploration for petroleum beneath the oceans along continental margins. The result was the acquisition of abundant data of very high quality, which showed that Sloss's unconformities (and many lesser ones as well) clearly extend beneath continental margins. Moreover, large international petroleum companies have data from several continents, which suggest approximate synchroneity of the major unconformities. Therefore, the major sequences are now thought to reflect worldwide sea-level changes (Fig. 4.22). The most obvious possible causes of such changes are large-scale glaciation and warping up or down of the floor of an entire ocean basin (see Fig. 1.9). The global cycles of sea-level fluctuation had variable periods on the order of 10 to 80 million years. The suggested maximum rise of sea level was about 350 meters (1,150 feet) above present level near the end of Cretaceous time, and the maximum fall was about 200 meters (650 feet) below present in late Paleogene time. For comparison, Pleistocene fluctuations were about 200 meters. Clearly such large worldwide fluctuations should have had profound effects upon the stratigraphic record of shallow marine and nonmarine regions all over the earth.

By documenting the geometric relationships on continental margins of the unconformities and wedge-shaped sequence packages of strata between them, the petroleum industry has found some profound general patterns of deposition, summarized in Fig. 4.23. The single most important conclusion is that, when sea level is high relative to a landmass, deposition of sediments is concentrated in the shallow marine areas called the continental shelf. When sea level is relatively low, more deposition occurs in the deeper marine realm. If a high level persists long enough, however, sedimentation may fill the shallow shelf and spill over into deeper water (Fig. 4.23C).

Additional Relative Time Scales

Index fossils are not the only things that provide worldwide, synchronous punctuations of the rock record with potential for correlating and subdividing that record. Several alternatives have been developed in recent years, alternatives that provide relative-age scales or chronologies with special applications.

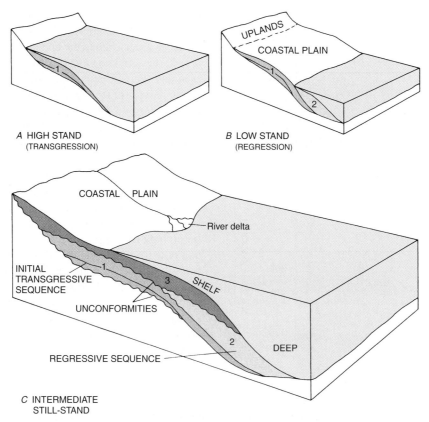

Figure 4.23 Effects of a sequence of sea-level changes upon sediment accumulation and unconformities on a continental margin. *A:* High stand of sea level, or transgression, causes most sediment derived from the continent to be trapped on the shallow continental shelf (sequence 1). *B:* Low stand of sea level, or regression, allows much sediment to accumulate in the deeper ocean (sequence 2). *C:* Intermediate long pause, or still-stand, of sea level, which causes sequence 3 to accumulate on the shelf and to spread seaward over the number 2 regressive sequence. Note the positions of the major unconformities that bound the three sequences. *(Adapted from* The Lamp, *Exxon Corporation.)*

The sequence unconformities found on continental margins gave rise to the worldwide sea-level curve shown in Fig. 4.22. Because the sequences seem to reflect global sea-level changes, the major unconformities may be useful for correlation to unstudied regions, especially on continental margins. The curve has become a special kind of relative time scale.

The fact that the polarity of magnetized minerals, or "fossil magnets," in some Cenozoic volcanic rocks is opposite that of the present earth's magnetic field has been known since the early twentieth century. In the late 1950s, it was demonstrated through numerical rock dating that these reversals were synchronous worldwide. The earth's magnetic field apparently reverses itself occasionally, and this provides the basis for yet another relative-age scale. Magnetic polarity is determined for a sequence of rocks, and then the positions of polarity reversals are correlated to the standard polarity reversal scale. Although the earth's reversal history is best known for only the past 5 to 10 million years, results have been obtained from Cretaceous rocks (Fig. 4.24). The older reversal history is only poorly known, however. We shall learn more about polarity reversals in Chap. 7.

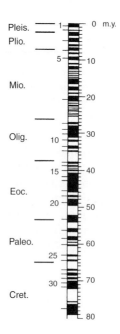

Figure 4.24 Magnetic reversal time scale for the past 80 million years (m.y.) *(From Heirtzler et al., 1968,* Journal of Geophysical Research, *v. 73, p. 2123; © American Geophysical Union.)*

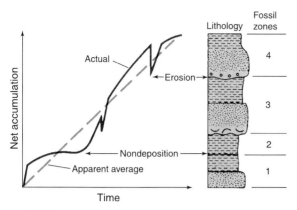

Figure 4.25   Graph of rates of deposition of a hypothetical sequence of strata. The continuous *apparent* average (linear) rate is determined by simply dividing the time interval by the thickness of strata shown. In marked contrast is the episodic "actual rate." The latter accounts for changing rates of accumulation as well as removal by erosion; it represents the more common situation.

Chemical analyses for rare or trace elements and for isotopes in sedimentary rocks also provide potential for relative geochemical time scales. For example, the analysis of ratios of two oxygen isotopes in marine fossils has provided a scale of changes in seawater composition extending back for at least the past 1 million years. These changes reflect temperature-related fluctuations during the Pleistocene ice ages as gigantic glaciers expanded and melted many times. The pattern of climatic fluctuations recorded chemically in the skeletons in deep-sea sediments is now well enough dated that it provides a seawater oxygen isotope scale that is used widely in oceanic studies. We shall refer to this isotopic scale again in Chap. 16. Other chemical elements may, in the future, provide additional scales.

A Sense of Time and Rates

Let us conclude this chapter about stratigraphic concepts by reconsidering the catastrophic versus uniform views of the earth.

The sequence-bounding unconformities suggest that Lyell missed the mark with his extreme gradualism and that the catastrophists were closer to reality with their spasmodic view. The sequences between the major unconformities represent many millions of years of sedimentation, but the strata within these intervals show a punctuated history on a finer scale as well. Familiar events, such as river floods and hurricanes, are recorded again and again in ancient sediments, and the so-called rare event, such as the 500 year flood mentioned in Chap. 1, is the rule rather than the exception. As British geologist D. V. Ager put it, "The stratigraphic record is like the life of a soldier—long periods of boredom separated by brief periods of terror." If spasmodic events are so important, did they recur in a regular, periodic fashion, or were they irregular, or episodic? As we shall see both types of phenomena are recorded, and in some cases transient episodic events were superimposed upon a record of continuing background, or ambient, periodic events. Sorting out such differences is both challenging and fascinating.

Geologists have long dreamed of determining quantitative rates of evolution, sedimentation, mountain building, and the like. They have attempted rather naive approaches, such as equating relative thickness of formations with rates, and have made simple linear extrapolations of poorly determined rates across great thicknesses of strata. A slightly more sophisticated approach is to compute average rates of sedimentation (or evolution) by dividing the total thickness of some interval (or number of fossil zones) by the total time represented. Although instructive, the result is only an *apparent* average rate, at best, for it takes no account of the many breaks in the sequence due to fluctuations in or cessation of deposition or erosion, as shown in Fig. 4.25. We are still a long way from realizing the dream of exact quantitative evaluation of the stratigraphic record. At present, we must be satisfied with statistical estimates of rates, which involve shaky assumptions and inadequate treatment of the episodic character of most strata. In Chap. 5, we shall learn more about numerical dating technique, which eventually may become refined enough to allow much more accurate determinations of rates.

Summary

Modern Relative Time Scale

- Application of **Steno's principles of stratigraphy, Hutton's corollaries to superposition,** and **fossil correlation** led to the modern relative time scale. For the Phanerozoic record, major divisions are based upon the larger patterns of change among fossil marine animals.

Stratigraphic Classification

- Stratigraphic classification today requires the distinction of the tangible but incomplete rock record from the abstraction time (e.g., Cambrian System versus Cambrian Period). This is because there are many gaps (**unconformities**) in the rock record.

- **The formation** is the basic local rock unit and is distinguished by one or more characteristic properties—that is, by **lithology.** A formation is defined formally at, and generally is named for, a type locality.
- Recognition of lateral variations of lithology within a formation (generally involving **intertonguing**) led to the concept of **sedimentary facies**—namely, variations related to depositional environments. As environments shift through time, patterns of facies are produced, which by the application of **Walther's Law,** allow us to interpret major past changes of geography.
  1. A **transgressive facies pattern** reflects a shrinkage of a land area due to the encroachment of the sea. A **regressive facies pattern** is the opposite.
  2. **Worldwide transgressions and regressions** are caused by the expansion and melting of large glaciers or by large-scale warping up and down of the sea floor.
  3. **Local transgression and regression** may be caused by subsidence, or uplift, of the crust; by rapid coastal sedimentation (as by deltas); or by a combination of these.
- **A fossil zone** is the interval of strata characterized by a particular index fossil. Environment as well as evolution have influenced the distribution of index fossils. Therefore, the best index fossils are those that evolved rapidly *and* were not very sensitive to sedimentary environments—in other words, organisms that lived or were dispersed in the air or water above the sediment surface.
- Three major types of unconformities can be distinguished: **angular unconformity, disconformity,** and **nonconformity.** Like formations, unconformities can be mapped and can be shown to vary in age and in time duration from place to place; they may disappear laterally into conformable strata.
- **Unconformity-bounded sequences** of strata comprising many formations and covering large areas provide new insights into the histories of continental margins and interiors. Their recognition on different continents shows a considerable degree of global synchroneity, which suggests a worldwide cause due to sea-level fluctuations.

Additional Relative Time Scales

- Besides the **worldwide sea-level curve** and **worldwide magnetic polarity reversals,** chemical phenomena may provide useful **geochemical time scales** that have special applications, especially in oceanic research.

Time and Rates

- Some regularly repeating, or **periodic, changes,** such as the advance and retreat of glaciers during an ice age, are reflected in the stratigraphic record. Irregular, or **episodic, changes** are more common, however, and they have punctuated the record with large-scale unconformities, which bound the major rock sequences, as well as with lesser breaks and **rare-event** deposits. The stratigraphic record is very punctuated, rather than being either continuous or gradual.

Readings

Ager, D. V. 1980. *The nature of the stratigraphical record.* 2d ed. New York: Wiley.

Berry, W. B. N. 1987. *Growth of a prehistoric time scale.* Palo Alto, Calif.: Blackwell.

Cande, S. C., and D. V. Kent. 1992. A new geomagnetic polarity time scale for the Late Cretaceous and Cenozoic. *Journal of Geophysical Research,* 97(B10):13917–52.

Carozzi, A. V. 1965. Lavoisier's fundamental contribution to stratigraphy. *Ohio Journal of Science,* 65:72–85.

Dott, R. H., Jr. 1983. Episodic sedimentation—How normal is average? How rare is rare? Does it matter? *Journal of Sedimentary Petrology,* 53:5–23.

Eicher, D. L. 1968. *Geologic time.* Englewood Cliffs, N.J.: Prentice-Hall.

Gignoux, M. 1955. *Stratigraphic geology.* San Francisco: Freeman.

Gretener, P. E. 1967. Significance of the rare singular event in geology. *American Association of Petroleum Geologists,* 51:2197–2206.

Jacobs, J. A. 1984. *Reversals of the earth's magnetic field.* Bristol, UK: Hilger.

Payton, C. E., et al. 1977. *Seismic stratigraphy.* Tulsa, Okla.: American Association of Petroleum Geologists, Memoir 26.

Prothero, D. R. 1990. *Interpreting the stratigraphic record.* New York: Freeman.

———, and F. Schwab. 1990. *Sedimentary Geology.* New York: Freeman.

Rudwick, M. J. S. 1985. *The great Devonian controversy. The shaping of scientific knowledge among gentlemanly specialists.* Chicago: Chicago University Press.

Schoch, R. M. 1989. *Stratigraphy: Principles and methods.* New York: Van Nostrand Reinhold.

Secord, J. A. 1986. *Controversy in Victorian geology: The Cambrian-Silurian dispute.* Princeton, N.J.: Princeton University Press.

Shaw, A. B. 1964. *Time in stratigraphy.* New York: McGraw-Hill.

Wilgus, C. K., et al., eds. 1989. Sea level changes—An integrated approach. *Society of Economic Paleontologists and Mineralogists Special Publication* 42.

(D. R. Prothero.)

Chapter 5

The Numerical Dating of the Earth

MAJOR CONCEPTS

Some drill and bore

The solid earth,

and from the strata there

Extract a register,

by which we learn

That He who made it,

and revealed its date to Moses,

was mistaken in its age!

William Cowper (late eighteenth century)

▶ Radioactive decay is the only process in geology that operates at a steady, statistically predictable rate. Examining the products of decay of radioactive elements trapped in the crystal structure of minerals is the only method of establishing the numerical age (in years) of geologic events.

▶ Because the radioactive decay products must be trapped in a crystal to be measurable, radioactive dating is useful only in crystalline rocks containing some radioactive elements, such as igneous or metamorphic rocks. Except for volcanic ashes and glauconites, no sedimentary rock can be dated directly. Instead, sedimentary rocks must be dated indirectly by their relative age with respect to datable igneous rocks, such as ash layers, lava flows, or intruded dikes.

▶ Any process that disturbs the radioactive element ratios will affect the apparent age of the rock. Thus, reheating of a rock may reset the atomic ratios; similarly any weathering or alteration will allow some atoms to escape and others to contaminate the crystal.

▶ Radiocarbon dating is the only method that can directly date fossils and organic material. However, it is useful only for the most recent 80,000 years, so it is primarily relevant to archeology and late Quaternary geology.

It is difficult to overstress the revolutionary impact upon modern geology of **isotopic, or radiometric, dating** of minerals. Not only has it stretched our conception of the total age of the earth by nearly 1,000 times, but also it has allowed us to date meteorites and moon rocks and to show that they are roughly the same age as the earth. Isotopic dating has allowed us to treat time numerically rather than only in relative terms, as was done in Chaps. 1–4. This made possible at last the subdivision and long-range—even intercontinental—correlation of Cryptozoic rocks, which lack index fossils yet represent 75 percent of earth history. It has also given more precision to the subdivisions of the younger, fossiliferous record, especially in complex mountain belts where igneous and metamorphic rocks are widespread (Fig. 5.1). Finally, isotopic dating allows the exciting possibility of determining rates of processes acting in the past with an increasing degree of quantitative rigor.

During the past 50 years, geologists have realized that the entire earth is the result of one grand chemical evolution from a heterogenous, lifeless beginning to a complexly organized inhabited world. The mutation of different isotopes of an element from unstable, radioactive parents to stable daughter isotopes is an important example of inorganic evolution and is the theme of this chapter. We first shall note several early ingenious attempts at numerical dating and then review the chemical basis for isotopic dating. Next we shall discuss specific methods used for such dating of minerals, rocks, and archaeological materials, dating that has led to the current numerical isotopic time scale. Finally, we shall evaluate the accuracy as well as the critical assumptions and some complications underlying isotopic dating. In Chap. 6, we shall see that, besides providing the basis for a numerical time scale, the decay of radioactive isotopes has played a major role in the overall chemical evolution of our planet.

Figure 5.1 Example of cross-cutting relationships that establish relative ages. The igneous dike cuts through the red shales but is truncated by the overlying sandstone unit. The radiometric date on the dike would give a minimum numerical age for the shales and a maximum numerical age for the sandstone. *(D. R. Prothero.)*

Early Attempts to Date the Earth

You will recall from Chap. 1 that, on the basis of the Scriptures, it was long ago argued that the earth was about 6,000 years old (see Box 5.1). Buffon suggested in 1760 that it was 75,000 years old, and he specified numerical dates for each of six epochs of earth history (see Chap. 2). Hutton, Lyell, and their followers reasoned from the implications of their uniformity doctrine and the observable rates of most geologic processes that the earth must be far older than that. Finally, Charles Darwin needed as much time as possible for the evolution of the amazing array of life forms known by 1859. Millions of years seemed necessary before the appearance of human beings, but no one had yet provided a sound quantitative measure.

Near the end of the nineteenth century, a British geologist estimated that all the time since the beginning of the Cambrian amounted to about 75 million years. He based his conclusion upon the maximum known thickness of strata of that age span multiplied by an assumed average rate of sedimentation derived from study of modern depositional rates. Use of the accumulation rate of salt in the sea was attempted next by an Irish scientist named Joly in 1899. Using the then known average rate of delivery of salt to the sea by rivers, he found that it would have taken about 100 million years to develop the present salinity. Obviously neither of these "ages" could have been the total age of the earth, as their authors realized.

All early attempts at assessing the age of the earth in terms of numerical time were based upon shaky assumptions. For example, if we were to choose an element other than sodium or chlorine to estimate the age of seawater, we would find a different apparent age because each element accumulates at a different rate. Today it is impossible to accept the simple extrapolation back in time of present rates of practically any processes. Nonetheless, early attempts to date the earth showed an evolution of thought in one direction, toward older and older estimates. By the middle of the nineteenth century, most geologists thought of a total age on the order of several hundred million years, but they had no numerical method to prove this.

Kelvin's Dating of the Earth

In 1846, the great physicist Lord Kelvin launched a 50-year combat with geologists over the earth's age and the nature of its historical development. This was an outgrowth of his displeasure with Lyell's strict uniformitarianism because that concept seemed

Box 5.1

Absolute Measures of Time

There are several ways to count backward, year by year, that are both more accurate and extend further in time than counting the generations mentioned in the Old Testament. Annual growth rings in still-living trees take us back nearly 5,000 years, and, by correlating distinctive tree-ring sequences from living to nearby dead trees the chronology can be extended to at least 12,000 years ago. A similar technique can be applied to sediments deposited in some lakes that freeze every winter, especially those near glacier margins. Such sediments are characterized by conspicuous laminations (called varves) that consist of pairs of summer silt and winter clay layers (Fig. 5.2). These annual pairs can be counted, and by correlating modern lake sequences with those of nearby former lakes, a good chronology has been established for the past 12,000 years. A less certain one extends back to 20,000 years ago—more than three times longer than Archbishop Ussher's Scriptural chronology.

Glacier ice also shows layering as a result of new snow added yearly. The layering is expressed as subtle differences of ice density, crystal texture, and proportions of dust. The layers can be measured microscopically in ice cores drilled from glaciers. Ice crystals also contain trapped air bubbles, which contain chemical species whose atmospheric abundances vary seasonally; therefore, measurement of variations of these species also can provide a year-by-year chronology. The best documented ice-core record, which is from Greenland, shows annual layering extending back at least 250,000 years.

Some climatic fluctuations result from well-known periodic variations of both the earth's orbit and its axis, known as Milankovitch cycles. As noted in Chap. 4, oxygen isotope ratios in skeletons from deep-sea cores record such climatic fluctuations. Because the period of the changes seems well established from astronomic observations, we have an independent time calibration for the observed deep-sea oxygen isotope fluctuations. One can count peaks on the isotope curve much like tree rings; at present, this method yields a record for deep-sea sediments extending back at least 800,000 years (see Fig.

Figure 5.2   Seasonal varve laminations averaging about 2 cm thick in fine Pleistocene glacial lake sediments in France. Because each pair of darker and lighter layers represents a single year, these laminations can be counted like the growth rings of trees (see Fig. 13.41) to determine numerical dates in years. Darker layers are clay deposited during winter when lakes were frozen, and lighter layers are silts deposited during summer when lakes were open. (© John S. Shelton.)

16.16). This is still only a fraction of geologic time, however, so we must seek other means of obtaining numerical dates for most of the time scale. Isotopic dating provides such means.

---

inconsistent with physical principles, as noted in Chap. 2. A completely uniform, unchanging earth was, to Kelvin, impossible. Deep mines in Europe showed that temperature increases with depth in the earth; this was taken as proof that the earth is losing heat from its interior. Either the earth must be cooling after an initial, very hot stage, or it contains a continuing internal source of heat energy. Lyell appealed to chemical reactions in the interior to produce heat in an endless cycle, allowing for a steady-state earth. Kelvin considered Lyell's untiring heat engine to be perpetual motion and thus unacceptable. He discounted any renewing internal heat source and accepted the long-standing assumption that the earth was originally molten. Therefore, heat was being dissipated through time, and the earth was not unchanging. Kelvin had, in fact, endorsed the evolutionary view of the earth that eventually would displace Lyell's conception.

The gradient of increasing temperature as one goes deeper and deeper into the earth is today known to average about 30°C per kilometer. Using an old estimate that was about three times too great, Kelvin extrapolated the apparent rate of cooling of the earth backward to a time when the earth presumably was molten; this should be close to the total age of the earth, he reasoned. Kelvin's calculations satisfied him that the earth is between only 20 and 30 million years old. His arguments seemed flawless, for no other terrestrial or solar heat sources were known. In the eyes of most scientists, Kelvin had won his campaign decisively and left the geologists reeling from his dazzling mathematics. They were reluctant to accept such a modest age but were incapable of mustering a rigorous counteroffensive until American geologist T. C. Chamberlin in 1899 challenged Kelvin's assumption that the earth had begun as a molten body. He introduced an important new hypothesis (to be discussed in Chap. 6) that the planets formed by the accretion, or collecting together, of cold, solid chunks of matter. According to Chamberlin, the earth must have heated up sometime *after* its initial formation through some internal process quite independent of solar heat.

Radioactivity

The last great breakthrough, the one that set the stage for modern historical geology, occurred in 1896 in the Paris laboratory of physicist Henri Becquerel. The Frenchman and his assistants, the Curies, discovered the phenomenon of radioactivity through the exposure of photographic film tightly sealed from light but held next to a uranium-bearing mineral. Some kind of radiation invisible to the eye had exposed that film. Two hitherto unknown chemical elements were soon discovered, radium and polonium, which form from uranium by changes in the atomic nucleus that involve then unknown types of radiation. Demonstration that many elements have several nuclear species called isotopes followed quickly. Some of these species are unstable; that is, their atoms are not permanent, as had been assumed for all atoms. The nuclear isotopes of a single element differ in number of neutrons in the atomic nucleus, yet isotopic species of any given element have similar chemical properties, and it is for this reason that isotopes went undetected for so long.

Isotopes with unstable nuclei, called *parent isotopes,* undergo spontaneous change until a permanent, stable configuration is reached, called *daughter isotopes.* This process of change is called radioactive decay, and it results in one or more of three types of **emissions** from the nucleus at a fixed average rate characteristic of any particular isotope. It was such emissions that exposed Becquerel's film. The emissions were originally named *alpha* rays (which proved to be a helium nucleus—that is, a helium atom stripped of electrons), *beta* rays (a free electron), and *gamma* rays (similar to X rays). Radioactive decay can be written like a chemical reaction in the following manner:

Unstable parent isotope → stable daughter isotope
+ nuclear emissions + heat

Isotopes are designated by a number representing the total sum of neutrons and protons in the nucleus, called mass number, which is generally written at the top left of the chemical symbol for the element. Emissions from the nucleus of a radiogenic isotope change the identity of that isotope by increasing or decreasing the number of neutrons in the nucleus (Fig. 5.3). For example, when the uranium isotope with mass number 238 (^{238}U) decays, an alpha particle (^{4}He) is emitted from its nucleus. This has the effect of reducing the number of neutrons and protons in the nucleus by 2 each (the two neutrons and two protons are what make up the ^{4}He particle); therefore, the resulting isotope has a mass number that has been reduced by 4. This new isotope is thorium 234 (^{234}Th), which is also unstable. By emission of an electron, it decays to protactinium 234 (^{234}Pa) and so on through a complex series of intermediate, unstable daughter isotopes, ultimately ending up at stable lead (^{206}Pb), as shown in Fig. 5.3. Thus, the

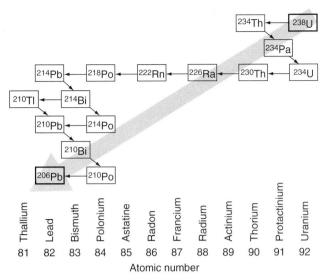

Figure 5.3 Diagrammatic portrayal of the decay history of the uranium 238 series. Rates of decay for the intermediate daughter products vary from less than 1 second (polonium 214) to 1,622 years (radium 226). *(From How old Is the earth? by P. M. Hurley. NY: Doubleday.)*

radium and polonium discovered in Becquerel's laboratory proved to be but intermediate daughter isotopes in the complex ^{238}U **decay series**—that is, a series of steps from an unstable parent to a stable ultimate daughter isotope. Uranium and thorium have the most complex series, whereas most isotopes used for dating involve simpler, direct decay from parent to daughter.

Further details of changes within the nucleus during decay are not essential to understanding the isotopic dating of minerals and rocks, and therefore we do not treat them further here. What is most important to realize is that radioactive decay is a statistical event. That is, neither the identity of the particular atom that will decay next nor the exact time at which the event will occur can be predicted absolutely. What we can predict is what fraction of the total parent atoms present will decay in a given period of time. Observations of many emission events from many atoms of a particular nuclear species over an extended time period provide a *statistical average rate of decay.*

In the laboratory, neither heating nor cooling nor changes in pressure or chemical state affect the average rate of spontaneous decay. In many rocks, dark rings are observed in crystals of minerals such as mica, rings that enclose tiny grains of radioactive minerals. These rings are radiation halos resulting from crystal damage by alpha particles propelled from decaying isotopes. In 1911, it was shown that the distance traveled by alpha particles is proportional to the decay rate of the parent isotope. The radii of halos in rocks of widely differing relative geologic ages was found to be constant for a given alpha-emitting isotope, a finding that supported the laboratory evidence that decay rates are constant through time. Halo radius is a function of decay rate rather than mineral age. Halo color is related to age, with older minerals having darker halos, but differences of color are too hard to discriminate to make radiation halos useful for rock dating.

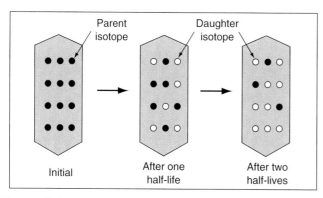

Figure 5.4   Diagrammatic portrayal of decay of a radioactive isotope incorporated within a mineral crystal at the time of initial crystallization. Note the changing ratio of parent to daughter isotope after one and then two half-life time intervals.

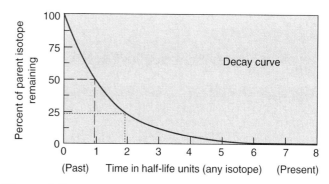

Figure 5.5   Simple arithmetic plot of a universal isotopic decay curve. After one half-life has elapsed, 50 percent of the original parent isotope remains; after two half-lives, half of that, or 25 percent; and so on.

First Dating of Minerals

Further knowledge of radioactivity came with breathtaking rapidity after the turn of the twentieth century. In 1902, then-Canadian physicists E. Rutherford and F. Soddy, after counting radium emissions for many days, reasoned that *total emission activity was proportional to the number of unstable parent atoms still present.* This meant that emission must decrease in some regular fashion through time and that decay could be expressed mathematically. They also suggested that the progressive accumulation of daughter isotopes might provide a basis for the numerical measurement of geologic time.

In 1905, a Yale chemist, Bertram Boltwood, proved that lead was among the disintegration products of uranium. This is an ironic twist of the alchemists' ancient dream of making precious gold from lead; instead, apparently common, ugly lead was formed from a rare element considered to be almost as precious as gold. In 1907, Boltwood, following Rutherford's suggestion, developed the idea that a decay series could provide a means of dating the time of crystallization of minerals that contained a radioactive element. Boltwood reasoned that in unaltered minerals

> of equal ages, a constant proportion must exist between the amount of each disintegration product and the amount of parent substance . . . and . . . the proportion of each disintegration product with respect to the parent substance must be greater in those minerals which are the older.

To compute the ages of minerals, Boltwood needed to know the rate of decay of uranium, a rate so slow that it had not yet been measured, so he estimated the rate of decay indirectly from the known rate for radium, a rapidly decaying intermediate member of the uranium series (Fig. 5.3). He then calculated the ages of minerals from their lead-uranium (Pb-U) ratios as

$$\text{Age} = \frac{\text{amt. of daughter isotope (Pb)}}{\text{amt. of parent isotope (U)}} \times 10^{10} \text{ yr}$$

Boltwood's results for ten localities on three continents ranged from 410 to 2,200 million years. In reality, his dates were about 20 percent too large because the mathematics of the decay

process was not fully understood until 1910 and because, unknown to him, more than one decay series—each with a differing decay rate—were involved in the minerals he studied. Even allowing for this error, the oldest apparent age listed by Boltwood was nearly 100 times greater than Kelvin's figure and 10 times the wildest guess of any geologist for the total age of the earth!

Concept of Decay

To analyze many different isotopes whose decay rates vary widely, it is convenient to have a generalized expression applicable to all decay series. Remember that what is dated is the *time of crystallization* of a mineral—that is, the time of incorporation of an unstable isotope into that mineral (not the time of origin of that isotope in the universe, which in most cases was earlier than the origin of the earth). With exceptions to be discussed later, it is assumed that, once a mineral crystallized, a closed chemical system was formed such that any daughter product then present in the mineral was formed only from the decay of the original parent isotope therein (Fig. 5.4). It is like the closed system of an hourglass with trapped sand grains. The grains in the top of the glass are like parent isotopes, whereas those falling to the bottom are like daughter isotopes; the ratio between them at any moment is a measure of time elapsed.

Dating is a matter of determining the ratio of parent to daughter atoms and then calculating a mineral's age by multiplying by the decay rate of the parent. More atoms decay early in the history of a mineral, and emissions then decrease in a regular way. Fig. 5.5 shows a decay curve with time expressed in half-life units. Note that decay, like so many other processes in nature, is nonlinear (see Fig. 1.12). The **half-life,** which is the time required for one-half of a given amount of any particular unstable nuclear species to decay, is a more convenient expression of decay rate than is the decay constant (λ). Their mathematical relationship is explained in Box 5.2. Half-life can be better envisioned by considering an analogy. Imagine a piece of cake to represent all the parent isotopes in a mineral crystal. This is the last piece of cake left as you sit with friends around the

Box 5.2

Mathematics of Radioactive Decay

Any scientist who studies radioactivity—especially a geochronologist, who determines mineral and rock dates—requires a more complete mathematical treatment of radioactive decay. The following is the expression used for calculating dates:

$$\text{Age} = \ln\left(\frac{\text{daughter atom}}{\text{parent atom}} + 1\right) \times \frac{1}{\lambda}$$

Being exponential decay can be conveniently expressed in logarithmic form such that

$$\lambda t = \ln\frac{N^\circ}{N^t}$$

where λ = decay constant (probability of a decay event within some unit of time)

N° = number of atoms of parent isotope at time zero (when cooling reaches the blocking temperature)

N^t = number of atoms of parent after time t of decay

$\ln$ = natural logarithm (base 2.78)

It is convenient to have some fixed unit of decay time that expresses concisely the decay characteristic of a particular isotope. Such a unit is the half-life ($t_{1/2}$), the time required for one-half an original amount of a nuclear species to decay. At one half-life, the N°/N^t ratio equals 2; therefore,

$$\lambda t_{1/2} = \ln 2$$

$$t_{1/2} = \frac{\ln 2}{\lambda} = \frac{0.693}{\lambda}$$

Table 5.1	Principal Decay Series Used for Mineral and Total-Rock Dating		
Parent Isotope	**Half-Life**	**Ultimate Stable Product**	**Effective Age Range**
Samarium 147	106 billion years (b.y.)	Neodymium 143	> 100 million years (m.y.)
Rubidium 87	48.8 b.y.	Strontium 87	> 100 m.y.
Thorium 232	14.0 b.y.	Lead 208	> 200 m.y.
Uranium 238	4.5 b.y.	Lead 206	> 100 m.y.
Potassium 40	1.25 b.y.	Argon 40	> 100,000 years
Uranium 235	0.70 b.y.	Lead 207	> 100 m.y.
Carbon 14	5.730 years	Nitrogen 14	0–80,000 years

Source: After Odin, 1982.

table, and you are still hungry but do not want to appear greedy. Therefore, you politely cut the piece in half and eat only one half. After awhile you are still hungry, so you cut the remaining piece in half and eat one of the two pieces. Ten minutes later you are still hungry, but no one else has eaten a piece, so you repeat the process. The diminishing size of the cake left on the plate is like the decrease of parent isotopes as decay proceeds in a mineral.

By using half-life as a time scale unit, our curve in Fig. 5.5 can apply to the decay of any isotope, regardless of the fact that the actual numerical values of half-lives vary enormously, as shown in Table 5.1. Undoubtedly some short-half-life isotopes contained in the early earth have so completely decayed that we cannot detect them, yet their daughter products must remain. For example, two isotopes of the gas xenon are thought to be decay products, respectively, of iodine 129 and plutonium 244, which do not occur naturally on earth (although they have been pro-

duced by nuclear explosions). The half-lives of [129]I and [244]Pu are so short that, if they were present in the early earth, as is supposed, after only a few hundreds of millions of years they had diminished to unmeasurably small amounts.

Modern Isotopic Dating

Decay Series Used

During the mid-twentieth century, isotopic dating of minerals increased tremendously and the subscience of geochronology developed. By 1930, uranium-lead dating had progressed to the point that it could be said that the end of the Cambrian Period was about twice as old as the Permian. In 1934, the first isotopic time scale was published, and even text-books were beginning to recognize isotopic dating. Discovery about 1940 of the rubidium

87-strontium 87 series and the proof in 1948 that radiogenic potassium 40 decays to the inert gas argon 40 initiated the use of other isotopes as geologic clocks in addition to those of well-known uranium and thorium. This was a great advance, for Rb and K, unlike U and Th, occur in several common rock-forming minerals.

The most important decay series in current use were shown in Table 5.1, but others continue to be developed. Differences in decay rates impose certain limitations on the relative usefulness of each series for minerals of different relative ages. Carbon 14, for example, decays so rapidly that after about 80,000 years there is not enough of it left to be measured accurately with present analytical techniques. Therefore, this isotope is useful only for late Pleistocene and Holocene times; it is especially useful in archaeology, climatology, and studies of the movement of ground water and ocean circulation. Until recently, the interval from about 100,000 years to 1 million years ago was the least datable, but refined K-Ar techniques, the use of certain intermediate daughter products of uranium, and fission-track dating (described in the section "Fission-Track Dating") have helped close this gap.

An ingenious method has been developed for dating Pleistocene deep-sea clays using the ratio of protactinium 231 (half-life 34,300 years) to thorium 230 (half-life 80,000 years). Both are intermediate daughter products of uranium decay, as is uranium 234, which also is used. All these isotopes are taken from seawater by clay minerals, and their ratios in sediments become a function only of the time since absorption from the water. This method allows dating of sediments as much as 175,000 years old, which is double the range of carbon 14 dating.

The long-half-life isotopes are of no use for dating minerals less than about 100 million years old because no perceptible change will have occurred in such a short time. Uranium, thorium, samarium, and rubidium are especially applicable to Cryptozoic rocks (i.e., over 700 million years). The potassium-argon series is slightly less useful for very old rocks. This is not because of the decay rate of the series but because the ultimate stable daughter product argon, like helium, being a gas, is lost from crystals more easily than are other daughter products. Argon *leakage* is a problem if a mineral has been reheated after initial crystallization through burial or metamorphism to temperatures exceeding 150° to 200°C (Fig. 5.6). In short, potassium-bearing minerals can have their isotopic clocks reset more readily than other minerals or decay series, but this is not necessarily bad, because dating metamorphic events is often just as important as dating the original crystallization of a rock. Moreover, it is now possible to correct for argon loss by recalculating the age of a sample using two K-Ar series with very different decay rates. Artificial irradiation converts some ^{39}K to ^{39}Ar, and age can then be calculated from the ^{40}Ar/^{39}Ar ratio as well as from the ^{40}K/^{40}Ar ratio, thus providing a cross-check. This is called argon/argon dating and has become one of the most important techniques.

Isotopic dates for igneous and metamorphic rocks are, in reality, dates of cooling to a critical threshold. At this threshold, called **blocking temperature,** a given mineral becomes a closed chemical system for a particular decay series (Fig. 5.6). *The*

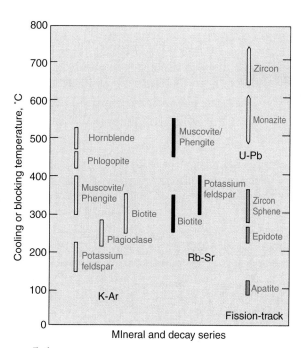

Figure 5.6   Cooling or blocking temperatures for several minerals and decay series. As a rock cools, a given mineral becomes a closed chemical system for a particular decay series at the latter's blocking temperature. If temperature is later raised above that level, then the parent-daughter ratios for that decay series may be altered, resulting in resetting and discordant dates. *(Figure by Odin, from Episodes 1982, p. 5. Reprinted by permission of the International Union of Geological Sciences.)*

parent-daughter ratio measured for dating relates only to the time of blocking. For igneous rocks that crystallized rapidly at or near the earth's surface, blocking temperature is reached immediately, so a calculated isotopic date coincides with time of crystallization—the *primary age.* For deep-seated igneous and metamorphic rocks, however, cooling occurs so slowly that blocking temperature is generally reached long after crystallization is completed. This necessitates thorough knowledge of a rock derived from both field and laboratory studies in order to derive geologic meaning from any isotopic date.

Both *delayed cooling* and *resetting of isotopic clocks* by metamorphic reheating above blocking temperatures can be detected by studying at least two different isotopic series that have responded differently during a rock's history (Fig 5.6). Resetting is discussed more fully in the section "Discordant Dates from Resetting."

Daughter Lead Ratios

Today geologists also establish dates by comparing ratios of daughter products. Comparison of different daughter lead isotopes, such as ^{206}Pb and ^{207}Pb (as in Table 5.2), provides some of the most accurate rock dates possible for the entire uranium series. Because these daughter isotopes result from series with different decay rates, proportions of the lead isotopes change through time. Their changing ratio provides a special isotopic clock.

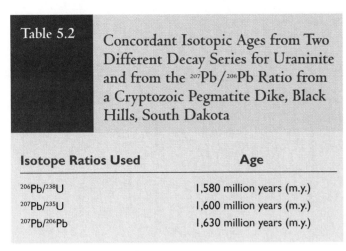

Table 5.2	Concordant Isotopic Ages from Two Different Decay Series for Uraninite and from the ²⁰⁷Pb/²⁰⁶Pb Ratio from a Cryptozoic Pegmatite Dike, Black Hills, South Dakota

Isotope Ratios Used	Age
²⁰⁶Pb/²³⁸U	1,580 million years (m.y.)
²⁰⁷Pb/²³⁵U	1,600 million years (m.y.)
²⁰⁷Pb/²⁰⁶Pb	1,630 million years (m.y.)

Source: After Wetherill et al., 1956, *Geochimica et Cosmochimica Acta,* v. 9.

Isotope ratios for lead minerals (such as galena, PbS) can be used for dating by comparing the proportions of different lead isotopes present. One isotope, ²⁰⁴Pb, is not known to be forming today by any radioactive decay process now operating on earth. Therefore, its total abundance has not changed since the origin of the earth's crust. Significantly it is more abundant in old lead minerals, whereas geologically younger minerals contain progressively more of the radiogenic lead isotopes produced by uranium and thorium decay (review Table 5.1). In other words, as decay has created radiogenic lead, some of that new lead has been freed from its parent uranium-bearing minerals by melting or weathering and has become recombined with other leads in later-formed lead minerals. Such dilution of the earth's original nonradiogenic ²⁰⁴Pb has increased through geologic time in direct proportion to the continuous generation of new radiogenic lead isotopes through decay. The systematic change in the ratios among the lead species in younger and younger minerals provides a special isotopic clock because each species is formed at a different rate (Fig. 5.7). Extrapolation of lead ratios back through time, as in Fig. 5.7, also provides an estimate of the total age of the earth (see the section "Memorable Dates").

Carbon 14—a Special Case

The value of ¹⁴C for dating was demonstrated in 1947 by American W. F. Libby. Unlike most other isotopes used for dating, an unstable isotope of carbon is produced in the upper atmosphere. High-energy cosmic particles shatter the nuclei of oxygen and nitrogen atoms there, releasing neutrons, which in turn produce ¹⁴C when they collide with other nitrogen atoms (Fig. 5.8). Unstable ¹⁴C decays eventually to stable ¹⁴N. Most of the radiocarbon produced is oxidized to carbon dioxide (CO_2), some of which enters the hydrosphere. Production of ¹⁴C in the atmosphere is assumed to be at a nearly constant rate; therefore, ¹⁴C in the atmosphere should have long since reached a steady state. That is, a constant balance should exist between its production and its decay such that the ratio of radiocarbon to stable carbon is assumed to be essentially constant through time.

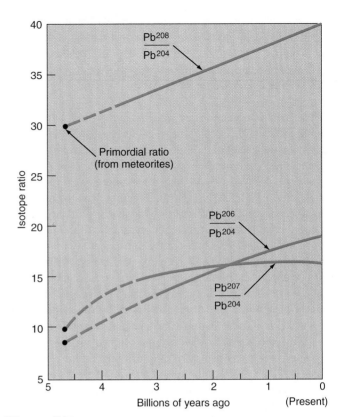

Figure 5.7   Use of daughter lead isotopes for dating. The ratios of each of three radiogenic lead isotopes to nonradiogenic lead 204 (which has no known parent on earth) have all changed through time, but at different rates. Comparison of such ratios for a mineral provides a means of dating. In addition, extrapolation of the solid curves derived from many lead minerals of differing ages back to *primordial ratios* found in meteorites provides an estimate of total age of the earth. *(Based upon work of Nier, Patterson, and Tilton.)*

From either the air or the water, organisms acquire small amounts of ¹⁴C along with the stable isotopes ¹²C and ¹³C. After death, the organisms acquire no new carbon, so that over time the ¹⁴C content decreases progressively by decay. Any dead organic matter, such as wood, charcoal, some bone, and human artifacts such as cloth, as well as shells and young limestones are all potentially datable (Fig. 5.8). The number of ¹⁴C and ¹⁴N atoms present is so small that direct measurement is difficult. Instead, dating is generally done by use of a radiation counter to compare the emission activity of the specimen with that of a standard sample whose activity is known very accurately. The number of emissions, or "counts," in a given time period is proportional to the number of atoms of ¹⁴C present; the older the specimen, the lower its activity level. Since the industrial revolution, burning of fossil fuels has diluted, whereas atmospheric nuclear testing has increased, the ¹⁴C content of the total carbon reservoir. These changes have necessitated using wood from a tree that died in the early nineteenth century for the standard sample. The rather short ¹⁴C half-life of 5,730 years, together with the limit of sensitivity of counting devices, imposes an upper limit of measurable age of about 50,000 to 60,000 years by conventional methods. A new approach using a reactor to count atom ratios has pushed the limit back further.

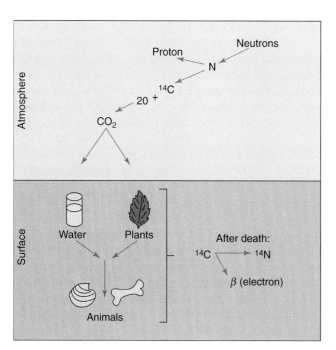

Figure 5.8   Constant generation of carbon 14 in the upper atmosphere by cosmic particle bombardment that produces neutrons, which convert nitrogen atoms to ^{14}C by expelling a proton. Some ^{14}C is ultimately incorporated into organisms whose remains can be dated.

Figure 5.9   Microscopic view of fission tracks in a zircon crystal after etching. Each track is the result of crystal damage by spontaneous fission of an atom of uranium 238. The number of tracks within a given area is a function of age. (Black bar at upper right is 0.02 mm long.) *(Courtesy of J. D. Macdougall.)*

Fission-Track Dating

The newest method of isotopic dating is based not upon spontaneous radioactive emissions but upon a rarer effect of decay characteristic of a few isotopes: spontaneous fission of an atomic nucleus. In ^{238}U, the principal fissionable isotope in nature, for approximately every 2 million decay emissions of *alpha* particles (helium nuclei), one atom undergoes fission instead. Its atomic nucleus splits, forming two energetic particles that repel one another. As these particles speed apart, they strip electrons from adjacent atoms. The havoc produced leaves minute imperfection lines in minerals and glasses. When etched, these lines, called **fission tracks,** become visible under a microscope as the tracks of fleeing nuclear fragments (Fig. 5.9). One can count the density of tracks to obtain an estimate of the number of atoms that have already undergone fission and then determine by conventional analytical procedures the amount of ^{238}U still present. In principle, we can then calculate the age, as with other types of dating, from the ratio of original total ^{238}U atoms to the number of unfissioned atoms still remaining, times the known average rate of fission events for this isotope. The half-life of ^{238}U fission is about 10^{16} years.

Fission-track dating is relatively simple and inexpensive, can be used for very small samples, and is almost as precise as other dating techniques for geologically young materials. For minerals older than about 100 million years, however, an accuracy of only about $\pm$ 10 percent is possible. The initial successes of this method were in dating archaeological and geologic materials ranging from several hundreds to a few millions of years in age,

that least accessible range between ordinary ^{14}C and potassium-argon dating. Manufactured glass colored by uranium-rich minerals as young as 30 years has been dated, as has pottery as young as 700 years.

Discordant Dates from Resetting

Daughter isotopes are usually trapped within the crystal structures of the minerals in which they originated. But metamorphic recrystallization of a rock tends to purge minerals of many daughter products because they are chemically dissimilar from their parent species and may not form true chemical bonds in the same minerals. Helium and argon may diffuse and leak completely out of the original rock, whereas other daughter products become incorporated in other minerals where they are more chemically "at home." Meanwhile, remaining parent isotopes begin a new decay history in *purged crystals* (Fig. 5.10). The latter grains will yield isotopic dates of the metamorphic, clock-resetting event. If none of the premetamorphic daughter products were lost from the rock during recrystallization, the parent-daughter ratio for the total rock will still indicate the time of original formation of the rock.

Fission-track clocks also may be reset because heating causes healing of the tracks. Thermal energy excites atoms sufficiently to allow them to recover from fission damage and restore order within crystals. The result will always be ages that are too young (as with K-Ar resetting). As a consequence, fission-track dates commonly are discordant with other isotopic dates, but this can generally be detected and used to advantage in discerning

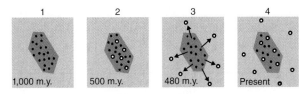

Figure 5.10   The effect of metamorphism in redistributing daughter isotopes (circles) in a rock that (1) crystallized originally 1,000 million years (m.y.) ago; (2) after 500 million years, some parent isotopes (dots) in a feldspar crystal had decayed; (3) metamorphism 480 million years ago drove the daughter atoms out of the crystal, but they were retained in the surrounding rock; (4) present dating of the feldspar would reveal the metamorphic event, whereas a total-rock date would reveal the original crystallization 1,000 million years ago.

complex rock histories. The tracks themselves may give a clue to resetting by showing two distinct groups, one of older, partially healed, short tracks and another of younger, normal-length ones formed after a heating event.

Refined dating today requires several isotopic dates from different minerals as well as from the total rock. Different isotope series in the same specimen provide cross-checks because of their different responses to metamorphism. For example, consider a granite that was reheated to 350°C and then cooled again. As shown in Fig. 5.6, a *metamorphic date* will be yielded by fission tracks and both K-Ar and Rb-Sr in feldspars and biotite mica, but a primary date will be yielded by U-Pb in zircon as well as by total-rock analyses of Rb-Sr and U-Pb. Such painstaking multiple analyses give the maximum possible information about both a rock's isotopic history and its geologic history.

Accuracy of Isotopic Dates

For many years, the numerical results from measurements of radioactive isotopes were called "absolute ages." Most of them were neither. As we have seen, the numbers are *dates* of geologic events, which may or may not correspond to the real primary age of a rock. Moreover, because of such complications as leakage, delayed cooling, and thermal resetting, the resulting numbers are certainly not absolute. None of this discredits either the accuracy or the value of numerical dating but merely recognizes reality somewhat belatedly. The results should be called isotopic dates, whose meaning must be interpreted in terms of everything that is known geologically about a given rock.

Several factors limit the accuracy of isotopic dates. First, the statistical nature of decay means that, even under the best of conditions, only average decay rates are determinable. This introduces a small uncertainty, though for very old minerals it is negligible. A second source of uncertainty originates in the laboratory analyses of the isotope ratios. The mass spectrometer, introduced in 1929, provides the most sensitive analyses of isotope abundance, with from $\pm$ 0.2 to 2.0 percent accuracy possible.

Because of analytical limitations, isotopic dates generally are accompanied by an uncertainty figure, such as 100 $\pm$ 5 million years. This means that analytical limitations do not allow one to say more than that the true date lies between 95 and 105 mil-

lion years. Under ideal conditions of rock freshness, analytical care, and relatively old material (with accurately measurable amounts of daughter material present), the analytical error for isotopic dates can be as low as $\pm$ .2 percent. It is important to realize that some uncertainty exists for every isotopic date, but in an introductory book such as this, the uncertainty factor is of little consequence, so it will be omitted from most dates quoted.

Isotopic Time Scale

Because most datable minerals containing radioactive elements originate in igneous rocks, the bulk of isotopic dating has been confined to such rocks. Detrital minerals containing unstable nuclear species occur in sediments, to be sure, but to date them is to date not the time of sedimentation but, rather, the time of crystallization of the mineral in a parent igneous or metamorphic rock or mineral vein prior to erosion, transport, and deposition in a sediment.

Because the relative geologic time scale is based upon fossiliferous sedimentary rocks, obviously it is urgent to establish numerical dates for the sediments. Direct isotopic dates of sedimentary rocks are possible for only a few minerals that contain unstable species *and that crystallize in the environment of deposition*. Such a material is "glauconite," a heterogeneous mixture of green, micalike silicate minerals containing potassium and iron. It apparently forms only in marine environments where deposition is slow. The potassium in glauconite consists in part of ^{40}K, therefore, the mineral can be dated by the K-Ar method.

Potassium-bearing feldspars form in certain sediments soon after deposition, and these, too, are potentially datable. Unfortunately, neither glauconite nor such feldspars are very common, and they give rather unreliable results. The uranium method has been used for black shales containing minute amounts of uranium-bearing minerals associated with carbon compounds. The uranium compounds formed in the sediments, so they provide dates of sedimentation. ^{14}C can be used to date shells, bone, or plant material in very young sediments.

Because most dating must be done on igneous or metamorphic rocks, it is necessary to relate such ages indirectly to the relative geologic time scale in order to establish a numerical scale. Here the basic principles of stratigraphy come into play. A volcanic rock interstratified with fossiliferous sediments and containing datable minerals is clearly of the same geologic age as those sediments and provides an ideal point on the isotopic time scale. An intrusive igneous body, however, is younger than all rocks through which it cuts. If it cuts fossiliferous strata, we have an older geologic age limit for the intrusion. If younger fossiliferous sedimentary rocks rest unconformably upon the intrusion, or at least contain pebbles derived from it, then a younger geologic age limit can also be established (Figs. 5.1, 5.11), for these sediments must be younger than the intrusion. If the age difference between the older and younger fossiliferous sediments is small, the intrusion is closely dated in the relative geologic scale, and its isotopic date provides another precise control point on the numerical time scale. The scale is constantly being refined as the techniques for analysis of isotopes are improved.

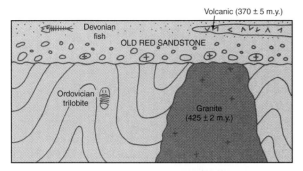

Figure 5.11   Scottish rocks studied by Hutton, showing how key dates are established for an isotopic time scale. The granite is post-Ordovician and pre-Devonian, thus is of Silurian age. A date for it establishes the Silurian Period as about 425 million years (m.y.) old. The date for a lava flow within the Old Red Sandstone establishes that part of the Devonian Period is about 370 million years old. (Compare Figs. 4.5 and 4.11.)

Memorable Dates

At last we can inquire intelligently of our planet's true birthdate. The chemical composition of the solar system must be relevant to its age as well as to its mode of origin. It is thought that most of the natural chemical elements in the universe evolved through thermonuclear reactions in stars; additional elements also developed outside the stars (as in the earth) through radioactive decay of unstable isotopes. The chemical elements in our part of the universe apparently evolved from the lightest and simplest atoms, hydrogen (75 percent) and helium (24 percent), to the heaviest and rarest ones, such as uranium and lead. The continued presence in the earth of moderate-half-life isotopes, such as ^{40}K ($t_{1/2}$ = 1.25 b.y.), places an upper limit on the date of formation of the earth. If the earth were more than 6 or 7 billion years old, all of those isotopes would have decayed.

No individual isotopic date represents the time of origin of a parent isotope, for most of them originated before the earth was formed (the majority of elements in the solar system are thought to be from 7 to 10 billion years old). The dates instead indicate times of incorporation of a given isotope into a mineral crystal. Isotopes had been decaying ever since their extraterrestrial origin, but only after incorporation into the closed system of a crystal could the decay products be trapped with their parent isotope to provide a measurable isotopic ratio useful for dating. Uranium, lead, and rubidium dating have revealed that igneous rocks as old as 3 to 3.5 billion years occur in South Africa and northwestern Russia near Finland, and over 3.7 to 3.8 billion years in southwestern Greenland and in Minnesota. Recently rocks over 3.96 billion years old have been dated from northern Canada, Wyoming, and China. Zircon grains from a Cryptozoic metamorphosed sandstone in Australia yielded dates of 4.3 to 4.4 billion years, but the source rock from which the grains were eroded has not yet been identified. Thus, some of the outer crust of continents is more than 3.9 billion years old. Some of the moon's crust appears to be of about the same age, according to isotopic dating of the Apollo mission specimens (3.5 to 4.2 billion years).

The oldest rock dates still do not represent the total age of the earth, for it must have taken the crust a while to form, as we shall see in Chap. 6. An ingenious way to estimate the elusive total age was suggested through work with lead isotopes in the 1930s. Today there is about one part of "new" radiogenic lead to two parts of original "common" lead in the crust. Certain meteorites contain leads but no uranium or thorium parents. The proportions of different lead isotopes therein are assumed to have remained fixed since the meteorites formed, and lead ratios in these meteorites are taken as a standard representing the approximate **primordial lead ratios** in the early earth (Fig. 5.7). By comparing ratios of the four lead isotopes in minerals of various ages, we can extrapolate back to ratios identical to the assumed primordial leads of meteorites. Such extrapolation indicates that the primordial ratio should have existed on earth about 4.6 to 4.7 billion years ago. This date apparently represents the time when elements in the earth first became arranged into definite chemical compounds, and it is taken as the beginning of truly geologic time. It must be close to the planet's birthdate.

How much earlier the very first assembly of the earth began is more difficult to fix. Astronomers believe, from theories of stars and galaxies, that the Milky Way galaxy is about 10 billion years old and that our sun formed about 6 billion years ago. Meteorites have been dated at between 4.3 and 4.6 billion years old, nearly identical both with the date indicated by lead ratios for the initial formation of earth minerals and with dates from moon rocks. Assuming that meteorites, as well as the planets and moons, have always been as closely linked to the sun as they are today, they can hardly be older than that star. Therefore, it appears that the earth as a solid sphere is not more than 5.0 billion years old, and that the outer crust of continental areas started forming between 3.8 and 4.0 billion years ago.

The first bona fide records of fossil life come from rocks that are about 3.6 billion years old and contain structures and chemical compounds formed by bacteria and possibly primitive algae. The first recognizable animal fossils are a mere 1,000 million years old or so, but the first abundant animal record began only about 600 million years ago; humans are a mere 2 to 3 million years old. Therefore, for almost one-fourth of earth history, there is no recognized record of any life form, and only the last quarter, Phanerozoic time, contains an abundant record, particularly of index fossils useful for geologic dating and correlation of strata. If we were to proportion this book strictly according to the numerical time scale, 15 chapters should be given over to the Cryptozoic, 1.25 to the Paleozoic, 0.6 to the Mesozoic, and only 0.25 to the Cenozoic. However, we actually do the opposite because most is known about the latest, brief part of our planet's history.

From isotopic dating, we can assess the lengths of the geologic periods and derive a numerical calender (see endpapers). From the results of extensive analyses, the dates of most period boundaries are now fairly well established, as summarized in the endpapers. The average duration of index fossil zones can be determined for periods whose boundaries are accurately dated isotopically. One simply divides the total time span of the period by the number of zones recognized for that period, and this computation shows that many zones represent as little as 1 million years. We can then use the average zone durations for a period to interpolate approximate dates of events for short time intervals that are below the level of resolution of isotopic dating. Thus, the study of earth history is becoming ever more quantitative, for we can now also compare rates of change, such as mountain building, erosion, sea-level changes, and organic evolution. Throughout this book, we shall examine many events in terms of numerical, as well as relative, geologic age and rates.

An Assessment

We have seen that estimates of the age of the earth have increased up to the present-day figure of 4.7 to 5.0 billion years. The maximum figure is not expected to increase much more, however, because of the limit on the age of the sun imposed by astronomical data. Early dating methods, such as those based upon increasing salinity of the seas, total accumulation of stratified sediments, and Kelvin's cooling of the earth, were all ingenious but suffered from erroneous assumptions and inaccurate knowledge of true rates of the processes involved. The most brilliant argument is no better than its weakest assumption. Kelvin's hypothesis was doomed by the incorrect notion that all the earth's internal heat was residual from a hot origin. Radioactivity was discovered near the end of Kelvin's life, so he had no way of knowing that the earth contains its own internal heat-generating mechanism. Isotopic dating has exonerated Hutton and Lyell by showing the earth to be many times older than even their wildest estimates.

Now, what about the assumptions behind geochronology? Those neocatastrophists who believe that the earth can be no more than 6,000 to 10,000 years old question the constancy of decay rates and doubt that any mineral could be a closed chemical system. Their criticisms seem to have been answered by (1) the sophisticated detection of, and correction for, either excess or diminished daughter products; (2) the detailed knowledge that different blocking temperatures provide about closed systems; and (3) both experimental evidence and radiation halo evidence for constant decay rates. Most convincing, however, is the consistency of results obtained from several different decay series having *markedly differing rates of decay.* It is extremely improbable that such consistency could be obtained if the basic assumptions behind dating were invalid.

Isotopic dating of minerals revolutionized geology in the twentieth century. Besides numerical dating, radioactivity provides an internal source of earth heat, the chief form of terrestrial energy. The discovery of radioactivity also provides a dramatic example of inorganic evolution—the irreversible change of one element to another. Modern thermonuclear theories of stars and of the origins of the elements provide a compelling evolutionary conception for the entire solar system, as well as for the earth. We explore this more fully in Chap. 6.

Summary

Isotopic Dating

- **Isotopic,** or **radiometric, dating** is based upon the formation of one isotope from another. **Radioactive,** or **radiogenic, isotopes** have unstable atomic nuclei. A **parent isotope** undergoes spontaneous **radioactive decay,** which results in the **emission** of one or more kinds of radiation from its nucleus; a **stable daughter isotope** results. The parent-to-daughter transformation is called a **decay series.**
- **Decay rate,** or **emission activity,** for any given radioactive isotope is constant through time.
- Decay rate is most conveniently expressed as **half-life,** which is the time required for the decay of one-half of a given amount of unstable parent.
- **Blocking temperature** is the cooling threshold below which a given mineral becomes a closed chemical system for a particular decay series. The parent-daughter ratio dates the time of such closure.
  1. **Slow** or **delayed cooling** may yield dates that are significantly younger than the **primary age** or **time of crystallization.**
  2. **Resetting of isotopic clocks** results from reheating above the blocking temperature and allows dating of **metamorphic events.**

- **Daughter-isotope ratios** provide a special isotopic clock. Different daughter isotopes are formed at different rates; therefore, their ratios change in a mathematically predictable way through time. The separate ratios of three radiogenic lead isotopes to nonradiogenic (thus unchanging) ^{204}Pb, when extrapolated back to their presumed **primordial ratios,** provide the best estimate of the total age of the earth. The resulting date of 4.6 to 4.7 billion years is thought to represent the first organization of elements into chemical compounds.
- **Radiocarbon dating** of organic material is unique because radiogenic ^{14}C is formed continually in the atmosphere; its half-life is only 5,730 years. It is important for dating archaeological and geologic events of the past 80,000 years.
- **Fission-track dating** is based upon the relative density of tracks resulting from radiation damage by passing fission particles. The older the material containing the tracks, the denser the tracks, but tracks can be healed by heating, which results in clock resetting.
- **Discordant dates** result from some disturbance of isotope ratios, such as metamorphic reheating. **Leakage** of daughter gaseous Ar or He, differential **purging** of daughter isotopes from certain minerals and their

redistribution in other minerals, and **healing** of fission tracks result in two or more dates from the same sample. A **metamorphic date** will be indicated by any decay series that was heated above its blocking temperature, but a **primary date** will still be recorded by any decay series whose blocking temperature was not exceeded and/or by a **total-rock** analysis.

- A combination of techniques using several decay series provides **isotopic dates** that, when integrated with geologic data, provide meaningful numerical calibration of the relative geologic time scale. Because only a few

sedimentary minerals can be dated isotopically, most calibration must come from igneous rocks, whose relative geologic ages are well established.

- **Memorable dates** (millions of years) are

Meteorites	4,300–4,600
Moon rocks	3,500–4,200
Pb ratio dating of earth	4,600–4,700
Oldest known crustal rocks and minerals	3,900–4,400
Beginning of Phanerozoic Eon	540
Paleozoic-Mesozoic boundary	250
Mesozoic-Cenozoic boundary	66

Readings

Berggren, W. A., D. V. Kent, M.P. Aubry, and J. Harden bol. 1995. *Geochronology time scales, and global stratigraphic correlation.* SEPM Special Publication, 54.

Brush, S. G. 1982. Finding the age of the earth by physics or faith? *Journal of Geological Education,* 30:34–58.

Burchfield, J. D. 1975. *Lord Kelvin and the age of the earth.* New York: Science History.

Carr, P. F., et al. 1984. Toward an objective Phanerozoic time scale. *Geology,* 12:274–77.

Dalrymple, G. B. 1991. *The age of the earth.* Stanford, Calif.: Stanford University Press.

———, and M. A. Lanphere. 1969. *Potassium-argon dating.* San Francisco: Freeman.

Harland, W. B., et al. 1989. *A geologic time scale.* Cambridge, UK: Cambridge University Press.

Holmes, A. 1931. Radioactivity and geologic time. In *Physics of the earth IV.* Washington, D.C.: National Research Council Bulletin, 80.

Hurley, P. M. 1959. *How old is the earth?* New York: Doubleday-Anchor.

Joly, J. 1909. *Radioactivity and geology.* London: Constable.

Lewis, C. 2000. The dating game: *One man's search for the age of the earth.* Cambridge, UK: Cambridge University Press.

Macdougall, J. D. 1976. Fission-track dating. *Scientific American,* December, 114–22.

Odin, G. S. 1982. *Numerical dating in stratigraphy* (2 volumes): Chichester and New York: Wiley.

Snelling, N. J., ed. 1985. *The chronology of the geological record.* Memoir 10 of The Geological Society of London.

(Apollo 15 surface photo; NASA Photo 71-HC-1277. Courtesy of
Paul D. Lowman and National Aeronautics and Space Administration.)

Chapter 6

The Origin and Early Evolution of the Earth

The eye sleeps until the mind wakes it with a question.

Arabian Proverb

There is something fascinating about science. One gets such wholesale returns of conjecture out of such a trifling investment of fact.

Mark Twain, *Life on the Mississippi*

▶ The earth formed from spinning matter in a solar cloud, or nebula, about 5 billion years ago. As the planet condensed, it picked up further matter from the solar gas cloud, and by 4.6 billion years ago gravitational attraction had condensed a small protoplanet.

▶ Once the earth had formed, it was heated by gravitational condensation, by the rapid decay of abundant radioactive elements, and by numerous asteroid impacts. As the interior melted, the dense iron and nickel derived from the original solar nebula settled to the core of the planet, and the silicate-rich portion remained as the mantle.

▶ By 3.8 billion years ago, the earth was no longer constantly bombarded by meteorites, and its surface was cool enough for bodies of liquid water and primitive sedimentary rocks to form. Chemical differentiation of materials in the upper mantle produced the first stable oceanic and continental crust, although the early protocontinents were constantly being metamorphosed and remelted.

▶ Much of the lighter gases (hydrogen and helium) inherited from the solar nebula escaped the atmosphere because the earth's gravity was too weak to retain them. The rest of the atmospheric gases came from the mantle by volcanic eruptions or by sunlight breaking down other gases. Free oxygen did not exist in the early atmosphere but was added later by the photosynthetic activities of plants and bacteria.

In an age of space exploration when humans have collected rocks from the moon (Fig. 6.1), unmanned devices have sampled Mars and Venus, and fly-by vehicles have analyzed comets, asteroids, Jupiter, Saturn, Uranus, Neptune, and their moons, we all have many questions about the solar system in which we reside. To explain the origin of planets and the formation of their solid spheres, oceans, and atmospheres is among the greatest challenges for science. This chapter is devoted to the important, though speculative, question of origins and earliest histories of planets, a question that provides a unifying focus for geology, geophysics, and astronomy. First we shall summarize ideas about the origin of planets, then we shall review the evidence available for interpreting the present interior makeup of the earth, and next we shall examine hypotheses to explain that makeup and origin of the earth's crust, oceans, and atmosphere. We shall see that radioactivity is fundamental to these hypotheses. Finally, we shall discuss the important chemical and thermal regulatory feedback relationships between seawater and the atmosphere, as well as their implications for climate. This chapter provides a background for all the subsequent chapters, which deal with more strictly geologic aspects of earth history.

Distribution of the Elements

Before proceeding, let us compare the overall chemical composition of the earth with that of the solar system. Only about one-quarter of the 100-odd known elements are at all common, over 95 percent of the universe being composed of only the two lightest, hydrogen and helium. In a general way, the heavier elements tend to be progressively less abundant. With the development of nuclear physics in the 1930s, it was concluded that most of the elements evolved successively from hydrogen through a series of complex thermonuclear reactions in stars. Only very large celestial bodies develop sufficient internal gravitational pressure and temperature to trigger such reactions and form stars, which is why the planets are just planets.

Practically all the atoms of various elements making up our earth originated in an exploding star before our solar system formed. Only the daughter elements from radioactive decay of initially incorporated unstable isotopes have evolved within the earth.

A mere fifteen to twenty elements can be called major constituents of the earth, and some of these are not very conspicuous. To the fifteen listed in Table 6.1 must be added the light, more volatile hydrogen, carbon, nitrogen, neon, and argon. Helium, though second only to hydrogen in the universe, is almost totally missing from the earth because it has been lost to space. But why is there any hydrogen, which is even lighter than helium? Because much of the earth's original hydrogen supply was trapped by combining with oxygen to make relatively heavy water molecules. The abundance of water is the most unique chemical characteristic of the earth, the characteristic that justifies the nickname "blue planet," coined by astronauts peering down at it from space.

Figure 6.1 Footprints on the moon. Mount Hadley rises 4,400 meters above the dust-covered plain. Note apparent planar structures on the mountainside inclined downward to the left, possibly indicating an outcrop of rocks tilted by meteor impact or volcanic eruption. *(Apollo 15 surface photo; NASA Photo 71-HC-1277. Courtesy of Paul D. Lowman and National Aeronautics and Space Administration.)*

The average density of rocks exposed at the earth's surface is about 2.7 times that of water (i.e., 2.7 grams per cubic centimeter), with the very densest ones having a density of slightly over 3.0 grams per cubic centimeter. The densities of these rocks were easily measured long ago, and in the late eighteenth century it was shown that the overall bulk density of the entire earth is about 5.5 times that of water. Therefore, there must be very dense material in the earth's interior. Although early cosmogonists postulated differential settling of matter according to density from an assumed chaotic early stage (resulting in a concentric arrangement of solids, liquids, and gases), determination of bulk density was the first confirmation that the earth's interior is somehow zoned.

Probable Origin of the Earth

Hot Origins

Early explanations of planetary births had in common the assumption of a hot origin from a sun that was assumed to be older than the planets. More than 200 years ago, Buffon popularized the long-held hot origin with a vivid hypothesis that the gravitational pull of passing comets had torn away hot masses from the sun, which then cooled to form planets (Fig. 6.2).

Buffon's hypothesis was abandoned because so many near-collisions are improbable and because comets have very weak

Table 6.1	Estimated Abundances of Principal Elements in the Solid Earth and in Meteorites		
Element	Earth's Crust, %	Total Earth, %	Meteorites, Av. %
Iron (Fe)	5.0	36.0	27.2
Oxygen (O)	46.6	28.7	33.2
Magnesium (Mg)	2.1	13.6	17.1
Silicon (Si)	27.7	14.8	14.3
Sulfur (S)	0.03	1.66	1.9
Nickel (Ni)	0.008	2.0	1.6
Calcium (Ca)	3.6	1.7	1.3
Aluminum (Al)	8.1	1.3	1.2
Cobalt (Co)	0.008	0.09	0.1
Sodium (Na)	2.8	0.14	0.6
Manganese (Mn)	0.045	0.05	0.3
Potassium (K)	2.6	0.02	0.1
Titanium (Ti)	0.442	0.08	0.1
Phosphorus (P)	0.105	0.21	0.1
Chromium (Cr)	0.010	0.47	0.3

Source: Adapted from B. Mason and C. B. Moore. 1982. *Principles of geochemistry,* (4th ed): New York, Wiley.

Note the similar ratios of most elements for "Total Earth" and "Meteorites."

Figure 6.2 Artistic rendition of Buffon's hypothesis of planetary origin by passage of a comet near the sun. *(From de Buffon, 1749, Histoire naturelle.)*

gravity fields. After Buffon, most hypotheses called for condensation of the planets from a hot, gaseous cloud, or **solar nebula,** surrounding the sun, rather than their being derived from the sun itself. During the twentieth century, more complex nebular hypotheses were proposed to overcome the difficulties of those earlier ones.

A Cold Beginning

After 1900, scientists preferred hypotheses that called for the aggregation of cold clouds of dust and gases. According to this scheme, the internal heating of the earth occurred *after* its aggregation. The energy produced by such heating continues to disturb the earth, as evidenced by modern earthquakes, volcanoes, and mountains.

The first important suggestion of a cold origin of the planets was the *planetesimal hypothesis* formulated by geologist Chamberlin and astronomer Moulton early in the twentieth century at the University of Chicago. Their hypothesis called for close passage of another star past our sun. The pull of the intruder's gravity field presumably extracted solar gaseous materials, which first condensed in space to form small, solid bodies called **planetesimals.** Then countless numbers of these cold, asteroid-like bodies tens to hundreds of kilometers in diameter aggregated to form

planets. As the planets grew, their strengthening gravity fields attracted still more particles.

Solar Nebula

In the mid-twentieth century, astronomers von Weizäcker and Kuiper modified the planetesimal concept. They tried to explain the simultaneous origin of the entire solar system in a unified *nebular hypothesis,* which considered both the differences among the planets and the recent discoveries of nuclear physics. From the revolution of planets in practically the same plane around the sun, as well as from the thermonuclear theory for stars, von Weizäcker and Kuiper reasoned that, about 5 or 6 billion years ago, a gigantic, disc-shaped interstellar cloud of gases and "dust" was spinning in our part of the galaxy. It consisted chiefly of hydrogen and helium, with lesser amounts of oxygen, nitrogen, silicon, calcium, aluminum, sodium, potassium, magnesium, carbon, sulfur, and iron and still less of the heavier metallic elements. Rotation was induced by motion of the galaxy as a whole. Gravity concentrated more mass at the center of the disc, and the resulting compression raised the temperature there to several million degrees. At a critical temperature, thermonuclear reactions began, providing the sun's heat energy through hydrogen fusion. In other words, the sun was born.

Spin of the embryonic solar system would inevitably induce turbulence in the envelope of gas and dust—the solar

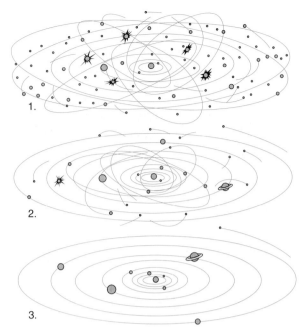

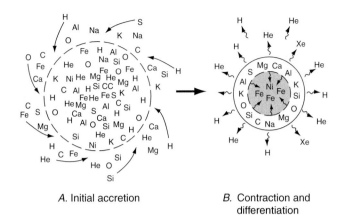

A. Initial accretion

B. Contraction and differentiation

Figure 6.4   Two hypothetical stages in the evolution of the early earth from *A* a large, low-density, homogeneously heterogeneous protoplanet to *B* a smaller, denser, internally differentiated planet. Note loss of light hydrogen (H) and helium (He) to space.

Figure 6.3   Three stages in the early evolution of the solar system through the accretion of the planets by the incorporation of planetesimals and gases. Many random orbits gradually were reduced to a few within a single plane as planetesimals were incorporated into a few planets. *(After Levin, 1972, Origin of the earth: Tectonophysics, v. 13, pp. 7–29. Reprinted with permission from Elsevier Science.)*

nebula—surrounding the growing and heating sun. Planetesimal bodies were thought to have condensed from the nebula first. Then, in regularly spaced eddies in the spinning cloud, the cold planetesimals, dust, and gases concentrated and collided rapidly to form nine or ten *protoplanets* of similar composition (Fig. 6.3). The planetary orbits that resulted are those that least disturb each other, in other words, an equilibrium was achieved. Comets were considered to be surviving planetesimals, with very elliptical orbits tossed out of the inner solar system by Jupiter's gravity. Moons either had formed in like manner as sisters in secondary eddies associated with the planetary ones or were captured later as slaves to their respective host planets.

As their gravity fields strengthened, the protoplanets enlarged by sweeping up still more material from the dust cloud. After millions of years, the protoplanets were essentially complete but were much larger than now (Fig. 6.4). Although their masses were much greater than today, their densities would have been small, for they consisted primarily of hydrogen and helium, with very minor amounts of heavy elements. As their gravity fields strengthened, the protoplanets contracted and became more dense. Apparently they also experienced a separation, or **differentiation,** of most heavier elements toward their centers. Although the composition of all protoplanets was assumed to be about the same initially, hydrogen and helium were so light that they could escape the modest gravity fields of the smaller planets. Accordingly, the four *inner, terrestrial planets* (densities 5.1 to 5.5 g/cm^3) presumably lost far more of their lighter elements

than did the larger, *outer, gaseous planets* (densities 0.71 to 2.47 g/cm^3). Further modification of compositions presumably resulted from a relative depletion of the light gases from the inner solar system by the sun's radiation, the "solar wind," literally blowing any remaining interplanetary nebular gases to the outer part of the system.

Recent Modifications

In recent years, it has been argued that the planetesimals and protoplanets accreted with initially different compositions due to a compositional gradient across the nebula. Meteorites are thought to provide clues to the composition of the planetesimals (Table 6.1), at least for the inner planets, and chemical theory provides an idea of the probable condensation temperatures (Fig. 6.5) and order of condensation of the different components. The most refractory elements—calcium, aluminum, and titanium—would condense into chemical compounds in the range 1,500 to 1,300°C. Elements such as iron, nickel, cobalt, magnesium, and silicon would condense and form compounds in the range 1,300 to 1,000°C, whereas most remaining elements would condense between 1,000 and 300°C; water would condense below 100°C. From the limited data available, the relative iron content of the planets seems to decrease with distance from the sun, suggesting that within the solar nebula there was a temperature gradient such that, more heavy elements (such as iron) condensed in the hotter neighborhood of the sun.

Some geochemists believe that there is too much Fe and Ni left in the earth's mantle for one-stage accretion and differentiation of the protoplanet to be valid. They propose instead a two-stage accretion by which about 80 percent of the earth first accumulated and the metallic core differentiated. Then, after the nebula had cooled somewhat, the outer 20 percent of the earth was accreted to form the mantle, rich in lighter elements but still containing considerable iron.

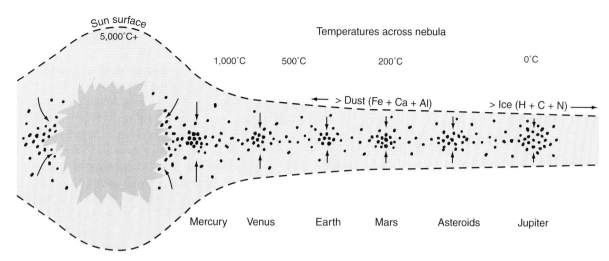

Temperatures across nebula

Figure 6.5   Cross section of spinning, disc-shaped nebular cloud showing formation of the inner protoplanets by concentration of planetesimal bodies. Temperatures shown are temperatures of initial condensation in different parts of the nebula estimated from physical chemical theory applied to the mineral phases observed in meteorites. (Not drawn to scale.)

The large, low-density, outer, gaseous planets presumably condensed from the cool outer portion of the nebula wherein ices of water (H_2O), ammonia (NH_3), and methane (CH_4) were prevalent. These planets, of which Jupiter is the best example (Figs. 6.5 and 6.6), are notably depleted in heavier elements. Apparently condensation was most efficient in the neighborhoods of Jupiter and Saturn, so that growth was fastest and greatest there.

Planetary satellites probably formed from the nebula as the protoplanets did, but some formed within present orbits of regular systems of Jupiter and Saturn, and others were captured later by the planets. Mars's two small moons may be captured asteroids, but our own is essentially a small planet. Recent space probes show an amazing variety among Jupiter's thirteen moons. For example, Ganymede, Callisto, and Europa have densities of only 2 to 3 g/cm^3 and probably consist mostly of ice. Ganymede and Callisto have many impact craters, but Europa does not, since it has been tectonically resurfaced. Io lacks impact craters but has a number of large volcanic craters; no fewer than six apparent plumes from active eruptions were photographed by *Voyager 2* in 1979! Very likely, the volcanoes erupt pure sulfur magmas and gaseous sodium. The eruptions may be caused by a strong gravitational tug-of-war between Jupiter and one or two or the other satellites, with Io caught in the middle (Fig. 6.6).

Because the revised nebular hypothesis accounts for many seemingly diverse phenomena and provides simultaneous origin of sun and planets *as part of one great evolutionary process,* it is considered the best available working model of the solar system. Beware, however, for popularity does not prove correctness. Table 6.2 summarizes the probable time scale for solar system evolution from the initial contraction of an interstellar cloud to the end of the protoplanet stage about 4 billion years ago.

Figure 6.6   Jupiter viewed from the *Voyager 1* spacecraft when it was 28.4 million kilometers from the planet, February 5, 1979. The dark moon Io can be seen near the right edge of Jupiter's disc and bright Europa to the right. Colorful bands with giant turbulent eddies, especially the Great Red Spot (several times the earth's diameter), are clearly visible in Jupiter's atmosphere. Jupiter's larger satellites and the Great Red Spot were first observed by Galileo with his crude telescope 400 years ago. *(NASA photo 79-HC-65, courtesy Jet Propulsion Laboratory.)*

Table 6.2	Possible Time Scale for Early Solar System Evolution during the First 1 Billion Years, Beginning with Contraction of an Interstellar Cloud. (Time Zero Is Assumed to Have Been about 5 Billion Years Ago.)

0—Spinning nebula of dust, gas, and ices

10,000—Protosun separating from nebular cloud

100,000—Asteroid-sized planetesimals accreting from nebula, Jupiter and Saturn forming in outer nebula

1,000,000—Inner four terrestrial planets forming from nebula, separation of cores of terrestrial planets

1,000,000,000—Last planetesimals swept up and large craters formed by the last major impacts, great volumes of basalt erupted in lowlands of terrestrial planets

Source: Adapted from J. A. Wood. 1979. *The solar system:* Englewood Cliffs, N.J.: Prentice-Hall.

Planetology of the Inner Solar System

Space exploration has added greatly to our knowledge of the solar system through manned and unmanned landings as well as fly-by missions. The study of other members of the solar system, which is called planetology, is relevant to hypotheses about the origin and early history of the earth, especially because the surfaces of the other terrestrial bodies are still pristine; they lack earth-style weathering and erosion processes. Moreover, geologists have been very involved in recent studies of planetology, which is causing us to look up as well as down at our rocks.

We know the most about our own moon, which is almost a planet, being nearly as large as Mercury. Like the earth, the moon is differentiated into three zones—crust, mantle, and possibly a small core—and it is as old, too. The lowland (maria) areas of the moon have rocks most like terrestrial basalt but richer in titanium, zirconium, yttrium, and chromium and relatively depleted in sodium, potassium, and rubidium. Their densities are 3.1 to 3.5 g/cm^3. The lunar highlands contain less-dense rocks rich in plagioclase feldspar (anorthositic gabbro). These lunar "continents" are rich in aluminum like the earth's continents but are depleted in sodium, potassium, silicon, rubidium, and more volatile constituents. The highlands also have many more impact craters than do the lowlands. Isotopic dating of lunar rocks suggests that extensive melting of the moon's surface occurred 4.5 to 4.2 billion years ago. This was followed by differentiation of its interior to produce the basalts, which were erupted from about 3.8 until about 3.2 billion years ago (Table 6.2).

For over a century, different theories have been proposed for the origin of the moon. Most of these were discredited when the first samples of moon rock were brought back in the 1970s. Some astronomers thought that the moon condensed alongside the earth as a "sister" satellite, but the lack of iron in the moon shows that it did not condense from the same undifferentiated materials as the earth (the moon has no iron-nickel core). The second theory suggested that the moon spun off the earth as a "daughter" blob from the early earth. However, calculations show that the earth-moon system doesn't have enough angular momentum to make this happen, and the moon doesn't lie in an orbit over the equator. The third idea was that moon was a "pickup" from space, a rock that flew in and was captured by the earth's gravity. However, the moon has the same oxygen isotopic composition as the earth, so it had to be derived from the early earth in some way.

The lunar samples brought back by space missions, instead, showed that the moon has a composition very similar to the earth's mantle and must have formed after the iron core had condensed and separated from the mantle. In the late 1970s, several scientists showed that the impact of a body the size of Mars could have blasted some of the earth's early mantle material into space, which later condensed to form the moon about 4.5 billion years ago. In recent years, numerous geochemical studies have strengthened this theory.

The lowlands of the terrestrial planets other than the earth are also thought to contain basalt, and their highlands, like the moon's, seem to be less dense because of their higher elevations. Highlands are consistently more cratered, as well; by analogy with the moon, most such cratering is thought to have been completed by 3.9 billion years ago. Circumstantial evidence suggests that the intense bombardment by asteroid-size bodies, which produced all the craters, may also have triggered the large-scale melting that resulted in basaltic eruptions. Thus, it is thought that the processes of differentiation and cratering were universal during the first 1 to 2 billion years of solar system history, but their record has been obscured by erosion here on the earth.

Although the similarities strongly suggest some common history for the terrestrial planets and our moon, there are important differences, too. The others show very different styles of structural activity than Earth, and, except for Venus, they are greatly depleted in volatile constituents. Neither Mercury nor the moon has any atmosphere at all, and Mars has only a very tenuous one of carbon dioxide with minor amounts of nitrogen, oxygen, and water vapor. Venus has an atmosphere ninety times denser than Earth's and with a very different composition (97 percent CO_2). Earth is unique in having abundant H_2O, most of which is present in the liquid state most of the time. Mars is the only other planet with any significant water, and most of that is now locked up in polar ice caps or permafrost. During formation, Earth's size was such that its gravity could retain enough hydrogen to combine with oxygen to form water, and the distance from the sun was just right for most of that water to remain liquid. Venus, although nearly the same size as Earth, apparently lost all but a trace of its hydrogen. Presumably its runaway greenhouse effect caused more complete depletion of hydrogen, so that all its oxygen combined with carbon instead to form abundant carbon dioxide. Mars, on the other hand, is so much smaller than Earth that its gravity could hold only a little volatile hydrogen and oxygen. Being farther from the sun, what water is present remains mostly frozen. The early Earth also was large enough to generate sufficient internal heat to cause not only differentiation but also profound structural deformation, which persists even today of a type never equaled by any other planet. Ours is a very special world, indeed!

Leftovers—Asteroids, Meteorites, and Comets

As the protoplanets grew, or accreted, from the solar nebula, they literally cleaned up their respective neighborhoods by accretion or ejection of smaller bodies. Eventually the most mutually stable orbits became established, much like the particle orbits in Saturn's rings. Accretion apparently occurred very rapidly during the first million years or so (Table 6.2). Things were not completely tidied up, however, for today we still have meteorites, asteroids, and comets left over as clutter to be explained. All three of these are considered members of the solar system, and they may be closely related to one another.

The **asteroids,** which constitute upwards of a million objects from about 1 to 950 kilometers in diameter, circle the sun like planets between the orbits of Mars and Jupiter (Fig. 6.5). They are irregular in shape and apparently in composition. We have no direct samples of asteroids as yet, but remote sensing suggests metallic ones nearer the sun, followed by silica-rich and carbon-bearing ones. This arrangement is consistent with the inferred radial compositional gradient across the solar nebula.

We know a lot more about the composition of meteorites, which are simply miniature asteroids or fragments of large asteroids that have fallen to Earth. Indeed, it was long assumed that all meteorites came from the asteroid belt until several recently studied ones showed features suggesting an origin on the moon and Mars; apparently a few are not from the asteroid belt. Regardless of origin, two main compositional types are known: metallic iron-nickel ones with densities about 7.5 g/cm^3 (25 percent) and nonmetallic, or stony, ones with densities of about 3.5 g/cm^3 (75 percent). Their combined overall composition approaches that of the total earth (Table 6.1), and the *ratios* among major constituents are sufficiently similar to that of the terrestrial planets to support the long-held view that they have some kinship. If, as the nebular hypothesis states, all protoplanets underwent gravitational differentiation more or less simultaneously, then the two classes of meteorites may be samples of the metallic core and stony mantle of accreting bodies large enough to differentiate. Not enough material remains in the solar system to make another planet.

The asteroids are clearly planetary in their behavior, and meteorites are small pieces of asteroids, planets, moons, or all three that were let loose in the solar system and then captured by the Earth's gravity field. But why do such aberrant objects exist? Two hypotheses require mention. First is a long-standing dogma that asteroids and meteorites represent a disrupted planet whose debris continues to orbit in the asteroid belt. A second alternative suggests that huge Jupiter's gravity field prevented the accretion of a planet in the main belt of asteroid orbits, so a horde of planetesimals resides there and still reflects initial compositional differences within the solar nebula. This second hypothesis is now preferred by most planetary geologists.

Comets are not much better known than asteroids, Halley's faithful one notwithstanding. They are objects of low density, and most are less than 10 kilometers in diameter. An apt characterization is "dirty snowballs" because their cores seem to be composed mainly of ice crystals and some stony particles. If one of them approaches the sun, the solar wind produces a gaseous tail on the comet. Some asteroids may be comet cores whose volatile constituents have been baked away by many such passages. Studies of Halley's comet on its 1986 passage support this view. Most comets seem to originate in the outermost solar system, where ice and dust are thought to orbit in the Oort Cloud, which is thought to consist of material "blown" from the inner solar system 4 to 5 billion years ago by the solar wind. The more than 100 periodic comets known to humans, such as Halley's, which penetrate the inner solar system and approach the sun, are exceptional and may have been deflected into their distorted orbits by gravitational interactions with (1) Jupiter and other planets, (2) a passing star, or (3) interstellar gas clouds when the solar system periodically passes through the plane of the Milky Way galaxy.

As we shall see in chapter 14, there has been much publicity associated with the hypothesis that impacts by asteroids and/or comets caused mass extinctions of organisms such as the dinosaurs. This is another reason that geologists today must look upward as well as downward for clues about the earth.

Nature of the Earth's Interior

Early Speculations

An early suggestion that the earth's interior contains iron followed the discovery of the magnetic field in 1600. Seemingly the planet must be a gigantic, spherical, iron lodestone. Other internal compositions have also been suggested, however, including hydrogen and silicon. Speculations about the physical makeup of the interior have included all-liquid, all-solid, partly solid–partly liquid, and even gaseous. Twentieth-century geophysical data were required to constrain these speculations and to move us toward a more sophisticated—and perhaps more accurate—understanding of the earth's interior.

Analogies from Meteorites

Recall that the earth's overall density of 5.5 g/cm^3 requires that the interior must contain denser material than the rocks of the crust, whose average density is only 2.8 g/cm^3. More than a century ago, it was suggested that the interior may be composed of materials like those that make up meteorites. This suggestion has continued to be accepted, in spite of uncertainties about the origin of meteorites and asteroids, and it is strengthened by chemical arguments, such as the fact that radiogenic ^{143}Nd is produced in the earth at the same rate as in stony meteorites. The fact that meteorites have five times as much iron but only three-fourths as much oxygen and silica as crustal rocks (Table 6.1) is explained by the postulated early gravitational differentiation of the protoplanets, which presumably concentrated heavier elements in interiors and lighter ones in crusts and atmospheres.

Seismological Evidence of Internal Structure

Below the deepest mines and wells lies indirect geophysical evidence of internal zonation. Results from seismology, especially, led to recognition of the zones in the earth's interior as shown in Fig. 6.7 and Table 6.3.

A near-surface discontinuity in earthquake shock-wave transmission discovered in 1909 is named the **Mohorovicic discontinuity,** or **Moho**; the velocity of wave propagation increases below it. The Moho generally has been taken as the base of the earth's crust. It lies an average of 5 kilometers below the sea floors and 33 kilometers below continental surfaces but is as deep as 70 kilometers under some portions of continents (Fig. 6.8).

A zone of relatively low seismic velocity and low rigidity—probably partly fluid or plastic—is present below the Moho within the upper mantle (between 60 and 250 kilometers deep). It may be of great importance in major structural adjustments. Indeed, it is now realized that this *low-velocity zone* is more fundamental than the Moho. Everything above this zone is today called **lithosphere** ("rocky sphere"). Thus, the lithosphere comprises all of the crust, including the Moho, plus the top part of the mantle.

A second great seismic discontinuity at about 2,900 kilometers below the surface separates the solid mantle from the core (Fig. 6.7). In 1926, it was demonstrated that the core material directly below the mantle had the physical properties of a fluid. Then, in 1936, a solid inner core was recognized.

Further Evidence from the Magnetic Field

The earth's magnetic field must be related to overall earth physics, and it has a history that, as we shall see in chapter 7, has a profound bearing upon interpretations of the history of the crust. The earth's field is similar to one surrounding a two-poled bar magnet roughly aligned with the earth's axis. Precise observations have indicated that the field was not completely fixed; in five centuries, the poles have migrated in a crudely circular path spanning approximately 20 degrees of longitude. Also, as noted in Chap. 4, the field has reversed its polarity many times (see Fig. 4.24). The period of reversals has varied considerably, as we shall see in Chap. 7.

The cause of the magnetic field has not been fully explained, though interesting hypotheses have been proposed. Several features must be explained: (1) the field has two poles located near the geographic poles; (2) it shows irregular variations in both position and polarity; (3) these variations bear no relation to the crust and, so, must have their origin deep in the earth.

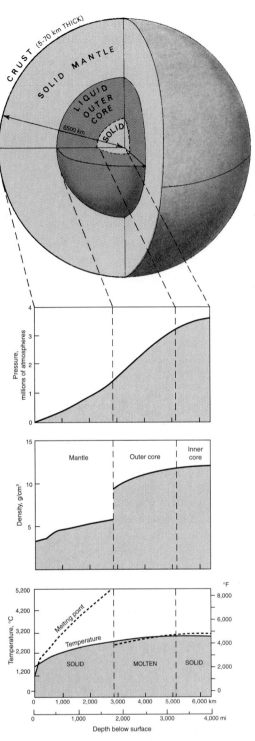

Figure 6.7   The earth's interior. *(Adapted from A. N. Strahler, 1960, Physical geography, with permission of John Wiley & Sons; and A. N. Strahler, 1963, The earth sciences, with permission of Harper & Row, Publishers.)*

The most widely accepted view, which was first suggested by the Frenchman Ampere in 1820, is that internal electric currents produce a magnetic field much like that formed around a wire transmitting a current. The assumed electrical currents require some driving mechanism to maintain them, however. A

Table 6.3	Major Zones of the Earth					
	Thickness or Radius, km	**Volume, 10^{27} cm³**	**Volume, %**	**Mean Density, g/cm³**	**Mass, 10^{27} g**	**Mass, %**
Atmosphere	(Outer atmospheric limit indefinite)			0.12×10^{-8}	0.000005	0.00009
Hydrosphere	3.80 (av.)	0.00137	0.1	1.03	0.00141	0.024
Crust	17	0.008	1.4	2.8	0.024	0.4
Mantle	2,883	0.899	82.3	4.5	0.016	67.2
Core	3,471	0.175	16.2	11.0	1.936	32.4
Total earth	6,371	1.083	100.0	5.52	5.976	100.0

Source: After B. Mason and C. B. Moore. 1982. *Principles of geochemistry,* 4th ed.: New York: Wiley.

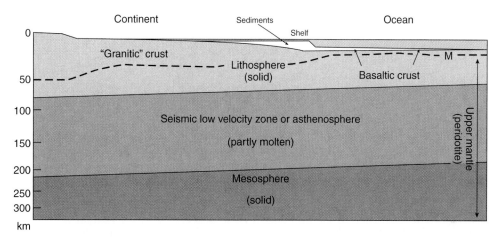

Figure 6.8   Structure of the upper 300 kilometers of the earth, showing relations among the crust, upper mantle, and lithosphere. The M seismic discontinuity (or Moho) long chosen to define the crust-mantle boundary now seems to have been overemphasized. Although M probably represents a compositional contrast (see Fig. 6.11), the contrast between the lithosphere and the low-seismic velocity zone is physically more profound. The zonation shown is based almost entirely on seismic velocities; compare Table 6.3 for other properties. (Not to exact scale.)

core rich in iron and nickel, similar to the metallic meteorites, would be a good electrical conductor, and a fluid outer part of such a core would allow mechanical motion of electrical charges. Physicist W. M. Elsasser suggested in 1939 that electrical currents in the outer core could generate and sustain the magnetic field much the way an electric dynamo does (Fig. 6.9). Near-coincidence of the field and spin-axis orientations are consistent with this explanation.

Elsasser assumed that the outer core is an electrical conductor and is in motion as a result of heat convection therein. Recently it has been suggested that wobble (precession) of the earth's rotational axis, coupled with the deflection caused by the spin of the earth, may drive the dynamo instead. It is well known that, if a conductor moves within a magnetic field, an electric current is generated inside the conductor. An analogy can be made with a conventional electric generator. The earth's core is the conductor moving within the earth's magnetic field; it is as

though the current being generated were fed back through the magnets to maintain the field.

According to Elsasser's dynamo hypothesis, the Earth's magnetic field results directly from core motions, and rotation of the earth affects both orientation and strength of the field. Core motions, which otherwise might be random in direction, are preferentially oriented by the earth's spin, producing a strong field whose axis on the average over time closely parallels the spin axis. The reversals of polarity are not fully understood but probably result from occasional instabilities of motion within the fluid outer core. It has been suggested that large asteroids hitting the earth may perturb the relative motion between mantle and core sufficiently to cause the reversals. The reversals are completed suddenly in a geologic sense; that is, they take only a few thousand years, as shown by careful studies through a series of young volcanic rocks. During a reversal, the magnetic poles migrate and the strength of the field varies erratically; it probably weakens to near zero for brief spells.

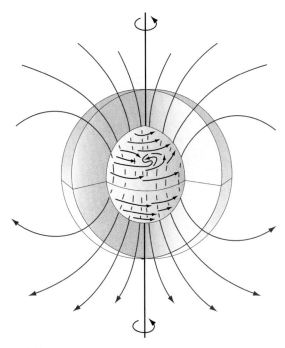

Figure 6.9 Cutaway diagram illustrating Elsasser's dynamo theory of the earth's magnetic field. Differential rotation of solid mantle (faster) and fluid core (slower) presumably induces eastward flow of electrically charged, ferromagnetic material in the outer core. Flow of electrical charges induces a strong field (large curved arrows), which is maintained by the earth's rotation. Large anomalies may be caused by complex eddies in the outer core (small curved black arrows).

The earth's magnetic field shields living organisms from damaging cosmic radiation. That the field is old is suggested by presence of a record of shallow marine life on earth for over 3,000 million years and of land life for at least 400 million years. Measurements of remanent magnetism frozen in very old rocks suggest approximately constant average intensity of the field for at least 2,700 million years—except, of course, during the brief reversal episodes.

Chemical Composition of the Deep Interior

Besides analogies with meteorite compositions and evidence from the magnetic field, there is other evidence that bears upon the important question of the composition of the earth's interior. First we assume that the interior consists of relatively common elements (Table 6.1) arranged in mineral phases compatible with seismic evidence, temperatures, and pressures for different depths. Observation of heat flow from the interior indicates that radioactive isotopes cannot be very concentrated below the crust (if they were, the earth would be much hotter). Finally, analogies long have been drawn from the densities and seismic properties of known surface-rock materials to allow us to speculate about conditions at depth. We shall now pull together and summarize the diverse evidence.

From all evidence available, it was reasoned years ago that the bulk composition of the crust beneath oceans is close to that of **basalt** (density 2.9 g/cm^3), that of the continental crust is close to that of **granitic rocks** (density 2.7 g/cm^3), and that of the upper mantle (the part that is lithosphere) seems to be closest to **ultramafic rocks** (density about 3.3 g/cm^3), which occur locally within structurally complex mountain belts. Ultramafic rocks, such as **peridotite,** consist almost exclusively of dark so-called *mafic minerals* (*ma-,* magnesium and *-fic,* ferric or iron). Important examples are pyroxene and olivine, which make up the rock peridotite. Ultramafic rocks lack feldspar and quartz and, so, are relatively poor in potassium and silica (see Box 6.1). They are similar in composition to the stony meteorites.

An ultramafic upper mantle of density 3.3 to 4.5 g/cm^3 still would not yield an overall earth density of 5.5 g/cm^3. Therefore, the core must be much denser because it makes up only 16 percent of the earth's volume (Table 6.3). Only the seismic properties of the core can be measured, and they give information primarily about rigidity. It has been shown by comparing density and seismic characteristics of the core with experimental results for many chemical elements that the transition metals (e.g., iron, nickel, cobalt) satisfy practically all known conditions. Of these, only iron and nickel are abundant enough in nature to form the major part of the core. But if the core were pure iron-nickel, it would be even denser than observed; therefore, lighter elements, such as sulfur and silica, are thought to be present in some iron-nickel alloy.

Large masses of peridotite are seen in many mountain belts, where they have been exposed by erosion following profound structural movements that carried the peridotite up from the mantle. Some oceanic basalts also contain chunks of peridotite carried to the surface intact. These dense, silica-poor rocks provide the most direct evidence of the makeup of the upper mantle.

Unusual small bodies of peridotite represent samples of the mantle punched up through the crust by streams of gas and fluid under high pressure. A variety of peridotite (called kimberlite) contains the form of pure carbon called diamond. Like other minerals characteristic of the ultramafic clan, diamond is very dense (3.5 g/cm^3) as well as hard, which reflects the high-pressure environment of the mantle at depths greater than 70 kilometers. (The less dense form of carbon called graphite, as in a "lead" pencil, forms near the surface at much lower pressures.) Minute inclusions of other minerals, which crystallized simultaneously within host diamond crystals, have yielded surprisingly old ^{147}Sm-^{143}Nd dates of 3.2 to 3.3 billion years. This suggests that at least some diamonds—perhaps your own—may date from the early formation of the mantle.

Eruptions of magma derived from below the crust should give us direct samples of mantle material. Indeed, earthquakes occurring at depths of 40 to 60 kilometers during volcanic eruptions indicate that lavas on Hawaii come from the upper mantle. These lavas yield basalt, however, which is more siliceous than peridotite. The volcanic equivalent of peridotite is olivine-rich **komatiite** (see Box 6.1). Although extremely rare today, it can be found among rocks older than 2.5 billion years. How is it that lavas derived from the mantle over later geologic time have a composition different from that of their parent peridotite? The answer is a complex chemical fractionation of elements called **partial melting,** which gives rise to basaltic magma richer in silica

Box 6.1

Major Minerals and Rocks of the Earth's Crust and Upper Mantle

The following tables provide a summary of the major constituents of the outer solid earth. They are to be used for reference as needed.

The Major Igneous-Rock-Forming Minerals (See also Table 6.4)

Name	Composition	Color
Quartz	SiO$_2$	Colorless
Potassium feldspars: orthoclase and microcline	K-Al silicate	White to pink
Plagioclase feldspar series: albite, oligoclase, andesine	Na-Ca-Al silicate Na > Ca	White to light gray
labradorite, by-townite, anorthite	Na < Ca	Dark gray
Iron-magnesium silicates:		
biotite mica	K-Mg-Fe-Al silicate	Black or brown
hornblende (amphibole group)	Ca-Mg-Fe-Al silicate	Black to dark green
pyroxene group	Ca-Mg-Fe silicates	Green, brown, black
olivine	Mg-Fe silicate	Green

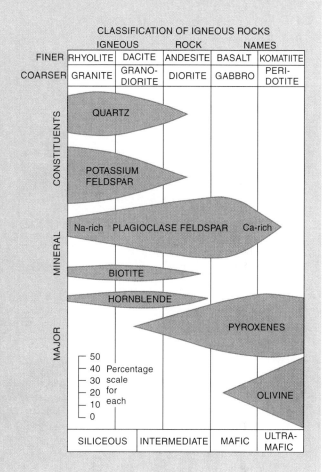

than the average mantle. Besides silica, other relatively light elements, such as aluminum, calcium, and sodium, are preferentially mobilized to become concentrated in such magmas. Variation in combinations of temperature, pressure, and water content in the area of melting accounts for this partial distillation, which can be reproduced in the laboratory. Whereas erupting basaltic magmas have temperatures of about 1,100°C, komatiite magmas require temperatures of at least 1,600°C. Therefore, the upper mantle apparently was hotter during early Cryptozoic time.

Dawn of Earth History

Chemical and Thermal Evolution

The early earth must have undergone a great deal of heating, which would have facilitated differentiation; intense heat may even have melted the planet completely for a short while. Initial heating caused by gravitational contraction of the protoplanet would have raised the temperature at the center to about 1,000°C, and the early earth must have had about five times as much radioactive heat production as now (Fig. 6.10), which would have raised the temperature another 2,000°C or so. A third source of heat would have been large meteorite impacts prior to 4 billion years ago, for which the moon, Mercury, and Mars provide abundant evidence.

Regardless of whether accretion of the protoplanet occurred in one or two stages, it seems clear that a distinct core and mantle existed by at least 4.5 billion years ago, which is the minimum date for the formation of distinct mineral systems in the earth as indicated by lead isotopes. Persistence of a liquid outer core beneath a solid mantle may seem puzzling, but magnesium and iron-rich silicate minerals typical of the mantle have higher melting points than iron at any pressure and temperature combination. Present temperature in the outer core is estimated to be near

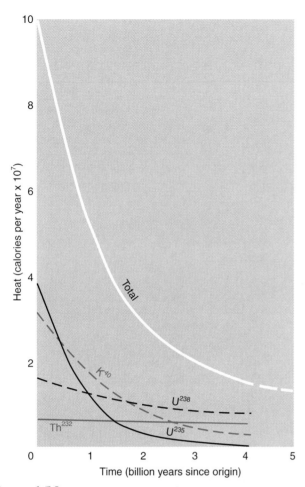

Figure 6.10   Estimates of changing heat production from major radioactive isotopes since the origin of the earth. Note that most heat results from decay of ^{235}U and ^{40}K. *(Adapted from B. Mason, 1966, Principles of geochemistry, 3d ed.: by permission of John Wiley & Sons, Inc.)*

2,500°C, which is above the melting point of core material but below that of mantle silicate minerals at that depth.

Is the earth today heating up, cooling off, or remaining in a thermal steady state? Volcanoes, hot springs, and high temperatures in deep mines and wells provide proof that heat is still flowing from the interior, and although much of the original heat-producing radio-active decay has ceased (Fig. 6.10), insulation by the crust and mantle impedes heat dissipation so much that the interior may be only midway in its cooling history. In any event, the interior is still evolving, and consequences of this chemical and thermal evolution are important because the heat energy causes great structural disturbances. We shall see that mountain building, the most conspicuous of such disturbances, has gone on throughout the earth's history. Eventually structural turmoil must decline, as it did long ago on the moon, when the thermal energy reservoir is exhausted. We can only speculate about where in its energy dissipation history the earth is today.

Origin of the Crust

We now turn our attention to the earth's crust, our chief concern for the remainder of the book. The crust must have formed by chemical differentiation of light elements from the mantle. N. L. Bowen suggested in 1928 that the crust of continents, whose average properties approach those of granite, originated by chemical fractionation from an ultramafic-rich mantle.

Many lines of evidence suggest that the crust derived from the denser mantle, but the mechanisms are more complex than Bowen's laboratory model suggested. As the mantle differentiated, relatively light silicon, oxygen, aluminum, potassium, sodium, calcium, carbon, nitrogen, hydrogen, and helium and lesser amounts of other elements rose to the surface to form the crust, seawater, and atmosphere (Fig. 6.11). The most familiar rock-forming minerals are produced in the proportions shown in Table 6.4. Atomic mass is not the only factor to determine their ultimate residence, however, because atomic size and electrical charge are more important than atomic mass for heavy elements, such as uranium, thorium, and the rare earth elements. Atoms of these elements are too large to fit into the closely packed crystal structures of dense silicate minerals found in ultramafic rocks but are comparable in size to potassium or calcium. It is not surprising, therefore, to find them preferentially concentrated in the more open structures of the minerals found in crustal rocks.

For oceanic crust, the melting of upper mantle material and extrusion at the surface of the resulting magma seems straightforward. Komatiite was produced early by simple, complete melting when the mantle was very hot, but basalt was produced instead by partial melting after cooling had progressed for a couple of billion years.

Origin of the continental crust is less well understood and apparently has been more complex. If the entire earth contained as much radioactive material as the continents do, the globe would be entirely molten. Even though uranium is not exactly abundant in granites, the 0.0006 percent uranium in granite is 10 times as much as in basalt and 1,000 times more than in ultramafic rocks! Isotopic dating suggests that continental crust, for whatever reasons, was not stable until about 4.3 to 4.4 billion years ago, about 200 million years after the formation of the core and mantle. For decades, the oldest crustal rocks known on earth were from southern Greenland, where dates of about 3.8 billion years came from gneisses that had once been sediments. In 2001, a group of scientists found detrital zircons in Australia that yielded dates of 4.3 to 4.4 billion years, although these minerals were reworked into much younger sediments. These zircons are the oldest dated crustal *minerals,* although they are not found in the oldest *rocks.* In 1999, dates of 4.03 billion years were obtained from detrital zircons in the Acasta Gneiss in northwestern Canada (Fig. 6.12). In 1992, rocks dated at 3.8 to 3.96 billion years were discovered in Wyoming and China. These rocks are currently the oldest known crustal fragments on the earth's surface. All these dates indicate (1) that some sort of granitic or dioritic protocrust was being weathered and eroded from the earth's surface around 4.3 to 4.4 billion years ago and (2) that sedimentary rocks were accumulating from weathered crustal debris.

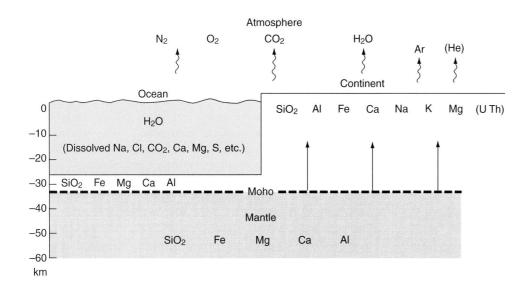

Figure 6.11   Present distribution of major chemical elements and uranium, thorium, helium, and argon, which reveals the elemental differentiation of the atmosphere, seawater, and crust from the earth's mantle. For each division, elements are listed in approximately decreasing abundance from left to right. (Seawater depth is greatly exaggerated to accommodate lettering.)

Table 6.4	Proportions of Commonest Rock-Forming Minerals in the Earth's Crust	
		Range of Crustal Abundances (Percent)
Alkali feldspar (KAlSi$_3$O$_8$ and NaAlSi$_3$O$_8$)		9–21
Plagioclase feldspar (NaAlSi$_3$O$_8$↔CaAl$_2$SiO$_8$		31–41
Quartz (SiO$_2$)		12–24
Pyroxenes (Ca [Mg, Fe] Si$_2$O$_6$)		0–11
Iron and titanium oxides (magnetite, hematite, ilmenite)		2
Biotite mica (complex K, Mg, Fe, Al, Ti hydroxyfluorosilicate)		4–11
Olivine ([Fe, Mg]$_2$ SiO$_4$)		0–3
Muscovite (complex K, Al hydroxyfluorosilicate)		0–8
Other minerals		0–3
		100

Source: After D. L. Anderson. 1989. Theory of the earth. Boston: Blackwell.

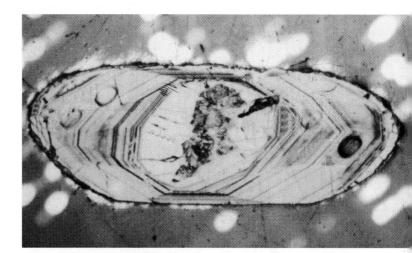

A

B

Figure 6.12   A: Microscopic view of one of the 4.03-billion-year-old zircon grains extracted from the Acasta Gneiss, Slave Province, Northwest Territories of Canada. The grain is 0.5 mm long. The polished surface has been etched with acid to highlight crystal growth zones. (Courtesy of Samuel A. Bowring.) B: Outcrop of the Acasta Gneiss, one of the oldest dated rocks on earth. (Courtesy of Samuel A. Bowring.)

With the early earth roughly five times more radioactive than today plus abundant meteorite impacts, there was ample heat to keep the crust unstable through almost continual volcanic activity between 4.5 and 4.4 billion years ago. Crust probably formed, then was reassimilated, then reformed repeatedly. Not until about 4.3 billion years ago did heat and volcanism decline to the point at which large chunks of crustal rock could be preserved. The proposed stages for the early evolution of the earth are summarized in Table 6.5.

Table 6.5	Proposed Stages for the Early Evolution of the Earth

Billions of Years Ago	Stage
4.7–4.5	The solar nebula condensed, and the sun and protoplanets began to form; the earth's core began to condense from the mantle.
4.5–4.4	Intense heating from gravitational contraction, radioactive decay, and meteorite bombardment; liquid iron-nickel core mostly separated from mantle; loss of gases from interior of the earth with lighter hydrogen and helium escaping, and other gases forming the primitive atmosphere.
4.4–4.0	First crustal rocks formed, including silicic crustal materials, although they were largely remelted and recycled; meteorite bombardment reached a climax; continued release of atmospheric gases from mantle, surface conditions are cool and stable enough for abundant water, sedimentation, and possibly life.
3.9	Meteorite bombardment declined abruptly as most loose nebular debris has been swept up into planets and the sun; formation of first stable continental crust.
3.8	Rapid buildup of silicic crustal material to form protocontinents.
3.6–3.5	Stable but thin continental crust grouped into protocontinents with "microplate" tectonics; first fossil evidence of cyanobacteria and stromatolites.

The new dates of 4.3 to 4.4 billion years are derived from zircons in Australia. The isotopic chemistry of these zircons also suggests that the earth was cool enough for water and oceans on its surface 4.4 billion years ago. This leaves only 100,000 to 200,000 years for the surface to cool—much faster than previously thought.

Origin and Evolution of the Atmosphere and Seawater

The Problem

The atmosphere and oceans must have somehow also originated by chemical differentiation. Even the origin of life seemingly was linked with differentiation, for it required certain unique chemical characteristics of the early atmosphere and oceans (see Chap. 9). Once formed, life influenced the further development of both the atmosphere and seawater to produce a complex global feedback system.

The atmosphere has undergone important changes through time and is evolutionary like the solid earth. To investigate the possible nature of its development, we first compare our present atmosphere with that of Jupiter because the latter apparently has retained original gases from the solar nebula. We also compare our atmosphere with the gases of meteorites because they probably resemble the early terrestrial protoplanets. From Table 6.6, it is clear that these compositions are markedly different from each other, so no simple, obvious explanation of the earth's atmosphere is immediately forthcoming from either of them.

Two major hypotheses exist to explain the evolution of the present atmosphere, and there is no reason that both mechanisms could not have contributed. There is one common feature: the assumption that considerable molecular hydrogen and helium were lost to space because of their very small atomic masses. What hydrogen remained was mostly oxidized to form seawater. The greatest single problem facing any hypotheses of atmospheric development is to explain the abundance of molecular oxygen (O_2), which is missing from all other planets except Mars. It has long been assumed that the early atmosphere had practically no molecular oxygen and that most of it has accumulated slowly through geologic time. There is now growing evidence, however, that a small amount of oxygen did exist as early as 3 to 3.5 billion years ago. In the following sections, we examine two hypotheses for the origin of our atmosphere and the photosynthesis hypothesis for oxygen.

Outgassing Hypothesis

A widely accepted explanation, first urged by geologist W. W. Rubey in 1951, is that most atmospheric gases reached the atmosphere from the interior by the process called **outgassing,** which means gaseous transfer to the surface through igneous processes. It had been suggested in 1910 that the trace amounts of helium found in our present atmosphere originated from radioactive decay of uranium in the crust, and in 1937 it was suggested that atmospheric argon, which is surprisingly abundant compared with the other rare gases, was derived similarly from decay of potassium 40 in the earth. Both the present abundance of argon 40 and the decay rate of potassium 40 are in good agreement with this hypothesis.

Volcanoes and hot springs are known to expel steam, carbon dioxide, nitrogen, and carbon monoxide (Table 6.6). By assuming outgassing to be a normal part of overall density differentiation, we might explain all the atmospheric nitrogen, helium, argon, and water vapor. In addition, the overwhelming preponderance of steam in the expelled gases provides a ready source of seawater. Oxygen had a separate origin through photosynthesis.

Photochemical Dissociation Hypothesis

The second hypothesis, also proposed in the 1950s, assumes an early atmosphere like that of Jupiter today, with methane, ammonia, and some water vapor (Table 6.6). The atmosphere would be devoid of any molecular oxygen and therefore of **ozone** (O_3). Ozone filters out most of the lethal, short-wavelength ultraviolet radiation from the sun, making the earth habitable (Fig. 6.13). An ozone layer occurs at the upper part of the molecular-oxygen-rich lower atmos-

Table 6.6	Comparison of Present Atmosphere with the Gases of Other Planetary Bodies and Volcanoes; Gases Listed in Approximate Decreasing Order of Abundance					

Earth's Present Atmosphere			Jupiter's Atmosphere	Meteorites (Average)	Volcanoes (Average)	Geysers and Fumaroles
Major						
Nitrogen	(78%)	Stable	Methane	Carbon dioxide	Water vapor (73%)	Water vapor (99%)
Oxygen	(21%)	Unstable (reacts with Fe and C)	Ammonia	Carbon monoxide	Carbon dioxide (12%)	Hydrogen
Argon	(0.9%)	Stable	Hydrogen	Hydrogen	Sulfur dioxide	Methane
Carbon dioxide	(0.03%)	Unstable (reacts with silicates)	Helium	Nitrogen	Nitrogen	Hydrochloric vapor
Water vapor	(Variable)	Unstable (reacts freely)	Neon	Sulfur dioxide	Sulfur trioxide	Hydrofluoric vapor
Minor	(Traces only)			Methane*	Carbon monoxide	Carbon dioxide
Ozone		Unstable (dissociates)				
Neon		Stable		Nitrous oxide*	Hydrogen	Hydrogen sulfide
Helium		Stable		Carbon disulfide*	Argon	Ammonia
Krypton		Stable		Benzene*	Chlorine	Argon
Xenon		Stable		Toluene*		Nitrogen
Hydrogen		Unstable (reacts with oxygen to form water)		Naphthalene*		Carbon monoxide
Methane		Unstable (dissociates)		Anthracene*		

*Only present in nonmetallic carbonaceous chondrites.

phere (about 15 to 30 kilometers). Ozone is manufactured in the atmosphere by high-energy ultraviolet radiation from the sun. Oxygen molecules are broken, and the free oxygen atoms then combine with other molecules to form ozone. Ozone is unstable, however, so the third atom soon splits off to combine with another free O to form O_2 again. The process is a steady-state one; that is, ozone is constantly forming at the same rate at which it is breaking up.

High-energy ultraviolet radiation can trigger other photochemical, or light-induced, reactions. Such reactions are known to occur in the upper atmosphere today and presumably were more common throughout the atmosphere before ozone accumulated. It is estimated that ozone began to form when the molecular-oxygen level reached about 10 percent of its present abundance. Changes resulting from such reactions in the primitive atmosphere might have been

1. Primeval water vapor dissociated into hydrogen and oxygen, with most hydrogen escaping to space:

$$2H_2O + uv \text{ light energy} \rightarrow 2H_2 \uparrow + O_2$$

2. Newly formed molecular oxygen reacted with methane to form carbon dioxide and more water vapor:

$$CH_4 + 2O_2 \rightarrow CO_2 + 2H_2O$$

3. Oxygen also reacted with ammonia to form nitrogen and water:

$$4NH_3 + 3O_2 \rightarrow 2N_2 + 6H_2O$$

4. After all CH_4 and NH_3 were converted to CO_2 and N_2, *then* O_2 could accumulate as more water vapor dissociated.

In this way, the present nitrogen–carbon dioxide–oxygen atmosphere might have formed.

Oxygen from Photosynthesis

Like lifeless Venus today, the early earth probably had a great deal of CO_2 in its atmosphere. Here on Earth, however, photosynthetic organisms require CO_2 and release O_2. A close balance between the abundance of carbon in sedimentary rocks and molecular oxygen

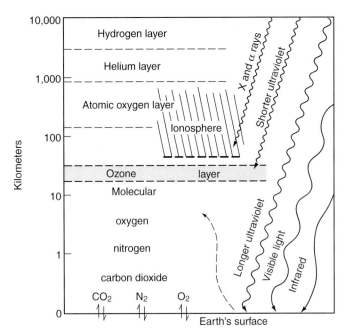

Figure 6.13   Profile of the atmosphere, showing stratification, types of radiation, radiation shields (ionosphere and ozone layers), and surface-atmosphere interchanges.

in the atmosphere seems to confirm that most O_2 has been released through photosynthesis, with much carbon temporarily stored in organisms and sediments. Any subsequent oxidation of that stored carbon, as by burning or respiration, returns CO_2 to the atmosphere, from where it is recycled by photosynthesis. Thus, there is a continuous steady-state flux of carbon and oxygen among the earth's surface, atmosphere, and life.

Today photosynthesis is universally regarded as the major source of atmospheric oxygen. Primitive photosynthetic cyanobacteria, which could produce O_2, appeared no later than 3.5 billion years ago, and red-colored strata containing strongly oxidized iron appeared by at least 2.5 to 2.8 billion years ago. Once molecular oxygen was present, ozone could form and provide an ultraviolet shield for further life development. There is an important interrelation between organisms and the atmosphere resulting from feedback: ozone provides a shield for all land life, yet apparently organisms themselves produced most of the oxygen from which it forms.

A plausible atmospheric evolution is summarized in Fig. 6.14. If the atmosphere is indeed evolutionary, then we must inquire if it is still changing and, if so, at what rate. Although organisms generally are remarkably adaptable, it is difficult to conceive of any drastic changes of atmospheric or oceanic composition having taken place since the appearance of highly organized marine animal forms nearly a billion years ago. Complex interactions between life and its total ecological environment provide a strong argument for relative stability of the atmosphere and oceans for at least the past billion years or so. Moreover, most Cryptozoic sedimentary rocks are not greatly different from those of younger ages, suggesting that the chemistry of the seas and atmosphere has not changed drastically for at least 2 to 2.5 billion years (see also Chap. 8).

Origin of Seawater

Seawater is not difficult to explain because any hypothesis for the atmosphere also provides abundant water. The origin of the oceans becomes largely a question of the beginning and the rate of accumulation of water and of dissolved salts.

Following the outgassing hypothesis, the rate of accumulation of seawater is tied directly to atmospheric production of water vapor and therefore to chemical fractionation of the solid earth. Rubey reasoned that the volume of seawater has grown in direct proportion to the increase in volume of continental crust through time. Did atmosphere and seas (and crust) accumulate slowly at a more or less uniform rate, or did they accumulate early and rapidly? A rough index might be obtained from the rate of release of helium and argon 40 to the atmosphere, assuming the outgassing hypothesis to be correct. If water were released at a comparable rate, it would suggest that the oceans accumulated over a long period of time, but most rapidly early in history when greater abundance of radioactive elements produced about five times as much heat as now. Furthermore, recall that there is isotopic evidence that the earth had an ocean as early as 4.4 billion years ago. This implies that most of the outgassing of the atmosphere and seawater probably was completed by then as well (Fig. 6.14).

Ocean-Atmosphere Regulatory Systems

Global Chemostat

Seawater undergoes chemical exchanges with the atmosphere, the solid earth, and life, resulting in a complex chemical regulatory system, or *global chemostat* (Fig. 6.15). Gases such as O_2 and CO_2 dissolve in seawater in proportion to their abundances in the atmosphere, helping to stabilize or buffer the composition of both—another example of feedback. The proportions of salts in seawater remain nearly constant through precipitation of any overly concentrated salts as sediments. Conversely if seawater becomes depleted in something, that element will be redissolved from sediments on the ocean floor. Examples of important ocean-sediment exchanges include storage of calcium and carbon in limestone ($CaCO_3$), calcium and sulfur in anhydrite and gypsum ($CaSO_4$), and sodium and chlorine in common salt ($NaCl$).

Important exchanges also occur between life and the atmosphere. Bacteria are important in overall cycling of nitrogen between the earth and the atmosphere; some release it, whereas others fix it in nitrogenous compounds. Remember that plant photosynthesis extracts carbon dioxide and releases oxygen, whereas animal respiration consumes oxygen and releases CO_2. O_2 is also used in large quantities for the oxidation of minerals at the crust surface, a weathering process of major importance for at least the past 2.5 to 3.0 billion years. Carbon is temporarily stored in coal and in carbonate rocks (limestone and dolomite). It is estimated that over 600 times as much CO_2 is so stored as now exists in the atmosphere, hydrosphere, and biosphere combined. Earth would have a CO_2-dominated atmosphere like that of Venus if it were not for the abundance of these carbonate rocks.

Through feedback processes—the continual chemical flux among atmosphere, seawater, rock weathering, and life metabo-

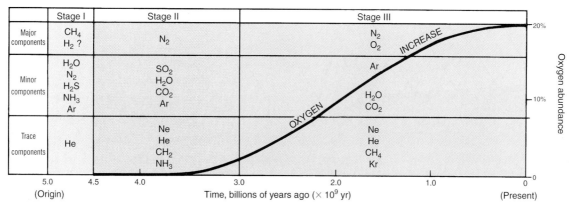

Figure 6.14   Probable evolution of composition of the earth's atmosphere. *(Adapted from H. D. Holland, 1964, in* The origin of the atmosphere and oceans; *by permission of John Wiley & Sons, Inc.)*

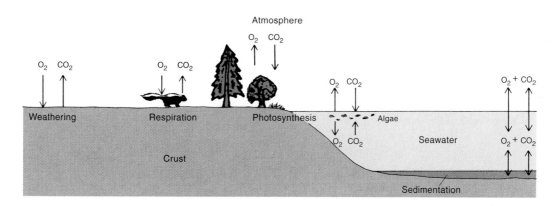

Figure 6.15   The global chemostat system exemplified by the cycling of oxygen (O_2) and carbon dioxide (CO_2) among crust, atmosphere, seawater, and organisms. Other important elements, such as nitrogen, phosphorus, sodium, sulfur, calcium, and potassium, cycle in a similar manner.

lisms—a global chemical equilibrium, or steady state, has been maintained for most of geologic time. Any perturbation in the composition of one realm tends to be damped out by a compensating change in another (Fig. 6.15).

Global Thermostat and Climate

Average global temperature, which controls overall climate, depends upon many factors, some of which are terrestrial and some extraterrestrial in origin. Most important are solar output, orbital variations, reflectivity of the earth, location of poles and oceans, and transparency of the atmosphere. Changes of solar output obviously would affect global temperature, but if such changes do occur, they have yet to be observed; therefore, we shall consider only the other factors.

Any change that shifts the ratio between incoming and outgoing radiation must affect the global heat budget, and reflectivity of the earth's surface is one important factor (Fig. 6.16). Viewed from space, the earth presents a variable face, with dark oceanic and forested areas and light clouds, snow, and deserts. The magnitude of radiation reflected from different surfaces expressed as a percentage of the incoming radiation is termed **albedo.** Measurements show the importance of relative albedo (Table 6.7). Land areas have far less heat retention capacity than

water (Fig. 6.16). Surface seawater absorbs and is heated by solar radiation to form the **photic zone,** and oceanic circulation distributes that heat widely over the earth. Therefore, *the larger the ratio of ocean to land surface, the warmer will be the overall global temperature, and vice versa.* It follows also that, if the earth's poles are located in open oceanic areas, they will be warmer than if they are landlocked. As we shall see, the relative coldness of polar areas in turn greatly affects overall climate. Yet another geographic factor of secondary importance is the configuration of continents as it affects atmospheric and oceanic circulation (e.g., mountains or isthmuses that block such circulation).

Transparency of the atmosphere is affected mainly by cloud cover and dust, which inhibit incoming radiation, and by certain chemical compounds, which trap outgoing radiation. Clouds affect albedo as well as transparency (Table 6.7). Fine dust and smoke particles are such effective inhibitors that a mere 7 percent increase of overall atmospheric "dustiness" would lower the average annual temperature approximately 1°C. The 1982 eruption of El Chichon in Mexico produced one of the largest ever observed discharges of sulfur dioxide (SO_2), which combines with water vapor to form radiation-absorbing sulfuric acid droplets; that single eruption reduced temperature at least 0.3°C.

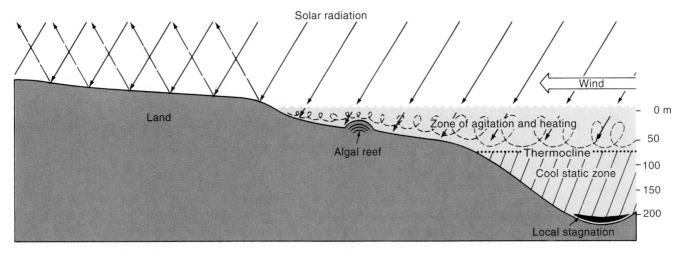

Figure 6.16 The global thermostat system exemplified by the contrasting heat retention of land and water. Shallow, agitated water is heated by solar radiation to form the earth's most important heat reservoir. This photic zone above a temperature boundary (thermocline) is also the habitat of algae and microscopic plants (phytoplankton), which form the base of the aquatic food chain. Below that boundary, water is cool and less agitated; the deepest zones may become stagnant and depleted in oxygen.

Table 6.7	Relative Earth Albedo (Percentage of Total Incoming Solar Radiation Reflected)
Snow or ice	45–85%
Clouds	40–80%
Grasslands	15–35%
Bare ground and/or light-colored rocks	15–20%
Forests and/or dark-colored rocks	5–10%
Water	2–8%

Trapping of long-wavelength infrared radiation in the atmosphere causes the well-known *atmospheric greenhouse effect.* If all long-wave radiation from the earth's surface could escape to space, the earth would be cold and uninhabitable. Such escape is inhibited, however, by molecules of CH_4 (methane), CO_2, water vapor, and ozone. Over 100 years ago, British physicist John Tyndall suggested that variations of atmospheric CO_2 content might explain major climatic changes, such as that which caused glaciation. In recent years, interest in his hypothesis has been renewed because of the realization that continued burning of fossil fuels, which began with the industrial revolution 200 years ago, may be increasing atmospheric CO_2 sufficiently to cause significant warming of the climate. A cooling of climate would be expected due to a reversed greenhouse effect during periods when very large amounts of carbon were removed from the atmosphere and stored in sediments such as limestone, coal, or carbon-rich shales. Measurements from cores of old glacier ice show that the atmospheric CO_2 content was considerably lower during the last glacial episode. The possible influence of CO_2 upon climate is explored further in later chapters.

Summary

Solar System

- The **entire solar system** is thought to have condensed from a spinning, disc-shaped **solar nebula,** which had both temperature and compositional gradients from the sun outward. The small **inner, terrestrial planets** condensed at high temperatures and were enriched in heavier elements, whereas the large **outer, gaseous planets** condensed at lower temperatures and are composed of light, volatile elements. Only Earth has abundant free oxygen and liquid water.
- The **terrestrial planets** all seem to have heated up rapidly as asteroid-like **planetesimals** accreted to form **protoplanets,** which **accreted** under their own gravity. It is not clear if the protoplanets melted completely, but their interiors did. This facilitated internal **gravitational differentiation** about 4.5 billion years ago, which resulted in relatively **iron-rich cores** surrounded by **silicate-rich mantles** and crusts divided between basaltic lowlands and less-dense highlands.
- **Impact craters** were formed by intense bombardment of planetary surfaces by planetesimals until about 3.8 billion years ago. The orbiting **asteroids** are probably leftover planetesimals that were prevented from accreting into a planet by Jupiter's strong gravity. Rare subsequent asteroid and/or comet impacts have been suggested as causes of major extinctions of organisms. **Meteorites** may be fragments of asteroids, whereas comets are composed mostly of ice.

Earth

- **Earth's core.** A dense core rich in iron and nickel is inferred from Earth's bulk density, the magnetic field, and an analogy with metallic meteorites. The dimensions and liquid character of the outer core are established from seismology. Interaction between spinning solid mantle and liquid outer core is thought to drive the magnetic field, much like an **electric dynamo.**

- The **earth's mantle** is composed chiefly of **ultramafic peridotite,** which is rich in magnesium and iron; its bulk composition approximates that of stony meteorites. **Oceanic crust** in early Cryptozoic time was composed of **komatiite,** a volcanic rock formed by complete melting of peridotite. As the mantle has cooled, however, later oceanic crust has been composed of **basalt,** a more silica-rich volcanic rock formed by **partial melting** of peridotite. A **low seismic velocity** zone of low rigidity defines the base of the **lithosphere,** which is 60 to 250 kilometers thick.

- The **Moho discontinuity** defines the base of the earth's crust. **Continental crust** has resulted from more complex, poorly understood processes of differentiation from the upper mantle, processes driven by thermal energy resulting in large part from radioactive decay. Only the earth has experienced the distinctive form of crustal turmoil and differentiation that resulted in plate tectonics. Ultimately such activity must cease as thermal reservoirs become depleted. Lighter elements, such as K, Na, Ca, Al, Si, and O, have been concentrated together with traces of the heavier elements U and Th, which stay in the melt until the very final stage of cooling. Continental crust has the average composition of **granitic rocks.**

- The **atmosphere** and **seawater** also must have formed by global differentiation, but their present compositions have resulted from several processes. **Outgassing** of the lightest elements from the interior could have given rise to N_2, CO_2, H_2, H_2O, and He. **Photochemical dissociation** of early compounds, such as CH_4, NH_3, and water vapor, also could have given rise to N_2, CO_2, H_2O, and some O_2. Although small amounts of oxygen apparently were present in the early atmosphere, most of it has been generated by **photosynthesis.**

- **Global chemostat, thermostat,** and **ozone shield.** The ocean-atmosphere system provides important regulatory functions. Feedback interactions among the atmosphere, seawater, crust, and life have maintained a near steady-state chemical and thermal equilibrium through most of geologic time. If heat or a particular chemical element increases in one realm, the others take it up to maintain overall balance. The **ozone layer** in the lower atmosphere provides an important shield from ultraviolet radiation.

- The **global heat budget** depends upon the ratio of incoming to outgoing radiation. The most important factors are

  1. **Reflectivity of the earth** or **albedo:** greater relative oceanic area means warmer climate because of the greater heat retention of the aquatic **photic zone** versus land. Moreover, ocean currents transport heat to colder latitudes.

  2. **Location of poles:** landlocked poles are colder than oceanic ones, and they affect mid-latitude climate profoundly.

  3. **Atmospheric transparency:** clouds and dust inhibit incoming radiation, resulting in cooling. CO_2, CH_4, water vapor, and O_3 inhibit the escape of outgoing radiation, resulting in warming (**atmospheric greenhouse effect**).

Readings

Ahrens, L. H. 1965. *Distribution of the elements in our planet.* New York: McGraw-Hill.

Anderson, D. L. 1989. *Theory of the earth.* Boston: Blackwell.

Beatty, J. K., and A. Chaikin, eds. 1990. *The new solar system.* Cambridge, Mass.: Sky.

Birch, F. 1965. Speculations on the earth's thermal history. *Bulletin of the Geological Society of America,* 76: 133–54.

Bowring, S. A., and I. S. Williams, 1999. Priscoan (4.00–4.03 Ga) orthogneisses from northwestern Canada. *Contributions to Mineralogy and Petrology,* 134: 3–16.

Carr, M. H., R. S. Saunders, R. G. Strom, and D. E. Wilhelms. 1984. The geology of the terrestrial planets (SP-469). Washington, D.C.: National Aeronautics and Space Administration.

Compston, W., et al. 1985. The age of (a tiny part of) the Australian continent. *Nature,* 317: 559–60.

Elsasser, W. M. 1958. The earth as a dynamo. *Scientific American,* May, 1–6.

Frakes L. A. 1979. *Climates through geologic time.* Amsterdam: Elsevier.

Jacobs, J. A., R. D. Russell, and J. T. Wilson. 1973. *Physics and geology.* 2d ed. New York: McGraw-Hill.

Lewis, J. S., and R. G. Prinn. 1984. *Planets and their atmospheres.* New York: Academic Press.

Liu, D. Y., A. P. Nutman, W. Compston, J. S. Wu, and Q. S. A. Shen. H. 1992. Remnants of $\geq$ 3800 Ma crust in the Chinese part of the Sino-Korean craton. *Geology,* 20: 339–43.

Lowman, P. D., Jr. 1972. The geological evolution of the moon. *Journal of Geology,* 80: 125–66.

Mason, B., and C. B. Moore. 1982. *Principles of geochemistry.* 4th ed. New York: Wiley.

Mason, B., and W. G. Melson. 1970. *The lunar rocks.* New York: Wiley-Interscience.

Mueller, P. A., J. L. Wooden, and A. P. Nutman. 1992. 3.96 Ga zircons from Archean quartzite, Beartooth Mountains, Montana. *Geology,* 20: 327–30.

Ringwood, A. E. 1979. *Origin of the earth and moon.* Heidelberg: Springer-Verlag.

Robertson, E. C., ed. 1972. *The nature of the solid earth.* New York: McGraw-Hill.

Schopf, J. W., ed. 1983. *Earth's earliest biosphere: Its origin and evolution.* Princeton, Princeton University Press.

Schwarzbach, M. 1963. *Climates of the past.* London: Van Nostrand.

Valley, J. W., W. H. Peck, E. M. King, and S. A. Wilde. 2002. A cool early earth. Geology, 30: 351–354.

Wilde, S. A., J. W. Valley, W. H. Peck, and C. M. Graham. 2001. Evidence from detrital zircons for the existence of continental crust and oceans on the earth 4.4 Gyr ago. *Nature,* 400: 175–181.

Wood, J. A. 1979. *The solar system.* Englewood Cliffs, Prentice-Hall. (Paperback, Foundations of Earth Science Series.)

York, D. 1993. The earliest history of the earth. *Scientific American,* 90–109.

JOIDES Resolution

(Courtesy of the Ocean Drilling Program, Texas A & M University.)

Chapter 7

Mountain Building and Drifting Continents

MAJOR CONCEPTS

What matters is how far we go?

His scaly friend replied,

There is another shore, you know,

Upon the other side.   The Mock Turtle, *Alice in Wonderland*

They [plates] can't curl down; they

must curl up

To form a kind of dish

To stop the oceans spilling out

And losing all the fish.   B. C. King and G. C. P. King (1971)

Nature, 232:37.

▶ Many different explanations for the origin of mountain belts and their associated thick wedges of sediment have been proposed since 1857. All these models assumed that the continents were fixed in their present position. In the late 1800s, many scientists tried to explain the compression of mountain belts by crustal cooling and contraction of the earth.

▶ Evidence that the continents had drifted apart was accumulating until presented in detail by Alfred Wegener in 1915, but most geologists still rejected the notion of moving continents until the 1950s, when evidence of ancient magnetic fields recorded in the rocks were discovered. These fossil magnets showed that continents had been located at other latitudes in the geologic past.

▶ In the 1950s and 1960s, exploration of the previously little-known deep-sea floor showed that the mid-ocean ridges were the sites of sea-floor spreading and the generation of new oceanic crust. Eventually this discovery led to the scientific revolution of plate tectonics.

▶ Since new crust is created at mid-ocean ridges, it must be destroyed somewhere else on the planet, or the earth would have to expand. Most crust is recycled in subduction zones, where one oceanic plate goes down beneath another plate (either oceanic or continental) and causes melting, which results in a volcanic arc.

▶ Plate tectonics provides a modern explanation of why sedimentary basins occur where they do, why they subside and accumulate thick packages of sediment, and why the same basins are often caught up in later mountain belt collisions.

When German meteorologist Alfred Wegener died in Greenland in 1930, he was, to paraphrase the Mock Turtle, exploring "another shore upon the other side" of the Atlantic Ocean from his native Europe. It was Wegener who, more than any other person, had championed the dramatic theory that past continents had been torn asunder and drifted apart to form the present continents with shores "upon the other sides" of new oceans.

The origin of continents and ocean basins remains a puzzle not fully resolved even today. For example, was continental or oceanic crust the more primitive? A century ago, it was suggested that the Pacific Ocean may be a scar from which the moon was torn. Following this postulated catastrophic event, the remaining continental crust presumably was broken and redistributed as we see it today. An alternative suggestion made about 1925 had the earth expanding to about double its original size. In the process, the ocean basins formed as cracks between the ruptured continents. Most other early speculations also assumed an original, uniform continental crust, but for the past 50 years or so, it has been generally believed that continental crust evolved slowly through geologic time. The hypothesis of growth, or *accretion of continents* fits the overall chemical differentiation of the earth discussed in Chap. 6, for it postulates the gradual concentration of potassium, aluminum, silica, oxygen, sodium, and other light elements from the earth's interior (see Fig. 6.11).

Although early advocates of continental accretion assumed that each continent grew where it is now located, subsequent rearrangement by displacements is not incompatible with accretion. Long ago, scholars observed the curious parallelism of outlines of the opposing coasts of the South Atlantic, suggestive of separated pieces of a jigsaw puzzle. In 1859, the first map fitting South America against Africa was published. Then in the early part of the twentieth century, the concept of large-scale rearrangement of continents, or **continental drift,** began to receive considerable attention, beginning about 1910, especially through the writings of Alfred Wegener.

This chapter discusses the large-scale structures of the earth's crust. First we shall examine the early ideas about how mountain chains and sedimentary basins were formed. Then we shall review important hypotheses on mountain building, which leads us back to Wegener's theory of continental drift. Next we shall examine the data that brought continental drift back into the limelight, data mostly from geophysics and marine geology (Fig. 7.1). Finally, we shall examine the modern theory of plate tectonics and see how old notions of geosynclines can be explained in modern terms.

Figure 7.1 The ocean drilling vessel *JOIDES Resolution* (or SEDCO/BP 471) operated by a consortium of a dozen nations for scientific drilling into the ocean floor. This ship is equipped to drill cores from sedimentary and igneous rocks from as far as 8.2 kilometers (5 miles) beneath the sea surface. In 1985, the *Resolution* replaced the smaller and less sophisticated *Glomar Challenger,* which during the preceding 15 years had traveled 600,000 kilometers and drilled 1,092 holes at 624 sites around the world. The information from the international Deep-Sea Drilling Project (DSDP) helped to confirm the hypothesis of plate tectonics and to unravel the history of oceans and climate for the past 100 million years. *(Courtesy of the Ocean Drilling Program, Texas A & M University.)*

Orogenic Belts and Mountain Building

The Geosynclinal Concept—Made in America

Beginning in 1857, American geologist and paleontologist James Hall of New York laid the groundwork for one of geology's early generalizations. For many years, geologists in both Europe and America had been speculating about the crumpled nature of strata seen in mountain ranges (Fig. 7.2). Hall noted that, in the Ap-

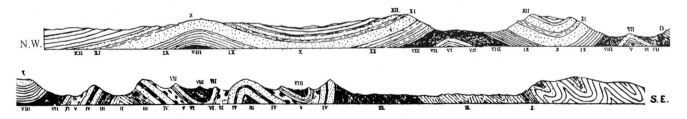

Figure 7.2 One of the first accurate cross sections of a mountain range showing folds in strata in the Appalachian chain across Pennsylvania from the northwest *(top)* to the southeast *(bottom)*. Roman numerals distinguish different formations. *(From Rogers and Rogers in* Transactions of American geologists and naturalists for 1840–1842, pp. 474–531.)

Table 7.1	Major Structural or Tectonic Elements of the Earth's Crust	

Stable Elements	**Unstable Elements**
Continental cratons—Topographically subdued continental plains with little seismicity or volcanism; granitic crust *Oceanic abyssal plains*—Deep ocean plains with little seismicity or volcanism; basaltic crust	*Orogenic belts (mountain belts and island arc-trench systems)*—Andesitic volcanoes; shallow to deep seismicity, compressional folding and thrust faulting, granitic batholiths, regional metamorphism and low heat flow *Oceanic ridges*—Basaltic sea-floor mountains with shallow seismicity; extensional faulting and abnormally high heat flow

palachian Mountains—today classified as an orogenic (or mountain) belt—the strata of the Paleozoic Systems are not only more deformed but are also ten times thicker than are their counterparts in the Mississippi Valley region farther west—today classified as part of the **craton** (Table 7.1). He suggested a simple cause-and-effect such that the greater load of sediments in the Appalachian belt depressed the crust as deposition occurred. Finally, the crust could bend down no farther, so it failed, and the strata were crumpled (Fig. 7.3).

In 1873, another leading American geologist, J. D. Dana, coined the term *geosynclinal* (*geo* = earth, *syncline* = downfold) for the old zone of thick strata. He rejected Hall's simple explanation of the great subsidence by sediment-loading alone because less dense sediments could not depress more dense crust nearly so much as was required by the great thickness of strata. The geosyncline was inferred by Dana to be a result—not a cause—of the fundamental structural instability of mountain belts. He argued that buckling of the crust beneath geosynclines was the primary cause (1) of subsidence and thick sedimentation and (2) of mountain building (Fig. 7.3).

European geologists quickly adopted the basic concept of thick sediments as the ancestors of mountains. Unlike the Americans, who believed that sedimentation always kept the subsiding geosyncline full of shallow-marine and nonmarine deposits, the Europeans believed that geosynclines began as deep troughs on the sea floor and that much of the thick sediment contained therein was of deepwater origin. By 1930, the Europeans had gone further by suggesting a close similarity between geosynclines and deep-marine trenches and volcanic islands of the Pacific. This began an important series of proposed links between ancient orogenic belts and the modern island arc-trench systems. From these early ideas developed one of the great unifying concepts in geology, which brought for the first time a recognition that the stratigraphy and structure of mountain belts are closely related and that such belts may generate new continental crust.

A **geosyncline** was defined as an elongate belt of thick strata that was linked closely with the formation of mountains. Ancient geosynclines characteristically show tightly compressed folds and low-angle overthrust faults, and most contain granitic batholiths and regionally metamorphosed rocks. Thus, geosynclines have developed within structurally unstable orogenic belts located between stable regions. Modern oceanic island arc-trench systems have many of the same structural qualities as ancient

A. Hall's theory
Warping due to sedimentation

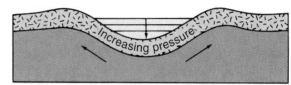
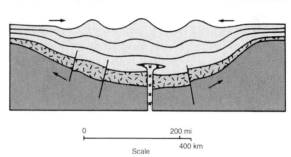

B. Dana's theory

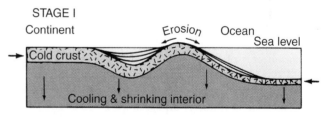
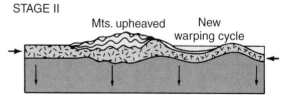

Figure 7.3   *A:* Hall's theory of mountain building postulated that thick sedimentation along the edge of the continent depressed the crust. The downbending crust was stretched and faulted, overlying sediments were compressed and folded, and topographic mountains were raised at the surface. *B:* Dana's geosynclinal theory postulated bending of crust as the interior cooled and shrank. Bending was concentrated at continental margins, where erosion of an upraised geanticline provided sediment to an adjacent geosyncline. Finally, the crust failed, mountains raised, and the geosyncline was welded to the continent. Repetition of this process caused growth of the continent through mountain building.

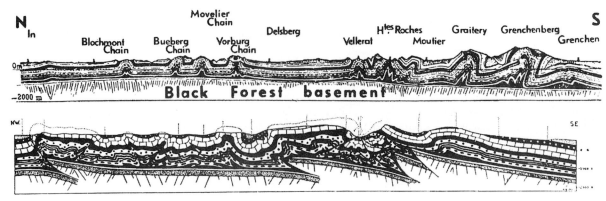

Figure 7.4   Two contrasting interpretations of different parts of the Jura Mountains of the French-Swiss border. *Upper:* Slippage of strata over the basement along a flat shear surface *(décollement)*, requiring no crustal shortening. *(From Buxtorf, 1908, Naturelle carte geologique Suisse.) Lower:* Faulting of the basement as well as superficial strata along steep faults, which does require some lateral shortening of the crust. *(From Aubert, 1949; by permission of Geologische Rundschau.)* Apparently both interpretations are at least partly correct.

mountain belts, *but with very little sedimentation.* The modern systems are either too young or too small to have generated the huge volumes of sediment characteristic of ancient geosynclines.

As the study of mountain belts progressed, it became apparent in the early twentieth century that their development involved all facets of geology. Besides early recognition that granitic batholiths were formed in orogenic belts, it was also realized that volcanic rocks make up a sizable proportion of the stratified geosynclinal sequences. Therefore, volcanism somehow must play an important role in the evolution of such belts. These discoveries further linked modern volcanic arcs with ancient orogenic belts, as the Europeans had suggested in the 1930s.

Unfortunately, *geosyncline* has acquired almost as many meanings as there have been geologists using the term. One is reminded of Humpty Dumpty's pronouncement in Lewis Carroll's *Through the Looking Glass:* "When I use a word, it means just what I choose it to mean—neither more nor less." The term has little value today except in the historical context just discussed. Modern concepts of plate tectonics involving both continental drift and spreading oceans have so dramatically revolutionized our thinking that the geosynclinal terminology has little relevance. Thick strata formerly all classified as geosynclinal may have several very different tectonic origins. Therefore, it is confusing to try to adapt this old term to the new concepts—revolutionary new hypotheses require new terminology. Nonetheless, the linkage between thick sedimentary sequences and mountain belts must be explained by any successful tectonic theory.

Vertical versus Horizontal Tectonics

The recognition of tightly folded strata in mountain belts during the late 1700s and the recognition of overthrust faults in southern Germany in 1826 naturally cried out for explanation. Plutonists such as James Hutton had explained folding as wrinkling due to a simple vertical uplift of mountain ranges by the rise of molten granite into them; the superficial strata either were pushed or slid to the sides (see Fig. 2.13). Beginning about 1840, the crumpling was envisioned by many as the result of viselike compression within the mountain belts, a view that implied a shrinking of the crust. By 1900, an intense debate had developed between those who believed that mountains were formed by lateral squeezing and those who believed that they were formed by vertical upwarping alone. Most geologists (such as Dana) thought that the folds and thrust faults proved that the earth's circumference had been shortened across mountains. Several physicists and a few geologists, however, following Hall, argued instead that isostatic upwarping could create topographic mountains and cause the superficial sediments to deform under the influence of gravity without disturbing the more rigid basement rocks beneath. It would be like a rug wrinkling when pushed across a floor. One of the reasons for this suggestion was the apparent weakness of the rocks involved in over-thrust faults. It seemed easier to imagine them sliding than being pushed for several kilometers. According to this gravitational sliding hypothesis, only the superficial strata must be deformed; according to the lateral compression hypothesis, the rigid basement rocks also must be faulted and deformed. This provides a test of the hypotheses. Practically all mountain belts show significant deformation of their basement rocks (as in Fig. 7.4, lower). Therefore, it is accepted today that mountain belts are primarily the results of profound lateral compression across relatively narrow zones; clearly, immense forces must be required to accomplish such impressive work.

A Cooling and Shrinking Earth

Dana proposed a simple and appealing explanation of geosynclines and mountain building, a view that accounted for compression due to gradual cooling of the earth. In his day, the universal assumption was that the earth was cooling from a molten origin (see Chap. 6). Because it was cooling, the earth must also be contracting, and its cold, rigid crust must bend or rupture somehow as the interior shrinks. The wrinkling skin of a drying apple provided a favorite analogy. Dana argued that most of the compressive adjustment of the crust would occur at the boundaries between continents and ocean basins—exactly where North American mountains occur now. First the edge of the continent bent downward, allowing thick (geosynclinal) strata to accumulate there. Further bending of the crust then caused compression of these

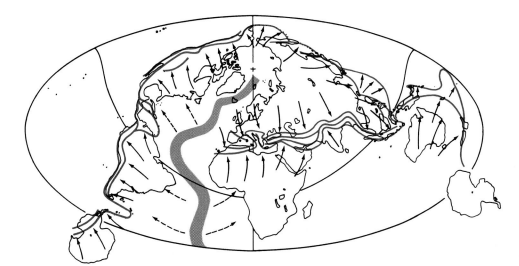

Figure 7.5   F. B. Taylor's and A. Holmes's conceptions of formation of Cenozoic mountains by continental drifting. Note parallelism of mid-Atlantic submarine ridge (gray) to opposing continent margins. *(Adapted from Taylor, 1910,* Bulletin of the Geological Society of America; *Holmes, 1929,* Mineralogical Magazine.*)*

strata and the upheaval of mountains (Fig. 7.3). It was a beautifully simple hypothesis, but by 1910 the recognition of radioactive heating of the earth's interior discredited contraction. Moreover, unlike North America, many of the earth's orogenic belts are not found at continental margins. A new hypothesis for mountain building was needed.

Continental Drift as a Cause of Mountain Building

The general idea of large-scale continental displacements began to be discussed seriously in 1908, when F. B. Taylor, an American geologist, suggested that drifting of the continents had caused wrinkling of the crust to produce all the great Cenozoic mountain systems (Fig. 7.5). He suggested that the leading edges of the moving continents depressed the oceanic crust ahead to form troughs in which thick sediments could accumulate. Further movement then compressed and upheaved those strata to produce the present lofty mountains rimming the Pacific. Upheaval of the Alpine-Himalayan chain carried the process a step further by culminating in the direct collision of continents (Fig. 7.5). For a mechanism to drive his continents, Taylor speculated that catastrophic tidal action was induced in the solid earth by capture of the moon during Cretaceous time. Although his explanation is unacceptable, Taylor had given birth to the important new idea of linking mountain building with resultant compression of the crust by drifting continents.

The first detailed reconstruction of the continents was attempted almost immediately by another American, H. Baker. He postulated an original, single, huge landmass, or **supercontinent,** that suddenly split to form the present Arctic and Atlantic Oceans at the end of Miocene time (Fig. 7.6). Baker was at least as imaginative as Taylor in devising a cause. He speculated that variations in orbits brought Earth and Venus close enough together to produce such severe tidal distortion that a large portion of original continental crust was torn from the Pacific to form the moon, with the remaining continental crust rupturing and slipping toward the Pacific void. There is, however, no record in Cenozoic strata of a catastrophe of such magnitude; indeed, the last major displacements of continents began in early Mesozoic time.

Figure 7.6   H. Baker's replacement globe reconstruction of continents prior to supposed separation. Note rotations and squeezing together of islands and peninsulas to improve fit. *(After Baker, 1912; with permission of Michigan Academy of Sciences.)*

Other people soon began to favor continental drift, the most important of whom was Wegener. He drew upon evidence from geology, geophysics, biology, and climatology in developing the first complete and influential statement of the drift theory in 1912. He believed that drifting apart of the continents occurred over a long period during the Mesozoic and early Cenozoic Eras (Fig. 7.7). He became the first to attempt reconstruction of a former supercontinent by fitting edges of continental shelves rather than present coastlines, Wegener's way being a more geologically realistic approach. Besides the parallelism of continental margins, Wegener appealed to apparent paleoclimatic indicators in late Paleozoic rocks, including glacial, desert, and tropical rainforest deposits, to reconstruct Permian climatic zones (Fig. 7.8).

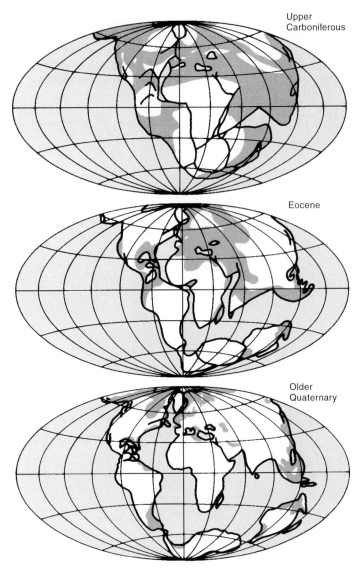

Upper
Carboniferous

Eocene

Older
Quaternary

Figure 7.7   Famous reconstruction of the continents by Alfred Wegener, showing three inferred stages of continental drift. Color represents extensions of the sea at the respective times. *(Adapted from Wegener, 1929,* Die Enstehung der kontinente und Ozeane; *by permission of F. Vieweg and Son.)*

South African geologist A. L. Du Toit completely consumed by drift fever, began in 1921 to strengthen Wegener's concept with a more detailed assessment of supporting geologic and paleontologic evidence. Near-identity of many fossils on now widely separated continents also lent support to drift. How could land plants with large seeds incapable of wind transport have been dispersed across wide oceans? And how could heavy land-dwelling reptiles have swum across? Even the earthworm was used to support drift. Modern worms of Madagascar resemble the worms of India more than they resemble the worms of nearby Africa, and those of eastern North America resemble the worms of Europe more than they resemble those of the American Pacific coast.

The first phase of the drift theory culminated in 1937 with publication of Du Toit's *Our Wandering Continents,* in which he scorched the conservative orthodox antidrifters (mostly American) with an attack upon their "groundless" worship of the dogma of fixed continents and ocean basins. He ridiculed, for example, the compromise of inventing countless narrow land or island bridges between continents to explain the undeniable similarities of many fossil and modern land organisms on opposite sides of the Atlantic.

Thermal Convection—Panacea for Mountain Building and Drift?

Wegener postulated that, in spite of apparent high viscosity of subcrustal material, small forces acting over very long periods, could cause that material to yield and allow crustal blocks to flow slowly across the upper mantle. Though this theory seemed incredible, we have since learned to be cautious about shouting "impossible."

Modern knowledge of material behavior proves that slow plastic flow does occur in the mantle. The question then becomes, What force might have caused the drifting? Wegener considered centrifugal effects of the earth's spin, tidal effects in the solid earth, and wobble of the axis as possible contributors, but all were rejected by physicists as inadequate.

The possibility of **thermal convection** within the earth was suggested as early as 1839, and in 1881 it was proposed as a possible mechanism both for localizing volcanoes and for

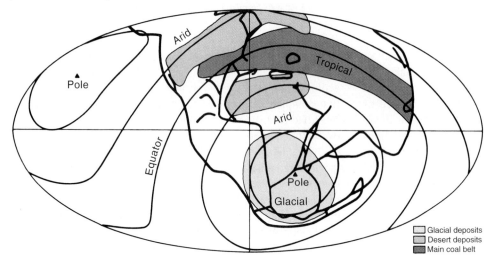

Arid

Tropical

Pole

Arid

Equator

Pole
Glacial

Glacial deposits
Desert deposits
Main coal belt

Figure 7.8   Wegener's reconstruction of the Permian continents and paleoclimate zones inferred from peculiar rock assemblages. Based upon paleoclimatic data, Wegener deduced paleoequator and pole positions, which are remarkably similar to positions indicated by paleomagnetic data four decades later. Because the map was drawn on present coordinates, paleolatitudes seem distorted. *(After Köppen and Wegener, 1924,* Die klimate der Geologischen Vorzeit; *by permission of G. Borntraeger Co.)*

Figure 7.9   The convection-current mechanism for rupturing and drifting continents as well as forming new ocean basins and mountains as conceived by Arthur Holmes in 1928. *Top:* Convection begins to stretch an overlying continental block. *Bottom:* Block ruptures and two fragments move apart. *(Adapted from A. Holmes, 1928, Physical geology.)*

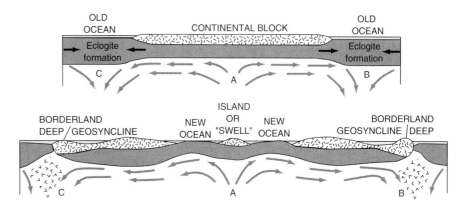

producing mountains by dragging and wrinkling the crust. In 1928, a brilliant British geologist, Arthur Holmes, who very early became interested in the implications of radioactivity, presented the first serious suggestion of thermal convection in the mantle as a possible driving mechanism both for mountain building and for continental drift (Fig. 7.9). The basic idea is simple. If more heat were generated in one portion of the deep mantle due to irregular distributions of radioactive isotopes, there would tend to be a very slow plastic flow of the hotter material upward. As it rose, this hot material would cool and flow laterally beneath the lithosphere; finally, when cooler than the average mantle below, it would sink. Thus, Holmes proposed a slow convective overturn of the mantle, like the convective circulation of warm air in a room heated by a radiator. He then assumed that, where the convective flow was upward, it might raise and even rupture the crust; where it was lateral, it might drag the crust along conveyor-belt fashion; where mantle motion turned downward, the light continents, being isostatically buoyant like a cork, would refuse to sink and, so, would buckle up to form mountains (Fig. 7.9).

Drift in Eclipse

After a hot controversy over drift during the 1930s, the matter cooled in the 1940s, with geologists around the world tending to throw their lot with either the pros or the cons and getting on to other tasks. Former president of the American Philosophical Society, W. B. Scott, expressed the prevalent American view in the 1920s by describing the drift theory as "utter, damned rot!" It was downright un-American to be "soft" on drift even as late as 1960. Reasons for strong opposition were twofold. First, the idea of such dramatic shiftings of continents, and seemingly only once in history, seemed counter to a lingering, narrow uniformitarian doctrine. Certainly continents had changed, but it seemed more uniform for them to do so in place. If they had to move, then they "should have" moved uniformly (i.e., continuously) or at least several times; a single great jump smacked of catastrophism! Second, no theoretical mechanism was known that could drive the large but very thin continents—slabs 100 times longer than they are thick—over the earth's surface. Geophysicists asked how weak continental rocks could be pushed over equally weak oceanic ones. It seemed like trying to push huge sheets of wet tissue paper. Zealots such as Du Toit, however, kept

the faith and "accepted the inescapable deduction from the wealth of geological evidence available to the unfettered mind." They were not overwhelmed by the protestations of the physicists because they remembered all too well the error of Lord Kelvin's dazzling mathematical arguments about the age of the earth (see Chap. 5). Here is an important illustration of divergent attitudes. The field observer's seat-of-the-pants feeling is that if "something *did* happen, it *can* happen," whereas the theoretician tends to disbelieve anything as dramatic as drift unless he or she can conceive of a sound mechanism and can describe it with equations. As in any other game, each side eventually wins some and loses some.

Paleomagnetism—Drift's Renaissance

Rebirth of interest in drift had to await new approaches, which came, ironically, largely through physics in the 1950s and 1960s. The study of rock magnetism gave the first new impetus. Physicists had begun measuring the magnetic properties of rocks about 1850 and had discovered that many very young lavas showed magnetization parallel to that of the earth's present field. About 1900, studies of ancient bricks and pottery showed that magnetization acquired during firing could be retained for at least 2,000 years. Next the contrasts between magnetic orientation in lavas, dikes, and adjacent baked sediments and orientations in the unbaked older sediments showed that igneous rocks tend to acquire and retain magnetic orientations parallel to the earth's field at the time of their cooling. Further studies showed that the magnetization is stable indefinitely unless the rock becomes heated to about 500°C, at which temperature the magnetization is lost.

In the late 1920s a French physicist, Mercanton, suggested that, because orientation of the magnetic field now bears a close relation to the earth's rotational axis, it might be possible to test the theory of continental drift with the magnetic characteristics of certain ancient rocks. This fertile suggestion was not followed up until after World War II, when rock paleomagnetic data began to be gathered on a large scale by a group of British physicists.

Several iron oxide minerals have properties that make them susceptible to a magnetic field. As such minerals crystallize in a

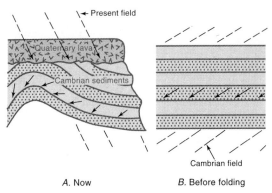

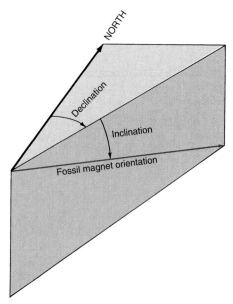

Figure 7.10   *A:* Orientation of remnant magnetism with respect to the present magnetic field in two different-aged rocks. *B:* Restoration of the Cambrian magnetic field orientation after correcting for post-Cambrian folding.

Figure 7.11   A hypothetical pie slice of the earth's crust showing the three-dimensional orientation of a fossil magnet. The declination angle is relative to present north and is measured in a horizontal plane; it provides an indication of paleolongitude. The inclination angle is relative to the horizontal earth's surface; it provides evidence for paleolatitude.

cooling magma, they become oriented parallel with the earth's magnetic field. If grains of such minerals are deposited gently in sediments, they will assume a similar orientation. These *magnetically susceptible* minerals retain their fossil, or **remanent, magnetism** unless later heated above their demagnetization temperature. Thus, many rocks, as well as pottery, containing these minerals today retain evidence of the orientation of the magnetic field when and where they were formed. If the field later shifted its position, we might find a record of such shifts "fossilized" in the rocks. We could, in fact, study the history of changes of relative orientation (and of changes of polarity) of the magnetic field, and at least one branch of geophysics would gain a historical dimension that it had lacked before.

Paleomagnetic studies based upon the early findings have been conducted at an ever increasing pace since 1950. The data show undeniable evidence of great relative changes between present geography and the magnetic field, but the results were at first difficult to interpret. If we assume that the field was always two-poled, as it is today, and that it owes its orientation to the spin of the earth (see Fig. 6.9), *then the two magnetic poles should always have been more or less coincident with the geographic poles of the earth's spin axis.* If these assumptions are correct, then determination of ancient magnetic pole positions from the measurement of "fossil magnets" in rock specimens should also reveal the approximate geographic *paleopole* and *paleoequator* positions. Fig. 7.10 shows the orientation of magnetization in a hypothetical modern rock in relation to the earth's present field (A) and in a Cambrian rock (B). Fig. 7.10B illustrates how different from the present field the *relative* orientation of the ancient field must have been when the Cambrian rock formed.

Both the **inclination angle** and the *declination angle* (Fig. 7.11) of the remanent magnetism for a rock are determined in the laboratory by using a sensitive magnetometer. The orientation of the specimen in the field must be established and corrections of orientation must be made if the rocks have been tilted after they formed. A simple rule is that *steep inclination of "fossil mag-*

nets" indicates formation of the rock at high latitude, whereas a very low angle indicates a location near the equator (Fig. 7.12). What a powerful paleogeographic tool this offers for determining the **paleolatitude** of an area through time (Fig. 7.13). *Paleolongitude* is not defined uniquely by angle of declination for a single locality because the same angle could exist at any longitude. Declination data for several widely scattered samples *of the same age,* if coupled with additional clues, such as shapes of shorelines or structural trends, may provide a unique solution, as shown in Fig. 7.14.

At first, geologists gathered paleomagnetic data chiefly from Europe and North America. As older and older rocks from each continent were studied, it appeared that the relative positions of these two continents with respect to the magnetic field, and therefore presumably also the earth's rotational axis, had changed markedly. But it was not a random change, for the apparent North Pole seemed to have been in the mid-Pacific Ocean 1 billion years ago and had migrated to its present position, which it reached in mid-Cenozoic time (about 25 million years ago). By 1955, it was thought that somehow the earth's rotational axis had shifted with respect to the crust, so that continents indeed had occupied different latitudes through time. After 300 years of speculation, drastic changes of latitude now had gained favor.

Although at first it was assumed that the magnetic field had shifted relative to a fixed crust, comparison of apparent pole migration paths for two or more continents revealed unexpected discrepancies. Paleopole positions determined from rocks of the

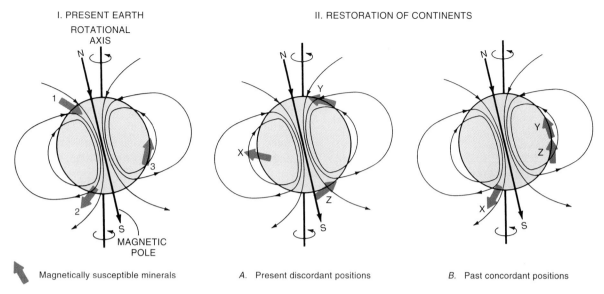

I. PRESENT EARTH

II. RESTORATION OF CONTINENTS

◥ Magnetically susceptible minerals

A. Present discordant positions

B. Past concordant positions

Figure 7.12   The relationship of the earth's magnetic field to remanent magnetism in rocks. *I:* Present earth showing magnetically susceptible mineral orientations (arrows) in modern rocks on three different continents. Note that the inclination of each is concordant with the magnetic field at its respective latitude. *II:* Restoration of paleolatitudes for three continents. *IIA:* The positions of fossil magnets in rocks of the same age on each continent. Each is *discordant* with the field at its present position. *IIB:* Their predrift positions are found by rotating each to a paleolatitude that is *concordant* with the magnetic field. Continents Y and Z may have been part of a single supercontinent before they drifted to their present positions.

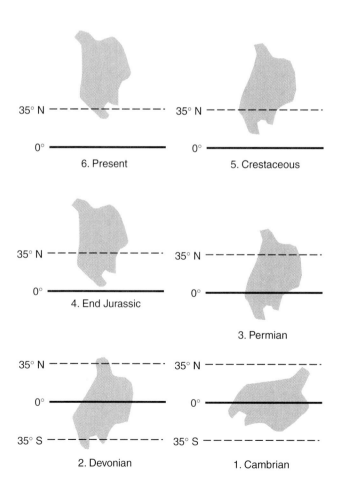

Figure 7.13   Different positions of North America relative to the equator from Cambrian *(1)* time to the present *(6)* according to paleomagnetic evidence. Note the progressive counterclockwise rotation and northward shift through time. Paleogeographic maps in will compare geologic and paleontologic data bearing upon paleolatitude and paleoclimate with these paleomagnetic restorations.

same age but from different continents were *discordant*—that is, they did not coincide. Only movements of continents could make them coincide. Assuming that the field always had been a simple two-poled one, as today, then no hypothesis of polar wandering alone was adequate, for, if the only motion taking place was that of the poles and field relative to the crust, *the apparent ancient pole positions should have been the same for all continents at any given time in the past.* Geophysicists soon concluded that, if the magnetic field had always been of the same sort, as today, then they must revive some hypothesis of continental drift in order to explain their data, no matter how painful that might be. *Seemingly the continents had to have moved relative to each other.* This was a curious irony, for recall that previously it was physicists who most vehemently protested against drift because no adequate physical mechanism seemed available. But the tables were turned, as expressed eloquently by American geophysicist Walter Munk at the height of renewed interest in polar wandering (1956):

> In this controversy between physicists and geologists, the physicists, it would seem, have come out second best! They gave

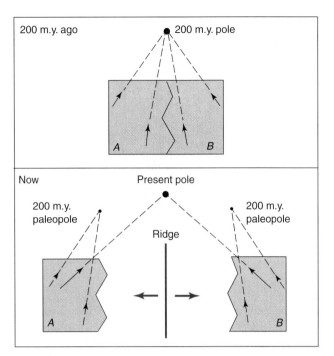

Figure 7.14   The method of restoration of continental positions of longitude in the past using paleomagnetic data. The declination angle (between a "fossil magnet" shown by arrow and a north line) points to a former magnetic pole position. Continent positions must be adjusted until ancient pole positions for two continents coincide, as shown in upper diagram. Outlines of continental margins also help, of course. (m.y. = million years.)

> decisive reasons why polar wandering could not be true when it was weakly supported by paleoclimatic evidence, and now that rather convincing paleomagnetic (i.e., physical) evidence has been discovered, they find equally decisive reasons why it could not have been otherwise. (From *Nature,* 77: 553–54.)

Paleomagnetic studies, as envisioned by their originator, Mercanton, not only provided a compelling test of the drift theory, but once displacements seemed proven, paleomagnetism also furnished an independent, nongeologic tool for reconstructing the predrift continents and for helping to date drift history. Most resulting reconstructions show impressive consistency with independent geologic evidence. For example, the paleomagnetic reconstructions of the continents are almost identical with those of Wegener and Du Toit (Fig. 7.7), which were based upon quite different evidence.

General Nature of the Sea Floor

Paleomagnetic studies have also forced geologists to rethink their model of the earth beneath our oceans. Until the 1960s, most geologists assumed that the crust beneath ocean basins was very old, topographically featureless, structurally tranquil, and essentially fixed in place. All these assumptions appear to be incorrect, for nowhere in deep oceanic sediments has extensive drilling dis-

covered fossils older than Jurassic. To our utter amazement, the continents turned out to be much older than the sea floor!

Profiles established by reflection of low-frequency sound waves from the sea floor and from buried layers beneath have shown that the ocean floors are anything but smooth. Broad oceanic ridges, deep trenches, escarpments, and countless submerged seamounts characterize it instead. Indeed, the pristine surface of the oceanic crust is more rugged than most continental areas, although sedimentation has smoothed the oceanic topography in some areas by burying original irregularities.

Rates of Continental Erosion and Oceanic Sedimentation

North America is now being eroded at a rate that could level it in a mere 10 million years, or to put it another way, ten North Americas could have been eroded since middle Cretaceous time 100 million years ago. If all present continents were reduced to present sea level and their refuse spread uniformly over the abyssal plains, a layer of sediments about 300 meters thick would result. The observed average total thickness of deep-sea sediments is only about 600 meters, an amount equal to the erosion of the present-sized continents *only twice during the past 200 million years.* It is clear that, in the past, the rate of erosion and/or the volume of land above sea level has, on average, been much less than now.

The apparent youthfulness of the entire present deep-sea floors came as a great shock, for, as we saw in Chap. 6, basaltic oceanic crust is chemically more primitive than granitic continental crust. Now we discover that not only do continents apparently contain the only record older than 200 million years but also there is a lot less deep-sea sediments than what we expected after such long erosion of continents. The evidence forces us to think that both deep-sea sediments and oceanic crust have been removed over the eons of geologic time.

Ocean Ridges

The most striking features, especially of the Atlantic and Indian Ocean floors, are submarine ridges (Fig. 7.15). Their symmetrical positions invite speculation that they are scars of a predrift configuration of the crust; for example, the Americas moved west, whereas Europe and Africa moved east, away from the mid-Atlantic ridge.

Oceanic ridges are made up mainly of basaltic lavas, as evidenced by islands along their crests, by samples dredged from their submerged portions, and from underwater photographs (Fig. 7.16). They have several important characteristics that make them unique, major structural features of the crust (Table 7.1). In addition to volcanic activity, ridges display only shallow seismicity beneath their axes (unlike trenches, where quakes can be either shallow or deep). Ridges are also characterized by greater than average heat flow through the crust along their axes. They clearly are zones where much subcrustal thermal energy is released, but the stress conditions that cause this energy release are different from the stress conditions that prevail beneath *magmatic arc-trench systems.*

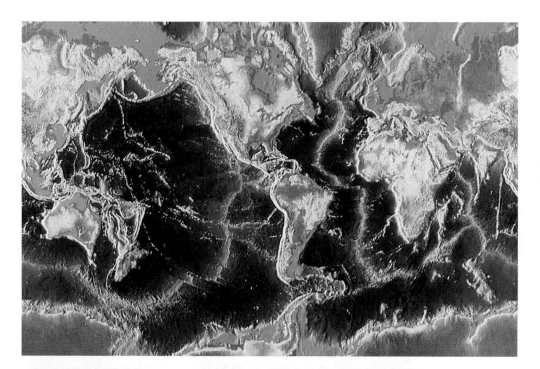

Figure 7.15 World's ocean-floor topography, revealing the roughness of much of the sea floor. Note the ocean ridges, linear escarpments, and deep trenches. Because it is less masked by sediments and vegetation than the continents, this topography reveals much more about the underlying structure of the crust. *(Satellite image courtesy Woods Hole Oceanographic Institution.)*

Figure 7.16 Submarine ellipsoidal ("pillow") lavas in the Galapagos rift zone off the coast of Ecuador. *(Courtesy Woods Hole Oceanographic Institution.)*

Another peculiarity shared by most ridges is a narrow depression that extends along their axes for thousands of kilometers. In Iceland, the largest exposed oceanic ridge area on earth, prominent elongate downfaulted structures called **rifts** are conspicuous across the center of the island from north to south (Fig. 7.17), and thirty active volcanoes occur along this zone. About 1960, American B. C. Heezen noted a close parallel of scale and morphology with African rift structures and suggested a similar origin. The Carlsberg Ridge of the northern Indian Ocean even passes into the continental African rifts (Fig. 7.18). In the 1970s, small research submarines allowed direct observation of many fissures and hot springs along many ridge crests. Prevalence of topographic rifts, together with seismicity and volcanism, all suggest—by analogy with structures on land—that

oceanic ridges are zones of extension along which the crust is being torn open. Ridge patterns may be complicated by branching. Any juncture of three separate rifts is called a **triple junction** (Fig. 7.18). The simplest case involves three branching ridges, but the arms of triple junctions also may be trenches or major faults.

The East Pacific Ridge is unique in being asymmetrically located in its ocean basin. It extends northward to or beneath the western edge of the North American continent (Fig. 7.15). Significantly the crust and mantle there are geophysically anomalous and display evidence of unusually widespread extensional structures in the western United States, suggesting that the ridge has disturbed the edge of the continent there much as the Indian Ocean ridge system has affected eastern Africa.

Figure 7.17   Thingvellier rift, or graben, in southern Iceland, marking the surface expression of the rifting of the mid-Atlantic ridge. Scarp is about 40 meters high. Volcanic eruptions and earthquakes are common in Iceland because the entire island is a subaerially exposed segment of the mid-Atlantic ridge (see Fig. 1.5). (© *John S. Shelton*.)

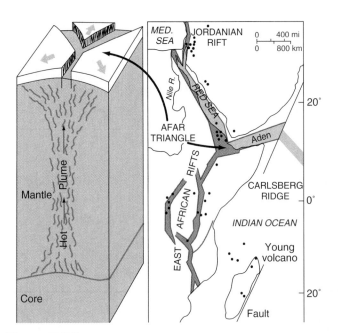

Figure 7.18   The East African rift system. The Afar Triangle is a triple-junction intersection of Red Sea, Aden, and East African rifts. It is probably the surface expression of a hot mantle plume. Such a system of diverging rifts initiates new continental drifting episodes. Dots indicate young volcanoes. (*Adapted in part from Levin,* The earth through time, *copyright © 1978 by W. B. Saunders; by permission of Holt, Rinehart and Winston.*)

Sea-Floor Spreading—a Breakthrough

Recall that, in 1928, British geologist Arthur Holmes presented a hypothesis of convection in the mantle as a cause of mountain building (Fig. 7.9), and in the same year Alfred Wegener accepted convection as one possible mechanism for continental drift. Since

that time, convection has been the most widely endorsed mechanism to explain large-scale tectonic features. In 1962, Princeton University geologist H. H. Hess proposed a bold new hypothesis that two opposing thermal convection cells rising beneath ocean ridges cause the abnormal heat flow observed there and produce tension in the crust, resulting in rifting. As rifting occurs, earthquakes are generated beneath ridges, and both hot water and new crustal material are erupted at ridge axes. Hess envisioned that, finally, the slow, convective flow laterally away from ridge axes carries older oceanic crust along as if on a conveyor belt, causing the spreading of sea floors through time.

The *hypothesis of sea-floor spreading* postulates a relatively recent origin of the Atlantic and Indian Ocean basins (and their crusts), born when former continents sitting atop rising convective cells were disrupted by rifting. Then the dismembered continental fragments separated from each other as juvenile oceanic crust was generated between them (Fig. 7.19). Eastern Africa and Arabia are today experiencing the beginning of a new phase of such disruption in response to the shift of mantle convection patterns that brought rising cells beneath that region. Spreading along the extension of the western Indian Ocean or Carlsberg Ridge tore open the Aden–Red Sea rift in the crust (Figs. 7.18 and 7.20). This African rifting during the past 25 million years typifies the initial breakup of huge supercontinents—the earliest stage in plate tectonics.

Rifts and Hot Mantle Plumes

A recent hypothesis of continental breakup involves postulated *hot mantle plumes* rising from the lower mantle (Fig. 7.18). New, sophisticated seismological studies are revealing complex heterogeneities in the deep mantle, which are thought to reflect a deep, global convective system. Plumes seem to originate from that system and are reflected at the surface by volcanic centers that are associated with domed uplift areas. These domes erupt peculiar alkaline-rich lavas. Plumes are not associated with mag-

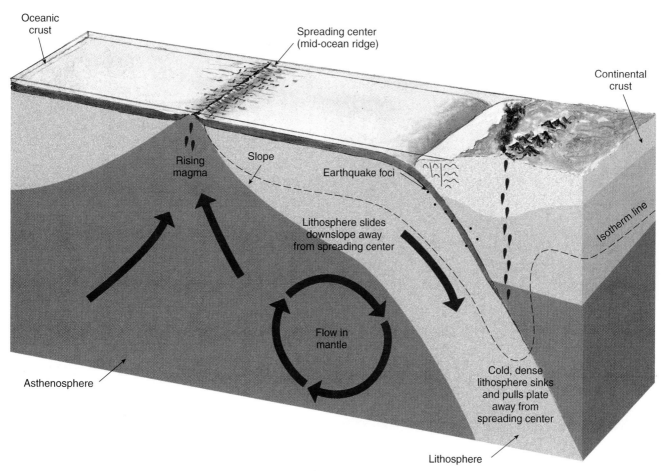

Figure 7.19   The hypothesis of sea-floor spreading showing its combined explanation of ocean ridges (divergent) and arc-trench (convergent) systems. Three lithosphere plates are shown moving over the weak low-velocity zone of the upper mantle. Magmas are produced in arcs by heating along the subduction zones where earthquake foci (dots) are concentrated in the downgoing, relatively cool slab. Magmas are generated at shallower depths under hot ridges where foci also are shallower. The 1,000°C contour illustrates contrast between hot upper mantle beneath ridges and the cooler region beneath arcs.

matic arcs and may or may not be associated with oceanic ridges. About 200 possible late Cenozoic plumes have been identified for the earth as a whole, with Africa being especially well endowed. The Hawaiian Islands are the best known example in oceanic crust.

The suggested mechanism by which rising hot mantle plumes drive rifting and continental drifting is as follows (Fig. 7.18). First, a continent is situated over a deep mantle "hot spot." Second, the uplift of the flat, brittle continental plate results in the formation of a triple junction with three radiating cracks; swarms of radiating igneous dikes are commonly associated. Third, as two of the three main cracks get wider, two arms of the trio tend to open up to form a new ocean basin. The third (or failed) arm, which is nearly perpendicular to the opening seaway, becomes a sediment-filled trough, called **aulacogen** by Russian geologists. It gradually becomes stable and is buried.

The southward-trending East African rift system seems to be an aulacogen of the Red Sea–Aden spreading new ocean. Although the East African rifts are tectonically active, they have not opened up as has the Red Sea–Aden system, because they have

continental crust beneath, rather than oceanic. Many great rivers flow down failed-arm troughs—for example, the Mississippi, Rhine, and Amazon. The aulacogen troughs commonly hold important petroleum resources because thick sediments accumulate rapidly in them and organic-rich deposits are commonly contained within these. Ore deposits also may be formed where hot solutions rise along the faults bordering the troughs.

Age Distribution of Oceanic Islands and Seamounts

About the same time that Hess first stated his concept of sea-floor spreading, Canadian geophysicist J. Tuzo Wilson postulated that oceanic islands, which are practically all volcanic in origin, tend to be symmetrically distributed as to relative age outward from submarine ridges. Youngest islands tend to be at or near the axes of ridges, whereas those more distant are progressively older. Moreover, still farther removed from ridge axes are submerged volcanic seamounts at progressively deeper levels. Many of these mountains were islands in the past, as evidenced by dredgings of

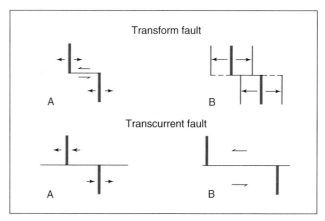

Figure 7.21   Contrast between motion on oceanic transform faults and transcurrent (or strike-slip) faults. Wide red lines are spreading ridge axes. A and B show two different stages for each case. *Upper:* Plates are spreading away from the ridge axis, and the transform fault connects two offset segments of that axis. Segments of adjacent moving oceanic crust slide past one another along the transform in the sense shown by arrows *while spreading proceeds.* *Lower:* In this case, sea-floor spreading ceased before A and then a transcurrent fault cut across the ridge. Between times A and B, the dead ridge was offset, as shown by arrows—in the *opposite* sense of displacement from that of the transform fault.

Figure 7.20   Gulf of Aden and southern Red Sea from a *Gemini* spacecraft. Note how well Africa *(below)* could fit against Arabia *(above),* suggesting that Africa was torn away to form the new ocean basin between them and begin a new phase of continental rifting. *(Courtesy of NASA; photo No. 66-54536.)*

shallow-marine fossils from their tops. Obviously the fate of oceanic islands, once they have ceased to be volcanically active, is to be eroded and to subside beneath sea level. Wilson argued that sea-floor spreading accounts for the symmetrical age distribution of these mid-ocean islands. He noted that sea-floor spreading also accounts for different types of tropical coral reefs, first described by Charles Darwin over 150 years ago. Active volcanic islands have fringing reefs ringing them, whereas doughnutlike coral atolls have a central lagoon instead of a volcano. Darwin had postulated that, when the central volcano became extinct, it subsided beneath sea level, but its reef kept growing upward to form an atoll. Wilson showed that the distribution of youthful fringing and older atoll reefs correlates well with the age distribution of volcanoes relative to oceanic ridges, as predicted by sea-floor spreading.

Major Oceanic Escarpments and Transform Faults

Exploration in the northeastern Pacific basin following World War II revealed a group of long, east-west-trending submarine escarpments explicable only as great zones of faulting (Fig. 7.15). Such zones have been found elsewhere, too, and seem to be character-

istic of sea floors. In general, they are perpendicular to oceanic ridges. In the mid-Atlantic, for example, the ridge crest has been offset along many such zones. The Indian Ocean floor displays an amazing array of escarpments, suggesting wholesale fragmentation of the crust along largely north-south fractures that look like railroad tracks, along which India traveled northward (Fig. 7.15).

With important exceptions, such as the San Andreas fault of California, such long, nearly straight fractures are rare on continents. Indeed, most of the Pacific fault zones terminate at the continental margins; thus, they reflect structural processes almost entirely restricted to oceanic crust. That the dominant movement along these zones is lateral, or transcurrent (that is, blocks move past each other horizontally rather than vertically), is indicated by offsets of the ocean-ridge crests. From the apparent offsets of the mid-Atlantic ridge near the equator, it appeared at first that the motion there was all toward the left (westward) because the north side of each fault seems to be displaced left (westward) as viewed from the south. Cumulatively such movement seemed to account for the great mid-latitude bend of the ridge.

Some of the faults terminate abruptly at the ridge crest, however, and a different interpretation of movement became possible in light of the sea-floor spreading hypothesis. In 1965, J. Tuzo Wilson conceived of a mechanism that produces what he termed **transform faults.** By this scheme, the faults formed during spreading have movement *opposite* that implied by mere apparent offset of ridge crests (Fig. 7.21); the faulting accompanies spreading, and lateral motion is "transformed" at ridge crests. Apparently such faults owe their origin to the fact that spreading rates cannot be uniform all along a spreading ridge on a sphere (Fig. 7.22).

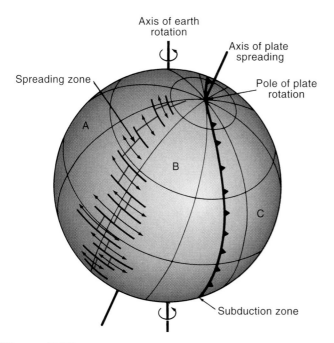

Careful studies of seismic records for earthquakes along ridge fractures confirm that the motions are indeed transform in nature. Besides connecting ocean-ridge segments, transforms may also connect ridge and volcanic-arc segments, or two arc segments. Furthermore, they may be members of triple junctions.

Confirmation of Sea-Floor Spreading

Sea-Floor Magnetic Anomalies

Parallel linear magnetic anomalies (deviations from normal magnetic intensity) unique to the oceanic crust were discovered in 1961. These were first mapped in the northeastern Pacific Ocean and then across the mid-Atlantic ridge south of Iceland (Fig. 7.23). Surveys elsewhere show a consistent pattern of narrow, alternating anomalies associated with all ridges. Near Iceland, the stripes show two striking features: first, they parallel closely the ridge axis; second, they have a remarkable bilateral symmetry such that those on one side of the axis tend to mirror those on the other (Figs. 7.23 and 7.24). The same symmetry has been found elsewhere.

What do these anomalies mean? They were mapped by ships or planes towing magnetometers along repeated traverses that crisscross the ridges. Resulting data recorded variations in total intensity of the magnetic field across the sea floor. Two interpretations suggest themselves. First, the stripe patterns might represent alternating zones with extreme contrasts of magnetic susceptibility and remanent magnetism. The magnitude of the anomalies, however, would require improbable alternations of very narrow, deep blocks of nonmagnetic and very strongly magnetic materials; the necessary geometry and susceptibility contrasts are so unlikely

Figure 7.22   Geometric relationships of a spreading ridge axis, transform faults, and subduction zones on a spherical earth. The lithosphere plates A, B, and C move with respect to an imaginary pole of plate rotation or spreading, rather than the earth's rotational pole. Transform faults are perpendicular to the spreading axis (i.e., parallel to imaginary lines of latitude about the plate rotation pole). Rate of spreading, indicated by relative length of arrows, increases from the pole to the equator of spreading. *(Adapted from Heirtzler, 1968, Sea-floor spreading, copyright © by Scientific American, Inc. All rights reserved.)*

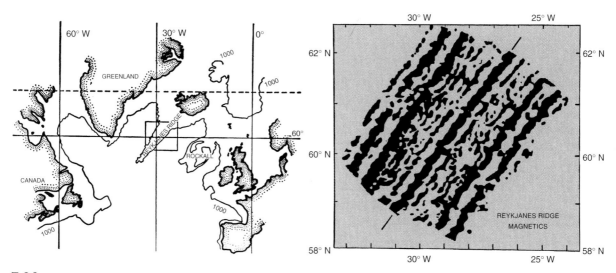

Figure 7.23   "Zebra-stripe" magnetic anomaly map showing alternating black (high magnetic intensity) and tan (low magnetic intensity) bands for oceanic crust along the mid-Atlantic ridge southwest of Iceland (in line with the rift of Fig. 7.17). It was the strikingly symmetrical relation of such anomalies to ridge axes that led to the hypothesis that these anomalies reflect successive reversals of polarity of the magnetic field while the sea floor was spreading away from the ridges. *(From Heirtzler et al., 1966, by permission of Deep Sea Research.)*

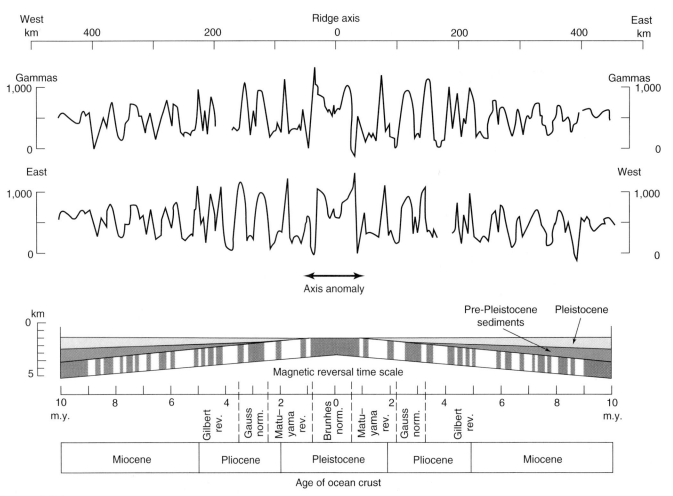

Figure 7.24　Ocean crust magnetic anomalies plotted as profiles of measured magnetic intensity along straight-line crossings of the South Pacific–Antarctic oceanic ridge. Note symmetry of anomalies on either side of the broad ridge-axis anomaly (upper curve is same as middle, but reversed to underscore the symmetry). Note also sharp contrasts of magnetic intensity peaks, which represent alternating "zebra-stripe" patterns of anomaly maps such as Fig. 7.23. Polarity-reversal episodes recorded in crust are idealized at bottom in gray-and-white bar with sediments shown above. *(From Pitman and Heirtzler, Science, v. 154, December 2, 1966, pp. 1166–1171; copyright © 1966 by the American Association for the Advancement of Science.)*

as to boggle the mind. The second, preferred, alternative is that the anomalies represent polarity reversals in successive bands of rock materials of uniform magnetic susceptibility. Thus, the dark bands in Fig. 7.24 are thought to represent normal polarity relative to the present earth field, and the blank ones reversed polarity. This explanation is consistent with the independent evidence of polarity changes of the earth's field discussed in Chaps. 4 and 6.

Magnetic-Polarity-Reversal Time Scale

As early as 1905, it was found that polarity in magnetically susceptible minerals in some late Cenozoic lavas was opposed to that of the present magnetic field. In the 1920s and 1930s, examples of reversals were documented in Cenozoic lavas from several continents. The last reversal, which produced the present field configuration, occurred in Pleistocene time nearly 1 million years

ago (Fig. 7.25). In thick sequences with many Cenozoic lavas, a number of successive reversals have been documented, and the rocks now can be dated isotopically, thus providing numerical calibration for the geomagnetic-reversal time scale described in Chap. 4. Similarly polarity studies have been made of deep-sea sediments from submarine cores, which can also be dated with fossils, and the results compare well with those for lavas (Fig. 7.25). As magnetically susceptible sedimentary minerals settled to the sea floor, they assumed a remanent magnetism imposed by the earth's magnetic field; thus, as in lavas, they preserve clues to past polarity.

Reversals of polarity of the earth's field have been important phenomena, with universal effects on minerals forming anywhere in the world in various types of rocks over at least the past 400 million years (see Fig. 4.24) and, we assume, much longer. The period between reversals has varied enormously. During late

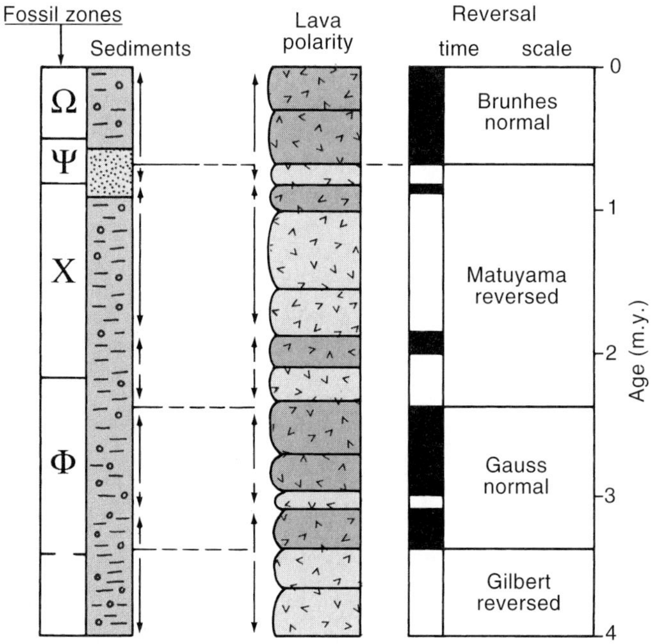

Figure 7.25   Diagrammatic representation of magnetic polarity reversals observed in late Cenozoic lavas and sediments in many parts of the world. Reversals in lavas can be dated by isotopic methods and in deep-marine sediments by fossils (fossil zones designated by Greek letters). Note the close correspondence between reversals in each. Polarity studies have led to compilation of a formal magnetic-reversal time scale, part of which is shown at right (compare Fig. 4.24). *(Adapted from Cox et al., 1963, Nature, v. 198, p. 1049; Opdyke et al., Science, 1966, v. 154, p. 349.)*

Cenozoic time, for which we have the best data, reversals have been occurring with frequencies between 10,000 and 1 million years. Relatively short polarity episodes also characterized earlier Cenozoic time, but late Mesozoic and late Paleozoic times had much less frequent reversals. Although the reversing process, which is instantaneous on the scale of geologic time, is not fully understood, we presume that this behavior results from temporary instability in the fluid outer core of the earth (see Fig. 6.9).

Discovery of an apparently universal polarity-reversal history for the earth's magnetic field represented a major breakthrough in geologic knowledge. It was suggested that symmetrical oceanic magnetic anomalies represent fossilized polarity episodes associated with the reversal history. Their pattern greatly strengthened the sea-floor-spreading hypothesis, for the anomalies apparently recorded each polarity-reversal event as the sea floor was spreading: As new crust formed in the central ridge rifts and cooled to the temperature at which magnetism becomes fixed, the crust acquired and retained the polarity of the earth's field at that time. The distance of each anomaly pair from a ridge axis would, therefore, be proportional to age (i.e., the time since the two members of the pair lay at the axis). The hypothesis that the anomalies are linked with known magnetic polarity reversals was strengthened by the fact that deep drill holes encountered fossiliferous latest Cenozoic

sediments resting upon oceanic basement on the central ridges but progressively older strata resting upon basalt outward toward the continents (Fig. 7.24).

Agreement between different kinds of evidence always strengthens a scientific hypothesis. Thus, the agreement between age patterns of volcanic islands, coral reefs, and magnetic anomalies as well as the existence of progressively younger sediments deposited on basaltic oceanic crust as one approaches ridge axes, greatly strengthened the argument for the sea-floor-spreading hypothesis.

Spreading History

The oceanic magnetic anomalies were likened by F. J. Vine to a tape recorder attached to a sea-floor conveyor belt. With D. H. Matthews in 1963, Vine proposed the relationship between the anomalies and polarity reversals and spreading. These British workers showed that the anomalies can be correlated between several different ridge systems. Vine inferred that a worldwide reversal history has been "taped" with great fidelity, and by comparing anomalies outward from ridge axes (Fig. 7.24) against the known polarity-epoch time scale established on land (Fig. 7.25), both average rates and an estimate of duration of spreading can be determined for different ridges. Resulting apparent rates of total spreading vary from about 2 centimeters per year south of Iceland to more than 18 centimeters per year in the South Pacific, with the average being 4 to 5 centimeters per year. Apparently the Pacific is now spreading almost ten times faster than the Atlantic.

Knowing the polarity reversal history from lavas on land, we can extrapolate out from ridge-crest regions, which represent only the past few million years, and estimate the total duration of spreading recorded in the anomalies across a given ocean basin (Fig. 7.26). Such analysis provides an estimate of the minimum age of the ocean basin. For example, spreading from the mid-Atlantic ridge apparently began between 150 and 200 million years ago, that from the northwestern Indian Ocean ridge began between 80 and 100 million years ago, and Australia and Antarctica did not separate until 65 million years ago. These figures match closely estimates of the time of continental separation that are based upon other evidence. They also suggest that little, if any, of the present oceanic crust can be as old as the Paleozoic Era! Much of the western Pacific and some of the northern Atlantic crust is Mesozoic, but most of the rest of the present oceanic crust is Cenozoic, which is confirmed by fossil evidence now available from the deep seas.

Extrapolations of the kind just mentioned are approximations only. In reality, spreading patterns vary somewhat along even the same ridge. Authorities also believe that the ridge-axis spreading sites themselves shift laterally as spreading occurs; sometimes they jump suddenly. On the East Pacific Ridge, only Pleistocene sediments have been found on the crest, whereas Pliocene and Miocene deposits occur on the flanks. A late Mesozoic period of active spreading apparently was followed by a mid-Cenozoic slackening, in turn superseded in Miocene time by the present spreading cycle.

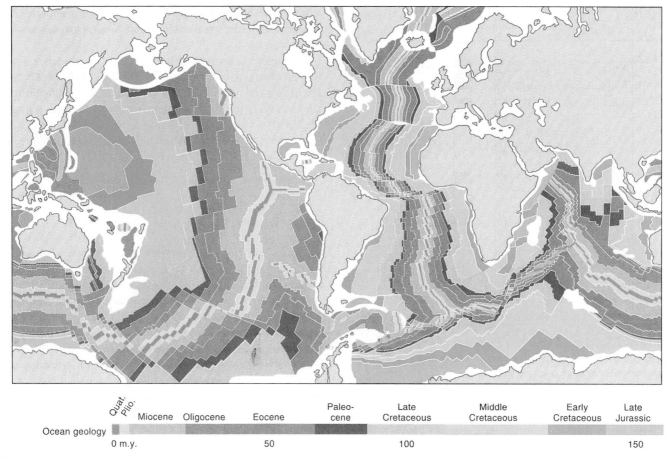

	Quat. Plio.	Miocene	Oligocene	Eocene	Paleo-cene	Late Cretaceous	Middle Cretaceous	Early Cretaceous	Late Jurassic
Ocean geology									

0 m.y.   50   100   150

Figure 7.26   Age of the oceanic crust based on magnetic anomalies and deep-sea drilling. Notice the intense red, orange, and yellow colors that mark the youngest rocks recently erupted along the mid-ocean ridge crests. The oldest rocks (blues and greens) are found at the edges of the Atlantic and along the western Pacific. Note how the rocks of similar age (= color) are offset by transform faults. *(From R. L. Larson, W. C. Pitman III, et al.,* Bedrock geology of the world, *1984. NY: W. H. Freeman.)*

Plate Tectonics

On a global scale, we find a surprisingly simple contrast in the major tectonic features of the crust. Oceanic ridges and the volcanic arc-trench systems are overwhelmingly the most seismically active zones today (Fig. 7.27). In the mid-1960s, it was argued that earthquakes define boundaries of huge, rigid plates on the outer earth. The structure of island arcs and ancient orogenic belts had long suggested great zones of dominant compression, and the oceanic ridges suggested extension; in the 1960s, such motions were confirmed by earthquake motions. The broad areas between these zones of compression and extension are relatively stable (Fig. 7.27 and Table 7.1). As the sea floors spread, continents apparently were swept along passively, as if carried on conveyor belts. Where continents are firmly coupled to adjacent spreading oceanic material, the edge of the continent is termed a **passive margin** because it is not being actively deformed. The Atlantic edge of the Americas is such a margin (Figs. 7.19 and 7.28).

In 1968, seismologists at Columbia University drew together sea-floor spreading and continental drift into a broad, unifying theory called **plate tectonics.** Based largely upon new seismic data from all over the world, this theory argues that the seismic zones (Fig. 7.27) define six large (and at least six small) **lithosphere plates** comprising the earth's outer 200 kilometers or so (Fig. 7.28). Each plate seems to behave as a rigid unit—moving away from a neighbor at an ocean ridge—a *divergent plate boundary*—and toward another plate along a volcanic arc and trench—a *convergent boundary.* Different plates also may meet either at triple junctions or at transform faults.

Not only do motions during earthquakes as well as folds and thrust faults seen in island arcs and young mountains suggest convergence, but since the 1930s it has been known that the deeper quakes (>100 kilometers) associated with trenches define a zone dipping 30 to 45° beneath an arc (Fig. 7.19). A recently detected speedup of seismic waves beneath this inclined seismic zone indicates the presence there of a huge slab of more rigid, apparently colder rock material extending downward into the mantle to a maximum depth of 700 kilometers. A 50-year-old idea of an inclined zone of shearing as much as 100 kilometers thick beneath volcanic arcs was thus confirmed, and the acknowledgment of underthrusting—called **subduction**—of a slab of oceanic material approximately 100 kilometers thick beneath the arc became inescapable. Thus, divergence of oceanic plates at ridges is balanced by subduction of the front of the plate at its convergent

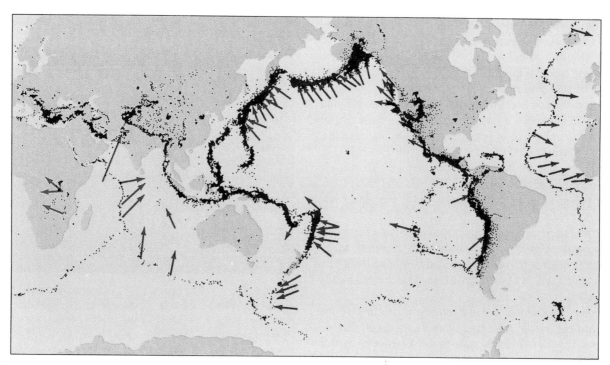

Figure 7.27 Distribution of earthquake epicenters between 1961 and 1967. Most active volcanoes also lie along the narrow zones of seismicity shown here. Arrows indicate direction of horizontal motion during earthquakes as determined from seismograph records. *(After M. Barazangi and J. Dorman, February 1969, World Seismicity Maps compiled from ESSA, Coast and Geodetic Survey, Epicenter Data, 1961–1967: Bulletin of the Seismological Society of America, vol. 59, no. 1.)*

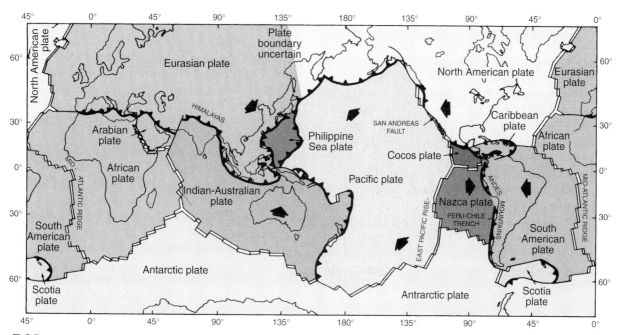

Figure 7.28 Major lithosphere plates defined by the zones of active seismicity shown in Fig. 7.27. Arrows summarize directions of plate motion, which confirm the hypothesis of sea-floor spreading by showing divergence or extension from ocean ridges and convergence or compression at volcanic arc-trench (subduction) zones. (Compare Fig. 7.15.)

margin. The subducted cold material is gradually heated and probably is very slowly assimilated into the mantle below 700 kilometers. Heating causes local partial melting in the crust above the subduction zone (Fig. 7.19). The resulting magma rises through the crust; some crystallizes below the surface, and some is erupted from the volcanoes of the arc. Most of the arc magmas are not basaltic, the way they are on ocean ridges, however, for the arc magmas are richer in potassium, sodium, and alumina. The resulting volcanic rock is called **andesite** (see Box 6.1). The andesitic arc magmas are "continental" in average composition, so that the accretion of new continental rock material really *is* occurring in young orogenic belts, just as old-timers postulated years ago.

Rare metallic elements also tend to be concentrated by plate-tectonic processes. Economically valuable ores of metals, which represent rare enrichments of elements that constitute but a trivial fraction of the earth as a whole, have been concentrated locally at plate margins.

Seismology has shown that the moving plates involve not only the crust but, rather, the entire lithosphere. Recall that the lower zone of the lithosphere, because of the decrease of seismic velocity within it, must be of lower strength than the plate above or the deeper mantle below. We must now think not of the crust as the most fundamental physical unit of the outer earth but, rather, of much thicker lithosphere plates, all in motion relative to one another on a worldwide scale.

What Drives Them?

The driving of the plates is attributed primarily to convection—the old panacea—within or below the low-seismic-velocity zone of the lithosphere. The two-times-greater heat flow observed at ridge crests as compared with deep-sea trenches supports the idea of convection in the mantle. An illustrative analogy for plate motions due to convection is provided by lava lakes in Hawaiian craters (Fig. 7.29). Dense, chilled lava crusts form on the surface and tend to sink, allowing less dense, still molten convecting lava to rise as ridges. Lava in these ridges becomes chilled, is pushed aside, and sinks slowly into the underlying molten pool to be reassimilated.

Geophysical studies of the stress across the interiors of the relatively rigid lithosphere plates indicate that the plates are now experiencing internal compression. The large but thin plates composed of rather weak rocks are like huge sheets of tissue paper, so it seems improbable that they can be simply *pushed* outward from ridges. Rather, they seem to be *dragged* passively along, piggyback fashion, by convection within the underlying mantle and then *pulled* by gravity downward along subduction zones (Fig. 7.30).

Plate Collisions

Recall that present rates of erosion should be able to reduce continents to sea level in a mere 10 million years; therefore, the persistence of continents requires some rejuvenation mechanism.

Figure 7.29 The lava lake in Kupaianaha crater near Kilauea, Hawaii, is a good analogue for plate tectonics. As fresh molten lava rises by convection and chills, similar to a mid-ocean ridge (foreground), slightly older chilled lava (darker) is shoved aside to sink in a compressive zone (background), analogous to a subduction zone. *(Courtesy of Peter Kresan.)*

The *principle of isostasy,* first proposed 100 years ago, provides the answer. Based upon gravity measurements, it was reasoned that high mountains are buoyed up at depth hydrostatically, according to Archimedes' principle. The gravitational buoyancy is thought to arise in the same way as the support of an iceberg by water displaced by the great mass of ice below the water surface. Isostasy postulates buoyant equilibrium among the larger units of the earth's crust according to their total mass. Thinner and/or more dense units (such as ocean basins) stand lower, whereas thicker and/or less dense ones (such as continents and mountains) stand higher. Furthermore, if a large load of sediment or ice were added, the crust would sink, whereas if some rock were removed by erosion, the crust would rise.

Isostasy predicts that the elevation of any large segment of the crust is directly proportional to the thickness and inversely proportional to the density of the crust. From seismic evidence, it is clear that continental crust is six to eight times thicker than oceanic crust (30 to 40 kilometers versus 5 kilometers) and is slightly less dense (2.7 versus 2.8 g/cm^3). Sure enough, thicker and less dense crust *does* stand higher than thinner, more dense crust. Further confirmation of isostasy comes from measured subsidence of the crust of nearly 200 millimeters (8 inches) under the load of impounded water after the building of Hoover Dam in 1935, as well as from the evidence of the depression of the crust by a huge load of Pleistocene ice sheets followed by measurable uplift or rebound of 250 meters after melting (see Fig. 1.9). Recent, very accurate measurements of elevations by lasers carried in orbiting satellites detect continuing isostatic rebound of a mere centimeter per year in areas formerly covered by ice sheets. From isostasy, we can see that a plate with a thin but denser oceanic

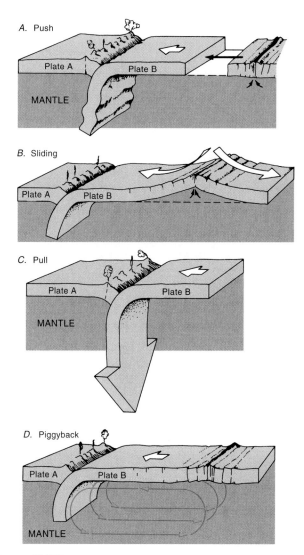

Figure 7.30 Four possible mechanisms that have been suggested to drive lithosphere plates: *A:* Push from an ocean ridge. *B:* Sliding down the slopes of an ocean ridge due to gravity. *C:* Gravitational pull of colder downgoing plate down a subduction zone. *D:* Carried piggyback on convection cells. *(After R. C. Dietz and J. S. Holden, 1973, Bulletin of the AAPG, vol. 57, pp. 2290–2296. Tulsa, OK: AAPG).*

crustal rind will sink more readily along a break than will a plate with a thick but less dense continental rind. Therefore, the subduction of oceanic material beneath continental material can be readily understood, as can the subduction of one plate with a thin oceanic rind beneath another of the same type, as is happening in the western Pacific. When two plates with thick continental rinds impinge, however, an awesome collision results. Their great isostatic buoyancy presumably would prevent either one from being subducted beneath the other.

The relative age of a subducting plate and the rate of sea-floor spreading also influence the nature of convergent plate margins. Because older plates have cooled more, they are denser and, so,

will subduct more easily than younger, warmer, more buoyant ones. Where faster rates of spreading bring younger plates to a convergent margin, the rate of subduction is faster and the angle of subduction is relatively less than with older plates (Fig. 7.31A). This results in much greater structural interaction or stress between the downgoing and overriding plates, which causes deformation and heating of the overriding plate across a very broad zone. The behavior of the leading edge of the overriding plate may also vary for reasons that are not yet fully understood. In some cases, rock is torn from this edge and subducted; it may then be underplated to the overriding plate. In other cases, ocean-floor sediments and igneous rocks are *scraped off* the subducting plate to be plastered onto the snout of the overriding one.

In its perverse way, nature has defied simple logic in several cases. There is clear evidence that thick slices of heavy oceanic material have been shoved up over lighter continental crust in a number of areas. Because this process is the opposite of subduction, it has been called **obduction** (Fig. 7.31B). Its cause is not clear. In a few other cases, continental material has been subducted beneath a continent (Fig. 7.31C). The best example of this is southern Asia, where seismic evidence indicates a doubling of the usual thickness of continental crust beneath the high Tibetan Plateau and the Himalayan Mountains. Apparently some of India's crust was driven beneath Asia, causing the unprecedented uplift of this "roof of the world."

Collisions between two continental masses are of special importance because such collisions are the major causes of mountain building. Along any old orogenic belt that now lies between two cratons, it appears that two continents collided—for example, the Himalayas (Fig. 7.31C). In these cases, intensely compressed and metamorphosed ultramafic rocks and basalts commonly define a **suture zone** where two plates collided. Such zones are all that is left at the earth's surface of a former ocean that separated the two continents before collision. Except where local obduction has occurred, *these narrow, mangled zones in continents are the only relics we can expect of any sea floor older than about 200 million years.* We find, then, that *convergent plate boundaries are destructive* and that staggering areas of the earth's surface of the past have completely disappeared. It is inescapable that several hundred billion cubic kilometers of lithosphere have been consumed in the past 3 to 4 billion years.

Besides collisions between two continents, which create supercontinents, there may also be collisions between two arcs, between an arc and a continent, or between ocean ridges and either arcs or continents. Also, relatively small fragments of continental character called **microcontinents,** such as Cuba or Madagascar, which were rifted away from large continents, may later collide with another continent or arc. There is increasing evidence that ancient orogenic belts tend to contain a complex collage of *tectonic terranes* consisting of microcontinents and arcs all jammed together with suture zones between to produce what may be called *collage tectonics* (Fig. 7.31D). Recognizing the boundaries of these terranes, dating their respective collisions, and determining the origin of each terrane are challenging, to say the least. It guarantees work and controversies for geologists for many years

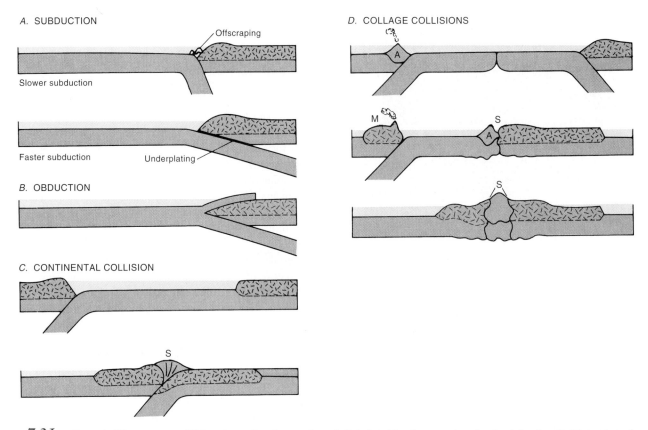

Figure 7.31   Several different types of lithosphere plate interactions. A: Relationship of rate and angle of subduction. B: Obduction of overthrusting of a slab of oceanic over continental crust. C: A continent-continent collision resulting from shrinkage of an intervening ocean as subduction proceeds. D: Collage tectonics resulting form multiple collisons: an arc-continent collision resulted from shrinkage of an intervening ocean by subduction at both of its margins followed by collision of a microcontinent (S—collision suture; A—volcanic arc; M—microcontinent).

to come. Collage tectonics is discussed and illustrated further in Chaps. 11 to 15.

Sea-floor magnetic anomalies provide an excellent record of lithospheric plate motions for the past 200 million years or so. For pre-Jurassic times, however, no such record exists (Fig. 7.26). Therefore, the reconstruction of more ancient plate positions involves jigsaw puzzle solutions based on paleomagnetic data for ancient latitudes of different plates and on study of orogenic belts with suture zones in order to delineate ancient convergent plate margins, all supplemented by paleontologic and sedimentary evidence of ancient geography (e.g., land versus sea, climatic belts, ancient oceanic currents, and wind directions).

Subsidence of Plates and the Accumulation of Sediments

The accumulation of great thicknesses of sediments over large regions is one of the prominent features of earth history and must be satisfactorily explained by any successful tectonic theory. Areas of exceptionally thick sediments vary greatly in size, shape, and tectonic setting but are collectively referred to as **sedimentary basins** (Fig. 7.32). The accumulation of thick sediments requires either that sea level rose or that the underlying lithosphere subsided during deposition—or both. Although im-

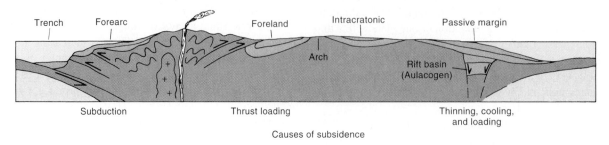

Figure 7.32   Six major types of sedimentary basins shown in their plate-tectonic settings. The major physical cause or causes of subsidence for each case are shown below the diagram.

portant world-wide rises of sea level can be demonstrated, many sediments are too thick to be explained entirely in this way. Therefore, we must consider several possible causes of subsidence, all of which are at least indirectly related to plate tectonics. Later we shall look at examples of these mechanisms from the geologic record.

Causes of Subsidence

Subduction subsidence is caused by the profound forcible depression of one lithosphere plate as it is subducted beneath another. This generally results in the formation of a long, narrow, deep topographic trench on the sea floor (Fig. 7.32). If, however, a large enough supply of sediment can reach the depression to keep it filled as subduction proceeds, then a *topographic* trench will not be present. Such is the case offshore from Washington and Oregon today, where there has been abundant sediment supplied by the Columbia River for millions of years.

Thermal, or *cooling, subsidence* is caused by the cooling of a lithosphere plate as it moves away from a hot, spreading ridge (Fig. 7.19). The initially hot material becomes more dense as it cools, and the resulting increase of density causes isostatic subsidence. This creates space for the accumulation of sediments on top of the plate. In a typical ocean basin, there is far more space below sea level than there is sediment to fill it, so the water remains deep. On a newly rifted continental margin that was initially near sea level, however, cooling subsidence may be accompanied by thick sedimentation during the 100 to 200 million years typically required for cooling to run its course (Fig. 7.33). Prisms of strata thousands of meters thick have accumulated in this way on most passive continental margins as such margins have moved away from spreading ridges (Fig. 7.34).

Erosion of the continental interior may provide much sand and clay to such a subsiding, passive continental margin, but under favorable tropical conditions, shallow-marine limestones may accumulate instead. Plate cooling and the resulting subsidence occur rapidly at first, but at a decreasing rate thereafter, which can be calculated from the principle of isostasy. Sediment accumulation decreases at the same rate unless sea level also was changing worldwide during deposition. Obviously a rise of sea level would allow still greater thicknesses of sediment to accumulate, whereas the opposite would be true for a fall of sea level.

Crustal-thinning subsidence can provide space for thick sediment accumulation as a result of isostatic subsidence of any crust that has been thinned. A phase change in the lower crust from normal to denser material is one way that thinning may occur. Stretching of crust beneath a newly formed spreading center also may thin the crust (Fig. 7.33). Thinning and cooling subsidence typically go hand in hand.

Another cause of subsidence is *loading* of a plate from above. We know that huge glaciers have depressed continents on a large scale (see Fig. 1.9). For example, ice caps 3,000 meters thick have depressed the crust 1,000 meters. A thick accumulation of sediments can weigh down a lithosphere plate in a simi-

lar way, but the amount of *sediment-loading subsidence* is limited by the initial depth of water and the ratio of densities of the loading material and of the crust and mantle beneath. From the principle of isostasy, we know that such subsidence will cease when a gravitational equilibrium is reached between the mass of the load and that of the underlying plate. If deposition of sediment began in deep water, a considerable thickness could accumulate before such equilibrium is reached. If, however, it began near sea level, then a lesser accumulation would be possible because waves and currents would rapidly spread the sediments laterally and prevent a thick vertical accumulation above sea level—unlike the case with glacier ice. Where an enormous supply of sediment is available, however, as at the mouth of the Mississippi River, sufficient mass of shallow-marine sediment may accumulate over a large enough area of the sea floor to cause widespread downwarping of the crust as the sediment is spread laterally (Fig. 7.35). Sediment loading provides a good example of feedback, for the deposition of the sediment helps to create more space for the accumulation of still more sediment. The rate of subsidence and deposition decreases through time, however, until isostatic equilibrium is achieved.

Another mechanism of loading subsidence is *thrust loading,* which involves the lateral shoving of huge slabs of rock in a series of overlapping, low-angle overthrust faults—like a bunch of cards sheared across one another. One common result of the collision of two lithosphere plates is the pushing of such slabs hundreds of kilometers long, tens of kilometers wide, and hundreds of meters thick away from a collision zone onto the edge of an adjacent craton. Such thrust loading causes isostatic downbending of the edge of the craton, which again creates a space for the accumulation of sediments to thicknesses of several thousand meters. The effects of such loading of the cratonic margin in front of a new mountain system may extend across a width of hundreds of kilometers from the mountain front. Downflexing adjacent to the mountains produces a **foreland sedimentary basin** with a complementary bulge, or *cratonic arch,* and a *cratonic basin* beyond the arch (Fig. 7.36). Isostatic modeling of thrust loading allows one to predict the rate of subsidence and the dimensions of these flexures: As with sediment loading, the rate of subsidence—thus, also of sediment accumulation—will be nonlinear—that is, fast early and progressively slower through time until isostatic equilibrium is reestablished. The depth of subsidence adjacent to the mountain front will be roughly one-half the height of the total load of thrust slabs. The cratonic flexures will be both wider and shallower the more rigid is the underlying lithosphere, and vice versa. Therefore, an initially warm plate may first develop a deep, narrow basin with a pronounced peripheral bulge. As the plate cools and becomes more rigid through tens of millions of years, the basin will broaden and the bulge will migrate and become more subdued (Fig. 7.36).

Plate Tectonics and Sedimentary Basins

The different types of sedimentary basins shown in Fig. 7.32 are distinguished by their particular plate-tectonic settings. Each type

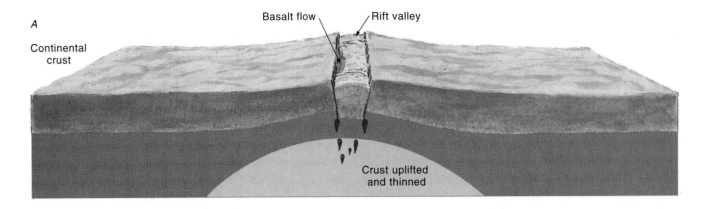

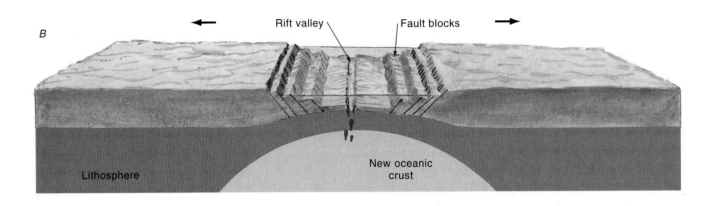

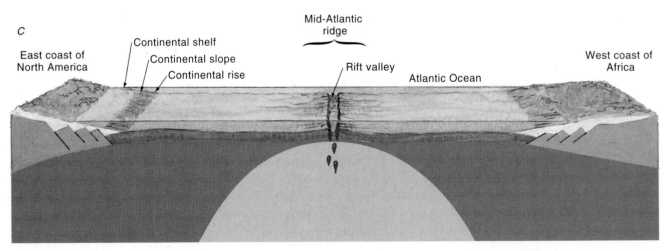

Figure 7.33   Steps in the development of a passive margin. A: Upwelling of mantle material along hot spots causes crust to bulge upwards and eventually break into a series of linear continental rift valleys. These rift valleys, like the modern East African rift (see Fig. 7.18), are filled with continental sediments and volcanics, which flow up through the fissures. B: As spreading continues, the rift valley widens until seawater flows in, forming a proto-oceanic gulf. Shallow-marine sediments are deposited on the floor of the gulf, as is happening today in the Red Sea. C: Continued spreading eventually pulls the two halves of the ancient continent farther and farther apart and produces an ever widening ocean between them. Meanwhile, the ancient rift valley and proto-oceanic gulf sequences sink down on the continental margins and are buried by deposits of the continental shelf, slope, and rise.

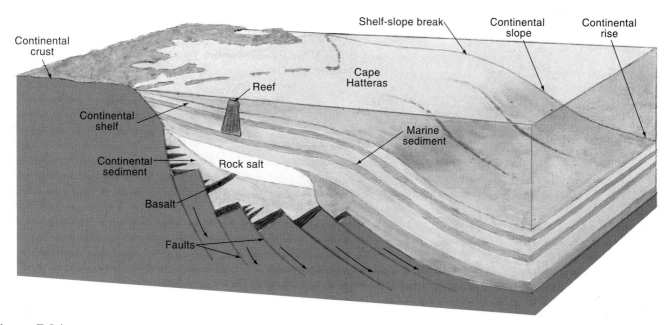

Figure 7.34   Detailed cross-section of a passive margin. In the Atlantic margin, the basement is composed of Triassic rift valley sediments in fault blocks and is overlain by marine sediments and rock salt from the Jurassic proto-oceanic gulf. During the Cretaceous and the Cenozoic, this entire package was buried by thick sequences of shelf, slope, and rise sediments. The present-day Atlantic coastal plain is still sinking and accumulating sediments.

Figure 7.35   Regional crustal subsidence produced by lateral as well as vertical accumulation of shallow-marine and nonmarine alluvial plain sediments. Through sedimentation over a large area, the crust is warped down a limited amount; the locus of active accumulation constantly shifts, spreading the load over an increasingly wide region and producing further subsidence. This has occurred on the Gulf of Mexico shelf in the region of the Mississippi delta during the past several million years.

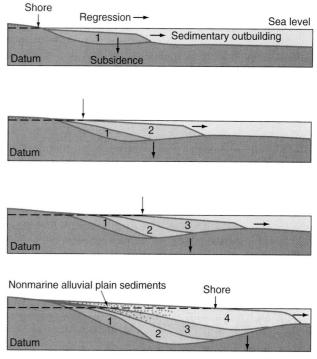

Accumulation of thick shallow-marine
and alluvial plain sediments

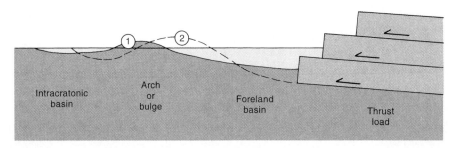

Figure 7.36   Thrust-loading susidence of a craton margin to form a foreland basin and an accompanying intracratonic basin and arch. Initial flexure forms profile 1. With time, the load approaches isostatic equilibrium and evolves to profile 2. Both the arch (or peripheral bulge) and the intracratonic basin become more accentuated vertically but also shift closer to the load. *(Modified from Quinlan and Beaumont, 1984, Canadian Journal of Earth Sciences, v. 21, pp. 973–996.)*

Table 7.2	Major Types of Sedimentary Basins	
Basin Type	**Characteristic Sediments**	**Depositional Environments**
Trench	Fine sediments overlying ocean-floor basalts	Deep-marine
Forearc	Heterogeneous gravels, sands, and muds derived from erosion of volcanic, metamorphic, and granitic rocks of the adjacent orogenic belt	Nonmarine to marine
Foreland	Heterogeneous gravels, sands, and muds derived from the orogenic belt and shed onto the continental craton; may be coal-bearing	Mostly river and deltaic
Intracratonic	Homogeneous quartz-rich sands and limestones, but may include muds, evaporites, or coal at certain times	Mostly shallow-marine, with some deltaic
Passive-margin	Quartz-rich sands and limestones passing seaward to muds	Shallow-marine shelf to deeper-marine
Rift or aulacogen	Earliest rocks volcanic overlain by thick gravel and sand; younger rocks may include evaporites and limestones	Rivers and lakes changing to shallow-marine

tends to have some identifying characteristics, although there can be considerable overlap (Table 7.2). We can now close the circle begun wherein we discussed early ideas about global tectonics and mountain building. We have shown how the theory of plate tectonics unifies many diverse aspects of geology, and we can now see how the old geosynclinal concept relates to our modern view of mountain building. It is clear that many passive or divergent continental margins have been converted to active or convergent margins through the commencement of subduction and the occurrence of various collisions. Thick strata deposited under passive conditions may be deformed and then buried beneath very different sedimentary and volcanic rocks deposited in trench, forearc, and foreland basins while active mountain building is occurring (Fig. 7.37). Therefore, the "geosynclines" of preplate-tectonic days turn out to comprise a heterogeneous mixture of very different types of sedimentary basins, which were superimposed upon one another as major changes of tectonic behavior occurred. Passive-margin basins can be followed by active-margin ones, and these in turn may be succeeded again by passive-margin deposition after mountain building ceases.

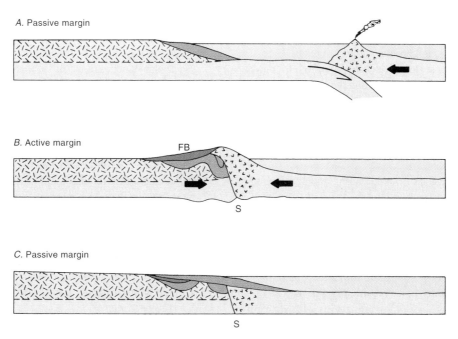

Figure 7.37   Conversion of a passive continental margin (A) into an active convergent margin by collision of an arc (B); thrust loading and erosion of mountains produce a foreland basin (FB). Cessation of tectonic activity and subsequent deep erosion of the mountains produce a new passive margin (C).

Summary

Mountain Building

- The **geosynclinal hypothesis** originally called for contraction of a cooling earth to buckle the crust at the edges of continents. A downwarped zone filled with sediment eroded from an adjacent uplifted area was called a **geosyncline**. As cooling and buckling continued, the geosynclinal sediments were intensely deformed by lateral compression and upheaved to make a mountain range. The mountain belt then became stabilized and was gradually eroded deeply. The geosyncline is today of historical interest only.

- **Continental drift** provided a compressional mechanism for mountain building both by the wrinkling of the leading edges of moving continents and by collisions of continents. **Supercontinents** resulted from such collisions; predrift reconstructions were done in jigsaw-puzzle fashion by matching similar geologic features.

- **Thermal convection** due to uneven heat generation in the mantle has been the most popular mechanism for buckling the crust since the cooling-shrinkage hypothesis of mountain building was nullified by the discovery of radioactive heating. Advocates of continental drift also have long appealed to convection to move continents. In spite of much circumstantial supporting evidence, drift was rejected by most earth scientists until **paleomagnetism** revived it in the 1950s.
  1. **Magnetically susceptible** minerals become fossil magnets when frozen parallel to the magnetic field when a rock is formed.
  2. **Remanent magnetism** can be measured and used to locate **paleopole** positions for a specimen if we assume that the present two-pole type of field has existed throughout history. **Inclination angle** of fossil magnets indicates **paleolatitude.**
  3. **Discordant paleopole positions** for rocks of the same age from different continents require continental drift to make them concordant.

Sea-Floor Spreading

- **Ocean ridges** are zones of extensional faulting or **rifting** characterized by basaltic volcanism. The Indian Ocean ridge connects with the East African rift system at a **triple junction** where three rifts meet. **Aulacogens** are sediment-filled troughs associated with initial rifting of continents at such junctions. **Hot mantle plumes** rising from the lower mantle seem to cause initial breakup at triple junctions.

- **Sea-floor spreading** postulates that new ocean floor is added as spreading occurs away from ridges. Continents may be rifted apart to drift passively as new oceans spread between. Distributions of heat flow, ages of volcanic islands, and **fringing reefs, atolls,** and **volcanic seamounts** all support the slow cooling and subsidence of ocean floor as it moves away from spreading centers.

- **Transform faults,** which offset segments of ocean ridges, move continually as the adjacent segments spread. They help to accommodate the greater widening of spreading ocean basins in equatorial than in polar regions.

- **Linear magnetic anomalies** in the oceanic crust, which are symmetrical and parallel to the ridges, record reversals of the polarity of the magnetic field as the sea floor spreads. The **polarity-reversal time scale** provides the basis both for dating spreading histories and for reconstructing the shapes of ocean basins and positions of continents during the past 200 million years while the present ocean basins have been forming. Other criteria must be used for earlier reconstructions.

Plate Tectonics

- **Plate tectonics** unifies sea-floor spreading and continental drift to provide the first truly global theory. **Lithosphere plates** are defined by the earth's major earthquake zones and plate motions seem to be driven by thermal convection in the upper mantle. Plates move apart at ridges or **divergent plate margins** and toward one another along **magmatic arcs** or **convergent plate margins.** Where the edge of a continent is not on a **tectonically active** convergent margin, but is fused to and moves with adjacent oceanic crust, we call this a **passive continental margin.**
- **Subduction** is the process of underthrusting one lithosphere plate beneath another at a convergent margin; much ocean floor is lost by subduction. Magmatic arcs dominated by **andesitic** eruptions are formed at the surface above subduction zones. Once a plate breaks and begins to be subducted, gravity helps to pull it downward into the mantle, where it is slowly heated and assimilated.
- **Obduction** is the process of local overthrusting of part of one plate over another plate at a convergent margin. Obduction is not as extensive as subduction.
- **Plate collisions** are the main causes of mountain building. Collisions may involve either two continents or varying combinations of magmatic arcs, continents, or **microcontinents.** Many orogenic belts now appear to be made up of **collages** of diverse **tectonic terranes** jammed together by successive collisions. **Suture zones** mark the boundaries of such collisions; they are characterized by crumpled mafic and ultramafic oceanic rocks quite different from adjacent ones. The collision of two or more continents forms a **supercontinent.**
- **Subsidence of plates** provides room for accumulation of thick strata in **sedimentary basins** of various kinds during periods of tens to hundreds of millions of years. Plate-tectonic processes that cause such subsidence include the following:
  1. **Subduction,** which creates oceanic trenches by forcible downbending
  2. **Cooling subsidence,** which allows thick sedimentation on a passive continental margin as it moves away from a ridge, cools, becomes denser, and sinks isostatically
  3. **Crustal thinning,** which causes isostatic sinking of a thinned crust
  4. **Sediment loading,** which involves isostatic downwarping due to the weight of sediment; this is most significant if deposition begins in deep water but is more limited if it begins near sea level
  5. **Thrust loading,** which occurs when a collision causes overthrusting of thick slabs of rock onto the edge of a craton; thick sediment can accumulate in the **foreland sedimentary basin** formed by the resulting isostatic subsidence
- **Sedimentary basins** are distinguished by their plate-tectonic settings and characteristic rock types. They include trench, forearc, foreland, intracratonic, passive-margin, and rift or aulacogen basins.

Readings

Condie, K. C. 1967. *Plate tectonics and crustal evolution:* London: Pergamon Press.

Cook, F. A., L. D Brown, and J. E Oliver. 1980. The southern Appalachians and the growth of continents. *Scientific American,* 156–68.

Cox, A., ed. 1973. *Plate tectonics and geomagnetic reversals.* San Francisco: Freeman.

———, and R. B. Hart. 1986. *Plate tectonics: How it works.* Palo Alto, Calif.: Blackwell.

Dana, J. D. 1873. On some results of the earth's contraction from cooling, including a discussion of the origin of mountains, and the nature of the earth's interior. *American Journal of Science,* ser. 3, 5:423–43; 6:6–14, 104–15, 161–72.

Dewey, J. F., and J. M. Bird. 1970. Mountain belts and the new global tectonics. *Journal of Geophysical Research,* 75:2625–47.

Dott, R. H., Jr. 1974. *The geosynclinal concept, in modern and ancient geosynclinal sedimentation.* Society of Economic Paleontologists and Mineralogists Special Publication 19, 1–13.

Du Toit, A. L. 1937. *Our wandering continents.* Edinburgh: Oliver & Boyd.

Glen, W. 1975. *Continental drift and plate tectonics.* Columbus: Merrill.

———. 1982. *The road to Jaramillo.* Stanford, Calif.: Stanford University Press.

Hall, J. 1859. *The natural history of New York, part 6—paleontology, J.3.* Albany: Van Benthuysen, Introduction, pp. 1–85.

Hallam, A. 1973. *A revolution in the earth sciences: From continental drift to plate tectonics.* New York: Oxford University Press.

Hess, H. H. 1962. History of the ocean basins. In *Petrologic studies: A volume in honor of A. F. Buddington.* Boulder, Colo.: Geological Society of America.

Isacks, B., J. Oliver, and L. R. Sykes. 1968. Seismology and the new global tectonics. *Journal of Geophysical Research,* 73:5855–99.

Menard, H. W. 1986. *An ocean of truth—A personal history of global tectonics.* Princeton: Princeton University Press.

Moores, E., ed. 1990. *Plate tectonics: Readings from* Scientific American. New York: Freeman.

Uyeda, S. 1978. *The new view of the earth: Moving continents and moving oceans.* New York: Freeman.

Vine, F. J. 1966. Spreading of the ocean floor: New evidence. *Science,* 154:1405–15.

———, and D. H. Matthews. 1963. Magnetic anomalies over ocean ridges. *Nature,* 199:947.

Wegener, A. 1966. *The origin of continents and oceans.* New York: Dover. (Paperback, English translation of 4th ed., 1929.)

Wilson, J. T. 1965. A new class of faults and their bearing on continental drift. *Nature,* 207:343.

Windley, B. 1984. *The evolving continents.* New York: Wiley.

Wyllie, P. J. 1976. *The way the earth works.* New York: Wiley.

(From: A Chesley Bonestell Space Art Chronology, by Melvin H. Schuetz,
Universal Publishers/uPUBLISH.com 1999)

Chapter 8

Cryptozoic History

An Introduction to the Origin of Continental Crust

M A J O R   C O N C E P T S

Mente et Malleo

By thought and dint of hammering

Is the good work done whereof I sing,

And a jollier crowd you'll rarely find,

Than the men who chip at earth's old rind,

And often wear a patched behind,

By thought and dint of hammering.

Andrew C. Lawson, formerly of the Geological Survey of Canada and the University of California

▶ The Cryptozoic Eon (the time before abundantly fossilized life) spans the interval between the origin of the earth 4.6 billion years ago to about 600 million years ago, or over 80 percent of earth history. Because Cryptozoic fossils are rare, correlation must be done by field criteria and isotopic dating, so our resolution of Cryptozoic events is relatively crude.

▶ Archean Era rocks (between 4.0 and 2.5 billion years old) are the products of the early differentiation of small protocontinents, which then collided and assembled into larger continents by accretion. These protocontinents are made of highly metamorphosed gneisses intruded by later granites. Between them are belts of metamorphosed proto-oceanic crust (greenstones) and their sedimentary cover of immature clay-rich sandstones (graywackes).

▶ In the Proterozoic Era (between 2.5 and 0.6 billion years ago), there were major changes in the earth's crust as full-sized continents assembled and typical plate tectonics began to operate along their margins. Thick wedges of mature quartz sandstones replaced abundant immature graywackes in many marine basins, and stromatolitic limestones and gypsum evaporites became more common.

▶ From an essentially anoxic atmosphere at the beginning, the earth gradually acquired free oxygen released by the photosynthetic activities of cyanobacteria and algae. Late Archean and early Proterozoic deposits of reduced iron in thick banded iron formations is widely regarded as evidence of no free oxygen. Other scientists argue that there is evidence that the earth's atmosphere had at least low levels of oxygen in the late Archean, about 3 billion years ago.

▶ Except for little free oxygen and frequent CO_2-caused greenhouse atmospheres, climates of the Cryptozoic were like those of the modern world. Major glaciations occurred in the early and late Proterozoic; the latter event left subtropical glacial deposits, which suggests that the earth nearly froze completely about 600 to 700 million years ago.

What did our planet look like 3 or 4 billion years ago when the first bits of permanent crust were trying to form? No people were around to bear witness—only a few mute bacteria. If you could use a time machine to fly past in a space vehicle, you would see a planet covered with countless volcanoes belching forth gases and steam all over the planetary surface, and you might see some spectacular collisions of giant meteorites with the earth's juvenile surface (Fig. 8.1). Continents, oceans, and mountain belts would not be clearly differentiated as yet, and the scattered, small areas of land would be covered with barren rocky plains and sand dunes rather than forests or grasslands. In short, there would have been precious little hint of what was to become the earth's surface as we know it today.

Cryptozoic, or Precambrian, time includes about 80 percent of total earth history—that is, from nearly 4.5 billion years to about 700 million years ago, yet for the first 500 million years or so, there is no decipherable geologic record, and it now seems unlikely that much will be found. Apparently the crust had not developed sufficiently to become permanent until about 4.4 billion years ago. In Chap. 6, we outlined the probable earliest history of our planet and some of the consequences of such a history. Beginning in this chapter, we shall concentrate attention upon North America as a model for testing hypotheses of continental development. First, we shall examine in some detail the preserved geologic record for the interval from about 4.4 billion to 0.5 billion years ago. It seems logically more sound to refer to the first three-quarters of earth history as Cryptozoic, but occasionally we shall use the more familiar term *Precambrian* (see Box 8.1).

We shall begin by discussing the Great Lakes region, where the first detailed chronology for Precambrian rocks was established more than 50 years ago. This region is used to illustrate

Figure 8.1 Artist's conception of the scenery during most of the first half billion years of the earth's existence. Much of the surface was molten, with a few cooling crustal fragments beginning to form microcontinents. Meteorite bombardment was intense, and the moon was almost twice as close, exerting enormous tidal pull on the earth's surface. The atmosphere had no free oxygen, but may have been loaded with nitrogen, methane, ammonia, carbon dioxide and water. *(From: A Chesley Bonestell Space Art Chronology, by Melvin H. Schuetz. Universal Publishers/uPUBLISH.com 1999).*

Box 8.1

How to Define the Base of the Phanerozoic?

Besides the uncertainties described in the text for defining the lowest Cambrian strata, uncertainties that have nagged geologists for years, we now have a new issue to consider regarding the Cryptozoic-Phanerozoic boundary. Beginning in 1947, fossil impressions of a remarkably diverse assemblage containing twenty-six species of animals were found in the Ediacara Hills of Australia and subsequently on most other continents. These impressions occur in strata that underlie fossiliferous Cambrian rocks without any significant break between. For the past three decades, many geologists have searched worldwide for still older animal fossils, but to no avail so far. Because Paleozoic rocks had always been distinguished from Cryptozoic ones by the presence of abundant animal fossils in the former *and* because the Ediacara strata are lithologically similar to and gradational with overlying Cambrian ones, it seems eminently logical to incorporate them within the Paleozoic. This would recognize their much greater kinship with the Cambrian System than with Cryptozoic rocks.

American P. Cloud and Australian M. Glaessner suggested in 1982 that a new Ediacarian System be formally recognized and classed as the basal division of the Paleozoic. Its base is approximately 700 million years old and directly overlies important glacial deposits. However, in 1952, Russian geologist Sokolov proposed the term *Vendian Period* for the same interval, and it has since become widely accepted by geologists working on the late Proterozoic. The Ediacarian now refers to the more recent of two epochs within the Vendian Period.

Because the Vendian is the first period of "visible life" perhaps it should be placed in the Phanerozoic Eon. If so, then we need a term comparable to Phanerozoic Eon for the three-quarters of earth history that preceded the Vendian. Earlier editions of this text advocated "Prephanerozoic" to replace the outdated concept of "Precambrian." However, this usage has not caught on. For lack of a better term, we refer to the 3.5 billion years of earth history prior to the first megascopic organisms as the Cryptozoic Eon, because this term has been widely used for over a century. Eventually stratigraphic commissions may decide to exclude the Vendian from the Proterozoic Era and Cryptozoic Eon, of which they are now a part.

methodology with no expectation that you should learn the details of the local geology. After showing the methods of developing that chronology in complicated rocks lacking any index fossils, we shall then examine the more fundamental problem of the early development of the earth's crust based upon evidence from Cryptozoic igneous and metamorphic rocks and the patterns of isotopic date provinces. Then we shall digress to consider some basic principles for the interpretation of ancient conditions from sedimentary rocks. These principles will apply not only to the Cryptozoic but also to the Phanerozoic record. Next we shall consider the sedimentary evidence bearing upon crustal development as well as upon the evolution of the early atmosphere and seawater. Finally, we shall consider the scant evidence of Cryptozoic climates. We defer the discussion of Cryptozoic life to Chap. 9.

The economic importance of Cryptozoic rocks is almost inestimable. Thousands of years ago, Native Americans mined copper in Michigan and traded it widely over the continent. Since the industrial revolution, we have derived untold wealth from these old rocks. The major source of iron ore is from peculiar *banded iron formations,* which are unique to the Cryptozoic. Other major metallic sources of gold, silver, copper, nickel, chromium, and uranium, to list but a few, also are enclosed in these ancient rocks. Perhaps half of the world's metallic mineral resources come from them.

At the outset, we emphasize that the Cryptozoic record is far more obscure than that for subsequent time. To be sure, many of the old rocks are severely deformed, metamorphosed, and deeply eroded (Fig. 8.2), but others are almost as youthful-appearing as Cenozoic ones. The most important single characteristic of the older record is its lack of index fossils, although microscopic cyanobacterial fossils now show promise for dating. As we saw in Chap. 2, the first use of index fossils for correlation and mapping was the major breakthrough that allowed the construction of a valid geologic time scale. But in Cryptozoic rocks, we have only very limited use of this important tool.

Considering the handicaps, it is remarkable what the pioneers of Cryptozoic geology accomplished in unraveling a chronology that had to be based solely upon the physical criteria of relative age first formulated by Hutton (see Box 4.2). It was chiefly those pioneers who perfected the use of such criteria to a high order. More recently, isotopic dating has revolutionized the study of the Cryptozoic even more than of younger eras. In general, it has confirmed much of the basic chronology established before by field geologic methods alone.

Development of a Cryptozoic Chronology

Sedgwick in Wales

You may recall from Chap. 4 that Adam Sedgwick first suggested that the eras be named for stages of life development. He also recognized clearly the relationships between older, unfossiliferous rocks and fossiliferous early Paleozoic ones in Wales. One of the first names proposed for Precambrian time was Azoic, meaning "without life"; subsequently Eozoic and Archeozoic ("ancient

Figure 8.2   Banded high-grade metamorphic rocks, typical of ancient shield regions, exposed by glaciation along Sondre Stromfjord, southwestern Greenland. Some of the oldest dated rocks in the world (3.8 billion years) occur near here. *(R. H. Dott, Jr.)*

life") and Cryptozoic ("obscure life") were suggested when presumed fossils were found. The most amusing term was Agnotozoic, proposed by a Wisconsin geologist who doubted that alleged fossils were truly organic.

Sedgwick recognized that Precambrian rocks in Wales are somewhat more deformed and metamorphosed than overlying Paleozoic ones, from which they are separated by an unconformity. In many areas of the world, however, this is not so, and the two are virtually identical in appearance except for a near absence of fossils in the older. Because upper Precambrian strata are not different from those of the lower Paleozoic, designation of a boundary is somewhat arbitrary where no unconformity is present. Traditionally the lowest stratigraphic appearance of Cambrian index fossils has defined that boundary, but no human being was around to paint a stripe on the rocks for us. In continuous or conformable sequences of strata, what assurance is there that Cambrian index fossils appear at a position exactly synchronous with their lowest position in Sedgwick's Welsh sequence? What if environmental or ecological factors were unfavorable for the organisms in certain areas at the precise moment when the curtain rose on Act 1, the Cambrian Period, in Wales? Moreover, in strongly deformed or metamorphosed strata, precise location of this boundary is almost hopeless. Today this boundary is selected by isotopic dating in such areas.

The Canadian Shield

Though Cryptozoic rocks were first recognized formally in Britain, the development of a chronology for them occurred largely in the Great Lakes region of North America. Later it was worked out in Scandinavia as well.

The Cryptozoic rocks on all continents are most widely exposed in the stable continental cratons. The region of more or less

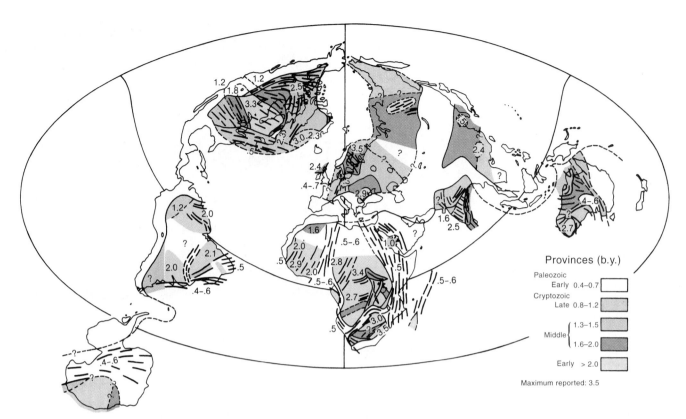

Figure 8.3   Cryptozoic and earliest Paleozoic isotopic-data provinces of basement rocks of continents (middle Paleozoic and younger orogenic belts shown blank, emphasizing their general overlapping relationship with older orogenic belt patterns). Note complexity of discordances between provinces. Quality of data varies enormously, being least complete in South America and Asia. *(From many sources, e.g., Cahen and Snelling, 1966,* The geochronology of equatorial Africa; *Compston and Arriens, 1968,* Canadian Journal of Earth Sciences; *Gastil, 1960, 21st International Geological Congress; Holmes, 1965; Hseih, 1962; Hurley et al., 1967,* Science, *v. 157, pp. 54–542; Jenks, 1956,* International Geology Review.*)*

Figure 8.4   Early working conditions on the Canadian Shield. Geologic party in large bark canoe on Lake Mistassini, Quebec, 1885. *Reproduced with permission of the Minister of Public Works and Government Services Canada, 2001 and Courtesy of the Geological Survey of Canada.*

uninterrupted Cryptozoic exposures in North America occupies primarily the eastern two-thirds of Canada, the United States margins of Lake Superior, and most of Greenland (Fig. 8.3). It is called the Canadian Shield because on a geologic map it has a shieldlike appearance. Younger strata covered most shields in the past; therefore, shields are largely accidents of erosion where the stripping off of later deposits has exposed the ancient basement. The tectonic term *craton* is more useful than *shield* because it defines the overall relative structural stability of a large portion of the earth's crust through a long time interval regardless of the age of the rocks exposed there today.

The Great Lakes Region

The first pioneer to probe the geologic secrets of the Canadian Shield was Sir William Logan, who in 1842 established the Geological Survey of Canada. Working conditions demanded heroic efforts, for much of the country was barely penetrable. In bush areas, transport was largely by canoe or on foot (Fig. 8.4); occasionally boats had to be fashioned on the spot. Logan's successor as Director of the Survey even suffered the heartbreaking indignity of having a faithful, but hungry, horse eat an entire field notebook at the end of a summer season.

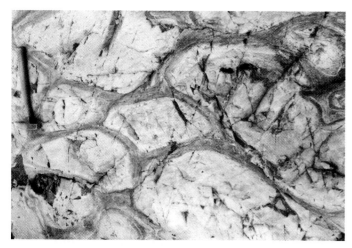

Figure 8.5   Ellipsoidal, or "pillow," structures in Archean greenstones 15 km west of Marquette, Michigan. The lavas are now vertical, but pimplelike protrusions on the lower side of several ellipses indicate the original bottom direction. These protrusions were produced by the weight of overlying pillows during cooling of the lavas. (Compare with Fig. 7.16.) *(R. H. Dott, Jr.)*

Logan and his successors found and named several divisions of sedimentary, metamorphic, and igneous rocks. The oldest rocks there stand in vertical positions, making it difficult to determine their original superpositional sequence (Fig. 8.5).

In 1882 a young man named Andrew C. Lawson (author of the poem at the beginning of this chapter) joined the Geological Survey of Canada and established a chronology for the Ontario-Minnesota border area. He proved conclusively that the oldest recognizable rocks there are metamorphosed basaltic volcanics (named Keewatin; Fig. 8.6). Contrary to what had been assumed since pre-Wernerian times, the most primitive crust was not granitic.

As mapping progressed in Minnesota, younger rock divisions were recognized and named (Fig. 8.6). The work was stimulated by discovery of iron ore near Marquette, Michigan, in 1844 (Fig. 8.7) and the start of iron mining in northeastern Minnesota in 1884. The Minnesota ore-bearing rock occurs within a sequence of metamorphosed lavas called "greenstone." Outcrops are very discontinuous in the swampy, forested country around Lake Superior; therefore, tracing of peculiar rock types, which serve as marker datums, facilitated mapping. Much of the iron ore

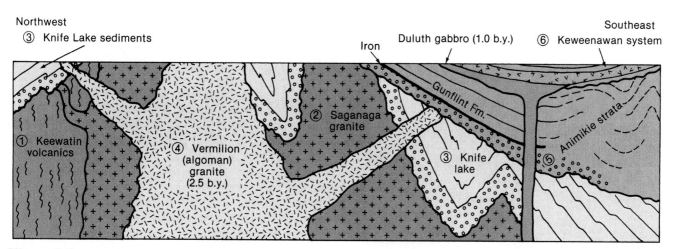

Figure 8.6   Diagrammatic cross section from the north shore of Lake Superior to northern Michigan. Stratigraphic relationships studied by early geologists are shown, as well as isotopic dates. Numbers indicate relative ages (1 = oldest). (Cryptozoic formations have been richly endowed with memorable—if unpronounceable—Indian names.)

Figure 8.7   Contorted banded iron formations near Jasper Nob, Ishpeming, upper peninsula of Michigan (not far from the pillow lavas in Fig. 8.5). Note the characteristic alternating bands of chert (red) and iron (gray). *(R. H. Dott, Jr.)*

A

Figure 8.8   A: Close-up of graded bedding (note the upward decrease in grain size in each of the gray beds) in an Archean graywacke from Ely, Minnesota. The grains settled from the fluid in order of decreasing mass. Where strata are overturned or vertical, the gradation of size indicates the original direction of top. *(R. H. Dott, Jr.)* B: Thick succession of dark Archean graywackes, showing multiple graded beds, interstratified with shaly limestones; from east of the Great Slave Lake, Northwest Territories of Canada. *(Courtesy of Paul Hoffman.)*

B

Figure 8.9   Large-scale cross-stratification formed by wind deposition about 1.75 billion years ago in the Big Bear Formation, Coppermine River, Northwest Territories, Canada. (See the rock hammer for scale in center.) This structure typifies sands transported by either vigorous wind or water currents; it represents the internal laminae of migrating dunes. Each set of cross-strata was deposited by a separate, large dune. The inclined laminations tend to be cut off sharply at their tops but tangential at their bottoms. Laminations are inclined in the direction of current flow; thus, the lowest set was deposited by wind blowing from left to right. These early Proterozoic strata are among the oldest recognized wind deposits and were formed on a landscape totally devoid of vegetation. *(Courtesy of Gerald M. Ross, Geological Survey of Canada.)*

is strongly magnetic and was first located because of the erratic behavior of compasses; for the same reason, iron-bearing strata were easily traced beneath concealed areas.

Strata containing iron ores south of Lake Superior soon were correlated with those of Minnesota, although in the former area the strata have been much more severely disturbed. Several features were important in determining superposition in the rocks; among these were graded bedding (Fig. 8.8), cross stratification (Fig. 8.9), and pillow structure in volcanic rocks (Fig. 8.5). Studies in this region were carried out largely by an army of United States Geological Survey workers under the guidance of C. R. Van Hise, later to become president of the University of Wisconsin. Copper, which has been mined in Michigan longer than iron, was an additional incentive for their work. It occurs in a thick succession of gently northward-dipping basalts and red-colored clastic sediments in the Keweenaw Peninsula of Michigan. A similar succession occurs on the north shore of Superior, where it dips south, thus forming a broad syncline beneath the lake (Fig. 8.6). Simple structure and distinctive rock types made correlation of these Keweenawan rocks across the lake simple.

Great Lakes Correlations

From stratigraphy alone, four major divisions of sedimentary and volcanic rocks were recognized. They were known to be separated by unconformities that reflected at least four major mountain-building episodes (called orogenies) during which extensive granites were formed (Fig. 8.10). These orogenies, which for convenience are named for some geographic region, punctuate the rock record in such a way as to provide several natural divisions (Table 8.1). But there were some thorny problems of correlation that could be resolved only through isotopic dating.

Isotopic dating has helped immensely to anchor age relations all around the Great Lakes. The numbers in Table 8.1 indicate the most important dates and make possible a workable standard Cryptozoic chronology for North America.

Correlation Beyond the Great Lakes

In the past, geologists commonly have used relative degrees of deformation and of metamorphism to argue that certain rocks are very old or very young. Great caution is required in exercising such criteria, however, because many misinterpretations are on record. Recall the neptunian error in regarding all metamorphosed

Figure 8.10   Intrusion of Archean granite (light) into Archean metavolcanic sediments (dark), near Nestor Falls, Ontario. Isotopic dates indicate that the granite is about 2.5 billion years old, formed during the Algoman orogeny. *(Courtesy of L. Gordon Medaris.)*

Table 8.1   Standard Chronology of Cryptozoic Events

CLASSIFICATION FOR GREAT LAKES STATES			Billions of Years	CLASSIFICATION FOR CANADA		Worldwide (Tentative)
Phanerozoic		Paleozoic Era	0.7		Hadrynian Era	Phanerozoic
Cryptozoic or Precambrian Eon	Proterozoic Era — Late	Grenville orogeny		Helikian Era	Grenville orogeny	Neoproterozoic
		Keweenawan System	1.2		Keweenawan System	Mesoproterozoic
		Unnamed orogeny			Elsonian orogeny	
		Unnamed system			Unnamed	
	Proterozoic Era — Early	Penokean orogeny	1.8	Aphebian Era	Hudsonian orogeny	Archeoproterozoic
		Animikean and Huronian Systems			Animikean and Huronian Systems	
	Archeozoic or Archean Era — Late	Algoman orogeny	2.5	Archaean Era	Kenoran orogeny	Archean II
		Timiskamian System			(Diverse local sequences)	
	Archeozoic or Archean Era — Early	Saganagan orogeny	3.0			Archean I
		Keewatian	3.8+			

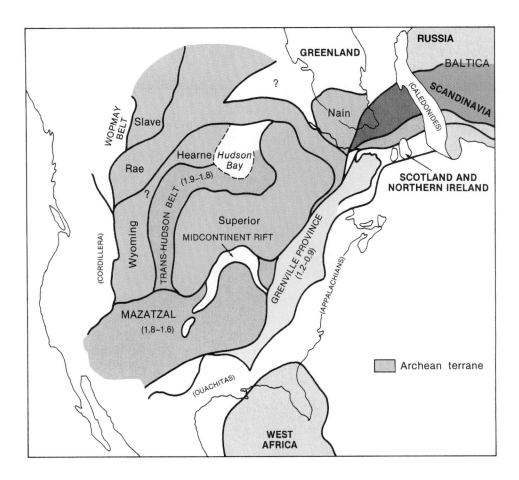

Figure 8.11   Geologic provinces of the core of North America that were assembled by the late Proterozoic. Numbers give the typical isotopic dates in billions of years (b.y.) found in each province. The oldest Archean microcontinents were the Superior, Nain, Slave, Rae and Hearne-Wyoming terranes, with dates older than 2.6 b.y. All of these terranes collided about 1.8 to 1.9 b.y. ago, trapping the Trans-Hudson belt between them. The terranes of the central United States were also accreted then, possibly by the collision with the Archean core of Africa to the south. The Grenville Province was sutured to North America about 1 b.y. ago, when the Baltic Platform collided. *(Modified from P. E. Hoffman,* Ann. Rev. Earth Planet. Sci. *16:543–603, 1988.)*

rocks as Primitive, when, in fact, some of then turned out to be Mesozoic and even Cenozoic (Chap. 4). The assumption that severe disturbance is necessarily related to great antiquity is too simple, for geographic or structural location may be more important. Cenozoic rocks in the Swiss Alps are far more deformed than are the late Cryptozoic ones around Lake Superior. And Logan's "old" gneisses near Ottawa proved to be younger than mildly folded late Cryptozoic strata on the shores of Lake Superior. More significant than age is whether a given group of rocks was formed within an orogenic belt or within a craton.

Before the advent of isotopic dating, it became popular to think of the Cryptozoic rocks as falling broadly into two groups: an older, intensely metamorphosed Archean and a younger, less disturbed Proterozoic. Far beyond the Great Lakes region, rocks were identified with one or the other of these groups solely on the basis of metamorphism and structure. This crude classification resulted in many errors, which could be corrected only through isotopic dating, which provides the best basis for long-distance time correlations of Cryptozoic rocks.

With the increase in isotopic dating, it has become clear that the North American continent contains several distinct *isotopic date provinces* (Figs. 8.3 and 8.11)—that is, large regions within which a certain range of dates predominates. Such provinces are by no means simple, and they typically yield a considerable total range of ages, but statistically each has a discrete cluster of dates falling within about a 0.3-billion-year interval. It is important to

remember that most of the dates upon which Fig. 8.11 is based are dates of orogenic events; thus, each province is a mosaic of related orogenic belts. The rocks found in these old belts are composed of volcanic materials, sediments, and large masses of granitic and high-grade metamorphic rocks, all of which attest to long-continued tectonic mobility. Many of the rocks have been metamorphosed more than once; thus, their "isotopic clocks" may have been reset several times. Dates obtained from such rocks generally record only the last readjustment of isotopes during an episode of heating. Some rocks retain two or more discordant dates, which reveal a complex history of multiple orogenies.

Most of the oldest isotopic dates (2.5 to 4.0 billion years) tend to occur in the center of the continent from western Ontario southwest to Montana and Wyoming. But at least two other nuclei of similar antiquity also occur, one in the far northwestern corner of the shield and the other in southern Greenland. Surrounding each nucleus are more or less concentric orogenic belts of younger ages (Fig. 8.11).

As isotopic dating progressed, it became clear that the dates from all continents tend to cluster in a roughly similar way (Fig. 8.3). The groups seem to represent crudely synchronous major granite-forming and metamorphic episodes; such peaks may eventually provide a basis for a worldwide standard Cryptozoic time scale (Table 8.1). Individual orogenic belts seem to have existed for about 0.8 to 1.2 billion years, during which time several periods of thick sedimentation, deformation, metamorphism, and

Table 8.2

Comparison of Chemical Compositions of Granitic Rocks* with Sedimentary and Andesitic Ones

Rock Type	COMPOSITION, %										
	SiO_2	TiO_2	Al_2O_3	Fe_2O_3	FeO	MnO	MgO	CaO	Na_2O	K_2O	H_2O
Granite	70.5	0.4	14.1	0.9	2.4	0.06	0.6	1.6	3.6	5.4	0.5
Quartz diorite	63.2	0.6	17.7	1.8	3.2	0.1	1.9	4.8	4.2	1.9	0.6
Average Proterozoic sediments	65.2	-	14.1	1.7	2.9	-	2.3	3.1	2.8	2.6	-
Average Phanerozoic sediments	58.8	-	13.6	3.5	2.1	-	2.7	6.0	1.2	2.9	-
Andesitic lava	60.1	0.5	17.8	2.0	3.4	-	3.5	6.3	4.2	1.3	0.3
Basaltic lava	48.4	2.7	13.2	2.4	9.1	0.1	9.7	10.3	2.4	0.6	-

*True granites are much less common than diorites, which have less quartz and potassium feldspar. Therefore, the broader term *granitic rocks* is used to include the whole assemblage of coarse, crystalline-textured, light-colored, silica-rich igneous rocks.
Sources: After Engel, 1963; Turner and Verhoogen, 1960.

granite formation occurred. The individual tectonic events, with periods of from 0.2 to 0.4 billion years, have been blended together to make up each of the Cryptozoic age provinces.

Evidence of Crustal Development from Igneous and Metamorphic Rocks

Importance of Granite

In the light of isostatic uplift and the long erosional history of continents, some rejuvenation, or addition, of continental crust through time seems inescapable. Partly from these considerations, but also from the igneous rock types of the early Cryptozoic, together with the arrangements of old orogenic belts, there has grown the *theory of continental accretion,* introduced in Chap. 7. Geologists in the early twentieth century realized that rock rich in silica, aluminum, and potassium—notably granites—seemed to be unique to continents. But field evidence suggested that granitic continental crust was not original and, so, must have increased in volume through time. Granitic continental material, being of slightly lower average density than the remainder of the crust and mantle, stands topographically higher than the ocean basins in accord with isostasy (see Chap. 7).

Apparently the original crust was thin and composed largely of komatiite and basalt. (Komatiite is the volcanic equivalent of ultramafic mantle rock.) It was postulated long ago that, by later redistribution of lighter elements through weathering, erosion, and igneous activity, some of the original crust was converted to granite to form embryonic continents. A century ago, it was also suggested that granite might form at depth in the crust simply by recrystallization—extreme metamorphism or granitization—of sedimentary-volcanic sequences to produce new granitic material without involving any liquid magma (Table 8.2). It has been sug-

gested that the banded gneiss of Fig. 8.2 had such an origin. Through time, more and more granitization of volcanic and sedimentary rocks presumably occurred, and the continents grew thicker and larger as a result. Granitization was a controversial idea during the early half of the twentieth century, but it is now accepted as one of several complex melting mechanisms.

Today geologists think that granites have formed from magmas produced by partial melting deep in the crust, a process that causes enrichment of such lighter elements as silica, aluminum, potassium, and sodium. Some of the magmas seem to have arisen by partial melting of predominantly sedimentary rocks (S-type) already rich in those elements, whereas a second group seems to have formed by partial melting of igneous rocks (I-type). Both these types form within orogenic belts, but a rare third variety (A-type) appears within cratons. Each type shows subtle but consistent differences of mineralogy, chemistry, and texture. When melting occurs, the liquid magma is less dense than adjacent rocks, so it rises. If the magma then crystallizes slowly within the upper crust, coarse granitic rock results. If the magma reaches the surface, however, a fine-grained volcanic rock, such as rhyolite, forms.

Obviously if only older continental material has been partially melted to form a granite (or its extrusive cousin, rhyolite), then no increase in volume of continental crust would be possible, only a recycling. But rocks such as andesite are derived by partial melting of the upper mantle near subduction zones and then erupted in volcanic arcs. Such arc rocks then can be accreted tectonically to continents and eventually be reconstituted to form new granitic material, because the composition of andesite is close to that of the most abundant granitic rocks, the diorites (Table 8.2). It also appears that sedimentary and volcanic rocks relatively rich in potassium, aluminum, sodium, and silica may be accreted or underplated to the bottom of continents from subducting plates carrying such rocks down with them. The increase of heat and pressure presumably causes partial melting of these

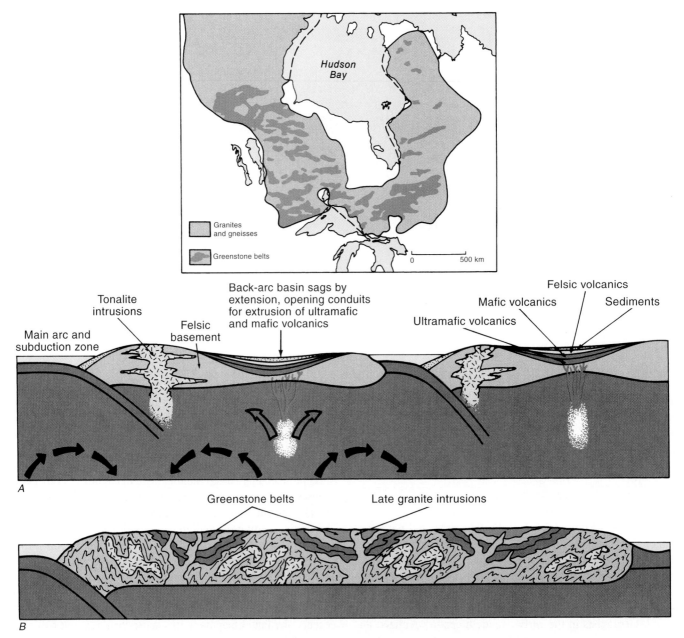

Figure 8.12   Evolution of greenstone belts. *Top:* Map of greenstone belts in the Superior province. A: Protocontinents made of gneisses jostle over the convecting mantle. Basins between the protocontinents are floored by oceanic pillow basalts and filled by graywackes eroded from the uplifted areas. *B:* When the protocontinents collide, they collapse the oceans filled with basalts and graywackes, forming greenstone belts. Finally, the entire assemblage is intruded by granitic rocks. (Tonalite = silica-rich diorite; felsic = feldspar- and quartz-rich igneous rock.) *(After B. F. Windley,* Evolving continents, *1977).*

materials, and the resulting magma is incorporated into the continental crust above. Thus, it appears that continental crust does continue to grow by a variety of accretion modes.

The chemistry of granitic rocks shows important trends through time. For example, younger granites are richer in potassium and radiogenic daughter isotopes. This seems to reflect increasing chemical differentiation of the earth as a whole, as discussed in Chap. 6, and to provide evidence of repeated accretions and a trend of chemical maturing of continental crust through time.

Significance of Isotopic Date Patterns

The patterns of isotopic date provinces appear to have an important bearing upon the hypothesis of continental accretion. The pattern of more or less concentric, younger-outward zones suggests that the continent may have accreted from a few old nuclei by the successive formation and ultimate consolidation of orogenic belts around them (Fig. 8.12). The nuclei were knitted together into a single craton, which then grew through addition of newer belts (what Paul Hoffman calls "the United Plates of America"). It is thought that, after long histories of orogenies, new

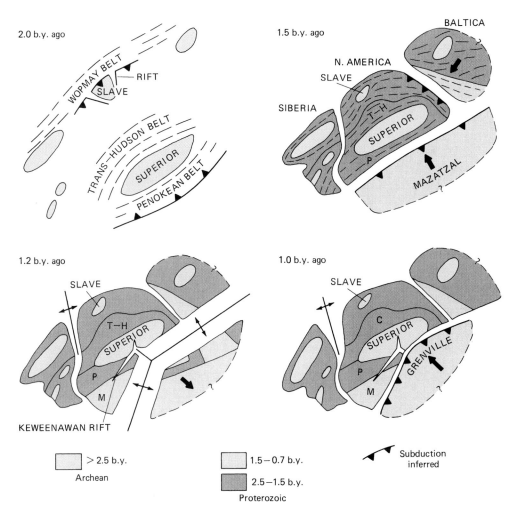

Figure 8.13 Hypothetical scenario for the assembly of North America during Proterozoic time based upon the tectonic and isotopic date patterns of Fig. 8.11 and the principles of restoration of ancient lithosphere plates outlined in Chap. 7. These restorations are based upon the logic of plate tectonics, and, although poorly constrained by field evidence in many places, they convey the collage nature of continents.

granitic material was formed in each successive orogenic belt, thus stabilizing and transforming the behavior of each. Deep erosion accompanying isostatic uplift reduced the mountains to low plains, exposing the bowels of the old mountain belts that we see today (Fig. 8.2). Shields, then, are mosaics of complexly interlaced ancient orogenic belts now stabilized within cratons—all deeply eroded. It is estimated that the original depth of formation of some of the great granitic batholiths was as much as 20 to 30 kilometers.

The hypothesis of accretion presents an extremely appealing and relatively simple generalization that unifies many types of geologic data and offers a logical explanation of crustal development through enlargement and thickening of the continents. The orogenic belt concept, igneous and metamorphic petrology, and overall geochemical differentiation of the earth are explained in a related manner. But nature rarely is simple! Note that some orogenic zones of very different isotopic ages are almost wholly superimposed, or **overprinted,** upon one another rather than simply being side-by-side. The Algoman orogeny around Lake Superior seems to have been so overprinted upon the much older Saganagan orogenic belt (Fig. 8.6). In other cases, younger belts cut off older ones at high angles rather than being neatly concentric. For

example, ghosts of reworked early Proterozoic trends can be discerned within the younger Grenville belt, where the rocks have been almost completely reconstituted.

Now let us apply the concepts of plate tectonics discussed in Chap. 7 to see how the Cryptozoic underpinnings, or "basement," of North America, shown in Fig. 8.11, may have been assembled. Fig. 8.13 presents a speculative scenario. Remember that ancient orogenic belts represent convergent plate margins where combinations of subduction and collisions cause mountain building accompanied by metamorphism and granite formation. Conversely where structural trends are truncated or cut off sharply, we suspect rifting along ancient divergent margins. Using clues from the structural trends together with the isotopic date patterns, we have indicated four possible stages in Proterozoic history. Two billion years ago, several small Archean granitic continents were accreting together as a result of the formation of convergent orogenic belts between them. Structural patterns suggest that, by 1.5 billion years ago, Siberia and possibly Australia and Antarctica had been fused to western North America and that Baltica (the old Scandinavian nucleus of Europe) was converging with eastern North America (the two collided about 1.2 billion years ago). Another continent, of unknown size, called Mazatzal, was also con-

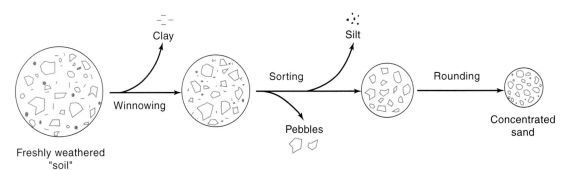

Figure 8.14
Idealized development of textural maturity of sand through abrasion and separation or sorting of different-sized grains.

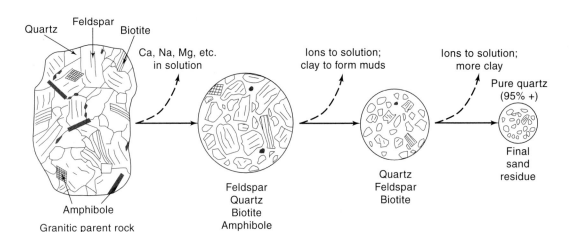

Figure 8.15   Typical changes through time of the mineral composition of sand that was derived from erosion of a granitic source. Less stable minerals are broken down both physically and chemically to leave a residue of most resistant mineral grains.

verging from the south to make a large supercontinent about 1.3 to 1.4 billion years ago. Then large-scale rifting began about 1.2 billion years ago, and finally the Grenville orogenic belt was formed along the southeastern margin about 0.8 to 1.0 billion years ago.

Interpretation of Crustal Development from Sediments

Bias of the Record

Further insight into crustal history can be gleaned from Cryptozoic sediments. The Lake Superior region contains thick, deformed sedimentary and volcanic accumulations that, for the most part, were deposited within ancient orogenic belts. They are typical of shield areas. Deep downfoldings protected sediments from subsequent erosion. Conversely flat-lying, easily eroded cratonic sequences are practically unknown in the Archean record. To further confound us, many of the orogenic belts suffered so much metamorphism that little can be said about their original sediments. By Proterozoic time, however, there were sizable stable cratons surrounded by orogenic belts. In the section "Terrigenous Versus Nonterrigenous Sediments," we shall digress to introduce some important principles for the interpretation of such tectonic elements from sediments. Keep in mind that the same principles apply equally to younger rocks and to Cryptozoic ones.

Terrigenous Versus Nonterrigenous Sediments

The composition of any clastic, or fragmental, sedimentary rock reflects, more than anything else, the sources from which it was derived. Of course, climatic conditions may modify composition through weathering, as may chemical changes after deposition, but for the present we shall ignore these complications. It is necessary to distinguish between two types of sediments: (1) *terrigenous clastic sediments* contain chiefly silicate minerals, such as quartz, and are derived from erosion of older rocks in land areas; (2) *nonterrigenous sediments,* formed within aqueous depositional environments where terrigenous material was not abundant, include *chemically precipitated* sediments, such as the evaporites (salt and gypsum), and carbonate rocks composed of fossil skeletal debris or precipitated calcareous particles. The latter are most characteristic of shallow, warm waters. In this chapter and Chap. 9, we shall consider primarily the terrigenous fragmental sediments; the nonterrigenous types are discussed in Chap. 12. The terrigenous clastics show trends of both textural and compositional changes that provide important clues for interpreting tectonic history (Figs. 8.14 and 8.15).

Textural Maturity

Clastic textures (Box 8.2) reflect primarily the rates and intensities of physical sedimentary processes. Maximum size or coarseness reflects the power of transporting agents, such as running water and wind. Wind normally moves only sand and silt, whereas moving water can carry gravel as well. Mudflows and

Box 8.2

Geology's First Important Instrument

Scientific revolutions that resulted from the invention of instruments such as the telescope in the sixteenth century and the compound microscope in the seventeenth century are well known. Less known, but of equal impact to geology was the invention in 1829 by a Scot named William Nicol of a device for polarizing light made from two crystals of calcite. He used his "Nicol Prism" to study thin slices of petrified wood, but English genius Henry Clifton Sorby saw the wider applicability of Nicol's invention for the study of optical, or light-transmitting, properties of minerals and rocks. Most minerals, if sliced very thin, are transparent, and their compositions and textures can be studied when polarized light is passed through the slices. In 1850, Sorby adapted Nicol's prism to construct the *petrographic microscope,* which is the single most important geologic instrument ever invented. He then proceeded to demonstrate the revolutionary value of his microscope with detailed microscopic descriptions of various sedimentary rocks. It was German and French, rather than British mineralogists, however, who quickly capitalized upon this revolutionary device for the systematic study of minerals and igneous rocks. Figure 8.16 shows photographs of two Cryptozoic sandstones taken through the polarizing petrographic microscope.

glaciers carry immense blocks for long distances because of their greater density and viscosity. In a general way, size tends to decrease with time and distance of transport. The reason is twofold, including a tendency for decrease of carrying power with distance for most agents of transport and reduction of particle size by abrasion. Degree of rounding of sharp corners of fragments also is related to intensity and duration of abrasion as well as to toughness of the materials themselves. The range of sizes in a given clastic sediment, generally described as size sorting, reflects primarily the total time of transport and constancy of physical energy of transportive agents. A sediment subjected to long and constant agitation (e.g., beach sand) tends to be well sorted because there is maximum opportunity for the early dropping out of large particles and for the removal or winnowing away of fine materials. The latter are deposited ultimately in less agitated environments. In this way, separation of different-sized materials occurs, with a final result being deposition of gravel near the source, well-sorted sand in another place, and well-sorted particles of fine silt and clay in still a third place. Generally, clastic sediments become finer as they are moved farther from their source (see Figs. 4.9 and 4.12).

From these considerations, we can formulate a useful generalization about the ideal textural evolution of terrigenous clastic sediments that will help us to interpret the history of any sample. Obviously with greater abrasion and winnowing by currents or waves, size of particles will be reduced, sorting of sizes will improve, and rounding will increase, as shown in Fig. 8.14. This idealized clastic evolution can be summarized as the degree of *textural maturity.* Figure 8.16A shows a texturally immature Cryptozoic sandstone (loosely termed graywacke) from an old German mining term. Such a texture generally implies rapid deposition, so that abrasion and winnowing were ineffective. Figure 8.16B shows a texturally mature Cryptozoic sandstone with well-rounded grains, which imply a much longer history of abrasion and winnowing.

Compositional Maturity

The mineral composition of a clastic sediment will change as its particles are subjected to repeated physical crushing and chemical destruction of the less stable minerals. Rock fragments tend to be ground down rapidly to their separate mineral grains, and dark (mafic) minerals, such as pyroxenes and amphiboles, suffer rapid chemical breakdown. This leaves a dominantly sand-sized (0.125 to 2.0 millimeters) ultimate residue of the more resistant material: quartz, chert, feldspar, and some mica, as well as very rare, but durable, accessory grains of minerals with high specific gravity, such as zircon, garnet, and magnetite. Of all these separate mineral grains, quartz is overwhelmingly the most abundant. Feldspar, though more abundant in parent igneous rocks, is of intermediate durability and, so, runs second to quartz in sediments. The others, though more durable than feldspar, are simply far less abundant in source materials. Chert (tough, fine-grained sedimentary rock composed of silica, SiO_2) is the most durable material that originates in sedimentary environments and is common in many conglomerates and sandstones.

An ideal evolution of *compositional maturity* exists (Fig. 8.15), with a progressive loss of the less stable and a concentration of more stable constituents as physical and chemical decay take their toll. Obviously the ultimate product of such an evolution is very pure quartz- or chert-bearing sand and gravel. Fig. 8.16 shows examples of the extremes of relative compositional (as well as textural) maturity.

There is a close relationship between destruction of the less stable coarse particles and the increase of fine residue products from that destruction. For example, the most abundant sedimentary rock is shale, which is composed largely of fine clay particles derived from weathering of feldspars, the most abundant minerals in the crust. Within shales, there is a spectrum of compositional maturity such that the clay species present reflect the relative thoroughness of chemical decay of the parent minerals.

Broad grouping into clans of sands and gravels according to composition is very useful as a simplification of knowledge. The clans recognized in this book are the quartz sandstones, the feldspathic clan, and the rock-fragment (or lithic) clan. These categories provide convenient summary nicknames.

Stratification

Stratification also provides important clues about depositional processes. The origins of different kinds of stratification have

A

B

Figure 8.16   Photographs taken through a polarizing microscope of two contrasting Proterozoic sandstones. *A:* An immature graywacke sandstone composed of a mixture of quartz, feldspar, and diverse rock fragment grains surrounded by a fine, dark clay; note the angularity and poor sorting of the grains (Tyler Graywacke, Wisconsin). *(R. H. Dott, Jr.) B:* Pure quartz sandstone composed almost solely of well-rounded quartz grains (note lack of fine, dark material); this is both compositionally and texturally mature (Baraboo Quartzite, Wisconsin). (Larger grains are about 1 mm in diameter.) *(R. H. Dott, Jr.)*

been studied both in modern environments and in the laboratory for many years. Early water tank experiments in Germany in 1825 used brandy and mercury as well as water as fluids. Since then, hundreds of artificial water channels (called flumes) and wind tunnels have been built to study erosion, transport, and deposition of sediments by currents and waves.

Most thin, horizontal lamination in very fine sediments (e.g., Fig. 5.2) is formed by slow settling of clay and silt particles from suspension within a transporting fluid. A minimum of current agitation of the depositional sediment surface is implied; therefore, a still environment, such as a deep lake or sea bottom, favors such stratification. (Such sediments are commonly called pelagic.)

Cross-stratification is the most common feature of sands and fine gravels deposited by either wind or water (Fig. 8.9). Its scale varies greatly, from layers only a few millimeters thick to sets of cross-strata tens of meters thick. Most cross-stratification

is formed by moderately strong, turbulent *traction currents* that roll and bounce particles over a loose sediment surface corrugated by small-scale ripples (Fig. 8.17) or larger-scale dunes (the boundary between these two categories being approximately 10 centimeters in height). Grains carried to the crests of the ripple or dune roll or slide down the steeper (lee) side. Therefore, the resulting inclined cross-stratification reflects the lee face of the migrating ripple or dune, and it dips toward the downcurrent direction (Fig. 8.18). Such factors as volume and coarseness of sediment, constancy of flow, and patterns of turbulence control the three-dimensional form of ripples and dunes and therefore control the internal geometry of cross-stratification.

It was long thought that sandstones with large-scale cross-stratification must represent wind-dune sands simply because such dunes were the only features known to produce such stratification. Beginning about 1960, however, it was recognized that large rivers and strong tidal currents also can form dunes at least a few meters high. Therefore, other properties, such as size, shape, and sorting of grains (Fig. 8.19) as well as the characteristics of ripples or any other associated features, must be studied together with the cross-stratification. The diagnosis of depositional processes is discussed further in Chap. 10.

Graded bedding (Fig. 8.8) represents episodic introduction of abnormally coarse material into a usually still environment generally characterized by delicately laminated fine materials. Such spasmodic deposition occurs principally where infrequent currents of denser water flow beneath less dense water and carry coarser debris to normally tranquil depths. **Turbidity currents** are the most important of such agents. They derive their driving energy from the presence of fine sediments thrown into suspension by earthquake shocks or by severe storm activity. Sediment is kept in suspension by turbulence as the slightly denser, muddy water mass flows downslope beneath less dense, clear water (Fig. 8.20). Such a current maintains its kinetic energy and continues to flow until the turbid water encounters a topographic depression or until it becomes diluted by mixing with clear water. In Lake Mead, where natural turbidity currents have been studied, muddy Colorado River floodwater (specific gravity 1.003) sinks beneath clear lake water (specific gravity 1.001) and flows 100 kilometers along the bottom to Hoover Dam, where the currents are halted. Very fine sand and silt with graded bedding are deposited from these turbidity currents along the lake floor. Such currents were not anticipated when Hoover Dam was built, and they are filling the reservoir faster than expected—much to the disappointment of the engineers.

About 1930, a British geologist, E. B. Bailey, noted that conspicuously cross-stratified sediments tend to occur apart from those with much graded stratification. He interpreted this dichotomy as reflecting fundamentally different sedimentary environments and depositional processes. He also extended his interpretation to a fundamental tectonic distinction. Cross-stratification was deemed restricted to shallow, agitated water characteristic of seas over relatively stable cratonic regions, whereas graded bedding was considered characteristic of the most structurally active regions, where deep, usually less-agitated water prevailed but into which poorly sorted coarser material occasionally was dumped. This was a very important generalization.

The Cryptozoic Sedimentary and Volcanic Record

Ancient Archean Rocks

No record is known for the first 100 million years of earth history. We do know, however, that, because of the extreme heat of the interior, the earth's entire surface must have been volcanic; it also suffered intense bombardment by large meteorites. As a result, the earliest crust must have consisted of ever changing small plates that had no chance of survival. As heat began to dissipate and large meteorite impacts ceased about 3.9 billion years ago, local crustal blocks began to survive.

Figure 8.17   Ripple marks in early Proterozoic (Huronian) quartzite 30 miles east of Sault Ste. Marie, Ontario. Ripples are exposed on three different stratification planes: those on the middle surface (just left of center) are perpendicular in trend to the others, indicating a 90° shift of wave and current directions between the times of deposition of these strata. *(R. H. Dott, Jr.)*

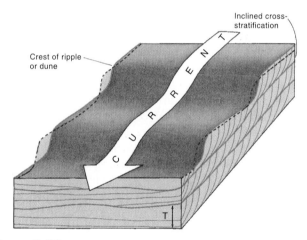

Figure 8.18   Origin of cross-stratification by the migration of ripples or dunes produced by vigorous bottom (tractional) currents. Grains roll and bounce over the dune crests, coming to rest on the lee faces. Successive inclined or cross-stratification forms as the lee face migrates; each lamina is a buried fossil lee face. Cross-stratification reveals not only the original top and bottom of steeply tilted strata but also ancient current directions. (T indicates thickness of cross-stratification.)

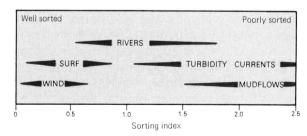

Figure 8.19   Comparison of relative sorting of sand-grain sizes by different sedimentary processes (as measured by the sorting index standard deviation). Sorting helps in determining the origin of an ancient sandstone; for example, note the great difference of sorting by surf and turbidity currents.

Figure 8.20   Experimental turbidity current of more dense, muddy, turbulent water flowing from right to left beneath less dense, clear water down a sloping laboratory flume approximately 50 centimeters deep. Such currents are driven solely by gravity because of the density contrast between water masses. *(Courtesy Gerald V. Middleton and Roger G. Walker, McMaster University, Ontario.)*

The oldest known rocks (2.5 to 4.0 billion years) comprise two different assemblages, both of which have been more or less altered: Best known are **greenstone belts,** which are made up of mildly metamorphosed volcanic rocks and associated sediments (Fig. 8.12). The second type includes high-grade metamorphic **gneiss belts** with very complex histories.

The greenstone belts (e.g., Keewatin in Canada, Fig. 8.6) were dominantly volcanic in character ("greenstone" derives from dark, green-colored minerals produced by metamorphism of mafic igneous rocks). Komatiites and basalts are most prominent, but andesitic and rhyolitic rocks also are present in the upper parts of some sequences. The komatiite greenstones are richer in magnesium than are younger basalts, which is interpreted as evidence of melting of mantle material at temperatures higher than those of today (Chap. 7). Again this is consistent with much greater radioactive heating. It has been suggested that komatiite lavas covered the Archean earth like water. Pillow structures (Fig. 8.5) indicate eruption of lavas underwater, but abundant ash deposits suggest that some volcanics were erupted from vents above water. Unusually large percentages of chromium and nickel in South African sediments older than 3.0 billion years suggest large, exposed sources of ultramafic rocks in the early crust.

Applying the concepts of sandstone maturity, we find that the clastic sediments of the greenstone belts are very *immature* graywackes. Conglomerates and sandstones composed chiefly of poorly rounded and poorly sorted volcanic fragments and feldspar predominate. Clearly, volcanic material was being eroded and deposited very rapidly after only slight weathering and abrasion. In the words of E. B. Bailey, such sediments look "poured in or dumped." Prevalence of graded bedding (Fig. 8.8) suggests deposition in relatively deep water between volcanic vents.

By analogy with younger times, a more or less oceanic setting is implied for the greenstones; however, associated andesitic and rhyolitic rocks suggest at least local evolution of arc conditions.

Although their origins are warmly debated, geologists generally agree that the Archean gneiss assemblages include plutonic (granitic), volcanic, and sedimentary parent rocks. Granitic pebbles in some Archean conglomerates and quartz-bearing sandstones also attest to the presence of some early granitic crust; zircon sand grains in Australia have yielded isotopic ages of 4.4 billion years. Selective concentration of lighter elements to form continental-type crust clearly had begun, but the chemistry of the granitic components of the gneiss belts differs in subtle ways from younger granites. Such things as isotope ratios and rare-earth-element distributions suggest that the mantle sources of the old granites were chemically more primitive than were the mantle sources of the younger granites. We must remember that there was nearly three times as much heat generation through radioactive decay 3.5 billion years ago than now; thus, widespread high-temperature metamorphism is not surprising. Embryonic continental crust must have been thin and unstable; partial melting of sedimentary and volcanic rocks seems to have been the rule. Recognizable sediments associated with the gneissic complexes include schists containing graphite, quartzites, banded iron formations, and minor marbles. The schists probably were deposited as shales, but whether the graphite reflects organic or inorganic carbon is unknown. Rare cross-bedded quartzites suggest deposition in agitated (shallow?) water under temporarily stable, local tectonic conditions.

What are the relationships between greenstone belts and gneissic regions? One view is that they differ only in degrees of metamorphism and reconstitution. Other explanations, however, assume that the gneisses are remnants of incipient continents, whereas the greenstone belts are remnants of oceans and volcanic arcs formed between continents.

Were plate tectonic processes as we recognize them in the younger geologic record operating in Archean time? Many geologists think not and argue that the Archean was unique. But the more we learn of Archean rocks, the more they seem to resemble younger ones after all; differences are more of degrees. There seem to have been some lithospheric plates, but apparently they were smaller, thinner, and less rigid then, so that relatively little material has survived in a recognizable form. A plausible scenario is that heat was still so great that volcanism occurred nearly everywhere. Orogenic belts as such were not well defined—indeed, most of the earth's surface probably was in turmoil much of the time due to intense mantle convection. Crust formed, was remelted, reformed, and remelted again, as is scum on a boiling pot or crust in a volcanic crater (Fig. 8.21). Finally, in late Archean time (about 2.5 to 2.6 billion years ago), a major period of granite formation occurred more or less worldwide (Algoman or Kenoran orogeny of the Canadian Shield). Apparently much of the present volume of continental crust accreted then by the aggregation, enlargement, and thickening of many minicontinents. This led to plate tectonics as we now know it (Fig. 8.13).

Early and Middle Proterozoic Sediments

Proterozoic sediments show some striking contrasts when compared with most Archean ones. Poorly sorted graywackes still formed, but light-colored, well-sorted pure quartz sandstones now became abundant. Although the latter have been recrystallized to metamorphic quartzites, there is still evidence that they were texturally as well as compositionally *mature*.

Proterozoic sandstones are of two principal types, differing chiefly in texture: (1) well-sorted quartz sandstones and (2) quartz-rich graywackes (Fig. 8.16). Feldspar is sparingly present in the graywackes, but other constituents, such as the volcanic rock fragments so abundant in Archean graywackes, are less common. The pure quartzites show cross-strata and ripple marks. Where conglomerates are associated with quartzites, they contain well-rounded, durable pebbles of quartz or chert. Interstratified with the quartzites are thick sequences of mudstone or slate with zones of dark graywackes and prominent banded iron formations consisting of alternating iron and chert bands (Fig. 8.7). Associated with some of the pure quartzites are limestones, many of which contain distinctive wavy laminated structures called **stromatolites** formed by marine, bottom-dwelling cyanobacteria (formerly called "blue-green algae") (Fig. 8.22). All these rock types are prominent south and east of Lake Superior (Animikean and Huronian systems).

What conditions could produce the first appearance of significant limestones associated with mature sandstones? A low or very distant stable land providing little or no clastic sediment and deposition in shallow, strongly agitated water are most likely. The wide extent of distinct limestone formations suggests, by analogy with younger strata, a marine origin. Stromatolite-forming cyanobacteria fixed on the sea bottom cannot grow in water deeper than about 150 meters—the photic zone—because at deeper levels the water absorbs the sunlight needed for photosynthesis. Evidence of strong agitation over wide areas also argues for a very shallow sea susceptible to regular stirring by winds. This type of sequence was deposited on a continental shelf—that is, a Cryptozoic craton flooded by a shallow sea.

Abundance of quartz sand indicates long and profound weathering of very large volumes of granitic or rhyolitic rocks. The volume of pure quartz sandstone provides a rough minimum index to the volume of material that must have been weathered and eroded to provide this concentrate. As an example, to produce

Figure 8.21   What the unstable early Archean surface of the earth probably looked like. Miniature plates and compressional zones are visible on the chilled surface of basalt in Keanaki Crater, Hawaii, following a 1974 eruption. The confused pattern resulted from constant turmoil due to convection below and the formation and crumpling of a surface crust (compare Fig. 7.29). Area of view approximately 100 meters wide. (R. H. Dott, Jr.)

Figure 8.22   Part of a large stromatolitic "reef" structure from 1.6-billion-year-old Proterozoic carbonate strata in the Wopmay orogen, Northwest Territories of Canada (see Figs. 8.11 and 8.24). (Courtesy of Paul Hoffman.)

the total volume of quartz residue represented by only one of several prominent quartzite-bearing formations would have required the complete weathering and erosion of at least 10,000 cubic kilometers (2,500 cubic miles) of granitic rock that contained a volumetric maximum of 25 percent quartz. To account for all of the incalculable amounts of quartz sand in the shield, this figure must

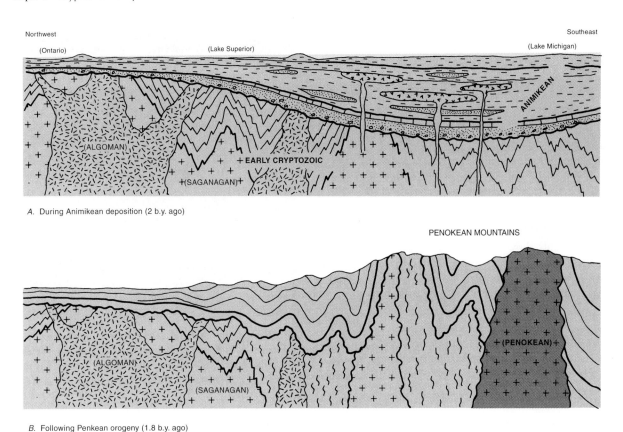

Northwest
(Ontario)   (Lake Superior)

Southeast
(Lake Michigan)

A. During Animikean deposition (2 b.y. ago)

PENOKEAN MOUNTAINS

B. Following Penkean orogeny (1.8 b.y. ago)

Figure 8.23   Simplified cross sections summarizing the history of the Penokean orogenic belt of the Lake Superior region. Penokean mountain building about 1.8 billion years (b.y.) ago apparently resulted from subduction just south (to right) of these sections.

be multiplied at least a hundredfold. The staggering volume of early Proterozoic quartz sand found both in well-sorted sandstones (now quartzites) and in ill-sorted ones (graywackes) implies several cycles of weathering, erosion, and concentration within a few hundred million years immediately following the Algoman orogeny. In each successive cycle, more and more unstable mineral grains would have been removed and quartz would have been gradually distilled. It is an inescapable conclusion that the early Proterozoic continent already was large and contained a great deal of granitic rock. Much of it was relatively stable, and it had been deeply eroded by 2.0 billion years ago. The sedimentary evidence, therefore, supports the concept of rapid early growth of continental crust at that time.

Early Proterozoic strata in the Great Lakes area become thicker from north to south, and the currents that deposited them flowed in the same direction. Volcanic rocks are present south of the Great Lakes, and the degree of deformation and metamorphism increases markedly in the same direction (Fig. 8.23). A stable craton still lay north of that area, and these relatively thin and less disturbed sandstones, shales, some limestone, and iron sediments were deposited. The clastic debris was derived from the weathering and eroding of that craton. These strata thicken southward across what was then a passive continental margin. Then volcanism developed and the margin was deformed by the Penokean orogeny about 1.8 to 1.9 billion years

ago. It has been suggested that a microcontinent with active volcanism collided from the south to cause the orogeny, but the evidence is very obscure.

Clearer evidence of modern-style plate-tectonic behavior has been recognized by Canadian geologist Paul Hoffman in northwestern Canada, where Cryptozoic rocks are much better exposed, thanks to Pleistocene glaciation. There the northwestern margin of the continent experienced rifting about 1.9 billion years ago (Fig. 8.13). Volcanic eruptions accompanied the rifting, but the transgression produced marine passive-margin and aulacogen sedimentation (Fig. 8.24A). Carbonate strata contain spectacular stromatolites (Fig. 8.22), which, together with associated tidal flat deposits, indicate very shallow deposition on a continental shelf. Over these strata were deposited deeper muds and sandy turbidity-current deposits derived from beyond the craton to the west. These reflect the conversion of the region to a convergent margin, the onset of subduction having caused uplift and erosion of an inferred microcontinent to the west and deepening of the continental margin (Fig. 8.24B). By 1.85 billion years ago, the entire region had been rapidly upheaved, crumpled, and thrusted toward the craton during the Wopmay orogeny (Fig. 8.24C). Coarse gravel and sand eroded from the Wopmay Mountains were shed onto the craton. Today Hoffman finds parallel tectonic zones here like those typical of Phanerozoic orogenic belts. A granitic-metamorphic zone passes eastward into a folded and

WEST   EAST

A. Rifting (1.9 b.y. ago)   AULACOGEN

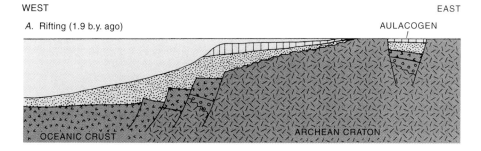

OCEANIC CRUST   ARCHEAN CRATON

B. Subduction (1.88 b.y. ago)

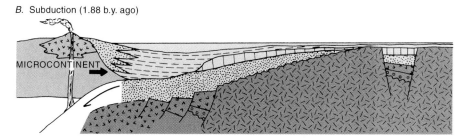

MICROCONTINENT

C. Collision (1.85 b.y. ago)

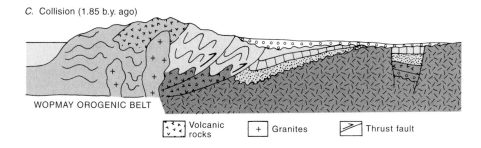

WOPMAY OROGENIC BELT

Volcanic rocks   + Granites   Thrust fault

Figure 8.24   Evolution of the Wopmay region in northwestern Canada in early Proterozoic time (see Figs. 8.11 and 8.13). This region shows the same structural characteristics as Phanerozoic orogenic belts, suggesting that plate-tectonic processes were operating much as in later times. *(Adapted from Hoffman and Bowring, 1984,* Geology, *v. 12, pp. 68–72.)*

thrust-faulted zone, which lies next to the nearly undeformed craton (Fig. 8.24C). Apparently the Wopmay region is an example of collage tectonics, with at least one "foreign" microcontinent tectonic terrane accreted to the Archean craton.

Late Proterozoic Rocks

Both the Penokean and Wopmay belts were upheaved, in their turn, into vast mountain ranges and, like older mountains, were reduced by erosion. Late Proterozoic strata deposited over these old, eroded mountains are nearly flat-lying. In the Lake Superior region, this interval was the lull before a storm, for sedimentation gave way abruptly to extensive outpourings of Keweenawan basaltic lavas (Fig. 8.25) accompanied by the intrusion of vast sills and the Duluth gabbro mass (1.0 to 1.15 b.y. ago). Following this igneous episode, sedimentation resumed, but now it produced widespread bright red–colored sandstones and shales. These strata are generally considered to be river and lake deposits, though some may be shallow-marine in origin.

Subsurface data from gravity surveys and scattered deep drill holes indicate that a belt of Keweenawan mafic, high-density rocks extends southwest from Lake Superior to northeast Kansas, probably the world's largest such mass (Fig. 8.26). Such rocks also underlie central Michigan in a band extending southeast from Lake Superior. Magmas were erupted along these trends, which cut across older structures. Beyond the rifts, some A-type granites

Figure 8.25   Columnar-jointed lavas from the Keweenawan eruptions, now exposed on the north shore of Lake Superior just north of Duluth, Minnesota. *(D. R. Prothero.)*

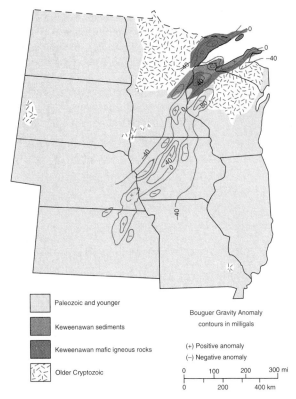

Figure 8.26   The mid-continent gravity anomaly, one of the largest such features known. Such a positive anomaly is the result of abnormally dense rocks, which cause a greater than average intensity of gravity. This anomaly is the result of Keweenawan basalt and gabbro, which extend from Lake Superior southward beneath Paleozoic strata. Apparently they occupy an old rift. *(Adapted from E. Thiel, 1956, Bulletin of the Geological Society of America, v. 67, pp. 1079–1100; by permission of Geological Society of America.)*

Legend:
- Paleozoic and younger
- Keweenawan sediments
- Keweenawan mafic igneous rocks
- Older Cryptozoic

Bouguer Gravity Anomaly
contours in milligals

(+) Positive anomaly
(−) Negative anomaly

0   100   200   300 mi
0   200   400 km

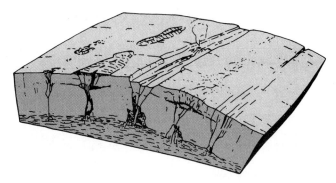

Figure 8.27   Extensional rifting and outpouring of lavas along a broad upwarp in cratonic crust as envisioned for the late Proterozoic Keweenawan of north-central United States and the Cenozoic East African rifts. *(After H. Cloos, 1939, Geologische Rundschau, v. 30, p. 401; by permission of Geologische Rundschau.)*

were intruded into the old cratonic crust. The Keweenawan rocks are only mildly deformed, unlike those of the old orogenic belts; therefore, the former represent a hitherto unprecedented condition in the central North American crust. Basalts are rock types more characteristic of oceanic areas and, so, are anomalous within cratons. Profound rifting of the continental crust occurred, allowing mafic magmas to rise to the surface from below the crust (Fig. 8.27). The lavas spread out on the surface of what was otherwise a stable, passive craton. Better-known late Cenozoic rifting and volcanism of similar scale are known in East Africa and in the Pacific Northwest of the United States.

The lavas in all three cases rose through fissures and flooded immense areas, burying everything in sight and gradually leveling the landscape; hence, they are called *flood basalts.* Collectively they represent an important class of tectonic features of the earth's crust. Flood basalts differ from andesitic volcanic rocks typical of orogenic belts in that (1) they are almost exclusively basaltic; (2) with few exceptions, they occur in cratonic areas; and (3) they are only mildly deformed. Clearly they require a structural explanation different from that for orogenic belts. Perhaps they are sites where huge crustal plates ruptured apart, as

mentioned in Chap. 7. The East African rift valleys, especially the Red Sea, are sites of incipient continental drifting. In North America, splitting of the continent apparently occurred 1.2 billion years ago, with this Keweenawan rift possibly representing a failed arm from the larger rift postulated in Fig. 8.13.

Almost immediately after the Keweenawan rifting, convergence along the southeastern margin of North America caused unusually severe compression and metamorphism, called the Grenville orogeny (Fig. 8.13, Table 8.1). This event is mysterious but is tentatively interpreted to be the result of a collision by another continent or microcontinent, which produced unusual heating of older crust across a belt 700 to 800 kilometers wide (Fig. 8.11).

The Cryptozoic Ocean and Atmosphere

The Orthodox View

During the 1950s, biochemists developed a theory for the origin of life that has had a profound influence upon geologic thinking ever since. That theory, which is outlined more fully in Chap. 9, required an oxygen-free, or **anaerobic,** ocean and atmosphere. Limited experimental results seemed to confirm the inference that complex protein molecules could not have formed if free molecular oxygen (O_2) existed at the earth's surface. Naturally the biochemists then asked if the geologic record supported this apparent requirement. Geologists quickly asserted that, "yes, indeed, Archean rocks bear evidence of anaerobic early conditions." As noted in Chap. 6, both chemists and geologists then hypothesized how our present oxygen-rich, or **aerobic,** atmosphere might have developed slowly.

What was the evidence cited for anaerobic Archean conditions? First, many Archean sediments are dark-colored due to the presence of unoxidized carbon (C), fine iron sulfide (FeS_2), and iron carbonate ($FeCO_3$) minerals. These constituents seemed to be more abundant in Archean than in younger strata, and if free

O_2 had been present, they should have been oxidized to other forms because both carbon and iron have strong affinities for oxygen. Moreover, it was argued that Archean rocks contain not only iron but also several other metals with affinities for oxygen, including manganese, copper, zinc, vanadium, and uranium, which are present widely in their least oxidized states. Sulfur, too, which has an affinity for oxygen, was thought to be present only in its unoxidized or reduced state, occurring especially in the mineral pyrite (FeS_2). Finally, differences between sulfur isotopes in Archean and Proterozoic sediments were thought to reflect a contrast in atmospheric oxygen content.

Conventional wisdom for the past 30 years has postulated that several Archean chemical "sinks" consumed any available early oxygen. These included the oxidation of hydrogen (H_2O) and carbon (CO_2). Only after these sinks were fully saturated with oxygen could sulfur, iron, and other elements begin to be oxidized. It was inferred that this saturation occurred sometime after photosynthetic organisms appeared and began to release O_2 into the atmosphere. Appearance of widespread, red-colored strata, called "red beds," among Proterozoic rocks seemed to confirm a gradual accumulation of free O_2. Their red color is due to a small percentage of completely oxidized (ferric) iron, which is a potent coloring pigment. The presence of oxidized sulfur ($CaSO_4$) in Proterozoic evaporite deposits also was cited as supportive evidence (Fig. 8.28).

Distinctive Cryptozoic banded iron formations ("BIFs" to the specialists) together with stromatolites have played a central role in the theory of gradual accumulation of atmospheric oxygen. Banded iron formations are made up of bands of metallic iron and chert, or cryptocrystalline silica (also known as flint, jasper, or novaculite). Chert is chemical in origin rather than clastic. It forms in two ways: by primary precipitation from water or by secondary replacement of original sedimentary grains. The primary type may be precipitated from seawater to produce bedded chert. The red chert (called jasper) interlaminated with pure iron in the Cryptozoic banded iron formations is an example (Fig. 8.7). This type forms only in the absence of clastic sand or clay. Red chert occurs especially near submarine springs and volcanic vents, which provide abundant silica in solution.

The second type of chert replaces original sedimentary rocks, such as shale or limestone, when silica is precipitated from solutions percolating slowly through the sediment. Irregular lenses, or chert nodules, form where silica concentrates, filling pores and chemically replacing preexisting material, such as the calcite in limestone. This silica is derived mostly from solution of the siliceous skeletons of certain microscopic organisms (e.g., diatoms, sponges, and radiolarians—Fig. 8.29) or from volcanic ash buried in the sediments. Both skeletons and ash contain silica in a chemically unstable form, which readily dissolves after deposition to be redistributed and precipitated by pore fluids as stable quartz.

Banded iron-chert deposits occur chiefly in Late archean and early Proterozoic rocks, which suggests some unique chemical condition. They have long been a puzzle not only because of their apparent confinement to Cryptozoic time but also because iron and silica show different chemical behaviors today. Iron tends to be transported in acid and precipitated in alkaline conditions, whereas silica does just the opposite. Because the peculiar association of iron with silica seemed to predate widespread red beds, it appeared to be related chiefly to oxygen-poor conditions, although there may have been some unknown biochemical influence, too. With CO_2 perhaps ten times more abundant than now, seawater would have been less alkaline and could have carried more calcium in solution. Abundance of CO_2 would have caused acidic rain and ground water (H_2CO_3), which would have dissolved much iron. If the entire early ocean and atmosphere were anaerobic, iron also could have been abundant in solution throughout the less alkaline seas. Depending upon local chemical conditions when it was deposited, the iron would combine with sulfur, carbonate, or silica (see Fig. 8.28 and Box 8.3). Because the atmosphere presumably was anaerobic, iron oxide minerals could form only locally where oxygen was present. American paleontologist Preston E. Cloud suggested that cyanobacterial colonies created such local, oxygen-rich oases around the stromatolites. Once free O_2 became generally abundant, however, most iron was deposited in oxide form. Iron was fickle, for it ceased its love affair with silica when the attractive newcomer O_2 appeared.

An Alternative View

Geologic evidence cited for Archean anaerobic conditions may not stand up well to critical scrutiny, and hindsight suggests that perhaps geologists "found" what the biochemists wanted. Some biologists long ago were puzzled by the geologists' slow-accumulation argument. These scientists reasoned on biological grounds that, once photosynthesis began, it should have generated free O_2 almost instantaneously in a geologic sense. Geologists countered that it took at least 1 billion years to "soak up" all available oxygen in the vast carbon and iron sinks of the Archean ocean.

A growing number of geologists have challenged the conventional wisdom on the following grounds. (1) The unoxidized carbon in Archean sediments may not be significantly different, either in total abundance or in isotopic makeup, from carbon in younger ones; moreover, free carbon as well as iron sulfide and iron carbonate minerals have been preserved in oxygen-poor muddy environments (such as swamps) right up to the present day. (2) Unoxidized manganese, copper, zinc, vanadium, uranium, and the like, although widely scattered, are not very common in Archean sediments and, so, like free carbon and iron sulfide, might be explained by local anaerobic conditions of deposition. (3) Some Archean red beds and oxidized sulfate evaporites have now been discovered, as well as some oxidized Archean soils. Therefore, challengers to the conventional wisdom believe that significant free O_2 existed even in the Archean atmosphere.

What about the iron/stromatolite argument? Here again, the evidence is ambiguous at best. Archean examples with red iron bands just as strongly oxidized as any younger ones have been known for over a century. And the many Cryptozoic stromatolites (as well as their associated shallow-marine sediments) today lack carbon, although they must have contained a great deal of it when

Box 8.3

Possible Highlights in the Chemical Evolution of the Cryptozoic Atmosphere-Ocean System

The following table suggests some of the generalized chemical reactions that probably occurred among the key elements hydrogen, oxygen, carbon, iron, sulfur, and silicon. "Early" and "Later" are relative only. Figure 8.28 contrasts the two competing views about the timing of the accumulation of sufficient oxygen to cause the "Later" reactions shown in the table.

Early: Any original oxygen quickly consumed by hydrogen and carbon

$2H_2 + O_2 \rightarrow 2H_2O$ (water)

$C + O_2 \rightarrow CO_2$ (carbon dioxide)

$Fe + CO_3 \rightarrow FECO_3$ (siderite)

$Fe + 2S \rightarrow FeS_2$ (pyrite)

$Fe + SiO_2 \rightarrow$ various iron silicate minerals

Later: With increasing oxygen available from photosynthesis

$4Fe + 3O_2 \rightarrow 2Fe_2O_3$ (hematite)

$Ca + S + 2O_2 \rightarrow CaSO_4$ (anhydrite evaporite)

$Ca + CO_3 \rightarrow CaCO_3$ (limestone)

$FeCO_3$, FeS_2, and iron silicates continued to form in oxygen-poor environments.

Figure 8.28   Two competing views of the timing of oxygen accumulation.

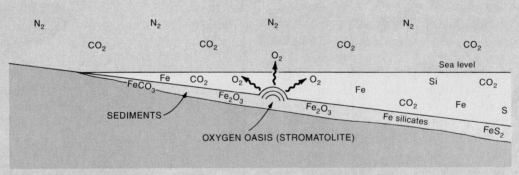

A SLOW ACCUMULATION OF OXYGEN

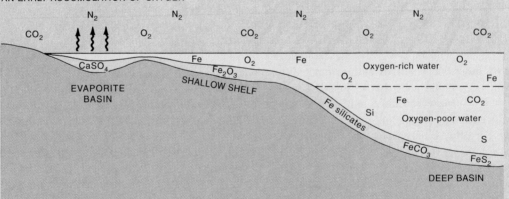

AN EARLY ACCUMULATION OF OXYGEN

Figure 8.29 Scanning electron microphotograph of siliceous rock known as chert, made up of the needlelike spicules secreted in the skeletons of sponges. Image is magnified 160 times. *(Courtesy of Richard Hammes.)*

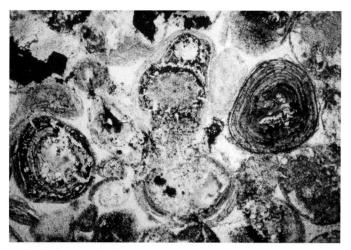

Figure 8.30 Microphotograph of concentric oolitic texture in chert, which is interstratified with iron-rich layers in a Proterozoic banded-iron formation (northern Michigan). The circular oolitic grains average about 0.5 mm in diameter. Black laminations and specks are fine-grained iron oxide. The fine, concentric laminations characteristic of oolite were precipitated from chemically saturated and agitated shallow water as grains were rolled vigorously by waves and currents (see Chap. 10). Such texture is characteristic of carbonate rocks rather than chert in the Phanerozoic record; therefore, we infer that Proterozoic oolitic limestone was deposited in an iron-precipitating environment and soon thereafter was replaced by silica to form iron-bearing chert. *(Courtesy of L. Gordon Medaris.)*

algae and bacteria formed them, implying that their carbon was oxidized. The challengers' view of iron deposition holds that much of the present character of these peculiar rocks is the result of chemical changes *after* deposition. As with soils, the chemical environment at the sediment surface may differ greatly from that at even a shallow depth. Decay of buried organic matter, activity of microbes living within the sediment, and slow percolation of waters having a composition different from that above the surface are some of the ways that sediment chemistry changes soon after burial. Minerals deposited at the sediment surface may become unstable after burial, causing both composition and texture to be altered to conform to the new conditions.

The banded-iron deposits imply quiet environments, but there are oolitic iron deposits (Fig. 8.30) that formed in agitated environments. Some may have formed as limestone, such as the stromatolites. Profound alteration later has added iron and silica, which have completely replaced the original carbonite minerals. Where the chemical environment remained oxidizing, iron oxide and iron silicate minerals are present, but if the postdepositional environment became anaerobic, then iron carbonate and iron sulfide formed together with silica. It still must be the case that some banded-iron deposits formed originally in stagnant, anaerobic environments where $FeCO_3$ and FeS_2 were precipitated. Some of these deposits have survived as such, but many have been oxidized by later weathering. Thus, whether one is orthodox or not, it seems inescapable that iron deposits have had very complex histories, from deposition in a variety of chemical environments

through variable early-burial changes and even later metamorphism and weathering to produce their present characteristics. They may have little to tell us about the atmospheric oxygen after all!

The most recent argument has come from calculations of the rates at which oxygen was absorbed by the carbon and iron sinks. According to Kenneth Towe, the amounts of carbon and iron in early Archean rocks are insufficient to account for all the free O_2 released by organisms, even assuming a global primary productivity only 1 percent of present levels. To account for this discrepancy, Towe argues that there must have been aerobic organisms breathing in the excess oxygen. Aerobic organisms apparently kept the free O_2 levels very low (about 0.2 to 0.4 percent) for most of the Archean, balancing the output of O_2 by photosynthesizing cyanobacteria and plants.

According to Towe, the abundance of organic carbon found in the oldest Archean rocks from Greenland also argues for efficient carbon cycling aided by some free O_2. How does Towe's argument account for the banded iron formations? Analysis of the concentrations of rare earth elements in banded iron formations has become a powerful tool for determining the oceanic conditions under which the formations were formed. Cerium, in particular, is a good depth indicator because it tends to be enriched in shallow waters and depleted in deeper waters. The low concentrations of cerium found in BIFs suggest that the bands were formed at some depth below the surface of the ocean. If some of them formed below the oxygenated surface waters of the ocean, they would not prove that the shallow surface waters or the atmosphere were anaerobic.

Where does the controversy leave us? Similarities in the composition of evaporites suggest little change in seawater composition for the past 1 billion years, but the chemistry of soils older than 1 billion years suggests more CO_2 and less O than now. Once all the smoke clears, it will probably turn out that the issue is simply one of timing and rates, rather than of dramatically different mechanisms. Whereas thought as recently as it was 1981 that atmospheric oxygen had reached about half its present abundance by 1.5 billion years ago, it now appears that the process occurred earlier—perhaps about 2.0 billion years ago (see Fig. 6.14). Photosynthetic organisms may have worked faster than geologists thought!

Cryptozoic Climate

Meager Evidence

Evidence about Cryptozoic climates is scant, but there is nothing that indicates conditions different from later geologic time. We have little inkling of the "typical" Cryptozoic climate—if there was such a thing. Chemical arguments suggest more atmospheric CO_2, a condition that should have caused a warmer average global temperature (greenhouse effect) and more acidic rains.

There is some direct evidence of climatic extremes, however. Evaporite deposits and mud cracks (Fig. 8.31) attest to dry and probably hot conditions sufficient to completely evaporate seawater locally. Proterozoic wind deposits also are known, and they suggest large desert dune fields. Evidence of cold periods is even better known.

Early Proterozoic Glaciation

Near the base of the upper Huronian sediments in Ontario, a peculiar, widespread assemblage of rocks called the Gowganda Formation occurs. The most distinctive rock type is a massive, almost completely unstratified and unsorted jumble of large boulders, pebbles, sand, and fine clay surrounded by a finer matrix of dark material. Another striking type of deposit in the formation is delicately laminated mudstone that resembles Pleistocene glacial lake clays containing laminae interpreted as seasonal layers; some of these layers contain scattered pebbles (Fig. 8.32). A widespread unconformity marks the base of the Gowganda Formation. At several localities where the surface is exposed, fine, parallel scratches are visible, which strongly resemble striations made by glaciers.

The peculiarities of the Gowganda Formation argue for glacial tills and river deposits, but some of the strata doubtless accumulated in the sea. Today many glaciers extend into the sea along the southern Alaska, Greenland, and Antarctica coasts. Drifting icebergs carry all manner of frozen-in rock debris offshore to be deposited helter-skelter over large areas of the sea floor as the icebergs melt. The pebbles dropped into laminated Gowganda mud-

Figure 8.31   Mudcracks in red shale from the Chuar Group, in the bottom of the eastern Grand Canyon. Rocks such as these, along with salt crystals, show that hot, dry conditions were common 1.8 billion years ago and that enough free oxygen was present in the atmosphere to turn the sediments rusty red. *(D. R. Prothero.)*

stones probably had a similar origin. A major episode of continental glaciation apparently occurred in Canada about 2.2 billion years ago. A large ice sheet spread out from the craton into the edge of the sea, much as Pleistocene continental ice sheets spread across the North Atlantic continental shelf from New England and eastern Canada. Recognition of ancient glaciation is significant in showing that the well-known Pleistocene refrigeration was not unique in history.

Late Proterozoic Glaciation

The most remarkable trademark of the late Proterozoic record is the presence of unsorted boulder-bearing glacial deposits slightly below fossiliferous Vendian strata. In our division of time into Phanerozoic and Cryptozoic eons (with the Vendian included in the Phanerozoic; see Box 8.1), these glacial deposits conveniently mark the end of the Proterozoic. What more fitting termination than a worldwide glaciation? Known as the Varangian (or Varanger) glaciation, these deposits were first described in northern Norway in 1891. Because similar deposits have been found on all but one continent (Fig. 8.33), their record is much better than that for older glaciations. Peculiar textures, wide distribution, and local scratched surfaces

Figure 8.32   Laminated mudstone with scattered pebbles and sand grains dropped from above (Gowganda Formation, near Blind River, Ontario). Association with glacial tills suggests dropping of stones from drifting icebergs. (R. H. Dott, Jr.)

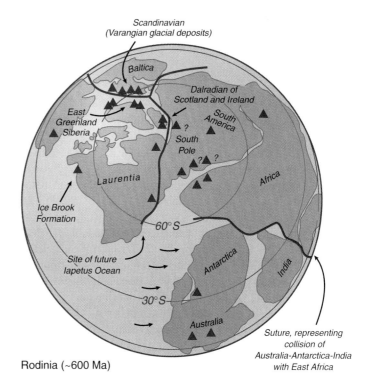

Figure 8.33   Global distribution of late Proterozoic (Varangian) glacial deposits (triangles), showing their occurrence in ancient equatorial regions. In some places, these glacial deposits are interbedded with marine limestones, further proving their low-latitude origin. Such evidence leads some scientists to suggest that the earth may have barely avoided freezing over completely in the Varangian. (After Eerola, 2001.)

Rodinia (~600 Ma)

A   *B*

Figure 8.34   Examples of Varangian glacial deposits. A: Large rocks showing glacial striations that were apparently dropped from a melting ice sheet into fine-grained marine sediments, southeast Avalon Peninsula, Newfoundland. *(R. H. Dott, Jr.)* B: Unsorted glacial till from the Kingston Peak Formation, south of Death Valley, California. *(D. R. Prothero.)*

beneath the conglomerates long ago led geologists to regard them as tills (Fig. 8.34), representing a major episode of continental glaciation between 700 and 800 million years ago. Continental ice sheets spreading onto shallow-marine shelves deposited till, which was more or less reworked by gravity sliding and currents, while icebergs dispersed and dropped boulders into normal marine sediments beyond glacier margins.

Late Proterozoic tills occur over an exceptionally broad latitudinal range, both of present latitudes and of apparent Vendian paleolatitudes (Fig. 8.33). The postulated position of the paleoequator is based upon rock magnetism and upon the distribution of younger, presumed tropical Lower Cambrian fossils called archaeocyathids. The very broad latitudinal distribution of the glacial deposits has been rationalized by assuming such a complete chilling of climate that glaciers could form at practically all lat-

itudes, or at least that icebergs could carry debris into very low latitudes.

For years, geologists tried to explain away these apparently equatorial glacial deposits by suggesting that the glaciated land masses had drifted rapidly away from the poles after the Varangian glaciation. However, some units, such as the Rapitan Group of the Canadian Cordillera, have glacial deposits between thick limestones, which are known to form in tropical or subtropical latitudes today. If this does not indicate subtropical glaciers, then either the process of limestone formation was radically different in the late Proterozoic or else this region danced rapidly from north to south and back in just a few million years!

This low-latitude glacial evidence can no longer be dismissed, because new paleomagnetic data from glacially varved sediments in the Elatina Formation of South Australia clearly show that the sediments were formed by sea-level continental glaciers only a few degrees from the late Proterozoic equator. It is now clear that the earth nearly became an icy snowball about 700 million years ago and might have ended up a frozen planet, like Mars. But what could have caused such a deep freeze? Some workers suggest that the earth's orbit became exceptionally elliptical, causing much greater temperature contrasts between winter and summer. Others have proposed that the earth's axial tilt with respect to its orbital plane became so extreme that the poles became warmer than the equator. Like Uranus, the earth's axis may have lain within its orbital plane. Both of these ideas can be ruled out by looking at the banding pattern found in stromatolites. Because cyanobacteria are highly sensitive to daily light fluctuations, they form daily bands that allow us to determine how many cycles there are in a month or year. All the stromatolites examined show that the earth's orbital cycles were not much different than those of today.

Joseph Kirschvink suggests another hypothesis for the snowball earth. He points out that plate reconstructions for the late Proterozoic place most of the continents in low and middle latitudes, with no landmasses over the poles (Fig. 8.33). This configuration has never occurred in earth history before or since. Today most of the earth's solar heat is trapped in tropical oceans, which were relatively scarce in the late Proterozoic. If shallow epicontinental seas had retreated from the Proterozoic continents, there would

have been a change from large areas of energy-absorbing oceans to large areas of highly reflective land surfaces, bouncing back the light energy and rapidly cooling the planet. As the process continued, the polar oceans would be covered by ice caps, and eventually glaciers would grow even near the equator.

However, the presence of pack ice in the oceans and glaciers on land would inhibit both silicate weathering and photosynthesis. Both of these processes tend to take CO_2 out of the atmosphere. In a glaciated world, the excess CO_2 would soon end up in the atmosphere, contributing to greenhouse warming. Thus, the growth of glaciers ultimately leads to their own destruction by rapid greenhouse warming, which melts back the glaciers. Kirschvink points out that this could produce a highly unstable planet, rapidly fluctuating between greenhouse and icehouse extremes. Such conditions might explain the fluctuations between tropical limestones and glacial tills.

The global snowball has important implications for the history of life. As we shall see in Chap. 9, for almost 3 billion years prior to the Varangian glaciation, life was entirely microscopic. Some geologists speculate that life might have disappeared entirely from this planet if the freezing had proceeded much further. After the Varangian glaciations, we see the first multicellular animals, the Ediacara fauna. Perhaps the refrigeration triggered some critical factor in evolution or opened up new ecological niches in regions that were once ice-covered.

Summary

- **Cryptozoic, or Precambrian, time** includes 80 percent of earth history, but the record of it is much more obscure than for Phanerozoic time. The greatest difference is the lack of animal index fossils older than 600 to 700 million years. Correlation must be done by physical field criteria and by isotopic dating.
- **Isotopic date provinces** within continents are based upon statistical clusters of dates from igneous and metamorphic rocks formed within various orogenic belts during several widespread **orogenies**. The patterns of such provinces generally support the hypothesis of **continental accretion.** The new granitic additions to continental crust seem to originate by **partial melting** of either older andesitic or sedimentary rocks. Most of the total volume of continental crust had formed by 2.5 billion years ago, but small accretions have continued to the present. Some old crust has been **overprinted** by younger tectonic patterns.
- The **Archean** record is dominated by two assemblages of rocks:
  1. **Greenstone belts** with mafic igneous rocks (e.g., komatiite and basalt) associated with heterogeneous immature clastic (fragmental) sediments (graywackes) rich in feldspar and volcanic rock fragments. These represent oceanic and arc material.
  2. **Gneiss belts** with varied metamorphic rocks, which probably represent remnants of embryonic, small continents. Radioactive heating was so great that little permanent crust could survive.
- By **Proterozoic** time, heat generation apparently had declined sufficiently to allow much larger masses of continental crust to survive. Clearly recognizable cratons bordered by well-defined orogenic belts suggest that plate-tectonic processes were operating much as today.
- Preserved **early Proterozoic** sediments different from Archean ones include the following:
  1. Widespread texturally and compositionally **mature terrigenous clastic** strata rich in well-sorted and rounded quartz sand are associated with **nonterrigenous chemical** carbonate and evaporite strata; **stromatolites** formed by algae or bacteria occur widely in the carbonate rocks. All these were deposited in broad, shallow epicontinental seas on stable cratons.

2. Features such as **ripple marks** as well as **cross-stratification** formed by **dunes** indicate transport of sand by **tractional currents** whose directions of flow can be deduced from such features. **Graded bedding** indicates deposition from **suspension** as from a sediment-laden **turbidity current.**

- **Late Proterozoic time** was characterized by important large-scale rifting, which was accompanied by eruption of widespread **flood basalts** in the central United States. The Proterozoic was closed by the Grenville orogeny on the southeastern margin of the continent.

- The **Cryptozoic atmosphere and ocean** were long thought to have been **anaerobic** until about 2 billion years ago, but there is abundant evidence of fully oxidized iron in Archean "red beds," **banded iron deposits,** and soils; oxidized calcium sulfate evaporites are also known. Other nonoxide sediments long thought to support the anaerobic hypothesis can be explained in terms of variable chemical environments during deposition and soon after burial. Photosynthetic organisms apparently had rendered the earth's surface at least slightly **aerobic** by 3.0 billion years ago.

- **Cryptozoic climates,** insofar as they can be deduced, seem not have been dramatically different from later ones, except that more CO_2 should have caused acid rain and warmer global temperatures due to the greenhouse effect. There is clear evidence of extremes of both glaciation and aridity.

- The end of the Proterozoic was marked by a great global glaciation event, the **Varangian glaciation,** which produced continental glaciers at nearly equatorial latitudes. The exact causes of this extraordinary glaciation are controversial, but the planet nearly became a lifeless frozen world, like Mars. When the Varangian glaciers retreated, multicellular life emerged for the first time after nearly 3 billion years of single-celled life on this planet.

- The more the Cryptozoic sedimentary record is studied, the more it seems to resemble that of later times. The absence of animal skeletons and the presence of banded-iron deposits are the most conspicuous differences. The apparent uniqueness of the Archean record seems to have been accentuated by a preservation bias, which destroyed any cratonic sequences that may have existed and badly mangled all other rocks in orogenic belts.

Readings

Cloud, P. 1972. A working model of the primitive earth. *American Journal of Science,* 272:537–48.

———. 1988. *Oasis in space: Earth history from the beginning.* New York: Norton.

Dimroth, E. 1977. Facies models 6. Diagenetic facies of iron formations. *Geoscience Canada,* 4:83–88.

Grieve, R. A. F. 1980. Impact bombardment and its role in protocontinental growth on the early earth. *Precambrian Research,* 10:217–47.

Harland, W. B., and M. J. S. Rudwick. 1964. The great infra-Cambrian ice age. *Scientific American,* 28–36.

Hartmann, J., and R. Miller. 1991. *The history of earth, an illustrated chronicle of an evolving planet.* New York: Workman.

Hoffman, P. 1984. Wopmay orogen: A Wilson cycle of early Proterozoic age in the northwest of the Canadian Shield. In *The continental crust and its mineral deposits.* Edited by D. W. Strangway. Geological Association of Canada Special Paper 20, pp. 523–49.

———. 1988. United plates of America, the birth of a craton: Early Proterozoic assembly and growth of Laurentia. *Annual Reviews of Earth and Planetary Sciences,* 16:543–604.

Hoffman, P. F., and D. P. Schraq. 2000. Snowball earth. *Scientific American,* January, 68–75.

Holland, H. D., B. Lazar, and B. McCaffrey. 1986. Evolution of the atmosphere and oceans. *Nature,* 320:27–33.

Kirschvink, J. L. 1992. Late Proterozoic low-latitude global glaciation: The snowball earth. *The Proterozoic biosphere, a multidisciplinary study,* Edited by J. W. Schopf and C. Klein. Cambridge, UK: Cambridge University Press.

Kirschvink, J. L., E. L. Oaidos, L. E. Bertani, N. J. Beukes, J. Gutzmer, L. N. Maepa, and R. E. Steinbager. 2000. Paleoproterozoic snowball earth: Extreme climatic and geochemical global change and its biological consequences. *Proceedings of the National Academy of Science,* 97:1400–1405.

Lowe, D. R. 1980. Archean sedimentation. *Annual Reviews of Earth and Planetary Sciences,* 8:145–67.

Medaris, L. G., ed. 1983. Early Proterozoic geology of the Great Lakes region. *Geological Society of America Memoir* 161.

Nisbet, E. G. 1987. *The young earth, an introduction to Archean geology.* Boston: Allen & Unwin.

Rankama, K. 1963–1968. *The Precambrian* (4 vols.). New York: Interscience.

Schopf, J. W., ed. 1983. *Earth's earliest biosphere: Its origin and evolution.* Princeton, NJ:Princeton University Press.

———, and C. Klein, eds. 1992. *The Proterozoic biosphere, a multidisciplinary study.* Cambridge, UK: Cambridge University Press.

Sims, P. K., and G. B. Morey, eds. 1972. *Geology of Minnesota: A Centennial Volume.* Minnesota Geological Survey.

Society of Economic Geologists. 1973. Precambrian iron-formations of the world (a symposium). *Economic Geology,* 68: 913–1179.

Walter, M. R. 1977. Interpreting stromatolites. *American Scientist,* 65:563–71.

Windley, B. F. 1984. *The evolving continents,* 2d ed New York: Wiley.

Chapter 9

Early Life and Its Patterns

Now in vast shoals beneath the brineless tide,

On earth's firm crust testaceous tribes reside;

Age after age expands the peopled plain.

The tenants perish, but their cells remain;

Whence coral walls and sparry hills ascend from pole to

pole, and round the line extend.

Erasmus Darwin, "Temple of Nature," 1803

▶ The oldest known fossils consist of filaments of cyanobacteria ("blue-green algae") in rocks about 3.5 billion years old. Because the heavy meteorite bombardment of the earth did not cease until 3.8 billion years ago, life must have originated in less than 300 million years early in the planet's history.

▶ Because the simple early fossils do not preserve details of biochemistry, we must simulate the origin of life through laboratory experiments. Forty years of laboratory work have shown that most basic chemicals of life and even simple "cells" with many of the properties of life can arise by natural processes.

▶ Life remained as simple, unicellular bacteria and cyanobacteria for over 1.5 billion years; multicellular animals did not arise until about 0.7 million years ago. For almost 3 billion years (almost two-thirds of earth history), the planet supported nothing more complicated than simple unicellular organisms and their colonies.

▶ The first multicellular animals were large, soft-bodied creatures known from rocks 600 to 700 million years old. The earliest animals with skeletons were tiny creatures that secreted simple caps or tubes of calcite or phosphate in the earliest Cambrian, about 550 million years ago. Slightly higher Cambrian rocks demonstrate an explosive radiation of trilobites, archaic molluscs, brachiopods, and echinoderms, as well as many experimental types of organisms with no living descendants.

When the *Apollo* astronauts returned from their voyage to the moon, they were asked what their most thrilling experience was. They were unanimous in saying that viewing the earth from space, and seeing how astonishingly beautiful and alive it is, was the most thrilling event. All the flybys of the solar system by explorer satellites show the other planets to be sterile landscapes—only the earth is a living entity. Organisms, principally through the process of photosynthesis, have created this one and only beautiful planet as we know it today from what was a dull, lifeless body over 3 billion years ago.

In this chapter, we shall be concerned with how life came to be and examine some of its early patterns. We shall investigate the origin of life by looking at the physical conditions necessary for its development. We shall address such questions as how the basic building blocks of organic molecules arose and how they became complex organic molecules, such as DNA. We shall also try to understand how the first cells came into existence. The most important event in life's history, which resulted in a new atmosphere, was the development of photosynthesis and the evolution of plants that provide animals with both food and an oxygen-rich atmosphere (Fig. 9.1).

The second part of the chapter explores the earliest record of life, from the single-celled plants that existed for several billions of years of the Cryptozoic up through the beginning of Vendian time, when the first multicellular animals appeared in the fossil record. We shall then look at the beginning of the Cambrian Period, when the first shelled, multicellular organisms appeared, and trace through the Cambrian the development of integrated and increasingly diverse faunas. Those unfamiliar with the classification of organisms should consult Appendix I.

The Origin of Life

Early Concerns

Ever since humans began to think, we have wondered where we came from and how all other living things came to be as they are. Before the theory of evolution was developed, people thought in terms of multiple origins. They observed that higher animals mated to produce offspring, though the exact process of conception eluded understanding until the invention of the microscope. One important early concept was that, because the female produced the offspring, only she was involved with heredity. With the development of the microscope, the sperm was discovered, a discovery that led to the curious idea that the sperm contained a complete miniature adult and the female provided only the environment for growth of the fetus. A more vexing problem was that of small living things that seemed to spring to life from inorganic matter in a putrefying environment. Experiments by Pasteur on putrefaction, along with the use of the microscope, demonstrated that even the lowest forms of life produce offspring by transferring a portion of their cell nuclei.

Thus, prior to the microscope, people confused reproduction with the actual origin of life. During the Middle Ages, many

Figure 9.1 Large, turban-shaped algal stromatolites exposed at low tide in Shark Bay, western Australia. The stick just above center is approximately 1 meter long. These concentrically layered, domed structures are formed when cyanobacteria (formerly known as "blue-green algae") trap sediment with their mucous coating, then grow upward through the sediment layer each day. Today large stromatolites are a very unusual occurrence, because most are heavily grazed by a variety of invertebrates. However, this scene must have been typical for over 3 billion years (b.y.) because stromatolites are the only common megascopic fossils known from rocks over 3.6 b.y. old until the Early Cambrian, about 540 m.y. ago. *(Courtesy of Paul Hoffman.)*

scholars were preoccupied with the idea of **spontaneous generation,** believing that putrefaction somehow produced a miraculous metamorphosis of nonliving to living matter. Father Athanasius Kircher (1602–1680), professor of science in the College of Rome, for example, divided animals into two groups, one that reproduced sexually and another that formed continuously by spontaneous generation. "It is obviously pointless to give these latter forms a place in the already encumbered ark," he wrote in discussing the passenger list for Noah's voyage. Even Buffon believed that organic molecules, released by putrefaction, came together to form simple organisms. It was Pasteur who finally proved by his famous sealed-flasks experiment in the early 1860s that there is no spontaneous generation.

Pasteur was asked by the wine growers of southern France to find out why their wines were turning to vinegar. He believed that airborne microbes may have infected the wine and then converted it. He designed balloon-shaped flasks made with long, drawn-out, wavy necks. The airborne bacteria entering the flasks were caught in the bends of the necks, and the liquid, sterile nutrients he had placed in the chamber of the flasks remained clear and sterile. Other flasks, with the necks broken off, containing the same sterile liquid nutrients rapidly putrefied as the entering bacteria colonized the nutrients. Thus, he showed that putrefaction caused the wine to go bad and, more important, that it was the product of an infection of bacteria and not newly created life ("spontaneous generation").

But if life does not arise from nonlife today, how did it arise in the first place? At first, this seems a paradox, but one critical factor had not been mentioned: spontaneous generation does not occur *under present earth conditions* because the earth's modern atmosphere is too corrosive and too oxidizing to allow it. (Just think how quickly iron objects rust when exposed to air and moisture.) The early atmosphere had almost no free molecular oxygen and was rich in the right kinds of chemicals to produce life. Under these conditions, the odds on life beginning change from impossible to highly favorable. Once life arose, it would use up all the nutrients and change the environment so radically that a second spontaneous origin would be impossible.

"A Warm Little Pond?"

In a letter to a friend, Charles Darwin suggested that "in some warm little pond, with all sorts of ammonia and phosphoric salts, light, heat, electricity, etc. present . . . a protein compound was chemically formed, ready to undergo still more complex changes." Darwin was prophetic, but it took 60 years before Russian biochemist A. I. Oparin and British geneticist J. B. S. Haldane followed up on Darwin's idea. In the 1920s, each man independently proposed that an earth with a reducing atmosphere and abundant methane (CH_4) and ammonia (NH_3) would have been the ideal "primordial soup" for the origin of life. Building upon these initial ideas, several generations of molecular biologists have contributed hypotheses about life's origins.

What factors constrain these biochemical theories of life's origin? As we shall see in the next section, well-preserved microfossils show that life had originated by at least 3.5 billion years ago. Because the earth is at least 4.5 billion years old, it was more than a billion years old before the first fossils were preserved. For the first half billion years, the planet was probably too hot for much water to condense. Until 3.9 billion years ago, the earth's surface was constantly churned and vaporized by extraterrestrial impacts. The atmosphere was rich in CO_2, methane, ammonia, and N_2, but not free O_2. There was no ozone layer yet, so the earth's surface was subjected to intense bombardment by ultraviolet radiation. The earliest sedimentary rocks indicate that modern sedimentary processes were operating by at least 3.9 billion years ago. This implies liquid water and surface temperatures that were close to those of the modern earth. Sometime between 3.5 and 3.8 billion years ago, life must have begun.

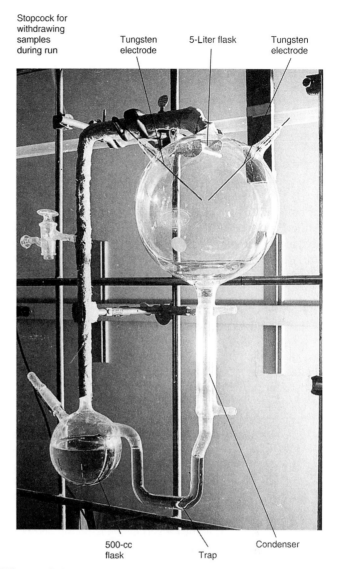

Stopcock for withdrawing samples during run · Tungsten electrode · 5-Liter flask · Tungsten electrode

500-cc flask · Trap · Condenser

Figure 9.2   An apparatus like this was used by Stanley Miller and Harold Urey to replicate the origin of organic compounds on Earth. The large flask had an "atmosphere" of ammonia and methane, and sparks from electrodes simulated lightning. Products of these reactions then went down through the condenser and flowed into the lower flask, where the primordial soup was brewing and steaming, returning gases to the upper flask. After about a week, the clear "soup" turned murky brown with organic compounds, including several amino acids. *(Courtesy of Stanley L. Miller, University of California, San Diego.)*

The exact details of this process, however, do not fossilize. Reconstructing the origin of life remains highly speculative, and there are numerous competing theories for the various steps. The first breakthrough came in 1953. Stanley Miller, a young graduate student at the University of Chicago, heard about Oparin's hypotheses from his graduate adviser, Nobel Prize–winning chemist Harold Urey. Miller set up a simple apparatus (Fig. 9.2), which contained a primitive atmosphere of hydrogen, methane, and ammonia in one chamber and a "warm little pond" that would catch the rain of condensed precipitates in a second chamber. For an energy source, he simulated lightning with sparks in the atmosphere. (Besides lightning, the early earth had other energy sources available, such as cosmic and solar radiation, meteorite bombardment, and volcanic heat.) After only a week, Miller's warm little pond turned into a muddy brown primordial soup. When he analyzed the soup, Miller found small quantities of four **amino acids,** the building blocks of proteins, the basic substance of life, plus many other organic molecules, such as cyanide (HCN) and formaldehyde (H_2CO). Finding these was remarkable, because amino acids are much more complex than the methane and ammonia in Miller's original mixture. Later experiments produced the twelve most common amino acids of the twenty known to occur in life. Another experiment with a dilute cyanide mixture produced seven amino acids. Apparently the abiotic synthesis of amino acids is not difficult, requiring only a source of the chemicals and a reducing environment. At least seventy-four amino acids have now been found in chondritic meteorites, so we know they are formed elsewhere in the solar system. Some biochemists have argued that life on earth was seeded by amino acids and nucleic acids supplied by meteorites. Whatever the source, the early solar system was clearly a rich primordial soup.

The next step in creating life is to link simple organic molecules into complex chains called **polymers.** Proteins, nucleic acids, sugars, fats, and enzymes are all complex organic polymers. Stringing together amino acids forms proteins. In the 1950s, Sidney Fox showed that, by splashing amino acids under hot, dry conditions, most of the proteins found in life polymerized instantly. Other experiments have shown that cyanide, clays, and heat are also capable of triggering condensation and polymerization of amino acids into proteins. Certain sugars polymerize quite easily by the condensation of formaldehyde. The *nucleotide bases* of nucleic acids (see Box 9.1) can be synthesized in several ways. Some of Miller's early experiments produced them. Heating an aqueous solution of ammonium cyanide produces the nucleotide base adenine, and dilute hydrogen cyanide bombarded by ultraviolet radiation produces adenine plus guanine. Along with naturally occurring phosphate, these components are essential to produce nucleic acids such as RNA and DNA.

In addition to nuclear material, the simplest cells have an outer membrane formed of **fatty acids** (the basic organic components of fats). Fatty acids are easily synthesized and have even been found in meteorites. Combining fatty acids and alcohol forms **lipids,** the chemical building block of most fats and oils. Lipid molecules are polar, with an alcohol "head" and a "tail" composed of strings of fatty acids (Fig. 9.4). The charged alcohol end is attracted to water, and the uncharged fatty acid tails are repelled by water, so lipids all line up with their alcohol heads on the outside and fatty acid tails on the inside when surrounded by water; the result is a two-layered structure. (This is why oil and water do not mix, but one "beads up" in the other.) When fatty acids are dried and concentrated, and then wetted again, they spontaneously condense into spherical balls, which also trap any DNA present. In one experiment, David Deamer found that the presence of simple proteins caused the amount of DNA trapped inside the lipid balls to be multiplied 100 times! In other words, the presence of certain proteins makes it very easy to concentrate nucleic acids and surround them with a lipid membrane.

Deamer's experiment addresses the next step beyond simple polymers—organized structures that approach cells. In addition to synthesizing proteins, Sidney Fox's experiments spontaneously produced droplets of protein about 2 microns (0.002 mm) in diameter, which he called **proteinoids** (Fig. 9.5). Oparin's experiments produced similar droplets, which he called "coacervates." These self-contained structures behave much as bacteria do. The coacervates maintain their organization when conditions change, increase in size, and bud spontaneously. Even more surprising, they selectively absorb and then release certain compounds in a process similar to bacterial feeding and excretion of waste products; some even metabolize starch! It seems that many of the properties of life are inherent in such simple abiotic structures as proteinoids and coacervates.

The Chicken-or-Egg Problem: RNA or Proteins?

The next step is the most difficult—coding and replicating complex information so that reproduction can take place. Without some kind of system for coding information and passing it on, natural selection is impossible, and the protocells of Fox and Oparin cannot be called living organisms. In modern living cells, the information is carried by a long sequence of nucleotides in RNA or DNA, which codes for a chain of proteins. Some proteins, in turn, act as enzymes to catalyze the replication of DNA. Protein synthesis and DNA replication are both essential processes, and each depends on the other, yet each is so complex by itself that it is highly unlikely that the entire system evolved at one time; it is much more likely that either each system appeared independently or one evolved first and stimulated the other. This is where the greatest controversy lies, reminiscent of the old chicken-or-egg problem. What is the most plausible means of building this predecessor of the genetic code? Which came first, the chicken (protein) or the egg (nucleic acids)? Or are there other alternatives?

Fox and others argue that the proteins came first. As we have seen, proteinoids have many of the properties of living cells, and Fox postulates that, after proteinoids formed, they tended to collect nucleic acids, which could then polymerize. The biggest difficulty with Fox's model has been the problem of replication. Today only nucleic acids can replicate, and we know of no examples of replication in a protein-based system. If proteinoids could copy themselves successfully, then why did they hand over their replication to nucleic acids?

The other major school of thought argues that nucleic acids (probably as single-stranded RNA, the genetic material of viruses) must have come first. After all, they are the only repli-

Box 9.1

Communities Without Sunlight: Deep-Sea Hydrothermal Vent Faunas

The processing of oxygen and carbon dioxide by photosynthesis to produce food involves, directly or indirectly, all but a tiny fraction of living organisms. Recently deep-undersea dives of the explorer *Alvin* found, in a series of hot springs on the crest of the East Pacific spreading ridge near the Galápagos Islands, a community of organisms that do not depend on photosynthesis. The springs, called black smokers, discharge hot, sulfur-enriched waters heated by basaltic lavas. Around the springs are rich, dense animal communities of 12-inch clams, 6-foot-long tube worms, crabs, and many other invertebrates, all living without any sunlight or even any connection to the sun's energy via photosynthesis (Fig. 9.3). Instead, the community's basic food is sulfur compounds spewing from the black smokers. Unusual sulfur bacteria harness energy from these compounds by means of chemosynthesis in a way similar to that by which plants process sunlight and CO_2 in photosynthesis. The energy for the community depends solely on these bacteria.

Similar communities have been found in many cold, deep-sea areas such as off the Florida west coast and the Oregon coast, but these communities derive their energy from methane-sulfur–enriched waters, which stem not from hot springs but from other sources, such as oil seeps. All these communities show that sunlight is not the only ultimate energy source possible in living systems.

A

B

Figure 9.3   *A:* The black smokers, chimneys made of Iron sulfide minerals, which precipitate around vents in the East Pacific Rise when mineral-rich seawater that has been heated by the upwelling magma rises upward with great force. (*Dudley Foster/Woods Hole Oceanographic Institution.*) B: The "black smoker" area in the mid-ocean rift valley is rich with life feeding on sulfur-reducing bacteria. They include 12-inch clams, strange crabs, and these 6-foot-long "tube worms." (*J. Frederick Grassle/Woods Hole Oceanographic Institution.*)

cating material found in living organisms. In this scenario, the nucleic acids function as a sort of "naked gene." Here the primary objection is one of complexity. Nucleic acids are far rarer and more difficult to synthesize than are protein polymers. Years of experiments have produced nucleic acids only a few dozen nucleotides in length, yet hundreds are needed for any true genetic code. In addition, all modern living cells require enzymes to cat-

alyze the replication reaction. How did "naked genes" copy themselves without enzymes?

This dilemma has been partly answered by the recent discovery of a type of RNA *that can also act as an enzyme.* Known as **ribozymes,** these RNA molecules are found in the **ribosomes,** the knots of RNA and protein in the cell that decode the DNA and assemble the protein polymers. In the naked-gene scenario,

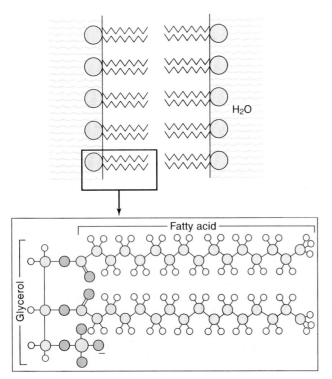

Figure 9.4   The basis for all living membranes is the lipid bilayer. Lipids, the building blocks of all fats and oils, are polar molecules with a head made of glycerol, which is attracted to water, and tails made of two fatty acid chains that are repelled by water. When lipids encounter water, they line up with the glycerol facing the water and the fatty acid tails pointing away from it. Lined-up lipids then form a bilayered membrane. Examples include the boundary enclosing oil droplets in water and the walls of living cells.

Figure 9.5   In experiments by Sidney Fox, hot (140°C) solutions of proteins that cooled slowly to about 70°C spontaneously formed these microspheres (about 1 to 2 microns in diameter) known as proteinoids. Fox argued that life may have originated by a similar process, and indeed proteinoids have been found in pools of water adjacent to the vents of Hawaiian volcanoes. *(Scanning electron micrograph by Steven Brooke, courtesy of Sidney W. Fox.)*

nucleic acids (RNAs) that catalyze themselves evolved first, producing an RNA world. Eventually the strands of nucleic acid began to assemble protein polymers, which took over the enzymatic function and developed into our present system. Currently this scenario is the most widely accepted.

Mud, Kitty Litter, and Fool's Gold

But are these the only possibilities? Graham Cairns-Smith argues that both scenarios are too complex. We need to look for something far simpler than complex organic polymers if we want to build a protocell that could replicate. Cairns-Smith proposes yet another alternative. The key to replication is a **template,** or some kind of scaffold or guide that could line up organic molecules in close order, allowing them to polymerize until they could do so on their own. Just as God is said to have created Adam from the dust of the earth, Cairns-Smith argues that life began as clay. At first, this seems preposterous. Life is based on carbon, not on silicate minerals. But clays are natural catalysts, causing some chemical reactions to occur hundreds of times faster than they do in sterile test tubes. Cairns-Smith points out that clay minerals have an open structure, which can act as a template and absorb organic molecules. The basic units of the clay mineral repeat again and again with slight imperfections, and these imperfec-

tions could cause "mutations" in the sequence. Clays could absorb molecules from the environment and catalyze their breakdown and synthesis into other substances. In short, clays would grow, modify their environment, and replicate—a very low-tech version of life. If they became enclosed in an organic membrane and trapped nucleic acids, clays would have all the components for life. Then Cairns-Smith postulates a *genetic takeover* event. High-tech nucleic acids would take over from the low-tech clays because nucleic acids can hold more information and replicate it more efficiently. The result would be fully organic life, taking over the world from the silicate-based "replicators," which were not really "life" as we understand it.

At the moment, Cairns-Smith's ideas are very controversial, but testing them in a laboratory should be possible. As long as we're considering inorganic templates, we should look at zeolites, complex silicate minerals that form by the breakdown of volcanic glass or collect in holes left by gas bubbles in lava. These silicates also have a complex, repeating structure, like clays, with surfaces that catalyze organic reactions. In fact, that is their primary use today. Most petroleum refining uses zeolites as catalysts, and they are such powerful absorbers of organic molecules that they are commonly used in filters and kitty litter.

The most recent and daring template hypothesis has been proposed by Gunter Wächtershäuser. Instead of clays or zeolites, Wächtershäuser suggests that iron disulfide, or pyrite (FeS_2, also known as "fool's gold"), was the original template. Pyrite occurs in great abundance in the reducing environments of hot volcanic vents of the deep oceanic spreading ridges (see Box 9.1). The crystal surfaces have a positive charge, which could attract negatively charged molecules, such as the phosphate backbones found in many organic molecules (especially nucleic acids). If a series of simple organophosphate compounds were lined up, all with their phosphatic ends attached to the pyrite, they would bump against one another, like guests at a crowded party, and begin to polymerize via organic bonds. Then they could unzip from their pyrite template, becoming free organic molecules. This scenario works not only for nucleic acids but also for cell membranes, for the origin of metabolic cycles, and for an astonishing number of biochemical reactions that involve phosphate groups and/or sulfide catalysis.

Wächtershäuser's hypothesis is so new that little experimental work has been done to test it. However, there are several powerful supporting lines of evidence. Deep-sea hot springs (Box 9.1) would be excellent environments for early life because they provide energy and nutrients in an environment shielded from meteorite impacts and cosmic radiation. More important, these springs are the home of primitive bacteria that use hydrogen sulfide (H_2S, the source of "rotten egg" smells), rather than water, as their source of hydrogen. Carl Woese has studied the genes of a number of types of bacteria and finds that some of them—those that live on hydrogen sulfide, on methane, or in extremely hot, salty springs—are the most primitive organisms alive (Fig. 9.6). Woese calls these organisms **Archaebacteria** and puts them in a separate kingdom, one much more primitive than the **Eubacteria** familiar from petri dishes. If the most primitive life was sulfide-based, rather than oxygen-based, then perhaps it arose on sulfide templates in submarine volcanic vents some 3.8 billion years ago. Rather than Darwin's "warm little pond," many scientists now suspect that life may well have originated in hot springs or sulfide-rich volcanic vents.

The Record of Early Life Before the Paleozoic

These ideas are interesting, but where is the evidence? Scenarios do not fossilize, so it will be difficult to rule them out by clear-cut geologic data. The oldest known fossils are thought to be the remains of single-celled **cyanobacteria.** Like Archaebacteria and other eubacteria, cyanobacteria are **prokaryotes**—single-celled organisms whose genetic material is not organized into a discrete nucleus (Fig. 9.7). Cyanobacteria are blue-green because they have chlorophyll, the chemical that allows plants to convert light, water, and carbon dioxide into complex organic substances during **photosynthesis.** (Green, red, and brown algae also photosynthesize, but their cells have a discrete nucleus—they are **eukaryotes.** That is why cyanobacteria are not really algae.)

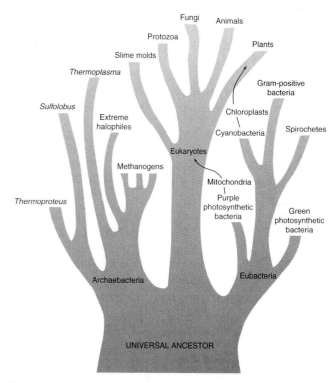

Figure 9.6 Family tree of life based on molecular similarities of RNA. About 3.5 billion years (b. y.) ago, the oldest fossils show, life had already split into two groups, the true bacteria (including most of the familiar bacteria, plus cyanobacteria and spirochaetes) and the Archaebacteria (mostly organisms that can live in extremely hot, saline, or anoxic waters; these include sulfur-reducing and methane-loving microbes). By 1.1 b.y. ago, there was a major radiation of the third major group, the eukaryotes, which include all single-celled organisms with nuclei and all plants, animals and fungi. *(Modified from Woese, 1984, The origin of life, Carolina Biological Supply Co.)*

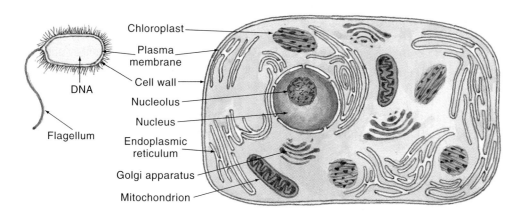

Figure 9.7 Prokaryotes, such as Archaebacteria and true bacteria, are small cells (only a few microns in diameter), and their genetic material (DNA) is not enclosed in a discrete nucleus. Eukaryotes (all other living organisms) have much larger cells (tens to hundreds of microns in diameter), with a discrete nucleus containing their genetic code in the form of DNA. They may also have a number of other organelles, including mitochondria for energy processing and possibly chloroplasts for photosynthesis.

Photosynthesis was an important breakthrough for the earth's first organisms. It represents a significant step because the earliest forms of life extracted their energy and nutrients from the biochemicals in the "soup" around them. Photosynthetic organisms, on the other hand, are **autotrophs** (literally, "self-feeding"). They get their energy directly from the sun by converting carbon dioxide and water into complex carbohydrates. The chemical reaction for photosynthesis is

$$CO_2 + H_2O + light \rightarrow (CH_2O) + O_2$$

Chlorophyll facilitates this reaction by acting as an "antenna," which traps photons of light. These photons, in turn, drive the reaction with their energy. Photosynthesis not only provides a mechanism for life to trap more nutrients and energy in the biosphere but also releases oxygen into the atmosphere as an important byproduct.

The current record holder for the oldest fossils occurs in rocks approximately 3.5 billion years old from the Warrawoona Group in northwestern Australia (Fig. 9.8A–E). Similar microfossils about 3.4 billion years old are found in the Onverwacht and Fig Tree Groups in the eastern Transvaal, South Africa. Both of these localities yield simple spherical microfossils, about 5 microns in diameter, arranged in strings that closely resemble cyanobacterial filaments. Although these tiny microfossils are small and poorly preserved, they are the oldest clear evidence of life we possess. Organic carbon from older rocks (such as the 3.8-billion-year-old rocks of Greenland) is alleged to be biogenic, but the evidence is not conclusive.

Much more impressive than their microfossils are the megascopic structures produced by these cyanobacteria. Today cyanobacteria form thick, spongy mats in shallow-marine lagoons. These mats grow rapidly when light permits photosynthesis. Their sticky filaments trap sediment but soon grow through the sediment to form a new layer. Eventually these layered mats can form cabbagelike domes or even laminated pillars (Figs. 9.1 and 9.8F), known as stromatolites ("layered rock" in Greek). Stromatolites are the commonest of Cryptozoic fossils; in some places, they form columns many meters tall. They are the only megascopic fossils in rocks from 3.5 billion to 700 million years in age. More important, they are found in the Warrawoona, Fig Tree, and Onverwacht Groups, along with the oldest microfossils (Fig. 9.8F).

It is remarkable that, after the appearance of these first fossils, *life showed almost no visible signs of change for 1.5 billion years!* Virtually all the fossil localities from the later Archean and early Proterozoic produce only single-celled prokaryotes: cyanobacteria plus fossils that suggest other types of bacteria. Stromatolitic mats became abundant about 2.8 billion years ago, and colonial bacteria and double-walled spheroids are known from about 1.7 billion years ago. The pond scum found today in salty lagoons is made up of true "living fossils," representing the typical form of life on this planet for most of its history! J. W. Schopf suggests several factors for such slow evolution. Cyanobacteria and other prokaryotes are extremely hardy and very flexible ecologically, so they can thrive under harsh conditions. No catastrophe ever completely wipes them out. Cyanobac-

teria are known to live in every possible setting, from extreme cold (near absolute zero) to extreme heat, in the driest deserts, in fresh water and extremely salty water, in acidic and alkaline conditions, and even in atomic blast zones. One specimen came back to life when it encountered tap water after 107 years in dry storage in a museum collection. They are the ultimate generalists, living in nearly every possible habitat on Earth. Another reason for their extreme conservatism is the fact that they reproduce asexually, so they cannot produce the same variability and potential for change typical of sexual organisms.

Gross cellular shape may have been unchanged for 1.5 billion years, but there is good evidence that considerable biochemical evolution was occurring. The earliest life was anaerobic, just as some primitive Archaebacteria that live in hot springs today. These earliest organisms are presumed to be **heterotrophs,** using outside nutrients (such as H_2S or methane) for their energy source. Recall that cyanobacteria are autotrophs and release molecular oxygen (O_2) into the atmosphere as a waste product. About 3 billion years ago, the shallow waters of the globe were filled with stromatolitic cyanobacterial mats, busily absorbing carbon dioxide from the atmosphere and pumping oxygen back in. As we saw in Chap. 8, however, most geologists argue that the atmosphere did not have much free O_2 until almost 2 billion years after photosynthesis started. The earth had several oxygen sinks to absorb oxygen, including banded iron formations and volcanic gases. In addition, some of these prokaryotes may have evolved the ability to use oxygen when it was available; in other words, they were aerobic organisms.

Between 2.0 and 1.8 billion years ago, an "oxygen holocaust" may have changed the earth. The geologic evidence reviewed in Chap. 8 suggests that the atmosphere had significant free molecular oxygen for the first time. Banded iron formations were replaced by oxidized red beds (sandstones and shales), and there were no more pyritic conglomerates or uraninite deposits (found only in reducing conditions). Oxygen from photosynthesis by cyanobacteria had reached a critical level in the atmosphere, estimated at about 1 percent of present, triggering global changes in climate and especially in life. Anaerobic organisms were driven below the surface by this corrosive "air pollution," which is now essential to most life, but some prokaryotes adapted to the new environment. Many of the more complex colonial bacterial fossils and double-walled spheroids found in rocks about 1.7 billion years in age must have been aerobic bacteria, living in fresh water and on the sea surface, where the oxygenated atmosphere could diffuse and form oxygen-rich oases.

The availability of free oxygen triggered another change. Microfossils from deposits about 1.75 billion years old and younger are much larger than the 10- to 60-micron size range typical of prokaryotes; such large cells are known primarily from eukaryotes. From 1.75 to 1.2 billion years ago, probable eukaryotic fossils were low in diversity, represented by simple unicellular structures known as **acritarchs** (Fig. 9.9). These spherical, organic-walled structures are thought to be the cyst stage of some kind of true algae, but their biological affinities are unknown. Whatever their origin, the large size (typically 60 to 200 microns, but some specimens up to 1 mm in size) suggests that

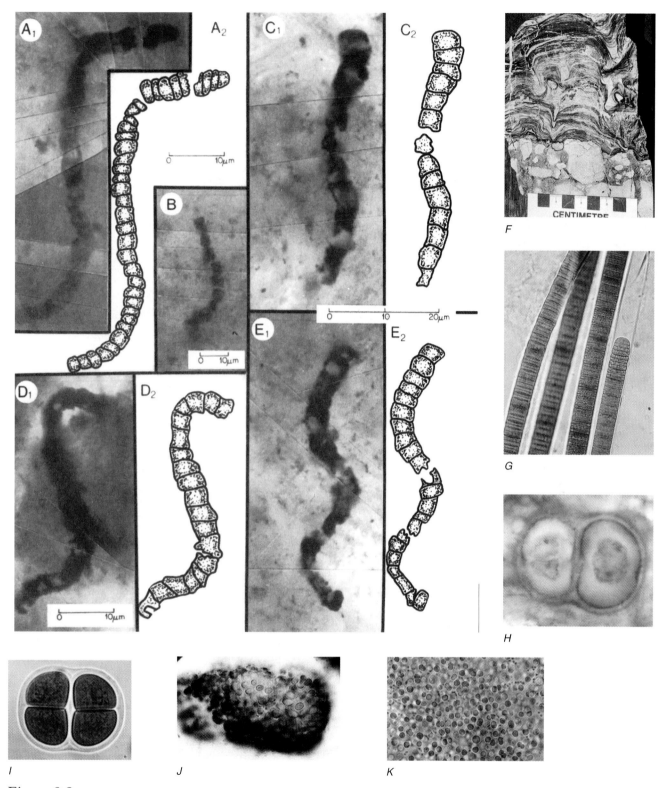

Figure 9.8   *A–E:* Some of the earliest known microfossils. These well-preserved filaments of cyanobacterial cells are from cherts at the Apex locality in the Warrawoona Group in western Australia, about 3.46 b.y. in age. *F:* One of the oldest known stromatolites, from the 3.4-b.y.-old beds in the Swaziland Supergroup of South Africa. *G:* A modern filamentous cyanobacterium, *Lyngbya,* whose cells are identical in size and shape to the cyanobacterial fossils from the Warrawoona Group. *H:* A four-celled colonial cyanobacterium surrounded by a thick sheath, from cherts in the Ural Mountains about 1.55 b.y. in age. *I:* Living *Gloecapsa,* a cyanobacterium identical in size and shape to the four-celled fossil. *J:* A colonial cyanobacterium called *Eoentophysalis* from cherts about 2.15 b.y. old in the Belcher Islands of Hudson Bay, Northwest Territories of Canada. *K:* Living *Entophysalis,* a colonial cyanobacterium identical to *Eoentophysalis* in size and shape. *(Photo F courtesy of Donald R. Lowe; photo J courtesy of H. J. Hofman; all other photos courtesy of J. William Schopf.)*

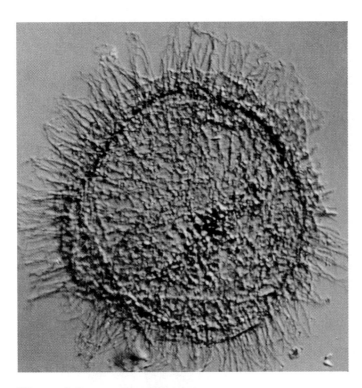 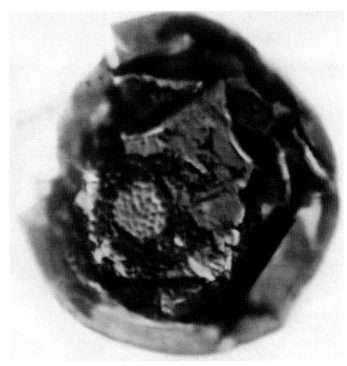

Figure 9.9   Microphotographs of acritarchs, organic-walled microfossils that may represent the cyst stages of early eukaryotic algae. Acritarchs were the commonest eukaryotic fossils in the late Proterozoic. Some were simple spheres, whereas others had complex wall structure or spines projecting outward. They ranged in diameter from 30 to over 100 microns. *(Courtesy of G. Vidal and M. Moczydlowska.)*

they are of eukaryotic origin. The world must have been colonized by eukaryotic algae and protozoans in the middle and late Proterozoic.

The origin of the eukaryotic cell from prokaryotes is not a trivial step. In addition to a discrete nucleus containing the genetic material, a characteristic that makes the cell a eukaryote, most eukaryotic cells have other internal structures (Fig. 9.7). All photosynthetic plants (including algae) and protozoans have *chloroplasts,* the structures that hold the chlorophyll needed for photosynthesis. Nearly all eukaryotic cells have *mitochondria,* the tiny organelles that are the site of energy conversion. Many protozoans also have a whiplike tail called a *flagellum,* which propels them through the water. It seems difficult to imagine how a simple prokaryotic bacterium could have subdivided itself to make these complex organelles. There is a simpler way: eukaryotes may be a living colony formed by **symbiosis.**

We are all familiar with symbiotic relationships, in which two organisms help each other to mutual benefit. The tick birds and oxpeckers riding on the back of a rhino get a free meal and protection, and the rhino is rid of parasites and gets a valuable lookout. There are many examples of symbiosis wherein one organism lives inside another. Our intestines are full of the bacterium *Escherichia coli* (the familiar *E. coli* from petri dishes). Bacteria actually do much of our digestion for us, providing us with nutrients in exchange for a home and food. Termites, sea turtles, and cattle

have bacteria in their digestive tracts that break down the cellulose in their plant diets. Most tropical corals and giant clams have algae in their tissues; these algae produce oxygen and help secrete the carbonate for skeletons. Lynn Margulis argues that the first eukaryotes arose from a symbiotic association of several types of prokaryotes (Fig. 9.10). If a large, thin-walled Archaebacterium were to incorporate cyanobacteria, they could evolve into chloroplasts. Anaerobic bacteria, such as purple nonsulfur bacteria, could become mitochondria. Spirochetes (prokaryotes responsible for diseases such as syphilis) have the same fibrous structure as the eukaryotic flagellum. Each of these once independent prokaryotes could have been colonized, providing energy, nutrients, or locomotion in exchange for protection.

Although this hypothesis once seemed far-fetched, it is widely accepted now, because some striking evidence supports it. Mitochondria and chloroplasts both have their own genetic material and can synthesize proteins on their own. They also divide independently of the cell in which they reside. Mitochondria and chloroplasts have their own ribosomes, which are the same size as prokaryotic ribosomes and smaller than ribosomes found elsewhere in a eukaryote. Mitochondria and chloroplasts are affected by antibiotics, such as streptomycin and tetracycline, just the way bacteria are; the rest of the eukaryotic cell is not. In addition, there are a number of examples of such symbiosis among the living bacteria. Some of the most primitive eukaryotes, such as the

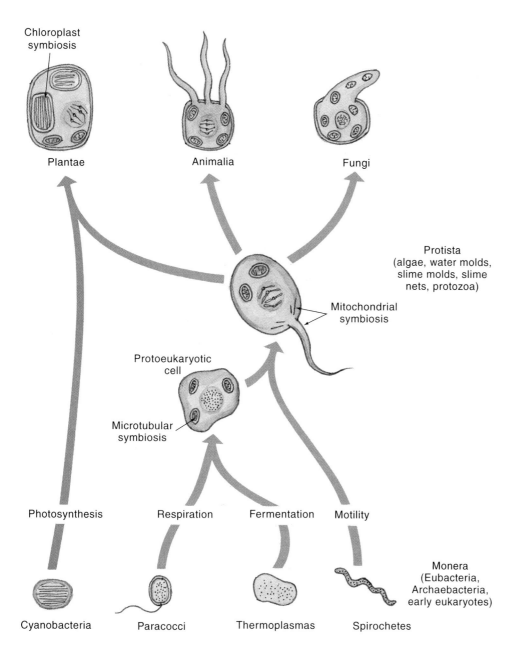

Chloroplast symbiosis

Plantae

Animalia

Fungi

Protista
(algae, water molds,
slime molds, slime
nets, protozoa)

Mitochondrial
symbiosis

Protoeukaryotic
cell

Microtubular
symbiosis

Photosynthesis

Respiration

Fermentation

Motility

Monera
(Eubacteria,
Archaebacteria,
early eukaryotes)

Cyanobacteria

Paracocci

Thermoplasmas

Spirochetes

Figure 9.10 Symbiotic origin of eukaryotes. According to Lynn Margulis, simple bacteria were colonized by purple nonsulfur bacteria, which became mitochondria. The whiplike flagellum that propels many protists may have originated from spirochetes. These associations could have given rise to either animals or fungi. If the supercolony incorporated photosynthetic cyanobacteria, these could have become chloroplasts, producing the first plants. Such symbiotic associations can be seen today in the primitive freshwater protozoan *Pelomyxa,* which has a nucleus (so it is a eukaryote), but its DNA is not arranged into chromosomes. It has no mitochondria but, instead, depends on a symbiotic bacterium living inside the cell.

freshwater amoeba *Pelomyxa,* lack mitochondria but contain symbiotic bacteria that perform the same respiratory function. In the laboratory, amoebae have evolved to become symbiotic with certain bacteria introduced into their tissues.

Although the symbiotic origin of eukaryotes must have taken place at least 1.75 billion years ago, eukaryotes did not really begin to diversify until 1.1 billion years ago. Schopf points out that these data correspond to the first appearance of cystlike structures, which may indicate the beginning of sexual reproduction. Why is sex so important? Asexual cloning or splitting produces only identical copies of the parent cell, so new variations can be propagated only very slowly. Sexual reproduction, however, allows organisms to exchange and mix genes, producing a new genotype with each offspring. This exchange allows evolution to

proceed much faster by providing a much wider spectrum of phenotypes for natural selection to operate on.

By 850 million years ago, acritarchs had reached a peak in diversity, with some specimens nearly a centimeter across. In the late Proterozoic, acritarchs declined in abundance and diversity, and about 675 million years ago, they reached their low point (although they did straggle on into the Ordovician). The cause of this great decline is uncertain, but Schopf suggests that it may have been related to late Proterozoic Varangian glaciations (discussed in Chap. 8), the reduction of atmospheric CO_2, and the increase in O_2, all of which hamper photosynthesis in micro-algae. Whatever the cause, the decline in eukaryotic algae set the stage for the next major event: the appearance of multicellular life.

A   *B*   *C*

Figure 9.11   Some of the complex Ediacaran metazoan fossils from the Ediacara Hills, South Australia. A: *Dickinsonia*, a segmented, wormlike creature that reached almost a meter in length. B: *Spriggina*, a more elongate, wormlike form. C: *Parvancorina*, a strange form that has been linked to arthropods. Other fossils are thought to resemble jellyfish or sea pens. *(Courtesy of R. L. Batten.)*

Metazoans, Vendozoans, and the "Cambrian Explosion"

Darwin's Dilemma

In *On the Origin of Species,* Charles Darwin wrote,

> Another difficulty . . . is the sudden appearance of species belonging to several of the main divisions of the animal kingdom in the lowest known fossiliferous rocks, namely those of the Cambrian.
>
> . . . If our theory be true, it is indisputable that before the lowest Cambrian stratum was deposited, long periods elapsed, and that during these periods the world swarmed with living creatures.
>
> . . . To the question why we do not find rich fossiliferous deposits belonging to these assumed earliest periods prior to the Cambrian system, I can give no satisfactory answer. . . . The case at present must remain inexplicable.

The fossil record of the origin of life presented great problems for Darwin in 1859 and remained problematic for almost a century. In most places, richly fossiliferous Cambrian strata full of trilobites were underlain by seemingly barren Precambrian rocks. What was the cause of this "Cambrian explosion"? Did life originate abruptly, or was there hidden somewhere an undiscovered fossil record of the transition to multicellular life?

The solutions to Darwin's dilemma began to appear about 50 years ago. In the early 1950s, Stanley Tyler of the University of Wisconsin began to look into a microscope at thin sections of cherts from the 2-billion-year-old Gunflint Formation, from the northern shore of Lake Superior. He found the first evidence of microfossils, such as those we have just discussed. In the 50 years since Tyler's discovery, the work of Elso Barghoorn, Schopf, and many others has documented hundreds of microfossil localities and has shown that single-celled organisms dominated the earth from 3.5 billion years ago until about 600 million years ago. The first answer to Darwin's dilemma is that life was present before the Cambrian, but we were looking in the wrong place—it was mostly microscopic (except for stromatolites). Another problem is that many places in the world have a profound unconformity between the local Cambrian and Precambrian rocks, so that the Precambrian-Cambrian transition is typically missing. In many of these places, the Precambrian rocks are also metamorphosed, rendering them unsuitable for the preservation of fossils.

The other breakthrough occurred in 1946, when Reg Sprigg found impressions of "jellyfish" in the Rawnsley Quartzite in the Ediacara Hills of South Australia. For decades after Sprigg's discovery, Australian paleontologists under the direction of Martin Glaessner collected more and more of these Ediacaran fossils, until there was clear evidence of a diverse assemblage of soft-bodied animals that predated the Cambrian. From about 1,500 specimens, there are about 30 species in 20 genera now known. Because the specimens are all impressions of soft-bodied animals preserved in a shallow-marine sandstone, they are often difficult to interpret (Fig. 9.11). However, there is no question that they represent multicellular animals, or **metazoans,** because most are several centimeters in size and some are much larger. Some frondlike animals and the flat "worm" *Dickinsonia* reached a meter in length.

Figure 9.12   Reconstruction of the Vendian sea floor, assuming the Ediacaran fossils are related to modern jellyfish, sea pens, and worms. *(Courtesy of U.S. National Museum, Smithsonian Institution.)*

Since the initial discoveries in Australia and Namibia, many other Ediacaran fossils have been found, in such places as the Yukon, Newfoundland, England, Scandinavia, Russia, Siberia, and China. Clearly this soft-bodied fauna dominated the world in the interval of time now called the Vendian. Many Vendian fossils vaguely resemble jellyfish, sea pens, and wormlike animals, but many more are clearly unrelated to anything living today (Figs. 9.11 and 9.12). Glaessner and most other paleontologists interpreted these as primitive relatives of living groups, such as jellyfish or sea pens. Critics have pointed out, however, that the Ediacaran fossils do not have the same symmetry or constructional details as modern organisms. Most of the "worms" do not have clear evidence of eyes, a mouth, an anus, locomotory appendages, or even a digestive tract.

Adolf Seilacher argues that the Vendian animals are not related to any living group but, rather, belong in a distinct group he calls the "Vendozoa." He sees the lack of detailed resemblances to living invertebrates as evidence that Vendozoans were built on an entirely different body plan. Many of the Ediacaran animals are either flat and leaflike or cushionlike. Seilacher interprets this as evidence that they were built in a quilted fashion, like a water-filled air mattress. Because there is no evidence for complex respiratory structures in such large animals, Seilacher argues that they must have used this quilting to increase their surface area for gas and food exchange. He suggests that, with their large surface area, they may have absorbed nutrients directly or maybe even had symbiotic photosynthetic algae in their tissues, as modern corals do. Seilacher's interpretations are very controversial. Research is hampered by the fact that all these "Vendozoans" are soft-bodied animals preserved as impressions only, so their interpretation is very difficult.

Whatever the zoological affinities of the Ediacara fauna, clearly metazoans were flourishing by the beginning of Vendian time, about 600 million years ago. Because preservation of soft-bodied animals is a rare accident, metazoans may have existed earlier in the late Proterozoic. Based on the molecular differences among the living invertebrate phyla, Bruce Runnegar argued that the precursors of the modern invertebrates may go back as far as 800 or 900 million years ago, when the only known fossils are stromatolites and microfossils, such as acritarchs. Other paleontologists do not find the molecular evidence convincing enough to extend the origin of metazoans so far back before their first occurrence in the fossil record.

Cambrian Explosion of Shelly Invertebrates

For times earlier than 600 million years ago, the fossil record yields only stromatolites and microfossils, such as acritarchs, which declined about 675 million years ago. From 600 to about 550 million years ago, the world was dominated by the Ediacaran soft-bodied forms. Except for some tiny, tube-shaped fossils made of calcite, there are no other animals preserved (Figs. 9.13 and 9.14). Early in the Cambrian Period (the Nemakit-Daldynian and Tommotian Stages, which are the first two stages of the Cambrian), Vendian animals disappeared completely and were replaced by little shelly fossils (Fig. 9.14). These tiny shells were so small and unimpressive that they were long overlooked, but recent research has shown that they are the harbingers of the abundant shelled invertebrates found later in the Cambrian. In addition to the "small shellies," earliest Cambrian strata yield the first evidence of significant burrowing or trace fossils in the sediments, including vertical burrows several centimeters deep. This shows that complex worms, with hydraulically stiffened, tubelike bodies capable of burrowing, must have been present.

After the first appearance of small, shelly fossils in the earliest Cambrian, the marine fauna grew richer. Together with deep

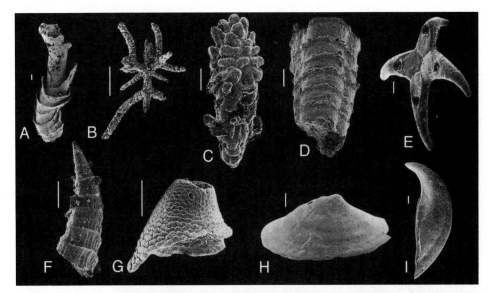

Figure 9.13   Fossils from the lowest Cambrian Tommotlan Stage are mostly microscopic phosphatic tubes, caps, and spicules from tiny animals, which were just beginning to form skeletons. They include the following. *(A) Cloudina hartmannae,* the earliest known animal fossil with a mineralized skeleton, from the late Proterozoic (Vendian) of China. *(B)* A spicule of a calcareous sponge. *(C)* A spicule of a possible coral. *(D) Anabarites sexalox,* a tube-dwelling animal with triradial body symmetry. *(E)* A spicule from a possible early mollusc. *(F) Lapworthella,* a cone-shaped fossil of unknown relationships. *(G)* A skeletal plate of *Stoibostrombus crenulatus,* another animal of unknown relationships. *(H)* Skeletal plate of *Mobergella,* a mollusclike fossil. *(I)* Cap-shaped shell of *Cyrtochites,* another mollusclike fossil. All scale bars are 1 mm. All specimens except A are from the Tommotian. D–G are from South Australia, and H–I are from Yunnan Province, China. *(Courtesy of Stefan Bengston.)*

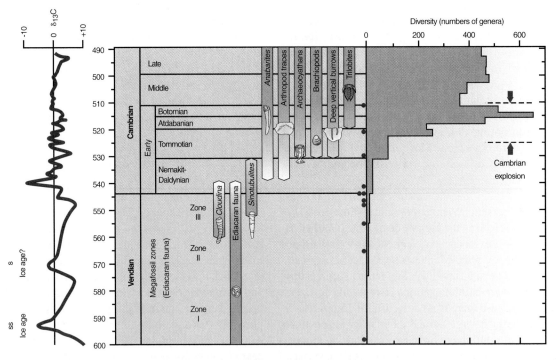

Figure 9.14   Vendian and Lower Cambrian strata document the transition from the simplest soft-bodied metazoans to complex skeletonized invertebrates. Soft-bodied Ediacaran animals arose first in the early Vendian, followed by tiny organisms with simple, tubelike skeletons, such as *Cloudina* and *Sinotubulites.* At the Early Cambrian (Nemakit-Daldynian and Tommotian Stages), many additional "little shellies" are found, along with archaeocyathans, brachiopods, and burrows and complex trace fossil indicating larger, soft-bodied, wormlike metazoans. In the Atdabanian Stage, there was a huge explosion of diversity with the arrival of trilobites, and this high diversity continued through the rest of the Cambrian (although archaeocyathans died out by the Middle Cambrian). In the center is the carbon isotopic curve, which shows several huge negative (to the left) events during the Vendian ice ages and another at the beginning of the Cambrian, which may reflect recycling of organic carbon from deep bottom waters as the continents moved rapidly apart and oceanic circulation changed. *(Modified from Kirschvink et al., 1997; McMenamin, 1987; and Knoll and Carrol, 1999.)*

vertical burrows, large invertebrates with hard skeletons first appeared. These include spongelike organisms known as archaeocyathids and the first group with two hard shells hinged together, the brachiopods. Shortly after these groups appeared, the third stage of the Cambrian (Atdabanian) marks the first appearance of the most typical Cambrian group, the trilobites.

What triggered the replacement of microfossils with large, soft-bodied animals in the Vendian? Why were the latter replaced by shelly fossils in the Cambrian? There are several hypotheses competing, but some points seem clear:

1. The extinction of the late Proterozoic acritarchs seems to correlate with a gigantic glacial event about 600 million years ago. As we saw in Chap. 8, the Varangian glaciation was so severe that glaciers flowed to sea level even in equatorial regions. The earth may have come perilously close to complete refrigeration and extinction of virtually all life. The abundant late Proterozoic microorganisms were nearly wiped out and never completely recovered after the glaciers retreated.

2. For years, scientists have suggested that megascopic metazoans did not evolve during the Proterozoic because atmospheric oxygen was too scarce for large animals to function. Recently Andrew Knoll and his colleagues have found evidence from carbon isotopes that the retreat of the Varangian glaciers was also marked by a major increase in atmospheric oxygen. Based on the size and surface area of some Vendian metazoans, some scientists estimate that Vendian oxygen levels had reached 6 to 10 percent. If their interpretation is substantiated, then perhaps limited oxygen was the critical factor preventing the evolution of multicellular life for almost 3 billion years.

3. The latest Proterozoic was clearly an episode of tectonic change. There is abundant evidence of rifting and volcanic activity caused by the breakup of a supercontinent before and after the Varangian glaciation. Most of the passive continental margins began to sink as their ocean basins rifted apart, producing a worldwide transgression. This tectonism had several important effects on life. Transgression expanded the area of shallow marine shelf available for life. Evidence from strontium isotopes suggests that the Vendian was also marked by a rapid increase in nutrients, such as calcium and phosphate, trapped in the deep ocean. A sudden increase in the abundance of carbon right at the Vendian-Cambrian boundary suggests that upwelling brought up rich sources of nutrients, which allowed explosive growth of organisms. Rifting and volcanic activity may have released many of these nutrients into the shallow-marine environment, where shell-building organisms could use them.

4. Increased nutrients provide the necessary materials for animals to build hard shells, most of which are made of calcite. It is natural to assume that the adaptive advantage of shells was for protection from predators and competitors, but there are alternative explanations. Calcium is a natural byproduct of metabolism, and organisms can maintain their calcium balance by secreting it as hard parts within their bodies. Living organisms cannot secrete calcium carbonate until the atmospheric oxygen level reaches a critical threshold. This may explain why the earliest shelly fossils are mostly either simple calcite or phosphatic tubes or small, platelike objects imbedded in their soft bodies (Fig. 9.14). In fact, one of the striking features of earliest Cambrian is the great abundance of organisms that made skeletons of phosphate or silica, which were very abundant and easier to mineralize at lower oxygen levels.

   Once calcite secretion became established, it allowed simple, wormlike animals to evolve into a much greater variety of forms, including colonial archaeocyathids, two-shelled brachiopods, and multisegmented trilobites. This may also explain why the great radiation of calcite-secreting organisms did not begin until the latest Cambrian and Ordovician, when atmospheric oxygen apparently reached modern levels.

5. The shallow marine world was once covered with thick cyanobacterial mats, but in the Early Cambrian they nearly disappeared. The appearance of the small shelly fossils and deep burrows are correlated with this decline in stromatolites. This is probably no accident. Before the appearance of small invertebrate animals, nothing fed on cyanobacterial mats. Some of these small shelly fossils must have been primitive molluscs that grazed and cropped stromatolites. Once they evolved, they would have a virtually unlimited food source and could easily cut the cyanobacterial mats and domes to pieces. Indeed, stromatolites survive today only in environments that are hostile to grazing invertebrates. These include lagoons too salty for grazing snails (such as Shark Bay, Australia—see Fig. 9.1) and shallow channels in the Bahamas where currents are too strong for clinging invertebrates.

   Opening up the mats would free up the ocean floor for burrowing organisms to feed, and these in turn would so disturb the substrate that cyanobacterial mats could not return to dominance. Steve Stanley suggests that this feedback between croppers and burrowers (known as the *cropping hypothesis*) opened up new ecological niches. Prior to the Cambrian, marine life lived in an almost two-dimensional world of stromatolitic mats and flat, quilted Ediacaran animals. With the advent of burrowing, the development of colonial reeflike archaeocyathids, and the opening up of the sea floor, organisms could exploit many niches on, above, and below the sea floor that were previously uninhabited. The development of multiple levels of biotic activity is known as **tiering,** and in later chapters we shall see its importance throughout the history of life. More than anything, the opening of niches and increase in tiering seem to explain the explosion of diversity during the Early Cambrian.

6. Diversification of invertebrates led to even more complex ecological relationships, such as predation. By Middle Cambrian time, there is evidence of large predators, which reached half a meter in length, as well as trilobites with

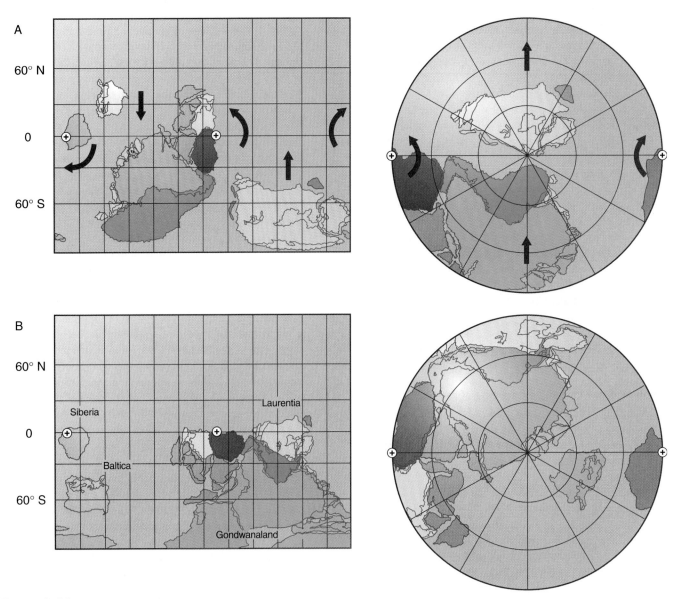

Figure 9.15   Paleogeographic reconstructions of the continents before *(A)* and after *(B)* the proposed inertial interchange or true polar wander event. Africa rotates counterclockwise, putting most of it near the South Pole, whereas Laurentia (mostly North America) shifts rapidly northward from the South Pole to the equator. *(From Kirschvink et al., 1997, Science, v. 277, Fig. 4. Copyright © 1997 American Association for the Advancement of Science. Reprinted with permission.)*

healed bite marks. A number of the Cambrian animals have spiky or platy armor, which suggests devices to thwart predators. The appearance of predators represents not only a brand new ecological niche but a more complex food chain as well, from primary producers (plants) to secondary feeders (grazing invertebrates and mud-grubbing trilobites) to third-level predators. The appearance of predators means that the prey species must have either developed armor or learned to burrow, leading to further selection for new evolutionary types.

In recent years, paleomagnetic poles from Vendian and Cambrian rocks have shown that the continents underwent enormous amounts of movement in Cambrian time. With the many new ra-

diometric dates that show that the Early Cambrian was much shorter than previously thought, it became apparent that these continental movements had to be extraordinarily rapid by modern plate-tectonic standards. North America moved from a position near the South Pole to near the ancient equator in about 10 to 15 million years, whereas Gondwanaland spun around a point in Antarctica, sending North Africa from the pole to the equator as well (Fig. 9.15). This is farther and faster than the race of India from Gondwanaland to Asia in the Cretaceous and early Cenozoic (Fig. 15.29), one of the fastest plate motions known. This rapid motion of the continents coincides with the peak of the Cambrian explosion (Fig. 9.13), so as discussed in point 2 on p. 195, it could have contributed to the diversification of life by releasing nutrients, by causing global sea levels to rise, or by placing more continents in the tropics, where shallow-marine conditions were ideal for the diversification of marine life.

In 1997, however, Joe Kirschvink, Robert Ripperdan, and David Evans proposed an even more interesting idea. According to their *inertial interchange true polar wander hypothesis,* the earth changed the direction of its spin axis 90° relative to the continents during the earliest Cambrian. Regions that were previously the North and South Poles were shifted to the equator, and the regions that had once been equatorial on opposite sides of the earth became the new North and South Poles. This rapid movement of the earth's surface relative to its interior may have occurred because of an imbalance in the mass distribution of the planet itself. Such a phenomenon may have already taken place on other planets. Mars, for example, has its largest volcanoes and huge mass concentrations near its equator. Many astronomers believed that those massive crustal elements may have formed at other latitudes but then shifted to the equator to balance the spin of the planet. This is analogous to the way a skater or a top spins on its axis—it is most stable when the mass is concentrated toward the center, and it wobbles uncontrollably if there are large, asymmetric massive protrusions at the top or bottom of the axis.

Kirschvink and colleagues argue that, once the late Vendian continents reached their highly unstable, asymmetric positions near the poles (Fig. 9.15), they made the earth's rotation unstable. To comply with the laws of conservation of mass and conservation of angular momentum, they shifted rapidly (in less than 15 million years) to equatorial positions, completely rearranging the positions of the continents. This hypothesis would explain not only why the ancient magnetic directions of North America and North Africa moved so rapidly from the pole to the equator but also some other phenomena as well. The rapid swings in the carbon isotope curve (Fig. 9.13) in the late Vendian and earliest Cambrian suggest that there were major changes in oceanic circulation, freeing up lots of buried organic carbon. Such rapid oceanic circulation changes might be caused by the shifts in the positions of the continents. Oceanographic and nutrient changes of this scale would have fragmented old ecosystems and may have triggered the Cambrian explosion.

When the Trilobites Roamed

By Middle Cambrian time, the shallow-marine environment had become filled with a great diversity of invertebrates. The most common of these were the trilobites (Fig. 9.16), which evolved so rapidly in Cambrian seas around the world that they are a primary tool for correlation. More than 600 species are known from the Cambrian, but by the Ordovician their dominance had ended. Trilobites are members of the Arthropoda, the "jointed legged" phylum of animals that also includes insects, spiders, centipedes, ticks, mites, scorpions, crabs, lobsters, shrimp, and many other kinds of animals. Like all other arthropods, trilobites had jointed legs covered by a skeleton composed of segments. Because their skeletons were composed of both calcite and organic chitin, they fossilized much better than do the shells of most modern arthropods, which have an external skeleton composed only of chitin.

The earliest trilobites, the olenellids (Fig. 9.16A), clearly show that trilobites are descended from arthropods with unspecialized segmentation. Olenellids had a well-developed head segment (the cephalon) like all other trilobites, but they also had a large number of segments in their body (thorax) and their small "tail" region (pygidium) is very spiny and segmented. By the Middle Cambrian, primitive olenellids had been replaced by more specialized trilobites (Fig. 9.16B–D). Some of these had either fused their tail segments into a single, platelike pygidium or had reduced the spines along their margins. Others evolved specialized eyes, which gradually moved away from the center; some were apparently blind and floated in the surface waters as plankton (Fig. 9.16E). Most of these trilobites spent their lives feeding on food particles in the sea floor, crawling along with their many legs. Trilobite burrows and feeding tracks are among the most common trace fossils of the Cambrian.

True corals and other reef-forming organisms had not yet evolved in the Early Cambrian. That ecological niche was occupied by the archaeocyathids, a strange group of organisms shaped like double-walled ice cream cones (Fig. 9.17A–I). Archaeocyathids were typically 25 mm in diameter and up to 150 mm high, but some giants reached over a meter in length. Perforated by pores, they were probably filter-feeders, as sponges are. Although some paleontologists think archaeocyathids are just aberrant sponges, other specialists argue that their anatomy is completely different from that of modern sponges or any other living group. Because archaeocyathids are now extinct and we have no soft tissues preserved, their relationships are controversial. They probably evolved from colonial protozoans, just as sponges did, and occupied the same ecological niche. Archaeocyathids were especially successful in shallow tropical seas, forming extensive "reefs," which were many meters thick in Early Cambrian deposits of Russia, Australia, Africa, and western North America. After worldwide dominance, they died out completely in the early Middle Cambrian, eventually being replaced by sponges and true corals.

The next most common fossils are the *brachiopods,* or "lamp shells." These animals have two clamlike shells joined for protection. Their anatomy, however, is very different from that of clams. A curled, featherlike device called a lophophore is used for filtering out food particles from the water that passes through. A long, fleshy stalk called a pedicle extends through the hinge area and helps the animal burrow or attach to hard surfaces. The earliest Cambrian brachiopods (Fig. 9.17J–K), a group known as the inarticulates, had no teeth and sockets in their hinge area to hold the shells together; instead, the shells were held together by muscles. Inarticulates had simple, disc-shaped shells and a long pedicle for burrowing. Some Cambrian inarticulates are barely distinguishable from the living brachiopod *Lingula* (Fig. 9.18A), a true "living fossil" unchanged in 500 million years. Most inarticulates have shells of chitin and phosphate, typical of many invertebrate skeletons in the earliest Cambrian before calcite became the dominant skeletal material.

A fourth major phylum of the Cambrian is the molluscs. Today they are represented by animals as diverse as octopus, clams, and snails, but in the Early Cambrian they were much simpler

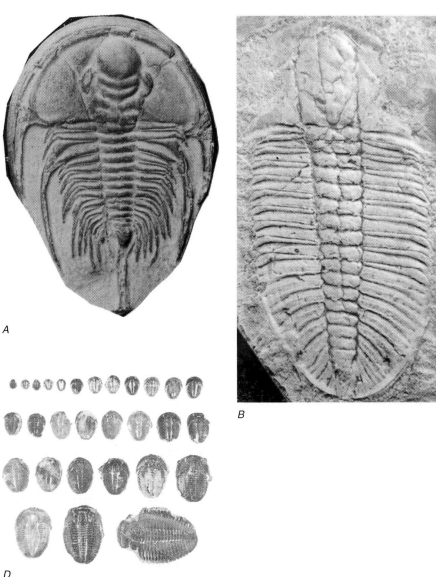

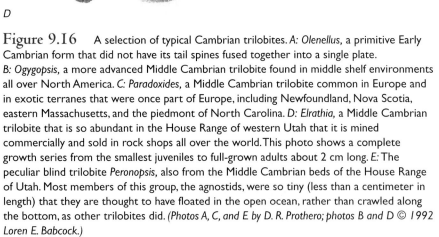

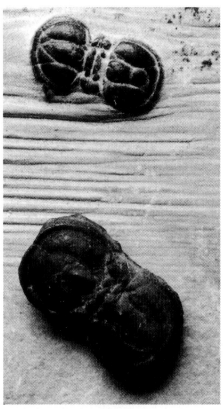

Figure 9.16   A selection of typical Cambrian trilobites. *A: Olenellus,* a primitive Early Cambrian form that did not have its tail spines fused together into a single plate. *B: Ogygopsis,* a more advanced Middle Cambrian trilobite found in middle shelf environments all over North America. *C: Paradoxides,* a Middle Cambrian trilobite common in Europe and in exotic terranes that were once part of Europe, including Newfoundland, Nova Scotia, eastern Massachusetts, and the piedmont of North Carolina. *D: Elrathia,* a Middle Cambrian trilobite that is so abundant in the House Range of western Utah that it is mined commercially and sold in rock shops all over the world. This photo shows a complete growth series from the smallest juveniles to full-grown adults about 2 cm long. *E:* The peculiar blind trilobite *Peronopsis,* also from the Middle Cambrian beds of the House Range of Utah. Most members of this group, the agnostids, were so tiny (less than a centimeter in length) that they are thought to have floated in the open ocean, rather than crawled along the bottom, as other trilobites did. *(Photos A, C, and E by D. R. Prothero; photos B and D © 1992 Loren E. Babcock.)*

(Fig. 9.17L–O). Many of the simple tube, cone, and cap-shaped shells of the Tommotian probably belonged to simple molluscs. Some Cambrian molluscs were simple cap-shaped forms that attached to rocks, much as modern limpets do. Living descendants of these early molluscs can be found in deep waters off South America. Their soft anatomy still retains segmentation of the muscles and gills, showing that molluscs evolved from segmented worms that developed a shell. By the middle Early Cambrian, true snails and clams had evolved, but they remained rare until the Ordovician.

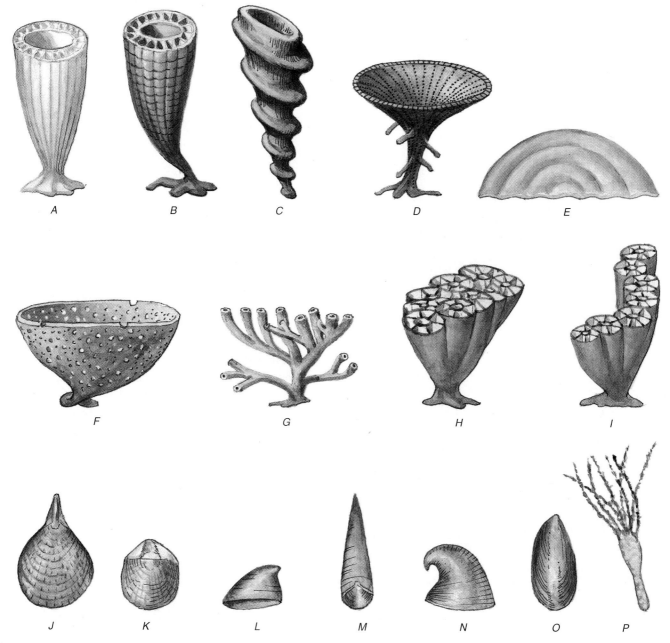

Figure 9.17 Selection of typical Cambrian animals. *A–I:* A variety of shapes demonstrated by the Early Cambrian archaeocyathids, spongelike organisms that formed the earliest reefs on Earth. Specialists still argue whether archaeocyathids were true sponges or simply an extinct early experiment in spongelike body form that left no descendants. Their typical "double-wall I-beam" construction is unique to the group and not found in any sponge. Most specimens were less than a meter in height, although some formed huge reefs, which spanned hundreds of meters. *J–K:* Inarticulate brachiopods, with simple phosphatic shells and no mechanical joint in the hinge. *L–O:* A variety of archaic molluscs, most with simple conical or cap-shaped shells, like the modern limpet. *P:* An eocrinoid, an early experiment in stalked, filter-feeding echinoderms. True crinoids replaced them in the Ordovician (see Chap. 11). Note that most of these animals are archaic or "experimental" members of their respective phyla, and most were rare after the Cambrian. *(Modified from Boardman et al., 1987, Fossil invertebrates, Blackwell; and A. R. Palmer, 1974, American Scientist, v. 62, pp. 216–225.)*

A fifth important phylum in the Cambrian is the *echinoderms,* or "spiny-skinned" animals. Today they are represented by starfish, sea urchins, sand dollars, crinoids ("sea lilies"), and sea cucumbers. Most echinoderms are built of calcite plates, and some are propelled by an internal hydraulic system of water canals and tube feet. All living echinoderms are built on a pattern of five-fold symmetry, such as the five arms on starfish or the five areas of tube feet on sea urchins. However, some Cambrian echinoderms are very bizarre by modern standards (Fig. 9.17P). They represent "experiments" in echinoderm design, with a wide variety of

A

B

Figure 9.18   Living brachiopods. A: The living inarticulate brachiopod *Lingula,* which has two phosphatic shells held together only by muscles. Today it lives in shallow intertidal muds, burrowing down with its long fleshy pedicle. (© *David Wrobel/Biological Photo Service.*) B: A living articulate brachiopod, with a calcite shell held together by a mechanically interlocking hinge. Note the delicate, featherlike lophophore, a filter-feeding device found in all brachiopods. (© *David Wrobel/Biological Photo Service.*)

Figure 9.19   *Helicoplacus,* an "experimental" echinoderm from the Lower Cambrian of California. Based on its spiral arrangement of plates, it apparently transported food up the spiraling grooves by means of tube feet. Original specimen approximately 10 cm long. (*Courtesy of J. Wyatt Durham.*)

shapes never seen again. Some were attached forms with tentacles, vaguely like the living crinoids. Others were disc-shaped forms with a three-armed spiral on the top. The oddest of all was *Helicoplacus,* a spindle-shaped form with a spiral arrangement of plates (Fig. 9.19). No one knows how these strange beasts lived, but apparently they sat upright, with the pointed part embedded in the sediment.

Several themes are apparent when we examine the Cambrian fauna. First, most of the animals were extremely primitive forms that either did not survive or are represented today by "living fossils." Many appear to be early experiments in the evolution of their phyla, stages that were pushed aside by later developments or went extinct. Second, the ecological communities were very simple. The stromatolite-cyanobacterial-archaeocyathid "reef" community, for example, included only a few burrowers, mostly worms and inarticulate brachiopods. Trilobites were busy sifting out food from the mud of the sea bottom, and the simple molluscs and echinoderms were found in rocky areas where they could attach. Missing was the tremendous variety of shallow and deep burrowers we find today, the great range of filter feeders (including reef-building corals and sponges), and—most striking of all—large predators.

These themes were reinforced by the discovery of one of the most remarkable fossil localities in the world, the *Burgess Shale.* First collected in 1909 from Middle Cambrian deposits in the Rocky Mountains near Field, British Columbia, Burgess Shale animals have since been found on several other continents. What is remarkable about the Burgess Shale fauna is the abundance of soft-bodied animals preserved in magnificent detail, giving us a rare window into life that seldom fossilizes and showing that the Cambrian world was much richer than what we imagined when we had only trilobite fossils (Fig. 9.20). Of course, trilobites are found there, but so are many different strange and unusual organisms without hard parts. These include at least twenty types of segmented, jointed-legged arthropods with body designs fundamentally different from those of any living arthropod, a number of wormlike animals, and at least fifteen other kinds of animals that cannot be fit into any living phylum of invertebrates. Each represents a different version of metazoan design that has not survived.

Some of these animals are truly bizarre. The appropriately named *Hallucigenia* (Fig. 9.21D) appeared to have simple, tubular legs and paired spikes on its tubelike body, but no one is sure which end was the head and which was the tail. *Opabinia* had five eyes, multiple segments, and a long "nozzle" on its front (Fig. 9.21B). *Wiwaxia* was covered by a complex armor of plates and spikes (Fig. 9.21C). The most impressive beast was *Anomalocaris,* a predator almost half a meter long with two armlike feeding appendages and a mouth that looked like a pineapple slice (Fig. 9.21A). This animal was one of the largest predators known

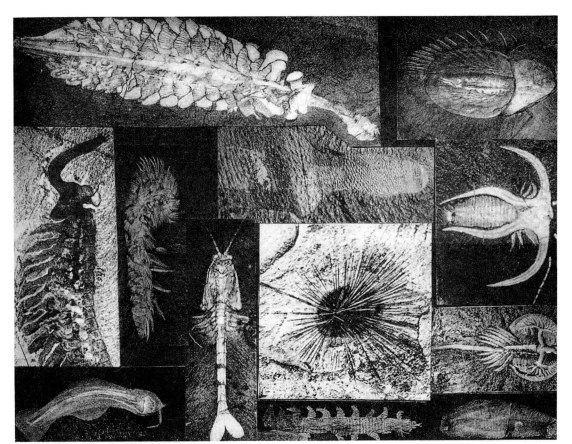

Figure 9.20
Examples of soft-bodied animals from the Middle Cambrian Burgess Shale at Field, British Columbia. Note the exquisite detail of the appendages. *(Courtesy of U.S. National Museum, Smithsonian Institution.)*

Figure 9.21 Reconstructions of some of the amazing Burgess Shale animals (see Fig. 9.20). *A:* The meter-long predator *Anomalocaris.* Originally found in disarticulated pieces, its distinctive "pineapple-slice" mouth was once reconstructed as a "jellyfish" in older Burgess Shale dioramas. *B:* The bizarre *Opabinia,* with five eyes and a front "nozzle"; the original fossil is shown at the top and middle left in Fig. 9.20. *C:* The armored crawler *Wiwaxia.* *D:* The appropriately named *Hallucigenia.* Once a complete mystery that was reconstructed with the spines pointed downwards, it is now thought to be related to the living wormlike creatures known as onychophorans. *E:* The wormlike creature *Amiskwia. F:* The stalked "goblet creature" *Dinomischus.* *G:* The tiny Burgess Shale arthropod *Sarotrocercus,* swimming on its back. *(Modified from drawings by M. Collins in S. J. Gould, 1989,* Wonderful life, *Norton.)*

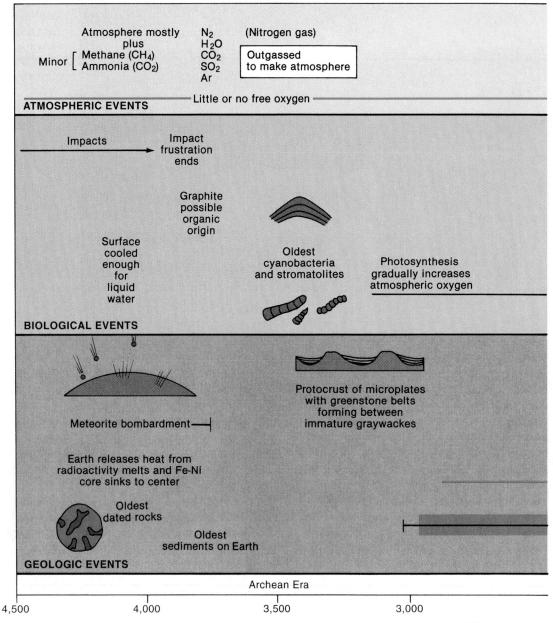

Figure 9.22   Time line of the early evolution of life during the Cryptozoic and Cambrian.

until the Late Cambrian, indicating that predators of the time were relatively small and rare and did not have jaws or pincers for crushing prey.

The Burgess Shale window shows us that Cambrian life was not as simple and monotonous as the hard-shelled faunas of trilobites, inarticulate brachiopods, and primitive molluscs preserved at most localities would indicate. The Cambrian had diversity comparable to that of later periods, but with an important difference: almost all this diversity is in experimental forms, each of which would represent a different phylum or class in a modern classification. By contrast, from the Ordovician onward, most fossils fit into about eight phyla, and just a few classes and orders within these phyla account for most species. In other words, the Cambrian organisms represented a wide range of evolutionary innovations, most of which are now extinct, and subsequent life forms have been constructed along a few basic body plans ever since the Ordovician. In his bestseller *Wonderful Life,* Stephen Jay Gould points out that this view is very different from our usual conception of life gradually diversifying from simple, primitive beginnings into all the body plans of living animals. Instead,

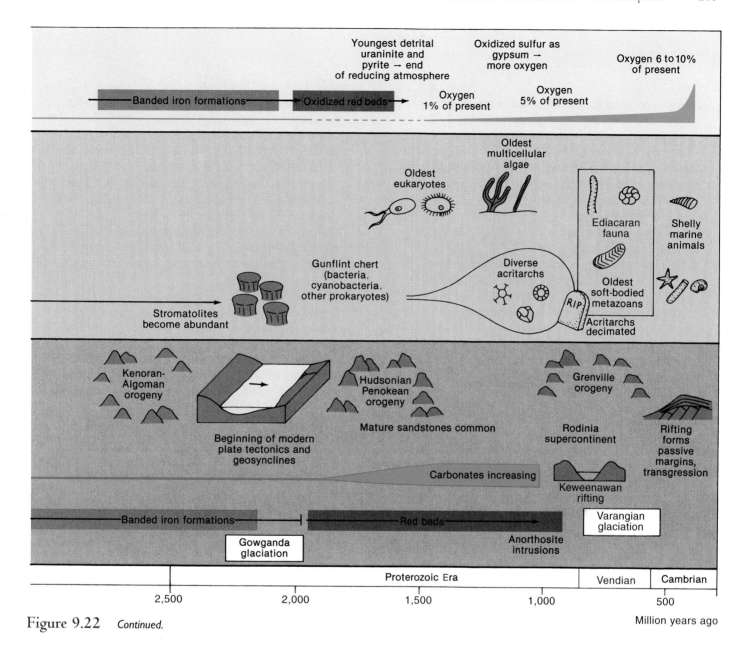

Figure 9.22 *Continued.*

the Cambrian was characterized by a wild profusion of diversity, which has subsequently been pruned by extinction into only a few successful types found today.

Because these soft-bodied animals rarely fossilize, it is difficult to tell when they became extinct. However, we do have a good record for the trilobites and other hard-shelled animals. By the late Middle Cambrian, a series of extinction events hit the trilobites particularly hard. Each extinction horizon was followed by a new adaptive radiation of trilobites, which was then hit by another extinction event several million years later. What caused these extinctions? Most of the victims were specialized, warm-water trilobites, and those that recolonized the shelf and established a new adaptive radiation were from deep, colder waters. This suggests that each extinction event was caused by a cooling of the oceans. Unlike later extinctions, however, we have no clear evidence of Late Cambrian glaciation that might explain it. Whatever the cause, by the Ordovician, trilobites were no longer dominant, and many other characteristic Cambrian groups (such as archaeocyathids and archaic echinoderms) were also extinct. In their place was a great explosion of new organisms that dominated the Ordovician and the rest of the Paleozoic.

Summary

Origin of Life

- Early concerns about the origin of life confused reproduction with the actual origin of life. Before the invention of the microscope, **spontaneous generation** was held to be the origin of new life.

- Original physical conditions for the origin of life involved an **anaerobic** atmosphere rich in water vapor and simple compounds of hydrogen, carbon, and nitrogen but poor in free oxygen in contrast to our present atmosphere of nitrogen, carbon dioxide, and oxygen (**aerobic**) atmosphere.

- The simplest building blocks of life, such as **amino acids** (which are linked to form **proteins**), **sugars,** and **fatty acids** (which link to form complex fats, or **lipids**) can easily be produced in laboratory experiments that simulate the original conditions of the early earth. Other experiments show that droplets of proteins can condense, surrounded by a lipid membrane, to form **proteinoids** with very lifelike properties.

- The central puzzle of research into early life is whether the genetic information was encoded first in strings of proteins or in nucleic acids, such as RNA. Recent discoveries of RNA that can also catalyze reactions the way a protein does suggest that there was once an RNA world of naked genes and that protein sequences were added later. Some researchers suggest that inorganic materials, such as clays, zeolites, or pyrite, were **templates** on which organic molecules could cling, then line up and polymerize.

Fossil Evidence of Early Life

- Earliest life in the form of single-celled autotrophic **cyanobacteria** (blue-green algae) is found in some of the oldest rocks known (in northwestern Australia, dated at about 3.5 billion years). From most of the Cryptozoic, until about 650 million years ago, only single-celled or filamentous cyanobacteria are known. They occur in structures called **stromatolites.**

- The first cells were **prokaryotes**—they had neither an organized nucleus nor **organelles** (sub-cellular structures) and reproduced by simple cell division. The first **eukaryotes**—cells with an organized nucleus and organelles—appeared in the late Precambrian.

- Multicellular animals (**metazoans**) are known first as **trace fossils,** usually either tracks or trails of wormlike creatures.

- **Ediacaran fossils** are the earliest animal fossils. They are apparently not related to later groups found in the Paleozoic and are distinct in that they lack **mineralized skeletons.**

- **Cambrian faunas** are first known from strata containing scattered phosphatic skeletons, which in turn are overlain by strata showing a dramatic increase in the volume and diversity of fossils preserved with calcitic skeletons. The **trilobites** were the first major group of organisms to leave an extensive record of very rapid diversification during the Cambrian.

- The **Burgess Shale fauna** is a remarkable window into a world of forms not known elsewhere because it contains the remains of soft-bodied organisms that are rarely fossilized. The fauna includes rare arthropods and many worm groups, as well as phyla that have no living relatives.

Readings

Barghoorn, E. S., and S. M. Tyler. 1956. Microorganisms from the Gunflint Chert. *Science,* 147: 563–77.

Bernal, J. D. 1967. *The origin of life.* Cleveland: World.

Brasier, M. D. 1992. Global ocean atmospheric change across the Precambrian-Cambrian transition. *Geological Magazine,* 129:161–68.

Cairns-Smith, A. G. 1985. *Seven clues to the origin of life.* Cambridge, UK: Cambridge University Press.

Cone, J. 1991. Fire under the sea: The discovery of the most extraordinary environment on earth—Volcanic hot springs on the ocean floor. New York: Morrow.

Conway, Morris, S. 1985. The Ediacaran biota and early metazoan evolution. *Geological Magazine,* 122:77–81.

———. 1987. The search for the Precambrian-Cambrian boundary. *American Scientist,* 75:156–67.

———, and H. B. Whittington. 1979. The animals of the Burgess Shale. *Scientific American,* (July) 122–35.

Cowen, R. 1990. *History of life.* Boston: Blackwell.

Cowie, J. W., and M. D. Brasier, eds. 1989. The Precambrian-Cambrian boundary. New York: Oxford University Press.

Dyson, F. 1985. *Origins of life.* Cambridge, UK: Cambridge University Press.

Fox, S. W., ed. 1965. *The origins of prebiological systems and of their molecular structure.* New York: Academic Press.

Glaessner M. F. 1984. *The dawn of animal life.* Cambridge, UK: Cambridge University Press.

Gould, S. J. 1989. *Wonderful life: The Burgess Shale and the nature of history.* New York: Norton.

Jones, M. L., ed. 1985. *Hydrothermal vents of the eastern Pacific: An overview.* Biological Society of Washington, Bulletin 6.

Kirschvink, J. L., R. L. Ripperdan, and D. A. Evans. 1997. Evidence for a large-scale reorganization of Early Cambrian continental masses by inertial interchange true polar wander. *Science,* 277:541–45.

Knoll, A. H. 1985. The distribution and evolution of microbial life in the late Proterozoic Era. *Annual Reviews of Microbiology,* 39:391–417.

———. 1991. End of the Proterozoic eon. *Scientific American,* October, 64–74.

———, and S. B. Carroll. 1999. Early animal evolution: Emerging views from comparative biology and geology. *Science,* 284:2129–37.

———, and M. R. Walter. 1992. Latest Proterozoic stratigraphy and earth history. *Nature,* 356:675–78.

Lipps, J. H., and P. W. Signor, ed. 1992. *Origin and early evolution of the Metazoa.* New York: Plenum Press.

Margulis, L. 1981. *Symbiosis in cell evolution.* San Francisco: Freeman.

———. 1982. *Early life.* Boston: Science Books International.

———. 2000. *Symbiotic Planet: A New Look at Evolution.* New York: Basic Books.

McMenamin, M. A. S. 1987. The emergence of animals. *Scientific American,* (June) 94–102.

———, and D. L. S. McMenamin, 1990. The emergence of animals, the Cambrian breakthrough. New York: Columbia University Press.

Miller, S. L. 1953. A production of amino acids under possible primitive earth conditions. *Science,* 117:528–29.

Narbonne, G. M. 1998. The Ediacara biota: A terminal Neoproterozoic experiment in the evolution of life. *GSA Today,* 8:1–6.

Oparin, A. I. 1938. The origin of life. London: Macmillan.

Palmer, A. R. 1974. Search for the Cambrian world. *American Scientist,* 62:216–24.

Runnegar, B. 1982. The Cambrian explosion: Animals or fossils? *Journal of the Geological Association of Australia,* 29:395–411.

———. 1992. Evolution of the earliest animals. In *Major events in the history of life.* Edited by J. W. Schopf. Boston: Jones and Bartlett.

Schopf, J. W. 1978. The evolution of the earliest cells. *Scientific American,* Feb. 111–38.

———, ed. 1983. Earth's earliest biosphere: Its origin and development. Princeton, NJ: Princeton University Press.

———. 1992. The oldest fossils and what they mean. In *Major events in the history of life.* Edited by J. W. Schopf. Boston: Jones and Bartlett.

———. 1999. *Cradle of life.* Princeton, NJ: Princeton University Press.

———, and C. Klein, eds. 1992. *The Proterozoic biosphere, a multidisciplinary study.* Cambridge, UK: Cambridge University Press.

———, and B. M. Packer. 1987. Early Archean (3.3-billion- to 3.5-billion-year-old) microfossils from Warrawoona Group, Australia. *Science,* 237:70–73.

Seilacher, A. 1989. Vendozoa: Organismic construction in the Proterozoic biosphere. *Lethaia,* 22:229–39.

Shapiro, R. 1986. *Origins, a skeptic's guide to the creation of life on earth.* New York: Summit.

Vidal, G. 1984. The oldest eukaryotic cells. *Scientific American,* (January) 48–57.

———, and A. H. Knoll, 1983. Proterozoic plankton. *Memoir of the Geological Society of America,* 161:265–77.

Wächtershäuser, G. 1988. Before enzymes and templates: Theory of surface metabolism. *Microbiological Reviews,* 52:452–84.

Walter, M. R., ed. 1976. *Stromatolites.* Amsterdam: Elsevier.

———, and G. R. Hays. 1985. Links between the rise of the metazoa and the decline of stromatolites. *Precambrian Research,* 29:149–74.

Ward, P. D., and D. Brownlee. 2000. *Rare earth: Why complex life is uncommon in the universe.* New York: Copernicus.

Wills, C., and J. Bada. 2000. *The spark of life: Darwin and the primeval soup.* New York: Perseus.

Woese, C. R. 1979. A proposal concerning the origin of life on the planet earth. *Journal of Molecular Evolution,* 13:95–101.

Zhuravlev, A. Y., and R. Riding, eds. 2001. *The ecology of the Cambrian radiation.* New York: Columbia University Press.

(D. R. Prothero.)

Earliest Paleozoic History

The Sauk Sequence—an Introduction to Cratons and Epeiric Seas

In the mud of the Cambrian main

Did our earliest ancestors dive;

From a shapeless, albuminous grain

We mortals are being derived.

Grant Allen, *The Ballade of Evolution*

▶ In the late Proterozoic, the global supercontinent began to break up. By the Early Cambrian, all the continental fragments were surrounded by slowly subsiding passive margins. Continuing subsidence and/or rising global sea levels produced great shallow seaways upon the continents themselves, so that by the Late Cambrian, the continents were almost completely drowned.

▶ Subsiding passive margins accumulated very thick and continuous packages of shallow-marine sediments, while the continental interior had only shallow basins into which relatively thin and discontinuous sedimentary sequences were deposited. Whereas the Cambrian sequences in the passive margin may be thousands of meters thick, the same amount of time is represented by a few tens of meters of sediment (and numerous unconformities) within the continental interior.

▶ Cambrian shallow-marine sediments are sheetlike deposits consisting mostly of mature quartz sandstones, with exceptional rounding and sorting; shales are relatively rare. It is thought that these sands were products of deep continental weathering and extensive wind erosion (because there were no land plants yet). Their geometry and structures were products of the very low relief on the Cambrian continental surface flooded by shallow seas of immense proportions, perhaps accentuated by strong tidal currents from a moon nearly twice as close as it is today.

▶ At the end of the Cambrian, carbonate-secreting organisms occurred in great abundance. They took over the vast shallow Cambrian epicontinental seaways and shut off the sand supply.

The beginning of Phanerozoic time represents the greatest punctuation mark in the entire record of earth history because of the appearance of complex and diverse animals, as discussed in Chap. 9. The soft-bodied Ediacara fauna was followed by the appearance of hard, more preservable Cambrian skeletons. For reasons presented in Chap. 8, we include poorly fossiliferous Vendian strata as the first division of the Paleozoic Era. These rocks, which span approximately the interval from 700 to 550 million years ago, are best preserved within early Paleozoic continental margins but are poorly represented within the cratons. For the most part, they were deposited as thick wedges of strata draping the newly rifted passive margins of fragments of one or more Proterozoic supercontinents.

This chapter emphasizes cratons and the nature of the shallow epeiric seas that have flooded them repeatedly. It treats the interval from 700 to 500 million years ago (Vendian, Cambrian, and Early Ordovician). Using a variety of maps and cross sections, we shall see how the geologist can interpret a craton's tectonic history and paleogeography. Similar illustrations are used throughout the remainder of the book.

Next we discuss and interpret the characteristics of both terrigenous clastic (Fig. 10.1) and carbonate rocks of the Sauk Sequence. Sauk rocks range in age from Vendian beyond the craton

Figure 10.1   Spectacular cliffs of the Lower Cambrian Potsdam Sandstone, Ausable Chasm, New York. Shallow marine sandstones like these are found at the base of the Cambrian transgression all over North America. *(D. R. Prothero.)*

Box 10.1

"SWEATing" It Out

Whatever the reason for the Varangian glaciations (Chap. 8), the planet was clearly warming up by the later Vendian toward a Cambrian greenhouse world. The causes for this warming are controversial. Many geologists think that one important factor was the breakup of the late Proterozoic supercontinents on which the Varangian glaciers had formed. Both North America and the Baltic Shield seem to have separated from the rest of the supercontinent at this time. The rapid separation of continents suggests renewed seafloor spreading and possibly increasing levels of greenhouse gases from the rapidly erupting mid-ocean ridges.

Most paleogeographic reconstructions explain the rifted passive eastern margin of North America as a result of the opening of the newly formed Iapetus Ocean east of North America. If the southwestern United States was also a rifted passive margin during the Vendian and Cambrian, where is the other half of the rift? In 1991, Paul Hoffman, Eldridge Moores, and Ian Dalziel independently proposed that East Antarctica may be a suitable candidate (Fig. 10.2A). Much of the Transantarctic Mountain basement rock is composed of Vendian rift margin sequences, overlapped by Cambrian shelf carbonates, forming a belt as long as the rifted margin of southwest North America. Moores coined the term *SWEAT hypothesis* (*South West U.S./East Antarctica*) for the notion that the Transantarctic Mountain belt is the other half of rifted western North America. Most authors pointed out that there is also a Grenville-type mountain belt in East Antarctica adjacent to the Transantarctic Mountains, which could be the southern remnant of the Grenville Province of eastern North America (Chap. 8). In this reconstruction, western South America adjoins the eastern North American margin, allowing the

Patagonia terrane to link up with the restored Grenville belt. The other margins of East Antarctica were in contact with the India and Australia as part of Gondwanaland. In Dalziel's 1992 reconstructions, the eastern margin of North America rifted away from South America, not the Baltic Platform (Fig. 10.2A)

In another paleogeographic reconstruction, Karlstrom and others have argued that Australia, not East Antarctica, was attached to the western United States; this is known as the "AUSWUS" (*AUS*tralia-*Western United States*) reconstruction.

All of these reconstructions remain controversial, since the data are so hard to obtain and interpret. Matching up distinctive rock types and mountain belts between regions leaves a lot of room for interpretation. The best evidence would from paleomagnetic pole positions. Unfortunately, the paleomagnetic data are not clear as to which pole position is best.

Fig. 10.2 shows one of the most recently published reconstructions for the positions of the continents in the late Proterozoic, about 750 million years ago (Fig. 10.2A) and at the end of the Proterozoic, about 545 million years ago (Fig. 10.2B). The late Proterozoic supercontinent (Fig. 10.2A) has been called Rodinia, and most geologists have come to recognize that this supercontinent was an important part of the late Proterozoic, even though there is still controversy about how to reconstruct it. The proposed latest Proterozoic supercontinent has been called Pannotia, although there are still many different interpretations of how to reconstruct it (compare Fig. 10.2B with Fig. 9.15). Clearly the art of reconstructing the positions of continents this ancient is still in its early stages.

margin through Early Ordovician. The sequence was defined by American L. L. Sloss and named for Sauk County, Wisconsin, where typical Upper Cambrian quartz sandstones are exposed. Profound unconformities mark both the base and the top of the sequence, making it a fundamental package of strata that record the Sauk transgression of the craton. That transgression, which was apparently due to a worldwide rise of sea level, culminated with the flooding of nearly all of North America about 500 million years ago in Early Ordovician time. A subsequent regression, to be discussed in Chap. 11, then reexposed the entire craton to erosion; this resulted in the upper unconformity.

Finally, we use the Sauk Sequence to show how one can interpret the conditions of deposition, paleoclimate, and paleogeography of a craton from the study of sedimentary rocks and fossils. Because no human was around as witness, our detective work must rely greatly upon analogical reasoning from the knowledge of modern processes and environments. Again, we shall apply this approach and build upon it in later chapters.

Vendian History

Tectonic Framework

North America as we know it was formed by the dismemberment of a Proterozoic supercontinent nearly 600 million years ago (Fig.

10.2). Evidence of such dismemberment is provided by Vendian basaltic rocks around the margins of the continent and fault-bounded troughs containing very thick Vendian strata within the margins of the craton (Figs. 10.3 to 10.5). As noted in Chap. 7, such troughs, or aulacogens, cutting into continental margins are thought to be characteristic of newly rifted continental margins (for example, the East African Rift Valleys of Fig. 7.18). Also characteristic is great subsidence of the passive margin of a newly rifted continent (see Fig. 7.33). As noted in Chap. 7, such subsidence is caused primarily by cooling of hot lithosphere as it moves away from a spreading ocean ridge, but this cooling is augmented by any thinning of lithosphere during rifting and by subsequent sediment loading (see Fig. 7.34). This is another example of feedback in that cooling-subsidence creates space for sediment accumulation, which in turn causes more subsidence to create still more space, and so on. These combined effects, however, occur fastest early and then at an ever diminishing rate.

Many of the present continents show the same simultaneous pattern of subsidence, so it is inferred that one or more large supercontinents existed and then broke up rapidly between 500 and 600 million years ago. The configuration of that supercontinent(s) is controversial (see Box 10.1).

As the Proterozoic supercontinent broke up and new ocean basins opened between its pieces, marine sediments were

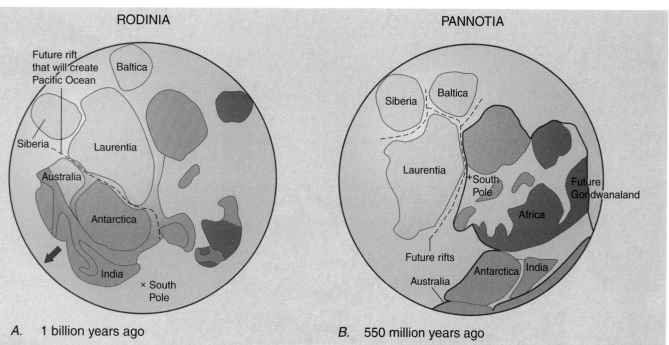

RODINIA

Future rift that will create Pacific Ocean

Baltica

Siberia

Laurentia

Australia

Antarctica

India

× South Pole

A. 1 billion years ago

PANNOTIA

Siberia

Baltica

Laurentia

+South Pole

Future Gondwanaland

Africa

Future rifts

Australia

Antarctica

India

B. 550 million years ago

Figure 10.2 Reconstructions of the Rodinia (1 billion years ago) and Pannotia (550 million years ago) supercontinents. Rodinia formed at the end of the Grenville orogeny, placing Laurentia (most of North America,) in the center of the supercontinent at the equator. In this reconstruction, Siberia adjoins northern Canada, whereas Australia ("AUSWUS" reconstruction) connected with the western margin of North America. Other reconstructions ("SWEAT" hypothesis) place East Antarctica in that position, but here it is farther to the south. The Australia-Antarctica-India block then rifted away from western North America, forming major late Proterozoic passive margins. By the Early Cambrian, they had collided with eastern Laurentia during the Pan-African orogeny to form Pannotia. Combined with South America and Africa, these regions would eventually combine to form Gondwanaland, whereas Laurentia would split away yet again and move to the equator during the Cambrian (see Fig. 9.15). *(Modified from Dalziel, 1992, GSA Today, v. 2, pp. 240–241, and from Unrug, 1997, GSA Today, v. 7, pp. 1–6.)*

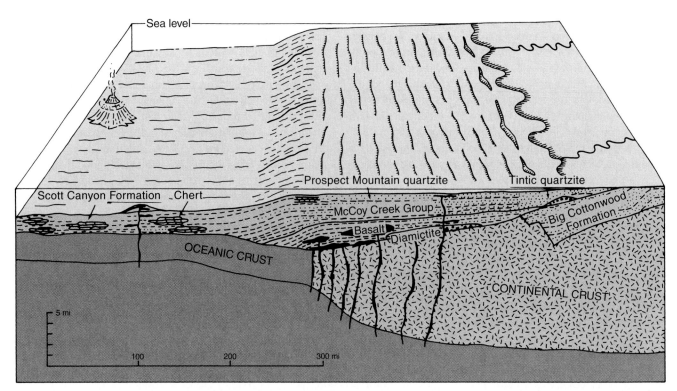

Figure 10.3   Diagrammatic restoration of Vendian and Lower Cambrian strata from California to Utah, suggesting that western North America was a passive continental trailing edge. Basalts suggest rifting of North America away from some other continent by sea-floor spreading. Typical of trailing edges is the accumulation of thick prisms of continental shelf sediments with mature quartz sandstones and carbonate rocks. (Compare with Fig. 7.34.) *(from J. B. Stewart, 1972, Bulletin of the Geological Society of America, vol. 83, pp. 1345–1360. Reprinted by permission of the Geological Society of America via Copyright Clearance Center.)*

deposited in thick prisms across the newly rifted margins. All margins of North America were passive during Vendian and Cambrian times. Although Vendian marine strata are thick and widespread all around the margins of North America and many other continents, they are poorly represented within cratons, which remained mostly above sea level as eroding lowlands. At the left end of Fig. 10.4 we see a typical example from the western United States of marine Vendian strata beyond the edge of the craton and passing conformably upward into Early Cambrian ones. Within the craton to the right, however, Vendian rocks are confined to narrow aulacogens, and most of these strata are nonmarine. By contrast, a thin sheet of marine Late Cambrian strata extends widely across the craton and is unconformable upon older rocks there (Fig. 10.6). The sea must have surrounded the continent at first and gradually spread over the craton during Late Cambrian and Early Ordovician times. Because this transgression affected most continents simultaneously, it was probably caused by a worldwide rise of sea level resulting from the breakup of the supercontinent(s). That

breakup represents one of the most fundamental global reorganizations of lithosphere plate configurations in the entire history of the earth. It created North America as we now know it.

The Cambrian Craton

Structural Modifications of Cratons

The North American Phanerozoic craton is defined by a profound three- to fourfold thickening of strata of many successive systems around the margins of the continent (Fig. 10.7). As pointed out in Chap. 7, cratons are only *relatively* more stable than other regions. Moreover, the cratonic surface in Cambrian time, though low in relief, was not perfectly flat. Cratons have undergone structural changes, largely of a mild (nonorogenic) sort, and these changes are reflected in the stratigraphic record.

Broad, gentle warpings of the cratonic crust, apparently in response to small changes of isostatic equilibrium due to density

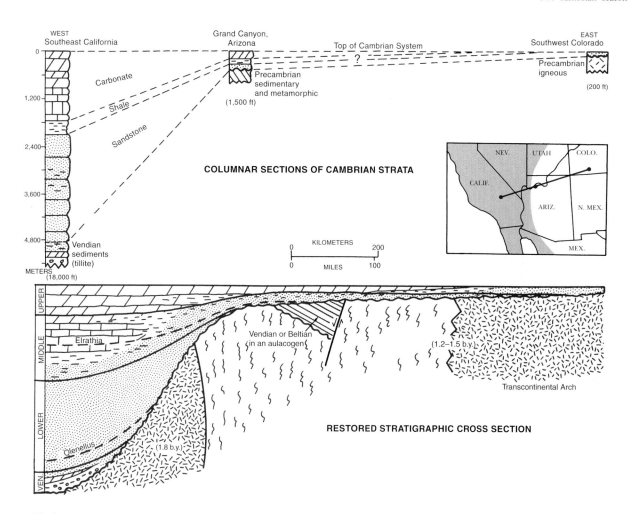

WEST
Southeast California

Grand Canyon,
Arizona

Top of Cambrian System

EAST
Southwest Colorado

Precambrian
igneous

(200 ft)

Carbonate

Shale

Sandstone

Precambrian
sedimentary
and metamorphic
(1,500 ft)

Vendian
sediments
(tillite)

METERS
(18,000 ft)

COLUMNAR SECTIONS OF CAMBRIAN STRATA

KILOMETERS

MILES

UPPER

MIDDLE

LOWER

VEN

Elrathia

Olenellus

Vendian or Beltian
in an aulacogen

(1.2–1.5 b.y.)

Transcontinental Arch

(1.8 b.y.)

RESTORED STRATIGRAPHIC CROSS SECTION

Figure 10.4   Cross sections showing Vendian and Cambrian strata in the southwestern United States. Columns *(top)* portray factually the character of preserved rocks exposed at several localities. The restored cross section *(bottom)* was constructed from the columns and other, scattered information in between.

changes in the underlying lithosphere, not only have raised areas called **arches** but also have depressed areas called basins. These structures may have been formed as by-products of the flexing of cratons by thermal subsidence of rifted continental margins or in response to compression of craton margins by approaching island areas or continents. Faults also are present in cratons, and a few show great displacement. Sedimentation on cratons has been profoundly influenced by such structural modifications. In turn, basins and arches have greatly influenced the accumulation of oil and natural gas, salt and gypsum, chemically pure limestone, and other economically important resources associated with sedimentary rocks. Therefore, the structure and history of cratons are of more than passing interest to humanity.

The largest single feature of the Cambrian craton is the so-called transcontinental arch extending from Lake Superior southwest toward Arizona (Fig. 10.7). It can serve as an example of how the geologist approaches questions about cratonic history. Along its sides, Cambrian sandstones lap unconformably against the Cryptozoic basement (Fig. 10.4). In many places, such as the Grand Canyon, Cambrian strata lap against old hills on the Cryptozoic surface (Figs. 10.9 and 10.10). But along the arch axis, especially in the subsurface beneath the Plains states, Cambrian strata are entirely missing and younger rocks rest upon the Cryptozoic there. Are these patterns original, or do they merely reflect post-Cambrian erosion? Finally, what are the age and history of this great arch? Fig. 10.8 reveals further clues not

B.

Figure 10.5   Typical outcrops of late Proterozoic deposits from western North America. *A:* late Proterozoic shales of the Belt Supergroup, Glacier National Park, Montana. These deposits were apparently formed during an early phase of continental rifting as the passive margin began to subside. *(D. R. Prothero.)* *B:* Thick quartzites and shales of the Uinta Mountain Group, Flaming Gorge, Utah. These deposits were accumulated in an aulacogen that formed perpendicular to the ancient Proterozoic continental margin of western North America. *(R. H. Dott, Jr.)*

A.

apparent from the thickness map alone. We see that dominantly quartz sandy facies border much of the arch, suggesting that it was a high feature supplying some of that sand during Cambrian time; perhaps it was an old Cryptozoic drainage divide present as a range of low hills converted to islands when the Late Cambrian sea flooded the craton. The presence of several major post-Cambrian unconformities over the transcontinental arch indicates, in reality, a long history of periodic flooding alternating with erosion. Although it influenced sedimentation in the Cambrian, much of the upwarping occurred in several later spasms. This is typical of cratonic arches.

Development of Arches and Basins

Let us now examine the warping of cratons in more detail. Compared with basins, arches typically contain thinner stratigraphic sequences interrupted by many unconformities. This implies that basins subsided relatively more rapidly and received greater and more continuous sedimentary accumulations than did

arches. The arches intermittently were shoal areas experiencing little or no sedimentation or were islands undergoing active erosion. In Fig. 10.11 (upper), note that thinning of strata over the arch is due both to originally thinner accumulations there *and* to erosion at unconformity surfaces. Interpretation of cratonic history rests heavily upon knowledge of the ages of warpings of arches and basins; therefore, it is crucial to make correct interpretations of the causes and timing of the thick and thin accumulations. In Fig. 10.11 (lower), we see an identical situation of thinner and thicker strata, but note that a different historical explanation is required in this case, for these structural features were formed entirely *after* deposition of the rocks. In the upper case, thinning over the arch was caused by lesser subsidence and smaller accumulation rates as well as by periodic erosion. But the lower case shows thinning caused solely by subsequent erosion or truncation (cut off) at an unconformity. The "basin" in this case is purely secondary and is the result solely of downfolding and truncation *after* deposition. If one is exploring for petroleum traps, this sort of analysis is of utmost importance,

B.

A.

C.

D.

Figure 10.6   Typical outcrops reflecting the great Cambrian transgression. A: Lower Cambrian sequence at the base of the Grand Canyon. The ledge-forming sandstones in the foreground are the Tapeats Sandstone, a shallow-marine shelf deposit. They are overlain by the poorly exposed Bright Angel Shale, representing quieter, deeper waters, and then by the cliff-forming Muav Limestone, deposited in offshore carbonate shoals. This sequence lies on the tectonically quiet craton, so it is only a few hundred meters in thickness. *(Courtesy of Peter Kresan.)* B: A similar sequence can be seen in the cliffs of the Panamint Range in Death Valley, California (visible in background). The prominent light band is the near-shore marine Zabriskie Sandstone, overlain by offshore shales and limestones of the Carrara and Bonanza King Formations (in the middle and top of the cliff). Here, the Lower Cambrian sequence is over 5,000 meters thick, since these rocks were deposited on the rapidly subsiding passive margin (see Figs. 10.3 and 10.4A). *(D. R. Prothero.)* C: Lower and Middle Cambrian deposits from Jasper National Park in the Canadian Rockies are about 5,000 meters in thickness, because they were also deposited on the subsiding continental margin. *(Courtesy of Stephen C. Porter.)* D: By contrast, the middle of the continent was not submerged until the Late Cambrian, at the peak of the Sauk transgression. These cross-bedded sandstones were formed by wind-blown sand dunes; they are now exposed in the Dells of the Wisconsin River. The total thickness of Cambrian rocks in this region is less than 200 meters. *(D. R. Prothero.)*

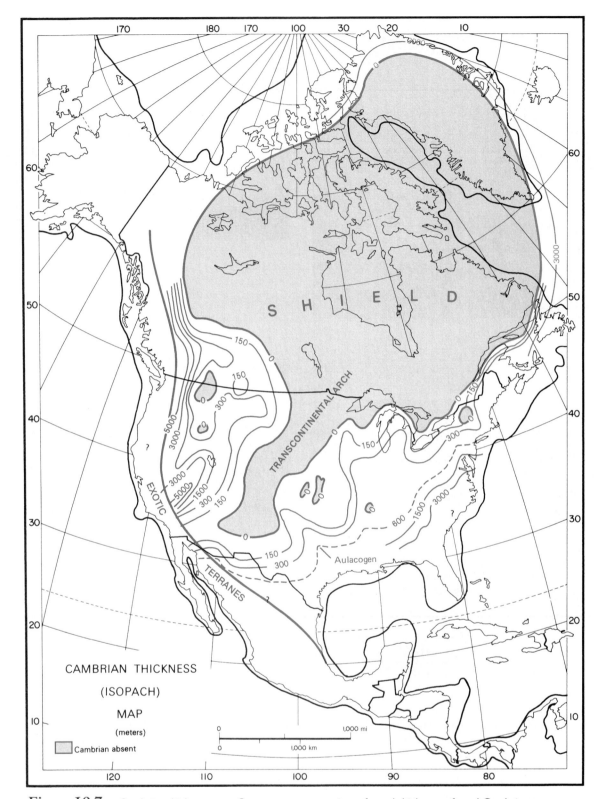

Figure 10.7 Cambrian thickness map. Contours connect points of equal thickness of total Cambrian strata. Marked contrasts in thickness are revealed (especially at the margin of the craton), but their causes must be interpreted carefully. For example, subsequent erosion has removed Cambrian strata from large areas. (See Box 10.2 for sources.)

Box 10.2

A Primer on Geologic Maps and Diagrams

Geologists use many kinds of diagrams to portray their information and interpretations. The principal kinds employed in this book are summarized in this feature, and the symbols used are explained in the accompanying legend.

Fig. 10.4 is a type of cross section—a vertical slice through the earth—used extensively to portray stratigraphic data and to help interpret patterns of thickness, sedimentary facies, and ages of strata. The upper *columnar* sections portray factually the character of preserved strata, exposed at several localities. The lower *restored cross section* was constructed from the columns above and from additional information from intermediate areas (other examples of restored cross sections appear in Figs. 8.23 and 8.24). Subsequent deformation and erosion are ignored. It is an interpretive graph in which the vertical axis is stratigraphic thickness and the horizontal axis is distance between data points. The geologist infers thickness and lithologic facies variations between separate columns of the upper diagram. Such cross sections are helpful in portraying regional relationships among strata in a vertical plane and are invaluable companions to various kinds of maps, which show relationships in the horizontal plane. The vertical scale is exaggerated to show detail. Note the transgressive facies, and how time lines overlap progressively eastward onto the basal unconformity (*Olenellus* and *Elrathia* are trilobite index fossils). Other kinds of cross sections are also important, especially those that show present structure and topography (see Figs. 7.2 and 7.4).

Maps portray various kinds of information in a horizontal plane and are important in conjunction with vertical cross sections. Figure 10.7 is a *thickness* (or *isopach*) *map*. Contour lines connect points at which a given age of strata has the same thickness. Such maps reveal major contrasts of thickness between different tectonic elements, such as cratonic arches and basins; they also delineate the margins of cratons beyond which thicknesses increase greatly.

Facies maps portray the geographic variations of sediment types of a given age, as in Fig. 10.8. Their interpretation is greatly facilitated by restored cross sections.

Paleogeographic maps show any of a variety of geographic phenomena. Most obvious are the past distributions of land and sea, mountains versus lowlands, ancient ocean currents, and winds, paleolatitude, climatic zones, and the like. These maps represent the ultimate product of historical interpretation so must be prepared using the information contained in all the other types of illustrations. Fig. 10.10 is one of the first paleogeographic maps ever published, and Fig. 10.9 is a modern example for the same period.

Principal sources for preparing these types of illustrations for this book include Alberta Society of Petroleum Geologists, 1964, *Atlas of facies maps of western Canada;* Clark and Stearn, 1960, *Geological evolution of North America;* Douglas et al., 1963, *Petroleum possibilities of the Canadian Arctic;* Eardley, 1961, *Structural geology of North America;* Kay, 1951, *North American geosynclines;* Levorsen, 1960, *Paleogeologic maps;* Martin, 1959, *Amer. Assoc. of Petroleum Geologists Bulletin;* Donald E. Owen, unpublished maps; Raasch et al., 1960, *Geology of the Arctic;* Rocky Mountain Association of Geologists, 1972, *Geologic atlas of the Rocky Mountain region;* Shell Oil Company, 1975, *Stratigraphic atlas;* Sloss et al., 1960, *Lithofacies maps;* Surlyk and Hurst, 1984, *Geological Society of America Bulletin.*

MAP SYMBOLS

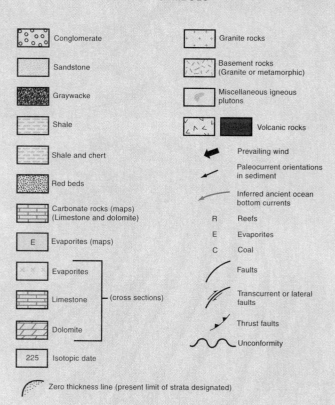

-Continued on page 216

Continued from page 215—

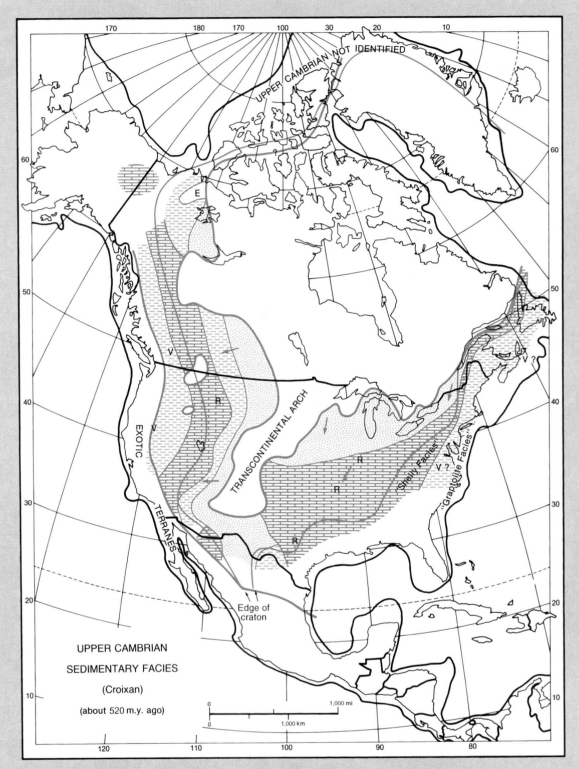

Figure 10.8   Upper Cambrian sedimentary facies map (note symbols in Box 10.2). Nowhere does the heavy zero thickness line represent a Cambrian shoreline; erosional nature of this boundary is proven, especially at the southeastern side of the transcontinental arch and in west-central Canada, where the zero line intersects facies boundaries nearly at right angles. Clearly the facies once extended beyond the zero line. Note southerly orientation of paleocurrent arrows.

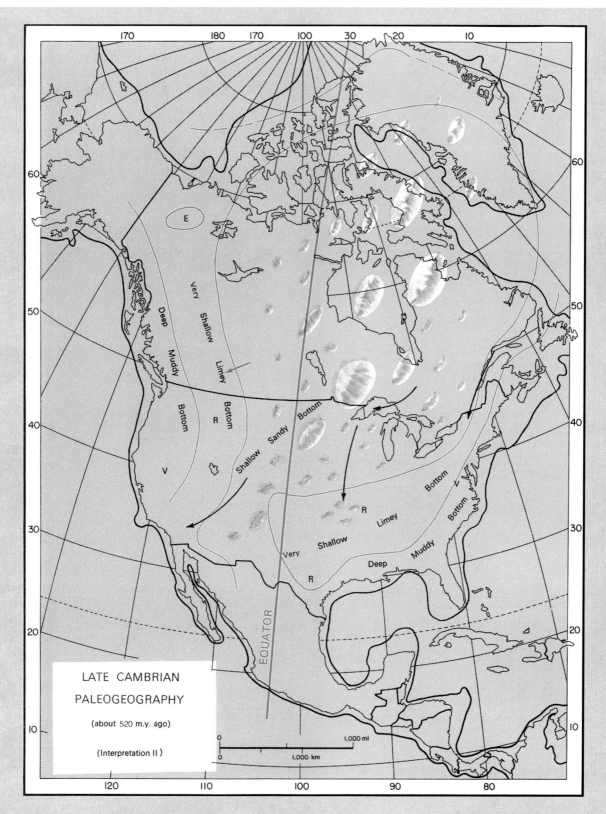

Figure 10.9   Late Cambrian paleogeography. Facies maps provide factual evidence for constructing paleogeographic maps, which inevitably are very interpretive. The latter may show land, sea, equator, poles, currents, winds, climatic zones, etc. Wherever we must restore geography beyond the zero limit of facies maps, there is much chance of error. For example, northeastern Canada and Greenland could have been one large, low island. (Compare Fig. 10.10.)

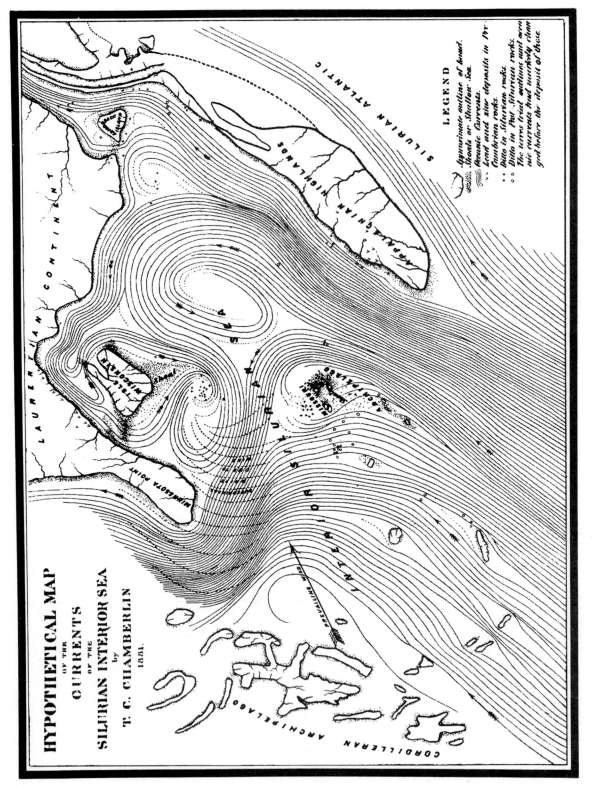

Figure 10.10   One of the first paleogeographic maps ever published shows the paleography of the eastern United States for early Paleozoic time. In 1881, the name Silurian often included all of what now comprises Cambrian, Ordovician, and Silurian. This map really refers to the Late Cambrian. Thin lines and arrows indicate supposed ocean currents. (Compare Fig. 10.9.) (*After* T. C. Chamberlin, Geology of Wisconsin, v. IV, p. 530.)

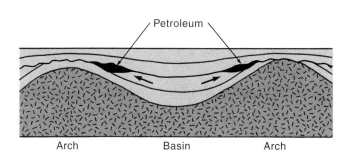

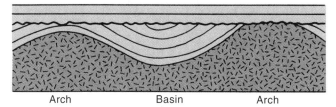

Figure 10.11 Contrasts of thicknesses and unconformities between cratonic basins and arches formed *during* versus *after* sedimentation. *Upper:* Differential warping *during* sedimentation caused thickening of strata in basins and thinning over arches. Note that timing of warping favored accumulation of petroleum, which migrated up-dip from the basin where it was generated. *Lower:* Warping *after* sedimentation produced thickness patterns superficially similar to the upper case, but here the contrast between basin and arches is due solely to differential erosion *after* warping had occurred. This warping was too late to trap petroleum.

because potential structural traps must have formed before petroleum began to migrate through permeable strata if any was to accumulate (as in Fig. 10.11, upper).

How do we acquire the information for studying broad and subtle cratonic structures? Much of the evidence for arches lies in surface outcrops, but only the edges of basins are exposed. Therefore, subsurface information is required to define basins and buried arches. Deep drilling for water and petroleum during the twentieth century revolutionized the study of regional stratigraphy by providing the third dimension to our observations. Where deep drillhole data are lacking, some insight into subsurface relationships can be gained with various geophysical devices. Magnetic and gravity surveys reveal information about special types and thicknesses of buried rocks, but seismology provides the greatest insight into thickness, structure, and, in some cases, even gross lithology of buried strata (Fig. 10.12).

The Sauk Transgression

General Setting

By the end of Cambrian time, at least three-fourths of the low surface of the North American continent had been flooded by the epeiric sea formed by the Sauk transgression, a sea that continued to spread in Early Ordovician time. This sea was like modern continental shelves except that it was much larger. The average

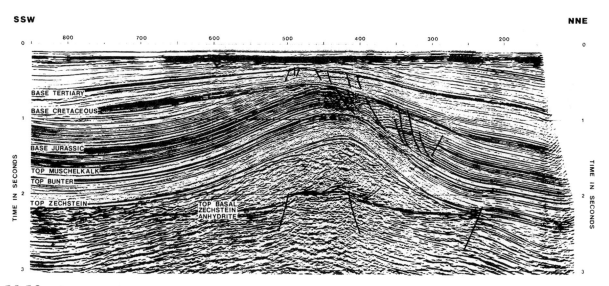

Figure 10.12 Seismic cross section (14 kilometers wide and about 3,500 meters deep) beneath the southern North Sea, which is a modern epeiric sea covering part of the European craton. Structure of the rocks is clearly revealed in this inaccessible subsurface setting. The large faulted anticline was formed in Late Jurassic time by rising Permian evaporite (Zechstein anhydrite), which is less dense than overlying sediments. Pre-Cretaceous erosion then produced a clear unconformity, after which further spasmodic upwarping by rising evaporite has influenced both Cretaceous and Tertiary sedimentation as evidenced by thinning of strata over the anticline. Depth expressed as "Time in Seconds" refers to the round-trip time required for sound energy to be reflected back to the surface from various strata. *(From A. W. Bally, ed., 1983, Seismic expression of structural styles, American Association of Petroleum Geologists, Studies in Geology Series, 15, p. 2.3.2–8; this cross section contributed by Owen and Taylor and British Petroleum Co.; with permission.)*

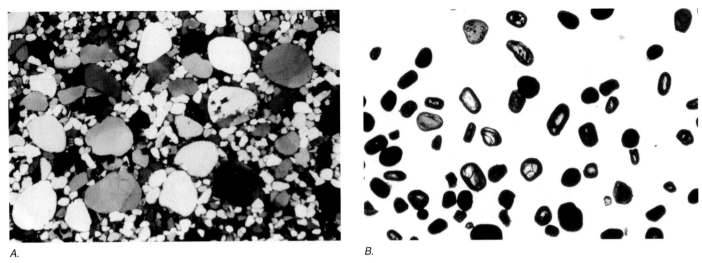

A.   *B.*

Figure 10.13   Microscopic photographs of typical Upper Cambrian sandstones, southern Wisconsin. A: Quartz grains averaging 0.25 millimeter diameter. Rounding is so exceptional for larger grains as to suggest an abrasive history equivalent in distance to rolling by water around the earth thirty or forty times. B: Concentrated fraction of well-rounded mineral grains of relatively high specific gravity (2.8 to 3.5) from a Cambrian sandstone like that in A. Minerals present average 0.15 millimeter and consist of very durable zircon, tourmaline, and garnet. Such minerals constitute less than 1 percent by weight of typical Cambrian sandstone. *(A:R. H. Dott, Jr.; B: Courtesy of John A. Andrew.)*

apparent rate of transgression was about 18 kilometers (10 miles) per million years (or about 0.05 feet per year). Actually the transgression was spasmodic, being interrupted by several partial retreats of the shoreline, as evidenced by local unconformities. Fig. 10.9 shows our interpretation of the paleogeography of North America near the end of Late Cambrian time and before the Early Ordovician maximum flood. Paleomagnetic data indicate that North America lay along the equator and was oriented 90° clockwise to present orientation. Such a map necessarily involves much interpretation, especially in areas where no Cambrian strata are present today, as in most of Canada and Greenland (Fig. 10.7). For example, other interpretations show a single lowland there, rather than many smaller islands.

Quartz-rich sand was the dominant Cambrian sediment on the craton with glauconite-bearing fine sand in some areas; shale was minor. Carbonate sediments replaced all of these during Early Ordovician times. The pattern of facies in Figs. 10.4 and 10.8 shows that the clastic sediments were derived from the craton because the sandy facies occur on the inner (craton) side of the shale and carbonate facies. The cratonic surface was weathered and eroded for nearly half a billion years before Cambrian sedimentation began; therefore, a huge volume of clastic material must have been available. Undoubtedly it was deposited, eroded, and redeposited countless times during this interval by rivers, wind, and glacial ice before most was redeposited in the epeiric sea. No definite fossil record of land plants exists in rocks older than Late Ordovician. Without a significant plant cover to hold soil in place, weathering processes must have been different from what they are today, and wind would have played a much greater role in erosion and transport of sediment. Therefore, the sand grains must have been acted upon by winds as well as rivers for long periods on the late Proterozoic and Vendian continental surface during the nearly half a billion years before their redeposition.

Like Sherlock Holmes, the geologist-detective is confronted by many mysterious questions. What was the origin of the unusually pure quartz sandstones? Were they spread in thin sheets by marine or nonmarine sedimentary processes? Why is there so little shale—the most abundant sedimentary rock of all? Why did carbonate deposition replace quartz sand in Early Ordovician time? How deep was the epeiric sea and was it calm or vigorously stirred? Was the climate warm or cold—wet or dry? Like Holmes, we must seek many different clues. We study the sediments and fossils by applying and adding to the basic concepts introduced in Chaps. 8 and 9. Keep in mind, as noted in Chap. 8, that we rely heavily upon a knowledge of modern sedimentary processes to provide keys for interpreting ancient conditions. By invoking the doctrine of actualism, introduced in Chap. 2, we constantly reason by analogy between modern and ancient sediments and organisms.

Mature Quartz Sandstones

Cambrian cratonic sandstones rank among the most mature in the world. They are notable for how perfectly the grains are rounded and sorted, as well as for possessing up to 99 percent quartz with only traces of other stable minerals (Fig. 10.13; see also Figs. 8.14 and 8.15). Together with a widespread, sheet-like distribution, this maturity implies lowlands and tectonic stability.

Sorting of the Cambrian sandstones narrows our options to wind, surf, and vigorous marine currents as the most probable agents of deposition. Rounding of the coarser Cambrian sand grains is quite remarkable. Experimental studies with the transport of sand by water in laboratory flumes and by air in wind tunnels indicate that wind is more than 100 times as effective as water for rounding quartz. This is because in water the impact between grains is cushioned by the greater viscosity of that fluid. Microscopic abrasion marks on the grain surfaces also point to

The Sauk Transgression 221

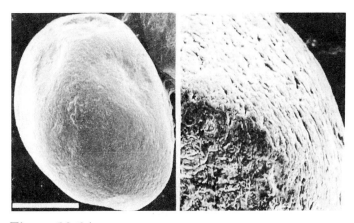

Figure 10.14   Quartz grains abraded by wind as viewed with very high magnification through the scanning electron microscope. *Left:* Pleistocene dune-sand grain about 0.25 mm long from Texas. The well-rounded shape is due to wind abrasion (compare Fig. 10.13). Bar is 100 microns long. *Right:* Enlarged view of a grain similar to the grain shown on the left from the Ordovician St. Peter Sandstone in Minnesota showing microscopic features characteristic of intense wind abrasion (sometimes called "upturned plates"). Similar features occasionally appear on glacially abraded grains, but there is no independent evidence for glacial influences in this case. *(Courtesy of James Mazzullo.)*

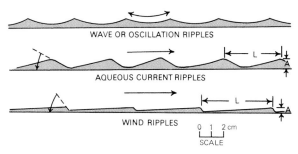

Figure 10.15   Profiles of ripple marks of different origins. Leeface angle of inclination, wavelength (L), and amplitude (A) vary for wind- versus water-formed current ripples. *(After E. H. Kindle, 1932, in* Treatise on sedimentation *[2d ed.], edited by W. H. Twenhofel; by permission of Williams & Wilkins Co.)*

wind erosion (Fig. 10.14). These facts, in addition to the lack of any fossil evidence of pre-Silurian land vegetation, suggests that the wind played a dominant role in the abrasion of the sand grains.

Next we have to decide if the *final deposition* of sand was accomplished by wind, water, or both. The presence of trilobites and other skeletons as well as trace fossils is taken to indicate final deposition of much of the sandstone in the sea because no pre-Silurian land animal fossils are yet known. But what about sandstones that lack even trace fossils yet seem superficially identical otherwise with the fossiliferous ones? Careful examination of stratification types has revealed subtle differences between the two. The shapes and sizes of ripple marks as summarized in Fig. 10.15 are especially important. Symmetrically shaped ripples are formed by the to-and-fro oscillation of waves on water, whereas most asymmetrically shaped ripples are formed by traction currents. Applying this knowledge, then, geologists interpret unfossiliferous Cambrian sandstones with very low, asymmetrical ripples as wind deposits, whereas unfossiliferous sandstones containing asymmetrical ripples with higher profiles are interpreted as river deposits. (These differences are related to the great contrast of viscosity of the two media.) Where the two types of unfossiliferous sandstones are interstratified, we infer deposition on wide, sandy river beds having wind dunes adjacent to them.

Cross-stratification larger in scale than ripples is the most conspicuous feature of the Cambrian sandstones all across the continent (Figs. 10.6D and 10.16). It owes its origin to the migration of dunes, as shown in Fig. 8.18. Cross-stratification provides an important indication of the direction of the wind or water current that formed it. The Cambrian paleocurrents shown in Fig. 10.9 were derived from the orientation of cross-stratification and

Figure 10.16   Geometrically complex cross-stratification typical of Upper Cambrian marine sandstones (Beartooth Mountains, Montana). Current flowed directly toward camera. Darker streaks are glauconite-bearing laminations. Notebook at bottom is 15 cm long. *(R. H. Dott, Jr.)*

ripples. But both water and wind currents can produce dunes, thus also similar-appearing cross-stratification. Wind more commonly forms large dunes a meter or more high, which generate large-scale cross-stratification (Fig. 10.6D), but large rivers and strong tidal currents can also form rather large dunes (Fig. 10.17). This characteristic ambiguity or perversity of nature may drive some people to distraction, but the tenacious geologist relies upon *associations* of other features (e.g., fossils, ripples, microscopic abrasion marks) to distinguish wind- from water-formed cross-stratification. Because a shallow ocean current is not unlike a very wide river, it may also be difficult to distinguish marine from river cross-stratification; environmentally sensitive

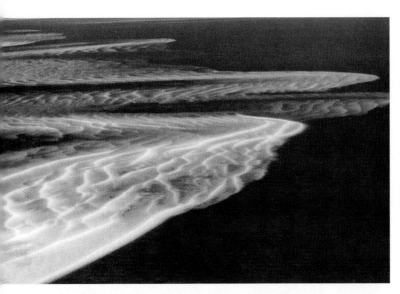

Figure 10.17   Air view of large submarine sand dunes formed by tide- and storm-driven currents on the Little Bahama Banks. Dune crests are only 1 to 2 meters below the water surface. The dunes consist of oolite. *(Photo By: M. O. Hayes, Courtesy of SEPM.)*

Figure 10.18   Trace fossils in the Upper Cambrian Jordan Sandstone, Mendota, Minnesota, caused by the burrowing action of wormlike marine animals with no hard, fossilizable skeleton. Deposition of the sediment must have been relatively slow to allow such extensive burrowing. Intensely burrowed intervals commonly alternate with unburrowed layers, which accumulated too rapidly for animals to survive. *(R. H. Dott, Jr.)*

fossils again provide the most important clues. Because there is no record of land life before Late Ordovician time, the mere presence of either skeletal or trace fossils in Cambro-Ordovician strata is taken to be diagnostic of marine deposition (Fig. 10.18).

We conclude from such detective work that wind and rivers first distributed the Cambrian sands in widespread sheets and that the Sauk transgression caused marine reworking of much of that sand. Tidal currents probably helped to spread the sand still farther and, together with waves, to winnow whatever clay may have been associated. Where is the abundant clay that must have formed by decay of the immense volumes of igneous and metamorphic rocks indicated by the pure quartz sand concentrate? More than half of the original source rocks should have weathered to clay, yet shale is relatively rare in Cambrian rocks of the craton. Apparently much of it was blown and washed from the craton, to find its way into deeper, less agitated zones of the sea beyond, for much shale does occur in Vendian and Cambrian strata beyond the Cambrian continental craton margins, especially on the down-current (present western) side of the continent (Fig. 10.3).

Change to Carbonate Deposition

As submergence of the craton progressed, the vast supply of weathered quartz sand was gradually redeposited. By the beginning of Ordovician time, so little land remained exposed that deposition of terrigenous clastic material virtually ceased. It is as though a great sand conveyor machine had been turned off after running for more than 100 million years. Meanwhile Upper Cambrian carbonate rocks had been forming already for millions of years on the craton margin away from the polluting effects of quartz sand (Fig. 10.8). Now, with final breakdown of the

"quartz-sand machine," a vast "carbonate factory" took over most of the shallow epeiric sea floor, which was otherwise favorable for carbonate deposition. Although these sediments presumably were deposited as limestone, most were converted later to dolomite.

Much limestone consists of shell debris and may display cross-stratification, ripple marks, and other features typical of sandstones formed of terrigenous silicate detritus. Such rocks are *clastic limestones.* Several conditions can be inferred from them. By analogy with living animals, most marine invertebrate fossil communities required agitated, well-oxygenated, clear water with plenty of sunlight for best growth. The textures of the shelly facies confirm accumulation in seawater with current activity sufficient to roll, abrade, and sort dead shells. Early Ordovician carbonate rocks imply a minimal influence of mud and sand from eroding lands and a shallow-marine environment where primitive algae or bacteria and carbonate-secreting invertebrate animals could thrive. Today such creatures are most abundant and most diversified in the shallow, clear, well-lighted, agitated, warm seas of the tropics and subtropics, although some local shelly sediments also form at higher latitudes. Normal-marine conditions are indicated for the outer rim of the epeiric sea, where a very shallow carbonate bank formed a ring around the craton and was bathed with nutrient-rich waters of the deep, open ocean that surrounded North America (Fig. 10.9). Over the central craton, however, low diversity of the fauna, complete dolomitization, and scattered evidence of evaporite minerals indicate greater than

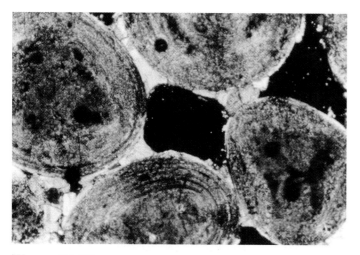

Figure 10.19   Microscopic photograph of modern oolite grains (average 1 to 2 mm in diameter) from Bahama Bank submarine dunes like those of Fig. 10.17. Each sphere consists of concentric laminations of carbonate precipitated around a nucleus grain. Cementation by carbonate has begun where the spheres touch. *(Photo by P. M. Harris; Courtesy of SEPM.)*

normal seawater salinity. Apparently circulation of water from the open ocean had become restricted here and a high rate of evaporation produced this hypersalinity.

Importance of Oolite

Fragmented fossils, scattered quartz grains, carbonate pebbles, and ripple marks in the Early Ordovician carbonate rocks point to considerable agitation during their deposition. Also common in these rocks are layers of crowded spherical, sand-sized grains called **oolites,** consisting of microscopic, onionlike, concentric carbonate laminae surrounding some nucleus, such as quartz or a shell fragment (Fig. 10.19). They are accretionary, and where they have been observed forming today (as in Great Salt Lake and on the tropical Grand Bahama Banks), precipitation of carbonate occurs as they are rolled. Agitation of the water causes loss of carbon dioxide gas to the atmosphere, producing saturation of calcium. The process, if facilitated by warm temperature which accelerates the escape of CO_2. By reasoning from the present to the past, we infer that ancient oolites also attest to strong agitation in shallow, calcium-rich waters. Oolitic carbonate rocks tend to have conspicuous cross-stratification, which also attests to agitation by currents strong enough to form submarine dunes like those of Fig. 10.17. On modern oolite banks, such currents are generally of tidal origin.

Depth of the Sauk Sea

Abundant evidence of wave and current agitation in both quartz sand and carbonate sediments suggests that the epeiric sea was shallow. By analogy with modern continental shelves, the Late Cambrian sandy bottom was probably less than 200 meters deep and had considerable variation of depth. Widespread stromatolites (Fig. 10.20) provide somewhat more precise depth limits

Figure 10.20   Late Cambrian stromatolites from Lester Park, New York. Pleistocene glaciers have scoured off their tops, revealing the intricate concentric layering of the domed columns (see Fig. 9.1). Although stromatolites were the commonest megascopic fossils for over 3 billion years, by the Cambrian they were disappearing, and they have been scarce since the Ordovician. Most scientists believe that this decline is due to the appearance of grazing invertebrates in the Cambrian. Hammer in lower left shows scale. *(D. R. Prothero.)*

for the carbonate sediments. As noted in Chaps. 8 and 9, these structures can form only in water shallow enough for photosynthesis. The light-bathed photic zone can be no deeper than 150 to 200 meters in the clearest water for the growth of the photosynthetic algae and cyanobacteria that create them. In Cryptozoic time, stromatolites appear to have formed over the full range of depths down to about 200 meters. Today, however, merciless grazing by snails tends to restrict most stromatolites to the uppermost intertidal and wave-splash zones, where these voracious animals cannot venture. Only where hypersalinity or very strong currents discourage the snails can large stromatolites form at greater depths today. Because the snails appeared in early Paleozoic time, we infer that Cambro-Ordovician stromatolites

Figure 10.21   Modern mudcracks, formed by shrinkage of muds during wetting and drying cycles. This photograph shows two generations of mudcracks, a large, deeply cracked set and later thin mud curls on the surface. *(R. H. Dott, Jr.)*

Figure 10.22   Flat-pebble conglomerate of fine, white sandstone chips in a dark green glauconitic matrix. Pebbles were partially cemented before being torn up by episodic storm waves (Upper Cambrian Tunnel City Group, Wisconsin). Such localized conglomerates are especially common as thin layers in both sandstone and carbonate facies of late Precambrian, Cambrian, and Lower Ordovician strata on most continents. *(R. H. Dott, Jr.)*

with associated fossil snails had depth ranges similar to modern ones. Therefore, the presence of stromatolites suggests the approximate position of sea level.

Fine dolomitic sediments associated with Cambro-Ordovician stromatolites commonly show polygonal networks of cracks formed by shrinkage of drying muds. This happens today on the highest exposed parts of muddy tidal flats and in dried-up mud puddles (Fig. 10.21). Such *polygonal cracks* also suggest very shallow depths with extended periods of exposure to the atmosphere. Commonly the polygonal mud chips were later stirred by storm waves and currents to produce dolomitic *flat-pebble conglomerates*. Sandy flat-pebble conglomerates (Fig. 10.22) apparently were formed by storm waves at greater depths, however.

To summarize, the widespread association of oolites, stromatolites, and polygonal cracking in dolomites points to very shallow depths for the late Sauk Sea. It seems clear that depth was greatest during Mid–Late Cambrian time (tens of meters) and then became much shallower during Early Ordovician carbonate deposition. One distinguished geologist has boasted that he could have walked all the way across this Ordovician sea bottom with his nose above water. Rapid upward accumulation of the carbonate sediments themselves apparently restricted the circulation of normal seawater from outside the craton as the worldwide rise of sea level, which had created the epeiric sea in the first place, ceased. Apparently a still-stand was reached before sea level fell again at the end of Early Ordovician time. In an example of sedimentary feedback, shallowing and evaporation produced widespread hypersaline conditions as a result of sedimentation during the last days of the Sauk Sea.

Modern Analogues for the Sauk Epeiric Sea

All the preceding discussion of Cambro-Ordovician sediments relies greatly upon knowledge gained from modern environments and from experimental studies of sedimentary processes. To make our analogies more explicit, let us now examine a modern continental shelf. The northern Gulf of Mexico provides well-studied examples of all the different facies discussed in this chapter, although they cover smaller areas than in the Cambrian epeiric sea. Indeed, even the largest modern continental shelves are less than one-fourth the size of the Sauk Sea.

The Gulf shelf is less than 200 meters deep, has moderately strong currents over it (Fig. 10.23), and has varied sediment sources. The Mississippi River supplies the bulk of detritus. On the outer shelf, much clay occurs, but deposition there since the last rise of sea level has been nearly negligible. In subtropical Florida to the southeast, by contrast, the land is entirely Cenozoic limestone and the climate is humid-subtropical. Therefore, no silicate clastic debris is supplied from it to the adjacent shelf. Environmental conditions off southern Florida, on the Bahama Banks, and off the Yucatan Peninsula of Mexico, favor carbonate sedimentation. Carbonate sands containing considerable oolite are prominent in both areas but are especially so around the Bahama Banks, where tidal currents pile the grains into dune-covered shoals (Fig. 10.17). The interior of the Banks is so shallow that circulation is restricted and evaporation causes hypersalinity—exactly the same conditions that we inferred for the Early Ordovician central epeiric sea on a much larger scale!

In all three areas, where nutrient-rich waters rise against the shelf edge, ideal growth conditions for organic reefs exist. Behind the Florida Keys reef zone, in the protected, very shallow waters of Florida Bay and on the inner parts of the Bahama Bank, fine limy muds are forming.

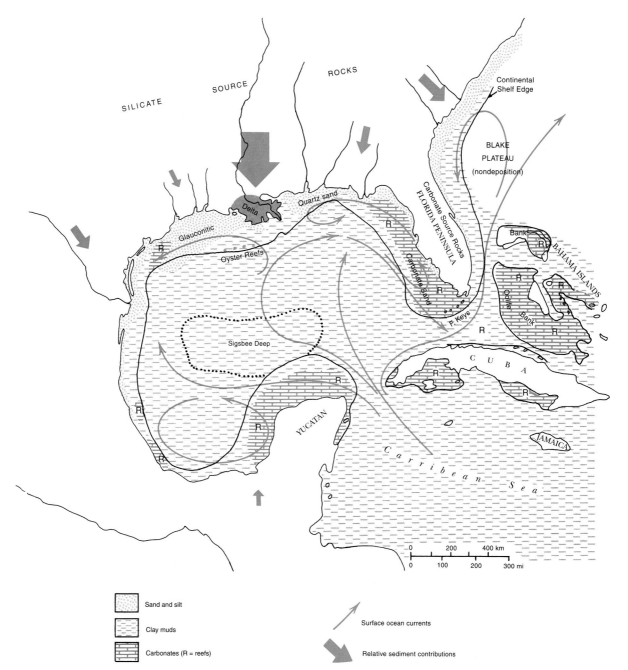

Figure 10.23 Recent sediments around southeastern United States for comparison with ancient epeiric sea deposits. Much of the sediment, particularly on the outer shelf, is relict from the last fall of sea level. On the average, sand on the outer shelf is moved by storm waves only about once very 5 to 10 years. Note differences among the sources of clastic sediments in the north and the presence mostly of carbonate rocks in the south. Practically all these shallow-water shelf sediments have counterparts in Cambro-Ordovician cratonic strata, but the deep-water muds do not. *(Adapted from F. P. Shepard et al., 1960,* Recent sediments of the northwest Gulf of Mexico; *with permission of American Association of Petroleum Geologists.)*

Quartz sands along the Alabama-Georgia coast are similar to those of the Upper Cambrian of the central craton. West of the Mississippi delta there is a large area of *nondeposition* where new sediment is not being added. The inner part of this area has a sandy bottom influenced only by major storms, whereas the outer part has fine glauconitic sediments. **Glauconite** forms on the sea floor where deposition is so slow that sluggish chemical reactions can convert clays and certain other minerals to this green, mica-like iron and potassium silicate mineral. Similar slow reactions can produce hard, mineralized surfaces of nondeposition called *hardgrounds* in a variety of other sediments, but especially carbonate ones. Iron, manganese, and phosphate are the usual mineralizing materials. These tend to form at times of deepest still-stands.

Figure 10.24 *Left:* Matagorda barrier island, central Texas coast, immediately after Hurricane Carla, 1961. *(Courtesy of Shell International Exploration and Production Inc.) Right:* The same coast a few weeks later. (Gulf of Mexico to left, Matagorda Bay to right in both views.) Hurricane-driven waves breached the island, spreading great volumes of sand both seaward and landward, but the breaches closed quickly, restoring equilibrium. *(Courtesy of Shell International Exploration and Production Inc.)*

Some features of the Gulf of Mexico do not have counterparts on the early Paleozoic craton. Ordovician rocks possess only small stromatolite reefs; large animal reefs such as those of the present Keys, Bahama, and Yucatan areas did not form on the craton until Silurian time. Likewise, nothing comparable to the Mississippi delta is known in early Paleozoic rocks, although deltaic deposits are prominent in younger strata. There are no early Paleozoic cratonic sediments like those of the deep, central Gulf, which include fine muds and some sands with graded bedding deposited by turbidity currents. Some post-Ordovician cratonic basins, however, were deep enough to have similar sediments.

Importance of Episodic Events

Hurricanes profoundly influence sedimentation on the Gulf shelf today. Their effects are most obvious along the shoreline, which is normally very stable but quickly modified when a violent storm strikes (Fig. 10.24) and sweeps much sand out onto the shelf. As noted in Chap. 4, events such as Gulf Coast hurricanes that may seem to be rare events on the ordinary human time scale (e.g., the "100-year storm") must be regarded as common on the geologic time scale, as shown by Fig. 10.25. By analogy, we must now ask if the early Paleozoic strata record primarily average, day-to-day conditions or extreme, episodic ones? Recall that vigorous wave and current agitation characterized much of the epeiric sea. Undoubtedly regular, more or less continuous currents (e.g., tidal currents) were present, but sporadic and more violent processes also are indicated. Most obvious are layers of coarse conglomerate interstratified with quartz sandstones near islands composed of Proterozoic rocks (Fig. 10.26) and sandy, flat-pebble conglomerates deposited farther from such islands, which must have been formed quickly during storms (Fig. 10.22). Other evidences of such events include thin shelly layers concentrated by winnowing away of

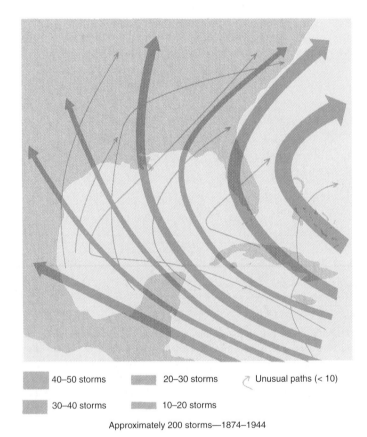

▨ 40–50 storms	▨ 20–30 storms	↗ Unusual paths (< 10)
▨ 30–40 storms	▨ 10–20 storms	

Approximately 200 storms—1874–1944

Figure 10.25 Relative frequency of various hurricane paths in the Gulf of Mexico and Caribbean region. An average of three storms per year strike the southeastern United States; since 1900, a minimum of two and maximum of twenty-one hit per year. Practically every portion of the coasts shown will average at least one storm per century. *(Data from I. R. Tannehill, 1950, Hurricanes, 7th ed: Princeton.)*

Figure 10.26   Thin layer of coarse conglomerate with well-rounded boulders between quartz sandstones. Boulders up to 1.5 meters in diameter, which are composed of resistant Proterozoic quartzite, were derived from nearby islands in the Cambrian epeiric sea. Tropical storm waves occasionally swept boulders away from sea cliffs to form many such local layers (Upper Cambrian near Baraboo, Wisconsin). *(R. H. Dott, Jr.)*

Figure 10.27   Hummocky stratification in fine sandstone (Upper Cambrian Tunnel City Group, Wisconsin). This distinctive, convex-upward structure was formed by rare vigorous stirring of fine sand on a marine shelf by unusually large waves, such as those caused by storms. Dark laminations are rich in glauconite, which was concentrated from other mineral grains by episodic storm wave agitation. Scale is in centimeters. *(Courtesy of Jane L. Sutherland.)*

sand, isolated layers with symmetrical wave-formed ripples, and strata with a distinctive undulatory lamination called **hummocky stratification** (Fig. 10.27). All of these were formed by unusually large waves that could stir the sea bottom briefly at depths greater than the depths reached by normal, fair-weather waves. Fig. 10.28 shows the spatial relationships among these features and their interstratification with fair-weather deposits. Episodic events are also recorded in the carbonate facies—for example, by dolomitic flat-pebble conglomerates and many thin zones of oolite.

At the opposite extreme from storms is nondeposition. As on the northern Gulf shelf today, large areas where sedimentation was very slow existed in the Sauk epeiric sea. The resulting slowly deposited fair-weather clastic sediments, represented by fine glauconite-rich strata thoroughly burrowed by animals, must have accumulated very slowly to allow time for both glauconization and complete destruction of stratification by burrowing animals (Fig. 10.18). In the carbonate facies, intervals of nondeposition are represented by mineralized hard-ground surfaces.

It is clear that much, if not most, of the sediments of the Sauk epeiric sea—be they terrigenous clastics or carbonate rocks—were transported by the less frequent, more violent events that punctuated the fair-weather boredom when nothing much was happening except burrowing, glauconization, and the growth of stromatolites. The history of Sauk sedimentation is well represented by the solid curve of Fig. 4.25; the total time represented by nondeposition and erosion by storms may be greater than the time represented by the preserved rocks!

Early Paleozoic Paleoclimate

Recall that paleomagnetic data indicate a tropical location for all of North America within ± 20° of latitude (Fig. 10.9). Whereas widespread Vendian glacial deposits indicate a cold climate about 700 million years ago, an equatorial location suggests much warmer conditions for Cambro-Ordovician time. Can we confirm this? Climatic evidence from organic reefs, land plants, coal, soils, and oxygen isotope data are not available in early Paleozoic strata, but the abundance of oolite, some very shelly limestones, and local evaporite deposits do tend to support a warm climate. Presence of wind deposits and the scattered evaporites make us think of arid deserts. The intimate association of river with wind deposits, as well as local evidences of a high water table, argue instead for at least a subhumid climate. Clearly the present is an imperfect key to the early Paleozoic past because there was no significant land vegetation. Therefore, the landscape would have been very foreign to our experience, for without vegetation, wind could have blown enormous quantities of clay, silt, and sand and built large sand-dune fields *even in a humid, tropical climate*. As in Cryptozoic time, the landscape may have resembled that of barren Mars today.

We have established that the Sauk epeiric sea was shallow and episodically stirred by violent storms. Given its tropical location, we would expect the steady trade winds to have rippled its surface on a daily basis. Current circulation in any shallow sea is driven chiefly by prevailing winds and the tides with a tendency for the flow over many shelves to parallel the coastline, as in the

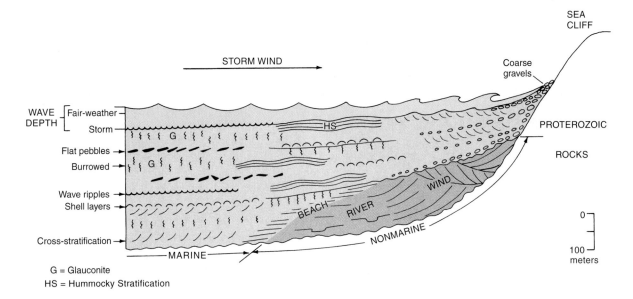

Figure 10.28　Composite diagram showing episodic storm deposits interstratified with fair-weather sandstones typical of Upper Cambrian clastic sediments in the central craton. Also shown are early nonmarine river and wind-dune deposits buried by the Sauk transgression. (Based upon strata in southern Wisconsin; compare Figs. 10.6D, 10.16, 10.22, 10.26, and 10.27.)

northern Gulf of Mexico (Fig. 10.23). Flow is also modified by the spin of the earth (Coriolis effect) and by large changes of air pressure. Thus, the flow along the sea bottom that affects sediment transport may differ considerably from the direction of the prevailing winds that first set the surface water in motion. Granted these complications, it is interesting to see a general conformance of the early Paleozoic paleocurrent pattern with the expected paleotrade winds. It is tempting to conclude that day-to-day circulation was entirely trade-wind driven, but we must beware of such circumstantial evidence, for modern trade-wind-driven currents flow at speeds too modest to move much sediment. Therefore, we believe that stronger tidal currents and storm-driven currents were responsible for moving most of the Cambrian sand.

Whatever the dominant control of day-to-day current circulation in the early Paleozoic epeiric sea, it is abundantly clear that tropical storms were superimposed upon fair-weather current and wave regimes (Fig. 10.28). They had an enormous effect upon the epeiric sea, just as they do today upon the continental shelves of southeastern North America.

Summary

Vendian History

- Rifting of a Proterozoic supercontinent formed North America about 600 million years ago. This was part of a profound reorganization of the global lithosphere. Post-rift thermal subsidence allowed the accumulation of thick Vendian and Cambrian marine strata on the passive continental margins. Only in Late Cambrian time did a rise of sea level cause marine transgression of the craton. The resulting epeiric sea reached its maximum size in Early Ordovician time before a fall of sea level caused regression.

Cambrian Craton

- Cratons tend to be differentiated into the following elements, whose histories were critical to sedimentation and to the formation of many economic deposits:
  1.　Arches with relatively thin and less complete sequences of strata
  2.　Basins with thicker and more complete sequences
  3.　Faults that may form aulacogens within the outer portions of cratons
- Sedimentary rocks and fossils of the Vendian to Early Ordovician unconformity-bounded Sauk Sequence provide

insight into conditions before and during transgression of the craton by the Sauk epeiric sea. North America would have seemed almost as foreign to humans then as Mars does today. It lay entirely within the tropical trade winds belt; the climate was warm and humid, but land areas had neither vegetation nor animals.

- The terrigenous clastic sediments are so mature that they are difficult to analyze. Marine ones look much like nonmarine ones except that the former contain skeletons, trace fossils, and the marine mineral glauconite. Complete lack of evidence of either land plants or animals is a striking feature of the nonmarine strata.
  1. Exceptional purity and sorting reflect redeposition of quartz sand already purified considerably during Proterozoic time. Clearly much of the sand produced by weathering through geologic time has not left the continents but has been recycled again and again.
  2. Exceptional rounding and microscopic abrasion features of grains and distinctive low-profile, asymmetrical ripples in many of the sandstones reflect the profound role of wind prior to the appearance of land vegetation in Late Silurian time.
- Cross-stratification, although abundant, is ambiguous for interpreting environments of deposition. It is an important indicator of the direction of flow during deposition of sand, however.
- Unusual rarity of shale apparently reflects extreme winnowing of fine material both by wind and by waves and currents over millions of years.
- The sheetlike geometry of the quartz sandstones is attributed to the spreading of sand widely over the low craton surface by wind, rivers, and tidal currents.

- Carbonate sedimentation, which had proceeded for more than 10 million years around the continental margin, finally spread across most of the craton as submergence gradually shut off the sand supply near the end of Cambrian time.
  1. Shelly, clastic limestones with a diverse fauna suggest normal-marine tropical conditions around the perimeter of the continent. All of the sea floor was now within the photic zone.
  2. Dolomite with a sparse fauna, much oolite, hints of evaporite minerals, and widespread stromatolites and polygonal cracks suggest very shallow water with restricted circulation and hypersalinity over the central craton in Early Ordovician time. This "stagnation" of the central epeiric sea reflects the cessation of rise of sea level—a still-stand. It also illustrates a sedimentary feedback because carbonate sediments "grow" so rapidly under optimal conditions in shallow water that they can choke the free circulation of normal seawater and thus alter their own further sedimentation.
- Episodic events are well illustrated by the Sauk Sequence. Storms such as tropical hurricanes produced local conglomerates; shell layers; undulatory, or hummocky, stratification; and symmetrical (wave) ripples. At the opposite extreme, nondeposition and very slow deposition are represented by glauconite-rich fine sandstones, mineralized hard-ground surfaces, and many intensely burrowed intervals lacking any stratification. This record speaks of long periods of fair-weather boredom punctuated by brief periods of stormy terror.

Readings

Dalrymple, R., G. M. Narbonne, and L. Smith. 1985. Eolian action and the distribution of Cambrian shales in North America. *Geology,* 13:607–10.

Dott, R. H., Jr. 1974. Cambrian tropical storm waves in Wisconsin. *Geology,* 2:243–46.

———, C. W. Byers, et al. 1986. Aeolian to marine transition in Cambro-Ordovician cratonic sheet sandstones of the northern Mississippi Valley, U.S.A. *Sedimentology,* 33:345–67.

Hayes, M. O. 1967. *Hurricanes as geologic agents.* University of Texas, Bureau of Economic Geology, Report of Investigations 61.

Holland, C. H., ed. 1971. *Cambrian of the New World.* London: Wiley-Interscience.

Kuenen, Ph. H. 1960. Sand. *Scientific American,* April,

Laporte, L. 1968. *Ancient environments.* Englewood Cliffs, N.J.: Prentice-Hall.

Middlemiss, F. A., P. F. Rawson, and G. Newall, eds. 1971. Faunal provinces in space and time. *Geological Journal Special Issue 4.*

Rocky Mountain Association of Geologists. 1972. *Geologic atlas of the Rocky Mountain region.* Denver: Hirschfeld Press.

Shell Oil Company. 1975. *Stratigraphic atlas—North and Central America.* Houston: Exploration Department.

Sloss, L. L., E. C. Dapples, and W. C. Krumbein. 1960. *Lithofacies maps: An atlas of the United States and southern Canada.* New York: Wiley.

Smith, G. L., C. W. Byers, and R. H. Dott, Jr. 1993. Sequence stratigraphy of the Lower Ordovician Prairie du Chien Group on the Wisconsin Arch and in the Michigan Basin. *Bulletin of the American Association of Petroleum Geologists,* 77:44–67.

Swett, K., and D. E. Smit. 1972. Paleogeography and depositional environments of the Cambro-Ordovician shallow-marine facies of the North Atlantic. *Bulletin of the Geological Society of America,* 83:3223–48.

(Courtesy Keene Swett.)

Chapter

11

The Later Ordovician

Further Studies of Plate Tectonics and the Paleogeography of Orogenic Belts

We crack the rocks

and make them ring,

And many a heavy pack we sling;

We run our lines and tie them in,

We measure strata thick and thin,

And Sunday work is never sin,

By thought and dint of hammering.

Andrew C. Lawson, formerly of the Geological Survey of Canada

and the University of California

▶ The typical Cambrian sea-floor fauna of trilobites and archaic and experimental animals was replaced in the Late Cambrian and Ordovician by a more typical Paleozoic fauna dominated by cephalopods, corals, brachiopods, bryozoans, and crinoids. These organisms demonstrated much greater ecological complexity, with several types of swimmers, floaters, attached filter feeders, and burrowers exploiting different levels above and below the sea floor. There were also several new types of predators, so their prey required greater defensive specialization.

▶ The great Cambrian epicontinental seaway retreated at the end of Early Ordovician time. In the Middle Ordovician, the seaway slowly returned to the continental interior, producing widespread, richly fossiliferous carbonate rocks over enormous areas of the continent.

▶ The eastern margin of North America had been a passive margin since the late Proterozoic. In the Middle Ordovician, it became a subduction zone when fragments of island arcs and other terranes began to approach. By the Late Ordovician, these had collided with North America, producing a high mountain range, which shed deltaic and nonmarine sediments onto the continental interior to the west.

In Chap. 10, the discussion of early cratonic history included the widespread deposition of carbonate sediments during Early Ordovician time. Chap. 11 concerns later Ordovician history;—from about 460 to 440 million years ago;—which began with a regression due to a worldwide fall of sea level for a few million years. The resulting cratonwide unconformity defines the base of the Tippecanoe Sequence, the second major package of Phanerozoic cratonic strata. The ensuing Tippecanoe transgression culminated 440 to 450 million years ago in Late Ordovician time, with one of the largest epeiric seas ever. Richly fossiliferous strata deposited therein attest to a new tropical marine paradise for very diverse invertebrate animals.

We begin by discussing Ordovician life, which is remarkably different from that of the Cambrian. The former represents the earliest of the faunas that characterized the later Paleozoic. After briefly discussing cratonic events of the later Ordovician, we then focus upon the main topic of this chapter, which is the first mountain-building episode to affect Phanerozoic North America. This was the Taconian orogeny of the Appalachian region, so named for the Taconic Mountains of southeastern New York, where an unconformity and thrust faulting were produced by this event (Fig. 11.1). We use the Taconian orogeny to illustrate further the sedimentary as well as structural effects of, and to elaborate upon the general theories about, mountain building introduced in Chap. 7. This will provide important background applicable to later chapters. Finally, as an extension of Chap. 7, we interpret Taconian mountain building in terms of plate tectonics, requiring consideration also of what was happening in other continents.

Ordovician Life

Explosive Radiation of the "Paleozoic Fauna"

Ordovician marine life was dramatically different from that of the Cambrian. As we saw in Chap. 9, Cambrian faunas were simple in structure and low in ecological diversity. Bottom-feeding trilobites dominated, along with suspension-feeding archaeocyathids, inarticulate brachiopods, and stalked echinoderms. Because there were few predators, there was not much selection pressure on the many experiments in animal design. During the latest Cambrian, this pattern began to change, and by the Middle Ordovician the sea floor looked radically different.

The most obvious difference is diversity. Any reconstruction of Ordovician life (Fig. 11.2) or sample of Ordovician sea floor (Fig. 11.3) is packed with animals, many of which first evolved during this period. Only about 150 families of animals are known from the Cambrian, but by the Late Ordovician there were more than 400 families, and that level was maintained through the rest of the Paleozoic. This diversity is a consequence of much greater ecological complexity. The simple Cambrian food chain of deposit-feeding trilobites and a few suspension feeders was replaced by a complex food chain (Fig. 11.4). Suspension feeders proliferated in several invertebrate phyla, and they grew larger so that they could filter food-bearing currents at different levels

Figure 11.1 Ordovician sandstones and shales in southwestern Newfoundland, near Port au Port Bay. The sandstones were deposited by deep-water turbidity currents in a marine trough in front of a volcanic arc but were then folded as the arc was thrust westward onto the continental margin during the Late Ordovician Taconic Orogeny. *(Courtesy Keene Swett.)*

Figure 11.2 Diorama of the Late Ordovician sea floor, based on fossils from the Cincinnati Arch region. The largest predator was the squidlike, straight-shelled nautiloid in the foreground. A large, well-armored trilobite swims in the lower left, and abundant brachiopods and horn corals cover the sea floor. Delicate crinoids *(right background)* and branching bryozoans *(left background)* occupy the higher levels of the sea floor. *(Image # K10276(4) American Museum of Natural History.)*

above the sea floor; in other words, ecological tiering increased (Fig. 11.5). In the Cambrian, only a few sponges and archaeocyathids protruded more than a few centimeters above the bottom, but by the Ordovician, there are several types of organisms that reached half a meter or more above the sea floor. Some animals attached to the sea floor, such as the brachiopods, fed on food particles in currents in the lowest few centimeters, whereas others,

Figure 11.3   Late Ordovician strophomenid and orthid brachiopods were so abundant that some limestones are literally paved with them. *(Courtesy of U.S. National Museum.)*

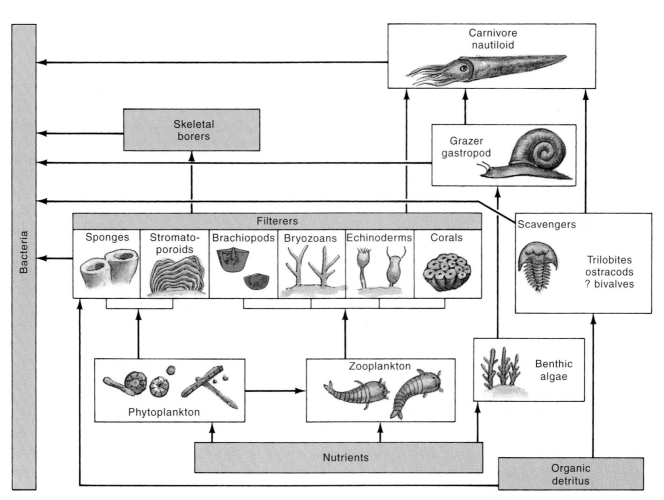

Figure 11.4   Ordovician life showed a much higher level of ecological complexity than any previous time in geologic history. The first complex food webs developed, with a base of primary producers (algae) grazed by primitive snails, and microscopic plankton fed upon by a wide array of filter feeders (sponges, stromatoporoids, brachiopods, bryozoans, echinoderms, and corals). Trilobites scavenged the detritus on the bottom. The top predators were the giant straight-shelled nautiloids, some of which had shells over 2 meters long! When these organisms died, their nutrients were recycled back into the food chain by bacterial decay.

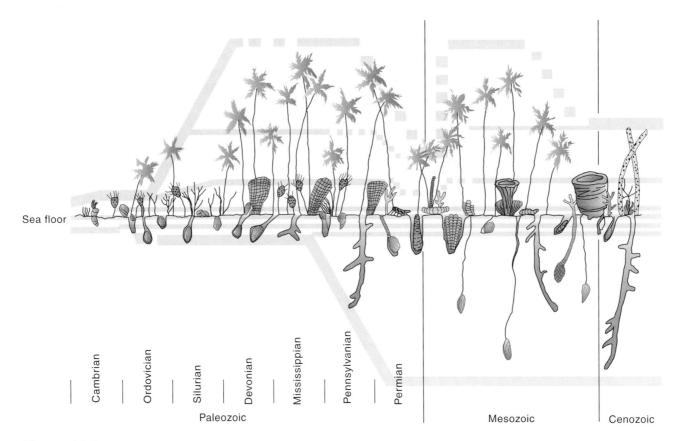

Sea floor

Cambrian | Ordovician | Silurian | Devonian | Mississippian | Pennsylvanian | Permian

Paleozoic

Mesozoic | Cenozoic

Figure 11.5   Another novel feature of Ordovician ecosystems was the increase in tiering, or multiple feeding levels above and below the sea floor. In the Cambrian, most invertebrates fed only a few centimeters above the sea floor or were very shallow burrowers. By the Ordovician, long-stalked crinoids and branching bryozoans began to take advantage of food-bearing currents up to 3 meters above the sea floor. In the late Paleozoic and Mesozoic, these high-level feeders had become even taller, and a new tier was added below the sea bottom by deep-burrowing clams. *(Modified from W. I. Ausich and D. J. Bottjer, 1991, Jour. Geol. Education, v. 39, pp. 313–318.)*

such as crinoids, fed on currents as far up as their long stalks would reach; some had stems longer than 3 meters.

By far the most common fossils in Ordovician rocks are brachiopods. Unlike the inarticulate brachiopods (Fig. 9.18A) of the earlier parts of the Cambrian, which held their shells together with muscles alone, the dominant group in the later Paleozoic fauna is the *articulate brachiopods* (Fig. 9.18B). Articulates had teeth and sockets in their hinge area to keep the shells better connected, and constructed their shells of calcium carbonate rather than the phosphate used by inarticulates. Articulate brachiopods covered the shallow sea floor in tremendous numbers, filtering out food particles from the water with their featherlike lophophores. Two groups of articulates are particularly characteristic of the Ordovician. The orthids (Figs. 11.3 and 11.6A) were very primitive articulate brachiopods, with relatively simple, finely ribbed shells. Although they originated in the Early Cambrian, their heyday was the Ordovician. The most abundant group, however, are the strophomenids (Figs. 11.3 and 11.6B), which during the Ordovician covered the sea floor like a pavement in some areas. Typical Ordovician strophomenids have a long, straight hinge that gives them a "D"-shaped outline. Many are extremely flat. Some have one shell nested inside the other, like stacked bowls, leaving very little room inside for the animal.

Next in abundance to the brachiopods are their close relatives, the *bryozoans,* or "moss animals." Bryozoans are colonial animals that form coral-like skeletons with thousands of tiny holes. Each of these pinhole-sized chambers houses a tiny filter-feeding animal with a lophophore like that of a brachiopod. Bryozoans were extremely diverse in the Paleozoic. Late Ordovician sea floors had a great variety of bryozoans, some with branching colonies, like certain living corals, and others with massive colonies formed in irregular, lumpy shapes (Fig. 11.6C, D). Calcified bryozoan fossils first appear in the Early Ordovician of China. Because all their close relatives occur in the Cambrian, bryozoans also probably evolved at that time but did not yet have a fossilizable hard skeleton.

Brachiopods and bryozoans filtered water a few centimeters above the sea bottom, but the true skyscrapers were the *crinoids,* or "sea lilies." Crinoids are echinoderms related to sea stars and sea urchins, but they are built in a different way: Their arms are used as filter-feeding fans, and their bodies are suspended on a long stalk rooted to the bottom (Figs. 11.2 and 13.49B). Cambrian stalked echinoderms were primitive animals known as eocrinoids (Fig. 9.17P). In the Ordovician, they were replaced by the major groups of crinoids typical of the rest of the Paleozoic. Some crinoid stalks were so long that the tentacles could feed several meters above the sea floor.

Figure 11.6 Representative Ordovician fossils. *A:* The primitive articulate brachiopod *Hebertella,* a common orthid. *B:* The "D"-shaped strophomenid brachiopod *Rafinesquina.* Both orthids and strophomenids were abundant on the Ordovician sea floor (see Fig. 11.3). *C:* The massive, lumpy bryozoan *Prasopora. D:* The delicate, branching bryozoan *Hallopora.* Ordovician reefs were built by members of several different phyla, including *E:* layered stromatoporoid sponges; *F:* the "honeycomb" tabulate coral *Favosites; G:* the solitary rugosid ("horn coral") *Streptelasma; H:* the "sunflower corals," or receptaculitids, now thought to be skeletons of dasycladacean algae. In response to predators, Ordovician trilobites became more specialized for burrowing, or protected by more spines and armor than their Cambrian predecessors. They included *I:* the foot-long "snowplow" trilobite, *Isotelus* (this specimen shows a healed bite mark on its right cheek); *J:* a mass-death assemblage of a similar trilobite, *Homotelus; K: Flexicalymene,* a trilobite that could roll up like a sowbug; *L:* the "lace-collar" trilobite *Cryptolithus,* with the delicate, lacy brim on the front of the head, long cheek spines, short body and "Jimmy Durante nose." *M:* The peculiar Ordovician gastropod *Maclurites.* This early experiment in snail evolution apparently carried the shell with the point directed down and forward, the opposite of most modern marine gastropods. Some macluritids were very large, reaching almost a foot in diameter. *(Photos I and J copyright 1992 Loren E. Babcock; all other photos by D. R. Prothero.)*

One of the most important innovations of the Ordovician was the appearance of the first true coral reefs. In the Cambrian, archaeocyathids and sponges formed reeflike mounds, but these mounds were never as large or diverse as coral reefs. Some Ordovician reefs exceeded 100 meters in length and 6 to 7 meters in height. A great variety of colonial organisms built these reef complexes. The earliest examples were built by bryozoans. In the Late Ordovician, two groups of typical Paleozoic corals became the dominant reef builders: the *rugosids,* or "horn corals," and the *tabulates,* or "honeycomb corals" (Fig. 11.6F, G). In addition to corals, the Ordovician reef complexes were built by an extinct group of sponges known as *stromatoporoids* (Fig. 11.6E). Stromatoporoids typically formed a thick, layered structure (*stroma* is Greek for layer and the root of the word *stromatolite,* the layered structures built by cyanobacteria; see Chap. 8). A fourth colonial reef builder typical of the Ordovician were the *receptaculitids* (Fig. 11.6H), or "sunflower corals," so-called because they resemble the head of a sunflower. Recent studies suggest that receptaculitids were probably formed by algae, so they were plants, not animals, like corals and sponges.

In addition to the proliferation of suspension feeders, the other great change in the Ordovician was the increase in predators. The largest predator in the Cambrian was the 45-cm long, soft-bodied *Anomalocaris* (Fig. 9.21A). By the Middle Ordovician, there were hard-shelled predators reaching 10 meters (30 feet) in length! These were the *nautiloids,* a group of molluscs related to the living squid and octopus (Fig. 11.2). Today their only surviving relative is the chambered nautilus of the South Pacific, but during the Late Cambrian, these predators began to diversify rapidly from other molluscan groups. They had a large head with tentacles, like an octopus or squid, but were encased in a long, conical shell that stuck out behind them. By the Early Ordovician, nautiloids were abundant worldwide, and their grasping tentacles and parrotlike beaks must have wreaked havoc on the complacent, unprotected Cambrian fauna. Certainly the primitive trilobites of the Cambrian must have been a favorite prey item. Not surprisingly, trilobites are much less common in the Ordovician than in the Cambrian. Nor is it surprising that the later trilobites also show specializations for protection against predators. Some had long spines (Fig. 11.6I–L); others could curl up into an armored ball; some had smooth shells on both front and back, like a plow for burrowing.

Although nautiloids were undoubtedly the top predators, another important predator group arose in the Ordovician: the sea stars, or "starfish." Modern sea stars use the suckers on their tube feet in their powerful arms to pull open clams, then turn their stomach inside out, insert it into the clam shell, and digest the clam. The earliest sea stars may not have been able to turn their stomachs inside out, but they apparently were important predators on the abundant brachiopods and few clams on the Ordovician sea floor.

Trilobites were the primary scavengers in the Ordovician, but not the only ones. In the Late Cambrian, simple, limpetlike molluscs evolved into the first snails, or *gastropods.* Ordovician snails had coiled shells, but these shells were peculiar: some were coiled in a plane and sat backward along the mid-line of the animal's back. Other shells were spiraled along an axis in a direction opposite that followed by the shells of living marine snails and

may have held their shell pointing down and forward on the animal's body (Fig. 11.6M). Clearly gastropods went through some experimentation before they settled into their later system of coiling. Some gastropods were predatory, a fact we know from the existence of Ordovician brachiopod shells that have been drilled through and fed upon, a feat performed by modern moon snails. The advent of herbivorous gastropods caused another dramatic event: the final decline of stromatolites (Fig. 10.20). Although these cyanobacterial mats were first grazed in the earliest Cambrian, they persisted until the end of the Ordovician, when the grazing pressure virtually wiped them out. Today they survive in a few refuges that are too salty, or whose currents are too strong, to allow grazing gastropods.

Most of the animals just described lived in the shallow-marine waters of the flooded continents. One group of fossils, however, is found in all environments, including the black shales of the deep Ordovician oceans. These were the *graptolites* (Fig. 11.7). Their name means "written on stone," and for a long time they were known only from mysterious flat, carbonized streaks on black shales that looked like pencil marks. In recent years, these animals have been found preserved in three dimensions in limestones, and this has solved the mystery of what they were. Surprisingly graptolites were distantly related to our phylum (the chordates), a fact we know from studying living animals (known as hemichordates) having the same detailed structure in their skeletons. Both living hemichordates and graptolites are colonial animals, with multiple cups on a long central rod, each cup housing a tiny filter-feeding animal. The earliest graptolites (from the Late Cambrian) were bushy, branched structures that apparently attached to a hard surface, but Ordovician graptolites were reduced to a few branches and floated on the open ocean. Some specimens even preserve little floats that allowed them to hang in the surface waters and filter out plankton. This explains why graptolites were fossilized in deposits of both shallow and deep waters. They floated on the ocean surface all over the world, and when they sank to the bottom in deep waters, they were preserved in the stagnant muds that became black shales. This circumstance also makes them the best possible kind of index fossils because they also evolved very rapidly, and they occur worldwide in all marine facies.

Another long mysterious group of fossils were the **conodonts** (Fig. 4.19). These microscopic, toothlike fossils were made of calcium phosphate (like our teeth and bones), so they were long thought to be the teeth of some extinct vertebrate. However, we now believe that conodonts were not teeth because they show no sign of wear on their tips, and growth rings show that they were enclosed in the body, not exposed on a jaw. Recently found specimens suggest that conodonts were part of a wormlike or eel-like organism distantly related to chordates, and they were used to support some kind of grasping or breathing structure in the mouth or throat region. Although primitive relatives of conodonts appeared in Cambrian time, true conodonts did not become diverse until the Ordovician. Conodonts, like graptolites, are excellent index fossils because they evolved rapidly and occurred worldwide in a variety of facies. They disappeared in the Triassic.

Another microfossil group found in the open ocean were the *ostracodes* (Fig. 11.8). These are microscopic, shrimplike

Figure 11.7 Graptolites arose in the Late Cambrian, but their abundance in the Ordovician and Silurian makes them important index fossils for these periods. Most graptolites apparently floated on the sea surface *(A, Glossograptus)* or hung down from floating debris or seaweed *(B, Syndograptus; C, Tetragraptus; D, Didymograptus; E, Cyrtograptus)*. *Lower left:* The complex budding pattern of the graptolite chambers closely resemble those of the living hemichordate *Rhabdopleura (lower right)*. Based on this evidence, paleontologists have concluded that each graptolite chamber contained a small, filter-feeding animal which was closely related to vertebrates. *(Modified from C. L. and M. A. Fenton, 1989, The fossil book: Doubleday.)*

A.

B.

Figure 11.8 Ostracodes are tiny crustaceans that have two kidney bean–shaped shells hinged over their backs. *A:* The earliest ostracodes are from the Early Cambrian. This specimen of *Vestrigothia* from the Late Cambrian of Sweden preserves the appendages in extraordinary detail; it is magnified 100 times. *(Courtesy of Klaus J. Mueller.)* *B:* By the Ordovician, some ostracodes had reached up to 3 centimeters in length. The shells of *Leperditia* were particularly abundant in Ordovician and Silurian limestones. *(D. R. Prothero.)* *C:* Later ostracodes were again very tiny. This specimen of *Calinocytheresis* from the Pliocene of Italy is magnified 25 times. *(Courtesy of Philip A. Sandberg.)*

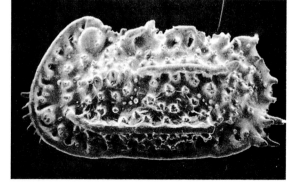

C.

crustaceans that lived inside two kidney bean–shaped shells hinged over their backs. The earliest ostracodes in the Cambrian may have used calcium phosphate for their shells, but later ones used calcite. Some Ordovician ostracodes reached almost a centimeter in length, and they were apparently suspension feeders living on the bottom in the stromatolitic mats. Since the Ordovician, however, ostracodes have gotten smaller, and all living ostracodes are microscopic (less than 1 mm). They live both in the surface plankton and on the bottom of marine and fresh waters.

Finally, the Ordovician marks the radiation of our own subphylum, the *vertebrates*. The earliest vertebrates were jawless fish with an internal skeleton made of cartilage (Fig. 11.9). These fish probably filtered food either out of the mud on the sea floor or directly from sea water; their simple, ringlike mouth opening could not bite. Their only hard parts were bony scales, so they rarely fossilize as complete animals. Most Ordovician fish are known only from isolated plates and scales in marine rocks, indicating that vertebrates first arose in the sea (not in fresh water, as scientists once thought). Small, bony scales that resemble fish scales were found in the Upper Cambrian Deadwood Sandstone of Wyoming, suggesting that vertebrates arose in the Late Cambrian (Fig. 11.9A). Recently, soft-bodied fossils from China have pushed the vertebrates all the way back to the Early Cambrian. From these simple ancestors, all later vertebrates (including humans) evolved.

Thus, the Ordovician marked a great radiation of many groups of animals, a radiation that increased both the diversity and the ecological complexity of shallow-marine communities. The Ordovician radiation had a cast of characters very different from typical Cambrian animals (such as trilobites, archaeocyathids, inarticulate brachiopods, and archaic molluscs and echinoderms). J. J. Sepkoski analyzed diversity data and found that the Ordovician radiation marked the beginning of the typical *Paleozoic fauna*. As shown in Fig. 11.10, the Paleozoic fauna was by articulate brachiopods, corals (tabulates and rugosids), crinoids, bryozoans, graptolites, and predators such as cephalopods (e.g., Ordovician nautiloids) and sea stars. Although the species of this cast of characters changed, these groups dominated the sea floor until the great Permian mass extinction ended their reign.

What caused the great Ordovician radiation that established the Paleozoic fauna? There is no clear answer to this question, but there are a number of suggestions. Once factor to notice is that the Late Cambrian and Ordovician marked the highest sea levels in earth history up to that point (Fig. 10.9). Every continent was almost completely flooded, forming enormous areas of shallow epeiric seas in which marine life could diversify. Second, we saw in Chap. 9 that oxygen levels in the Vendian and Cambrian were gradually increasing, and several lines of evidence suggest that they finally reached modern levels (about 20 percent of atmospheric gases) in the Ordovician. Modern organisms that use calcite in their skeletons cannot thrive unless oxygen levels reach 16 percent of the present level (or about 3 percent of atmosphere).

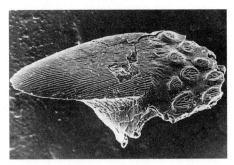

A.

B.

Figure 11.9   Fossils of the oldest jawless fishes. A: Bony armor plate of *Anatolepis*, from the Upper Cambrian Deadwood Formationof Wyoming. *(Courtesy of J. E. Repetski and the U.S. Geological Survey #543.)* letter B: *Athenaegis*, a nearly complete fish specimen from the Silurian of northern Canada. Note the simple, slitlike mouth for filter feeding, the asymmetric tail, and the lack of fins. *(Courtesy of Mark H.V. Wilson.)*

CAMBRIAN FAUNA

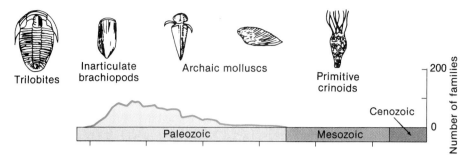

Trilobites

Inarticulate brachiopods

Archaic molluscs

Primitive crinoids

Cenozoic

Paleozoic   Mesozoic

Number of families

200

0

PALEOZOIC FAUNA

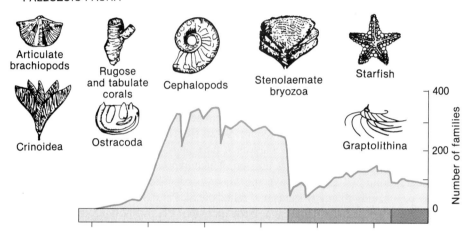

Articulate brachiopods

Rugose and tabulate corals

Cephalopods

Stenolaemate bryozoa

Starfish

Crinoidea

Ostracoda

Graptolithina

Number of families

400

200

0

MODERN FAUNA

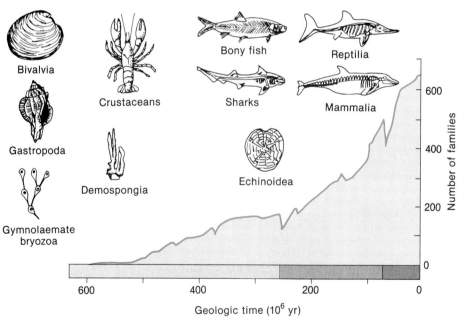

Bivalvia

Crustaceans

Bony fish

Reptilia

Gastropoda

Sharks

Mammalia

Demospongia

Echinoidea

Gymnolaemate bryozoa

Number of families

600

400

200

0

600   400   200   0

Geologic time (10^6 yr)

Figure 11.10 Sepkoskl's concept of three dominant faunas of the fossil record. The "Cambrian fauna" was dominated by trilobites and archaic brachiopods, molluscs, and echinoderms. In the Ordovician, the typical "Paleozoic fauna" took over. It was dominated by articulate brachiopods, bryozoans, crinoids, corals, nautiloids, graptolites and ostracodes. The "Paleozoic fauna" was nearly exterminated at the end of the Permian and replaced by the "Modern fauna," which has dominated the oceans for the entire Mesozoic and Cenozoic (although most of its members were present in the Paleozoic). These include molluscs, such as bivalves and gastropods, plus echinoids, crustaceans, modern groups of sponges and bryozoans, and a variety of swimming vertebrates, including sharks, bony fish, marine reptiles in the Mesozoic, and whales in the Cenozoic. *(After Sepkoski, 1981, Paleobiology.)*

Notice that most members of the Cambrian fauna had organic skeletons (such as the Burgess Shale arthropods) or phosphatic skeletons (such as the inarticulate brachiopods, the ostracodes, and the conodonts), and those that used calcite were only lightly calcified (trilobites and molluscs, for example). By contrast, Ordovician animals had massive calcite skeletons, and all Ordovician animals began to thrive about the same time. Finally, the appearance of efficient predators with crushing jaws, such as the nautiloids, put pressure on archaic animals to become specialized or go extinct. Trilobites and soft-bodied animals were affected and possibly others as well. G. J. Vermeij has argued that, any time a new predator appears, the ecological complexity and defense mechanisms must escalate, leading to greater diversity as well.

Ordovician Extinctions

The end of the Ordovician marked one of the first great mass extinction episodes in the history of life. By most measures, this event was the second most severe in earth history—worse than the event that killed the dinosaurs and second only to the great Permian extinctions. Worldwide, more than 100 families of marine animals did not make it into the Silurian. In North America, more than half the species of brachiopods and bryozoans died out. A third of all the brachiopod families, especially among the orthids and strophomenids, disappeared. The crinoid-stromatoporoid-tabulate-rugosid-receptaculitid reef community was decimated and did not recover until late in the Silurian Period. Nautiloids

were also decimated, and trilobites declined even further. The most striking fact about these extinctions is that they were concentrated in tropical groups, and the survivors and replacements were adapted either to deep waters or to cold waters from high latitudes. This suggests that there was a severe cooling event in the world ocean in the Ordovician. Indeed, a major glaciation of the southern Gondwana supercontinent, centered in what is now North Africa (then in a polar position), occurred in the Late Ordovician (Fig. 13.47). This chilled the world ocean enough that only the cold-adapted invertebrates could survive to repopulate the sea floor in the Silurian.

Ordovician History of the North American Craton

Mid-Ordovician Regression and the St. Peter Sandstone

After extensive Early Ordovician dolomite deposition for several million years, a widespread unconformity was produced over virtually the entire craton (see Box 11.1). The sea retreated at least to the margins of the craton, and the interior was subjected to extensive erosion, which exposed Cryptozoic rocks in some areas. Above the unconformity lies a very widespread pure quartz sandstone called the St. Peter Sandstone, the basal unit of the Tippecanoe Sequence (Fig. 11.11A). Though the sea had re-

A.

B.

C.

Figure 11.11   Typical Ordovician outcrops of the Tippecanoe Sequence. *A:* The basal transgressive sandstone in the Midwest is the St. Peter Sandstone, famous for its extremely mature, well-rounded sand grains. The rock is so porous that it forms the major aquifer in the region. In some places, the St. Peter is so weakly cemented that it can be mined with a fire hose for clean, pure sand to make glass. *(R. H. Dott, Jr.)* *B:* Above the St. Peter Sandstone is a series of Upper Ordovician limestones and dolomites, which produce abundant fossils. Here, a geology class collects fossils from outcrops near Dickeyville, Wisconsin. *(D. R. Prothero.)* *C:* The Cincinnati region is world famous for its abundantly fossiliferous Upper Ordovician shales and limestones (see Fig. 11.3). These typical outcrops show large coral heads protruding from the layer examined by the student. *(D. R. Prothero.)*

treated from the craton, the topography was subdued everywhere. Maximum relief observed in erosional channels is about 100 meters. Because of this, only a small relative vertical fall of sea level was needed to cause exposure of a tremendous area of the former shallow sea floor (see Fig. 4.14). As soon as the sea floor was exposed, weathering, solution, and erosion began to remove earlier Ordovician dolomites. In the center of the craton, erosion reexposed Upper Cambrian quartz sandstones. These were recycled and the sand again was spread out (for perhaps the dozenth time since the Proterozoic) from the craton core to form a veneer over much of the craton. The St. Peter Sandstone is virtually identical with mature Cambrian sandstones (see Fig. 10.13). The enormous sheet of St. Peter sand totals more than 20,000 cubic kilometers!

Fossils indicate that the sandstone is slightly older near its outer limits than in the center of the craton (Fig. 11.12). It is a classic transgressive deposit whose deposition spanned about 5 million years. River and wind dispersal distributed the sand all across the craton in a thin sheet, but then most of it was redeposited by the sea as transgression renewed during the Middle Ordovician. Although some wind and river deposits are still preserved locally, much of the preserved sandstone apparently represents near-shore deposits formed at the advancing edge of the sea. The sand was transported outward from the center of the craton. Pure quartz sand drifted southwest into Oklahoma and even as far west as central Idaho and Nevada. In the latter region, it is interstratified in an unusual association with black, graptolitic mudstones, which were deposited beyond the western margin of the continent in deep water. This interstratification occurred when sea level reached its lowest level during the regression, an event that brought the shoreline clear out to the craton margin and allowed some sand to be redeposited into deep water.

Later Ordovician (Tippecanoe) Epeiric Sea

After deposition of the St. Peter Sandstone, marine carbonate sediments formed across the entire craton. They contain a rich fauna that is better preserved than earlier ones. Many Upper Ordovician limestones are 80 or 90 percent shell material and, so, constitute a **shelly facies.** The Upper Ordovician strata of the Ohio-Indiana region are among the most fossiliferous rocks in the world, and they were the inspiration and training ground for a "who's who" of outstanding early American geologists (Figs. 11.3 and 11.11B, C). Upper Ordovician fossil-rich carbonate strata also are the most extensive marine deposits on the entire craton; a cratonic basin near Hudson Bay contains a large volume, and smaller patches occur elsewhere in the Canadian Shield (Fig. 11.15). The Late Ordovician epeiric sea represented one of the most complete floods experienced by any continent (Fig. 11.16); nothing quite like it exists today. Both the great diversity of species and paleomagnetic evidence indicate that the continent was still tropical. The character of some Ordovician sediments suggests that the Tippecanoe sea was somewhat deeper over cratonic basins than was the earlier Sauk sea. Hummocky stratification, shell-rich layers, and hard-ground surfaces of nondeposition provide evidence of the continuing importance of episodic deposition.

Shaly Deposits of Later Ordovician Time

In the Appalachian region, much of the later Ordovician sequence is different from that just described. It is characterized by dark mudstones with sporadic graywacke sandstones and chert layers. The top of the Ordovician sequence over much of the eastern craton also contains thin but persistent dark gray shale (Fig. 11.17). This pelagic sediment—the *graptolitic shale facies*—is an oddity for the craton and must record some important change of terrigenous source areas. To determine its source, we must examine facies maps and cross sections (Figs. 11.15 and 11.18). The shale becomes thicker toward the east, where practically the entire Ordovician sequence in the northern Appalachian region is composed of dark shales. Still farther east, important volcanic rocks occur in the Ordovician succession. Both lava and ash are found, indicating that important volcanic islands were formed here (Box 11.2). In summary, the facies patterns point to the elevation of islands above sea level in the Appalachian region, with the islands shedding sand and much fine mud, which was then carried westward. Most of this sediment was deposited within the Appalachian orogenic belt, but as the volume increased, mud

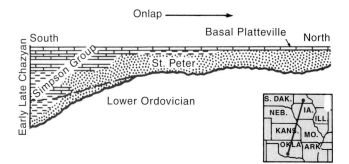
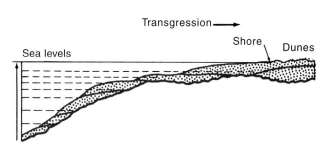

Figure 11.12   *Left:* Diagram of mid-Ordovician strata in the southern craton showing northward decrease in age of transgressive quartz sandstone facies (St. Peter). *Right:* Integration of countless successive near-shore deposits formed during transgression to produce a widespread, tabular mass of sandstone. Black dashed lines represent successive time datums. (Vertical scale is exaggerated.)

Box 11.1

Unconformity-Bounded Sequences Revisited

In Chap. 4, we introduced the concept, first developed by L. L. Sloss in the 1950s and 1960s, of regional rock units called *sequences,* which are assemblages of many formations bounded by exceptionally widespread unconformities. In Chap. 10 we illustrated the oldest sequence, the Sauk, and in this chapter we encounter the second one. With this added background at hand, further general consideration of sequences is in order.

Sequence boundaries vary in age from place to place, so that the sequences are not universal time divisions with strict age limits. Strata adjoining the bounding unconformities may vary in age several or even tens of millions of years. The unconformity beneath Cambrian rocks across North America and that beneath the St. Peter Sandstone together define the Sauk Sequence. The St. Peter Sandstone is the basal transgressive deposit of the second sequence the Tippecanoe. A total of six unconformity-bounded sequences are now widely accepted for North America (Fig. 11.13). Sloss chose American Indian names for the sequences to avoid confusion with the European names for the periods of the geologic time scale.

The sequence concept applies chiefly to cratonic strata and was conceived long before the plate-tectonic theory appeared around

1970. Increasingly we see evidence, however that the changes reflected by the great unconformities (and accompanying transgressions and regressions) are probably related to lithosphere plate motions to sea-floor warping, and to tectonic events in orogenic belts at plate margins. Cratons tended to be more emergent when plate activity (spreading and subduction) was modest, as in Vendian time whereas they tended to be submerged when plate activity was greater, as in the Late Ordovician. It has been suggested that this may be due to more heat being retained beneath continents and causing isostatic rise in the former case and the opposite when plates were more active. Another tectonic factor affecting sea level is the expansion and rise (or shrinkage and fall) of ocean ridges to raise (or drop) sea level worldwide. Any plate tectonic explanation of cratonic changes implies a worldwide effect, however. Can this be proved? Fig. 11.14 indicates that indeed different cratons were affected more or less simultaneously by several sea-level changes. This suggests that *worldwide* sea level changes account for most of Sloss's major sequences and implies some global plate-tectonic mechanism as the probable cause of those changes.

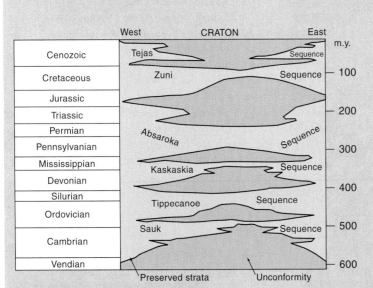

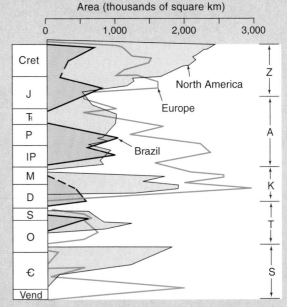

Figure 11.13 Six major unconformity-bounded sequences designated by L. L. Sloss for the North American craton. Dark areas represent major unconformities;—that is space time without any rock record present today. These major unconformities subdivide the cratonic stratigraphic record into widespread "packages" of strata. Note that the more complete deposition records are at the craton margins. The St. Peter Sandstone lies at the base of the Tippecanoe Sequence. Compare Fig. 4.22. *(After Sloss, 1963, Bulletin of the Geological Society of America, v. 74, pp. 93–114.)*

Figure 11.14 Comparison of stratigraphic records for three separate cratons expressed as area of preservation. Overall similarity of curves suggests worldwide sea-level changes as a major cause of transgressive and regressive events evidence here Local structural warping also affected the individual patterns especially after Ordovician time. North American sequence names are abbreviated at right (see Fig. 11.13). *(Adapted from L. L. Sloss, 1976, Geology, v. 4, pp. 272–276; and P. C. Seares et al., 1978, Bulletin of Geological Society of America.)*

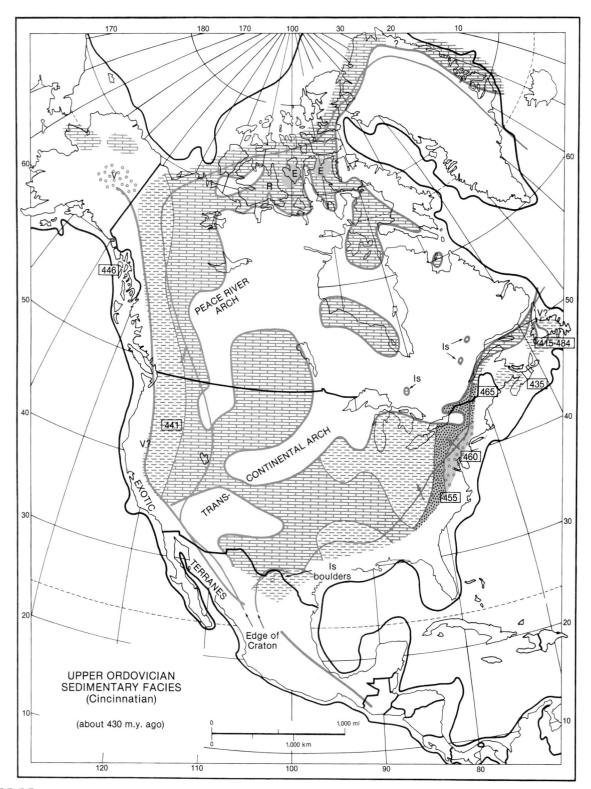

Figure 11.15   Upper Ordovician sediment patterns for North America. Widely scattered patches of sediments on the Canadian Shield prove the great extent of the Late Ordovician sea. Absence of Ordovician strata on several arches proves subsequent warping and erosion of these arches. Note the spread of red beds and marine shales westward from the Appalachian region, forming a clastic wedge. (See Box 10.2 for symbols and sources.)

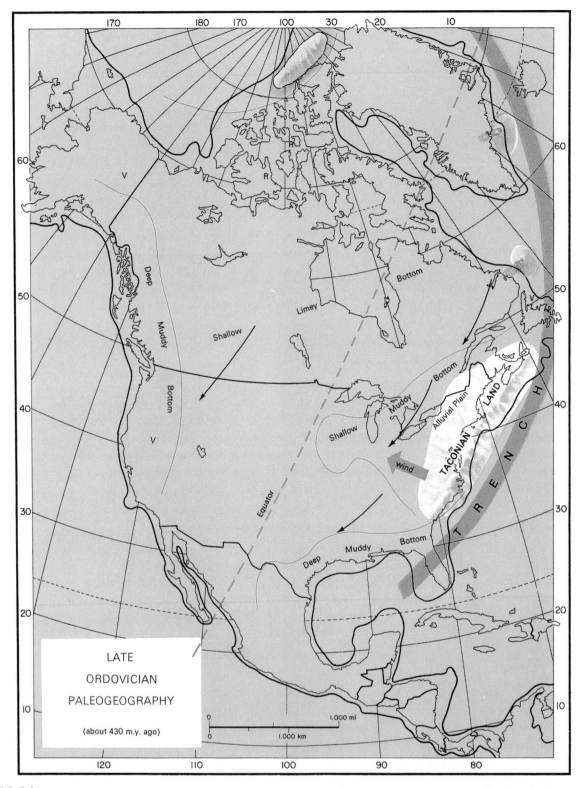

Figure 11.16   Late Ordovician paleogeography interpreted from Fig. 11.15. Note especially the Taconian land, widespread marine inundation, and position of the equator according to paleomagnetism. Possible wind direction is interpreted from volcanic ash deposits. (Compare Fig. 11.36.)

Figure 11.17   Ordovician black shales and thin white limestones (Utica Formation) along the New York Thruway, central New York. In this area, mud derived from the east spread beyond the Appalachian belt onto the craton. Differential flow under the influence of gravity on water-laden mud caused contortions after deposition. *(R. H. Dott, Jr.)*

literally spilled onto the edge of the craton in New York. Near the end of the period, clay and volcanic ash were carried as far west as Iowa (Fig. 11.15). The distribution of the shale facies suggests westerly flowing paleocurrents; ripple marks and other current-formed features confirm this. By latest Ordovician time, the continent appeared as shown in Fig. 11.16.

Cratonic Basins and Arches

Except for the transcontinental arch discussed in Chap. 10, basins and arches were not well defined during Sauk deposition. By Late Ordovician time, however, such features were beginning to become delineated and were destined to influence later cratonic sedimentation profoundly. For example, a basin was accentuated beneath Michigan. A broad arch extending from Ohio across Kentucky to Tennessee seems also to have formed during Late Ordovician time. The origin of cratonic arches and basins has long been a puzzle. There is growing evidence that they may owe their existence to reactivation of ancient structural zones within the Cryptozoic basement rocks. But what could cause such reactivation? Intuitively we might seek some plate-tectonic explanation, so we shall return to speculate about this question at the end of the chapter after investigating what was going on along the eastern margin of the continent.

Ordovician Mountain Building in the Appalachian Orogenic Belt

Evidence of Increasing Structural Mobility

Let us now consider the specific evidence for an Ordovician orogeny and for dating such an event. The reasoning involved applies equally to all younger orogenies.

In the northern Appalachian region, some spectacular, ill-sorted conglomerates with fragments from mixed Cambrian and Early Ordovician formations occur helter-skelter as local deposits within Ordovician black shale sequences; some of the blocks are many meters long (Fig. 11.20). These unusual deposits occur at several scattered localities along the mid-Ordovician continental margin. They represent submarine avalanches of debris derived from shallower-marine environments. The material became unstable and slid and rolled into the deeper, adjacent muddy environment. Such bouldery mudstones reflect disturbances of the sea floor—probably by earthquakes—heralding more profound unrest destined to produce major mountain building later in the period.

Volcanic ash (Box 11.2) and abundant dark, graptolite-bearing muds deposited widely within the Appalachian region and far out onto the craton also reflect unrest and uplift in the east. In the southern Appalachian region, local red shales appeared in mid-Ordovician time, reflecting the elevation of land there.

In the northeastern United States and southeastern Canada, black shale is succeeded eastward by Upper Ordovician red shale. This red-bed facies coarsens eastward in Pennsylvania to sandstone and some conglomerate (Figs. 11.15 and 11.18). This assemblage of black and red sediments is called a **clastic wedge** because of its shape in cross section (Fig. 11.18). Clastic wedges are characteristically associated with orogenies, being the debris eroded from the rising mountains and deposited in the adjacent foreland basin. The red deposits reflect a major change in depositional conditions from the marine dark shales. Because the red deposits lack fossils and show thorough oxidation of iron, it was suggested long ago that they were nonmarine river and deltaic deposits that bordered a large land being eroded somewhere to the east.

Box 11.2

Dividends from Volcanic Ash

Correlation Between Facies

In the Appalachian Mountain region, there is a major lateral facies gradation from dominantly *graptolitic shale facies* westward to a *shelly limestone facies* more characteristic of the craton (Fig. 11.15). Because of the differences of environment reflected by these facies, there is also a difference of fauna in each. In turn the differences in fossil assemblages make it difficult to correlate between the facies on the basis of index fossils; this Illustrates the limitation of facies fossils discussed in Chap. 4.

Volcanic ash layers are found in the Ordovician strata of the Appalachian belt (Fig. 11.19). They are represented today by thin, clay-rich layers called *bentonite* formed by the alteration of volcanic dust particles. Because each layer was erupted instantaneously in a geologic sense, individual ash strata provide *time datum markers*. Certain of the ash layers extend into both shaly and carbonate facies, so they constitute unique thin strata for correlation between the two facies (Fig. 11.18). Also, those layers that have not been entirely altered can be used for *isotopic dating* to provide a numerical date for a sequence

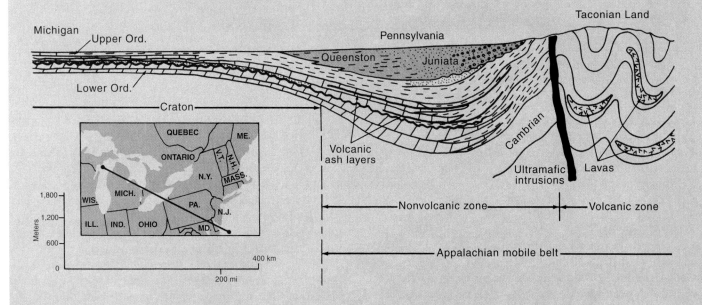

Section without vertical exaggeration

Figure 11.18   Restored cross section showing relations of Ordovician facies to the Taconian land. Note the volcanic ash layers that transcend facies boundaries to provide useful time datums for correlation purposes. Note mid-Ordovician unconformity and Upper Ordovician clastic wedge deposited in a foreland basin. *(Adapted from M. Kay, 1951, Geological Society of America Memoir 48.)*

Evidence for Dating the Mountain Building

At a number of localities within the Appalachian region, important unconformities are visible within the Ordovician sequence and between Ordovician and Silurian or Devonian strata (Fig. 11.21). Extending from the St. Lawrence River valley south into eastern New York is a zone of major overthrust faulting that originated near the end of the Ordovician. Farther east, in New England and maritime Canada, some isolated, small granite masses are overlain by Silurian strata and were intruded into older Cambrian and Ordovician ones. Isotopic dates confirm that some of these granites were formed during Late Ordovician and Early Silurian time.

Long, narrow ultramafic igneous rock masses also were emplaced along the orogenic belt (Fig. 11.21). You will recall that these unusual rocks represent mantle material faulted into the crust during severe deformation, commonly along suture zones between two lithosphere plates. Ordovician sandstones and conglomerates are heterogeneous in composition. This derivation from erosion of limestones, sandstones, shales, volcanic rocks, and rare granitic rocks indicates the erosion of complexly deformed rocks of many kinds and ages. Finally, Upper Ordovician rocks were themselves folded and metamorphosed (Fig. 11.1).

The evidence points to a major interval of mountainous uplift, erosion, and intrusion of some igneous plutons late in the of strata (as isotopic dating of glauconite can provide for Upper Cambrian strata of the craton).

Of the many Ordovician ash layers found in both Europe and North America, one is particularly impressive. Known as the Big Bentonite in northern Europe it is over 2 meters thick in some places (Fig. 11.18). Recently this bentonite was correlated with the Milbrig bentonite in North America, since the two have similar chemical composition and occur in the same biostratigraphic zone on both continents. The same bentonite has been found even in South Africa and South America. This gigantic ashfall has been dated with several different isotopes at about 454 m.y. Based on the geographic distribution and geochemistry of the ash, it must have erupted from an arc volcano in the southern part of the active Taconian belt. When the volume of this eruption is calculated, the amount is truly staggering. Over 1,140 cubic kilometers of ash blanketed eastern North America, Europe, and the still-closing Iapetus Ocean to a depth of 2 meters or more! This makes it the largest known eruption in all the Phanerozoic, dwarfing even the gigantic eruptions of the Pleistocene (Fig. 16.14).

Paleowind and Paleoclimate

Perhaps of even greater interest is the clue that volcanic ash distribution provides about the possible Ordovician wind pattern in eastern North America. This distribution suggests that the ash was derived from known volcanic centers of the Appalachian orogenic belt to the east (Fig. 11.16). The winds blew from the east (in terms of present geography), which is the opposite of present prevailing winds across North America. Marine currents would also influence ash distribution after it settled into the sea.

It is interesting to note the relationship between the ash distribution and the possible position of the Ordovician equator with respect to North America as indicated by paleomagnetic data. In Fig. 11.16, note that the equator extended from northeast to southwest—nearly bisecting the continent. This would place the Appalachian orogenic belt within the trade wind zone, and the ash distribution could have

Figure 11.19   Outcrop of the Middle Ordovician Deike bentonite (crumbly gray layer at chest level) in the midst of a thick sequence of marine limestone (resistant rocks), South Carthage, Tennessee. This immense ashfall from a volcano somewhere just off the present Atlantic Coast filled the Ordovician seas, forming a layer over a meter thick in the Appalachians, and extending all the way to Wisconson and Minnesota. (*Courtesy of Warren D. Huff.*)

been influenced in part by trade winds. All the richly fossiliferous Ordovician carbonate deposits of the craton would fall within 40° latitude of the equator, whereas evaporites and reefs in present Arctic Canada would lie within 20° of the equator. But if the equator then occupied its present position with respect to the continent, the carbonate rocks would span 70° of latitude. As will be shown in Chap. 12, such sediments today form chiefly in low, warm latitudes, so it would be less surprising to find them at 40° N than at 70° N.

Ordovician Period. It is called the **Taconian orogeny,** for the Taconic Mountains of southeastern New York, and it represents the first of several great Paleozoic mountain-building episodes in the Appalachian belt. All later orogenies produced similar effects.

Paleogeography and Sedimentation within Orogenic Belts

Background

Now that we have encountered the first Phanerozoic orogenic event, and because the Appalachian region was the cradle of early

ideas about mountain building, we shall now expand upon concepts introduced in Chap. 7. Reconstruction of the ancient geography of regions from their stratigraphic records is one of the highest goals of the earth historian. We have already demonstrated the reconstruction of cratonic geography for the Cambrian and Ordovician periods. Now we discuss the paleogeography of orogenic belts.

Early Ideas About Borderlands

More than a century ago, American J. D. Dana suggested that ridges of Cryptozoic or Precambrian igneous and metamorphic rocks lay along the borders of North America throughout much of

Figure 11.20   Outcrop of the Cow Head Breccia from the west coast of Newfoundland. Blocks of shallow marine limestone up to 3 meters in diameter have rolled and slid into a deep-marine trough filled with black shales, showing that extreme tectonic uplift and subsidence was taking place during the early phases of the Taconian Orogeny. *(R. H. Dott, Jr.)*

Phanerozoic time (Fig. 11.22). His postulated **Precambrian borderlands** were based upon two observations: (1) many clastic sediments within what he called the "Appalachian geosyncline" coarsen eastward (e.g., Upper Ordovician red beds) and (2) farther east in New England lie predominantly granitic and metamorphic rocks considered in Dana's time to be very old. These "Precambrian-looking" rocks naturally were assumed to be the sources of the clastic sediments that had been derived from that direction. It was also suggested that the present Atlantic Coastal Plain had been a part of the borderland, as shown in the old paleogeographic map of Fig. 10.10. When the adjacent Appalachian geosyncline ceased to subside at the end of the Paleozoic Era, the borderland had disappeared mysteriously, as had mythical Atlantis.

The inference that the borderlands were composed of Precambrian rocks resulted from an old fallacy of age correlation of rocks solely on the basis of degree of metamorphism and deformation. In 1930, Harvard University geologists discovered Silurian and Early Devonian marine brachiopods in New Hampshire mica schists long assumed to be Precambrian, so the borderland of rocks originally designated "Precambrian" actually was in large part composed of much younger rocks. But not only did the supposed borderlands contain mostly Paleozoic and younger rocks, to make matters worse, they could not have been permanent lands during Paleozoic and Mesozoic time, for the fossils found included forms known to have inhabited only marine environments. Intermittently seas must have occupied the borderland area as well as the adjacent geosyncline. Clearly borderlands were neither permanent nor composed wholly of Precambrian rocks. What, then, was the true nature of these extracratonic lands?

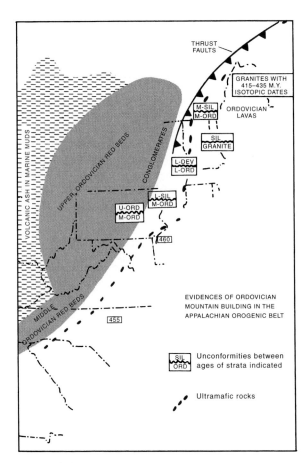

Figure 11.21   Summary of evidences of the Late Ordovician Taconian Orogeny.

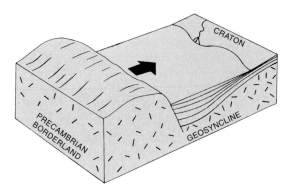

Figure 11.22   The old hypothesis of Precambrian borderlands as the sources of most geosynclinal sediments. Since 1940, this model, which calls for large highlands *outside* geosynclines, has been replaced by a model of lands raised *within* geosynclinal belts (see Fig. 11.23).

Modern Concepts of Borderlands

Soon after fossils were discovered in the borderland areas, a German, Hans Stille, and Americans Marshall Kay and A. J. Eardley independently drew the distinction between magmatic, or *igneous-bearing* ("eugeosynclinal"), and nonmagmatic, or *non-igneous-*

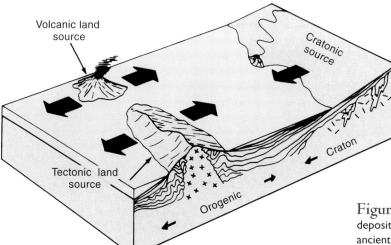

Figure 11.23   Three major source types for sediments deposited within subsiding parts of orogenic belts. Composition of ancient sediments indicates the influences of each type for different past times.

bearing ("miogeosynclinal"), *subdivisions* of orogenic belts. The important new observation that those ancient belts contain a distinct zone of volcanic and intrusive igneous rocks of diverse ages became the key to a reinterpretation of borderlands as a series of temporary islands periodically raised within the magmatic portion of the belts. Details of stratigraphy show, besides abundant volcanic outpourings, many angular unconformities, coarse conglomerates, varying ages of igneous plutons, and metamorphism, as we saw in our discussion of the Taconian orogeny (Fig. 11.21). In short, there is ample evidence of the persistence of great structural mobility throughout much of the histories of these belts. The rocks, moreover, attest to formation of both volcanic and nonvolcanic islands intermittently *within* the orogenic belt. Perhaps most compelling is the fact that today some orogenic belts extend seaward *into* still-active volcanic arc-trench systems. Clastic detritus deposited in the geosynclines was derived from such islands, not from borderlands wholly composed of Precambrian rocks and lying outside the geosyncline. The rocks exposed to erosion in the islands were mostly only slightly older deposits augmented periodically by additions of igneous extrusions and intrusions. Plate tectonics, of course, provides a straightforward explanation of these islands.

Prior to the beginning of mountain building in mid-Ordovician time, the Appalachian region was a broad, tranquil continental shelf on a passive continental margin receiving clastic sediments only from the craton (Fig. 11.23). Mature quartz sand continued to enter from the craton even when the region became an orogenic belt in later Ordovician time, but less mature sands from other sources soon became more important (see Fig. 11.24).

A *volcanic source* (Fig. 11.23) is exemplified by some of the Ordovician strata of the eastern Appalachian belt. As indicated earlier, during mid-Ordovician time, extensive volcanism occurred there, and volcanic islands were formed, as evidenced by volcanic-rich clastic sediments associated with lavas (Fig. 11.24) and by the fine volcanic ash blown far to the west (Fig. 11.16).

A **tectonic land** source (Fig. 11.23), which was emphasized particularly by Kay, is exemplified by Upper Ordovician and

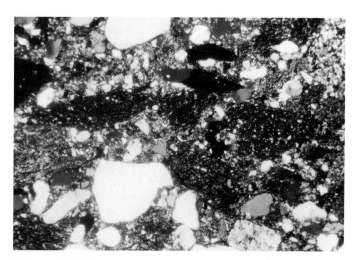

Figure 11.24   Microscopic photograph of a typical immature Ordovician (Normanskill) graywacke from eastern New York. Grains include quartz (white), feldspar (gray), and varied sedimentary rock fragments (dark) derived from erosion of earlier Paleozoic strata in the ancient Taconian tectonic land to the east. Note the very poor sorting and rounding (grains average about 2 millimeters in diameter). *(R. H. Dott, Jr.)*

Lower Silurian strata of the Appalachian region. As shown in Figs. 11.15 and 11.18, these sediments coarsen toward the east. Microscopic study shows that the sandstones belong to the immature (Fig. 11.24) feldspathic and lithic graywacke clans discussed in Chap. 8. The composition and great volume of these sandstones indicate that a large land composed chiefly of older, partly metamorphosed sedimentary and some igneous rocks was their principal source.

A Modern Analogue

How large might the Taconian land have been? The clastic wedge contained more than 500,000 cubic kilometers of sediment, which implies the erosion of a land 1,000 kilometers long by 200

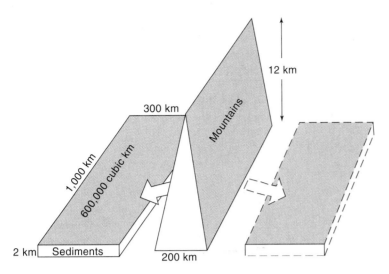

Figure 11.25   Approximate volume of Ordovician clastic sediments deposited west of the source area *(left)* plus an assumed equal volume shed eastward. Together these would require erosion of a mountainous mass about 1,000 × 200 × 12 kilometers high (or 7 kilometers higher than Mt. Everest). But erosion would proceed simultaneously with uplift; therefore, the mountains need never have been so high.

kilometers wide and several kilometers high (Fig. 11.25). New Guinea today is a close analogue. It is 2,000 kilometers long by 600 kilometers wide and averages nearly 2 kilometers in elevation (Fig. 11.26). New Guinea also is marginal to a craton, that of Australia, and an epeiric sea now floods the northern edge of that craton much as does (although smaller than) the Ordovician epeiric sea over the North American craton. Finally, sedimentation is also similar in that carbonate deposits are forming on much of the Arafura epeiric sea floor. Terrigenous clastic material is accumulating on advancing deltas along the south side of New Guinea, where large rivers drain the mountainous upland of the tectonic land and flow south across a broad, heavily forested alluvial plain. Southward regression of the sea toward the craton is happening today in southern New Guinea because of the advance of the shoreline by sedimentation, just as it did in North America 440 million years ago.

Controversy About Sea Level and Sedimentation

In Chap. 7, we noted that American geologists assumed from the outset that all "geosynclinal" sediments were deposited either in shallow-marine water or just above sea level. This view was based upon the clastic wedge of the western portion of the Appalachian Mountain belt, where the rocks contain ripple marks, stromatolites and other shallow-marine fossils as well as river and delta deposits. Meanwhile, however, European geologists were arguing that "their" typical geosynclinal sediments were deep-marine in origin—perhaps even trench deposits. They pointed to the fact that the base of the thick sequence of Mesozoic strata exposed in the Alps consists of fine-grained chert and black shale overlying thick basalts with ellipsoidal, or pillow, structure (Fig. 11.27), which in turn overlie ultramafic peridotite and serpentine. The Europeans called this assemblage the **ophiolite suite** and showed, by analogy with modern oceanic materials, that it represents an ancient deep sea floor (Fig. 11.28).

It is now clear that both groups were partly right and that many different environments of deposition existed in ancient

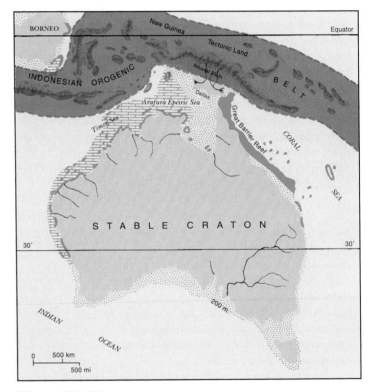

Figure 11.26   Relations of modern tectonic lands of New Guinea and Indonesia to the Australian craton, a close analogue to Late Ordovician conditions in North America. New Guinea is a complex tectonic and volcanic borderland with dimensions comparable to the Taconian land; Australia is about the size of the United States. Rivers carry clastic sediments southward from the mountains toward the craton, where they are deposited in deltas at the margin of an epeiric sea, which, like its larger Ordovician counterpart, has carbonate sedimentation over much of its area; reefs (darkest color) are also common around tropical coasts. Rotate page 90° clockwise for a closer comparison with Fig. 11.16. *(Analogy originally suggested to writers by T. S. Laudon.)*

Figure 11.27   Pillow lavas and pillow breccias from Cape St. Francis, Newfoundland. These lavas and associated chert were part of an Ordovician ophiolite sequence formed on the sea floor in deep water beyond the continental shelf. (Compare Fig. 7.16.) *(R. H. Dott, Jr.)*

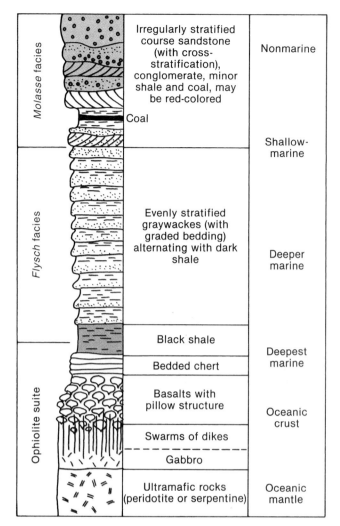

Figure 11.28   Idealized vertical sequence from oceanic igneous rocks and chert (*ophiolite* sequence) upward through graptolitic-graywacke (*flysch*) strata to red clastic (*molasse*) deposits. Such successions are commonly seen in orogenic belts and represent a progression from deep-marine to nonmarine conditions, as long ago argued by Europeans. Such a sequence is seen in the Ordovician record of the Appalachian belt, especially in Newfoundland.

orogenic belts, making the relationship of sedimentation to sea level a critical factor. As noted in our discussion of subsidence in Chap. 7, if deposition began in deep water, as the Europeans believed for *their* geosyncline, a considerable thickness of marine sediments could accumulate without requiring much subsidence. But if thick strata consist entirely of shallow-marine types, then sedimentation must have just kept pace with subsidence, which is exactly what early Americans believed for *their* geosyncline. This latter view requires one of the structural causes discussed in Chap. 7 to upset isostatic equilibrium and allow crustal subsidence during deposition. Based upon the fact that all the subsidence mechanisms involve a gradual decrease in subsidence rate, we might expect many thick sedimentary sequences to show upward-shallowing trends as sedimentation rate surpassed subsidence rate. Indeed, most clastic wedges associated with orogenies *do* show exactly such a trend from relatively deep-marine sediments upward through shallow-marine to nonmarine ones (Fig. 11.28). This trend results in the lateral spreading, or **progradation,** of younger river, delta, and shoreline deposits over slightly older, shallow-marine ones, as in Australia and New Guinea. We see that progradation, the "gradual moving forward" of a shoreline, can force a local regression *due only to sedimentation proceeding faster than subsidence* (Fig. 7.35).

The ophiolite suite of the Alps is overlain by evenly stratified, alternating sandstone and shale sequences called *flysch,* which has many of the same characteristics as the Ordovician graptolitic shale-graywacke sequence of the Appalachian region. The immature graywackes in both show conspicuous graded bedding (Fig. 11.29), which Europeans interpreted as early as 1930 as due to sediment being "rapidly poured into deep water." With the recognition in 1950 of the importance of turbidity currents in transporting relatively coarse sediments into otherwise quiet environments and the discovery of ripples on the deep-sea floor (Fig. 11.30), a modern reinterpretation of many sandy strata as

deep-water deposits became possible. The long controversy between the American and European views of geosynclinal sedimentation was finally resolved.

In Newfoundland, an ophiolite suite is overlain by Ordovician graptolitic *flysch* (Figs. 11.1 and 11.28), and we infer that this was the case throughout the entire Appalachian orogenic belt. Basalts extruded onto the deep-ocean floor were buried first by both cherts and black muds. By Middle Ordovician time, however, immature sands showing graded bedding were introduced rapidly by turbidity currents and were deposited on *deep-sea fans* (Fig. 11.31). At the turn of the twentieth century, an American paleontologist noted that many graptolites in eastern New York had been oriented by bottom currents. More recently, current-sculptured *sole marks* found on the bottoms (soles) of sandstone strata interstratified with shale (Fig. 11.32) have provided geologists with much additional paleocurrent data to help them to interpret the paleogeography.

In Switzerland, the dark-colored *flysch* strata pass upward into nonmarine, light-colored, irregularly stratified conglomerate, cross-stratified sandstone, minor shale, and coal. These deposits are called **molasse** in Switzerland, where they reflect the final uplift of the Alps above sea level. Some of the *molasse* is red and is in every way analogous to Late Ordovician–Early Silurian nonmarine red beds of the Appalachian belt (Fig. 11.28), which we take to reflect erosion of the highest Taconian mountains (Fig. 11.18).

Figure 11.29   Graded bedding with finer texture upward in Ordovician graywacke (Martinsburg Formation) near Middletown, New York. These sandstones, which are typical of *flysch*, were deposited by turbidity currents. *(Courtesy of Earle F. McBride.)*

Plate-Tectonic Interpretation of Taconian Mountain Building

Now we can analyze the Taconian orogeny in terms of plate tectonics. First, we shall interpret the conversion of the Early Ordovician passive margin to a mid-Ordovician active continental margin by the collision of a volcanic arc or microcontinent. This collision caused thrust loading of the continental margin and formation of a foreland basin. After we examine the clastic wedge deposited in the foreland basin formed in the Appalachian region by thrust loading, we shall examine the early Paleozoic record of other continents across the present Atlantic Ocean to discover something about the larger plate-tectonic context for North America.

Early Paleozoic Passive-Margin Shelf

From Vendian to mid-Ordovician time, there was no orogenic belt in eastern North America. Instead, a structurally passive continental shelf faced a proto-Atlantic Ocean basin, called the Iapetus Ocean, which was completely destroyed before the present Atlantic formed. As noted in Chap. 10, sedimentation kept pace with cooling subsidence following the rifting of a postulated Proterozoic supercontinent. Thick, shallow-marine Vendian and Cambrian sands followed by later Cambrian and Ordovician carbonate rocks accumulated on the shelf, while fine muds and occasional limestone slide blocks (Fig. 11.20) accumulated on the continental slope beyond. Because no lands existed nearby and the entire shelf was located within the subtropics, carbonate deposition was favored here as well as on the craton.

Figure 11.30   Deep-marine ripple marks on a fine sand bottom at a depth of 3,500 meters in the South Pacific Ocean. Prior to the development of submarine photography around 1950, it was assumed that the deep seas were static and currentless; therefore, ripple marks could form only in shallow water. *(Official NSF photo, USNS Eltanin, Cruise 15, courtesy Smithsonian Oceanographic Sorting Center.)*

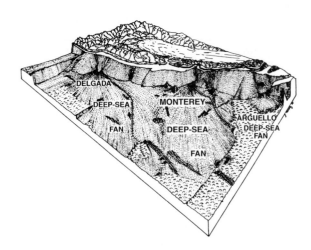

Figure 11.31   Diagrammatic portrayal of deep-sea fans west of California (view looking east toward continent; vertical scale exaggerated). Most sediment transported by turbidity currents is deposited on such fans. Note the gradual burial of topographic features. *(From H. W. Menard, 1960, Bulletin of the Geological Society of America, Vol. 71, pp. 1271–1278. Reprinted by permission of Geological Society of America via Copyright Clearance Center.)*

Figure 11.32 Flute structures on the bottom (sole) of an Ordovician graywacke (Martinsburg Formation), near Staunton, Virginia; the bulbous ends *(right)* point upcurrent. Turbulent bottom currents scoured elongate flutings in cohesive muds and deposited sand within them. Long after lithification, erosion exposed the structures on the now more resistant sandstone. *(Courtesy of Earle F. McBride.)*

Thrust Loading of the Continental Margin by Collision

Beginning in mid-Ordovician time, the eastern passive margin became a tectonically active one with the onset of Taconian mountain building. The first hint was the replacement of shallow-marine-shelf carbonate deposition by graptolite-bearing black muds and evenly stratified dark sandstones (graywackes). The organic content of these rocks implies rapid deepening to produce an oxygen-poor environment in which delicate graptolites could be preserved better than in shallow shelf environments. The cause of this abrupt subsidence to form a foreland basin was thrust loading of the continental margin by the westward obduction of oceanic material. The Ordovician thrust loading resulted from the closing of a narrow ocean by eastward subduction of that ocean's floor beneath a volcanic arc or microcontinent, which collided with North America and crushed the sediments of that basin against the rigid continent (Fig. 11.33). A long, narrow zone of ultramafic rocks along the Appalachian belt (Figs. 11.18 and 11.21) defines the suture for that collision.

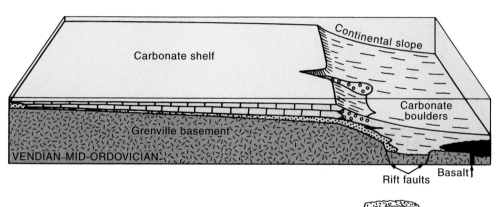

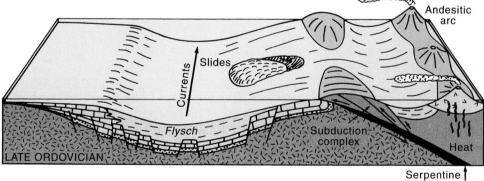

Figure 11.33 Restoration of eastern North America showing evolution from a passive to an active continental margin culminating in arc collision, which caused the Taconian orogeny. As the continental margin was downwarped in Late Ordovician time, carbonate sedimentation gave way to deeper-water *flysch* deposition. Finally, Taconian upheaval resulted in westward spreading of nonmarine *molasse*. *(Adapted from J. F. Dewey and J. M. Bird, 1970, Bulletin of the Geological Society of America, v. 81, pp. 1031–1061; and D. B. Rowley and W. F. Kidd, 1981, Journal of Geology, v. 89, pp. 199–218.)*

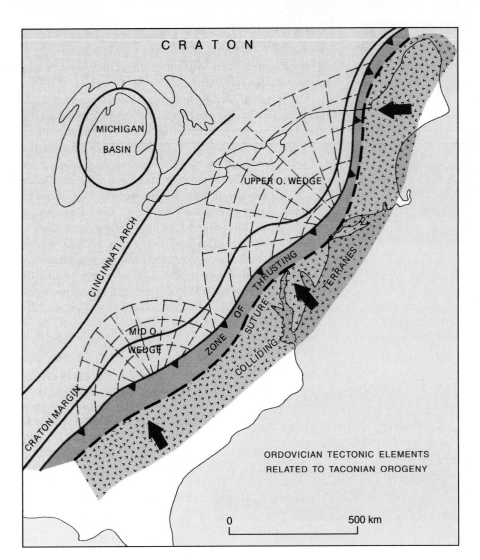

Figure 11.34 Relations of Taconian collisions and thrust loading of the continental margin to Ordovician clastic wedges and cratonic warping of arch and basin. *(Adapted from P. B. King, 1977,* The evolution of North America; *and Quinlan and Beaumont, 1984,* Canadian Journal of Earth Sciences.)

Throughout the foreland basin, facies patterns and paleocurrent criteria indicate that both the *flysch* and *molasse* deposits were derived from erosion of Taconian volcanic and tectonic lands that lay along the east margin of the basin (Figs. 11.18 and 11.33). The narrow ocean basin that lay between the continent and the approaching arc was similar to the modern Sea of Timor, which lies between the Indonesian island belt and northwestern Australia (Fig. 11.26). Like the Ordovician example, the Sea of Timor is closing today as sea floor is being subducted beneath Indonesia. In the future, this part of the Indonesian arc system will collide with Australia in the same manner as New Guinea already has done farther east.

A succession of thrust-loading events is suggested by the age distribution of easterly derived clastic wedges within the orogenic belt—that is, from mid-Ordovician in the south to Late Ordovician in the north (Fig. 11.34). Clastic wedges represent sensitive responses to plate-tectonic processes of collision and uplift within orogenic belts. According to mathematical modeling of crustal responses to thrust loading, the warping of cratonic arches and basins within the craton also may be a response to flexuring of the craton by the same loading (compare Figs. 7.36 and 11.34).

Relations Across the Atlantic

There is a long recognized similarity between the early Paleozoic faunas of eastern North America and those of northwestern Europe (Fig. 11.35), a likeness that long ago suggested some connection. The tectonic history of the two sides of the North Atlantic also suggests a former physical connection. The presence in Britain and Norway of an early and middle Paleozoic orogenic belt that has an almost perfect mirror image in eastern Greenland and Newfoundland is especially compelling. A few ardent advocates, such as Wegener, invoked continental drift many years ago to explain these similarities, but their restorations had Europe and North America joined throughout the Paleozoic Era (see Fig. 7.7). A modern plate-tectonic interpretation, however, postulates that Europe and North America were separated until middle Paleozoic time. During the Ordovician and Silurian Periods, these two continents and one or more microcontinents were moving toward each other as the intervening Iapetus Ocean was being consumed by subduction on both sides (Fig. 11.36). Recall that Taconian deformation is now interpreted as the result of the collision of an arc-trench system and microcontinents with eastern North America. The result is an example of collage tectonics, as discussed in Chap. 7 (see Figs. 11.33 and 11.37).

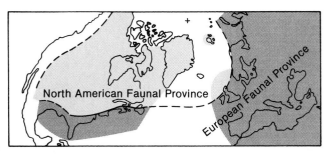

Figure 11.35 Cambrian faunal provinces on both sides of the North Atlantic Ocean, 2,000 kilometers apart. The "European" fauna is dominated by graptolite-bearing shales and trilobites such as *Paradoxides* (Fig. 9.15C), whereas the "American" fauna, including trilobites such as *Ogygopsis* (Fig. 9.15B), occurs chiefly in fossiliferous carbonate rocks. Long ago, it was thought that land barriers separated and isolated the two faunas, but they are found in close proximity in western Newfoundland and elsewhere. Today we think that most of Newfoundland, Nova Scotia, eastern Massachusetts, and parts of the southern Appalachians are actually part of Avalonia, an exotic terrane that was closer to Europe in the Cambrian than to North America. These terranes became sutured to North America during the Devonian, then remained attached to their new home when the rest of Europe pulled away during the Triassic and Jurassic (see Fig. 14.5).

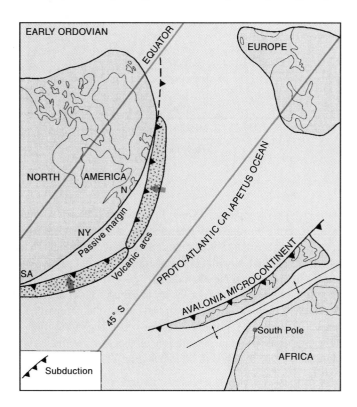

Figure 11.36 Early Ordovician paleotectonic map showing inferred relationships of North America to Europe and Africa. Volcanic arcs of the Piedmont terrane were impinging upon eastern North America and the Avalonia microcontinent apparently was rifted from northwestern Africa. N—Newfoundland, NY—New York State, SA—southern Appalachian region. *(Adapted from P. E. Schenk, 1971, Canadian Journal of Earth Sciences, v. 8, pp. 1218–1251; Cocks and Fortey, 1982, Journal of Geological Society of London, v. 139, pp. 465–478; W. S. McKerrow, 1988, Geological Society of London Special Publication 38, pp. 405–412.)*

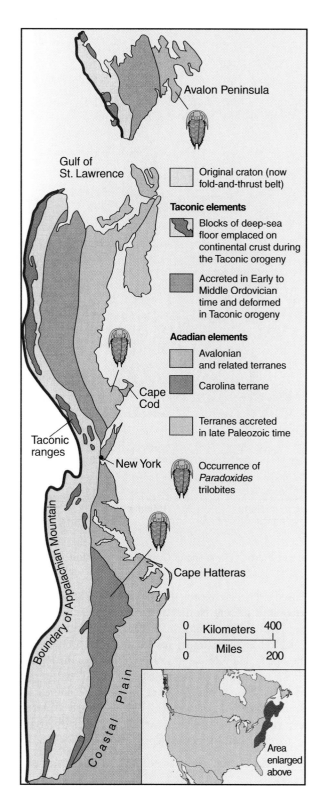

Figure 11.37 Exotic terranes of eastern North America. Distinctive fossils, including those of the Early Cambrian trilobite *Paradoxides*, characterize terranes accreted during the Acadian orogeny. *(Modified from H. Williams and R. D. Hatcher, Geology 10:530–536, 1982.)*

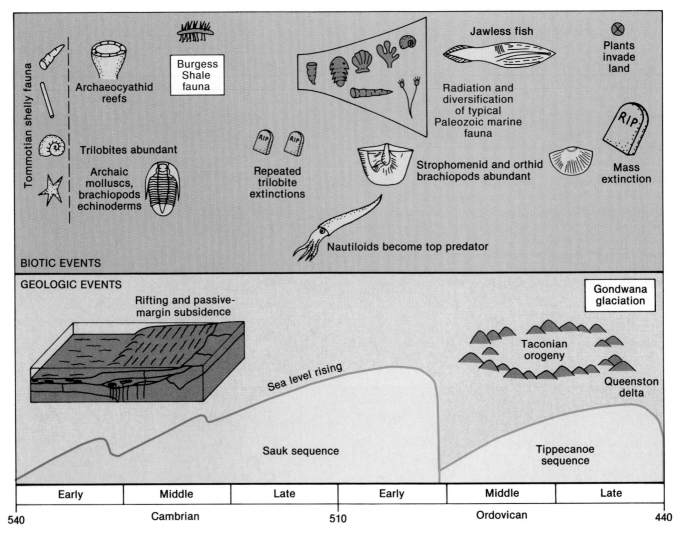

Figure 11.38   Time line of Cambrian and Ordovician events.

Based upon this concept, the map of Fig. 11.36 shows one interpretation of present North America, Greenland, Europe, and Africa to their inferred Ordovician positions. The resulting jigsaw puzzle has some fragments of different continents (e.g., Britain, Spain, and Florida) in strange places, but the relocations explain the geologic evidence much better than do the present positions of these fragments. A point of particular interest is that widespread glacial deposits discovered by French geologists in northern Africa are of latest Ordovician to earliest Silurian age. These, together with paleomagnetic evidence, show that Africa lay near the early Paleozoic South Pole, whereas North America was equatorial; Europe was intermediate. A more recent interpretation suggests that South America may have been next to the eastern United States, with northwestern Europe and Africa adjacent to Greenland and Newfoundland.

All the tectonic elements shown in Fig. 11.37 were crushed together by continental collision in Devonian time, as detailed in Chap. 12. When the present Atlantic basin began forming in Mesozoic time, irregular separation of the two continents left fragments of one attached to the other (Fig. 11.35).

Summary

Ordovician Life

- Ordovician life is drastically different from that of the Cambrian. During the Ordovician, marine ecosystems expanded as a uniform climate and high sea levels produced epeiric seas all over the world, providing a vast new series of habitats for bottom-dwelling marine invertebrates. The major groups of the typical Paleozoic fauna became established, including articulate brachiopods, bryozoans, tabulate and rugose corals, and crinoids. These diverse filter feeders established multiple tiers at different levels above the sea bottom, further subdividing their habitat and expanding their diversity.

- Ecological diversity was increased by the evolution of large predators, including gigantic nautiloids. In response to this escalation of predators, the unspecialized trilobites of the Cambrian were replaced by trilobites either adapted to burrowing or rolling up or else covered with spines for protection. The earliest swimming fish—and planktonic animals, such as graptolites and ostracodes—further increased the utilization of the marine realm.
- In the Late Ordovician, a mass extinction decimated many typically Ordovician groups, especially among the brachiopods, bryozoans, and nautiloids. This extinction seems to be related to cooling caused by glaciation in North Africa, which was then located over the South Pole.

Ordovician History of the North American Craton

- Sauk deposition was terminated by regression of the sea from the entire craton, probably due to a worldwide fall of sea level. The resulting unconformity marks the base of the Tippecanoe sequence, which is overlain widely by pure quartz sandstone largely recycled by erosion of Cambrian strata. Subsequent Tippecanoe epeiric sea deposition was dominated by richly fossiliferous carbonate rocks and, in the eastern craton, by dark-colored shales. A diverse fauna is consistent with a tropical paleolatitude indicated by paleomagnetic data.

Appalachian Margin of the Continent

- A passive continental margin with a shallow-marine carbonate shelf in Early Ordovician time deepened suddenly in mid-Ordovician time to form a foreland basin. This basin was filled rapidly by a clastic wedge whose sediments were derived from erosion of volcanic and tectonic lands that lay to the east.
- Eastern North America had become a tectonically active margin, as evidenced by the clastic wedge and by local unconformities and conglomerates, eruption of volcanic rocks, intrusion of granites, and emplacement of ultramafic

rocks. All of this constitutes evidence of the Late Ordovician Taconian orogeny, which was the first of several Paleozoic mountain-building episodes in the Appalachian region.

Plate-Tectonic Interpretation

- Mid-Ordovician closure of a narrow ocean basin between North America and a volcanic arc is postulated. The ocean floor was subducted eastward, and as the arc approached, oceanic ophiolite was obducted westward onto the continental margin. Resulting thrust loading depressed that margin to create the foreland basin. Successive episodes of thrust loading culminated with complete closure of the ocean basin and collision of the arc and possible microcontinents with North America along a suture marked today by crushed ultramafic rocks. The Taconian orogeny provides an example of collage tectonics, in which various tectonic terranes have been accreted to a continent by plate motions.
- Thrust loading accounts well for the origin of the foreland basin, and the clastic wedge sediments show the shallowing-upward trend during filling expected from subsidence theory. Deep-ocean ophiolites are succeeded by *flysch* with black shale and immature sandstones (graywacke) deposited by turbidity currents in moderately deep water and containing fossil graptolites; paleocurrent directions can be determined from sole marks on the bottoms of the sandstones. The turbidity currents built deep-sea fans in the basin. Finally, nonmarine *molasse* deposits composed of deltaic and nonmarine red beds filled the basin and caused the shoreline to prograde westward onto the craton as subsidence had virtually ceased.
- Thrust loading of the eastern continental margin apparently also affected arches and basins for several hundred kilometers across the craton to the west, for there is evidence of activation of some such features during Ordovician time.

Readings

Bird, J. M., and J. F. Dewey. 1970. Lithosphere plate-continental margin tectonics and the evolution of the Appalachian orogen. *Bulletin of the Geological Society of America,* 81:1031–60.

Bruton, D. L., ed. 1984. *Aspects of the Ordovician System: A Handbook.* Oslo: Universitets-forlaget.

Dapples, E. C. 1955. General lithofacies relationship of St. Peter Sandstone and Simpson Group. *American Association of Petroleum Geologists,* 39:444–67.

Dott, R. H., Jr., and R. H. Shaver, eds. 1974. *Modern and ancient geosynclinal sedimentation.* Tulsa, Society of Economic Paleontologists and Mineralogists Special Publication 19.

Huff, W. D., S. M. Bergström, and D. R. Kolata. 1992. Gigantic Ordovician volcanic ash fall in North America and Europe: Biological, tectonomagmatic, and event-stratigraphic significance. *Geology,* 20:875–78.

Kay, M., and E. C. Colbert. 1965. *Stratigraphy and life history.* New York: Wiley.

King, P. B. 1977. The evolution of North America. 2d ed. Princeton, N.J.: Princeton University Press.

McBride, E. 1962. Flysch and associated beds of the Martinsburg Formation (Ordovician), central Appalachians. *Journal of Sedimentary Petrology,* 32:39–91.

Quinlan, G. M., and C. Beaumont. 1984. Appalachian thrusting, lithospheric flexure, and the Paleozoic stratigraphy of the eastern interior of North America. *Canadian Journal of Earth Sciences,* 21:973–96.

Schenk, P. E. 1971. Southeastern Canada, northwestern Africa, and continental drift. *Canadian Journal of Earth Sciences,* 8:1218–51.

Sloss, L. L. 1976. Areas and volumes of cratonic sediments, western North America and eastern Europe. *Geology,* 4:272–76.

Stewart, J. H., et al., eds. 1977. *Paleozoic paleogeography of the western United States.* Los Angeles: Pacific Coast Section, Society of Economic Paleontologists and Mineralogists.

Williams, H., and R. D. Hatcher, Jr. 1982. Suspect terranes and accretionary history of the Appalachian orogen. *Geology,* 10:530–36.

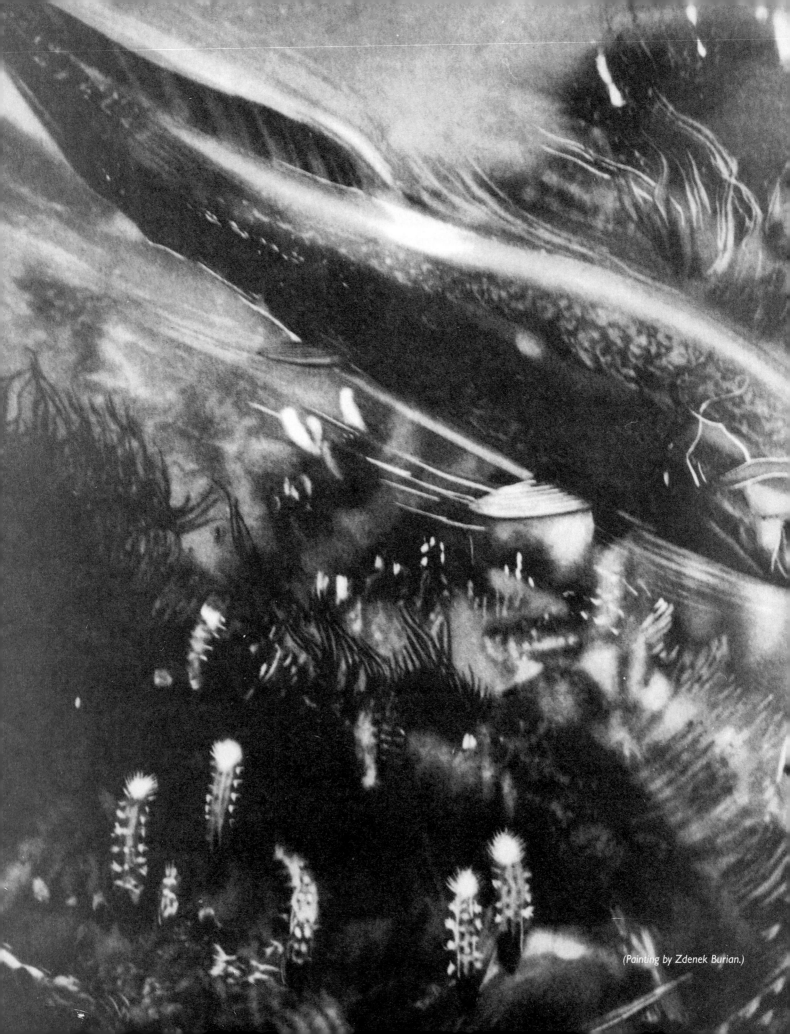

(Painting by Zdenek Burian.)

Chapter

12

The Middle Paleozoic
Time of Reefs, Salt, and Forests

MAJOR CONCEPTS

He who with pocket-hammer

smites the edge

Of luckless rock

or prominent stone, disguised

In weather-stains

or crusted o'er by Nature.

The substance classes by

some barbarous name,

And thinks himself enriched,

Wealthier, and doubtless

wiser than before.

William Wordsworth, *The Excursion* (1814)

▶ During Silurian and Devonian times, marine life on the sea floor diversified further. An unusually large number of great reef complexes formed in the shallow seaways, especially on the margins of epicontinental basins. Ammonoid cephalopods and jawed fishes were particularly diverse in the Devonian. A major extinction of tropical marine life occurred in the Late (but not latest) Devonian.

▶ The first definite land plants appeared in the Late Ordovician, and by the Devonian, there were forests made up of primitive vascular plants, including "club mosses," horsetails, true ferns, and seed ferns. In the Silurian, the vegetation was inhabited by the first land animals, including millipedes, spiders, scorpions, and primitive insects. In the Late Devonian, amphibians evolved from the lobe-finned fish and crawled out onto land for the first time.

▶ The great later Ordovician continental seaway persisted through the Silurian until the Early Devonian, when it retreated. It was soon replaced by a third great seaway, which covered the continent in the mid-Devonian. In North America, these Silurian-Devonian marine seaways were very shallow and tropical, and they accumulated great thicknesses of marine limestones. In many places, restricted marine circulation produced thick Silurian evaporites.

▶ After the Late Ordovician collision on the eastern margin of North America, the great mountain belt gradually wore down in the Silurian Period. However, the same region was deformed again in the mid-Devonian by collision with Europe and with another exotic terrane, Avalonia. This great collision built another, even greater mountain belt, which shed yet more deltaic and nonmarine sediments westward onto the continent. The same effects were seen on the other side of this mountain belt as it shed sediments eastward into Europe in the Silurian and Devonian.

▶ The western and northern margins of North America, which had been passively sinking in the early Paleozoic, experienced collisional tectonics for the first time in the Late Devonian.

259

In Chap. 10, we discussed the characteristics of clastic and, to a lesser extent, carbonate sedimentation on cratons. In Chap. 11, we emphasized the results of Ordovician mountain building in eastern North America and its bearing upon general mountain-building theory. In this chapter, we shall discuss middle Paleozoic (Silurian and Devonian) history, which spanned the interval from 440 to 355 million years ago.

First, we review important developments in organisms, including the great diversification of fish (Fig. 12.1), the first invasion of land habitats, and the achievement of large-scale organic reef building by marine organisms. Next, we investigate the origin of both carbonate and evaporite sediments on cratons and discuss briefly the origin of petroleum. Finally, we examine the effects of the second major Phanerozoic mountain-building event, the Acadian orogeny, which affected the entire eastern margin of North America between about 400 to 355 million years ago. This profound event resulted from the collision of the Avalonia microcontinent with southeastern North America. Meanwhile, the closely related (but slightly earlier) Caledonian orogeny resulted from the collision of northeastern North America (Greenland) with northwestern Europe. These events together produced much granite, metamorphism, and a very large mountain chain. Warping of arches and basins across the craton was more pronounced than in Ordovician time, and this warping influenced the location of organic reefs, evaporite deposition, and petroleum accumulation.

In spite of important differences, there was a broad similarity between the tectonic and sedimentary patterns of Ordovician North America and those of Devonian North America. Each had an early cratonwide unconformity followed by an epeiric sea covering the craton and receiving dominantly carbonate sediments; each was characterized also by orogeny along the eastern continental margin, which resulted in deposition of similar clastic wedges there.

The simultaneous invasion of land by plants and air-breathing invertebrate animals was followed about 80 million years later by an invasion by land vertebrates (amphibians) in Late Devonian time. These events indicate that an atmospheric ozone shield from deadly ultraviolet radiation existed by middle Paleozoic time, if not earlier. Because apparently suitable nonmarine habitats existed earlier, it is interesting to speculate about the possible cause-and-effect relationships between biological evolution and physical evolution of the solid earth and atmosphere, which combined to produce the geologically abrupt appearance of nonmarine fossils in the Silurian Period.

Middle Paleozoic Life

Marine Communities

At the close of the Ordovician Period, a mass extinction apparently caused by global cooling triggered by Gondwanan glaciation wiped out most of the warm-water invertebrates. The Silurian survivors were mostly cold-adapted animals, either from high latitudes or from deep waters. Consequently Early Silurian seas were populated by a low diversity of animals, and many of these species were found worldwide. By the Late Silurian and Devonian, life had

Figure 12.1 Typical underwater scene in the Late Devonian seas near Cleveland, Ohio. A gigantic 10-meter-long placoderm *Dunkleosteus* is pursuing archaic 3-meter-long sharks of the genus *Cladoselache*. In the middle Paleozoic, there was a great variety of fishes, including archaic armored jawless fishes, many other types of jawed placoderms and even lobe-finned fishes. Indeed, some authors call the Devonian "the age of fishes." *(Painting by Zdenek Burian.)*

Figure 12.2 Diorama based on a Middle Devonian fossil site in central New York. Note the great abundance of rugose corals, as well as stalked crinoids *(left foreground)*, large-eyed phacopid trilobites *(center foreground)*, and long-hinged spiriferid brachiopods *(foreground)*. Straight-shelled nautiloids *(right foreground)* were still important predators, but the first coiled ammonoids *(right center background)* had also appeared in the Devonian. (Image #: 10253(2) *American Museum of Natural History Library.*)

recovered from the Late Ordovician crisis, and the marine ecosystems that developed were as complex as those of the Ordovician.

The major phyla of the Paleozoic fauna (brachiopods, bryozoans, corals, crinoids, graptolites) returned, but different families and orders dominated (Fig. 12.2). Among the brachiopods,

260

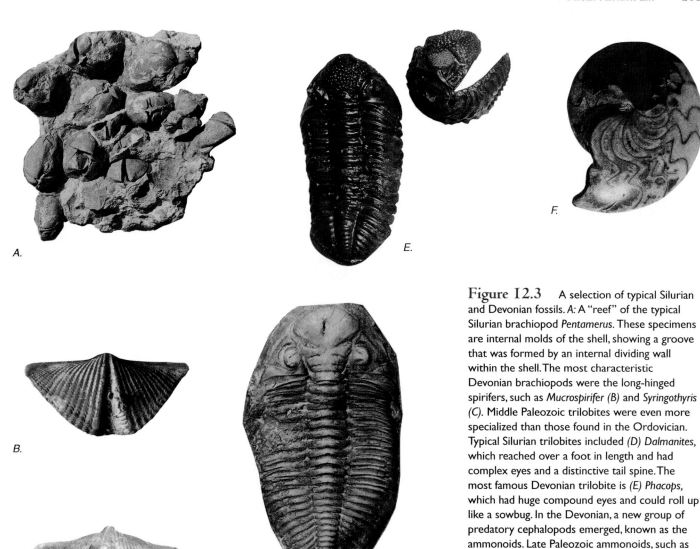

A.

B.

C.

D.

E.

F.

Figure 12.3 A selection of typical Silurian and Devonian fossils. *A:* A "reef" of the typical Silurian brachiopod *Pentamerus.* These specimens are internal molds of the shell, showing a groove that was formed by an internal dividing wall within the shell. The most characteristic Devonian brachiopods were the long-hinged spirifers, such as *Mucrospirifer (B)* and *Syringothyris (C).* Middle Paleozoic trilobites were even more specialized than those found in the Ordovician. Typical Silurian trilobites included *(D) Dalmanites,* which reached over a foot in length and had complex eyes and a distinctive tail spine. The most famous Devonian trilobite is *(E) Phacops,* which had huge compound eyes and could roll up like a sowbug. In the Devonian, a new group of predatory cephalopods emerged, known as the ammonoids. Late Paleozoic ammonoids, such as *(F) Manticoceras* from Devonian beds of the Tindouf Basin of Morocco, had a zigzag pattern of suture between the external shell and the internal chamber walls; this pattern is known as **goniatitic.** *(D. R. Prothero.)*

the thin-shelled, relatively flat orthids and strophomenids were greatly reduced in numbers. In their place were brachiopods with much thicker shells and more robust, elongated, deeper bodies. The most characteristic Silurian brachiopods were the *pentamerids* (Fig. 12.3A), which had robust, teardrop-shaped shells. Pentamerids had a set of internal walls that show up as clefts in the fossil molds of these animals. Pentamerids typically lived in dense communal clusters called pentamerid reefs. During the Devonian, pentamerids declined, and the sea floor was taken over by a different group of brachiopods, the *spirifers* (Fig. 12.3B, C). Spirifers got their name from their lophophore, which is arranged in a pair of spirals. Although spirifers varied in body shape, typical Devonian spirifers had very long hinges and, so, resembled a pair of wings. Spirifers continued to flourish in the late Paleozoic and survived until the Jurassic, but their heyday was the Devonian.

Bivalves and gastropods continued through the Silurian and Devonian Periods, although they were far outnumbered by bra-

chiopods. Bivalves expanded to freshwater habitats for the first time. Some animals were in decline. Trilobites were relatively scarce in the Silurian (Fig. 12.3D). Devonian trilobites included the unusual *Phacops,* which not only could roll up but also had huge compound eyes (Figs. 12.2 and 12.3E). Nautiloids were also declining, although they remained a major predator in the Silurian. During the Devonian, however, their role was taken by their descendants, the *ammonoids* (Figs. 12.2 and 12.3F). Unlike the nautiloids, which were mostly straight-shelled, ammonoids grew a shell that curled into a tight spiral. An interior "bulk-head," which separated the living cephalopod from the earlier uninhabited parts of its shell, is called the septum. As the cephalopod grows, it vacates old chambers and secretes the septum to close them off, and the closed hollow chambers provide flotation. When the outer shell is removed from a fossil, the septum is visible. The line between the septum and the outer wall is called the "suture." In nautiloids, the septa were mostly simple, curved

Figure 12.4   Diorama showing two kinds of "sea scorpions," or eurypterids; some reached almost 2.5 meters in length. Eurypterids had pincerlike claws on one set of appendages and a swimming paddle on another. They were the largest predators of the Silurian. *(© The Field Museum, #GEO8019c.)*

Figure 12.5   Reconstruction of typical Devonian fishes. In the upper right is an advanced lobe-finned fish, ancestral to amphibians and other land vertebrates. The rest are armored jawless fish, the last survivors of the earliest vertebrate groups that first diversified in the Late Cambrian and Ordovician (see Fig. 11.9). All of these groups, along with the jawed and armored placoderms (Fig. 12.1) were extinct by the end of the Devonian. *(Painting by Zdenek Burian.)*

structures, so that the sutures were always simple curves. However, ammonoids soon developed complex creases in their septa for reinforcement and structural strength. This produces a complex folded suture. Late Paleozoic ammonoids typically had a *goniatitic* suture pattern, which shows up as a series of zigzag folds on the side of the shell (Fig. 12.3F).

The top predators in the Silurian were not nautiloids or ammonoids, however, but the "sea scorpions," or *eurypterids* (Fig. 12.4). These arthropods, related to scorpions and horseshoe crabs, ranged through the entire Paleozoic but reached their greatest abundance during the Silurian. Most were 13 to 50 cm (5 to 20 inches) in length, but one giant reached 2.5 meters (9 feet)! Eurypterids had scorpion-like claws on their front appendages, and their rear appendages were shaped like paddles. Early eurypterids were marine, but during the Silurian they spread to brackish lagoons, fresh waters, and even swamps. The Silurian Bertie Waterlime of New York, a brackish-water limestone, is world famous for its abundance of eurypterids.

Planktonic organisms also thrived in the warm Silurian and Devonian seas. Acritarchs and ostracods continued to populate the microplankton, as they had since the Cambrian. Graptolites, which were nearly wiped out in the Ordovician extinction, again radiated. In Britain, the twelve species that survived into the Silurian evolved into about sixty species in the 5 million years of the Early Silurian. Some Silurian graptolites formed either intricate spirals or multiple branches radiating from a single spiral (Fig. 11.7E). By the Late Silurian, however, free-floating graptolites were in their final decline. The last known genus, *Monograptus,* got its name from its single row of tiny animals along a single branch. One species, *Monograptus uniformis,* is used to recognize the Silurian-Devonian boundary. Shortly thereafter, planktonic graptolites became extinct, leaving only the primitive attached bushy types (relics of the Cambrian) to hang on until the Pennsylvanian.

The Age of Fishes

The greatest innovations of the Silurian and Devonian occurred not in the marine invertebrates but in the vertebrates. Simple jawless fish had been around for almost 100 million years, since the Early Cambrian. Their isolated bony scales and plates (Fig. 11.9A) are the only fossils we have for them. During the Late Silurian, fish began to diversify, and by the Devonian, they were so abundant in both marine and freshwater deposits that the Devonian is often called "the age of fishes."

Jawless fish continued to be abundant, but they were no longer just simple forms armored only with bony scales. During the Silurian, they developed armored head shields and body armor (Fig. 12.5). They had a slitlike mouth with no jaws, an indication that they must have lived by filter feeding. Their cartilaginous internal skeleton could not fossilize, but their body armor covered them all the way to their asymmetrical tails. Some had a flat base on their head shield, suggesting that they fed along the bottom. Having no muscular fins on the side, they must have been clumsy swimmers.

The first fish with jaws, the *acanthodians,* incorrectly known as "spiny sharks" (Fig. 12.6), also appeared in the Silurian. They had large eyes and a strong bite, as well as a row of fins supported by spines along the side of the body. They must have been good swimmers and active predators. The Devonian saw many different types of jawed fishes. These included the first true sharks, whose skeleton is made entirely of cartilage (Figs. 12.1 and 12.7). Typical Devonian sharks had very broad pectoral fins; a long, flat skull; and spines in front of two dorsal fins. The most spectacular jawed fish of the Devonian, however, were an extinct group

Figure 12.6 A more successful middle Paleozoic fish group were the spiny **acanthodians.** With their large eyes and strong fins and jaws, they were active predators. Their many fins were supported by bony spines on the leading edge. Acanthodian spines are known from the Silurian, making them the earliest known jawed vertebrates, and the group lasted until the end of the Paleozoic. More important, all bony fishes (and amphibians and their land vertebrate descendants) are more closely related to acanthodians than they are to any other early fish group (see Fig. 12.7). *(Painting by Zdenek Burian.)*

Figure 12.7 Family tree of the living and extinct groups of fish. The Devonian witnessed a great radiation of archaic jawless fish and jawed placoderms, both of which did not survive into the Carboniferous. There were also archaic sharks (which have living descendants), acanthodians (the earliest jawed vertebrates), and lobe-finned fishes, including lungfish and rhipidistians, the group closest to amphibians. In the late Paleozoic and Mesozoic, most of these archaic groups disappeared, to be replaced by a great radiation of bony fish.

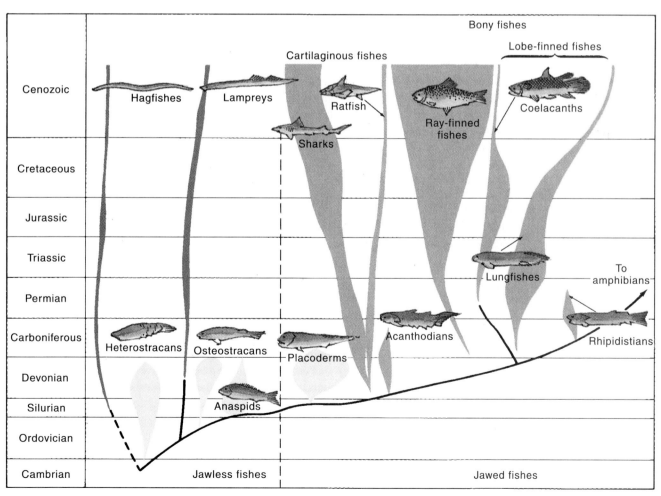

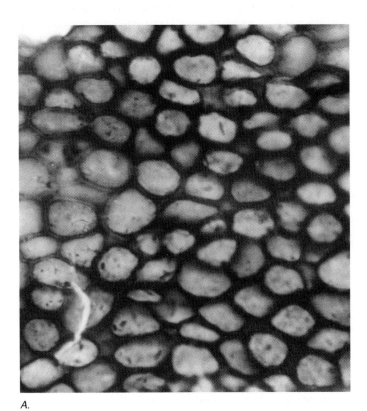

A.

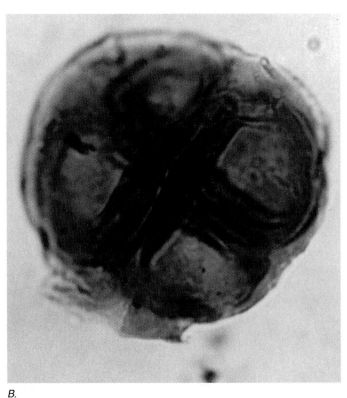

B.

Figure 12.8   Ordovician plant fossils. Although the evidence for land plants in the Ordovician is sparse, the presence of four-part spores ("tetrads") suggests that some form of vascular plant had invaded the moister habitats on land. These fossils include *(A)* plant cuticle (×600) and *(B)* spores from the Upper Ordovician of Libya (×1500). *(Courtesy of Jane Gray.)*

known as *placoderms* (Fig. 12.7). Like sharks, their internal skeleton was made of cartilage, but they developed extensive bony armor plating on the head and trunk. Some were heavily armored down to their fins and probably fed on the bottom. Others were lightly armored and flattened, and they lived much as do modern skates and rays. The most spectacular were the gigantic armored predators, such as *Dunkleosteus* (also called *Dinichthys*), which had sharp biting plates instead of teeth and reached 12 meters (40 feet) in length (Fig. 12.1). Clearly the largest predator in the Late Devonian seas, it must have terrorized even sharks, since the latter reached no more than 1.2 meters (4 feet) in length.

Except for sharks, most of these fish, including the armored jawless fish and the placoderms, did not survive the Devonian; only the acanthodians lasted until the Permian. Two other groups of fish that arose in the Devonian have living descendants. These groups had skeletons made of bone rather than cartilage. One group, the *ray-finned fish,* flourished during the late Paleozoic; today 99 percent of living bony fishes are from this group (Fig. 12.7). As the name implies, their fins are supported by bony spines radiating from the body. The other group, the *lobe-finned fish,* had a different fate (Fig. 12.5). These fish were characterized by a club-shaped fin supported by stout bones (homologous with our arm bones) that gave the fin better support; ultimately, this fin allowed the fish to walk on land (Fig. 12.13). Lobe-finned fish also had internal nostrils and eventually lungs, allowing them to breathe out of water. There were three main groups of lobe-finned

fish: lungfish, which live today in fresh water and can survive extended drought by burrowing in the mud of a drying lake bottom; the coelacanths, known only from fossils until a "living fossil" was discovered off the coast of Africa in 1937; and a third lineage that led to amphibians and other land vertebrates.

Invasion of the Land

Vertebrates were latecomers to the land. Up until the Ordovician, the land had only sparse plant cover. Soils were probably held in place by **microbiotic crusts** (also known as **cryptogamic soils**), a community of fungi, bacteria, and algae that today can be found in the arid regions forming a soft, spongy mat on barren clays and silts. In Upper Ordovician deposits, there are fossil spores (Fig. 12.8B) indicating that more advanced plants were present on land and burrows several centimeters in diameter suggesting that some kind of invertebrate animal (possibly a millipede) lived and fed among them.

For a semiaquatic plant to thrive on land, several requirements must be met. The plant must have a waterproof cuticle (Fig. 12.8A) to prevent desiccation in the dry air, a strong supporting structure to lift it off the ground because it can no longer depend on the buoyancy of water, and a means of passing the sperm to the eggs, which are no longer immersed in water. We find the first *vascular plants,* which have tubes for transporting water and nutrients through the tissues, in the Early Silurian. One of the best

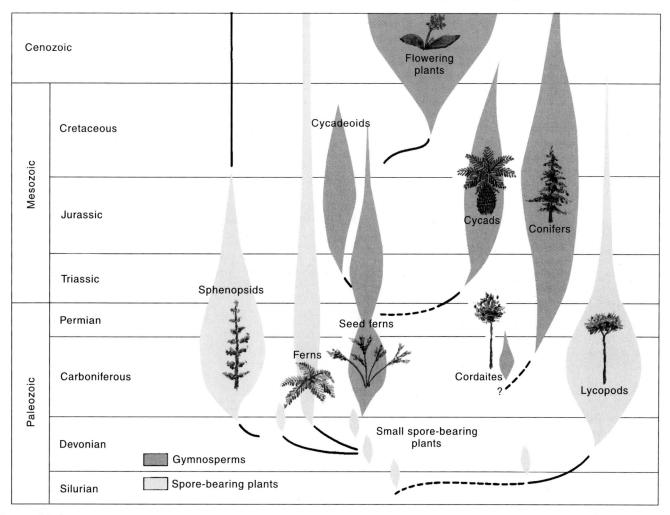

Figure 12.9 Evolutionary radiation of land plants. Spore-bearing plants were dominant in the Silurian, Devonian, and Carboniferous. By the Late Permian, seed ferns and other gymnosperms, such as conifers, had taken over the forests of the earth, and they continued to prevail through most of the Mesozoic. In the mid-Cretaceous, flowering plants radiated and soon became the most successful group of plants. (*Modified from A. H. Knoll and G. W. Rothwell, 1981,* Paleobiology, *v. 7, pp. 7–35.*)

known is the Early Devonian *Rhynia* (Figs. 12.9 and 12.10A), built of simple, leafless stalks with a waterproof cuticle and spore-bearing organs called *sporangia* at their tips; few specimens are more than a half a meter tall. Early Devonian plants lacked roots or leaves and were confined to creeping along the ground. By Middle Devonian time, the vascular bundles in a plant such as *Psilophyton* (Fig. 12.10B) occupied a large part of the stem, producing a much stronger stalk and more efficient water transport. These plants also evolved roots, both for support and for removal of nutrients out of the soil and water.

By the Late Devonian, lycopsids, including the living club mosses and "ground pine," covered the landscape (Fig. 12.10C, D). Most were low, creeping forms growing near water, but some reached a meter or more in height, with some Carboniferous lycopsid trees reaching 30 meters (100 feet) in height. Lycopsids had long, slender, very simple leaves, which issued directly from the trunk in a spiral arrangement. As a result, the bark of lycop-

sid trees has distinctive diamond-shaped leaf pad scars arranged in spirals. Lycopsids are more advanced than psilophytes in that the sporangia of the former are arranged into tight clusters called cones. In many lycopsids, the male and female cones are separate, so that cross-fertilization with other plants is favored over self-fertilization. This arrangement enhances genetic variability, giving an important evolutionary advantage (just as sexual reproduction is favored over cloning in animals).

A second important group of spore-bearing plants was the *sphenopsids,* or joint-stemmed plants. The only living sphenopsid is the horsetail, or scouring rush, *Equisetum* (Fig. 12.10E), commonly found along stream banks today. Sphenopsids have a long, hollow stem that is jointed, with leaves and sporangia clustered at the joints. The name "scouring rush" comes from the fact that *Equisetum* has crystals of silica in its tissues, making the plant good for scouring the cooking pots of early American settlers.

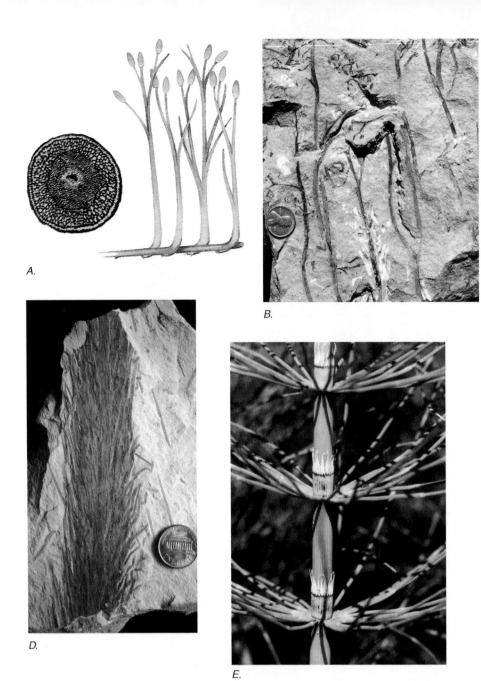

A.

B.

C.

D.

E.

F.

G.

H.

Figure 12.10   Land plants from the middle Paleozoic. *A:* Reconstruction of *Rhynia,* one of the most primitive land plants known, from the Lower Devonian Rhynie Chert beds of Scotland. The vascular tissue (central zone in stem) was relatively Inefficient in conducting fluids, so the stems were seldom more than a few centimeters long. *B:* Early Devonian plants, such as *Psilophyton,* had much larger vascular bundles, producing a stronger stalk with more efficient water transport, so the plants could grow larger. Living psilophytes are very similar to their Devonian ancestors. *C:* Living *Lycopodium,* or "club moss." The lycopsids, which arose in the Early Devonian, are normally small ground plants, but some were giant trees in the late Paleozoic. *D:* The Late Silurian or Early Devonian lycopsid *Baragwanathia,* from Victoria, Australia. Although it is one of the oldest known vascular plant fossils, it is more complex than *Rhynia* and some later forms. *E:* Living *Equisetum,* known as "horsetails," "scouring rushes," or sphenopsids. This group, with their distinctive jointed stems, also arose in the Devonian; some formed giant trees in the late Paleozoic. *F:* The first plants with seeds instead of spores were known as seed ferns. These are seeds of *Trigonacarpus,* a common seed fern from the Carboniferous. *G:* By the Late Devonian, there were tree-sized seed ferns up to 10 meters tall. These are seed fern stumps from the Late Devonian Gilboa Forest, in the Catskill Mountains of New York. *H:* These living tree ferns spore-bearing, rather than seed ferns) give a good approximation of Late Devonian forests. (*A from M. E. White, 1990, The flowering of Gondwanaland: Princeton University Press; G, by D. R. Prothero. All others courtesy Bruce Tiffney.*)

Figure 12.11   Artist's conception of the Late Devonian landscape. Tall seed fern and lycopsid trees are conspicuous, but most plants were low-growing psilophytes, lycopsids, sphenopsids, and ferns that clustered close to the water's edge. Against this backdrop, early land arthropods flourished, and eventually the first amphibians crawled out of the water. *(Painting by Zdenek Burian.)*

The third important group of spore-bearing plants is the true ferns, which are abundant today in any shady, damp area; there are more than 10,000 living species (Fig. 12.10H). These have large, complex leaves, and their sporangia occur in small clumps under the leaves. Devonian sphenopsids and ferns were small plants, but by Carboniferous time, both were common as large trees.

Finally, the Late Devonian saw a more important development: the first seed plants, known as *seed ferns* because of their fernlike foliage (Fig. 12.10F, G). Unlike spores, seeds do not require continuous moisture to survive because they have their own food stores and a waterproof cover. The sperm of seed plants do not have to swim through water to reach the egg (as they do in spore-bearing plants) but can fertilize by other means. The earliest seed plants soon developed into large trees up to 10 meters (33 feet) tall, and the Late Devonian landscape was covered with the world's first forests (Fig. 12.11).

Land plants created a new habitat, which was soon exploited by the first land animals, the arthropods. Late Ordovician burrows suggest that a millipede-like animal was already terrestrial at that time. The Lower Devonian Rhynie Chert of Scotland contains a number of fossil arthropods, including the oldest known scorpions, spiders, mites, millipedes, and wingless insects called springtails. By the Devonian, the swamps were filled with a variety of crawling, burrowing, and even flying arthropods (Fig. 12.12).

The first amphibians appeared on dry land at the end of the Devonian. Like land plants, land vertebrates had to cope with problems of desiccation, support, and land reproduction. Amphibians still reproduce in the water today, but they developed strong limbs and a semipermeable skin early in their evolution. From animals such as *Ichthyostega* (Fig. 12.13), we infer the transition between lobe-finned fish and typical amphibians. Its limbs were fully functional, yet they were basically similar to the bones of the lobed-finned fish. *Ichthyostega* retained fishlike vertebrae, bony gill covers, and a tail fin. The bones in the head were also fishlike, but the snout was long, like that of an amphibian. *Ichthyostega* had reinforced ribs to support its lungs when it was sprawling on dry land away from the buoyancy of water. Recently found specimens show that it had six to eight toes on each foot, so that it had not yet reduced to the typical five toes. *Ichthyostega* is a classic transitional form between fish and amphibian. By the Carboniferous, amphibians had lost their remaining fishlike features, except that they still returned to water to lay eggs.

What led vertebrates to struggle into this new, hostile, dry environment, with no water for support? Some scientists have suggested that they did so because their pools dried up and they had to wriggle across land to find another one. It is more likely that they left the water because competition and predation from other fish were much greater there, whereas the land represented an unexploited resource: plenty of arthropods to feed upon and no larger predators. In fact, the transition to land is not that difficult: a number of modern ray-finned fish can breathe air, and many have become semiterrestrial. For example, the mudskipper lives partly out of the water to catch insects and escape predators. If there were not already land vertebrates competing, the ray-finned fish might also have become completely terrestrial.

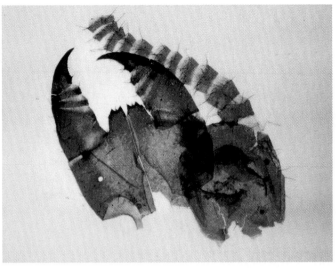

A.

B.

Figure 12.12 Fossils of some of the earliest known land animals. *A:* Head and poison claw of the earliest known centipede; the entire animal was about 10 mm long. These specimens come from the same deposits as the Late Devonian Gilboa seed fern stumps shown in Fig. 12.10G. *B:* Framework of compound eye from one of the earliest known insects, also from the Late Devonian Gilboa forest. *(A and B courtesy of William Shear.)* *C:* Millipedes are the oldest known land animals, with specimens found in Middle Silurian rocks. This extraordinary specimen is from the Late Carboniferous of West Virginia and shows excellent preservation of its long spines. *(Photo by Bruce Frumker,* © *Cleveland Museum of Natural History.)*

C.

Late Devonian Mass Extinctions

The Late Devonian is marked by another severe extinction event. It occurred between the last two stages of the Devonian, the Frasnian and Famennian (not at the very end of the period, as is the case with several other extinctions). In the thick Devonian sequence of New York State, 70 percent of the marine invertebrate species were wiped out. Pentamerids disappeared completely; only 15 percent of Frasnian brachiopods survived, and ammonoids were devastated. Trilobites and gastropods also declined, but the most severe extinctions occurred in the reef community. Tabulate-rugosid-stromatoporoid reefs (Box 12.1) were devastated at the end of Frasnian, and all three groups were rare throughout the rest of the Paleozoic. Among microplankton, the acritarchs, which had persisted since the late Proterozoic, were almost completely wiped out. Most of the typical Devonian fish, including the armored jawless fish and the placoderms, were eliminated.

As in the case of the Late Ordovician extinctions, warm-water marine invertebrates were the most hard-hit by the Devonian extinction. By contrast, polar marine organisms in South America were virtually unaffected. After the mass extinction, limestone-producing reefs were temporarily replaced by reefs of the glass sponge *Hydnoceras* (Fig. 12.14). Modern glass sponges live in cool waters, suggesting that *Hydnoceras* took over the vacant reef habitat during the colder Famennian. All these factors point to a global cooling event, and indeed there is some evidence of another Gondwana glaciation at this time. Since the Ordovician, Gondwanaland had moved off the pole and remained off during the Silurian, before shifting back in the Early Devonian. By Late Devonian time, the South American portion of Gondwanaland had moved over the South Pole, producing thick Devonian glacial deposits in northern Brazil (Fig. 13.47). Some scientists have suggested that the cooling triggered a massive overturn of oceanic waters, bringing cold, nonoxygenated, nutrient-rich bottom waters to the surface. The

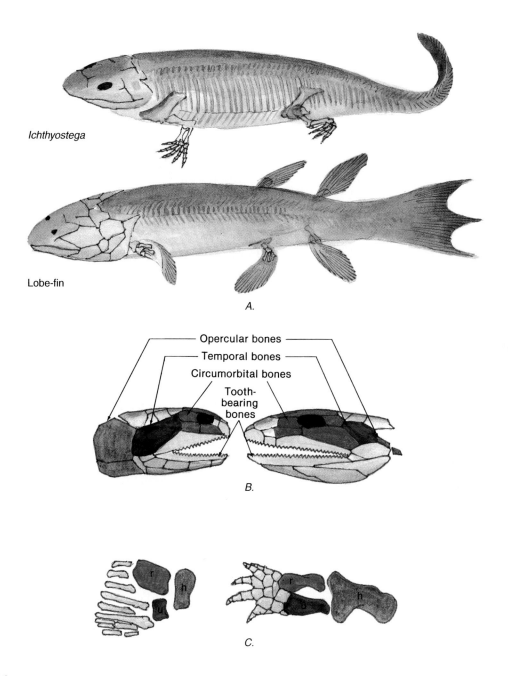

Ichthyostega

Lobe-fin

A.

Opercular bones
Temporal bones
Circumorbital bones
Tooth-bearing bones

B.

C.

Figure 12.13 The oldest known amphibians, *Ichthyostega* and *Acanthostega,* are found in Upper Devonian rocks of Spitsbergen and Greenland. *A:* In most anatomical features, they are very similar to lobe-finned fish, such as the rhipidistians (see Figs. 12.5 and 12.7). For example, they still retain the tail fin and the canals on their face for sensing motion in water. However, their rib cage is much more solid, so that they can breathe while lying on land. *B:* Although the proportions of the skull have changed in the earliest amphibians, they still retain fishlike features, including remnants of the opercular bones over the gill covers. *C:* The lobed fin was ideally preadapted for becoming the amphibian foot. The same basic elements (h, humerus or upper arm bone; r, radius; and u, ulna, lower arm bones) were modified for weight bearing. The fin supports became the toe bones. This illustration shows the standard five toes, but recently discovered specimens show that Late Devonian amphibians had as many as eight toes. Later amphibians eventually stabilized the toe count at five.

chief evidence for this idea is a sudden enrichment in carbon isotopes at the end of the Frasnian, indicating that much organic carbon from the deep ocean was suddenly brought to the surface waters. This may explain the widespread occurrence of black shales near the end of the Devonian.

Extraterrestrial impacts at the Frasnian-Famennian boundary have also been suggested, but the evidence is inconclusive. The extinction of warm-water animals took place over almost 5 million years throughout the Famennian, which seems more consistent with slow glacial cooling than with a sudden impact event.

Box 12.1

Fossils as Calendars

We have seen that isotopic dating gives us the best estimate of numerical dates in terms of years. But the smaller, units of time—the month and day—are so brief geologically that they cannot be resolved by any isotopic method known. As we have seen, fossils were used to establish the relative time scale, and now it appears that they will also prove useful for small-scale numerical chronologies. A very ingenious line of investigation initiated by John Wells of Cornell University about 1960 uses fossils to help bridge the gap between years and the smaller units of time.

Biologists have observed that modern corals deposit a single, very thin layer of lime once a day. It is possible, with some difficulty, to count these diurnal (day-night) growth lines and to determine the coral's age in days. More important, seasonal fluctuations cause the growth lines to change their spacing yearly, so that annual increments can also be recognized, much as in growth rings of trees.

Wells began looking for diurnal lines on fossil corals. He found several Devonian and Pennsylvanian corals that do show both annual and daily growth patterns. He was astonished to find that the Pennsylvanian forms had an average of 387 daily growth lines per year-cycle and that the Devonian corals had about 400 growth lines (Fig. 12.15). Counting between annual marks, Wells found an average of 360 growth lines per year on modern corals. He then constructed a graph (Fig. 12.16) based on the latest isotopic dating of the periods back to the beginning of the Cambrian, and this graph suggested a systematic decrease in the number of days per year through geologic time.

In 1968, Pannella and his associates, using many more organisms from Recent to Cambrian showed that, although a systematic decrease is probably valid, their graph formed an S curve rather than a straight line. This S curve is probably caused by the different effects of tidal action accompanying the changes of continents, ocean basins and epeiric seas. Studies using modern clams show that the growth lines on and within the shells record different environmental events that are also rhythmic, such as seasons and breeding periods. These lines may even record episodic events, such as storms, unusual temperature fluctuations, and attacks by predators. Much can be learned of past environments from such studies.

Figure 12.15   Devonian rugose coral *Heliophyllum,* showing a spectrum of growth lines. The white lines bracket the annual band, and the fine lines between are presumed to be daily growth lines: From evidence such as this, paleontologists have deduced that the days were shorter (about 22 hours) and the years had more days about (400). *(Courtesy of R. L. Batten.)*

Figure 12.14   The Late Devonian extinction was particularly severe on the great Devonian tabulate-stromatoporoid-rugosid reef community. In its place were low-diversity reefs built by the cold-water-loving glass sponge *Hydnoceras,* suggesting that the Late Devonian extinction was largely caused by global cooling. *(Courtesy of J. Keith Rigby.)*

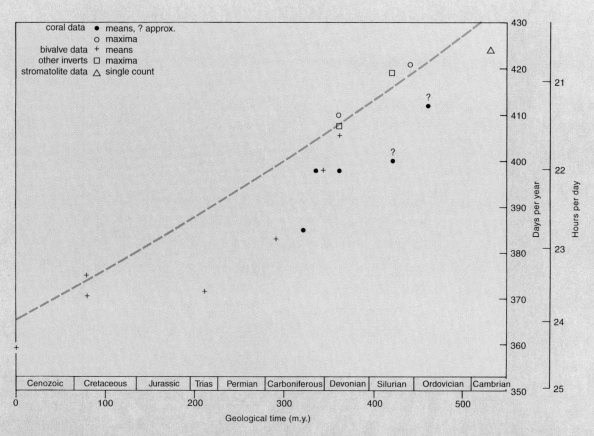

Figure 12.16   Changing length of day through Phanerozoic time based upon data gathered from corals, stromatolites, bivalves, and other invertebrates. The dashed line shows the gradual lengthening of the day and is based on a constant of 2 milliseconds per century. *(Muller and Stephenson, 1975.)*

Meanwhile, geophysicists and astronomers estimated that there has been a deceleration (presumably due to tidal friction) of the earth's rotational velocity in recent centuries amounting to 2 seconds per 100,000 years. If one extrapolates this rate (which may or may not be valid) the Cambrian day would have contained 21 hours and the Cambrian year 420 days. By the same reasoning, the length of a Devonian year should have been 400 days. Thus, the astronomic extrapolations and Wells's coral data for the Devonian are remarkably similar.

Modern stromatolites have also been shown to lay down daily increments, and studies of them in the fossil record show a similar line pattern; for example, stromatolites found in the Bitter Springs Formation in Australia (850 million years ago) show a 410-day year.

The Silurian and Early Devonian Continent

Aftermath of the Taconian Orogeny

In the Appalachian orogenic belt, erosion of islands raised during the Late Ordovician Taconian disturbance occurred during Early Silurian time. Deposition in the western part of the belt continued more or less uninterrupted from the Ordovician, but an unconformity marks the base of Silurian strata farther east. Gradual eastward encroachment of the sea over the erosional surface is clearly documented. Sandy and gravelly facies gradually shifted eastward as transgression proceeded.

Early Silurian clastic sediments were derived from the core of the orogenic belt east of New York. This derivation is clearly shown by the coarsening of sandstones and increase of conglomerate eastward (Fig. 12.17) and by the orientation of current-formed features. Silurian quartz sandstones are widespread in the Appalachian Mountains. They represent weathering, winnowing, and concentration from a tremendous volume of source rocks in the Taconian mountains.

Unusual Silurian iron-rich sedimentary deposits in the southern Appalachian Mountains provided ore for the important Birmingham steel industry; a similar ore used to be mined in Newfoundland. Apparently there was intense tropical weathering

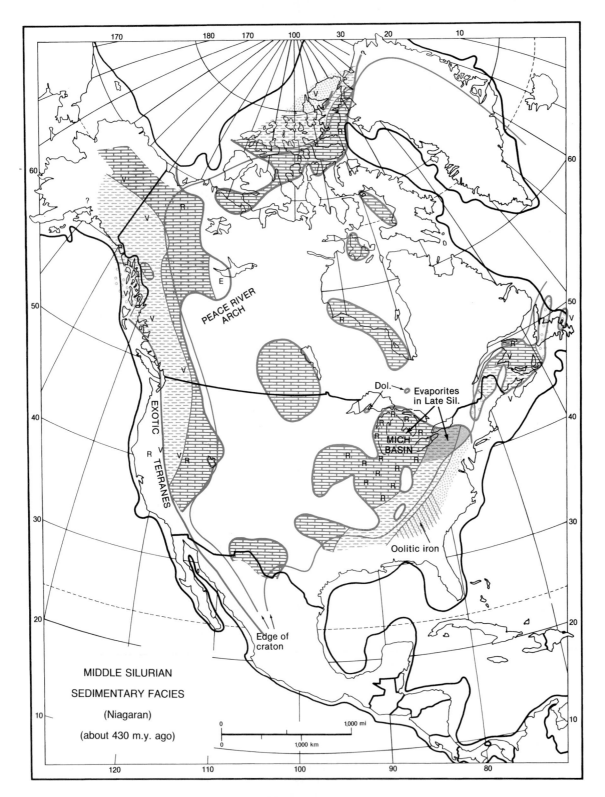

Figure 12.17
Middle Silurian lithofacies. Note large patches of carbonate rocks over the Canadian Shield region and widespread organic reefs, R. (see Box 10.2 for symbols and sources.) Some patterns on western margin of continent may not have been in their present locations yet.

Labels on map: 170, 180, 170, 100, 30, 20, 10, 60, 50, 40, 30, 20, 10, 120, 110, 100, 90, 80

PEACE RIVER ARCH

EXOTIC TERRANES

Dol.

Evaporites in Late Sil.

MICH BASIN

Oolitic iron

Edge of craton

MIDDLE SILURIAN

SEDIMENTARY FACIES

(Niagaran)

(about 430 m.y. ago)

0   1,000 mi

0   1,000 km

in the Taconian mountains. Rivers introduced unusually high concentrations of iron into a somewhat restricted marine environment. The sediments forming there, including marine shells and oolites, were replaced and cemented with red hematite.

A Carbonate-Rich Craton

Silurian and Devonian strata are scattered widely over the craton (Fig. 12.17). Their scattered distribution led to a long-held view that mid-Paleozoic epeiric seas were restricted in area and studded with many low islands. Such an interpretation was strongly influenced by the hazardous assumption that present limits of marine strata closely approximate their original distribution. When dealing with cratonic sequences, one must be cautious to evaluate the importance of unconformities that may account for a great deal of erosion of formerly more extensive strata. Such a discontinuity within the Devonian System is known to occur over all the craton

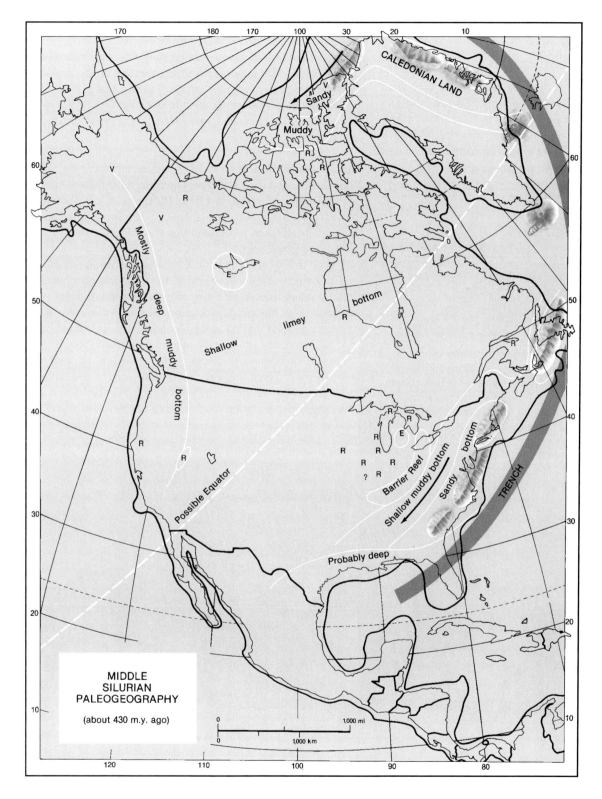

Figure 12.18
Middle Silurian paleogeography. Most of the continent lay beneath an epeiric sea. Western margin of continent cannot be restored with confidence.

and must account for much of the present distribution of Silurian strata. This unconformity marks the boundary of the Tippecanoe and the overlying Kaskaskia Sequence (see Fig. 11.13).

If we examine the sedimentary facies for the Middle Silurian, we see that most of the preserved rocks are marine carbonates and some shale that must have formed under a minimum influence of lands. Moreover, the deposits and their fossils suggest a former shallow sea with uniform conditions over an immense region rivaling that of the Late Ordovician (see Fig. 11.16). Therefore, Fig. 12.18 shows our preferred interpretation of Middle Silurian paleogeography, in which it is assumed that marine Silurian strata originally covered *all* the craton and were subsequently eroded to produce their present distribution. Similarly we shall see that marine Devonian strata (also dominantly carbonate) covered all the craton, too.

Modern Versus Ancient Carbonate Sedimentation

What can the modern seas teach us about Paleozoic carbonate deposition? Today carbonate sediments form in two very different settings (Fig. 12.19). The first of these is the shallow, warm, well-lighted sea with little polluting clastic material, which we call the "carbonate factory." The second is the deep ocean where,

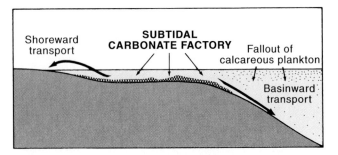

Figure 12.19   The two major origins of carbonate sediments: (1) the subtidal "carbonate factory," in shallow, (< 30 m) warm seas; (2) the settling of calcareous microfossils in the deep sea basins. Some carbonate sediment is also redeposited shoreward and basinward from the "factory" by waves and currents. *(Adapted from N. James, 1984, in* Facies models, *Geological Association of Canada.)*

since late Mesozoic time, floating microscopic planktonic organisms with calcareous skeletons have settled to the sea floor after death to form distinctive calcareous sediments called ooze. For analogies with ancient epeiric seas, we look to the carbonate factories on modern, shallow, tropical continental shelves (Fig. 12.20). It takes little imagination to translate the conditions on these shelves to the much larger Paleozoic carbonate-producing epeiric seas.

Lands were relatively small and low until late Paleozoic time, and shallow seas were large. Calcareous-secreting organisms thrived, and their skeletons contributed the bulk of carbonate materials. As noted in Chap. 10, the wide distribution of ancient carbonates with abundant and diverse invertebrate faunas long has been taken to indicate that the Paleozoic tropics were wider than at present.

Note that we assume relatively constant seawater chemistry and other ecological requirements for most marine organisms since Vendian time. With a few notable exceptions, fossil communities show the same associations of organisms and of sediments as do their nearest modern counterparts. It is improbable that *all* members of very complex communities could have changed requirements to the same degree and at the same rate through time. Therefore, the finding that ancient communities consisting of two or three dozen organisms showing mutual relations closely duplicating their modern counterparts argues for similar requirements through time (Fig. 12.21). Studies of isotopes of several elements found in ancient sediments (especially

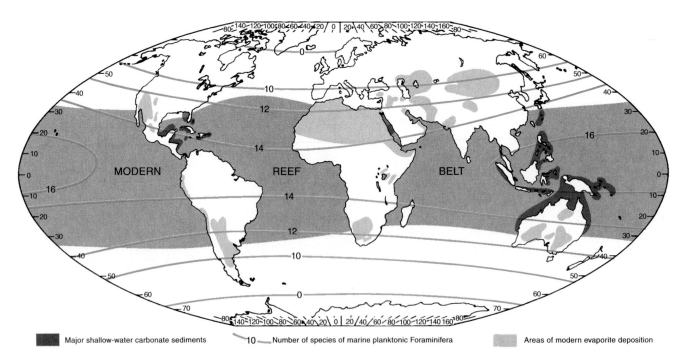

■ Major shallow-water carbonate sediments	10 — Number of species of marine planktonic Foraminifera	▒ Areas of modern evaporite deposition

Figure 12.20   Distribution of modern organic reefs, major shallow-marine carbonate deposition, and nonmarine evaporites. *(Adapted from Lowman, 1949, Geological Society of America Memoir 39; Rodgers, 1957, Society of Economic Paleontologists and Mineralogists Special Publication 5; Goode's world atlas, 1964.)* Also shown are contours indicating latitudinal diversity (number of species) of modern marine planktonic Foraminifera as a function of temperature. Most marine and nonmarine organisms show similar patterns of increasing diversity in warm, tropical latitudes. *(After F. G. Stehli, Science, v. 142, November 22, 1963, pp. 1057–1059; Copyright © 1963 American Association for the Advancement of Science. Reprinted with permission.)*

S and Sr) and of the relative abundance of two different mineral forms of $CaCO_3$ (calcite and aragonite) do suggest changes of seawater composition through time; however, the magnitude of such changes was not great enough to be devastating to marine life, thanks to the ocean-atmosphere chemostat system.

As is the case today, it appears that seawater has been near saturation with respect to Ca through most of earth history. During Proterozoic time, when CO_2 was more abundant in the atmosphere and in the seas, limestone and dolomite were frequently precipitated directly from seawater. During Phanerozoic time, however, organisms have been largely responsible for precipitating $CaCO_3$ biochemically to make their skeletons. Upon death, the skeletal particles become part of the sedimentary record—a kind of immortality. Complex chemical feedbacks among organisms, atmospheric CO_2, weathering of rocks, and temperature have controlled the abundance of skeletal marine organisms and subtle compositional variations of carbonate sediments through time.

Organic Reefs

General Characteristics

Although animals inhabit all portions of the seas, the greatest numbers of marine animals and plants and the highest level of diversity are found in warm, shallow waters within the photic zone. The floating plankton forms the beginning of the long marine food chain; microscopic plants begin the chain, followed by microscopic animals, which feed on the plants and on each other.

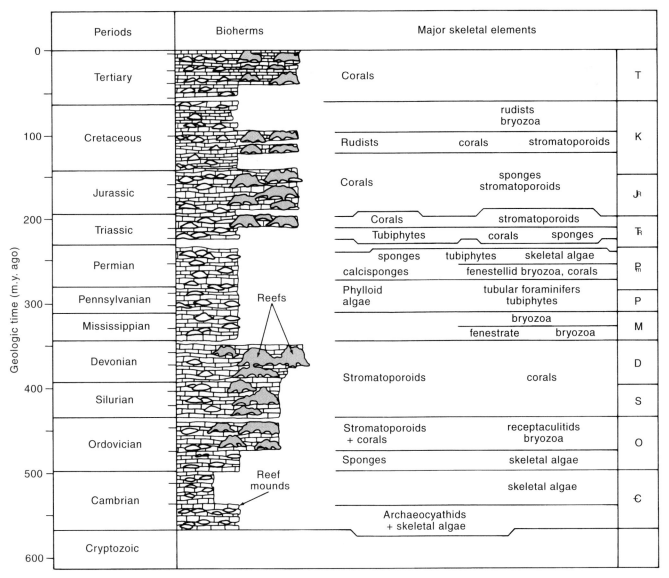

Figure 12.21 History of organic buildups (bioherms) and true reefs. Although the type of reef-building organisms changed through time, the ecological patterns remained very similar. Note the great abundance of reefs (widest profiles) during the middle Paleozoic, middle Mesozoic, and late Tertiary. *(From Noel P. James, 1984, REEFS, in Roger G. Walker (ed.) FACIES MODELS: Geological Association of Canada. Reprinted with permission.)*

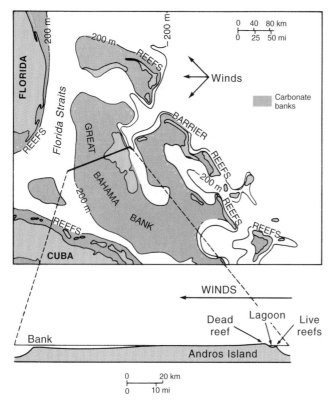

Figure 12.22   Modern reefs and carbonate banks of the Bahama Islands–Florida Keys–Cuba region. The Bahama Islands and the Keys are emerged dead Pleistocene reefs; living reefs lie seaward of them. Reefs have grown almost continually in this region since late Mesozoic time. Carbonate sands cover the bank (see Fig. 10.17). Contour line indicates 200 meters below sea level. *(Figure from Newell & Rigby, 1957. (SEPM) Society of Economic Paleontologists and Mineralogists Special Publication 5. Reprinted with permission.)*

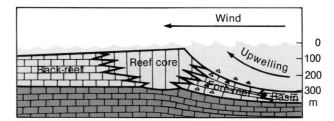

Figure 12.23   Diagram of typical ancient facies seen in many Silurian and Devonian reefs. Back-reef includes clastic skeletal sand and oolite; fore-reef includes blocks derived from windward side of reef.

Bottom-dwelling invertebrates feed upon both kinds of plankton and upon each other. Various swimming invertebrates, fishes, and marine mammals in turn are predatory on the other creatures. Because plants provide the ultimate basis of all animal diets, the food chain must begin in shallow, lighted water, where photosynthesis is possible. Further, agitation tends to make shallow waters well oxygenated for animal respiration, but nutrients must also be available.

Optimum growth conditions for many organisms exist where waves and currents impinge against submerged slopes or escarpments and are deflected upward, as along the northwestern side of the Florida Straits and the eastern, windward edge of the Bahama Banks (Fig. 12.22). The shallow, clear water is well lighted and oxygen-rich. At such locations in warm tropical and subtropical latitudes, crowded and diverse marine animal and algal communities develop, and the growth of skeletal organisms is so great that wave-resistant, moundlike masses of calcium carbonate may be built up to sea level. Cementation of such carbonate deposits can occur within a few thousand years. These are called *organic reefs* and are the most "urbanized" and most ecologically complex areas of the sea floor.

Organic reefs are geologically as well as biologically important, for large masses of carbonate rocks are built by reef-forming organisms. Many ancient reefs exist in Silurian and Devonian strata. They are of special interest because of their significance both environmentally and as traps for petroleum and even some metallic ores. The main reef-core rock commonly is characterized by a distinctive fabric of interlocking skeletal material. Typically it forms a massive, lenticular body surrounded by clearly stratified sediments (Figs. 12.23, 12.24 and 12.25).

Today symbiotic corals and calcareous algae are the most prominent reef builders, but in the past many other organisms contributed as well. These included algae alone prior to the Ordovician Period but stromatoporoids and other sponges, bryozoans, crinoids, brachiopods, and certain molluscs during later Phanerozoic time (Fig. 12.21).

Small, reeflike masses have been formed by algae since Archean time. Primitive filamentous bacteria and algae, most characteristic today of the intertidal zone, have built laminated stromatolitic structures throughout history (see Figs. 8.22 and 10.20), but beginning in early Paleozoic time, more complex organisms became important reef contributors (Fig. 12.21). Animals such as archaeocyathids and sponges began building small reefs in Cambrian and Ordovician times, but why there was such a sudden burst of highly complex animal and algal reef building in Silurian time is not clear. Probably it resulted from a combination of evolutionary changes and existence of shallow-marine conditions that gave some selective advantage to "urbanized" living.

Organic reefs, both modern and ancient, vary greatly in size and form. Some occur as long, linear *barrier reefs* at the edges of shelf or bank areas. Examples include the Florida Keys and Bahamian reefs (Fig. 12.22) and the largest of all modern examples, the 1,700-kilometer-long Great Barrier Reef of Australia (see Fig. 11.26). Also important are the more circular *fringing reefs* developed around islands and the Pacific **atolls** built on submerged prominences.

Reefs encompass many subenvironments characterized by differences in the organic community and sediments. The seaward side is generally rather steep-faced and is constantly battered by waves; it is the zone both of most active growth and of destruction. Fragments of reef rock periodically are torn loose to slide down the reef front into deep water (Fig. 12.23). Thus, an apron of coarse, poorly sorted, angular reef debris (or breccia) characterizes the *fore-reef facies* on one side of a massive reef

Figure 12.24   Underwater photo of a modern reef top near Key Largo, Florida. Living brain coral *(center)* and staghorn coral *(lower)* are surrounded by sand produced by wave erosion of the reef. The water is only a few meters deep, allowing photography under ordinary sunlight. *(Courtesy of George Lynts.)*

Figure 12.25   Aerial view of domelike reef core *(center)* over 100 meters thick in Upper Devonian Peechee Formation, now exposed in the Flathead Range of the Canadian Rockies, southern Alberta. Notice the sharp transition between the massive reef core and the sloping flank beds. *(Courtesy of Brian Pratt.)*

core. The debris shows crude stratification inclined as much as 40° away from the reef front. *Back-reef facies* on the opposite side of a massive core consist chiefly of stratified clastic carbonate sand derived from the reef, although oolite also may be present (see Fig. 10.19), as well as evaporite layers. Many ancient limestone reefs have been more or less converted to dolomite, causing considerable modification of original textures (Box 12.2).

Silurian and Devonian Organic Reefs

The most spectacular middle Paleozoic development occurred in reef-building organisms. The Silurian was characterized by abundant reef complexes reaching 10 meters above the sea floor and stretching up to 3 kilometers (2 miles). In North America, the reefs occurred from Tennessee and Ohio to Alberta to the Arctic (Fig. 12.17). By the Devonian, they were worldwide, and some reefs were immense, spanning tens of kilometers with heights over 100 meters (Fig. 12.25). Silurian reefs were built mostly by tabulate corals, such as the honeycomb coral, *Favosites* (Fig. 11.6F), and the chain coral, *Halysites* (Fig. 12.26C) and by rugosid corals plus stromatoporoids and a variety of unusual sponges (Fig. 12.26A, B). During the Devonian, a slightly different group of corals and sponges dominated. The most familiar were the beehivelike colonial rugosid *Hexagonaria* (Fig. 12.26E) and the lumpy, wrinkled solitary rugosid *Heliophyllum* (Fig. 12.26D), along with stromatoporoids and favositid tabulates.

A classic Devonian reef complex is exposed in Windjana Gorge in western Australia (Fig. 12.27). Part of a belt of reefs almost 350 kilometers long and 50 kilometers wide, the reef is located on the edge of the block-faulted Canning Basin, which had a relief of several hundred meters in the Late Devonian. By accident, the river gorge cuts a perfect cross section through the reef complex. There are dipping fore-reef beds with jumbled reef talus at the base, a massive reef core, and well-bedded back-reef lagoonal deposits (Fig. 12.27A). Even more striking is the zonation of organisms within various facies (Fig. 12.27B). In the open basin facies were free-swimming organisms, such as fish, cephalopods, crustaceans, and conodonts. The reef front was inhabited by animals with an ability to anchor to the unstable substrate, particularly crinoids, sponges, and brachiopods. The reef core was built of the wave-resistant colonial organisms, such as corals, stromatoporoids, and algae. The back-reef sheltered organisms that preferred quiet, protected conditions and could tolerate occasional extremes in temperature and salinity; these included gastropods, bivalves, corals, and stromatoporoids. In the far back-reef were stromatolites and cyanobacterial mats, formed in saline lagoons where no other organisms could thrive.

One of the most interesting aspects of organic reefs is that they preserve the *ecological succession* that produced them. This can be seen by analyzing their structure from base to top (Fig. 12.28). The shifting sands of the sea floor were first colonized by hardy, weedlike species, such as twiglike tabulates; small, solitary rugosids; and well-rooted crinoids. In the intermediate stage, broad, moundlike stromatoporoids, tabulates, and rugosids dominated. Finally, the reef reached maturity with a much

Box 12.2

Origin of Dolomite Rocks

Dolomite is especially common in Silurian and Devonian strata. Magnesium is about three times as abundant in seawater as calcium, yet calcium is added six or seven times faster by rivers (Table 12.1). At present rates of input, calcium abundance should double in only 1 million years and magnesium in 18 million years. Because the composition of seawater seems to have remained fairly constant, at least over later geologic time, both calcium and magnesium must be constantly removed from seawater so that some sort of balance—a dynamic equilibrium—is maintained.

Measurements also show that calcium is less soluble in water than magnesium and, so, is more readily extracted. As noted in Chaps. 6 and 10, the carbon dioxide content of seawater largely controls deposition of calcium but carbon dioxide is sensitive to temperature. With an increase of temperature, carbon dioxide escapes from water to the atmosphere. This causes dissolved calcium to become less soluble, and inorganic precipitation of calcium carbonate may ensue, especially in warm climates. The reaction is as follows, in which loss of CO_2 drives it to the right:

$$Ca^{2+} + 2HCO_3 \rightleftharpoons CaCO_3 + H_2O + CO_2 \uparrow$$

Given an abundant supply of calcium and magnesium from weathering of rocks on land (Table 12.1) the quantity of carbonate sediments that can be precipitated inorganically in the shallow seas is a function of the carbon dioxide content of seawater, which is controlled ultimately by the amount of that gas in the atmosphere. In reality, most limestone, or calcium carbonate, is precipitated by organisms that secrete calcareous skeletons. Thus, biochemical reactions more complex than the simple equation are involved.

A decrease of temperature accompanied by an increase of pressure, as in the deep seas, increases both calcium and magnesium solubilities. Therefore, many of the skeletons found more than 5,000 meters below sea level are composed of silica, which is less soluble there than is calcium carbonate.

Seawater is more nearly saturated with respect to calcium carbonate than more soluble magnesium carbonate even though magnesium is more abundant. Therefore, it follows that special conditions are required to precipitate magnesium carbonate to form dolomite.

It appears that some carbonate sediments could form by direct, inorganic precipitation from seawater through evaporation without the intervention of organisms. Dolomite $[CaMg(CO_3)_2]$ is the most common evaporitic carbonate mineral. Experimental as well as observational evidence indicates that dolomite commonly forms in an alkaline medium, such as seawater, with slight excess salinity and slightly elevated temperatures. Under such conditions, magnesium becomes insoluble and substitutes for some calcium atoms to form either impure calcite or dolomite.

Although a theoretical case can be made for direct precipitation of dolomite under evaporative conditions, there are immense volumes of dolomite rock that show no evidence of having formed that way. Most

| Table 12.1 | Relative Abundance of Four Chief Bases Dissolved in River Water and Seawater |

	River Water, %	Seawater, %
Calcium	73	3
Magnesium	11	10
Sodium	9	84
Potassium	7	3
	100	100

Paleozoic dolomites formed initially as accumulations of calcareous skeletons, yet no known organism secretes the mineral dolomite. Therefore, much dolomite rock must have been converted from limestone after deposition by the following reaction:

$$2CaCO_3 + Mg^{2+} \rightarrow CaMg(CO_3)_2 + Ca^{2+}$$

Full discussion of this reaction is beyond the scope of this book, but one of the reasons that calcium now becomes more soluble than magnesium has to do with relative concentrations of the two elements. If magnesium becomes more than eight to ten times as abundant as calcium, dolomite will form.

Recrystallization of limestone to dolomite rock is called *dolomitization*, and this process occurs in a variety of ways and at various times after initial deposition. Magnesium must have been introduced, and percolation of high-salinity brines through limestone is one important mechanism. In recent years, extensive search of modern sediments has shown that modern dolomite is common in supratidal *sabhka* muds. Evaporative increase of salinity in pore waters, as in the example of supratidal evaporites, causes magnesium to be precipitated. Continuous seepage of brines formed either in *sabhkas* or in restricted lagoons has been confirmed by experiments as well as by field observations of modern sediments. By pumping large volumes of magnesium-rich, hypersaline water through sediments, dolomitization of large volumes of limestone may occur.

Recently other mechanisms have been suggested. One involves repeated mixing of fresh water with seawater in the pores of limestones. Repeated changes of salinity affect relative solubility of magnesium versus calcium and favors dolomitization. Finally, in deep-marine sediments, dolomite forms where anaerobic bacteria reduce any sulfate (SO_4^{2-}); otherwise, sulfate seems to inhibit dolomite formation. Dolomitization remains somewhat mysterious because it is such a complex process.

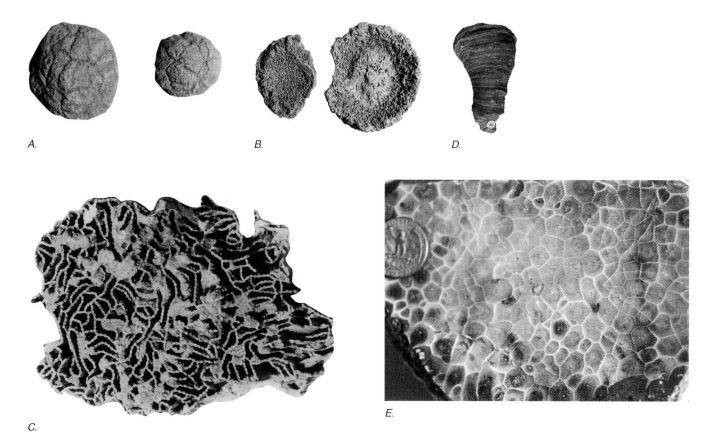

A.   *B.*   *D.*

C.   *E.*

Figure 12.26   Silurian and Devonian reefs were built by a wide variety of colonial organisms. In the Silurian, sponges such as *Astylospongia* (A) and *Astraeospongia* (B, with the star-shaped spicules which are responsible for its name) were particularly common, along with the tabulate "chain coral," *Halysites* (C). In the Devonian, reefs were built by layered stromatoporoid sponges (Fig. 11.6E); favositid tabulate corals (Fig. 11.6F); lumpy rugose corals, *Heliophyllum* (D); and the colonial rugose coral *Hexagonaria* (E). In this sliced and polished specimen, the close packing of the corallites gives them the characteristic hexagonal shape responsible for its name. Water-worn specimens of *Hexagonaria*, known as "Petoskey stone," are the state rock of Michigan. *(D. R. Prothero.)*

greater diversity of animals. A wave-resistant ridge of massive encrusting stromatoporoids protected the reef on the seaward side. Behind this ridge in quiet water lived a diverse assemblage of tabulates, rugosids, and crinoids. The protected lagoon sheltered not only crinoids and delicate bryozoans but also a diverse group of mud-dwelling brachiopods, clams, snails, and trilobites.

Because reefs grow in the surf zone only a meter or so below sea level, the depth of surrounding water can be estimated in some fossil examples by tracing a reef-front stratum down into basin deposits and noting the vertical difference in level. In Fig. 12.23, we saw that the depth was approximately 300 meters, which is reasonable for many of the Devonian barrier reefs of western Canada and for some Silurian examples in the midwestern states. The presence of fossil marine algae may also provide a depth indicator, because photosynthesis is possible only at depths less than 200 meters. Thus, we see that ancient reefs are unusually rich in clues to past environments. Besides providing indications of depth and wind direction, their regional distribution is also a clue to ancient warm latitudes, as we shall see shortly.

Marine Evaporite Deposits

Evaporites are important chemical sediments, which are commonly associated with carbonate rocks. These sediments require the evaporation of large volumes of water in order to concentrate brines enough to cause precipitation of various salts. Like carbonate rocks, evaporites of different ages provide clues about subtle variations of seawater composition through time. Evaporites are known from Proterozoic time to the present, but Silurian and Devonian strata contain the earliest extensive examples in North America. First we shall examine the two major environments in which evaporites form today. Then we shall discuss an important association of evaporites with Silurian reefs and dolomites in Michigan.

Restricted-Basin Evaporites

Today evaporites form in central Asia in the Kara-Bogaz Gol (Gulf), a restricted embayment of the Caspian Sea. The gulf's waters are replenished more or less continuously across a shallow

A.

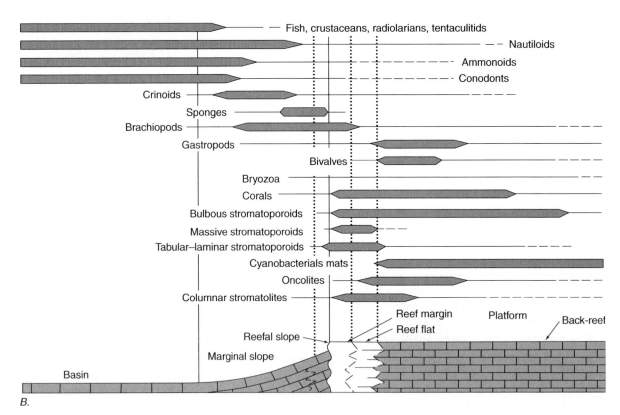

B.

Figure 12.27   A: Photograph of the spectacular Devonian reef complex on the southeast wall of Windjana Gorge in western Australia. Notice the steeply dipping flank beds on the reef front *(left)*, the massive reef core *(center)*, and the flat-lying beds of the back-reef lagoon *(right)*. (Compare with Fig. 12.23.) B: Ecological zonation of animals found near the reef. Free-swimming organisms, such as fish, crustaceans, nautiloids, ammonoids, and conodonts are found in the open waters of the basin. Attached animals, such as crinoids, sponges, and brachiopods, colonized the unstable reef slope. The massive reef core was built by corals and a variety of shapes of stromatoporoid sponges. The quiet, muddy waters of the back-reef lagoon were exploited by a variety of animals, including gastropods, burrowing bivalves, and delicate bryozoa, as well as stromatoporoid sponges. There were also occasional cyanobacterial mats, which formed stromatolites, as well as rolled balls of ripped-up algal mat known as oncolites. *(Phillip E. Playford.)*

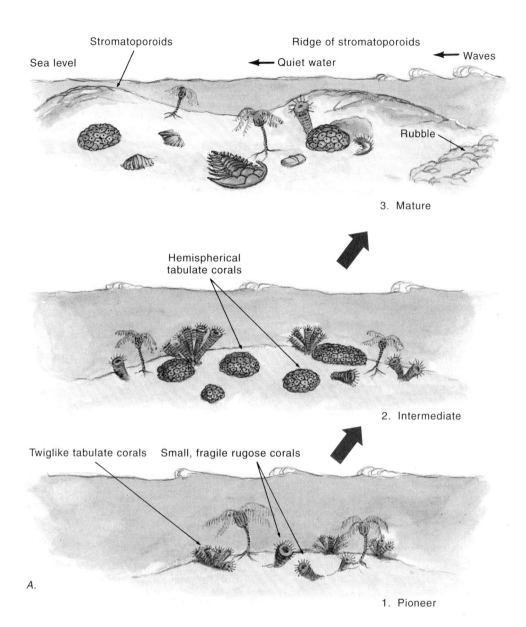

Stromatoporoids

Ridge of stromatoporoids

Sea level

← Quiet water

Waves →

Rubble

3. Mature

Hemispherical tabulate corals

2. Intermediate

Twiglike tabulate corals   Small, fragile rugose corals

A.

1. Pioneer

Figure 12.28 *A:* Typical example of ecological succession in a Devonian reef community. In the pioneering stage *(bottom),* organisms that can colonize the shifting sands, such as crinoids, bryozoans, small rugosid corals, and twiglike tabulate corals, are the first to become established. As they stabilize the bottom sediments and modify the currents around them, larger colonial corals can become established *(middle).* Eventually a mature reef community grows over the earlier colonizers, with massive corals and stromatoporoid sponges sheltering more delicate organisms from the pounding of the waves *(top). B:* Ecological succession can be seen in this cross section of a Silurian reef from Buckland, Ohio. Note that the lowest part of the reef core is mostly colonizers, such as crinoids, bryozoans, and brachiopods. These are succeeded in higher beds by stromatoporoids and eventually by corals. Meanwhile, crinoids become established on the reef flank once the corals and stromatoporoids have established a wave-resistant core. *(After R. H. Shaver et al., 1983,* Field trips in midwestern geology, *vol. 1, p. 159.)*

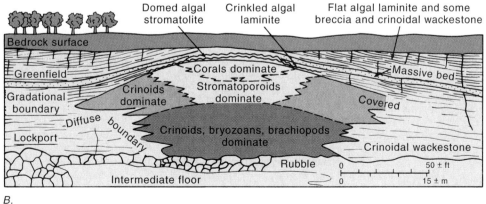

Domed algal stromatolite   Crinkled algal laminite   Flat algal laminite and some breccia and crinoidal wackestone

Bedrock surface

Greenfield

Corals dominate

Massive bed

Crinoids dominate

Stromatoporoids dominate

Gradational boundary

Covered

Diffuse boundary

Lockport

Crinoids, bryozoans, brachiopods dominate

Crinoidal wackestone

Rubble

Intermediate floor

0   50 ± ft

0   15 ± m

B.

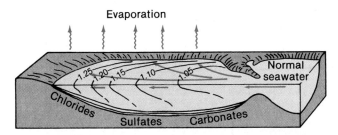

Figure 12.29   Equilibrium evaporite basin model illustrating precipitation due to restricted circulation. Normal seawater flows continually into the restricted basin at a rate closely balanced by evaporation; dense brine sinks and precipitates different evaporite minerals according to concentration. Numbers indicate water density. *(After Briggs, 1957, by permission of Michigan Academy of Science.)*

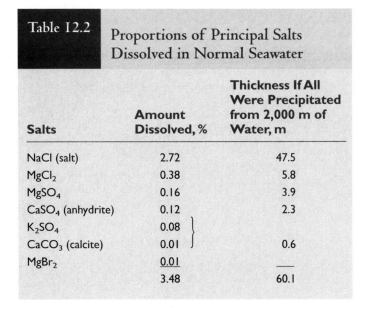

Table 12.2	Proportions of Principal Salts Dissolved in Normal Seawater	
Salts	**Amount Dissolved, %**	**Thickness If All Were Precipitated from 2,000 m of Water, m**
NaCl (salt)	2.72	47.5
MgCl$_2$	0.38	5.8
MgSO$_4$	0.16	3.9
CaSO$_4$ (anhydrite)	0.12	2.3
K$_2$SO$_4$	0.08	
CaCO$_3$ (calcite)	0.01	0.6
MgBr$_2$	0.01	—
	3.48	60.1

bar, but because the region is very arid, evaporation causes continuous precipitation of salts from very concentrated, dense waters at the bottom of the gulf (Fig. 12.29). These hypersaline brines cannot escape to the Caspian Sea because of the restricting bar. As a result, a steady state exists between precipitation of evaporite sediments at the bottom and replenishment and evaporation of water at the top. As long as this equilibrium persists, evaporite sedimentation continues. The Kara-Bogaz Gol provides a model to help us understand evaporite deposition in ancient restricted sedimentary basins and in various depths of water.

Seawater today contains about 3.5 percent dissolved salts, of which sodium chloride (NaCl) is the most familiar and most abundant. Theoretically the complete evaporation of seawater should produce sequential deposition of a series of evaporite minerals in reverse order of their relative solubilities in water (Table 12.2). Laboratory experiments performed as early as 1849 showed the theoretical salt precipitation sequence to be expected. It is the rule to find incomplete sequences, however, indicating that a variety of events may disturb the evaporation process. For example, cessation of precipitation or solution of earlier salts results from seasonal temperature or humidity changes or from destruction of circulation barriers to allow dilution of brines by normal seawater. Furthermore, a lateral sequence of evaporite mineral facies is common, as shown in Fig. 12.29, with a continuous flow of water undergoing constant evaporative precipitation of carbonate first, sulfates farther along, and chlorides at the farthest end of the flow. Under appropriate conditions, evaporite precipitation may occur very rapidly.

Supratidal *Sabhka* Evaporites

Supratidal deposition (slightly above high-tide level) also is of major importance. For example, salt flats several kilometers wide border intertidal lagoons of the Persian Gulf. The flats, called *sabhkas* in Arabic, are but a few centimeters above average high tide and occasionally are flooded during onshore wind storms. Seawater fills the pores of sediments beneath the salt flats and

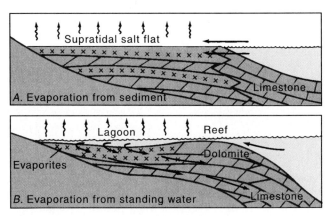

Figure 12.30   A: Evaporites and dolomite formed by infrequent washovers onto supratidal salt flats (called *sabhkas* in Arabic) and some seepage from normal marine waters at right. *(Adapted from Alling et al., 1965, SEPM Special Publication 13.)* B: Dolomization of limestone (CaCO$_3$) by seepage of very saline brines derived from restricted lagoon *(left)* where evaporation increases brine density. Dense, magnesium-rich brines are "pumped" through limestone toward right; calcium is carried away and replaced by magnesium to produce dolomite [CaMg(CO$_3$)$_2$]. *(Adapted from Adams and Rhodes, 1960, Bulletin of the American Association of Petroleum Geologists, v. 44, pp. 1912–1920; by permission.)*

then becomes abnormally saline due to evaporation in the arid climate. Gypsum and dolomite are then precipitated within the salt-flat muds (Fig. 12.30, Box 12.2). Under conditions of balanced subsidence and deposition, significant thicknesses could accumulate of such evaporite-bearing strata formed essentially at sea level. Lateral shifts of shoreline would cause transgressive or regressive migration of the evaporite facies.

Postdepositional Changes

Evaporites undergo many changes after deposition. These involve chiefly reactions with ground water because of the great solubility of the evaporite minerals. The most common change is simple solution of evaporite layers, causing subsidence or collapse of overlying and interstratified insoluble sediments into solution caverns. Collapse causes fragmentation of adjacent nonevaporite rocks, producing a jumble of breccia. Another typical change is the hydration of anhydrite to form gypsum near the surface—that is, the combination of two molecules of water with each molecule of calcium sulfate. As a result of such changes, we only rarely see original evaporite sequences in surface outcrops. Undisturbed, natural evaporite deposits are largely confined to the subsurface below the depth of penetration of fresh ground water. Indeed, because of near-surface solution, many evaporite deposits were unknown until deep drilling penetrated hitherto unexplored basins.

The Michigan Basin

Prominent Silurian sediments that include evaporites and reefs are found in New York, Ohio, Michigan, and western Canada (Fig. 12.17). These deposits are the basis for major plaster and chemical industries; potassium, important as fertilizer, is obtained from evaporites in Saskatchewan. Evaporites characterize chiefly Upper Silurian strata that formed immediately after the maximum reef development shown in Fig. 12.18.

In Michigan, especially, subsidence accelerated in a circular area called the Michigan Basin (Fig. 12.31). As suggested in Chap. 11, probably there was a direct relationship to thrust loading of the eastern margin of the craton. In the Michigan Basin, up to 1,500 meters of Upper Silurian sediment was deposited, chiefly dolomite rock $[CaMg(CO_3)_2]$, but with as much as 750 meters of rock salt (NaCl) and anhydrite-bearing ($CaSO_4$) strata. These evaporite strata required concentration of brines from an immense volume of seawater. If the Silurian sea were as saline as that of today, it would have required evaporation of the equivalent of a column of seawater nearly 1,000 kilometers deep (about 600 miles) to deposit 750 meters of evaporite strata! Certainly the water was never 1,000 kilometers deep over Michigan; rather, it was apparently fairly shallow at all times. Therefore, we postulate continual replenishment of water as evaporite sediments were precipitated over a subsiding basin floor—much like the modern Gulf of Kara-Bogaz.

In many ancient basins, circulation of seawater was restricted by a variety of possible causes, so that evaporative salts were precipitated. Though arid climate is necessary, excessively hot temperatures are not required, for dry winds could accomplish the evaporation effectively (evaporites are precipitating today in Antarctic lakes). But if we assume both moderately warm temperatures and dry winds, then evaporites can be explained readily wherever oceanic circulation in shallow seas was impaired. In Michigan, the fringing reef complexes that began developing

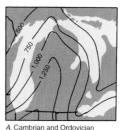 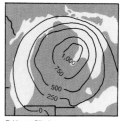

A. Cambrian and Ordovician   B. Upper Silurian

Figure 12.31   Two thickness maps showing that the circular Michigan Basin became well defined in Silurian (time contours in meters). The basin suddenly subsided as much then as it had throughout all of earlier Paleozoic time. *(Adapted from Cohee, 1948, U.S. Geological Survey Oil and Gas Chart 33; Alling and Briggs, 1961, Bulletin of the American Association of Petroleum Geologists, v. 45, pp. 515–547; by permission.)*

around the basin in Middle Silurian time apparently caused the restriction of circulation that led to later evaporite precipitation (Fig. 12.32). Also, a slight lowering of sea level apparently occurred near the end of Silurian time (Fig. 11.13), which would have produced islands and shoals around the basin margin and further restricted circulation. In New York, circulation was restricted not only by shoals on the west and south (Fig. 12.32) but also by land to the east. Red-colored fine clastic sediments derived from that land are found intimately interstratified with the marine evaporites there.

Paleoclimate and Paleogeography

An abundance of iron oxides and calcium sulfate evaporites in certain Silurian and Devonian sediments indicates a strongly oxidizing atmosphere. Important evaporite sediments prove a relatively high evaporation potential over large parts of North America, so one is tempted to conclude that the climate was warm. Over Devonian land in eastern North America, apparently there was moderate humidity, as suggested by forests that cloaked the lowlands there (Fig. 12.11). The presence of great organic reef complexes and rich, very diverse marine fossils in both Silurian and Devonian marine strata suggests warm, shallow, agitated seas by analogy with modern shallow tropical seas (Fig. 12.20).

Devonian land plants are similar the world over, suggesting that climate was essentially uniform. Wide distribution of richly fossiliferous middle Paleozoic marine carbonate rocks, and especially the great latitudinal spread of fossil reefs suggest subtropical conditions for North America, Europe, Siberia, and Australia (Fig. 12.33). Devonian evaporites also closely parallel the reefs. Based upon this reasoning, the average climate of the earth through time probably has been milder and more homogeneous than it is today. The present certainly is not a very good key to the past in terms of climate.

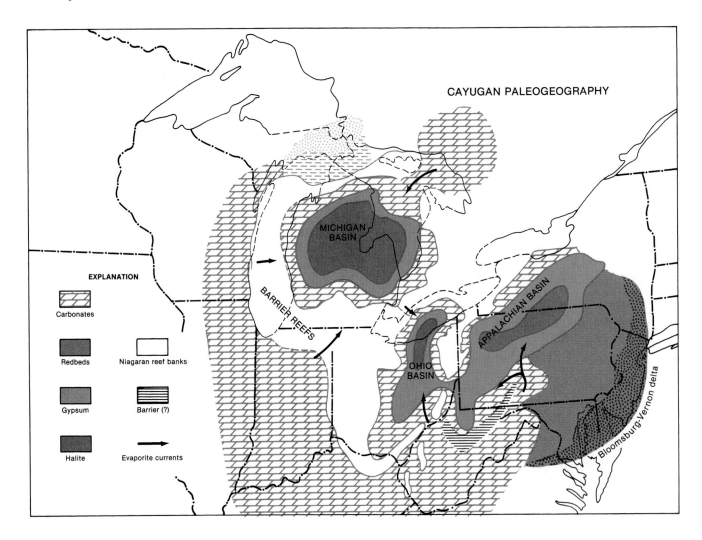

Figure 12.32   Late Silurian paleogeography of the Michigan–New York–Ohio evaporite basin. Barrier reefs restricted marine circulation into the basins; evaporites occur in basin centers. (*After Alling and Briggs, 1961*, Bulletin of the American Association of Petroleum Geologists, *v. 45, pp. 515–547; by permission.*)

Devonian Strata of the North American Craton

Regression and Transgression Again

In Late Silurian and Early Devonian times, marine deposition became restricted to a few basins connected by narrow seaways and by the marginal orogenic belts. In the latter regions, marine Upper Silurian and Lower Devonian strata are perfectly conformable. Over most of the craton, however, regressive facies developed, and finally most of the craton emerged as a lowland for a few million years to end the Tippecanoe Sequence. Younger, transgressive Devonian strata of the Kaskaskia Sequence rest unconformably upon a variety of older rocks, including even Cryptozoic ones on several arches.

The Early Devonian unconformity is a rare major cratonwide break, like that beneath the Ordovician St. Peter Sandstone, which is why it was chosen as the boundary between the second (Tippecanoe) and the third (Kaskaskia) major cratonic sequences illustrated in Figs. 4.22, 11.13, and 11.14. Pure quartz sandstone

south of the eastern Great Lakes (Oriskany of Fig. 12.39) resembles the lower Paleozoic quartz sandstones in every respect and, like them, was deposited at the margin of a transgressive sea. It was derived from erosion of those same older sands, reflecting the repetitive nature of early cratonic history.

Apparently the Ordovician and Silurian carbonate rocks originally deposited with the older sands were dissolved by ground water from the uplifted arches.

Cratonic Basin Deposition

Warping of the North American crust was widespread in Early Devonian time, and practically all the craton and even portions of the orogenic belts were above sea level and being eroded to produce profound changes on the continent. Basins and arches became more sharply delineated than ever before. The net result of warping and regression was much differential erosion. As the Kaskaskia transgression occurred near the middle of the Devonian Period, marine deposition resumed first in basins, which were most easily flooded, and gradually encroached upon arches (Fig. 12.34). As a result, Upper Devonian marine strata are more widespread than

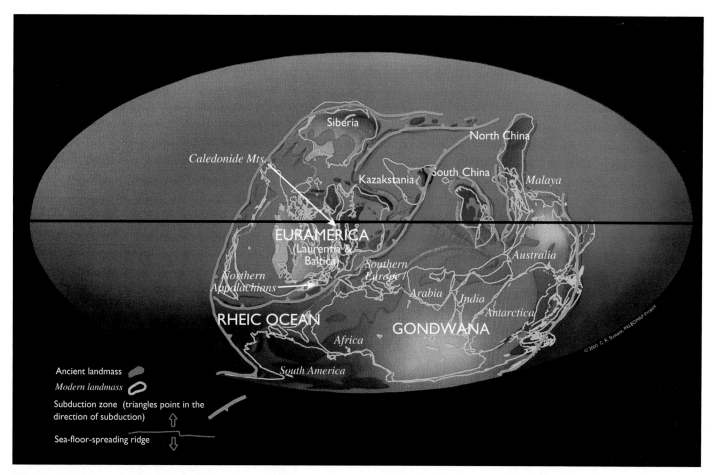

Figure 12.33   Map showing the distribution of continents and climatic belts in the Devonian. *(www.scotese.com)*

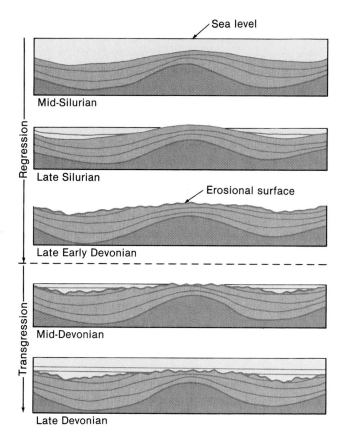

Figure 12.34   Effects of Devonian regression and transgression of cratonic arches and basins. Note differential erosion beneath the Kaskaskia unconformity and differential deposition above. Subsurface information is mandatory to make such reconstructions in basin areas.

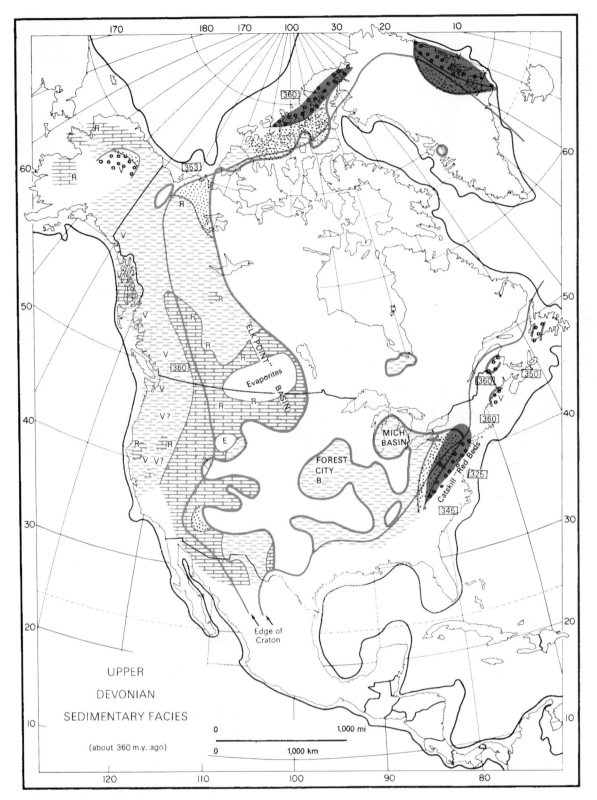

Figure 12.35
Upper Devonian sedimentary facies. Note importance of reefs and evaporites in western Canada and isotopic dates for widespread granite rocks. (See Box 10.2 for symbols and sources.)

UPPER
DEVONIAN
SEDIMENTARY FACIES
(about 360 m.y. ago)

Edge of Craton

Middle Devonian ones (Fig. 12.35). In many areas, younger carbonate sediments were deposited upon older ones, making recognition of the unconformity difficult. Besides this major cratonwide unconformity, as many as fourteen lesser breaks are recognized and are thought to reflect smaller sea-level changes.

The Michigan Basin continued to subside, and more evaporites were deposited. The Williston, or Elk Point, Basin farther west also received important evaporites. An immense barrier-reef complex developed around its margin and extended to the northwest in Canada (Fig. 12.36). Apparently it formed along a zone

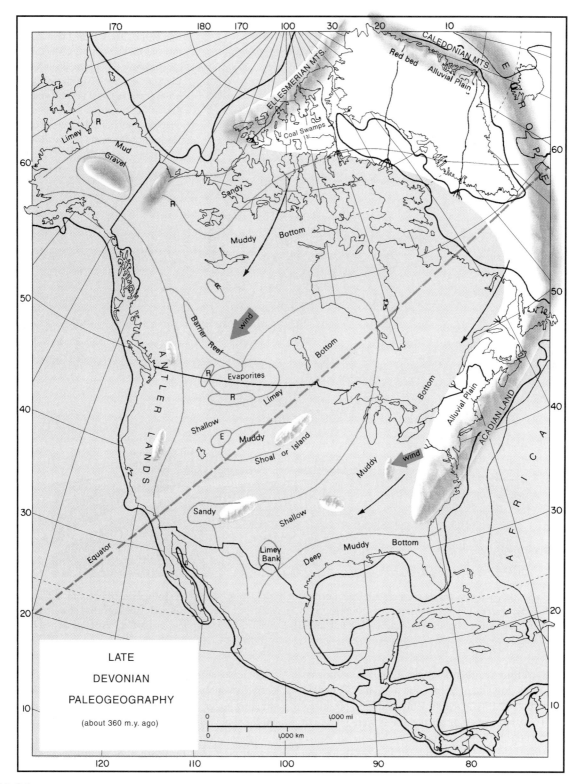

Figure 12.36   Late Devonian paleogeography. Note the importance of marginal tectonic lands in the east and north. Wind direction in the east is inferred from volcanic ash distribution; in the west, from reef facies. Note collision of Europe with northeastern North America and approach of Africa.

Figure 12.37   Outcrop on the northeast side of the Ozark Dome near Louisiana, Missouri, showing Ordovician (O), Silurian (S), and Mississippian (M) strata. C is Mississippian Chattanooga Shale, which overlies a subtle but widespread regional unconformity (top of the Kaskaskia sequence). Presence of the unconformity here is apparent only from the absence of Devonian fossils. *(R. H. Dott, Jr.)*

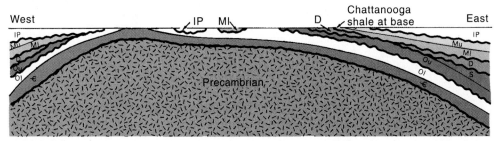

Figure 12.38   Section across a typical cratonic arch, the northern Ozark Dome, Missouri, showing several unconformities as they appear today (vertical scale exaggerated). Uniform thickness of Cambrian and Ordovician strata across the arch indicates that upwarping occurred after their deposition. For reasons listed in text, initial warping is dated as Early Devonian. Later warpings produced Mississippian and Pennsylvanian unconformities.

of deeper water upwelling westward against a shallow carbonate shelf. Devonian reefs long had been known in the Canadian Rocky Mountains (Fig. 12.25), and petroleum had been produced from some since 1920, but reefs beneath the plains were unknown until about 1947, when drilling encountered phenomenal petroleum reserves trapped therein. The discovery triggered one of the continent's greatest oil booms and provided a wealth of information about previously unknown buried rocks. Indeed, existence of the basin itself was hardly appreciated before that time.

Cratonic Arches

Because the stratigraphic record on arches is less complete than that on basins, and because some of the unconformities in the arch record are very subtle (Fig. 12.37), it is more difficult to date arch warpings than to date basin warpings. At least some Cambrian and

Ordovician strata are present on most arches, but Silurian and Early Devonian rocks are absent from all crests (Fig. 12.38). This situation suggests that warping and erosion began in Silurian time, but there are other clues to consider, too. If shoals or islands were present in Silurian time, sandy or muddy facies would be expected around them, yet only Silurian carbonate rocks are found (Fig. 12.17). Moreover, the paleogeologic map of the mid-Devonian unconformity reveals that the present distribution of Silurian rocks on the craton was controlled mainly by Early Devonian erosion. Therefore we conclude that many of the arches were not very important before Devonian time (for example, Fig. 12.38). Because of the number of arches truncated by Devonian strata, we conclude that this was a time of unusually severe cratonic deformation. We tentatively attribute such deformation to the side effects of mountain building in eastern North America.

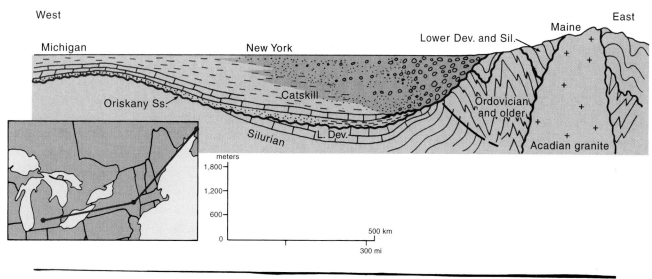

West

East

Maine

Lower Dev. and Sil.

Michigan

New York

Oriskany Ss.

Catskill

Ordovician
and older

Silurian   L. Dev.

Acadian granite

meters
1,800

1,200

600

0

500 km

300 mi

Section without vertical exaggeration

Figure 12.39   Restored cross section of Devonian rocks in the eastern United States showing the effects of the Upper Lower Devonian (pre-Oriskany) unconformity, the Acadian orogeny, and the Catskill clastic wedge. Oriskany sandstone was derived from erosion of older sandstones in the central craton. Note superimposition of Acadian on older Taconian folding *(upper right)*.

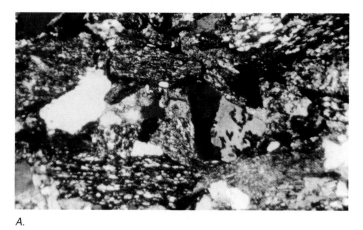

A.

B.

Figure 12.40   A: Microscopic photograph in polarized light of an Upper Devonian (Catskill) sandstone from southeastern New York. Immature sand composed largely of metamorphic rock fragments (e.g., quartzite and schist) derived from erosion of the Acadian uplift in present New England. Larger grains are about 2 mm in diameter. *(R. H. Dott, Jr.)* B: Thick sequence of black Devonian shales at Taughannock Falls, near Ithaca, New York, formed when muddy *flysch* deposits shed from the Catskill deltas covered most of western New York. *(D. R. Prothero.)*

The Acadian Orogeny in the Appalachian Belt

Evidence of Devonian Orogeny

Whereas the entire Devonian sequence averages less than 300 meters thick in the craton, it is nearly 1,800 meters thick in the Appalachian region of eastern Pennsylvania and New York (Fig. 12.39). In New England and southeastern Canada, Devonian rocks include considerable lava and volcanic ash, and throughout most of the Appalachian belt, later Devonian strata include red sandstone, conglomerate, and shale (Figs. 12.35 and 12.40)—part of a *Catskill clastic wedge* named for the Catskill Mountains of southeastern New York.

Like the older red clastic *molasse* facies of Late Ordovician age in the same region, Catskill sediments also coarsen toward the east and, so, must reflect elevation and erosion of another prominent land there, much like modern New Guinea (Fig. 11.26). We know that this land was composed mainly of slightly older fossiliferous Paleozoic rocks. The Devonian tectonic land encompassed the area of the older eroded Taconian land (compare Figs. 11.16 and 12.36) but also included some new microcontinent accretions. Thus, a major Devonian orogeny was superimposed, or overprinted, upon the older one. In the maritime region of southeastern Canada and in northern New England, granites of Devonian age and unconformities related to this orogeny are well displayed. Indeed, the famous granites of New Hampshire, which give the state its nickname, "The Granite State," are Devonian in age. This mountain-building episode has been named the **Acadian orogeny** for the old French colonial name for that region.

More granite and regional metamorphism (Fig. 12.41) developed during the Acadian event than during the Taconian event, and probably some thrust faults also were active. The Acadian orogeny, therefore, was a more severe disturbance of the earth's crust and represents the greatest orogeny for the Appalachian orogenic belt.

Dating the Orogeny

At several scattered localities in the Acadian region, angular unconformities with Mississippian strata resting variously upon deformed Lower Devonian or older rocks that were intruded by granitic plutons serve to date the orogeny. Also, Acadian granites have been dated extensively by isotopic methods (Fig. 12.35). In Nova Scotia and New England, many granites yield dates of from 360 to 330 million years ago. In Maryland a large mass known as the Baltimore Gneiss has been dated by the K-Ar method, using biotite mica, as from 350 to 300 million years (Early Mississippian). But zircon from the same rock yielded U-Pb dates of from 1,100 to 700 million years (late Proterozoic)! These were among the first **discordant isotopic dates** discovered for a single rock (see Chap. 5). Such discordances have proved to be so common as to be expected now in complex orogenic belts that have suffered several overprinted orogenies (Fig. 12.41).

Discordant dates reflect two major events in the history of the Baltimore Gneiss. This rock apparently was formed first as a granite during the widespread Grenville orogeny (1,000 to 700 m.y. ago), which affected the entire southeastern margin of present North America and produced the complex metamorphic and igneous rock basement beneath the Paleozoic Appalachian orogenic belt (see Fig. 8.11). The Maryland granite was again heated and deformed during the Acadian orogeny, at which time biotite's K-Ar isotopic clock was reset by argon leakage to 350–300 million years ago, which dates the time of cooling of the metamorphosed granite. Zircon is very resistant to temperature changes; therefore, it has retained U-Pb isotopic ratios that reflect the date of original crystallization of the granite. Biotite is very sensitive to heating in excess of 250°C during metamorphism, and ^{40}Ar, being a gas, tends to leak from the mica crystal lattice. The result is resetting of the mica's isotopic clock; all the ^{40}Ar now present in the biotite of the Baltimore Gneiss has accumulated from decay of ^{40}K since Acadian metamorphism and cooling of the rock below 250°C, the blocking temperature for K-Ar in biotite (Chap. 5).

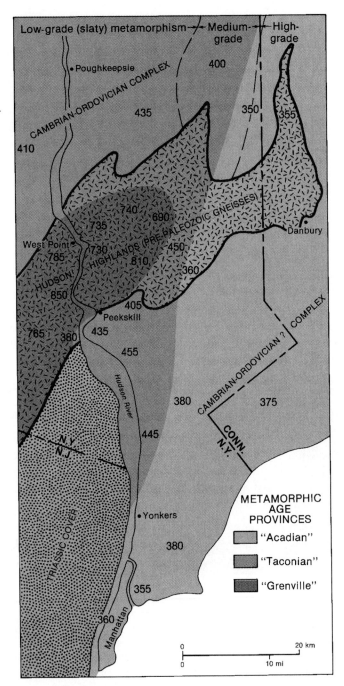

Figure 12.41   Superimposed, or overprinted, metamorphic events reflected in discordant K-Ar isotopic dates. Stratigraphic relations show that Hudson Highlands gneisses are all Cryptozoic; in the west, isotopic dates reflect their true age (850 to 730 m.y.). Fossiliferous Cambrian-Ordovician rocks to the north yield metamorphic dates reflecting the Taconian orogeny (435 to 400 m.y.), but some Cryptozoic gneisses also yield Taconian dates due to remetamorphism. Finally, *all* the rocks in the eastern half of the map, regardless of original age or type, yield overprinted Acadian dates (380 to 350 m.y.). *(After Long, 1962, p. 998; and Long and Kulp, 1962, Bulletin of the Geological Society of America, v. 73, pp. 980 and 983.)*

Isotopic evidence, unconformities, and the thick Middle and Late Devonian Catskill clastic wedge together indicate that Acadian mountain building occurred during Late Devonian and Early Mississippian time through an interval of perhaps 30 to 40 million years (370 to 330 m.y. ago), a significant span even to the geologist.

The Catskill Clastic Wedge

Depositional conditions for the Late Devonian clastic sediments in the eastern United States were virtually identical to those of Late Ordovician sediments of the same region. But the total volume and coarseness of the Catskill rocks exceed those of the Ordovician, and the pebbles and sand grains of the Catskill are chiefly composed of metamorphic and granitic rock fragments, feldspar, mica, and quartz (Fig. 12.40). The red color is due to the presence of a small percentage of iron oxide between the grains.

The Catskill sediments form a clastic wedge deposited as a vast alluvial coastal plain that sloped gently westward from the eroding Acadian mountains to the epeiric seashore. This plain was on the order of 300 to 500 kilometers wide at its maximum extent. At its eastern side, it was built of gravels and sands spread by rivers from the foot of the mountains. Many river channel deposits are evident in the eastern Catskill facies. Along the western shoreline, however, finer sediments were deposited on river deltas, and black marine muds and some thin sands were spread beyond the deltas onto the craton by turbidity currents to form a *flysch* facies (Fig. 12.35), as in Late Ordovician time (see Fig. 11.16). Volcanic ash also blew westward again over the eastern craton.

In southeastern New York, the red sediments contain remains of a spectacular buried fossil forest. Tree stumps 30 centimeters in diameter are found buried in their rooted positions (Fig. 12.10G), but some trees were rafted down rivers to the epeiric sea, there to be buried offshore in black shales. Identical land plants occur in Devonian red beds in eastern Greenland, and coal is present in Arctic Canada. The oldest known vertebrate land animals also occur in the Greenland strata (Fig. 12.13).

The Chattanooga Black Shale Enigma

Not all the clastic wedge was composed of nonmarine red beds. Black muds, which first began accumulating in the region west of the Catskill deltas in Middle Devonian time, spilled westward onto the craton in Late Devonian time (Figs. 12.35 and 12.38), just as in Ordovician time. By the end of the Devonian, such muds were deposited all across the craton from New York to Nevada and north into Canada. This unprecedented expanse of mud continued to accumulate on the craton for a few million years into Early Mississippian time, resulting in the Chattanooga Shale of Fig. 12.37 (so-named for exposures near Chattanooga, Tennessee). Erosion of numerous areas across the continent (Fig. 12.36) gave rise to this vast quantity of clay. The shale contains much organic carbon, traces of uranium and phosphate minerals, and iron sulfide. Slow deposition in an oxygen-poor environment uninhabitable by carbonate-secreting organisms is indicated, yet epeiric seas more typically were agitated and oxygen-rich; therefore, the Chattanooga is a great puzzle.

Many geologists have postulated unusually deep, stagnant water for the deposition of the cratonic black shale, as is indicated for the older graptolitic facies in orogenic belts. This idea may be supported by an apparent high sea level—possibly as much as 200 meters above present—as indicated by the worldwide sea-level fluctuation curve of Fig. 4.22. (Such a rise is also invoked as a cause of the disappearance of Late Devonian reefs.) Within only a few million years, deposition returned to more normal, shallow-water carbonate conditions (late Early Mississippian), which may reflect a subsequent fall of sea level.

Another possible explanation of the shale is that excessive worldwide salinity developed as a result of extensive evaporative conditions during Late Silurian and earlier Devonian times. This would have caused density stratification of the epeiric sea, resulting in a stagnant-water layer over the sea floor. We shall discuss causes of black shale deposition again in Chap. 15; however they formed, such organic-rich muddy sediments are important sources of petroleum (Box 12.3).

The Caledonian Orogenic Belt

A Paleozoic mountain belt, called Caledonian, has long been recognized in Britain and Norway along the northwestern margin of the European craton (Fig. 4.11A). It was from studies of this belt that James Hutton formulated his revolutionary eighteenth-century views of the earth. The Caledonian belt is a twin of the northern Appalachian–East Greenland system, and it contains similar rocks, which reveal a similar history, as noted in Chap. 11. Episodes of volcanism and deformation are recorded in the Caledonian belt at least as early as Late Cambrian time; more severe ones in the Ordovician correspond roughly to the Taconian, whereas the culminating **Caledonian orogeny** is Silurian—only about 10 million years before the Devonian Acadian event of America. Even major cratonic events show some correlation between the two continents. Basin and arch warpings on the two continents as well as major transgressions and regressions correspond to a remarkable degree (see Fig. 11.14), suggesting again that cratonic structure was also controlled by plate movements.

The first geologists to study eastern Greenland were Europeans, which was a happy circumstance, because what they found was so familiar that they might easily have forgotten they had left home. (It was in eastern Greenland that Alfred Wegener, father of the continental drift theory, lost his life in a blizzard in 1930.) We noted in Chap. 11 the great similarity of lower Paleozoic strata and fossils in America with those of northwestern Scotland (see Fig. 11.35). Devonian and Carboniferous sediments are also virtually identical in northwestern Europe and the Appalachian belt; many species of fossil land plants and fish in these rocks are the same on both continents as well.

Structural Symmetry

Structurally the two margins of the North Atlantic Ocean are almost mirror images. In eastern Greenland, as in the Appalachian belt farther south, thrust faulting carried thick orogenic-belt rocks westward onto the craton. In Norway, thrusting was eastward onto the European craton. In both cases, intense metamorphism

Box 12.3

Origin of Petroleum

The chemistry of petroleum leaves no doubt that organisms are the primary raw materials for oil and gas. The remains of practically any aquatic and land plants or animals can produce petroleum hydrocarbon compounds but the floating microscopic planktonic organisms are most important. It is the remains of such organisms that are largely responsible for the relatively high organic carbon content of fine-grained black sediments such as the Chattanooga Shale, which are the principal *source rocks* for petroleum. The origin of petroleum is therefore a special sedimentary process.

Either deposition must occur in anaerobic environments to avoid destruction of the organic compounds by oxidation or so much organic matter must be deposited that complete oxidation is impossible. Molecules in the living organisms are not the same as petroleum molecules; therefore, the former must undergo chemical changes after burial. These changes, which are collectively called maturation, are chemically so sluggish that temperatures of 80 to 140°C for liquid oil and 140 to 300°C for gas are required. Elevated pressure speeds the process and there is growing evidence that certain bacteria also can contribute to the alteration of the original compounds. Deposition of source sediments in environments with unusually high temperatures such as a newly rifted continental margin, will accelerate maturation.

Given appropriate conditions for maturation, petroleum next must migrate and then be trapped to provide an economically recoverable quantity. Migration occurs by the squeezing of new petroleum from compacting source muds into permeable reservoir strata. Because oil and gas are lighter than water, which is also present in the sediments,

they rise through the carrier stratum either to be intercepted by some trap or to escape to the surface through oil seeps (Fig. 12.42). Not only must there be appropriate structural and sealing conditions for entrapment, but the timing of formation of a given trap relative to time of maturation and migration is also critical. Many fine traps were formed too late and now contain only water.

Buried middle Paleozoic organic reefs are among the most prolific petroleum traps, but sandstones are also important. The famous Drake well drilled near an oil seep in 1859 at Titusville, Pennsylvania, tapped a petroleum-saturated Devonian sandstone that was folded during the Acadian orogeny.

As a by-product of practical oil seeking, science has been rewarded with a wealth of otherwise inaccessible subsurface geologic data, which adds greatly to the accuracy of our historical analysis, especially for deeply subsided basins. And such greater accuracy is essential if we are to find new petroleum reserves for our energy hungry world. Even with the best possible skill and luck, however, it is a losing battle in the long run, for petroleum—like all mineral resources—is a nonrenewable resource in terms of the human time scale. Our best estimates indicate that, 100 years from now, the world's petroleum reserves will have been largely exhausted. Even by the year 2050, we shall need other energy sources to supplement petroleum. Coal and nuclear fuels, with all their environmental hazards are most likely to be the first supplements, but they are nonrenewable, too, and will probably be exhausted by the year 3000. The problem of energy reserves for the future is discussed further in Chap. 17.

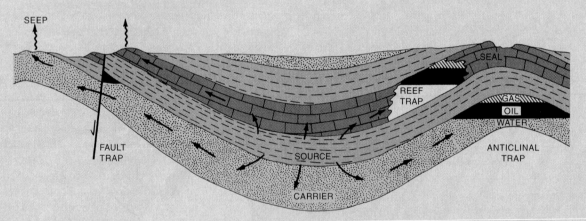

Figure 12.42   The origin and entrapment of petroleum in a sedimentary basin. Organic-rich source sediment was buried and heated sufficiently to enhance generation of petroleum, which then migrated upward through permeable carrier strata until it encountered an anticlinal, reef, or fault trap or escaped at the surface in a seep. (Compare Fig. 10.11).

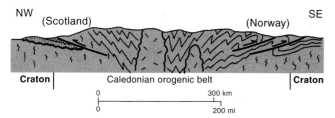

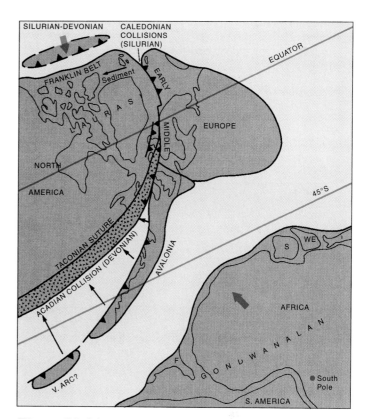

Figure 12.43 Simplified restored cross section of the Caledonian orogenic belt, northwestern Scotland to southern Norway. Rocks of the orogenic belt were thrust toward *both* margins; the belt apparently was symmetrical, being an *intercratonic* one formed between colliding Greenland and Europe.

Figure 12.44 Paleotectonic map showing final closure of the proto-Atlantic (Iapetus) Ocean in middle Paleozoic time. Successive north-to-south collisions of North America with Europe and Avalonia caused the Caledonian and Acadian orogenies. Note the approach of the African part of Gondwanaland from the south and a microcontinent from the north. F—Florida, S—Spain, WE—western Europe, I—Italy.) (Compare Fig. 11.36.) *(Adapted from W. S. McKerrow, 1988, Geological Society of London Special Paper 38, pp. 405–412.)*

and large granitic batholiths developed in what is now the seaward side of the orogenic belts. Locally a definite bilateral symmetry can still be seen across these belts today. In northwestern Scotland, Caledonian thrust faulting carried metamorphic rocks northwest over flat, lower Paleozoic carbonates that were part of eastern Greenland at that time; eastward thrusting occurred nearby in Norway (Fig. 12.43). In Newfoundland, a central zone of intensely deformed Ordovician and Silurian oceanic rocks lies between two zones of mildly deformed continent-margin rocks. From these clues, as well as from paleomagnetic data, we can reconstruct the continents as they must have looked before and after a Late Devonian collision (compare Fig. 11.36 with 12.44). This collision resulted in the final destruction of the old proto-Atlantic Ocean, as shown in Fig. 12.45.

Catskill–Old Red Sandstone Facies

The Devonian red clastic-wedge deposits of eastern Greenland and northwestern Europe (see Fig. 4.11A), like those of the eastern United States, reflect the rapid erosion of high mountains (Fig. 12.46). In Fig. 12.45, we see that the Old Red Sandstone of Europe was the mirror image of the American Catskill red beds. The sediments were transported by rivers in both directions away from a single great Caledonian-Acadian mountain range. Approximately equal volumes occur on both sides. The identity of land plant and fish fossils contained therein is readily explained as well by the collision.

Intercratonic Orogenic Belts

We now see more clearly that the *intercratonic* type of orogenic belt noted in Chap. 7 results from *continent-continent collision* (Figs. 12.44 and 12.45). Moreover, it follows that many presently *marginal orogenic belts,* such as the Appalachian, were intercratonic at sometime and have been subsequently dismembered by continental drift. It is only since the development of plate-tectonic theory, however, that such belts could be adequately explained.

Fig. 12.45 summarizes the probable evolution of the northern Appalachian-Caledonian system in terms of lithosphere plate movements; the scheme shown applies to any intercratonic belt in a general way. Of particular importance in unraveling these complex belts is the identification of suture zones of crushed oceanic

rocks (typically the mafic ophiolites), which mark the old boundary between converging plates. Such a zone is generally all that remains of the ancient ocean basin that was consumed between two converging continents. Besides the suture, the central zone characteristically has deformed volcanic arc rocks and granitic plutons, and shows high-temperature metamorphism. Bounding the central zone on both sides are intensely deformed sedimentary rocks thrust outward toward adjoining cratons (Figs. 12.43 and 12.45). Marginal orogenic belts formed between converging oceanic and continental crust do not show much bilateral symmetry (see Fig. 7.32).

Collision with Europe cannot account for the United States part of the Acadian belt. Apparently this part was compressed by the collision of one or more microcontinents and arcs, much as we inferred for the Taconian event (Fig. 12.44). The main microcontinent was the Avalon terrane, consisting of portions of coastal New England and the Canadian Maritime Provinces (Fig. 11.37), which rifted away from Africa in the Ordovician (Fig. 11.36). Its collision with North America during the Acadian orogeny is another example of collage tectonics.

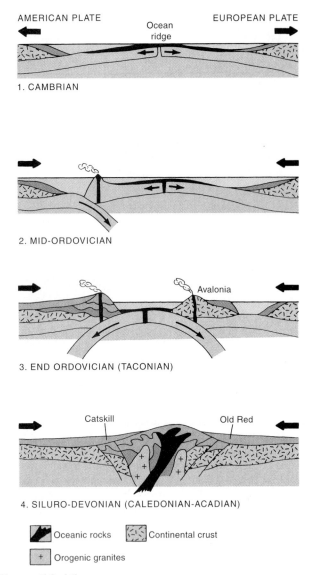

AMERICAN PLATE   EUROPEAN PLATE

Ocean
ridge

1. CAMBRIAN

2. MID-ORDOVICIAN

Avalonia

3. END ORDOVICIAN (TACONIAN)

Catskill   Old Red

4. SILURO-DEVONIAN (CALEDONIAN-ACADIAN)

Oceanic rocks   Continental crust

+ Orogenic granites

Figure 12.45   Hypothetical evolution of northern Appalachian-Caledonian orogenic system due to collisions of continents, arcs, and microcontinents. *(1)* Rifting had broken up a Proterozoic supercontinent to produce passive trailing edges on both sides of the proto-Atlantic (or Iapetus) ocean. *(2)* In mid-Ordovician time, an arc formed and approached North America. *(3)* In Late Ordovician, the arc collided to produce the Taconian orogeny. *(4)* In Silurian and Devonian times, collisions of Europe and the Avalonia microcontinent with North America produced the Caledonian and Acadian orogenies. Note symmetry of Catskill and Old Red clastic wedges on either side of an intercratonic orogenic system with an oceanic suture bisecting it. (Compare Figs. 4.11 and 12.44).

Figure 12.46   Coarse, angular Devonian ("Old Red") conglomerates, Solund District, western Norway, one of several areas of thick, nonmarine sediments deposited in downfaulted basins following the Caledonian orogeny. Note the heterogeneous composition representing varied older metamorphic rocks. *(Courtesy of Tor H. Nilsen.)*

Mountain Building in the Arctic and Western Cordillera

In the *Franklin orogenic belt* of Arctic Canada and in northern Alaska, Silurian and Devonian conglomerates and sandstones (Fig. 12.35) reflect uplift associated with Caledonian-Acadian mountain building. Caledonian thrust loading depressed northern Greenland in Silurian time (Fig. 12.44). Turbidity currents carried much sand and gravel derived from the Caledonian mountains westward along a deepwater (Franklin) trough (Fig. 12.47); they were deposited on a huge, elongate deep-sea fan. The Franklin and Caledonian belts even merge in northeastern Greenland, but structural relationships with Alaska are unclear, even though facies and history were similar there. Indeed, it is not even clear that most of Alaska was attached to North America yet, and the same is true for many parts of the present western edge of Canada and the United States. We shall learn more about this problem in later chapters.

Granitic rocks of latest Devonian and earliest Mississippian ages are widely scattered in the Arctic (Fig. 12.35). Local ultramafic and volcanic rocks of early and middle Paleozoic ages

Figure 12.47   Folds in Cambrian to Silurian shelf-margin shale and carbonate sequence *(dark)* and overlying Silurian turbidity-current deposits *(light)*. Kap Wohlgemuth, Naresland, northern Greenland, an extension of the Franklin orogenic belt of Fig. 12.44. Intense compression from north *(left)* toward the craton *(right)* occurred during the Ellesmerian orogeny. *(Photo: A. K. Higgins © Geological Survey of Denmark and Greenland.)*

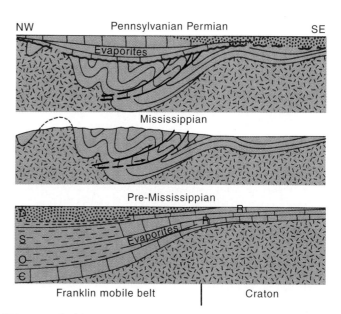

Figure 12.48   Sequential restored cross sections across the Franklin belt of Arctic Canada showing deformation produced by the Mississippian Ellesmerian orogeny. Note also distribution of reefs and evaporites. (Not to accurate scale.) See Fig. 12.44 for location.

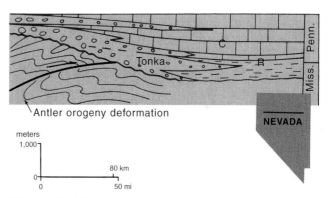

Figure 12.49   Evidence of the Antler orogeny in northern Nevada. Folding and thrusting began in Late Devonian and continued through Mississippian time; note westward overlap by later Paleozoic conglomerate as lands were eroded (C and R denote index fossil zones.) *(After R. H. Dott, Jr., 1964, Kansas Geological Survey Bulletin 169.)*

present on the northernmost tip of Canada apparently reflect oceanic or microcontinent rocks crushed against the northern edge of the continent. This crushing caused the folding and thrust faulting of Paleozoic strata southward against the craton during the Late Devonian–Early Mississippian **Ellesmerian orogeny** (Fig. 12.48). This event probably resulted from collision between either an arc or a microcontinent and northern North America, but proof is lacking (Fig. 12.44).

Disturbances also began to affect the Cordilleran region of western Canada and the United States for the first time. Important chert-bearing conglomerates were deposited in northern Alaska and northwestern Canada during Late Devonian and Early Mississippian times. In Nevada, at least 1,000 meters of Mississippian and Early Pennsylvanian sandstone and chert-rich conglomerate lies unconformably upon folded and thrust-faulted early Paleozoic oceanic sediments and pillow basalts (Figs. 12.49 and 11.27). Together with granites yielding Late Devonian ages in western Canada (Fig. 12.35), these rocks indicate an episode of mountain building called the **Antler orogeny.** Unlike the orogenies just discussed, however, Antler deformation and metamorphism were mild; there was little volcanism and no extensive nonmarine red-bed portion to the resulting clastic wedge. Here most of the sediments eroded from modest-sized tectonic lands were deposited below sea level by turbidity currents and at sea level on small deltas. Those sediments are characterized by an abundance of chert and quartz fragments resulting from erosion of uplifted early Paleozoic deep-marine cherts and sandstones deposited originally beyond the western edge of the continent (see Fig. 10.3).

What could have caused this peculiar Antler orogeny? There is no evidence of collision with another continent here, but in northwestern California there are Devonian rocks that appear to be remnants of a volcanic arc that collided with western North America to cause the orogeny. Their location during Devonian time is not yet satisfactorily proven, however.

Physical and Organic Evolution

The remarkable correlation between the explosive evolution of Devonian land plants and land animals on the one hand and the marked increase in land area as a result of mountain building on the other, hardly seems sheer coincidence (Fig. 12.50). It suggests an outstanding example of mutual feedback between physical and organic evolution. Epeiric seas reached their maximum size in later Ordovician, Silurian, and Middle Devonian times, but mountain building in eastern North America and in Europe provided diverse new habitats for invasion by neophyte land organisms, which were quick to fill the ecological vacuums. But why did these organisms not invade the

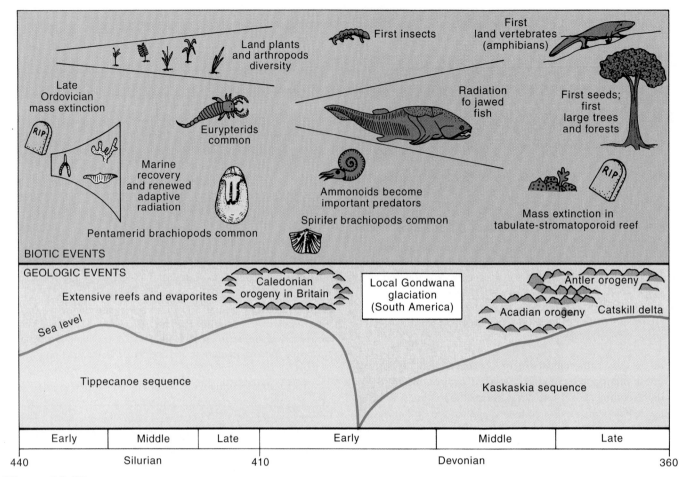

Figure 12.50   Summary time line of events of the Silurian and Devonian.

Late Ordovician Taconian land, which apparently was similar to Siluro-Devonian ones? It has been speculated that, for a considerable time after the atmosphere began to accumulate free oxygen, there was lethal ozone at ground level. Much more oxygen presumably had to accumulate before the ozone layer physically rose to a level safe for advanced forms of land life. Possibly this did not occur until mid-Paleozoic time, although that is hard to believe. In any case, it is clear that, from Silurian time onward, the development of land life was closely tied to physical history.

Another example of feedback between physical and organic history is provided by a comparison of Paleozoic fossils from North America and northern Europe. In Cambrian time, the marine invertebrates living on opposite sides of the proto-Atlantic (Iapetus) Ocean were so different as to comprise two distinct faunal provinces (Fig. 11.35). During middle Paleozoic times, the faunas began to show some mixing, and the faunas of Avalonia and northern Europe became homogeneous in Devonian time as the microcontinent merged with Europe (Fig. 12.44). After the Devonian collision of North America and Europe-Avalonia, the nonmarine floras and jawless fishes of the Catskill and Old Red *molasse* were virtually identical. Such decreases of diversity reflect decreasing provinciality, which is explained by the approach and final connection of the continents. We shall discuss even more dramatic examples of biogeographic feedback in Chaps. 13 and 14.

Summary

Mid-Paleozoic Life

- Ordovician diversification of marine animals carried into mid-Paleozoic time. Brachiopods, crinoids, and corals flourished, whereas trilobites and nautiloids declined. Animals such as stromatoporoids, colonial corals, and sponges became important reef builders in Silurian time and constructed the most widespread reefs ever in the Devonian Period.

- Explosive evolution of fish and invasion of land by plants and animals highlighted mid-Paleozoic time. Vascular plants and invertebrate animals conquered land no later than the end of the Silurian period, and lobe-finned fishes gave rise to amphibians in Late Devonian time. Forests covered coastal lowlands by the Late Devonian.

The Silurian and Early Devonian Continent

- Middle Paleozoic epeiric seas covered practically all the North American craton except during an Early Devonian regressive episode, which marked the beginning of the third major cratonic sequence (Kaskaskia).
- Carbonate sedimentation dominated the Silurian and Devonian epeiric seas.
- Large organic reefs were built widely by marine animals in warm, clear, agitated seas where upwelling of nutrient-rich water occurred. By analogy with the present oceans, all these rocks formed in shallow tropical seas.
- Evaporites [$CaSO_4$, NaCl, and some dolomite—$CaMg(CO_3)_2$] form where evaporation exceeds replenishment of normal seawater in two main settings—namely, restricted basins and supratidal *(sabhka)* flats.
- Dolomitization converts many limestones to dolomite rock by the replacement of Ca^{2+} by the Mg^{2+} ions carried in water percolating through them. Such replacement is facilitated by the flushing of pores with hypersaline brines, by repeated alternation from fresh to hypersaline pore waters, or by the presence of sulfate-reducing bacteria in pore water.
- Paleoclimate was warm, as indicated by evaporites and richly fossiliferous carbonate rocks with many reefs.
- Modern analogues suggest a tropical setting for middle Paleozoic diverse faunas, reef-bearing carbonate rocks, and evaporites; this is consistent with paleomagnetic evidence for North America. Present wide latitudinal distribution of these rocks is explained by later plate movements.
- Warping of basins and arches was more pronounced than ever before and apparently was related to extensive mountain building around the margins of the continent.

Orogenic Belts

- The Acadian orogeny (mid-Devonian–Early Mississippian) resulted in the Catskill clastic wedge, angular unconformities, widespread metamorphism, and extensive granites with isotopic dates of 330 to 360 million years. Several discordant dates record tectonic overprinting upon the effects of Taconian and older events. The Acadian event was caused by the collision of the Avalonia microcontinent with eastern North America—another example of collage tectonics.
- The middle Paleozoic Caledonian orogeny of East Greenland and northwestern Europe was linked closely with the Acadian event. It was caused by a continent-continent collision, which created an intercratonic orogenic belt in contrast to marginal orogenic belts, such as the Taconian.
- The Arctic Franklin belt experienced the Ellesmerian orogeny, whereas western North America had the mild Antler orogeny—both in Late Devonian to Early Mississippian time. Causes of these relatively milder events are unclear but probably are collisions of microcontinents or arcs with the northern and western margins of North America.
- Increasing tectonism created many and varied land areas. The first appearance of nonmarine fossil plants and animals during middle Paleozoic time suggests cause-effect feedback between new environmental opportunities and biological exploitation. Other organic results of physical changes include the decreasing provinciality of organisms as continents approached each other and Late Devonian marine extinctions, perhaps caused by widespread anaerobic conditions in epeiric seas.

Readings

Alberta Society of Petroleum Geologists. 1964. *Geological history of western Canada.* Calgary: Author.

Alling, H. L., and L. I. Briggs. 1961. Stratigraphy of Upper Silurian Cayugan evaporites. *Bulletin of the American Association of Petroleum Geologists,* v. 45, pp. 515–547.

American Association of Petroleum Geologists. 1973. *Arctic geology.* Tulsa: American Association of Petroleum Geologists Memoir 19.

Dinely, D. L. 1984. *Aspects of a stratigraphic system: The Devonian.* New York: Wiley.

Gray, J., and W. Shear. 1992. Early life on land. *American Scientist,* 80:444–57.

Haller, J. 1971. *Geology of the East Greenland caledonides.* New York: Wiley-Interscience.

House, M. R. 1979. *The Devonian system.* London: Palaeontological Association Special Papers.

Kahn, P. G. K., and S. M. Pompea. 1978. Nautiloid growth rhythms and dynamic evolution of the earth-moon system. *Nature,* 275:606–11.

Lovell, J. P. B. 1977. *The British Isles through geological time.* London: Allen and Unwin.

Oswald, D. H., ed. 1968. *Proceedings of the international symposium on the Devonian System.* Calgary: Alberta Society of Petroleum Geologists.

Pannella, G., C. MacClintock, and M. N. Thompson. 1968. Paleontological evidence of variations in length of synodic month since Late Cambrian. *Science,* 162:792–96.

Schreiber, C. B., and H. H. Lincoln, eds. 1985. *Sixth International Symposium on Salt,* v. 1. Alexandria, Salt Institute.

Scrutton, C. T. 1964. Periodicity in Devonian coral growth. *Palaeontology,* 7:552–58.

Sloss, L. L. 1972. *Synchrony of Phanerozoic sedimentary-tectonic events of the North American craton and the Russian platform.* Montreal: 24th International Geological Congress Proceedings, sec. 6, pp. 24–32.

Strand, T., and O. Kulling. 1971. *Scandinavian Caledonides.* New York: Wiley-Interscience.

Surlyk, F., and J. M. Hurst. 1983. Evolution of the early Paleozoic deep-water basin of North Greenland—Aulacogen or narrow ocean? *Geology,* 11:77–81.

Wells, J. W. 1963. Coral growth and geochronometry. *Nature,* 197:948–50.

Williams, H., and R. D. Hatcher, Jr. 1982. Suspect terranes and accretionary history of the Appalachian orogen. *Geology,* 10:530–36.

Wilson, J. L. 1975. *Carbonate facies in geologic history.* New York: Springer-Verlag.

(D. R. Prothero.)

Chapter 13

Late Paleozoic History

A Tectonic Climax and Retreat of the Sea

Here about the beach I wandered,

nourishing a youth sublime

With the fairy tales of science,

and the long results of time.

Alfred Tennyson

MAJOR CONCEPTS

▶ The great shallow seas of the Devonian persisted into Early Carboniferous time, producing the last great deposits of limestone (dominated by millions of crinoids) in the continental interior. After a mid-Carboniferous retreat, the seaways returned again, but now they accumulated coal-bearing deltaic deposits with a distinct cyclicity. Such global cycles were probably the result of sea-level changes due to geologically rapid fluctuations of glaciers in the Southern Hemisphere.

▶ The late Paleozoic was dominated by the collision of various continents to form the supercontinent of Pangea. In the Late Carboniferous and Permian, the eastern and southern margins of North America collided with Africa and South America, producing a great mountain belt from the Appalachians to the southern Rockies, which shed much clastic sediment into the interior of North America. By the Permian, the supercontinent was fully assembled and was uplifted so much that marine seaways virtually disappeared from the continental interior; nonmarine sedimentation (especially red beds, floodplain shales, dune sands, and evaporites) prevailed instead.

▶ The late Paleozoic sea floor continued to feature ammonoids, crinoids, bryozoans, and corals, although molluscs and brachiopods that favored the muddy bottoms of the Late Carboniferous were particularly abundant. The end of the Permian was marked by the greatest mass extinction in earth history, which may have wiped out 95 percent of marine species. A combination of extreme climatic fluctuations, runaway greenhouse warming, and massive volcanic eruptions may have been responsible.

▶ Land plants diversified further in the late Paleozoic, forming giant coal swamps in the Late Carboniferous. By the mid-Permian, conifers and other primitive seed plants adaptable to drier conditions replaced the lowlands coal forests and invaded uplands for the first time.

▶ Mid-Carboniferous time marked the appearance of the first vertebrates that could lay their eggs on land. They included two main groups: (1) the synapsid lineage, which dominated the forests of the late Paleozoic and eventually gave rise to mammals and (2) the reptiles, which eventually led to turtles, snakes, lizards, crocodiles, and dinosaurs during the Mesozoic.

We have seen in Chaps. 11 and 12 that, during early and middle Paleozoic time, the North American craton experienced repeated and widespread transgressions and regressions by epeiric seas. A Late Mississippian* regression followed by an Early Pennsylvanian transgression resulted in a cratonwide unconformity that defines the base of the Absaroka Sequence. This unconformity is recognized on other continents, so it probably had a worldwide cause. We have seen that, except at times of considerable mountain building, carbonate rocks formed widely on the craton (Fig. 13.1). During major orogenic episodes, as during Ordovician and Devonian times, clastic wedges engulfed orogenic belts and influenced cratonic sedimentation.

The late Paleozoic was a time of revolutionary changes (Fig. 13.2). Whereas the North American stratigraphic record had previously been dominated by marine conditions, during Carboniferous and Permian times (355 to 251 m.y. ago), it was increasingly influenced by tectonic disturbances that raised much of the craton above sea level. In this chapter, we stress the implications of this increasing tectonism in terms of structure, sedimentation, and life history.

Early Carboniferous strata represent a transition from the marine conditions of the middle Paleozoic to the more nonmarine conditions of Late Carboniferous and Permian time. We shall see that, by the end of the era, practically the entire continent stood above sea level and was inhabited by wide-ranging plants and animals. What had been a paradise for denizens of the sea was to become a land of swamp forests, lurking reptiles, and gigantic insects. Wholesale extinctions of many groups of shallow-marine invertebrates and explosive proliferation of land organisms accompanied the shrinkage of shallow seas.

A general increase in structural unrest also resulted in drastic changes in sedimentation. Near the end of the Mississippian Period, deposition of typical carbonate rocks almost ceased on the craton. Pennyslvanian and Permian strata contain great volumes of clastic sediments derived from peripheral mountains and from large uplands within the craton. Pennsylvanian strata contain our largest coal deposits, which formed under special climatic and topographic conditions favoring luxuriant growth of swamp forests. The coals occur in repetitive strata, showing a conspicuous pattern of geologically rapid alternations of rock types. Because this pattern is of worldwide significance, we believe it resulted from global fluctuations of sea level that were much more rapid than previous fluctuations.

Africa and South America, fused together as part of the supercontinent Gondwanaland, completed their collision with North America and Europe. This last compression resulted in extensive thrust faulting and uplift in the southern Appalachian belt—the Appalachian orogeny—and across Europe—the Hercynian orogeny. Simultaneously the collision of Siberia with eastern Europe formed the Ural Mountains, while several microcontinents collided and fused with southern Siberia to become a nucleus for China.

At the end of the Paleozoic Era, all the continents were more or less interconnected to form the super-supercontinent Pangea. Total land area was very large, and this situation greatly influenced the evolution of land life by facilitating dispersal of organisms and providing a great diversity of nonmarine ecological niches.

Mississippian Rocks

Last Widespread Carbonates

The Mississippian Period (355 to 322 m.y. ago) was characterized by the last widespread carbonate-producing epeiric sea in North America. Carbonate sediments spread over most of the craton and

Figure 13.1 This spectacular cliff of the Mississippian Redwall Limestone in the eastern Grand Canyon is typical of the widespread limestones representing enormous areas of shallow, warm, limey epeiric seas all over the world in the Early Carboniferous. Note the cavern, rafts, volleyball net, and river runners at the bottom for scale. *(D. R. Prothero.)*

---

*The Carboniferous Period named in Europe has been subdivided into the Mississippian and Pennsylvanian Periods in North America. Thus, when we discuss North American events, we use the latter terms, but for the rest of the world, only *Carboniferous* is appropriate.

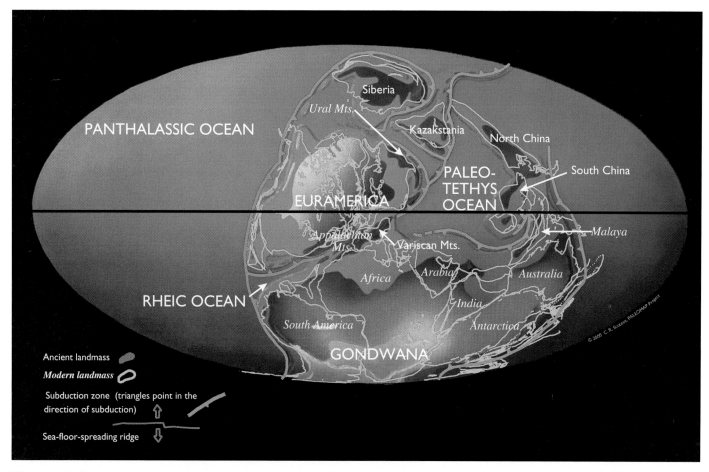

Figure 13.2   Global paleogeography for the Early Carboniferous. *(www.scotese.com)*

large portions of the orogenic belts (Fig. 13.3). Limestones rich in crinoid fragments predominated (Figs. 13.4 and 13.5). They display conspicuous clastic textures. Breakage and sorting of fossil fragments, oolite, cross-stratification, ripple marks, and scoured structures are common. These features point to close analogies with the modern Bahama Banks (see Fig. 12.22). Organic reefs occur widely in Mississippian strata from the southern United States to northern Alaska, but they were smaller than the great barrier reef complexes of middle Paleozoic time. Evaporite deposits were somewhat less important as well.

Sedimentary and Tectonic Changes

During Late Mississippian time, distinct changes in cratonic sedimentation began. There was a change in the composition of terrigenous sediments. Prior to Late Mississippian time, only very pure quartz sandstones and shales composed overwhelmingly of a single clay mineral (illite) were deposited. Beginning in later Mississippian time, heterogeneous sands bearing feldspar and mica, as well as quartz, began to appear on the southeastern part of the craton (eastern United States), and they became more wide-

spread during subsequent periods. Simultaneously, the mineralogy of associated shales became more complex.

At the end of the Mississippian Period, a major regression drained the entire craton, and the earliest Pennsylvanian seas were confined to marginal orogenic belts. As a result, a major unconformity occurs beneath Pennsylvanian strata over the craton (see Fig. 11.13). A great deal of warping and faulting occurred before the sea returned to the craton in mid-Pennsylvanian time. The transgression over this great unconformity began the next great cratonic succession—the Absaroka Sequence.

Most of the late Paleozoic sedimentary changes indicate exposure as a result of erosion of new, more heterogeneous source rocks. Facies and paleocurrent patterns, together with mineralogy, indicate that the source of much of the clastic material lay in eastern Canada and was composed chiefly of igneous and metamorphic rocks long buried beneath early Paleozoic strata; some material also was derived from lands in the Appalachian belt. Erosion during various earlier periods of regression gradually had stripped most of the Paleozoic veneer from eastern Canada, laying bare significantly large areas of old, crystalline Cryptozoic basement. Comparison of Mississippian and Pennsylvanian

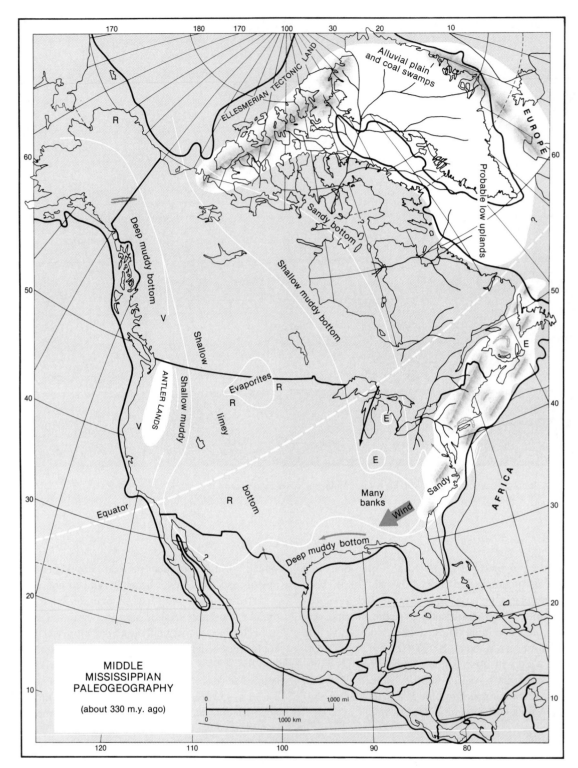

Figure 13.3   Middle Mississippian paleogeography. Note Acadian uplifts in eastern North America, Antler uplifts in the Cordillera, Ellesmere uplifts in the Arctic, and approach of Africa from the southeast. Compare Figs. 12.36 and 13.8 to appreciate the transitional nature of the Mississippian.

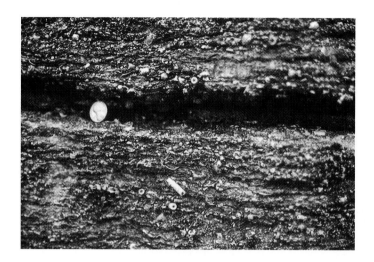

Figure 13.4   Densely packed crinoidal limestone from the Chester Group (Upper Mississippian) in southern Illinois. Note the abundance of partial stems, as well as disarticulated crinoid columnals with their star-shaped central hole. Some units, such as the Lower Mississippian Burlington Limestone, have a total volume of 300×10^{10} cubic meters, which represents the skeletal remains of approximately 28×10^{16} individual crinoid animals! After the crinoids died, their stems disarticulated and were moved around as coarse carbonate sand, just as seashell fragments litter many beaches today. *(D. R. Prothero.)*

A.

B.

Figure 13.5   The first thing that should come to mind when someone says "Mississippian" is "limestone." All over the world, most Mississippian deposits are typified by thick sequences of limestone. These include not only the amazing crinoidal limestones of the upper Mississippi Valley in Missouri, Iowa, and Illinois (Fig. 13.4), from which the Mississippian got its name, but also thick limestone cliffs all over the western United States, such as the Redwall Limestone of the Grant Canyon (Fig. 13.1), the Madison Limestone in the northern Rockies (*A,* shown here at Gates of the Mountains, Montana), or the Pahasapa Limestone (*B*) in the Black Hills of South Dakota, from which Wind Cave and Jewel Cave National Monuments were etched. *(D. R. Prothero.)*

Figure 13.6   Throughout the Rocky Mountains, there are thick sandstones with gigantic cross-beds, representing giant Pennsylvanian and Permian sand dunes. However, these are frequently interstratified with marine limestones, indicating repeated inundations by a shallow sea. It is as if a quarter of the Sahara Desert were flooded twenty-five to thirty times. This can be seen at the top of the Grand Canyon, where the Permian Coconino Sandstone is overlain by the Toroweap and Kaibab Limestones, or *(above)* in the Permian De Chelley Sandstone at Canyon de Chelley, Arizona. *(Courtesy of L. Gordon Medaris.)*

facies with earlier facies and with paleogeographic maps indicates gradual uptilting of eastern North America. As a result, epeiric seas of late Paleozoic time covered less and less of the eastern part of the craton and erosion cut deeper and deeper. Over the western portion of the craton and adjacent eastern Cordilleran orogenic belt, pure quartz sands were still being deposited intermittently throughout late Paleozoic time (Figs. 13.6 and 13.7). They were derived largely from erosion of earlier Paleozoic sediments on the west-central craton in Canada, where the Cryptozoic basement was not yet widely exposed. Lands raised in Late Devonian time in the Arctic and in the western Cordilleran belt continued to be eroded during Mississippian time (Fig. 13.3). By Pennsylvanian time, these lands had been mostly eroded away, but new ones appeared within the southern craton (Fig. 13.8).

Late Paleozoic Repetitive Sedimentation

Sedimentary Cycles

Beginning in Late Mississippian and continuing through Early Permian times, the strata deposited over the North American craton and inner parts of the orogenic belts displayed a striking repetitive pattern, which is present to varying degrees in late Paleozoic strata on other continents. Upper Mississippian deposits in the eastern states show sandstone-shale-limestone sets repeated several times vertically. Illinois geologists have shown

that the sandstones and shales represent deltaic deposits formed by river systems flowing from the southeastern Canadian region. In Pennsylvanian time, influxes also came from the Appalachian region (Fig. 13.8).

Practically all Pennsylvanian strata on the North American continent show some kind of repetitive pattern, but the most striking occurs in coal-bearing sequences. At least forty late Paleozoic cycles are known, many of which can be traced widely over the southern part of the craton. A typical cycle, or **cyclothem** (Figs. 13.9 and 13.10), begins at the base with cross-stratified sandstone and conglomerate resting unconformably upon older strata. Channel structures, fossil logs, and relatively poor grain sorting indicate that these coarse sediments were formed by river processes. The middle of the cycle contains coal and plant-bearing shales, whereas the upper part generally contains brackish water or marine fossiliferous shales and limestones. Many such cyclothems occur vertically, stacked upon one another. Lateral variations within them are just as important as vertical alternations, for both record drastic environmental shifts through time.

In the central states, the upper marine rocks of the cycles are best developed, whereas the coal-bearing, nonmarine rock types predominate in the east. Meanwhile, the western part of the craton was covered by a vast sand-dune desert (Fig. 13.6). At least twenty times, this desert was flooded by a shallow sea, which deposited thin but widespread marine limestones—again in a repetitive pattern, but involving different facies than farther east.

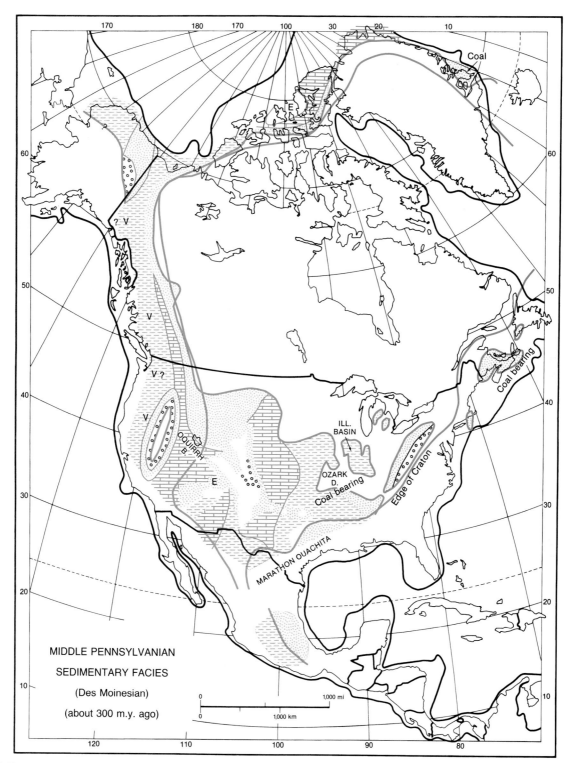

Figure 13.7 Middle Pennsylvanian sedimentary facies; note eastern coal-bearing regions and western coarse clastic areas reflecting uplifted regions (see Box 10.2 for symbols).

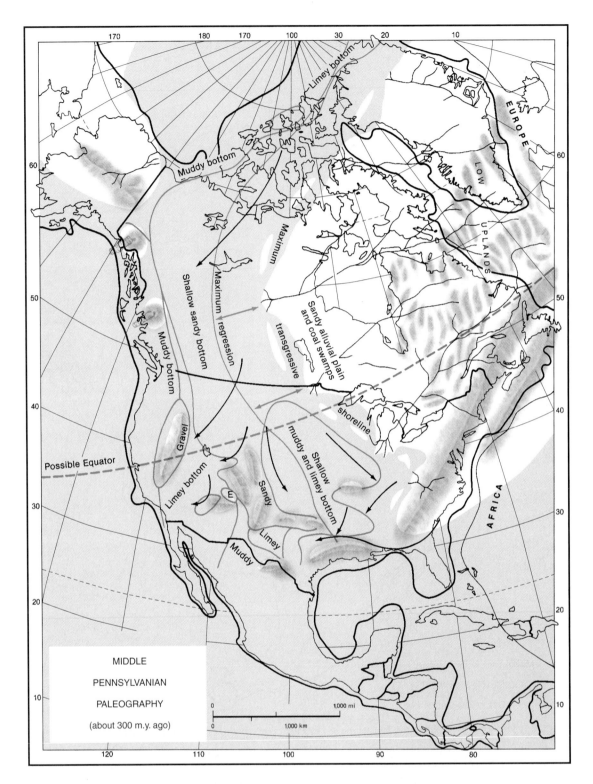

Figure 13.8 Middle Pennsylvanian paleogeography. Note enlargement of land area in eastern North America and many local uplifts in the southwest, as compared with the Mississippian. Note approach of Africa from the southeast, which caused upheaval of the Appalachian Mountains.

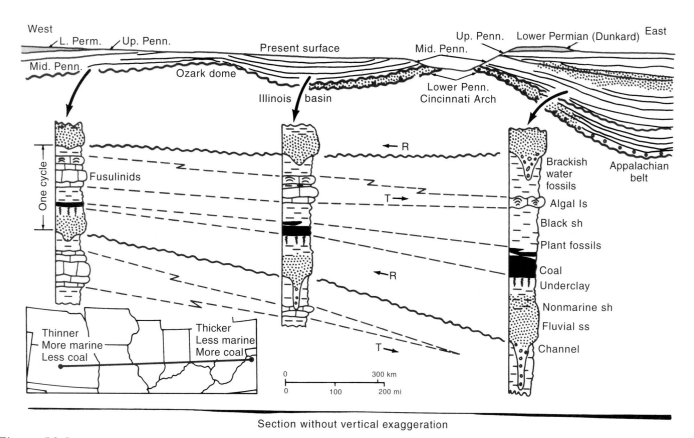

Figure 13.9   Idealized cross section showing lateral and vertical relations of one Pennsylvanian depositional cycle **(cyclothem)** reflecting major transgressions T and regressions R of the sea and shoreline.

A.

B.

Figure 13.10   *A:* A nearly complete cyclothem exposed along Interstate 24 in southern Illinois. At the base of the outcrop is the channel sandstone that marks the base of the cycle. It is overlain by tan floodplain shales and then a distinctive gray clay known as the underclay, since it lies under the black coal bed, and represents the anoxic muds in which the coal swamp grew. Above the coal is a series of marine shales and limestones, representing the maximum transgression before the sequence is truncated at the top by a channel sandstone from the overlying cyclothem. *B:* At Cagles Mill Dam, near Terre Haute, Indiana, the basal channel sandstone is a narrow lens. As a result, the overlying shales and coals have been compacted and draped over the bulge of the channel sandstone. *(D. R. Prothero.)*

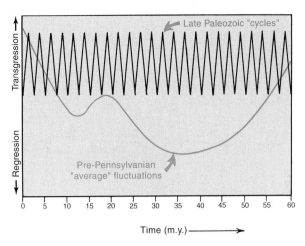

Figure 13.11   Relative average time period of early and middle Paleozoic transgressive-regressive episodes (10–30 m.y.) contrasted with the much shorter period late Paleozoic cycles (about 1 m.y.). New evidence suggests possible "rapid" Devonian fluctuations similar to late Paleozoic ones.

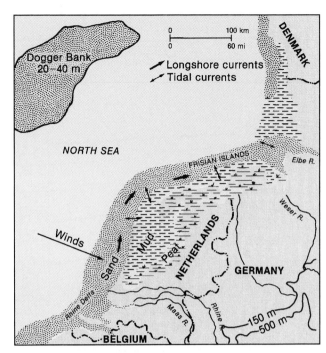

Figure 13.12   The Netherlands before diking began—model for ancient, low, swampy coasts. Note peculiar, nonprotruding delta of the Rhine River; vigorous tidal and longshore currents sweep sand from the shallow (epeiric) North Sea floor northeastward to form long sand spits and barrier islands. Contours at lower right indicate elevation of land surface.

Alternation of marine and nonmarine deposits points to many transgressions and regressions of the sea over wide areas. With possibly more than 100 cycles representing about 100 million years of Mississippian to Middle Permian time, it is apparent that these were geologically rapid oscillations averaging no more than a million years in length. These were much shorter intervals than were the major early Paleozoic transgressive-regressive episodes, which averaged tens of millions of years each (Fig. 13.11). Apparently the later episodes require a different explanation. Possible causes of the shorter oscillations include (1) *worldwide rise and fall of sea level* due either to advance and retreat of continental glaciers or to large-scale warping of the deep-sea floor; (2) *spasmodic tectonic up-and-down motions* of the entire continent; or (3) fluctuations of clastic sediment supplied to deltas, in response to *cyclic climatic changes* affecting erosion in the uplands; as more sediment was supplied, deltas advanced but transgression occurred when the supply decreased. The last cause, although an important process, is too localized to explain the large-scale Pennsylvanian oscillations. The second (a kind of Saint Vitus's dance hypothesis) overtaxes credulity because so many short-term oscillations were involved. We prefer the first mechanism, with changes in known late Paleozoic glaciers on Gondwanaland as the ultimate cause.

Glacially caused sea-level changes would affect coastlines everywhere on earth, which helps explain simultaneous repetitive sedimentation in both western and eastern North America and on other continents as well. It has been suggested that climatic oscillations caused by well-known variations in the earth's orbit and the tilt of its axis may be reflected in the cyclothems. The late Paleozoic record cannot now be subdivided with a fine enough time scale to test this idea fully, however.

Pennsylvanian strata provide clear examples of complex land-sea relationships. A *long-term tectonic rise of eastern North America* is evident (Fig. 13.8), but more *localized differential warping* of the craton also occurred on an intermediate scale, and this warping accentuated many arches and basins (Fig. 13.8). Finally, superimposed on both of those effects were the *short-term transgressions and regressions* presumably caused by sea-level oscillations. The unique rate of each process was superimposed upon the rates of the others to produce a very complex stratigraphic record.

Coal Swamps

Basal sandstones and shales of a Pennsylvanian cyclothem are interpreted as river and delta deposits, whereas the coals formed in vast coastal plain swamps containing junglelike vegetation. The *Dismal Swamp* of North Carolina (2,500 square kilometers), *Dutch lowlands* (15,000 square kilometers; Fig. 13.12), *Florida Everglades* (25,000 square kilometers), and south *Louisiana swamps* all provide useful modern comparisons, although none of these is as large as the late Paleozoic coal swamps must have been. Coal formed through the accumulation of vast quantities of plant material to make peat, which was transformed into coal by compaction (Fig. 13.13).

The ancient *coal swamps* were dominated by large, scaly-barked trees called lycopsids (Phylum Lycopsida; see Fig. 12.10C), which had evolved first in Late Devonian time. Other plants included ferns and tree ferns. Amphibians, primitive reptiles, air-breathing molluscs, and insects inhabited the soggy forests as well. Land floras and faunas of Europe and North

Africa were virtually identical with American ones. The presence of cold-blooded animals only and the nature of the vegetation over the three continents indicate that the climate must have been humid and warm. Carboniferous trees lack clear growth rings. Together with the presence of large, thin-walled cells in the tree trunks, this suggests both rapid growth and a lack of very distinct annual seasonal changes of either temperature or humidity, although some aridity is indicated by Late Pennsylvanian and Permian plants. The lycopsid flora is considered to have been a tropical one, based both upon its botanical characteristics and upon paleomagnetic evidence for paleolatitude (Fig. 13.8).

As trees and shrubs died and fell to the swamp floor, much of their debris was submerged and buried rapidly, thus excluding oxygen and preventing decay and attack by all but anaerobic bacteria. With time, the debris was compacted to peat by the weight of subsequent sediments. The typical thickness ratio of uncompacted peat and coal is about 10 to 1. To form a sizable commercial coal seam, therefore, required a fantastic quantity of vegetation. Through the gradual escape of volatile hydrocarbon compounds from plant tissues, peat changed to coal (which is more compact than peat and contains a higher percentage of carbon). Increased burial and compaction over time produced better grades of coal (that is, coal with greater heat values and less polluting soot). Principal areas of commercially important Pennsylvanian coal are the Illinois Basin, the central Appalachian Mountains, and the Maritime Provinces of Canada (Fig. 13.7).

Paleogeographic Reconstruction

Ultimately the sea flooded the low, coastal-plain swamps. The marine upper phase of each cylothem was formed during such transgressions, thus burying the peat. As with sediments previously discussed, the understanding of Pennsylvanian strata has been facilitated greatly by studies of modern sediments. Much of the North American craton must have been exceedingly flat and very near sea level. Therefore, the coal swamps then, like the present southern Florida Everglades, Dutch lowlands, and Louisiana swamps, were near sea level when the peat originally accumulated. Only a small rise of sea level or sinking of land could cause widespread inundation of former swamps (Fig. 13.14). Conversely a small drop in sea level would cause an equally widespread regression, enlargement of land area, and a great expansion of rivers, deltas, and swamp forests. These are exactly the conditions recorded in the cyclic strata. Between low mountainous uplands along the Appalachian orogenic belt and a

Figure 13.13   Diorama of a Pennsylvanian coal forest. Large trees with the scar patterns on trunks are lycopsids; single tree with the horizontal grooves in the right foreground is a jointed sphenopsid *Calamites*. *(Image #: K10234(3) American Museum of Natural History Library.)*

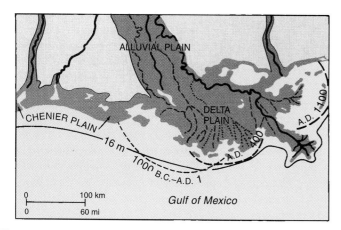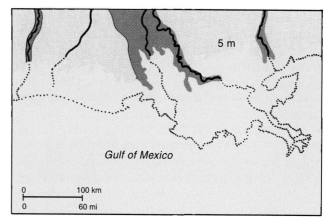

Figure 13.14   The Mississippi delta and swampy alluvial plain—model for certain Pennsylvanian deposits. *Left:* modern protruding or "birdfoot" delta (A.D. 1500 to present) and swampy deltaic plain integrated from three older deltas, now partially submerged. Sand from delta feeds chenier plain and barrier islands farther west (see Fig. 10.23). *Right:* Effect of a 5-meter rise of sea level for comparison with the transgression phase of Pennsylvanian cycles (no part of map area is more than 10 meters above sea level today). *(Adapted from R. Leblanc and H. Bernard, 1954, in* The Quaternary of the United States; *and H. N. Fisk and J. McFarlan, 1955:* Journal of Sedimentary Petrology.)

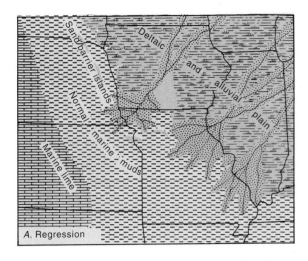

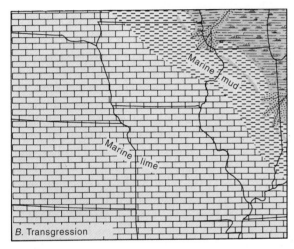

Figure 13.15   Regressive and transgressive fluctuations of environments for a typical Pennsylvanian sedimentary cycle in the central United States (compare Fig. 13.14). *(Suggested by data from H. R. Wanless et al., 1963: Bulletin of the Geological Society of America.)*

persistent epeiric sea over the western part of the craton, the shoreline oscillated across a swampy coastal plain 800 to 1,000 kilometers wide (Fig. 13.8). Overall sedimentary facies, paleocurrent patterns, and mineral composition of sandstones indicate that large rivers flowed west from the Appalachian region and southwest from eastern Canada to empty into the epeiric sea along its eastern margin (Fig. 13.15A).

Much of the Pennsylvanian sandstone and shale in the central United States represents environments associated with deltas (Fig. 13.14). As the shoreline oscillated, so did the deltas (Fig. 13.15). River-channel sandstones pass laterally into black shales, and coals formed in swamps between channels. Channel sands grade westward into silty shales containing brackish water and marine fossils. Frequent pulsations of deposition produced sedimentary features of great variety (Fig. 13.9).

Terminal Paleozoic Emergence of the Continent

Permo-Triassic Geography

The tendency of the eastern side of North America to tilt upward continued from late Paleozoic into Mesozoic time. Recall that changing facies and paleogeographic patterns indicated that normal marine deposits became progressively more restricted and westward encroachment of nonmarine detritus increased many-fold (Fig. 13.16). These tendencies culminated in Permo-Triassic time to produce more land area and of greater average elevation than ever before during the Paleozoic Era (Fig. 13.17).

By Early Permian time, a sea comparable in size to the modern Black Sea occupied the south-central craton, but final stands of the Permian epeiric sea were in present southwestern Texas and the northern Rocky Mountains. In the former area, a great Late Permian organic "reef" complex flourished around the edge of a basin that was more than 300 meters deep (Figs. 13.18 and 13.19), but as the sea retreated farther, the reef died. Thick evaporite and red-bed deposits rapidly filled the basin and buried it. The reef complex swarmed with an amazing variety of life. Brachiopods, including bizarre forms adapted to coral-like growth, were prominent, together with sponges, bryozoans, and advanced forms of algae; corals and crinoids were less important than in middle Paleozoic reefs.

At the western edge of the craton, phosphate and chert formed with black shales, as a result of the upwelling of nutrient-rich cold waters at the edge of the continent. Upwelling is common today along western coasts and is related to the way the spin of the earth affects marine circulation. Meanwhile, evaporites and red beds accumulated widely in the western United States, and evaporites and carbonate rocks accumulated in northern Canada, where middle Paleozoic mountains had stood (Fig. 13.16). Evaporites and red beds (Box 13.1) probably were deposited across central Canada, but erosion has removed them. The vast expanse of Permian evaporites records the slow drying up of epeiric seas as the land area of the earth enlarged to nearly an all-time maximum.

Permo-Triassic Paleoclimate

The position of the equator for late Paleozoic time, deduced from paleomagnetic evidence, is noteworthy in relation to the way various sediments, fossil types, and paleocurrents are distributed. Most significant is that coal-bearing deposits would lie at or within 20 or 25 latitudinal degrees of the apparent paleoequator (Fig. 13.8), thus in tropical to subtropical zones, where mild, humid, nonseasonal climatic conditions might be expected. Land apparently was low enough and land surface area small enough during Mississippian and Pennsylvanian time to allow moderately uniform rainfall over the continent. During Permian and Triassic time, the paleoequator position was not greatly different, but

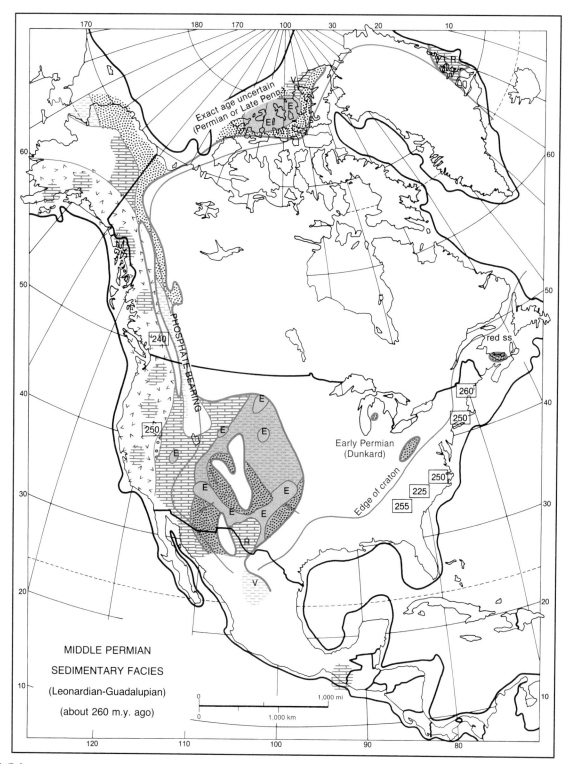

Figure 13.16   Middle Permian lithofacies and isotopic dates of granite plutons. Note the importance of evaporites and red beds in the western craton and phosphate and widespread volcanic rocks with associated limestones in the Cordilleran belt. (See Box 10.2 for symbols and sources.)

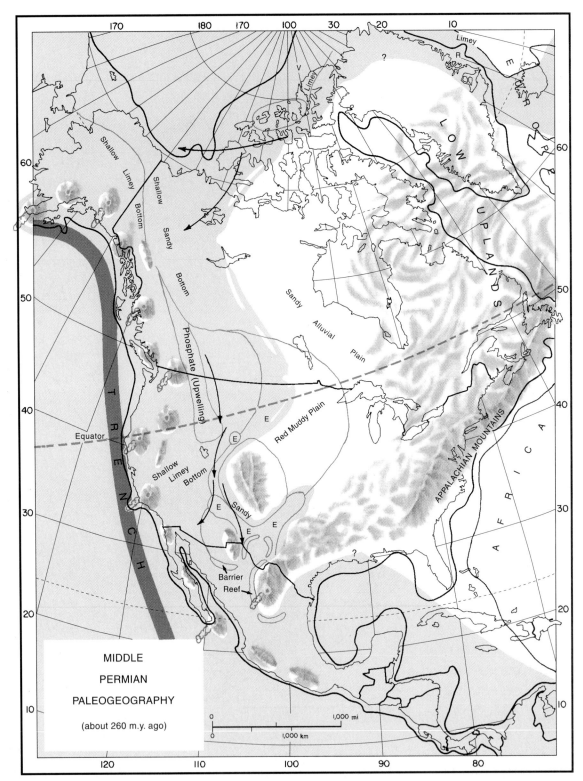

Figure 13.17   Middle Permian paleogeography. Note volcanic islands in the Cordillera and almost complete exclusion of the sea from the craton and eastern North America.

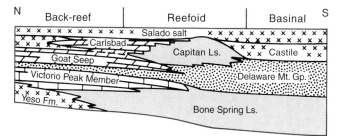

Figure 13.18 Complex Permian facies in the Guadalupe Mountains, New Mexico and Texas. *(After P. B. King, 1948: U.S. Geological Survey Professional Paper 215.)* "Those of us who grew up with West Texas Permian geology . . . learned facies the hard way . . . as we looked into the interior of the range, we saw all our fine units [formations] dissolve before our eyes, merging into a monotonous sequence of dolomite" (from left toward center in the figure; see also Fig. 13.19). *(P. B. King, 1949.)*

Figure 13.19 Outcrops of Permian rocks in the Guadalupe Mountains. El Capitan, formed of massive carbonate shoal limestones (formerly called the "reef core") overlying laminated basinal siltstones of the Delaware Mountain Group. *(D. R. Prothero.)*

Box 13.1

The Red Color Problem

Red beds are the trademark of Permian and Triassic strata on five continents. The vivid color adds to the beauty of many landscapes; especially throughout the western United States. Such deposits long have been considered to bear witness to unique paleoclimatic conditions, but the origin of their red color has long been a subject of controversy.

What can we learn from North American occurrences? The abundance of Permian evaporite deposits associated with many red beds indicates high evaporation rates over the epeiric sea and adjacent lowlands. Pennsylvanian-Permian wind-dune sands covering nearly a million square kilometers of the western part of the North American craton (Fig. 13.6) argue for a desert origin, and Permian fossil plants in Arizona confirm aridity because they have small, thick, hair-covered leaves and spines typical of modern dry-climate floras. Farther east, however, sporadic Permian and Triassic coals as well as aquatic plants, freshwater molluscs, fish, amphibians, and some aquatic reptile fossils indicate moderate humidity there. Late Triassic floras show climatic zonation from mild-temperate conditions in East Greenland through wet tropics in the Appalachian region, tropical savannahs (alternately dry and wet) in the southwestern United States, and wet tropics in southern Mexico (Fig. 13.20). All these areas also have Permian or Triassic red beds.

From paleontologic evidence, it would appear that red-colored sediments developed in both arid and humid areas, but this seeming paradox could be rationalized by relatively humid upland areas blanketed with abundant vegetation and more arid lowlands. Moreover, trees grow today even in true deserts along permanent rivers, and animal carcasses and logs float long distances from uplands down rivers through arid regions and even out to sea.

What red color indicates, first and foremost, is thorough oxidation of iron in sediments. Only a small percentage of iron is required because it is such a potent coloring agent and is so unstable in the presence of free oxygen. The chemical environment favoring red color must be strongly oxidizing and slightly alkaline. Though most examples appear to have been nonmarine, some red beds interstratified with marine limestone and gypsum were deposited in the sea. And red color seems to develop in both arid and humid regions. Red lateritic soils, for example, form today in the tropics, especially where there are alternating wet and dry seasons. Recent studies suggest that oxidation of iron minerals to produce red oxides disseminated among sediment grains occurs largely after deposition. Progressive breakdown of iron in sand grains has been observed in late Cenozoic sediments that are now turning red. In finer, clay-rich sediments, it appears that brown, hydrated iron minerals "age" to red hematite by dehydration after burial. The bulk of the evidence points to development of red color through intense oxidation of iron in sediments under relatively warm conditions in nonmarine and marginal evaporative marine environments. Optimum conditions seem to have included a stable land surface with well-drained soils and distinct wet and dry seasons. The large Permian and Triassic land areas seem to have met these conditions to an unusual degree.

Once established in sedimentary rocks, red color is practically indestructible, as evidenced by the present color of dirt roads and river waters in regions having ancient red beds. The dust that reddened skies during the "dust bowl" days of the 1930s also provides testimony; few distant recipients appreciated that their skies and eyes were being reddened by 200-million-year-old dust blown hundreds of miles. Red color even survives cooking, as pioneer vertebrate paleontologist E. C. Case found on collecting trips into Oklahoma and Texas around 1900. Occasionally Case mixed biscuits for his field parties, but they were so brightly colored by the local water that only the chef was enthusiastic about his ferruginous creations.

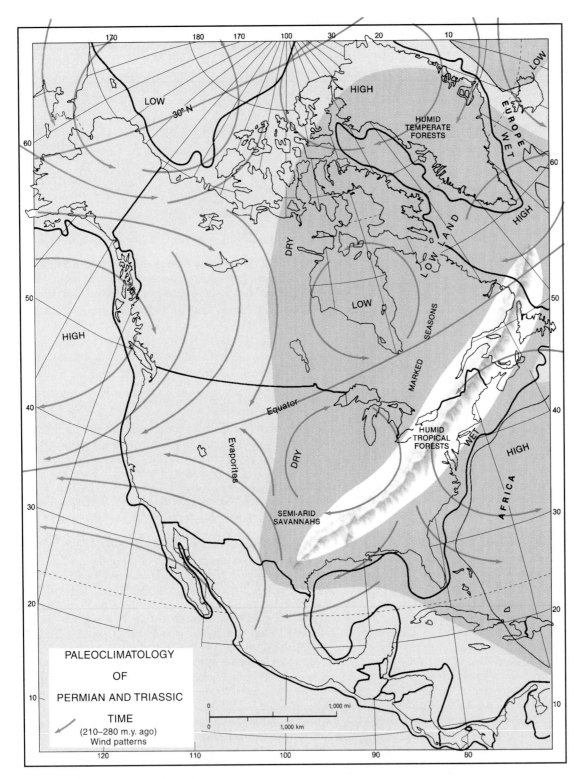

Figure 13.20
Hypothetical paleoclimatic map for late Paleozoic and early Mesozoic time (land is green). Accepting the paleomagnetically indicated equator position, we see that most of North America, Europe, and North Africa would have been in the Trade Winds belt, whereas today they lie chiefly in the Westerly Wind belt. The restoration explains fossil and sediment evidence very well. The Tethys seaway apparently supplied moisture to eastern North America.

both land area and elevation had increased. This change would cause greater differentiation of climate (Fig. 13.20); therefore, we postulate relatively humid Permo-Triassic uplands and drier lowlands. Much of North America would fall in the late Paleozoic tropics and thus in the Trade Winds belt. The western side of the continent consequently would be leeward of the Appalachian Mountains, a configuration that could account for the greater

aridity there. Fossil evidence summarized confirms these deductions and indicates a zonation of climate across North America. Evidence summarized in Chap. 14 shows even stronger zonation on a global scale.

The presence of land areas in Europe and Africa east of the old Appalachian Mountains would affect climate significantly. The subtropical high-pressure cell shown over Africa in Fig. 13.20

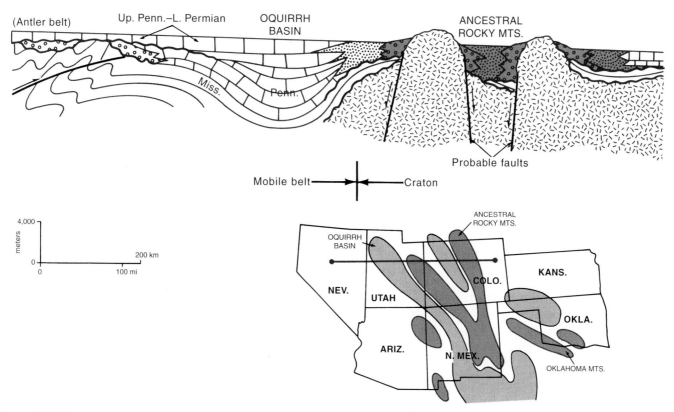

Figure 13.21   Mountains elevated by faulting (probably in part strike-slip in character) in Pennsylvanian time showing coarse, red facies deposited adjacent to them. Lower Paleozoic strata were so thin that erosion quickly exposed Cryptozoic basement. Farther west, the older compressional Antler orogenic belt became buried as the Oquirrh basin subsided abruptly and received 7,000 meters of Pennsylvanian strata. These features and other cratonic basins shown underscore the new instability of the crust.

would be drier than if over open ocean, and it would weaken and strengthen markedly with seasonal cooling and heating of the land. Some moisture, however, probably was available by evaporation from seas covering central Europe and northern Africa (Fig. 13.2). Such moisture would have been carried westward to the rising Appalachian Mountains, where Pennsylvanian humid-forest fossil plants have been found. The paleogeographic picture strongly suggests that the Appalachian region had a monsoonal climate (markedly seasonal rainfall due to seasonal reversal of prevailing winds) like that of India and southeastern Asia today. As Africa approached America and the ocean between them shrank (compare Figs. 13.2 and 13.20), the entire region would become drier. Moreover, as the mountains were raised higher by continent-continent collision, they would block moisture from the southeast and thus cast a rain shadow across western North America. These paleoclimatic predictions seem to be verified by the geologic record.

Tectonics

Cratonic Disturbances

As noted above, the North American craton suffered more severe deformation in Pennsylvanian time than ever before. Many basins and arches were warped and eroded again, and much petroleum and possibly lead and zinc ores were trapped by the structures

produced at this time. The most intense disturbances occurred in the southern part of the craton (Fig. 13.21). Both the Colorado and the Oklahoma Mountains were of the order of 1,000 meters high, and erosion rapidly removed Paleozoic strata to expose Cryptozoic igneous and metamorphic rocks. Thick, coarse red gravels, immature sandstones, and shales (Fig. 13.22) were deposited on alluvial fans and deltas between the mountain fronts and the nearby sea; these deposits grade abruptly into marine strata. As shown in Fig. 13.21, these mountains must have been large fault blocks, but it is very unusual for such high mountains to be raised within cratons. Their ultimate cause is uncertain, but we believe that they were internal cratonic by-products of the collision of Gondwanaland with southern North America.

Appalachian Orogeny

Until late Paleozoic time, the southern margin of North America was a passive stable shelf and continental slope, much like the Nevada region during early Paleozoic time (Fig. 10.3). Unlike the main Appalachian region farther northeast, tectonic disturbances did not affect the Marathon-Ouachita region (Fig. 13.7) until the Mississippian Period. Only thin, fine, deep-water shale and chert had accumulated there earlier. But now sand derived from the eroding mountains farther east spilled over the edge of the continent and was dispersed southwestward by turbidity currents (Figs. 13.23 and 13.24). Volcanic ash entered from the south, indicating

B.

A.

Figure 13.22   Evidence of Pennsylvanian uplift can be seen all around the Ouachitas and Rocky Mountains. A: Tilted red sandstones and conglomerates of the Fountain Formation at Red Rocks Amphitheater, Denver, Colorado. Concertgoers are surrounded by deposits of alluvial fans and rivers shed eastward off the Ancestral Rocky Mountains. Similar Fountain Formation outcrops are responsible for the "Flatirons" above Boulder and the Garden of the Gods near Colorado Springs. B: Outcrop of Collings Ranch Conglomerate in the Arbuckle Mountains of south-central Oklahoma. It unconformably overlies tilted Mississippian sediments, showing that the region was folded, uplifted, and eroded during the Pennsylvanian Ouachita orogeny. Steep mountains nearby must have shed these coarse boulders and gravels in huge flash floods and debris flows. (*D. R. Prothero.*)

that a volcanic island arc had formed, thus implying that subduction had begun. During the Pennsylvanian Period, the trough between the arc and the shelf to the north filled rapidly with clastic sediments, and by Permian time, the entire Marathon-Ouachita sector was uplifted and deformed (Figs. 13.23 and 13.24) together with the Appalachian belt.

Farther northeast, in the main Appalachian region, folding and thrust faulting of Pennsylvanian and older strata toward the craton also occurred (Figs. 13.25 and 13.26). Although results of this late Paleozoic deformation are most obvious, they were superimposed upon effects of the earlier orogenies discussed in Chaps. 11 and 12. Evidence for dating this final compression of southeastern North America is as follows:

1. In southwestern Texas and south-central Oklahoma, folded Pennsylvanian strata are overlain unconformably by Lower Permian ones.
2. In the central Appalachian Mountains, Pennsylvanian rocks were folded and faulted, together with older ones.
3. From Nova Scotia to Florida, nearly flat-lying Upper Triassic strata rest unconformably upon folded Paleozoic rocks (Figs. 13.26 and 13.27).

4. Isotopic dating of scattered granitic bodies formed during this final upheaval indicates an average age between 250 and 260 million years.

It follows that the final deformation of the orogenic belt, which is called the **Appalachian orogeny** (sometimes also called Alleghanian orogeny), occurred between Late Pennsylvanian and Late Triassic times, the exact timing varying somewhat with locality. Fig. 13.28 summarizes the Paleozoic history of the Appalachian belt in plate-tectonics terms.

Along the axis of the orogenic belt, Paleozoic and Cryptozoic rocks were most intensely deformed and metamorphosed by the overprinting of several orogenies. Carbonate rocks became marble, and other strata became either schists or quartzites. Older Paleozoic shales were converted to slate and some coals to anthracite (hard coal). Ultramafic (serpentine) rocks and basalts in the central zone represent sutures of old oceanic material crushed between colliding continents (Fig. 12.45).

Recall that the rocks of the eastern metamorphic belt of New England and the Atlantic Coast states long were considered to be entirely Cryptozoic because they "looked old." Later they were shown to include much Paleozoic rock and, so, are simply more

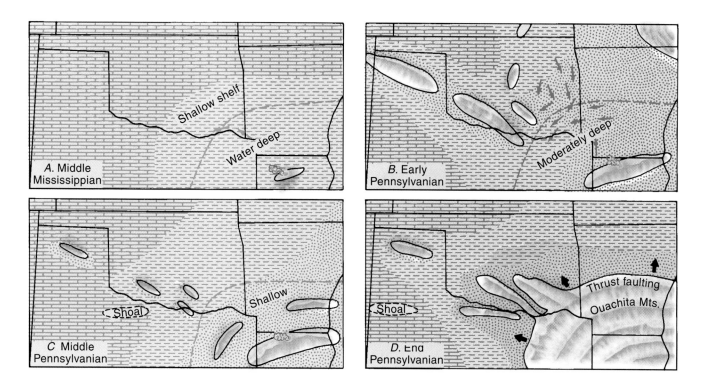

Figure 13.23   Changing late Paleozoic geography of the Oklahoma region. The Ouachita area was a deep-water area starved of sediments until Late Mississippian time. Turbidity currents then introduced large volumes of sand from the Appalachian region to the east, which gradually filled the deep troughs. By Late Pennsylvanian time, collision from the south had produced lands in the region, from which much red clastic sediment was shed. *(From published and unpublished work by L. M. Cline and associates.)*

Figure 13.24   Steeply tilted and folded turbidite sandstones and shales in the Ouachita Mountains of Arkansas. These *flysch* deposits (Fig. 11.28) filled a deep marine trough in the Mississippian and Early Pennsylvanian prior to the Ouachita orogeny, then were crumpled and uplifted by that Late Pennsylvanian collision. *(D. R. Prothero.)*

intensely disturbed equivalents of the strata of the Ridge-and-Valley region to the west (Figs. 13.25 and 13.26). As is typical of the deep root zones of most orogenic belts, it is difficult to differentiate the deformation and metamorphism produced by each successive orogeny because the effects have all been superimposed. Granitic rocks are also hard to differentiate; some are Cryptozoic, but some also formed during each episode of Paleozoic orogeny. Isotopic dating shows that the largest share of the granite and the metamorphism were of Acadian age.

Significance of Thrust Faulting

The Appalachian region was among the first orogenic belts to be studied in detail, and the fold structures long ago were taken to indicate a shortening of the circumference of the earth. The recognition of thrust faults lent much weight to the theory of crustal shortening across orogenic belts because lateral displacements on some thrusts could be shown to amount to several kilometers.

However, as noted in Chap. 7, there have been different ways of interpreting the thrust faults and folds. Some geologists suggested that the faults and folds flatten downward and merely represent superficial sliding of strata over a more rigid basement (see Fig. 7.4); this is the so-called thin-skinned hypothesis. Most workers, however, have considered the basement to be involved as well, in a more fundamental compression of the entire crust—the thick-skinned hypothesis. Maximum displacement on thrusts was envisioned in terms of tens of kilometers. With deeper

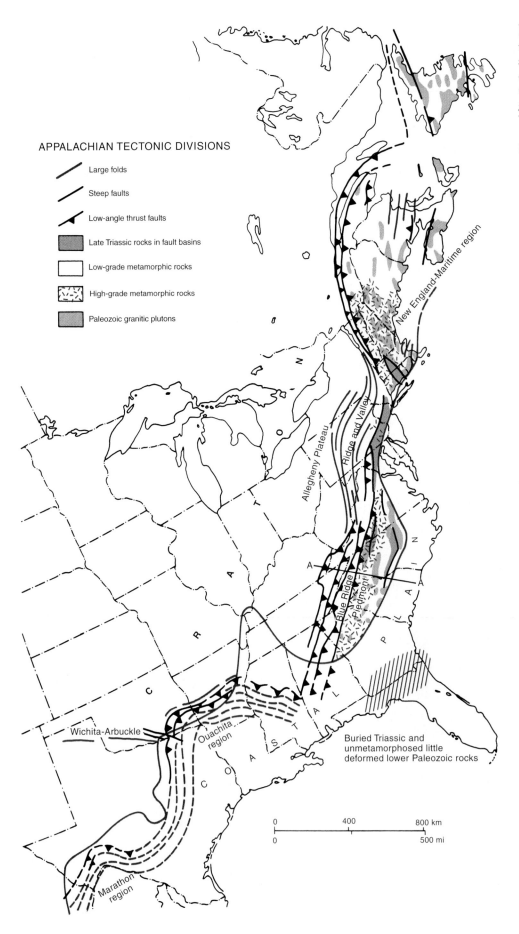

APPALACHIAN TECTONIC DIVISIONS

Large folds

Steep faults

Low-angle thrust faults

Late Triassic rocks in fault basins

Low-grade metamorphic rocks

High-grade metamorphic rocks

Paleozoic granitic plutons

New England-Maritime region

Allegheny Plateau

Ridge and Valley

Blue Ridge

Piedmont

Wichita-Arbuckle

Ouachita region

Marathon region

Buried Triassic and
unmetamorphosed little
deformed lower Paleozoic rocks

0   400   800 km

0   500 mi

Figure 13.25   Tectonic map of
the Appalachian orogenic belt. Note
apparent zonation across the belt from
obvious thrust faulting at the cratonic
margin eastward to a metamorphic and
granite pluton belt with superimposed
Triassic fault basins. A—A locates Fig.
13.26. (*Adapted from* Tectonic map of
North America.)

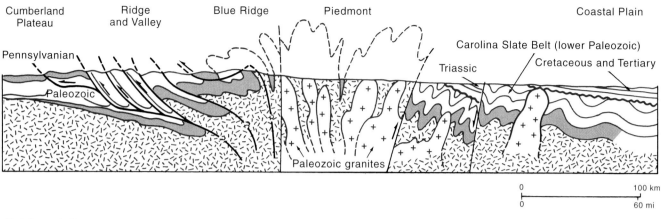

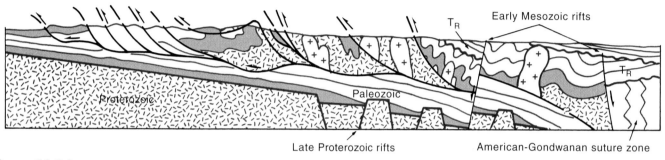

Figure 13.26   Cross sections of the Appalachian belt (see Fig. 13.25 for location). *Upper:* interpretation of 1950 implied a bilateral symmetry to the belt and showed thrust faults steepening downward. *Lower:* interpretation of 1979 based upon deep seismic reflections shows no symmetry and thrusts flattening downward to form one major fault zone at depth. Most dramatic is evidence that metamorphic and granitic basement rocks are part of this immense thrust sheet. *(Top adapted from P. B. King, 1950, Bulletin American Association of Petroleum Geologists. Bottom adapted from F. A. Cook et al., 1979, Geology.)*

Figure 13.27   Angular unconformity *(arrow)* between intensely folded Mississippian strata (Horton Group) below and Upper Triassic red-bed deposits (Wolfville Formation) at Rainy Cove, Nova Scotia. This unconformity reflects the late Paleozoic to Middle Triassic Appalachian orogeny. *(Courtesy of George D. Klein.)*

drilling and modern, high-resolution deep-seismic-reflection techniques, it is now possible to "see" to greater depths and to test these hypotheses. The lower section of Fig. 13.26 shows such a result that indicates thin-skinned thrusting on a breathtaking scale. It appears that a huge sheet 6 to 15 kilometers thick, *and apparently including Cryptozoic basement,* has been displaced westward at least 260 kilometers! This sheet is internally complex, with many subsidiary thrusts that flatten downward. Most dramatic is the evidence that apparently none of the rocks at the surface along this section are rooted in place (compare with the upper section). Rooted rocks all lie below the main thrust.

How does thrust faulting fit into the overall development of the Appalachian mountain belt? Fig. 13.28 summarizes a hypothetical evolution of the belt in terms of plate tectonics and shows what a collage of accreted terranes it is. After initial rifting of some ancient supercontinent between 600 and 700 million years ago, eastern North America was a passive continental margin characterized by shelf carbonate banks until mid-Ordovician time, when compression began due to closure of a small ocean between North America and the Piedmont arc (see Figs. 11.36 and 11.37). This compression culminated in the Taconian

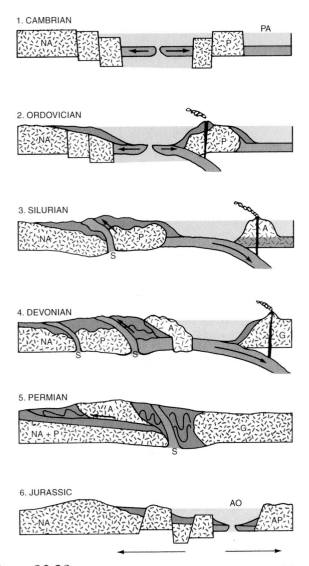

Figure 13.28 Hypothetical scenario for the evolution of the southern Appalachian orogenic belt. *(1)* Cambrian rifting formed proto-Atlantic Ocean (PA) and Piedmont microcontinent (P). *(2)* Ordovician subduction and closing of marginal ocean resulted in collision of microcontinent with North America (NA) by beginning of Silurian (**Taconian orogeny;** compare Figs. 11.36 and 11.37). *(3)* Subduction of proto-Atlantic plate then caused Late Devonian collision of Avalonia microcontinent (A) (**Acadian orogeny;** Fig. 12.44). *(4)* Continued subduction resulted in Permian collision of Gondwanaland (G) with North America, which in turn caused *(5)* large-scale overthrust faulting (**Appalachian orogeny).** *(6)* Early Mesozoic rifting Initiated present Atlantic Ocean (AO). Note that a narrow sliver of Gondwanaland crust was left attached to North America. (S—collision suture zones.) (Compare Fig. 12.45.)

orogeny because of collision (see Fig. 11.33). A second collision with the Avalonian microcontinent resulted in the Acadian orogeny in the central and southern parts of the Appalachian belt, whereas a continent-continent collision with Europe had occurred farther north. Finally, near the end of the Paleozoic Era, Gondwanaland, which had been approaching for some time (see

Figs. 11.36 and 12.44), collided dramatically with North America to produce the great folding and thrust faulting characteristic of the Appalachian orogeny. Significant thrusting has been produced by all kinds of collisions (and even by subduction alone), but clearly continent-continent collisions take the prize for the most spectacular thrust faulting!

Late Paleozoic Mountain Building in Eurasia

The end of the Paleozoic Era was a time of unusually widespread mountain building, the result largely of several continental collisions. These produced a gigantic, interconnected super-supercontinent called **Pangea.** Central Europe was the site of great mountain building in its Hercynian orogenic belt, an extension of the Appalachian belt (Fig. 13.29). The **Hercynian orogeny** was caused by the collision of Northern Africa with Europe. Similar-aged deformation around the southern perimeter of the Siberian craton may have resulted from the collisions of several small Chinese cratons (Fig. 13.2). Upheaval of the Ural belt at this time resulted from collision between lithosphere plates carrying Siberia and Europe. Late Paleozoic mountain building affected all other continents, too, but discussion of that is deferred until later in this chapter.

Both the Hercynian and Ural belts experienced earlier deformation, especially in Devonian time, after which Europe was widely covered by Early Carboniferous carbonate rocks. This was a lull before an orogenic storm, however, for the major Hercynian mountain building, characterized by widespread metamorphism and granitic plutonism, began in Late Carboniferous time. In northwestern Africa, as in Europe, erosion of tectonic lands produced large quantities of sand and mud, which were spread into the surrounding epeiric sea to form huge deltas and coal swamps like those of eastern North America. The resulting sediments show characteristics of repetitive sedimentation, too. Most of Europe's mineral wealth of coal and ores was formed at this time. For example, the ores mined in the ancient districts of central Germany and what used to be Czechoslovakia were of Hercynian origin. As in North America, the last epeiric sea deposition occurred in restricted basins of later Permian time. Evaporites and nonmarine red beds became more and more widespread, and at the beginning of the Triassic Period, all of Europe was land. Late Paleozoic plants, freshwater fish, amphibians, and reptiles, as well as the sediments, were almost identical to their counterparts in North America and northern Africa.

Siberia is much like North America in possessing a central craton completely encircled by orogenic belts. Marine Vendian strata are well represented on its margins, and most of the craton was inundated by Cambrian and later epeiric seas. Great volcanism characterized an early Paleozoic orogenic belt on the South, and Ordovician, Silurian, and Devonian rocks all include important red beds there. By Carboniferous time, coal-bearing deposits were widespread both in orogenic belts and upon the craton itself. Although it is generally believed that the Ural belt resulted from late Paleozoic collision of Siberia with Europe, it is at present difficult to relate the other Paleozoic belts of Asia to plate motions (but see Fig. 13.2). Fig. 13.29 shows the southern supercontinent Gondwanaland after collision with a northern supercontinent,

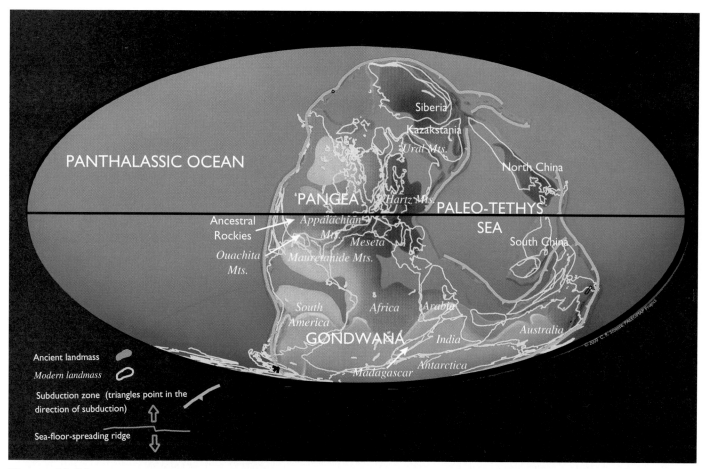

Figure 13.29   In the Late Carboniferous, Laurasia and Gondwanaland collided to form the supercontinent known as Pangea. The Appalachian-Ouachita collisional belt resulted from the suturing of African/South American portions of Gondwanaland to southeastern North America; a similar belt between southern Europe and Africa was known as the Hercynian belt. The Ural Mountain collision in central Russia formed by the suturing of the Baltic and Siberian plates. Additional collisions occurred along the eastern and southern margins of the Siberian plate as the various Chinese microplates began to accrete. Note the great extent of the Gondwana ice cap (light blue) at this point. (*www.scotese.com*)

which is called Laurasia (composed of North America, Europe, and Siberia). This great end-of-the-Paleozoic collision produced the super-supercontinent Pangea, which is our next topic.

Gondwanaland

In late Paleozoic time, Gondwanaland collided with North America and Europe, resulting in the final Appalachian and Hercynian orogenies (Fig. 13.29). Meanwhile, apparently a microcontinent made up of present southern South America, part of western Antarctica, and probably New Zealand converged upon southern Gondwanaland, where orogenic belts remained active throughout the Paleozoic Era. South African drift advocate, Alexander Du Toit, mentioned in Chap. 7, used these zones of late Paleozoic orogeny to make his now-famous reconstruction of Gondwanaland (Fig. 13.30). His so-called Samfrau geosyncline was finally crumpled either by collision of the postulated "Samfrau micro-

continent" with Gondwanaland proper or by very-low-angle subduction under this part of Gondwanaland (Fig. 13.30). The resulting Permo-Triassic mountain building was named the **Gondwanan orogeny** by Du Toit. Typical clastic wedges accumulated in front of the mountains raised by it.

Because much land area lay at high latitudes in late Paleozoic time, large ice sheets formed on up-land areas. Permo-Carboniferous glacial deposits are known on all five of the southern continents. Together with the similarity of present continental margins, the distribution of these deposits provided one of the criteria used for early reconstructions of Gondwanaland (Fig. 7.8). Their presence on either side of the modern equator and the absence of similar-aged tillites at high present latitudes in North America and Europe were baffling, to say the least, in terms of present geography. Moreover, in several places the ice seemed to have moved onto the continents from present deep-ocean basins (e.g., southeastern Australia and eastern South America; Fig. 13.30). The Dwyka Tillite of South Africa contains boulders with

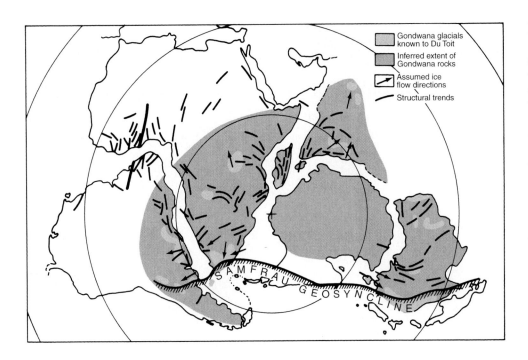

Figure 13.30   Du Toit's reconstruction of the Gondwanaland supercontinent during late Paleozoic time based upon congruence of shorelines, matching of structural features in ancient basement rocks (lines), and late Paleozoic orogenic belts—here combined in what Du Toit called Samfrau geosyncline (from the combination of letters from the names of three continents). Apparently a microcontinent compressed the Samfrau belt against main Gondwanaland in Permo-Triassic time. After reconstructing, Du Toit plotted distributions of Gondwana rocks and especially glacial phenomena. Heaviest lines at left represent same boundary on both continents between two major isotopic date provinces. (Compare Fig. 13.31.) (Adapted from Du Toit, 1937.)

Figure 13.31   Reconstruction of Atlantic border continents using best computer-fit of margins. Some late Proterozoic and Paleozoic orogenic belts defined by isotopic dating are added to show how well reconstruction could explain many disjunctive geologic features. Of great interest is the fact that long-supposed *marginal orogenic belts* become *bilateral intercratonic* ones in such a restoration (Fig. 13.29). Arrows indicate shedding of coarse, red clastic sediments from middle Paleozoic mountains. Colored squares indicate Grenville basement beneath Appalachian belt. *(After McElhinny 1967, UNESCO symposium of continental drift, Montevideo.)*

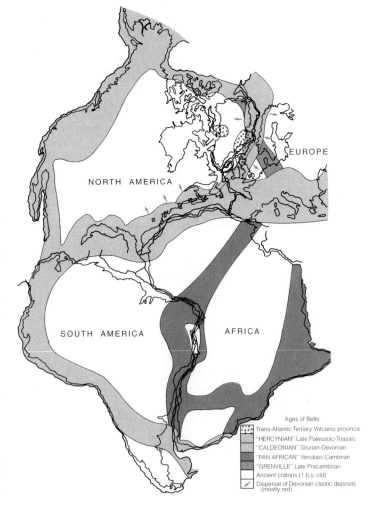

Lower Cambrian archaeocyathid fossils (see Fig. 9.17A–I) that are not known in Africa but that could have come from Antarctica. To these criteria, Du Toit added structural features (Fig. 13.30).

In the 1960s, the new clues for refitting the jigsaw puzzle provided by extensive paleomagnetic studies (see Box 13.2) and isotopic dating (Fig. 13.31) gave new precision and confidence to the restorations. Figure 13.32 shows how paleomagnetism can be used to trace the migrations of Gondwanaland. Deep-sea drilling has added yet more refinement, especially in discovering submerged continental basement that "fills" a former hole in the reconstruction between Africa, Antarctica, and South America (see area around "Frau" in Fig. 13.30). The greatest remaining uncertainty is the proper position of the Pacific side of Antarctica within restored Gondwanaland.

The Gondwana Rocks

Late Paleozoic and early Mesozoic strata of the southern cratons occur in a distinctive sequence that is amazingly similar on all five now widely separated continents. These rocks are collectively known as the **Gondwana rock succession.** (The name Gondwana is derived from an ancient tribe in India.) They have played such

Box 13.2

Paleomagnetic Reconstructions for Drifting Continents

Figure 13.32 illustrates the way in which paleomagnetic measurements on rocks from different southern continents can be used to reconstruct the predrift positions of the continents. It should be compared with Figs. 7.11 and 7.14.

A. This view shows the present positions of South America and Africa with the addition of the positions of the south pole *relative to each continent* as determined from rocks of different ages from those two continents. The two zigzag lines connecting the successive pole positions relative to each continent are called *apparent polar wandering paths*. Of course, it was actually *the continents that wandered with respect to the pole*, but it is more convenient to illustrate the relative motions in this way. It is equivalent to holding the continents fixed and seeing how the pole *appears* to have migrated.

B. Next we can reconstruct the Paleozoic positions of South America, Africa, and Australia by bringing together their apparent polar wandering paths, so that all Cambrian (Є), Silurian (S), Carboniferous (C), and Permian (P) pole positions match as closely as possible. When we do this "best fit," we find that several bends in the paths *as well as* the margins of South America and Africa fit together much as a jigsaw puzzle (compare Fig. 7.14). This paleomagnetic best fit must then be checked against independent geologic and paleontologic evidence (Ћ—Triassic; K—Cretaceous.)

C. This view portrays the early Mesozoic paleolatitude positions of all five Gondwana continents based upon paleomagnetic data as discussed in Chap. 7. The basis of this reconstruction is the *angle of inclination* of "fossil magnets" measured for rocks of the same age from each continent (see Fig. 7.14). The latitudes indicated for Australia and India do not agree fully with other geologic data (compare Fig. 13.30).

Apparent relative wandering paths appear also in other figures— for example, Fig. 13.31. These are to be read in the same way as *B*.

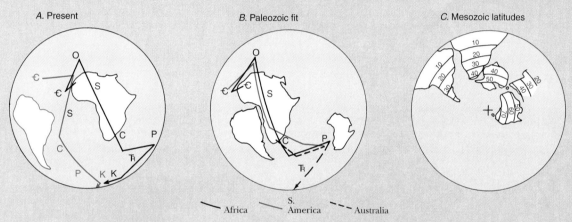

A. Present   *B. Paleozoic fit*   *C. Mesozoic latitudes*

—— Africa   —— S. America   - - - Australia

Figure 13.32 Paleomagnetic reconstructions of southern continents. *(Reconstructed base map adapted from Bullard, 1965, Royal Society of London symposium on continental drift.)*

a key role in the theory of continental drift that we shall examine them further.

When Gondwana rocks were first discovered a century ago, their age was unknown. The principal fossils belonged to a then strange and as yet undated flora. Because the relative geologic time scale was developed in Europe in largely marine strata, the Gondwana succession could be dated only by relating its largely nonmarine fossils to some more familiar, well-dated marine index fossils. This first became possible in northern India, where lower Gondwana tillites are interstratified with Upper Carboniferous and Lower Permian fossiliferous marine strata (Fig. 13.33). Elsewhere, upper Gondwana strata were found associated with fossiliferous marine Jurassic ones, so the total sedimentary succession was dated as Carboniferous through Jurassic. More recently, it has been shown that the top of the succession extends into the Cretaceous in some places. Once dated according to the marine index fossils of the European standard time scale, Gondwana fossils could then be used to correlate among different Gondwana localities on all five southern continents (Fig 13.34).

Figure 13.33 Diagrammatic cross section showing how nonmarine Gondwana rocks were dated in India by relating them to a few interstratified marine tongues of Himalayan (Tethyan) facies containing fossils referable to European standard geologic time scale, shown at right (UC—Upper Carboniferous).

On the five southern continents, the Gondwana succession has prominent glacial tillites in its lower part (Fig. 13.35). These distinctive, unsorted strata overlie scratched surfaces on old rocks abraded by boulders frozen in ice (Fig. 13.36). Many erratic boulders in the tillites also show scratches characteristic of glacial

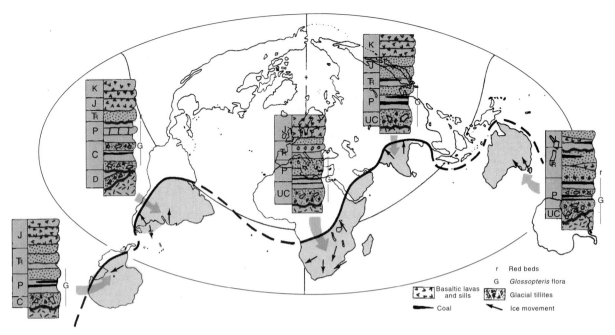

Figure 13.34   Distribution of Gondwana rocks showing their remarkable similarity on all five southern continents. *(Bartholomew's Nordic projection; used by permission.)*

Figure 13.35   Lower Gondwana (Dwyka) tillite, Port St. Johns, South Africa. This, the most famous ancient glacial deposit in the world, formed both on the African craton and in the north edge of the Cape orogenic belt. *(Courtesy J. C. Crowell.)*

Figure 13.36   Striated glaciated surface, or "pavement," beneath the Carboniferous Dwyka Tillite, Nooitgedacht, near Kimberly, South Africa. Surface was cut into pre-Cambrian metamorphosed volcanic rocks. *(Courtesy J. C. Crowell.)*

Figure 13.37   Giant "dropstone" released by a melting iceberg as it floated away from the Gondwana ice sheet during the Permian. Notice how the layered marine sediments were bent downward by the impact and then smoothly buried the dropstone once normal, quiet marine deposition had resumed. From the Parana Basin, Brazil. *(Courtesy J. C. Crowell.)*

Figure 13.38   Tongue-shaped leaves of the characteristic Gondwana seed fern *Glossopteris* (whose name means "tongue fern"). Found on all the southern continents (including India, Madagascar, and Antarctica) during the Permian, its seeds were too large to have been blown across the modern southern oceans, and apparently they did not float. Many geologists, including Wegener and DuToit, argued that these southern continents must have been united for this flora to be so widespread. *(Courtesy of Bruce Tiffney.)*

materials (Fig. 13.37). Large pebbles dropped into finely laminated sediments suggest rafting by melting icebergs (Fig. 13.37; compare Fig. 8.34). Such deposits in India and South Africa were the first pre-Pleistocene tills to be recognized—and only two decades after proposal of the theory of Pleistocene continental glaciation in Europe in 1840. Although some of the tillites occur within marine sequences, most of them are interstratified with nonmarine strata containing important coal seams (Fig. 13.34). Associated with the coals are many fossil plants belonging to an assemblage collectively referred to as the ***Glossopteris* flora,** named for a seed fern that bore distinctive, tongue-shaped leaves (Fig. 13.38). Several distinctive vertebrate animal fossils also are scattered widely in Gondwana strata on several continents. The younger Gondwana rocks include chiefly nonmarine sediments, such as conspicuous red beds and wind-blown sands. In some areas, such as eastern Africa, northern India, and South America, these intertongue with marine deposits.

The Gondwana rock succession is capped everywhere either by thick basalt flow rocks or by dikes and sills (Figs. 13.34 and 13.39). Isotopic dating shows that the basalts are Late Triassic or Jurassic in Australia, Jurassic in South Africa and Antarctica, chiefly Early Cretaceous in South America, and Late Cretaceous to Eocene in India. These rocks form immense basalt plateaus deep within the cratons, where basalt is not normally expected. Such flood basalts were fed by countless fissures, which must extend down to the mantle. Also of mantle derivation is the unusual diamond-bearing variety of ultramafic rock called **kimberlite** for Kimberley, South Africa; it also occurs in Arkansas, Colorado, Siberia, Greenland, Brazil, and Canada and generally in proximity to basalts. This ultramafic rock, which contains such high-pressure minerals as diamond and has a mixture of chaotic

Figure 13.39   Drakensburg Mountains in Royal Natal National Park, South Africa. The escarpment is capped by resistant basalts, which overlie wind-deposited white sandstone cliffs, both of which are Jurassic and constitute the upper Gondwana rock succession here (see Fig. 13.34). *(R. H. Dott, Jr.)*

textures, was emplaced by hot, explosive gases rising rapidly from the mantle. In southern Africa, kimberlites penetrated Upper Jurassic rocks, and diamond grains are found in Cretaceous sandstones, indicating intrusion during Early Cretaceous time soon after eruption of nearby basalts. Although there are exceptions, most known kimberlites are Mesozoic or early Cenozoic. Both they and the flood basalts provide important evidence of great crustal disturbances associated with the initial breakup of Pangea.

The *Glossopteris* Flora

The *Glossopteris* flora is surprisingly homogeneous, even though it occurs today on five scattered continents. Twenty species of leaves found in Antarctica are common also to India, now located across the equator and far away, yet *Glossopteris* seeds were too large to be windborne. How could a uniform land flora be dispersed across wide oceans? The obvious answer provided by continental drift is that those oceans did not exist when the flora lived.

The distribution of the *Glossopteris* flora does not overlap geographically that of the late Paleozoic lycopsid flora of North America, Europe, and North Africa, nor of the Asiatic flora (Fig. 13.40). Because of its intimate association with glacial deposits, the presence of well-developed seasonal growth rings in petrified tree trunks (Fig. 13.41) and the great abundance of leaves at most fossil localities, the *Glossopteris* assemblage is interpreted to have been a temperate-climate deciduous flora. The lycopsid

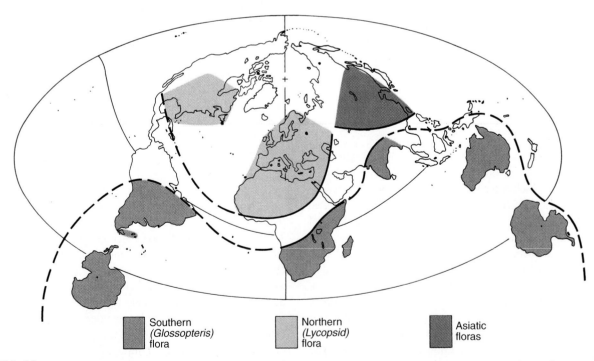

| Southern (Glossopteris) flora | Northern (Lycopsid) flora | Asiatic floras |

Figure 13.40   Present distribution of three late Paleozoic land floras of the world. Note the separation of lycopsid and *Glossopteris* floras. Climatic zones formed the barriers between the two floras in Africa. Their present arrangement resulted from their breakup and drift since the Triassic. *(Adapted from Just et al., 1959: National Research Council report on Paleobotany N22; and Gothan and Weyland, 1954, Lehrbuch der Paläobötanik.)*

Figure 13.41   Seasonal growth rings in petrified wood in the *Glossopteris* floral province (Permo-Triassic), Senekal, Orange Free State, South Africa. Such rings could reflect hot/cold, wet/dry, or dark/light seasons but probably resulted mainly from temperature changes. Diameter of log is about 40 centimeters. *(R. H. Dott, Jr.)*

Figure 13.42   The archaic reptile *Mesosaurus*, known from Permian freshwater deposits in both Brazil and South Africa. This fish-eating swimmer was less than a meter in length and clearly could not have swum across the modern Atlantic with its simple webbed feet, so it also demonstrated that South America and Africa had once been connected. *(Painting by Zdenek Burian.)*

flora, on the other hand, apparently was a tropical-climate one, consistent with paleomagnetic evidence for latitude (Fig. 13.29). The Asiatic flora apparently was a northern temperate one. The *Glossopteris* flora flourished over wide, swampy low areas on the perimeter of glaciated regions. Interlayering of *Glossopteris*-bearing strata with several tillites indicates considerable climatic fluctuation and implies that the flora was adapted to a wide range of ecological conditions. It persisted into Triassic time long after glaciation had ceased, but by the Jurassic Period, *Glossopteris* had been replaced by ginkgoes, cycads, and various new seed ferns (discussed in Chap. 14). This change in flora reflects a Mesozoic warming of climate as Gondwanaland moved away from the pole and began to break up. Besides these new plants, warming and drying of climate over much of Gondwanaland are suggested by evaporites, dune sands, red beds, and abundant reptiles.

Animal Fossils

An unsurpassed variety of Permo-Triassic amphibian, reptile, fish, and invertebrate fossils occurs in Gondwana strata. Some fossil leaves even show evidence of having been eaten by insects. A small fresh- or brackish-water reptile called *Mesosaurus* (Fig. 13.42) is unique to South Africa and Brazil and marks the Carboniferous-Permian boundary. Presumably it could not have crossed the present wide Atlantic and, so, was taken years ago as additional evidence for the existence of the reconstructed Gondwanaland. Although many other Gondwanaland vertebrates also are widespread in the southern continents, they also have much in common with their northern counterparts. Unlike the *Glossopteris* flora, vertebrates clearly could move easily between the northern and southern supercontinents. For example, African Permo-Triassic vertebrates have about as much in common with Eurasian, North American, and Siberian ones as with those of other Gondwana continents. And Jurassic dinosaurs in eastern Africa were identical to those of Wyoming! Clearly land animals managed to disperse north and south more readily than did plants, which perhaps is not surprising in view of the mobility of animals. Their wide distribution requires, however, that the herbivorous animals could adapt to a diet of either lycopsid or glossopterid vegetation.

Antarctica—a Triumph of Prediction

Two-thirds of Antarctica (the East) is a craton, whereas a youthful orogenic belt marks its Pacific (the West) margin. An early Paleozoic Trans-Antarctic orogenic belt along the western edge of the present craton has 500 million-year-old granites like those of the Pan-African belts. Early Paleozoic mountains were leveled by erosion, and then a Devonian epeiric sea flooded the craton, as in Brazil and Australia. Devonian marine fossils are now recognized to have close affinities to those of Australia, South America, and the Cape belt of south Africa, all of which make up the *Southern Faunal Realm*.

Figure 13.43   Gondwana strata (Beacon Sandstone) with black basaltic sills (Jurassic Ferrar Dolerite) exposed along glaciated walls of Finger Mountain near McMurdo Sound at west edge of the East Antarctic craton. *(Courtesy of R. F. Black.)*

Figure 13.44   Spectacular fold structures in Paleozoic strata, Sentinel Mountains, West Antarctic orogenic belt. This area represents westernmost known occurrence of Gondwana rocks in Antarctica; beneath them is a thick marine sequence much as in the South Africa Cape orogenic belt. Folding in both areas occurred during the Gondwanian orogeny. *(Courtesy of Campbell Craddock, University of Wisconsin.)*

A very widespread and distinctive sequence of cross-stratified sandstones with important coal seams overlies the Devonian rocks (Fig. 13.34). Important fossil plants were discovered in these sandstones only 100 kilometers from the pole during a 1901–1904 British expedition and by the ill-fated Scott South Pole expedition of 1910–1912. On the return down Beardmore Glacier from the pole, Scott's men collected many rock samples, which were found with their frozen bodies at their last camp. The plants were misidentified as lycopsids until reexamination years later proved them to be *Glossopteris.* Among the specimens were samples of basalt described as forming thick, black bands in the sandstones along the walls of Beardmore Valley (Fig. 13.43). Naturally these early revelations from the frozen continent were greeted with eager anticipation, especially by Southern Hemisphere geologists and by botanists, who were puzzled by the presence of coal and fossil trees at such hostile latitudes.

These clues led the ardent driftist Du Toit to conclude in the 1930s that Antarctica was a full-fledged "Gondwana Continent." Eventually it would be shown to possess *all* the Gondwana trademarks, he predicted. Besides *Glossopteris,* coal, and basalt, it also should yield tillites and Gondwana-type nonmarine vertebrate fossils. Moreover, it should have evidence of the Permo-Triassic Gondwanan orogeny (Fig. 13.30). It was 30 years before Du Toit's predictions were all fulfilled, however, because relatively little new exploration occurred until after World War II. The International Geophysical Year, or I.G.Y. (1957–1958), an imaginative venture in worldwide scientific cooperation, gave impetus to a new wave of scientific exploration on a scale so large that Antarctica appears in danger of overpopulation. As a result of this research, many more Gondwana plant localities were found, and Gondwana rocks were recognized beyond the craton where they

were caught in Gondwanan mountain building (Fig. 13.44; compare Fig. 13.30). Mesozoic basaltic rocks also were found to extend over an immense area, although their main concentration is around Beardmore Glacier. Isotopic dating indicates that these igneous rocks are about 160 million years old (Jurassic), or about the same age as igneous rocks on surrounding continents (Fig. 13.34).

The greatest Antarctic triumphs were two discoveries made in the 1960s. First, tillites were found within a complete sequence of Gondwana rocks containing coal and *Glossopteris* only 200 miles from the South Pole. Then in 1967, an amphibian jaw bone, and in 1969 synapsid bones, both of Triassic age, were found, proving that land animals as well as forests inhabited the continent 210 million years ago! At least one of the land-dwelling synapsids, *Lystrosaurus* (Fig. 13.45) was identical with Triassic forms long known on other southern continents. The kinship of Antarctica with the other four Gondwana continents was now fully established (Fig. 13.34).

Paleogeography and Paleoclimate of Pangea

Figure 13.46 shows a paleogeographic reconstruction of all of Pangea for Permian time. Early continental-drift theorizers postulated a single huge polar ice cap that spread radially outward from the center of Gondwanaland, as shown here (also Fig. 13.30). In Chap. 7, we noted also that the Wegener-Köppen late Paleozoic restoration of the continents placed the North Pole in the north Pacific Ocean, which would make it a warmer pole (Fig. 7.8). Probable glacial deposits of late Paleozoic age

Figure 13.45 When fossils of the synapsid *Lystrosaurus* were found in Antarctica in 1969, many skeptics of continental drift were finally convinced. This small (about half a meter in length) plant eater, with its short, squat body and almost toothless bill, clearly could not swim across the southern oceans as they exist today, yet it is now known from the Triassic deposits of South Africa, India, and Antarctica. *(Painting by Zdenek Burian.)*

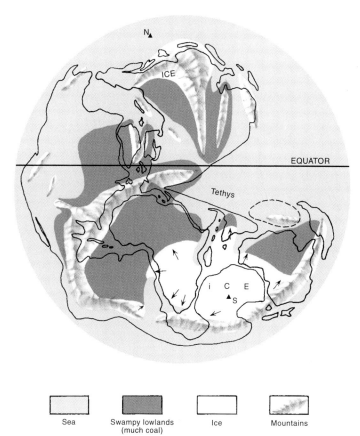

| Sea | Swampy lowlands (much coal) | Ice | Mountains |

Figure 13.46 Permian reconstruction of Pangea—a compromise of all available data. Marine embayments on Gondwanaland could have furnished moisture for glaciers. Glacial-marine deposits also exist in Siberia near the north paleopole (S—South Pole; N—North Pole).

recently recognized in northeastern Siberia suggest that there was, in fact, some northern glaciation at high latitudes symmetrical with that of Gondwanaland (Fig. 13.29). This is consistent with the temperate-climate fossil flora of northeastern Asia (Fig. 13.40). Wegener and Köppen drew the equator through the northern (lycopsid) coal belt, which already was considered tropical on paleobotanical grounds. Based upon evaporite deposits and presumed fossil dune sand, mid-latitude deserts like today's were postulated for both hemispheres (see Fig 7.8). Everything seemed to fit, and for drifters the most heartwarming development has been the almost perfect agreement of paleomagnetic results for ancient latitude with the old 1920s restorations (compare equators in Figs. 7.8 and 13.46).

American geologist A. A. Meyerhoff has pointed out that the total area covered by Gondwana glaciation was three times as large as the North American Pleistocene ice cap. A single landmass so large would have been too dry in its interior for the simultaneous nourishment of a huge cap as large as that in Fig. 13.46. In Pleistocene time, for example, Siberia was certainly cold enough but was too dry to have had a significant ice cap because it was located too far from moisture-providing seawater. In southwestern Africa and in Brazil, tillites occupy deep, ancient valleys. This indicates that considerable topographic relief existed, and Meyerhoff notes that some glaciers might have formed on high, cool plateaus even at low latitudes, just as glaciers are found today on high equatorial mountains in South America, eastern Africa, and New Guinea.

The location and age distribution of glacial deposits indicate that glaciation of all of Gondwanaland was not simultaneous (Fig. 13.47). It began first in northern Africa in Late Ordovician time. Then it started in South America in Silurian or Devonian time and continued there into Early Carboniferous time. It was

confined to the Carboniferous in Africa but spanned the Late Carboniferous and part of Permian time in India, Antarctica, and Australia (Figs. 13.34 and 13.47). Paleomagnetic evidence confirms also that Pangea was drifting as a unit across the South Pole (Fig. 13.31). As summarized in Fig. 13.47, the centers of glaciation shifted as this occurred. Therefore, we need not postulate a single ice cap at any one time that was larger than Pleistocene ones. The glacier nourishment problem is further alleviated by the recognition in recent years that epeiric-sea embayments existed within Gondwanaland (Fig. 13.46). These areas of water would have been important sources as moisture evaporated to feed the interior ice cap areas.

Northern Australia provides proof that parts of the northern fringes of late Paleozoic Gondwanaland also were cold. Permian brachiopod shells from a marine layer just above a tillite have been subjected to a geochemical technique for determining the temperature of the seawater in which the shells were formed. The analysis involved precise determination of the ratio of two isotopes of oxygen contained in the calcium carbonate of the shells. That ratio is related to the original water temperature, as explained more fully in Chap. 14. The results indicate a Permian seawater temperature of only 7°C, which supports the Gondwanaland restoration.

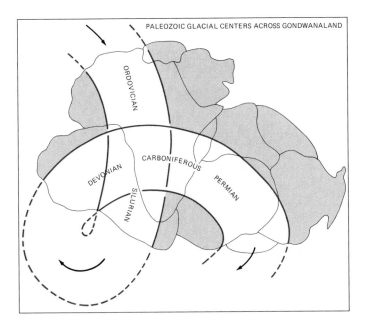

PALEOZOIC GLACIAL CENTERS ACROSS GONDWANALAND

Figure 13.47   Changing relative positions of Paleozoic glacial centers across Gondwanaland. For convenience, the path of migrating glacial centers is shown relative to Gondwanaland, but, in fact, it was the supercontinent that was moving relative to the South Pole. This caused the shifting of glacial centers as shown. *(Modified from Caputo and Crowell, 1985: Geological Society of America Bulletin.)*

Some parts of the supercontinent were warm, whereas glaciers persisted elsewhere. For example, Permian evaporites in northern South America and Permo-Triassic red beds containing some cold-blooded animal fossils in southern Africa both indicate warming of climate as those areas drifted away from the polar region (Fig. 13.47).

The *Glossopteris*-bearing coal deposits in Gondwanaland, like the northern lycopsid-bearing coals, formed largely in broad, coastal swamps, as indicated by the interstratification of these deposits with fossiliferous marine and brackish-water strata. Because of their higher latitude, the Gondwana swamps were more temperate and more variable in climate than the tropical lycopsid ones. Clearly the glaciers flowed down to sea level and out onto continental shelves at many places, as in northern Australia (Fig. 13.46). Presence of some warm-temperature indicators, such as evaporites and cold-blooded amphibians, need not be alarming, for today tree ferns grow within a mile of glaciers in New Zealand and ice extends down to areas with apple orchards in Norway. The point is that, if glaciers flow from their cool sources into warm areas more rapidly than melting occurs, they will persist there in spite of the warmth.

At least six distinct tillite zones are known in Africa, and more than two dozen have been claimed in Australia. Clearly the glaciers advanced and retreated repeatedly, but it is impossible to tell how many times. The *Glossopteris* flora was adapted to a variable cool-temperate climatic zone surrounding the ice caps, so, as the glaciers retreated, the flora expanded poleward, and vice versa. Considering the long time during which large glaciers existed on Gondwanaland (and, by analogy, with the Pleistocene), worldwide sea level must have fluctuated up and down many times as the ice shrank and expanded. All coastlines around the world—even in the tropics—would have experienced transgressions and regressions, and it seems more than coincidental that the most conspicuous **repetitive sedimentation** in the entire geologic record occurs in strata of Carboniferous and Early Permian ages in many parts of the world—exactly the span of the main Gondwana glaciation (Fig. 13.9). Whereas late Cenozoic glaciation in polar regions has lasted about 40 million years so far, late Paleozoic Gondwana glaciation spanned almost 100 million years.

Antarctica shows the most dramatic evidence of climatic changes through time of any place on earth. Today only lichens, algae, and a few puny grasses (and no truly land animals) grow there, but dense forests with sizable trees have grown there at several times in the past. Moreover, in Permo-Triassic time, vertebrate animals roamed the landscape only 100 miles from the present South Pole. Clearly the continent had a much milder climate during the past than now.

Major temperature fluctuations characterized late Paleozoic Gondwanaland, as indicated by the presence of cold-blooded animal remains and coal seams interstratified with tillites. The polar region apparently suffered greater oscillations of climate than did equatorial regions, just as in Pleistocene time. Whereas in much of Gondwanaland the tillites alternate with coal and reptile-bearing strata, in North America and Europe there is no evidence of significant climatic fluctuations—only a gradual, long-term increase of aridity of the land through Permian and Triassic times as all continents became larger.

Pangean Biogeography

With all the world's continents sutured together into Pangea, the oceanic and atmospheric circulation was far different from what we see today. Solar radiation and the reflection of that radiation from the earth's surface, together with those circulation patterns, affected the climate in late Paleozoic time. We have seen glacial evidence proving that both the southern and northern high latitudes were cold, while between them lay a tropical **Tethys Sea,** with a rich and diverse fauna (Tethys was a mythical Titaness, wife of Oceanus and daughter of Uranus). This sea separated Gondwanaland from Laurasia and nurtured a homogeneous fauna from Permian through Eocene time.

Permian marine faunas reflect a steep pole-to-equator temperature gradient (Fig. 13.46), with very cold waters in high latitudes. In the Asiatic part of Laurasia, faunas consist of very few species but individuals of large size. In Gondwanaland (e.g., Australia), we also find a typical cold-water fauna with few species. These two faunas are separated by the low-latitude Tethyan fauna containing hundreds of species, many of which can be traced east and west for thousands of kilometers. This warm-water fauna and its descendants existed through the Mesozoic Era until Eocene tectonic activity began to destroy the continuity of the seaway. Today only relics of that once glorious ocean and its fauna remain as the Caspian, Black, and Mediterranean Seas.

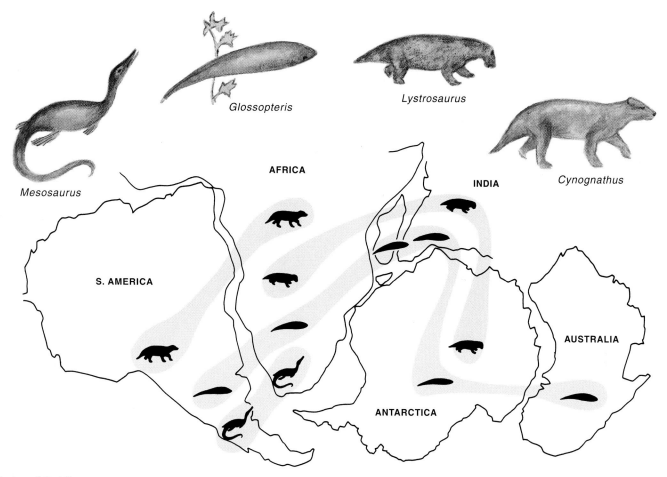

Figure 13.48   Reconstruction of Gondwanaland, showing the recorded distribution of key Permian and Lower Triassic genera, including *Lystrosaurus*. (From Edwin H. Colbert, 1973, figure 31, p. 72 from WANDERING LANDS AND ANIMALS. NY: Penguin Putnam Inc.)

Valentine and Moores summarized the relationship between the assembled continents and faunal diversity. When Pangea was assembled, the climate became zoned and strongly seasonal, with a steep latitudinal temperature gradient. In the oceans, there were few barriers, so that species were widespread, and hence there was moderate species diversity. The only barrier as such was the colder temperature of the water at higher latitudes, which excluded many species that could tolerate only warmer water, the result was lower polar diversity. Just as today, however, some species could tolerate a wide range of temperatures and even of salinity, and these species were more **cosmopolitan** (widespread) forms. As Pangea began to break up during the Mesozoic, atmospheric and oceanic circulation was altered, and the climate became more uniform and warmer. Distinctive biotas formed where separated by barriers. For example, widening deep-ocean basins between the separating continents were barriers to some marine organisms, just as they were to land-dwelling ones. Species diversity increased, particularly in shallow shelf seas, as stable but separated environments became established. Diversity of land organisms also increased on the now separated continents.

During Permo-Triassic time, vast stretches of lowlands on Pangea had homogeneous floras and faunas traceable over great distances. The diversity of land reptiles was quite low, compared

with the rich Cenozoic mammal faunas after the breakup of Pangea. *Lystrosaurus* (Fig. 13.45) illustrates the wide distribution of the land animals during this time, for it is known from many localities on three southern continents (Fig. 13.48). Its present disconnected, or disjunctive, distribution also serves as one of the most convincing pieces of evidence for an Antarctic-Africa-India connection. As mentioned earlier, the remains of *Lystrosaurus* were discovered in 1969 in Antarctica, and they proved to be identical to those of species found in South Africa and India. This reptile was a land dweller and could not have swum across any wide expanse of water. Previously discovered vertebrates, such as the small aquatic reptile *Mesosaurus* (Fig. 13.42), commonly had been discounted as having no relevance to the question of former land connections.

Disjunct faunas and floras can come about in two ways. One is that a living fauna becomes widely distributed but, due to some later environmental change along its range, the distribution is broken into disconnected patches. The second type of disjunction is caused by continents drifting apart to break up a former continuous range. Either the living organisms ride on a continent, which becomes a kind of **Noah's Ark,** or already fossilized remains ride on a continent, which is more like a **Viking Funeral Ship.** In the former case, the organisms may have adapted and

A.

B.

Figure 13.49   *A:* Undersea landscape of the shallow crinoidal shoals so typical of the Mississippian. Most of the stalked forms are crinoids or blastoids. Note the screw-shaped colonies of the bryozoan *Archimedes* in the left foreground. *(American Museum of Natural History.) B:* Modern example of a crinoid meadow. Although rare and smaller in scale today, such scenes capture the long vanished world of the Mississippian. Note how the crinoids are bent by the current, but their feathery arms curl back to trap food particles. These individuals live in 1,300 feet of water off the Little Bahama Bank. *(Courtesy of Charles Messing.)*

evolved as their ark sailed through different climatic zones, or they may have become extinct en route. Naturally the fossils floating on Viking Funeral Ship continents would not be affected by such environmental changes. In either case, the same genera or species will then be found on different continents and will be considered disjunct, as in the case of *Lystrosaurus.*

Late Paleozoic Life

In the Devonian Period, animals and plants developed complex communities on the land as well as in the ocean. Late Paleozoic time was marked by a continued expansion of terrestrial forests, as well as further evolution of the marine communities. By the Pennsylvanian, the great epeiric seas were beginning to withdraw, expanding the available habitat for terrestrial life. In the Permian, shallow-marine seas became very restricted and terrestrial deposits made up the bulk of the strata. We see new environments, such as vast coal swamps, never before encountered on this planet. Consequently our focus will shift away from a strictly marine emphasis to the innovations on land.

Marine Life

Although the Late Devonian extinctions affected the reef community and caused widespread extinction in brachiopods and ammonoids, the major phyla of invertebrates soon recovered and remained dominant for the rest of the Paleozoic. However, the relative abundance of certain groups changed dramatically from the earlier Paleozoic. In addition, several new groups arose during

the late Paleozoic, dominating the sea floor until the great extinction at the end of the era.

The differences can be seen by comparing the Mississippian sea floor (Fig. 13.49) with that of the earlier Paleozoic. As we saw earlier in this chapter, the Mississippian was characterized by the last great highstand of epeiric seas during the Paleozoic, flooding almost the entire continent under shallow, warm, tropical waters (Fig. 13.3). Consequently this enormous shallow-marine area was ideal for carbonate-producing organisms, especially *crinoids* (Figs. 13.49 and 13.50(A). All over North America, the most typical rocks of the Mississippian are crinoidal limestones, representing astronomical numbers of individual crinoid animals that lived, died, and then were scattered all over the sea floor! In these abundant crinoid meadows were another type of stalked echinoderm, the *blastoids* (Fig. 13.50B). A classic index fossil of the Mississippian is the blastoid *Pentremites,* whose head is shaped like a flower bud.

Crinoids and blastoids were gently waving on their long stalks above the sea bottom, but there were other important colonial filter feeders. Stromatoporoids and tabulate corals were very rare, and rugosids were scarce, solitary forms. There were a few small patch reefs in places but none of the great barrier reefs so typical of the Silurian and Devonian. Instead, the great crinoid-blastoid meadows featured another important group of filter feeders, the bryozoans. Unlike the lumpy or branching bryozoans of the Ordovician, however, Mississippian bryozoans (known as *fenestrate bryozoa*) formed lacy, fanlike structures (Fig. 13.51A). One of the most distinctive index fossils of the Mississippian is the fenestrate bryozoan *Archimedes* (Fig. 13.51B). Its name is derived from its shape, with the lacy fan of the bryozoan colony

B.

Figure 13.50   *A:* Densely fossiliferous slab covered with complete crinoids from the Mississippian of LeGrand, Iowa. This extraordinary locality has produced hundreds of slabs with many perfectly preserved specimens, so they are on display in museums all over the world. *B:* Flowerbud-shaped heads of the blastoid *Pentremites.* In life, these heads (about 1 to 2 cm in width) would have attached to long stems, like those of crinoids, and these animals would have had a similar filter-feeding mode of life, with tentacles extending from the five petal-like grooves on the top. Although blastoids were found throughout the late Paleozoic, *Pentremites* is so abundant and characteristic of the Mississippian that it is an important index fossil. *(D. R. Prothero.)*

arranged in a spiral around a corkscrew axis. This configuration reminded the scientist who named it of Archimedes' screw, a device invented by Hellenistic Greek scientist Archimedes (287–212 B.C.) to pump water up a tube. In many places, Mississippian rocks are largely composed of lacy sheets of fenestellids, and some of the Mississippian formations in Missouri, Iowa, Illinois, and Indiana were once known as the "*Archimedes* beds" for the great abundance of their "corkscrews."

Beneath the canopy of stalked crinoids and the lacy fenestellid bryozoans, the sea floor was still dominated by brachiopods. The spirifers had begun to decline, however, and they were replaced by a group of strophomenids that dominated the rest of the Paleozoic. Known as *productids* (Fig. 13.52), these brachiopods had one shell that was deeply cupped and sat on the sea floor, propped up by long spines, which either served as stilts on sandy or muddy bottoms or anchored the animals to hard bottoms. The other shell was a thin, flat lid, which the productid could raise

while filter feeding. By Pennsylvanian time, productids had become the most common brachiopods, and they remained so for the rest of the Paleozoic.

The great Mississippian crinoid-blastoid-fenestellid-productid communities had other familiar inhabitants as well. Most of the typical Devonian fish groups were extinct, but ray-finned fish continued to thrive and were the dominant vertebrate predators. In fresh waters, the largest predators were xenacanth sharks, which had distinctive, double-pronged teeth, a long spike on the head, and a strange symmetrical tail (Fig. 13.53). Some xenacanths reached 2.1 meters (7 feet) in length! Although severely affected by the Devonian extinctions, the goniatitic ammonoids recovered in the Mississippian and remained the dominant invertebrate predators. Other molluscs, such as clams and gastropods, continued to evolve but were scarce in the limy crinoidal carbonate banks.

Some typical early Paleozoic animals were no longer important. Graptolites were practically gone, and trilobites were extremely scarce. Among microfossils, acritarchs were virtually gone, but conodonts were still common, and another group, the amoeba-like protozoans known as *foraminiferans,* became abundant. Although protozoan fossils occur in the Cambrian, they were always rare until the end of the Devonian. In the Mississippian,

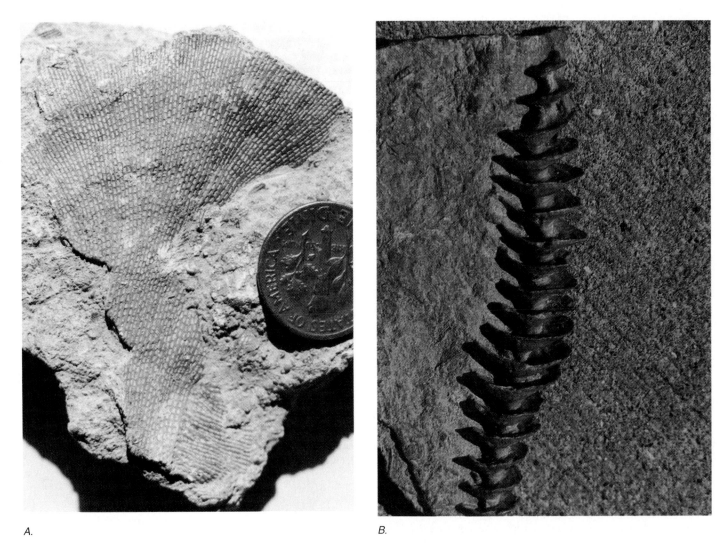

A.   B.

Figure 13.51   *A:* In addition to crinoids, the Mississippian witnessed a great abundance of other attached filter feeders, such as the bryozoans. Most were delicate, lacy forms called fenestellids, which had hundreds of tiny, filter-feeding animals along the latticework of their skeleton. *B:* The most characteristic Mississippian bryozoan was *Archimedes,* a fenestellid whose lacy skeleton was arranged around a central spiral "corkscrew" stem. It was named in honor of the great Greek mathematician who invented a water pump that had a screw inside a tube, reminiscent of the shape of the bryozoan fossil. *(D. R. Prothero.)*

they were among the many abundant carbonate-secreting organisms living on the sea bottom, feeding and creeping along, with their fingerlike blobs of protoplasm. In some places, they were extremely abundant. The famous Mississippian Indiana Limestone (Fig. 13.54) has been mined in central Indiana for facing the Empire State Building and many famous governmental structures. It is the single most widely used building stone in the United States, and the huge, abandoned limestone quarries were even featured in the movie *Breaking Away.* Indiana Limestone is made almost entirely of the shells of the foraminiferan known as *Endothyra.* For years, however, their tiny, spherical shells were mistaken for ooids, so this stone is often called the "Indiana oolite." In some areas, the Indiana Limestone is truly oolitic.

As we saw earlier in this chapter, the Pennsylvanian was marked by the partial retreat of the great Mississippian epeiric carbonate seas. There was a rapid fluctuation of carbonate and clastic deposits, especially in the deltaic regions of the Midwest. These muddy deltaic environments were not suitable for as great a variety of filter feeders as were the clear, tropical waters of the Mississippian. Consequently typical Pennsylvanian invertebrate communities have fewer corals, crinoids, blastoids, and bryozoans. Instead, they have a great diversity of mud-dwelling molluscs, especially clams and gastropods, and of productid brachiopods, propped up out of the mud on their spiny stilts. Bony fish and sharks must have also been important predators in these communities, since the spines of

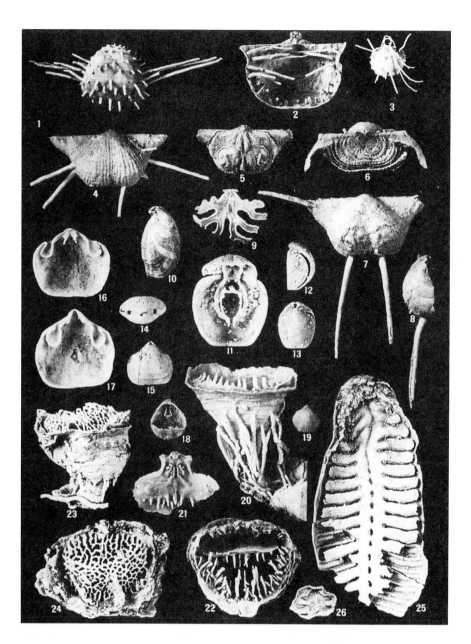

Figure 13.52 The dominant brachiopods of the late Paleozoic were the productids. Most had a deep, cup-shaped shell, which sat concave side up on the muddy bottom, supported by stiltlike spines (specimens 1–7); the other shell was a flat lid, which was raised when they filter fed. This plate shows Permian brachiopods, which include other productids that modified this basic shape in bizarre ways. For example, the richthofenids (specimens 20–24) stretched their lower shell into a long cone, and the upper shell was a tiny lid inside; over the mouth of the cone, they had a meshwork (specimens 22–24). In their conical shape, they converged on other filter feeders, such as rugose corals (see Figs. 11.6G, and 12.26D), some glass sponges (see Fig. 12.14), and later with rudistid clams (see Figs. 14.38 and 14.39A). The strangest productid of all was *Leptodus* (specimen 25), whose lower shell was a shallow "dish" and upper shell a "comb," which allowed water to flow in even when the shell was closed. *(American Museum of Natural History.)*

Figure 13.53 Once the primitive jawed placoderms of the Devonian (Figs. 12.1 and 12.7) became extinct, the dominant marine vertebrates were sharks and bony fish. The most impressive were the xenacanth sharks, which had a strangely symmetrical tail fin, double-pronged teeth, and an odd spike on its head; some were over 2 meters in length. This xenacanth is shown chasing some of the earliest ray-finned bony fishes, ancestors of nearly all living bony fish. *(Painting by Zdenek Burian.)*

Figure 13.54   Quarry of "Indiana Limestone," which produced the building stone for the Empire State Building and many other important structures, near Bedford, Indiana. During their heyday, these quarries produced 1.0 to 1.5 million tons a year, and today there are huge abandoned quarries, which cover many square miles. Abandoned quarries that were later flooded provided the swimming hole settings in the movie *Breaking Away.* "Indiana Limestone" is often called the "Bedford Oolite," but most of the "ooids" are actually shells of the foraminiferan *Endothyra.* (D. R. Prothero.)

acanthodians and the double-pronged teeth of xenacanth sharks are also common fossils.

The most important new addition to the late Paleozoic sea floor was a new group of foraminiferans called the *fusulinids* (Fig. 13.55). These were among the largest of all protozoans, growing in a spiral around a central axis to form a spindle-shaped shell. Although they were single-celled protozoans, some fusulinid fossils reached over a centimeter in length! Like many modern, large forminiferans, they probably had symbiotic algae living in their tissues, which enabled them to grow rapidly and secrete such a large calcite skeleton. The complex foldings in the walls of fusulinids make it possible to identify them accurately when they are cross sectioned. Fusulinids evolved rapidly in the late Paleozoic, with many species lasting only 2 to 3 million years. Because they were extremely widespread across the tropical shallow-marine belt on all continents, they are the best group of index fossils for the Pennsylvanian and Permian. Fusulinid specialists can correlate rocks from Siberia to central Europe to the American Midwest, subdividing late Paleozoic time into biostratigraphic zones of less than a million years. In some places, fusulinids were so abundant that the sea floor was packed solid with them, producing fusulinid limestones made of nothing but their skeletons.

By Permian time, shallow-marine environments in North America had become restricted to just a few places in the western and southwestern United States (Fig. 13.17). Marine invertebrate communities known from these areas are dominated by a few phyla. Fusulinids are found in great abundance in nearly every marine limestone, with hundreds of species known from the Permian alone. Productid brachiopods reached the climax of their evolution. In addition to simple cup-shaped forms, they evolved into several bizarre shapes. One kind of productid, the richthofenids (Fig. 13.52), were shaped like an ice cream cone on stilts, with one shell sitting in the mouth of the cone, like a lid. The most bizarre brachiopods of all, however, were productids such as *Leptodus.* These brachiopods had a shallow, leaf-shaped lower shell, and the "lid" shell was reduced to a pair of "combs," which allowed water to flow in even when the brachiopod "shut" its shell! These animals were glued to hard surfaces, and their irregular shape resembled that of modern oysters. There must have been very few shell-crushing predators in the Permian for such a delicate, unprotected animal to have thrived.

The best-known example of Permian marine communities comes from the famous "Permian reef complex" in western Texas and New Mexico (Figs. 13.18 and 13.19). Unlike true reefs of the earlier Paleozoic or present day, this one had no wave-resistant framework of corals (Fig. 13.56). Instead, these large accumulations were formed mainly by calcareous sponges and huge concentrations of bryozoans, productid brachiopods, calcareous algae, and fusulinids growing on the slope of a carbonate shoal, in this sense, it was not a true reef. The community was very diverse, with abundant clams and gastropods in muddier parts of the carbonate bank. Crinoids and solitary rugosid corals were also present but were much less important than the sponges, productids, bryozoans, and calcareous algae. Very rare, but still straggling on to the end of the Paleozoic, were the last of the trilobites.

Pennsylvanian Coal Swamps and Permian Coniferous Forests

In the Late Devonian, the first forests were emerging from the swamps, and diverse arthropods and the first amphibians were living in these forests. During the late Paleozoic, terrestrial life diversified into a much more complex range of communities, with a wide variety of new plants and animals.

The most striking developments occurred in land plants. lant life became so diverse and abundant that the most characteristic rock is coal. Indeed, the European name "Carboniferous" means "coal-bearing" in Latin, and the strata of this period were first called the "Coal Measures." In North America, the Pennsylvanian got its name because of the abundance of coal deposits of that age in the state of Pennsylvania; the Mississippian, named for the exposures in the central Mississippi River Valley, comprises the early portion of the Carboniferous in the United States. These ancient coal swamps were unprecedented in their scale, covering large areas of every continent for the first time in earth history. Coal never again accumulated to this extent.

The impact of the swamps has extended even to the present day. Carboniferous coal deposits were largely responsible for the industrial revolution, making industrial centers of the coal-producing

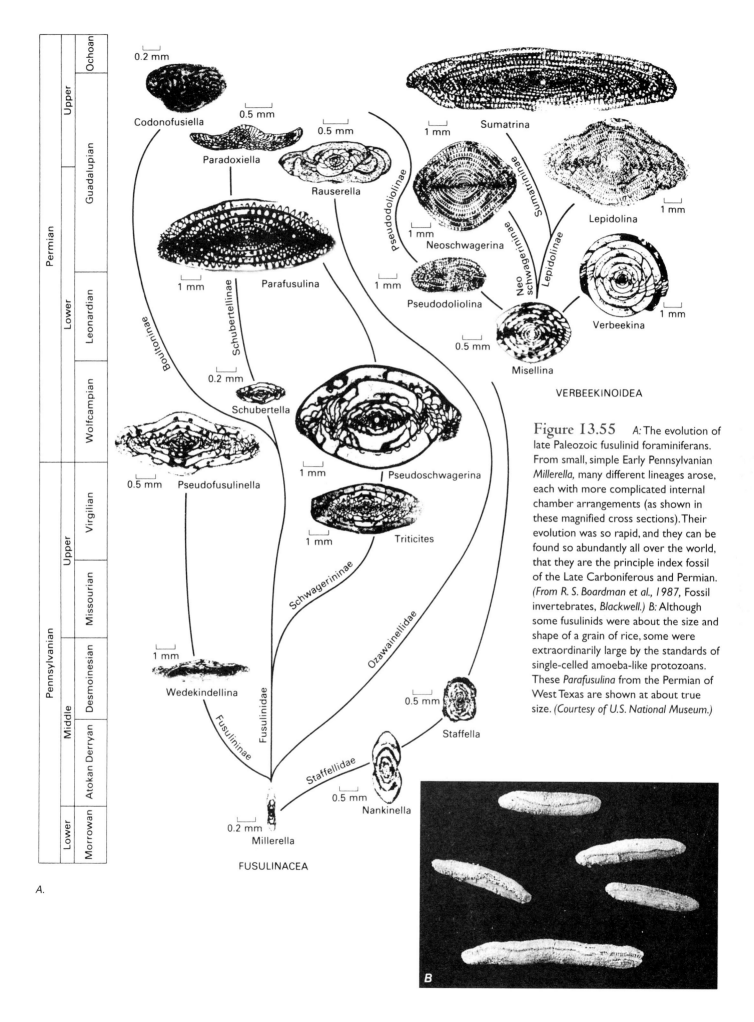

Figure 13.55 *A:* The evolution of late Paleozoic fusulinid foraminiferans. From small, simple Early Pennsylvanian *Millerella,* many different lineages arose, each with more complicated internal chamber arrangements (as shown in these magnified cross sections). Their evolution was so rapid, and they can be found so abundantly all over the world, that they are the principle index fossil of the Late Carboniferous and Permian. *(From R. S. Boardman et al., 1987,* Fossil invertebrates, *Blackwell.) B:* Although some fusulinids were about the size and shape of a grain of rice, some were extraordinarily large by the standards of single-celled amoeba-like protozoans. These *Parafusulina* from the Permian of West Texas are shown at about true size. *(Courtesy of U.S. National Museum.)*

Figure 13.56   Diorama of a Permian underwater scene represented by the fossils of the Glass Mountains of Texas. Prominent in the center are spiny productid brachiopods, with tall sponges in the background. The brown, beadlike strings in the left background are the calcareous sponge *Girtyocoelia*. In the front center are the leaflike brachiopod *Leptodus*, and the spiny coiled nautiloid *Cooperoceras* is shown in the left foreground. *(Image #: K10269(4) American Museum of Natural History.)*

Figure 13.57   Restoration of typical plants of the Pennsylvanian coal swamps. The tall, scaly tree in the center is the lycopsid *Lepidodendron*, and the tree with the bush of leaves at the tip of the stem is the lycopsid *Sigillaria*. The sphenopsid ("horsetail") *Calamites*, with the jointed stems in the right foreground, grows in front of tall tree ferns. The ground is covered by spore-bearing ferns. *(The Age of Reptiles, a mural by Rudolph F. Zallinger. © 1966, 1975, 1985, 1989, Peabody Museum of Natural History, Yale University, New Haven, Connecticut. Courtesy of the Peabody Museum of Natural History, Yale University.)*

regions of the England Midlands, several parts of Europe, and the American Midwest. Paleozoic plants have contributed directly or indirectly to the housing, production, and general well-being of Western civilization for over 200 years. They still constitute one of the earth's great reservoirs of stored energy.

During the Mississippian, the widespread epeiric seas left few habitats for terrestrial life, and the land plants were not much different from those of the Devonian (Fig. 12.11). However, the growth of floodplains and deltas in the Pennsylvanian marked a great expansion of coal swamps and other terrestrial habitats. These immense swamps were densely vegetated with a great variety of plants, many different types of which produced tree-sized vegetation for the first time. Among spore-bearing plants, the lycopsids produced gigantic trees reaching 30 meters (100 feet) in height, with their characteristic diamond-shaped, scaly bark. Two of the most common coal swamp trees were *Lepidodendron* and *Sigillaria* (Fig. 13.57). The joint-stemmed sphenopsids, or scouring rushes (Fig. 12.10E), also reached giant size. A common Carboniferous sphenopsid was *Calamites*, whose branches clustered at the joints between vertical-grooved stems. There were also tree-sized true ferns (which are spore-bearing), although true ferns were most abundant in the understory.

In addition to spore-bearing plants, seed plants began to flourish. The great trees of the Devonian Gilboa forest (Fig. 12.10G) were seed ferns, which had fernlike leaves but reproduced by seeds rather than spores. Seed ferns are among the most primitive gymnosperms, the plant group that now includes the conifers (pines, spruce, and redwood) and ginkgoes. Gymnosperms have both male cones containing pollen-bearing grains and female cones containing embryonic seeds. The pollen must be transported (usually by wind) to the female cone, where fertilization can take place and true seeds form (see Box 13.3). Seed

ferns, along with spore-bearing ferns, formed the middle story of vegetation, reaching heights of 3 to 4 meters (10 to 14 feet) below the canopy of lycopsids and sphenopsids.

There were other common gymnosperms, such as *Cordaites*, a tree with long, straplike leaves (Fig. 13.60C). *Cordaites* reached 30 meters (100 feet) in height but were typically found in drier ground, where their gymnospermous seeds gave them an advantage over water-loving lycopsids and sphenopsids.

In addition to seed ferns and *Cordaites*, there was a variety of other primitive gymnosperms in the late Paleozoic. One of the most famous was *Glossopteris*, the commonest plant of the southern Gondwana continent (Figs. 13.38 and 13.40).

With the cooling and drying of the Permian, the great coal swamps that so dominated the Pennsylvanian began to disappear. Land plants underwent yet another revolution comparable to their Devonian invasion of the land. This second plant revolution is so fundamental in plant history that some paleobotanists divide the last 600 million years differently than do paleozoologists. Our geologic time scale reflects the three great divisions in animal life ("Paleozoic," "Mesozoic," and "Cenozoic" mean "ancient animals," "middle animals," and "recent animals"). The divisions in plant history occurred at different times. Some paleobotanists use the term "Paleophytic" ("ancient plants") to refer to the Silurian through early Permian, dominated by psilophytes, lycopsids, sphenopsids, and seed ferns. The "Mesophytic" began in the Middle Permian, with the great drying event that devastated water-dependent swamp plants and

Box 13.3

The Advantages of Seeds

Seedless vascular plants require liquid water for successful fertilization. Their sperm must swim through water to the egg, and therefore plants such as ferns can grow only in habitats where at least temporary thin films of water are present (Fig. 13.58). This requirement restricted the earliest land plants to low, moist environments along sea coasts and rivers. Both their spores and fertilized eggs were dispersed haphazardly by water and wind currents.

Exploitation of higher and drier habitats required adaptations for protecting plant reproductive cells from desiccation. This was first accomplished by the gymnosperms ("naked seeds"). These first seed-bearing plants appeared in Early Carboniferous time in the form of a transitional group of seed ferns with fernlike foliage but having seeds for reproduction. Modern conifers are the most familiar gymnosperms, and their ancestors first appeared in Late Carboniferous time.

Gymnosperms avoid desiccation of their reproductive cells in a remarkable way. The male pollen grain which is carried to the egg by wind, possesses a moist tube, through which its sperm passes to fertilize an egg-bearing seed (Fig. 13.59). In other words, the pollen carries with it the portable moist environment essential for fertilization. After fertilization, the embryo develops from nutrients that surround it within the seed's protective cover.

Some seeds are winged for more efficient dispersal by wind, others are dispersed by animals.

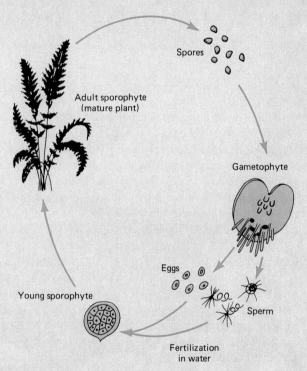

Figure 13.58 Generalized life history of a seedless vascular plant. The adult plant, called a sporophyte, produces spores. The few spores that survive grow into small, independent plants called gametophytes, which produce both eggs and sperm. Fertilized eggs give rise to the adult sporophyte plant. *(After McAlester, 1968, The history of life: Prentice-Hall.)*

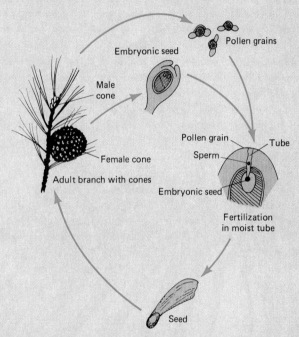

Figure 13.59 Generalized life history of a gymnosperm (a primitive seed plant). The adult plant produces both male cones (which make pollen grains) and female cones (which contain embryonic seeds). Pollen grains travel, usually by wind, to female cones, where a tube grows from the grain to fertilize the embryonic seed. A fertilized seed grows into a mature, cone-bearing adult. *(After McAlester, 1968, The history of life: Prentice-Hall.)*

A.   B.

Figure 13.60   Primitive gymnosperms began to appear in the Carboniferous and were dominant in drier habitats by the Late Permian. A: The primitive conifer *Walchia* from the Late Pennsylvanian of Kansas. *(Courtesy of G. Mapes.)* B: A Permian conifer with slightly broader leaves, very similar to the living gymnosperm *Araucaria* (see Fig. 14.45). *(Courtesy of Bruce Tiffney.)* C: Restoration of the primitive conifer *Cordaites*, one of the commonest late Paleozoic gymnosperms. Their distinctive, strap-shaped leaves and abundant cones are known from many coal swamp deposits, as well as from drier habitats. *(The Age of Reptiles, a mural by Rudolph F. Zallinger. © 1966, 1975, 1985, 1989, Peabody Museum of Natural History, Yale University, New Haven, Connecticut. Courtesy of the Peabody Museum of Natural History, Yale University.)*

C.

discriminated against spores in favor of seeds. The "Mesophytic" persisted until the Late Cretaceous, when flowering plants ushered in the modern era ("Cenophytic").

The Middle to Late Permian marked a great expansion of Mesophytic floras, especially gymnosperms. The drier uplands and floodplains were covered by a variety of gymnospermous trees, including the first conifers (Fig. 13.60). These primitive conifers had needlelike leaves arranged along their branches and small cones for their naked seeds. Lycopsids, sphenopsids, and seed ferns still persisted through the Permian (and survive today), but they became small, creeping forms restricted to watery habitats. True ferns flourished but were found mostly as low growth underneath the canopy of gymnosperms. We shall see this Mesophytic pattern of gymnosperms and ground-dwelling ferns through most of the Mesozoic, feeding the great dinosaurs of the Triassic, Jurassic, and Early Cretaceous.

Swamp Dwellers and Synapsids

In the Silurian, the first animal inhabitants of the newly vegetated land were arthropods, especially scorpions, spiders, and millipedes. A few primitive wingless insects were also found in the Early Devonian. By the Late Carboniferous, insects had undergone an explosive adaptive radiation, taking over the globe, which they still dominate today. Primitive insect groups with wings that cannot fold along the back, such as dragonflies and mayflies, were particularly common. One fossil dragonfly from the Late Carboniferous of France had a wingspan of almost a meter (28 inches)! By the end of the Carboniferous, insects with folded wings had appeared and they flourished in the Permian. Cockroaches, beetles, and relatives of grasshoppers and crickets were particularly common. Some cockroaches reached lengths of 8 to 10 centimeters (3 to 4 inches), and they were so abundant that the Pennsylvanian has been called the "Age of Cockroaches"— an amazingly successful creature, the cockroach! In the Permian, insects that could undergo complete metamorphosis from larva to

pupa to adult were also abundant. In addition to land-dwelling arthropods, we find the first land snails, and the fresh waters were inhabited by the first freshwater clams in the Carboniferous.

At the end of the Devonian Period, the first amphibians (represented by *Ichthyostega*) crawled out onto the land (Fig. 12.13). During the Carboniferous, amphibians were the largest animals in the great coal swamps, and they diversified considerably. From their originally fishlike bodies, amphibians soon developed into large predators, with flattened heads, long snouts, and short, sprawling limbs. Many had their eyes on top of the head, suggesting that they floated in the swamps, like alligators, and snapped up insects or fish. By the Permian, these large, flat-bodied *temnospondyl* amphibians had reached true crocodilian proportions. *Eryops*, from the Early Permian of Texas (Fig. 13.61), was over 2 meters (7 feet) long and weighed about 130 kilograms (285 pounds)! Other groups of amphibians specialized in different ways. Some were small, lizardlike animals; others were legless, like snakes; and one aquatic group of amphibians had bizarre sets of "horns" flaring out from the side of its head that looked like a boomerang (Fig. 13.61). No one knows what the horns were used for, although it has been suggested that they helped the animal swim up from the bottom to catch prey. Another major group of amphibians, the *anthracosaurs*, tended to have deep, rather than flattened, skulls and apparently were more agile and less sprawling than typical temnospondyls (Fig. 13.61). Some anthracosaurs had so many reptilian characteristics that it is difficult to tell where the amphibians ended and the reptiles began. Anthracosaurs continued to thrive well into the Permian, long after reptiles had become successful.

The earliest undisputed reptiles appear in the Middle Carboniferous, not long after the anthracosaurs. Their most important evolutionary innovation was the *amniotic egg* (Fig. 13.62). Amphibian eggs are porous, so that oxygen can enter and wastes

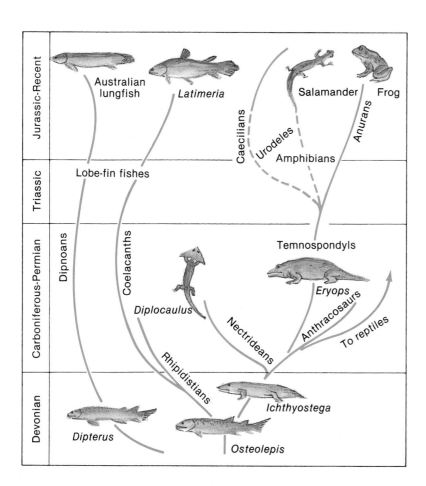

Figure 13.61 Family tree of late Paleozoic amphibians. Three typical groups of amphibians (all from the Early Permian of Texas) are shown. The commonest were the large, flat-bodied temnospondyls, such as *Eryops (top)*, which reached 2 meters in length. A second group were the nectrideans, including the strange *Diplocaulus (bottom)*, with its broad, "boomerang" head and salamander-like body. The function of the flanges on its head is unknown, but they may have helped the animal swim. The third group, the anthracosaurs such as *Seymouria (center)*, eventually gave rise to reptiles. Indeed, some anthracosaurs are so reptilian in certain features that paleontologists still argue about where to draw the line that defines "reptile."

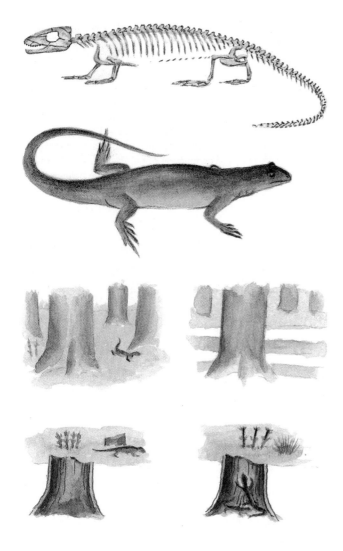

can escape, but they must stay immersed in water. No matter how successful they become on land, amphibians must return to the water to breed. However, the reptilian egg encases the embryo in an amniotic membrane, which allows oxygen to enter without losing water. The amniotic egg also has a yolk sac to feed the developing embryo and a sac for storing wastes. Most amniotic eggs have a shell to protect them. The great success of the amniotic egg allowed all *amniotes* (reptiles, birds, and mammals) to reproduce on land, a development that opened up many new habitats far from water.

Because we have no way of knowing whether some extinct anthracosaurs might have laid amniotic eggs, we use certain skeletal features to recognize the earliest amniotes. Among the earliest fossils widely recognized as amniotes is the little, lizard-like animal *Hylonomus* (Fig. 13.62), originally found inside fossilized rotted tree stumps of the lycopsid *Sigillaria* from an early Pennsylvanian locality near Joggins, Nova Scotia. Unlike anthracosaurs, these early amniotes were very lightly built, with slender limbs and strong, deep jaws.

By the Late Carboniferous, amniotes had split into two main lineages. One group, the reptiles, began with small, lizardlike animals such as *Hylonomus* and eventually diversified into turtles, marine reptiles, lepidosaurs (lizards and snakes), and archosaurs (crocodiles, dinosaurs, and birds). The other lineage is the **synapsids,** or "mammal-like reptiles." It is incorrect to call them "reptiles," however, because these animals diverged at the beginning of amniote evolution from a stock completely separate from *Hylonomus* and true reptiles. In fact, the earliest synapsids appeared side-by-side with the oldest true reptiles in the same Early Pennsylvanian fossilized tree stumps of Joggins, Nova Scotia. As the nickname implies, synapsids were the lineage from which mammals evolved.

In the Early Permian, both reptiles and synapsids evolved into a great variety of large land predators and herbivores. The famous Lower Permian red beds of Texas, Oklahoma, and New Mexico have produced a great variety of spectacular vertebrate skeletons,

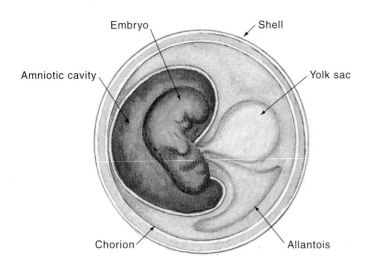

Figure 13.62   *Top:* Skeleton and reconstruction of one of the earliest known reptiles, *Hylonomus,* from the Early Pennsylvanian of Joggins, Nova Scotia. It was a small, lizard-like creature about a foot in length. *Middle:* The preservation of fossils at the Joggins locality is very unusual. Rotted stumps of the lycopsid *Sigillaria* were buried and formed open-pit traps, into which a number of small animals fell and were trapped and fossilized. *Bottom:* The defining feature of reptiles and all higher vertebrates is the amniotic egg. Adapted for being laid and hatched on land (rather than in water as for amphibians), it has a tough shell for protection and resistance to desiccation and several distinct membranes inside. The amnion surrounds and protects the growing embryo. The yolk sac provides the nourishment, and the allantois contains the waste products during development.

Figure 13.63   Early Permian synapsids and amphibians from the red beds of north Texas. The dominant predator, the finback *Dimetrodon* (center), was over 2 meters long, with long, stabbing teeth. Another finback was *Edaphosaurus* (light tan in foreground, with knobs on fin), a plant eater. The purpose of the fins on these synapsids is unknown, but many paleontologists think it helped them warm up and cool off quickly. The lizardlike synapsid *Varanosaurus* sprawls on the left, and the "boomerang-head" amphibian *Diplocaulus* (see Fig. 13.61) swims in the right foreground. *(Painting by Charles R. Knight/The Field Museum, #Ck45T.)*

revealing a diverse and successful terrestrial community. The members of this community ranged from small, aquatic, "boomerang-headed" amphibians to sprawling *Eryops* to reptile-like anthracosaurs, holdovers from the Early Carboniferous. Also on the scene were a number of archaic reptiles, including an ancient relative of turtles, which looked like beaked lizards. The dominant carnivorous role, however, was occupied by the primitive synapsids. Some were fish eaters, but the most famous of them all were the finbacked synapsids (formerly called "pelycosaurs," another unnatural "wastebasket" group). These included *Dimetrodon* (Fig. 13.63). It reached 2 meters (7 feet) in length and may have weighed 100 kilograms (220 pounds). Its ferocious mouth was filled with sharp, peglike teeth, including stabbing canines in the front. The fin on its back was apparently used as a heat collector and radiator. When *Dimetrodon* basked with its fin in the sun, it could quickly warm up its cold-blooded body to catch a slower-moving reptile or amphibian. Another finback, *Edaphosaurus*, was herbivorous. Along with the huge anthracosaur *Diadectes*, it was one of the few herbivorous vertebrates in the Early Permian. The slow-growing gymnosperms of the early Permian did not yet have to survive the destruction wrought by great browsing animals. Most of the food pyramid was apparently based on insect- and fish-eating amphibians and reptiles, which were in turn prey for carnivorous finbacks.

Late Permian terrestrial ecosystems are represented by spectacular vertebrate faunas in South Africa and Russia. These communities still had large, flat-bodied temnospondyl amphibians, but most of the other types of amphibians, including the anthracosaurs and "boomerang-head," were extinct. In their place was a spectacular radiation of synapsids, with more than 20 different families appearing in place of their finback ancestors (Fig. 13.64). There were more than 170 Late Permian synapsid genera and

Figure 13.64   By the Late Permian and Early Triassic, primitive finbacked synapsids had disappeared and had been replaced by more advanced synapsids. A pack of wolflike carnivorous *Cynognathus* are seen here, attacking the cow-sized plant-eating dicynodont *Kannemeyeria* (right). Like another dicynodont, *Lystrosaurus* (see Fig. 13.45), *Kannemeyeria* had a nearly toothless beak for biting off tough plants. *(Painting by Charles R. Knight/The Field Museum, #Ck22T.)*

only about 15 reptiles. Most of these synapsids were predators, and they began to look progressively more and more mammal-like. Their teeth have mammalian specializations, especially large, stabbing canines in front and multicuspid cheek teeth in back. Their skulls and jaws also underwent a number of evolutionary changes, which eventually led to mammals (Fig. 3.19). There is some evidence (such as pits on the snout for whiskers) that the most advanced Permian synapsids may have had hair or fur and other mammalian features. Some paleontologists have argued that these animals even kept their body temperature constant, as mammals do.

Although the large, carnivorous synapsids were on the main line of evolution, culminating with mammals in the Triassic, several groups of synapsids were herbivores. One group had a plant-cutting beak (Fig. 13.64), which in most species was toothless, except for large upper canines. *Lystrosaurus* (Fig. 13.45) was a small example of these plant-eating synapsids. Another group of huge synapsids had ugly faces capped by a thick, bony skull for head butting; one of them was over 2.5 meters (8 feet) long (Fig. 13.65). *Pareiasaurus*, a hippolike beast related to archaic Pennsylvanian reptiles, was even more grotesque, with bony bumps all over its face; it may have weighed 600 kilograms (1,300 pounds) and reached 3 meters (10 feet) in length. All these herbivores were undoubtedly eaten by the large, predatory synapsids, such as *Gorgonops*, which weighed 100 kilograms (220 pounds) and reached 2 meters (7 feet) in length. Hiding in the vegetation from these great predators were the earliest relatives of lizards and snakes and the earliest relatives of crocodiles and dinosaurs. In the Late Permian, however, both of these groups were represented only by small, lizardlike animals. Their spectacular future success was not yet apparent in a landscape dominated by synapsids.

The Late Permian Catastrophe

When mass extinctions became a trendy topic in the 1980s, almost all the attention was focused on the extinction event at the end of the Cretaceous because this was the event that wiped out the ever-glamorous dinosaurs. However, the Cretaceous extinction was only the second or third worst crisis in earth history. By far the most severe of all marine extinctions occurred at the end of the Permian. By some estimates, *it was so devastating that 90 to 95 percent of all marine species on earth died out!* If things had gotten much worse, life in the oceans might have been extinguished altogether. The Permian catastrophe literally marked the end of an era and radically rearranged the landscape of life. The phyla dominant since the Late Cambrian, the "Paleozoic fauna" (Fig. 11.10), were decimated, and many groups went extinct altogether. When life recovered millions of years later in the Triassic, a completely new cast of characters recolonized the sea floor. This was the "Modern fauna" (Fig. 11.10). As the name implies, this fauna still populates the oceans today.

The list of Permian victims is long and surprising. Some animals were declining throughout the Middle and later Permian, such as the trilobites, archaic molluscs, tabulate and rugose corals, and orthid brachiopods; the archaic molluscs and trilobites (Fig. 13.66) were extinct before the end of the period. However, other groups that thrived in the mid-Permian were decimated. The fusulinids literally covered the sea floor in the Late Permian, but they were all wiped out. Productid brachiopods were by far the most common shelly organisms, but they did not survive, either. Spirifers did survive in reduced numbers but faded out until their final extinction in the Jurassic. Modern brachiopods are mostly members of two other orders, the rhyn-

Figure 13.65 In addition to toothless, beaked dicynodonts (such as *Lystrosaurus* and *Kannemeyeria*), another important group of Late Permian herbivorous synapsids was typified by monsters such as these *Moschops*, which were over 2.5 meters long and had thickened bone in the roof of their skulls for head-to-head butting. *(Painting by Zdenek Burian.)*

chonellids and terebratulids, which were less important in the Paleozoic. Bryozoans, which built huge reefs in the Late Permian, were decimated. All the typically Paleozoic bryozoans, such as the lacy fenestellids and lumpy and branched Ordovician groups, went extinct. Most modern bryozoans belong to a group that arose in the Triassic. Crinoids were nearly wiped out; only one family survived to evolve into modern crinoids. The archaic nautiloids went extinct, and the ammonoids nearly died out; apparently only two or three ammonoid genera survived into the Triassic to begin another great Mesozoic radiation. Clam and gastropod genera suffered about a 30 percent reduction. In short, virtually every group of marine animals that was common in the Paleozoic was either wiped out or barely survived to straggle on, never regaining its former diversity or abundance.

Terrestrial extinctions were also severe. Permian floras show a gradual shift from *Cordaites* and ferns to conifers, cycads, ginkgoes, and other gymnosperms, but this change began in the Early Permian and was stretched out over 30 million years well into the Triassic. The floral change happened in different areas at different times, suggesting that it may have been largely due to climatic adjustment. Terrestrial vertebrates show several waves of extinction as more primitive synapsids and reptiles were replaced by successive waves of more advanced vertebrates. From the Late Permian into the Triassic, about 75 percent of the vertebrate families went extinct, including six families of archaic amphibians, a variety of primitive reptiles, and fifteen families of synapsids. Because the fossil record of these animals is very poor through the Late Permian and Early Triassic, it is hard to say whether these

Figure 13.66   Many marine invertebrate groups that were successful for most of the Paleozoic became extinct during the great crisis at the end of the Permian. In addition to the complete extinction of blastoids, rugosids, tabulates, and fusulinids and the near extinction of crinoids, bryozoans, most brachiopods, and ammonoids, many other groups suffered. These specimens of the trilobite *Anisopyge cooperi* from the Upper Permian of Texas are among the last of the trilobites known from the fossil record. *Anisopyge* was a tiny trilobite (about a centimeter in length). *(Courtesy of David K. Brezinski.)*

extinctions were concentrated at the Permo-Triassic boundary or spread out over 20 million years.

What could have caused such devastation in the ocean *and* on land? First of all, it is clear that the Permian extinctions were not a single catastrophic event. In fact, they were spaced out over the last two stages of the Permian, which spanned about 7 million years. In some animal groups, there were significant extinctions throughout the Permian. In addition, extinctions were concentrated on tropical, warm-water animals. For example, in the Late Permian, the fusulinids, bryozoans, rugosids, and a number of other animals became restricted to the warm waters of the Tethys Sea, suggesting that the higher latitudes were becoming too cold for them (although this restriction may be because most of the Upper Permian outcrops come from low-latitude regions).

From this evidence, some paleontologists attribute the Permian extinctions to global cooling, just as we saw in the great Ordovician and Devonian extinctions. Those previous extinctions occurred when the Gondwana continent had moved over the South Pole, triggering polar glaciation (Fig. 13.47). Not surprisingly, Antarctica was still over the South Pole in the Middle Permian. Unlike the earlier Paleozoic, however, the continental configuration of the Permian was unique. A single supercontinent, Pangea, stretched nearly from pole to pole, completely cutting off the oceanic circulation in the equatorial belt (Fig. 13.46). For the first time ever, we find evidence of an ice sheet in the Northern Hemisphere at the same time as the southern ice cap existed. Covering about half a million square kilometers (about 200,000 square miles), the Middle Permian glacial deposits in Siberia reach as much as a kilometer (0.6 mile) in thickness. These glacial deposits overlie marine limestones, indicating that the ice sheet rather suddenly followed the warm tropical ocean in the Northern Hemisphere.

Although cooling was indeed severe, recent research suggests that the main phase of glaciation was concentrated in the Middle Permian, long before there were any significant extinctions. By contrast, the latest Permian was marked by a rapid global warming episode, perhaps caused by a runaway greenhouse effect. This warming is strikingly demonstrated by the recent discovery of Late Permian fossil wood in Antarctica, which apparently grew in a warm, mildly seasonal climate, with no evidence of frost or cold winters at such high latitudes. Although the exact details of this warming are poorly understood, it is possible that such a rapid warming after the world had become adapted to cold, glaciated conditions might have severely stressed some marine invertebrates.

Instead of the cooling hypothesis, other scientists have argued that reduced shallow-marine habitat caused the Permian extinctions. Because several continents had fused together to form Pangea, all the continental shelves between them had been destroyed when the continents collided. This greatly restricted the available area of shallow seas, which could have affected many groups of marine invertebrates. Unfortunately for this explanation, the formation of Pangea and the destruction of the shallow shelves took place in the Early and Middle Permian, long before the great dying.

On land, there is evidence not only of severe cooling but also of significant drying. As the seas regressed, deserts stretched across much of temperate latitudes. Upper Permian deposits are dominated by thick sequences of dune sands and evaporites. This contrast between hot, dry conditions in the lower latitudes and glaciation in the high latitudes shows just how extreme the temperature gradient between pole and equator had become. Indeed, one of the best hypotheses for the Permian extinctions argues that climatic instability is the most critical feature of the Late Permian;

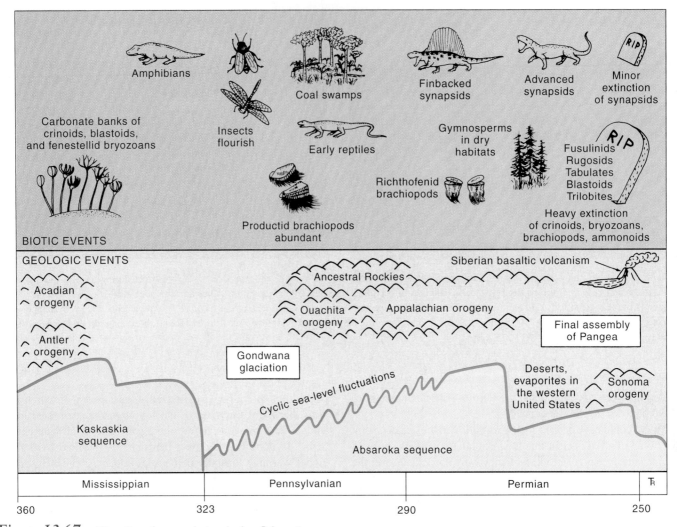

BIOTIC EVENTS

GEOLOGIC EVENTS

Figure 13.67   Time line of events during the late Paleozoic.

unlike cooling or loss of shallow-marine habitat, it took place at the time of the extinctions.

In addition to glaciation followed by rapid warming and severe climatic fluctuations, other events conspired to make the Permian crisis even worse. There was a dramatic change in carbon isotopes in the Late Permian and Early Triassic oceans, and this change is usually attributed to the marine regression that allowed oxidation of previously buried organic matter (particularly from the Carboniferous coal deposits). When this organic matter was oxidized, it depleted the nutrient supply available to organisms. Some scientists have suggested that this oxidation of organic matter may have depleted atmospheric oxygen levels from about 30 percent to less than 15 percent in the Late Permian.

The Late Permian also saw the largest episode of basaltic lava eruptions in the entire Phanerozoic, pouring out lavas in excess of 1.5 million cubic kilometers in Siberia. These basalts may have injected large amounts of sulfates into the atmosphere, further decreasing global temperatures. Recently these Siberian lavas have been redated, and their eruption occurred in a short interval of less than a million years right at the Permo-Triassic

boundary. Chinese strata, which preserve a detailed record of this boundary, give evidence of explosive silicic volcanic eruptions, which may have formed ash clouds around the world, further cooling the earth by blocking out sunlight. About the only factor that *does not* appear to have triggered Permian extinctions is extraterrestrial impact. Although iridium, the rare platinum-group metal associated with the Cretaceous extinctions (see Chap. 14), was once reported in the same Chinese rocks, further examination shows that iridium abundance was actually from a silicic volcanic ash layer, and there is no good evidence of a Permian impact.

According to A. Knoll, R. Bambach, D. Canfield, and J. Grotzinger, the dramatic change in carbon isotopes at the Permo-Triassic boundary suggests that carbon dioxide had accumulated in the deep waters of the oceans, then catastrophically surged to the surface. When marine organisms get too much carbon dioxide, they can die of **hypercapnia,** or carbon-dioxide poisoning. This hypothesis is supported by the fact that the extinction was worst in the more sluggish and immobile creatures (such as fusulinids, brachiopods, bryozoans, blastoids, crinoids, and tabulate and rugose corals) and less severe in the more mobile organisms with more sophisticated gill systems, such as the molluscs and arthropods. However, Paul Wignall and Yukio Isozaki argue that the carbon isotopes and abundance of black, organic-rich mudstones at the Permo-Traissic boundary in many sections suggest that the oceans were stagnant and **anoxic,** which would also kill off most of marine life.

The latest research on excellent Permo-Triassic boundary sections in China show that the extinction was much more severe and abrupt than has been previously appreciated. In these sections, the extinction is dated at 251.4 million years ago, occurred in less than 500,000 years, and wiped out 94 percent of the species known from the sections. It coincides precisely with the dates on the Siberian eruptions and the huge carbon isotope event. The boundary layer contains many microspherules, which are thought to be debris from the volcanic eruptions. According to Jin and others, these beds also do not support the global anoxia hypothesis but, instead, suggest that the Siberian eruptions produced a brief volcanic winter, followed by an extreme global warming and possibly by hypercapnia in the oceans. Jin and others note that the rapidity of the extinction might also be consistent with an impact as well, although no strong evidence to support such an event has been found. The debate over the "mother of all mass extinctions" is not over, but the accumulated recent evidence has ruled out some old favorite hypotheses, such as the coalescence of Pangea, or the global glaciation, and has shown that global hypercapnia and extreme greenhouse warming, possibly triggered by the Siberian eruptions, are the most likely causes.

In summary, so many factors coincided to make the Late Permian world hostile that it is not surprising that marine life nearly vanished altogether. The Northern and Southern Hemisphere glaciation and cooling, followed by runaway greenhouse warming, regression, climatic instability, depletion in nutrients (and possibly atmospheric oxygen) and massive volcanism, must have interacted in ways that made the latest Permian earth more inhospitable than at any time since the late Proterozoic glaciations nearly refrigerated the entire globe.

Summary

Late Paleozoic Geology

- Mississippian time was characterized by the last cratonwide carbonate deposition. Near the end of the period, immature sands appeared on the craton as a result of widespread exposure of Cryptozoic basement rocks on cratonic arches and as a result of the continuing erosion of older Paleozoic rocks in the Appalachian region.

- Late Mississippian regression ended the Kaskaskia Sequence and produced a cratonwide unconformity upon which the Absaroka Sequence was deposited.

- Repetitive sedimentation is conspicuous in late Paleozoic strata on several continents. As many as fifty cyclothems with both nonmarine and marine portions reflect transgressive-regressive oscillations across wide areas. These oscillations averaged about 2 million years in duration, which is but a fraction of the typical duration of older Paleozoic transgressive-regressive episodes. Global sea-level changes due to expansion and contraction of continental glaciers on Gondwanaland offer the best explanation of the late Paleozoic oscillations.

- Coal was formed in immense Pennsylvanian coastal plain swamps by the accumulation of vast quantities of vegetation. Burial in an anaerobic environment and compaction beneath transgressive deposits converted the plant material to coal. Because elevation of the swamps was so low, modest changes of sea level affected very large areas. Carboniferous climate was humid and tropical with indistinct seasons in North America and Europe but became somewhat drier during Permian time.

- Tectonic activity on all continents produced large areas of land during Permo-Triassic time and raised the average elevation of this land. Consequently epeiric seas almost completely disappeared for about 20 million years. Sediments colored red by small percentages of iron oxide together with evaporite deposits formed widely on most continents. The red color seems to have been favored by warm climate with distinct seasonal differences of rainfall; the evaporites indicate increasing aridity.

- Both fossils and sediments indicate stronger zonation of Permo-Triassic climate than for Carboniferous time. Evaporites and widespread sand-dune deposits suggest arid desert conditions on the western, leeward side of North America, whereas the geographic position and considerable elevation of the Appalachian Mountains suggest a strongly monsoonal climate in eastern North America.

- The Appalachian orogeny was the last mountain-building event for southeastern North America. It resulted from a continent-continent collision with northwestern Gondwanaland. This profound event produced much warping and faulting within the craton as well as the folding, thrust faulting, and elevation of the Appalachian-Ouachita-Marathon mountain system as an intercratonic orogenic belt.

- The Pangea supercontinent was formed by the aggregation of all continents during late Paleozoic time as a result of many collisions. Similarities between North America and other continents, such as the Hercynian orogeny of Europe and uplift of the Ural Mountains in Russia, are readily explained by this profound plate-tectonic reorganization.

- The Gondwana rock succession (late Paleozoic-Mesozoic), which is unique to the present five southern continents, is characterized by glacial tillites, nonmarine sediments containing coal, the *Glossopteris* flora, and a nonmarine vertebrate fauna. It is capped by basalts and is penetrated locally by diamond-bearing peridotites of mantle derivation.

Paleoclimate

- The paleoclimate of Pangea was clearly zoned. Glacial centers shifted as different parts of Pangea migrated across the South Pole. Glaciation began in northern Africa (Late Ordovician), then moved to South America (Silurian-Devonian) and eastward across Africa and Antarctica (Carboniferous) to Australia (Early Permian). It ceased as Pangea moved completely away from the pole (Late Permian). Meanwhile, the North Pole remained in an oceanic location, so that Laurasia was too warm for glaciation, except in northeastern Siberia. Waxing and waning of Gondwana glaciers provides the most probable cause of late Paleozoic repetitive sedimentation on many continents.

- The southern *Glossopteris* flora was adapted to a wide range of cool, temperate conditions. Two contemporary Laurasian floras were tropical (lycopsid) and northern-temperate (Asiatic), respectively. Nonmarine animal faunas, on the other hand, were cosmopolitan across all of Pangea and showed low diversity because of the accessibility of all land areas via land bridges.

- As Pangea drifted away from the pole in Permo-Triassic time and land area became larger, overall climate became

warmer and drier. All three late Paleozoic floras were replaced by new Mesozoic plants apparently better adapted to the new climatic conditions.

Tectonics and Life

- Many marine invertebrate groups, such as the crinoids, corals, brachiopods, and fusulinids, experienced major evolutionary changes. This development climaxed in the Permian, as seen in the rich reef developments in Southeast Asia and the southwestern United States.

- The Late Paleozoic global tectonic climax also greatly affected life, illustrating again an important feedback between the physical and organic evolution of the earth.

- Land life exploited the new, late Paleozoic conditions. Adaptations to avoid desiccation of embryos allowed both plants and reptiles to invade dry upland habitats as well as wet lowland ones already conquered in middle Paleozoic time by the seedless plants, invertebrate animals, and amphibians. The amniote egg made this habitat change possible for reptiles and synapsids.

- The greatest ever marine life crisis at the end of the Permian Period resulted in extinction of half the families of marine invertebrate animals and possibly 95 percent of the marine species in the ocean. The causes of this extinction are controversial, but regression, climatic instability, and massive volcanic eruptions in Siberia may have been important factors.

Readings

Alberta Society of Petroleum Geologists. 1964. *Geological history of western Canada.* Calgary: Author.

Anderton, R. A., et al. 1980. *A dynamic stratigraphy of the British Isles.* London: Allen and Unwin.

Bowring, S. A., D. H. Erwin, Y. G. Jin, M. W. Martin, K. K. Davide, and W. Wang. 1998. U/Pb zircon geochronology and tempo of the end-Permian mass extinction. *Science,* 280:1039–1045.

Douglas, R. J. W., et al. 1963. Geology and petroleum possibilities of northern Canada, in Proceedings Sixth World Petroleum Congress (sec 1). Frankfurt: 519–71.

Erwin, D. J. 1993. *The great Paleozoic crisis: Life and death in the Permian.* New York: Columbia University Press.

Fisher, G., et al. 1970. *Studies of Appalachian geology—Central and southern.* New York: Wiley-Interscience.

Gray, J., and A. J. Boucot. 1978. The advent of land plants. *Geology,* 6:489–92.

Hallam, A., and P. B. Wignall. 1997. *Mass extinctions and their aftermath.* Oxford, U.K.: Oxford University Press.

Hatcher, R. D., Jr. 1972. Developmental model for the southern Appalachians. *Geological Society of America Bulletin,* 83:2735–60.

Jin, Y. G., Y. Wang, W. Wang, Q. H. Shang, C. Q. Cao, and D. H. Erwin. 2000. Pattern of marine mass extinction near the Permian-Triassic boundary in South China. *Science,* 289:432–36.

Kemp, R. S. 1982. *Mammal-like reptiles and the origin of mammals.* New York: Academic Press.

Klein, G. D., ed. 1968. *Late Paleozoic and Mesozoic continental sedimentation, northeastern North America.* Boulder: Geological Society of America, Special Paper 106.

Knoll, A. H., R. K. Bambach, R. E. Canfield, and J. P. Grotzinger. 1996. Comparative earth history and the Late Permian mass extinction. *Science,* 273:452–57.

Logan, A., ed. 1974. *The Permian and Triassic systems and their mutual boundary.* Calgary: Canadian Society of Petroleum Geologists Memoir 2.

McKee, E. D., S. S. Oriel, et al. 1967. *Paleotectonic investigations of the Permian system.* U.S. Geological Survey Professional Paper 515.

Panchen, A., ed. 1980. *The terrestrial environment and the origin of land animals.* New York: Academic Press.

Rodgers, J. 1970. *The tectonics of the Appalachians.* New York: Wiley-Interscience.

Stewart, J. H., et al., eds. 1977. *Paleozoic paleogeography of the western United States.* SEM Pacific Coast Paleogeography 1.

Stewart, W. N., and G. W. Rothwell. 1993. *Paleobotany and the evolution of plants.* Cambridge: Cambridge University Press.

United States Geological Survey. 1979. *The Mississippian and Pennsylvanian Carboniferous) systems in the United States.* Professional Paper 1110.

Wanless, H. R., et al. 1963. Mapping sedimentary environments of Pennsylvanian cycles. *Bulletin of the Geological Society of America,* 74:437–86, also GSA Map MC-23.

Ward, P. D. 2000. *Rivers in time: The search for clues to earth's mass extinctions.* New York: Columbia University Press.

White, M. E. 1990. *The flowering of Gondwana.* Princeton: Princeton University Press.

Zen, E., et al. 1968. *Studies of Appalachian geology: Northern and maritime.* New York: Wiley-Interscience.

(Painting by John Gurche.)

Chapter 14

The Mesozoic Era
Age of Reptiles and Continental Breakup

MAJOR CONCEPTS

Pterodactyls and brontosauruses

Eyed with a wink,

Cheer up, sad world

It's kinda fun to be extinct.

Ogden Nash

▶ The Mesozoic was dominated by the tectonic effects of the breakup of Pangea. In eastern North America, the Atlantic Ocean began to open up, forming Triassic rift grabens, then Jurassic proto-oceanic gulf sediments (including evaporites), and finally a thick Cretaceous passive-margin shelf sequence. In the western interior of North America, the uplift of Pangea produced thick sequences of nonmarine sediments, including Triassic red beds, Lower Jurassic dune sands and floodplain deposits, and a Late Jurassic seaway followed by dinosaur-bearing floodplain deposits.

▶ As the Atlantic opened and the North American plate moved rapidly westward over the Pacific region, major collisional tectonics occurred on the western margin of the continent. In the Permo-Triassic, a major terrane collision occurred in northern California and Nevada. By the Jurassic, an Andean-style volcanic arc complex had developed all along the western margin, including the Sierra Nevadas. Throughout the Mesozoic, exotic terranes were accreted onto the western margin of North America.

▶ Unusually rapid sea-floor spreading due to the opening of the Atlantic in the Cretaceous produced high mid-ocean ridges. These displaced water out of the ocean basins to produce the largest continental seaways since the Paleozoic. High Cretaceous sea levels resulted in the accumulation of thick sequences of shale and chalk widely over the interiors of continents.

▶ Following the Permian extinctions, marine life underwent a profound change. Instead of typical Paleozoic invertebrates, the seas were dominated by bivalves, gastropods, and echinoids—much like the sea floor today. The appearance of several shell-crushing predators, including crabs and lobsters, reptiles, and fish, caused the decline of those invertebrates that could not burrow or resist these new predators. A variety of ammonites, marine reptiles, and bony fish swam in the seas.

▶ Synapsid vertebrates dominated the coniferous landscape in the Triassic, but in the Late Triassic they were replaced by dinosaurs and their relatives. Through the Jurassic and Cretaceous, dinosaurs continued to diversify and dominate the landscape, although mammals, birds, snakes, lizards, and frogs all evolved in the Mesozoic. By the mid-Cretaceous, flowering plants had appeared and triggered a great diversification of pollinating insects.

▶ The end of the Cretaceous was marked by a mass extinction, which wiped out not only the dinosaurs but also ammonites, marine reptiles, many marine invertebrates, and even most of the plankton. Both asteroid impacts and massive volcanic eruptions have been blamed for this extinction, although a combination of both is the likely cause.

The Mesozoic Era (66 to 250 m.y. ago) holds special interest because of dinosaurs (Fig. 14.1), of course, but also because of the colorful strata that highlight the stunning landscapes of the canyon and plateau country of the southwestern United States. Europe, too, has unique Mesozoic rocks in the famous chalk exposed in cliffs on either side of the English Channel (Fig. 2.6). Black shales are also important in Mesozoic sequences on several continents. Cretaceous strata are especially important economically because they contain the world's second largest coal reserves and a large share of its petroleum.

The Mesozoic Era is especially noteworthy for the breakup of Pangea (Fig. 14.2) and the profound effects that this global tectonic revolution had upon climate and the evolution of life—particularly nonmarine plants and vertebrates. For North America, the Mesozoic represents a marked change tectonically, a shift from compressional mountain building on the eastern and northern margins of the continent to major compressional mountain building on the western margin. The unconformable boundary between the Absaroka and Zuni Sequences punctuates the early Mesozoic record on the craton.

First, we shall examine the important Mesozoic tectonic changes, including the extensional tectonics in eastern North America and northern Europe that resulted in the opening of the Atlantic Ocean basin. Then we shall discuss the application of collage tectonics to mountain building in western North America, followed by a discussion of how these tectonic changes affected sedimentation and paleogeography. Next we shall consider the last great transgression in Cretaceous time, when one-third of the earth's total continental area was flooded; dark-colored mudstones prominent in many Cretaceous sequences will be discussed within the context of this transgression and its climatic effects. Last, we shall turn to a consideration of the important Mesozoic history of life. No treatment of the "Age of Reptiles" would be complete without a discussion of dinosaurs and the appearance of birds and mammals. We shall conclude by reviewing several hypotheses for the mass extinctions of dinosaurs and many other animals at the end of the era.

Extensional Tectonics—Opening of the North Atlantic

The Newark Rifts

The breakup of Pangea began near the end of the Triassic Period between present North America and Africa. The breakup was caused by a profound change of convection following a long period of heat accumulation in the mantle, due to the insulating effect of the supercontinent (Figs. 14.3 and 14.4). The breakup proceeded rapidly during later Mesozoic time with the ultimate

Figure 14.1   Herd of gigantic *Ultrasauros* pursued by predatory *Allosaurus. (Painting by John Gurche.)*

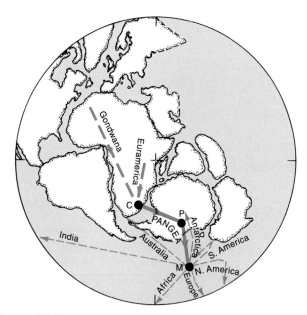

Figure 14.2   Map restoring relative past paleomagnetic South Pole positions for all continents except Asia. Medium-weight dashed arrows trace pole migrations *relative to continents* for Gondwana and Euramerica in early Paleozoic; C (Carboniferous) records their collision in late Paleozoic time (P—Permian). Heavy solid arrow traces pole migration for all of Pangea until breakup began in early Mesozoic time (M). Light dashed arrows record separation of the different continents since then. *(From Paleomagnetism and plate tectonics by Dr. M.W. McElhinny, 1973.)*

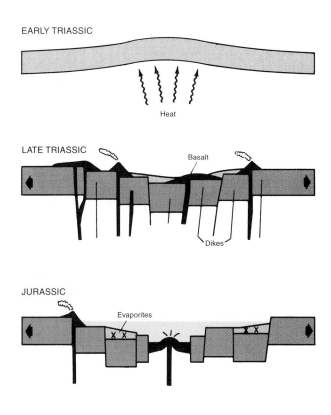

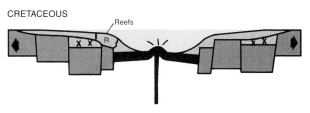

Figure 14.3   Initial breakup of a supercontinent by heating and doming followed by rifting and sea-floor spreading based upon the North Atlantic and Gulf of Mexico basins (compare Fig. 7.33). *Late Triassic:* Block faulting, swarms of basaltic dikes, volcanic eruptions (black), and nonmarine sedimentation. *Jurassic:* Initial sea-floor spreading and first marine deposition characterized by evaporites. *Cretaceous:* Normal marine passive-margin deposition including reefs (R); Tethyan fauna had become established in Gulf of Mexico by this time.

separation of Greenland and Europe from North America at the end of the Cretaceous Period (Figs. 14.5 and 14.6). This also opened a connection between the Atlantic and Arctic Oceans. Stretching of the Pangean crust first caused Triassic faulting (Fig. 14.4) accompanied by basaltic intrusions and extrusions, which continued into the Jurassic Period. The Palisades of the Hudson River, a thick basaltic sill exposed in northern New Jersey opposite New York City, is the most famous example of the latter (Fig. 14.7A). Rivers carved mountains from the upraised

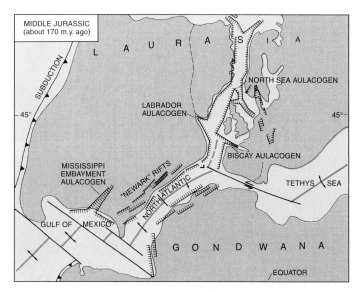

Figure 14.4   Paleotectonic map for Middle Jurassic time showing the initial opening of the North Atlantic and Gulf of Mexico ocean basins. Note several aulacogens along the new ocean margins and Newark rifts within eastern North America.

fault blocks and formed gravelly and sandy alluvial fans along the straight flanks of those mountains. Within the downfaulted rift basins, thick, mostly red-colored, nonmarine clastic sediments known as the Newark Supergroup accumulated rapidly (Figs. 14.7B and 14.8). Lakes formed in the central parts of the basins, and some were as large as the largest lakes in the modern East African rift valleys (see Fig. 7.18). It was along the margins of such Triassic lakes and adjacent riverbeds that wandering reptiles left their tracks.

Fossil plants suggest a humid subtropical climate, but evaporites and local wind-dune sands seem to suggest the opposite—aridity. How can these contrasting evidences be reconciled? Detailed studies of the lake deposits reveal that the climate must have changed repeatedly from humid to arid, causing the lakes to expand and shrink greatly. Moreover, fine laminations in the deposits, which are thought to be annual layers (*varves,* as in Fig. 5.2) can be correlated from basin to basin, counted, and then analyzed mathematically. The results suggest periodicities that match closely the climatic cycles known to result today from small variations in the earth's orbit (see Chap. 16).

Ocean Basins and Aulacogens

During the Middle Jurassic, the southern margin of North America began separating from South America to form the Gulf of Mexico basin (Fig. 14.4). Major rifting perpendicular to the margin of the North American craton initiated the Mississippi Embayment aulacogen beneath the present lower Mississippi River valley. In Chap. 7, we noted that such structures typically form at the triple junction where supercontinents first break up. This

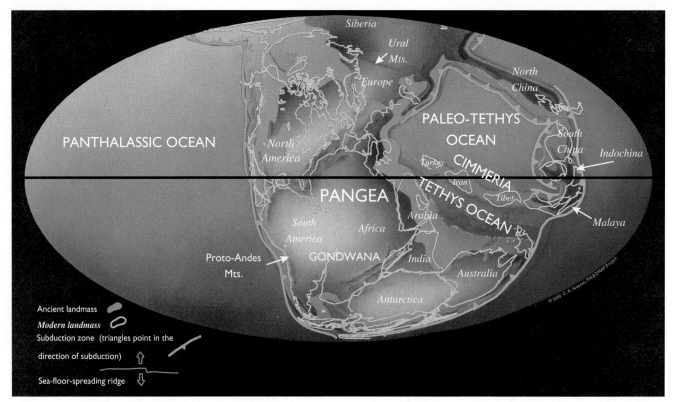

A.

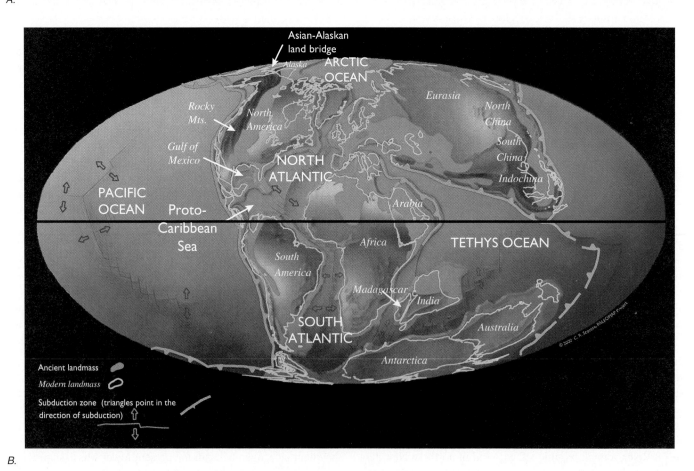

B.

Figure 14.5 Global paleogeography during the Mesozoic. A: Late Triassic, about 215 m.y. ago. Note that Pangea is just beginning to break up, with uplifts in the Atlantic margin as thermal bulges began the process of rifting. B: Late Cretaceous, about 95 m.y. ago. At this point, the Atlantic opened completely, and India began its movement away from the other Gondwana continents. Note that high global sea levels flooded the interiors of many continents, including the Great Plains region of North America, much of the Sahara and Arabian Peninsula of Africa, and most of Europe and the Near East (see Fig. 12.33 for color scheme). *(www.scotese.com)*

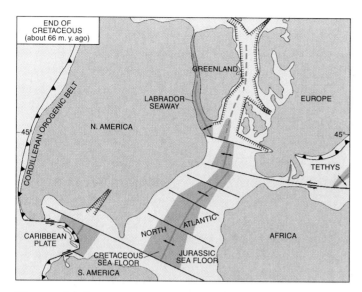

Figure 14.6 Paleotectonic map for end of Cretaceous time, showing the opening of the North Atlantic Ocean basin. Note initial separation of Greenland by opening of the Labrador Seaway and separation of Italy from the Iberian Peninsula. The Caribbean plate carrying Cuba also is shown forming between North and South America. Dark color shows new Cretaceous sea floor (compare Fig. 14.4). *(After B. C. Burchfiel and G. A. Davis, 1972, American Journal of Science, v. 272, p. 111.)*

material was converted to petroleum, which became the basis for the major North Sea oil boom of the 1970s and 1980s. Faulting as well as deformation by the upward intrusion of the low-density Permian evaporites (Fig. 14.6) formed many traps for the petroleum. During Cretaceous time, chalk was deposited widely over northern Europe and locally in southeastern North America at the time of maximum transgression (Fig. 2.6). This unusual rock consists of the microscopic calcareous skeletons of pelagic organisms (especially algae called coccoliths), which rained down from surface waters in sufficient abundance to form a type of limestone. Such **pelagic** sedimentation generally has been confined to the deep-ocean basins, but unusual rises of sea level have allowed it to spread into the deeper parts of the epeiric seas.

In Late Cretaceous time, the Atlantic widened rapidly (Fig. 14.6), and sea-floor spreading extended northward through the Labrador Sea to split Greenland from North America proper. In Paleocene time, spreading ceased there and shifted to a new spreading axis, which split Greenland from Europe. Iceland is a large outcrop of the ridge axis, which developed rapidly and extended into the Arctic Ocean basin. The North America–Eurasia Siamese twins were separated for the first time in more than 350 million years—that is, since their grafting together by the Caledonian-Acadian orogenies.

The Breakup of Pangea

Late Triassic to Early Jurassic rifting also disrupted the connection between northern Africa and Europe (Fig. 14.5). Meanwhile, several microcontinents were rifted from northern Gondwanaland in Triassic and Jurassic time to drift north and collide with southeastern Asia—and possibly even with North America (Fig. 15.29). Middle Jurassic transgression in eastern Africa and Madagascar heralded the initial opening of the Indian Ocean. Further widespread transgressions on all the Gondwana continents, together with sea-floor magnetic anomaly patterns, indicate that India separated from Africa and Australia in Late Jurassic time and that South America split from Africa in mid-Cretaceous time. The last separations were between Australia and Antarctica near the end of the Cretaceous and between Antarctica and South America in the Miocene.

aulacogen continued to subside and influence sedimentation throughout Mesozoic and Cenozoic time. Other aulacogens formed in the Labrador region as well as on the northwestern margin of the European craton, especially beneath the present North Sea (Fig. 14.4).

The Mesozoic aulacogens and associated rift basins along the margins of the new ocean basins initially received nonmarine clastic sediments. Unlike the Newark rifts, however, the marginal rifts were soon flooded by the sea as the oceans widened (Figs. 14.4 and 14.6). In warm regions, evaporites were typically the first marine sediments because the narrow, irregular rifts do not allow unrestricted inflow of normal marine waters at first (Fig. 14.3). Such was the case in the new Gulf of Mexico, where thick evaporites were deposited in Middle Jurassic time (Fig. 7.34); wind dunes formed along the adjacent arid coast. By the end of the Jurassic, normal marine conditions had developed, and the Tethyan fauna had migrated from Eurasia.

In Europe, Permian and Triassic evaporites had been deposited during an earlier phase of rifting. During Jurassic and Cretaceous time, the North Sea aulacogen became fully marine, with gravelly deltas deposited along the margins here and in eastern Greenland, where fault scarps were eroded rapidly (Figs. 14.4 and 14.8). In Late Jurassic time (and again in Late Cretaceous), organic-rich black muds were deposited widely. After subsequent burial and maturation, much of the organic

Triassic History of the Craton

Triassic strata on the North American craton are very similar to the Late Permian ones discussed in Chap. 13. Red beds continued to form widely over the craton as well as in the Appalachian and East Greenland regions during the Triassic Period (Fig. 14.9). Continental land area remained very large, even though average elevation was reduced. Mountains raised during the Pennsylvanian Ancestral Rockies orogeny in Colorado and New Mexico

A.

B.

Figure 14.7   A: Steep, columnar-jointed basalt cliff of the Palisades Sill, near Palisades, New York. This resistant igneous intrusion forced its way between the layers of Triassic sediments and now forms the western bank of the Hudson River (in background) from Hoboken and Jersey City, New Jersey, and many miles up the Hudson Valley. B: Exposure of Triassic fluvial sandstones and lake shales of the Newark Group in central New Jersey. These deposits filled the ancient rift valley during the early opening of the Atlantic. They are found all along the Atlantic margin, from North Carolina to Gettysburg battlefield in Pennsylvania to the Connecticut Valley to Nova Scotia and on the African side. *(D. R. Prothero.)*

MESOZOIC RIFT BASINS

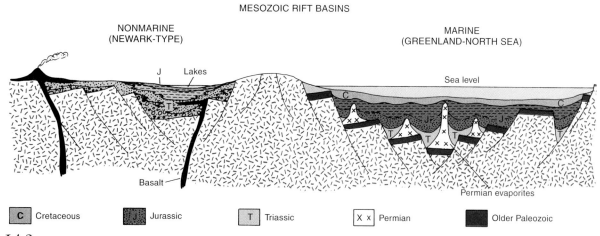

NONMARINE
(NEWARK-TYPE)

MARINE
(GREENLAND-NORTH SEA)

| C | Cretaceous | J | Jurassic | T | Triassic | X x | Permian | | Older Paleozoic |

Figure 14.8 Diagrammatic comparison of landlocked, nonmarine Mesozoic rifts of the Newark-type with the marine-type formed in eastern Greenland and northern Europe. The rise of evaporite intrusions in the North Sea aulacogen has formed many important petroleum traps. (Compare with Figs. 10.12, 14.3, 14.4 and 14.6).

A.

B.

Figure 14.9 All over the Rocky Mountain region, Triassic deposits are predominantly red beds formed in forested floodplains and rivers. They include the red beds of the Painted Desert of Arizona and Petrified Forest. A: Typical Triassic-Jurassic sequence of the Colorado Plateau, here seen in Flaming Gorge, Wyoming. Triassic red beds of the Chinle Formation are overlain by Lower Jurassic dune sands of the Navajo Formation. Similar sequences of Triassic red beds overlain by Jurassic dune sands can be seen throughout Utah, Arizona, western Colorado, and even near Las Vegas, Nevada. B: In some places, Triassic floodplains experienced drying episodes. Here the red shales of the Spearfish Formation in the Black Hills includes thick beds of white gypsum, formed when the area became a dry lake. Photo taken behind Evans Plunge, Hot Springs, South Dakota. (D. R. Prothero.)

were lower and became mostly buried by nonmarine Triassic sediments (Fig. 14.10). Sediments were transported by rivers flowing westward across an immense alluvial plain to the Cordilleran Sea; red beds accumulated on the plain. Forests covered at least the highlands, as evidenced by petrified logs (Fig. 14.11). Triassic red beds also have yielded important animal fossils, including remains of amphibians, early dinosaurs, and various other reptiles. Freshwater molluscs and fish point to some swamp and river environments.

Along the western margin of the craton, Triassic red sediments grade into the marine gray shales and limestones typical of the eastern part of the Cordilleran belt. Complex facies variations as well as an abrupt westward thickening of Triassic strata characterize the cratonic margin. Geologists long assumed that the widespread red beds to the east in the entire Rocky Mountain region were nonmarine deposits. In Wyoming and Utah, however, thin marine limestone and gypsum layers are interstratified in the red beds, forming tongues of western facies penetrating eastern

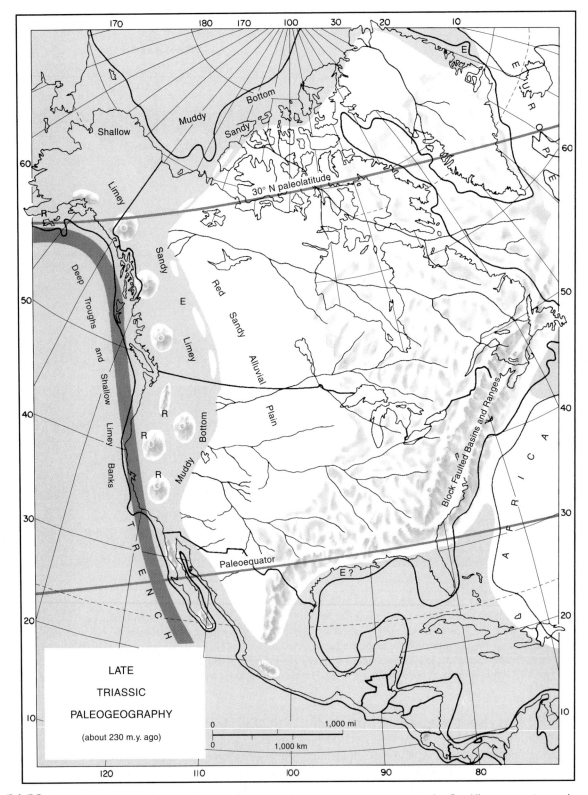

Figure 14.10   Late Triassic paleogeography. In spite of uncertainties about suspect terranes in the Cordillera, approximate character is portrayed. Note Africa and Europe still attached to North America as parts of Pangea.

Figure 14.11   Spectacular fossilized tree trunks from Petrified Forest National Monument in Arizona. These trees were once part of a large forest of conifers like the modern *Araucaria* but were buried in shales of the Upper Triassic Chinle Formation. These shales erode rapidly, exposing hundreds of trees as well as many fossils of land animals. (© *Biological Photo Service.*)

red strata. A considerable amount of the associated red strata represents marine lagoonal and tidal flat deposits. Marine conditions existed, at least intermittently, on the western part of the craton, where periodic transgressions of Triassic seas extended at least 600 kilometers eastward.

Jurassic History of the Craton

Ancient Navajo Desert

As the Triassic Period drew to a close, sedimentation changed markedly over western North America. A vast blanketing mass (approximately 40,000 cubic kilometers) of very-well-sorted sand was deposited along the entire western edge of the craton in the United States and even extended onto the eastern flank of the volcanic arc (Fig. 14.12). The Late Triassic and Early Jurassic sands consist of 90 percent quartz. They are strikingly similar to the widespread, mature Paleozoic quartz sandstones of the craton.

Where did so much sand come from? The relative purity and roundness of the grains indicate that they were derived from older sandy sediments through recycling. With much sandstone present in upper Paleozoic and Triassic strata of the western part of the craton, there is no problem of designating potential source rocks. Paleocurrent data indicate general southerly transport of the sand (Fig. 14.12). This points to derivation from the north, chiefly from the craton in the north-central United States and adjacent Canada, where a widespread pre-Middle Jurassic unconformity

indicates that upper Paleozoic and Triassic strata indeed were being eroded. Derivation from that region is consistent with the general pattern of tilting of the entire craton accompanied by a westward shift of most active sedimentary accumulation and the gradual westward stripping of strata that formerly covered most of the region.

Large-scale cross-stratification is a prominent trait of the Jurassic Navajo and other sandstones in Utah (Fig. 14.13). This spectacular feature was attributed long ago to the preservation of lee faces of dunes perhaps as much as 50 to 100 meters high; this origin is verified by other, more subtle features as well. We believe that sand derived from the craton was transported south parallel to an oscillating eastern shoreline (Fig. 14.12). Some of the sand apparently was deposited on the shelf, but onshore winds also produced coastal dunes, upon which dinosaurs left telltale footprints. Rare land plants and fossil burrows probably formed by beetles also are present. Interdune pond deposits and deformed cross-strata suggest a high water table in the dune field, and, with onshore winds, coastal fog may have characterized the region as it does the similar modern Namibian Desert of southwestern Africa (Fig. 14.14).

Return of the Epeiric Sea

Thin carbonate rocks containing marine fossils are interstratified locally with the upper sandstones, and all are overlain by widespread limestone, shale, and evaporite deposits loosely referred to as the *Sundance Formation* (Fig. 14.15). Beginning in Middle Jurassic time, the sea advanced widely over the western part of

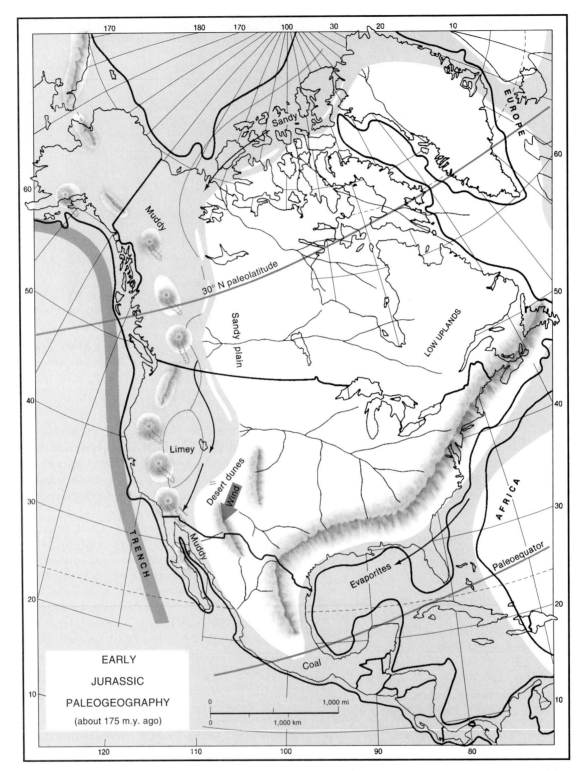

Figure 14.12 Late Early Jurassic paleogeography. Note beginning of encroachment of the sea along the western edge of the craton with deposition of Navajo dune sands next to the large embayment. Note Africa moving away but Europe still attached to North America. Evaporites were deposited in the newly opened Gulf of Mexico.

Figure 14.13   Large-scale cross-stratification approximately 5 meters thick in the Navajo Sandstone near Zion National Park, Utah. Details of the stratification indicate deposition in wind-formed dunes, some of which must have been of the order of 100 meters high. *(Courtesy of Gary Kocurek.)*

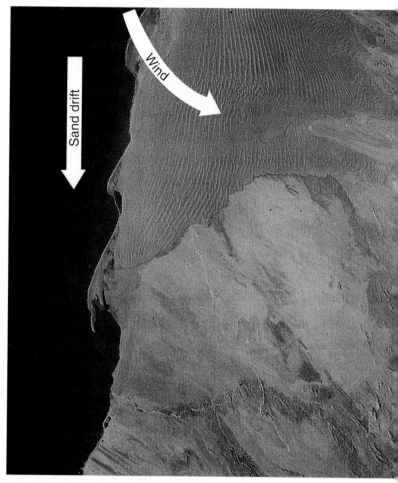

Figure 14.14   Coastal sand dunes of the Namibian Desert in southwestern Africa, produced by cool, dry onshore winds. This region is a close analogue for the setting of the Lower Jurassic Navajo Sandstone of Utah. Note that the dunes cease abruptly at a river because wind is unable to blow sand across the valley. Curved sand spits along the coast are produced by marine longshore drifts of sand from top to bottom. Photo inverted for clear comparison with Fig. 14.12. *(Gemini photo 65-45579 taken from about 200 miles above the earth; courtesy NASA.)*

the craton from the present Arctic and northern Pacific regions. Quartz sand deposition was followed in the northern Rocky Mountain–Canadian Plains region by conditions like those prevailing in Paleozoic epeiric seas.

Late Jurassic time marked the last significant carbonate and evaporite deposition anywhere on the craton. Middle and Upper Jurassic strata lap across a widespread major unconformity and are younger toward the center of the craton. This unconformity marks the base of the Zuni Sequence (Fig. 11.13). As transgression proceeded, normal marine conditions with better circulation prevailed and evaporite deposition practically ceased (Fig. 14.15). Nearly every type of epeiric sea deposit discussed for the Paleozoic formed in the Late Jurassic sea. Limestones rich in fossil fragments, oolites, algal material, fossiliferous shales, and cross-stratified glauconitic sandstones are all prominent, but complexity of the facies variations defies a simple summary.

Most of the Appalachian belt was above sea level and being actively eroded at this time, though areas around the present Gulf of Mexico were under water. Extensive, thick carbonate rocks as well as evaporites formed in Texas, Mexico, and Cuba (Fig. 14.15). The epeiric sea must have joined with waters encroaching northward from the present Gulf of Mexico region (Fig. 14.15) to produce a continuous seaway stretching to the present Arctic. There must have been an ecological barrier, such as water temperature, however, because the Jurassic faunas of the Pacific Coast and Canadian Rockies have affinities with Asian faunas and differ from those of the Gulf region. The latter are like the faunas of the tropical Tethys seaway of southern Europe and Africa (Fig. 14.5).

The Cordillera

Birth of a Mountain Belt

The western margin of North America is commonly called the **Cordillera** (Spanish for "mountain range"). Today it features such rugged mountain ranges as the Sierra Madres running down the spine of Mexico, the Rocky Mountains, the Sierra Nevadas, and the many ranges of western Canada and Alaska. It has a long and complex history of mountain building, which is just now beginning to be deciphered.

As we saw in Chap. 10, the western margin of North America began as a passive margin in the Proterozoic and Cambrian,

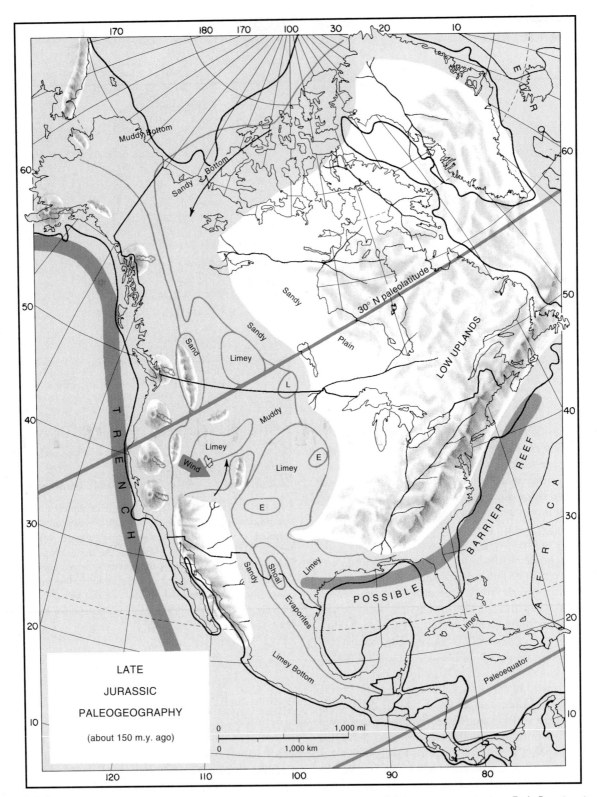

Figure 14.15   Late Jurassic paleogeography showing the first major transgression of the western craton since Early Permian time; note volcanism in the western Cordilleran belt. Africa has moved far enough away to open the Atlantic basin to normal marine circulation and allow the entry of the Tethyan fauna; note possible barrier reef along southeastern margin of continent (compare Fig. 14.3). Most of the continent now lay in the westerly wind belt.

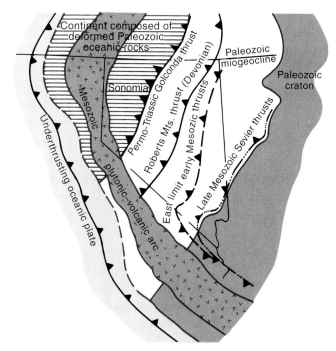

Figure 14.16 During the early Mesozoic, the old northeast-southwest trend of the ancient passive margin and the late Paleozoic collisions (such as the Roberts Mts. Thrust, formed during the Antler orogeny—see Fig. 12.49) was abruptly truncated by a new mountain belt trending northwest-southeast. The Mesozoic volcanic arc was intruded almost at right angles to the Paleozoic direction of compression, presumably because the subducting plate was moving northeasterly, rather than southeasterly, as it must have in the Paleozoic. The reasons for this abrupt shift are not entirely clear but must have been related to the global plate reorganization as Pangea broke up in the early Mesozoic. *(After B. C. Burchfiel and G. A. Davis, 1972, American Journal of Science, v. 272, p. 111.)*

sinking gradually as it rifted away from other continents and piling up a thick sequence of shelf sediments (Fig. 10.3). By the Devonian, the western margin had turned into an active subduction zone, as shown by the collision of a volcanic arc with the continent, which caused the Antler orogeny in central Nevada (Fig. 12.49). Through the Mississippian and Pennsylvanian, a thick sequence of sediments filled the Oquirrh Basin in Utah in front of the Antler highlands (Fig. 13.7). At the same time, the future southern Rocky Mountain region was disturbed by collisions to the south in the Ouachita region (Figs. 13.8 and 13.21). Both uplifts were apparently extinct and gradually eroding away in the Permian (Fig. 13.17).

Much more dramatic were the collisions that have shattered and deformed the Cordillera ever since the Permian. Although subduction along the Pacific Rim began in the Devonian, the pace of activity was much more intense in the Mesozoic. What could be the reason? Some geologists have suggested that Mesozoic subduction might have been more rapid because, as it pulled away from the spreading Atlantic, the entire North American plate was riding westward over the Pacific region (Figs. 14.4 and 14.6).

Another peculiarity of the Cordillera is that it changed orientation during the Mesozoic. During the early Paleozoic, the axis of the subsiding margin was oriented northeast-southwest, parallel to the Transcontinental Arch (Figs. 10.7 and 10.8). Cambrian shoreline deposits show this trend most clearly, with the near-shore facies belt running from eastern Montana to eastern California (Figs. 10.8 and 10.9). Even when the Antler collision occurred in the Devonian, the trend was also northeast-southwest. This orientation persisted until the mid-Permian (Fig. 13.16).

In the Triassic, however, this northeast-southwest axis abruptly shifted to new mountain belts that trended northwest-southeast (Figs. 14.10 and 14.16). The great Paleozoic structural belts that extended southwesterly through eastern California were abruptly sliced off in the Triassic and punctured by arc volcanoes running at right angles to the original margin. What could have caused such a dramatic shift? Again, the most likely explanation lies in the plate motions that began in the Mesozoic. Subducting margins usually form at approximately right angles to the motion of the downgoing plate, suggesting that the direction of subduction must have shifted in the Triassic. As discussed earlier, global plate reorganization took place when Pangea broke up in the early Mesozoic, and this shift in plate motions must have subjected the Cordilleran margin to a new stress regime.

Suspect Terranes

Long before the theory of plate tectonics was formulated, a striking faunal anomaly was recognized in the Cordillera where Permian Tethyan fusulinid foraminiferans characterize a zone of limestones associated with volcanic rocks extending from Alaska south to California (Fig. 14.15). Their nearest counterparts today lie in Japan and southeastern Asia. Before plate theory, this anomaly was explained by migration of the fauna from Asia via warm currents. This is still one reasonable hypothesis, as witnessed by penguins living today on the equatorial Galápagos Islands, which they obviously reached by following the cold current that extends up the western coast of South America. But plate tectonics forces us to consider the staggering alternative that a microcontinent containing these fusulinids—either a Noah's Ark or a Viking Funeral Ship, as described in Chap. 13—may have sailed thousands of kilometers from southeastern Asia or Gondwanaland to collide with North America in Mesozoic time. Biologists attempting to explain puzzling modern-day floral and faunal disjunct distributions around the Pacific also like this concept.

When plate-tectonic theory was applied to the puzzle of the Permian Tethyan fusulinids in central British Columbia, it forced a radical reassessment of the origin of these rocks. Paleomagnetic evidence from several areas in the western Cordillera showed that the rocks in these areas had come from much farther south. No longer could one assume that rocks in Idaho or Alaska had always been part of North America. Instead, geologists increasingly suspected that most of western North America is a collage of microcontinents from someplace else. A fault between two distinctive

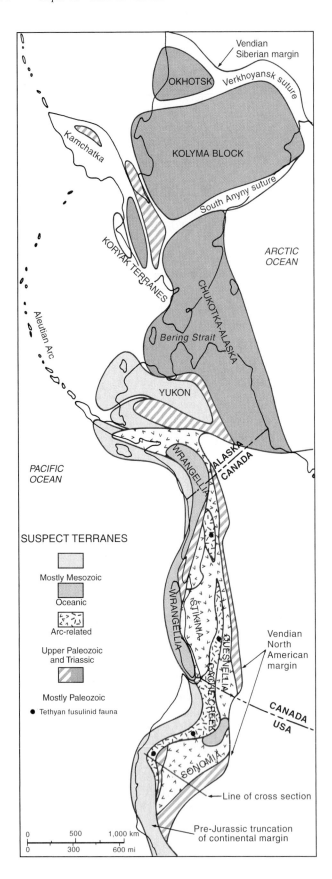

Figure 14.17   Suspect terranes of the Pacific margin of North America and adjacent Siberia, showing the complex collage character of the Cordilleran orogenic belt. Six suspect terranes are shown: two "mostly Paleozoic," three upper Paleozoic and Triassic (two arc-related and one oceanic), and one "mostly Mesozoic." Each named crustal block—an old arc or microcontinent—has distinct stratigraphy, fauna, and/or volcanic rock type. Colors indicate the dominant ages and types of igneous rocks in each terrane. All terranes have been displaced from their original locations and jammed together at different times, but most were together as shown by Cretaceous time. *(Adapted from J. W. H. Monger, 1975: Geoscience Canada, v. 2, pp. 4–9; R. A. Schweickert, 1976: Nature, v. 260, pp. 586–591; D. L. Jones et al., 1977: Canadian Journal of Earth Sciences, v. 14, pp. 2565–2577; M. Churkin, Jr., and J. H. Trexler, Jr., 1979: Geology, v. 7, pp. 467–469.)*

rock types might be more than just a fault contact; it might represent the suture between two continental fragments whose geologic histories were entirely different. Geologists began to match distinctive rock suites found in each coherent crustal block and soon saw that most had come from long distances. If the birthplace of a crustal block is uncertain, it is called a **suspect terrane;** those that are clearly from someplace else are called *exotic terranes.* As Fig. 14.17 shows, approximately 70 percent of western North America is suspect in origin, including almost all of Alaska, British Columbia, Washington, and Oregon and parts of Idaho, Nevada, California, and Mexico.

Striking similarities of stratigraphy coupled with paleomagnetic evidence suggest that two of the slivers now in westernmost Canada (Stikinia and an unnamed smaller one in adjacent southeastern Alaska), dominated by Paleozoic and/or early Mesozoic arc terranes, may have been moved north—presumably by strike-slip faults like the present San Andreas—from northwestern California through about 15° of latitude. Note at the bottom of Fig. 14.16 that the continent was cut off, or truncated, in early Mesozoic time. Slivers apparently were removed, of which Stikinia probably was one.

Even more dramatic is paleomagnetic and stratigraphic evidence that a terrane called Wrangellia (Figs. 14.17 and 14.18) may have been displaced between 35 and 65° northward from some equatorial region. This terrane has a unique sequence of Middle and Upper Triassic limestones overlying oceanic basalts (Fig. 14.19), which probably originated on an extinct ocean ridge far to the south. Wrangellia is bounded by other suspect terranes that contain Triassic and upper Paleozoic rocks with different stratigraphy and arc-type volcanics. These latter terranes formed in tectonic environments different from that of Wrangellia and are out of place. Wrangling over Wrangellia and its neighbors will doubtless go on for some time. One of the important issues is when these landmasses collided with each other and when each was emplaced as part of North America. Fig. 14.20 shows some of the clues used for such dating. Most of the Cordilleran terranes seem to have been in their present positions by no later than Cretaceous or early Cenozoic time.

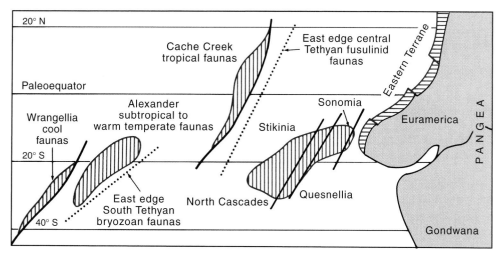

Figure 14.18 Reconstructed late Paleozoic positions of exotic terranes that were accreted to the Cordillera during the Mesozoic. Many terranes carry distinctive fossils, such as fusulinid foraminiferans or bryozoans that place them in terms of latitude or within the old Tethys seaway. Although the latitudinal position is well constrained, their relative east-west longitude cannot be pinpointed precisely. *(After C. Stearn and R. L. Carroll, 1989,* Paleontology, *John Wiley.)*

Figure 14.19 Chitistone Valley, Wrangell Mountains, Alaska, showing a large syncline with oceanic Triassic basalts overlain by Upper Triassic massive limestone and shale. These rocks are part of the suspect terrane Wrangellia. Glacier-shrouded peaks in background *(right)* include Pleistocene volcanoes. *(Courtesy of Gene L. LaBerge.)*

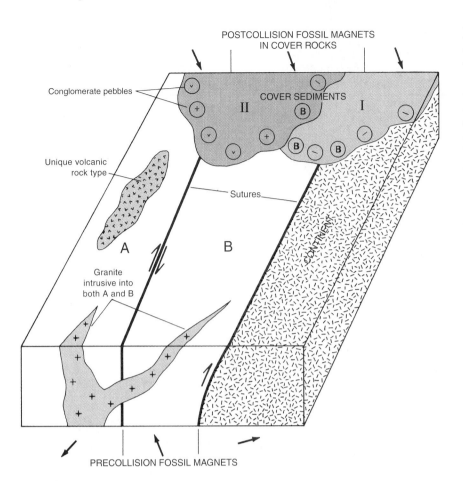

POSTCOLLISION FOSSIL MAGNETS
IN COVER ROCKS

Conglomerate pebbles

COVER SEDIMENTS

II   I

Sutures

Unique volcanic
rock type

A   B

Granite
intrusive into
both A and B

PRECOLLISION FOSSIL MAGNETS

Figure 14.20   Diagram showing application of superpositional principles for dating collision of two suspect terranes with a continent. Unconformable overlap and conglomerate pebbles derived from the continent and terrane B prove emplacement of B before deposition of I cover sediments. Terrane A must have been emplaced before granite intruded both A and B as well as before II cover sediments, which overlap I and contain pebbles of the granite and of a volcanic rock unique to A. Finally, paleomagnetic analysis shows different orientations of fossil magnets in rocks that predate the collisions but uniform orientations in postcollision cover sediments. Precollision fossil-magnet inclinations also provide evidence of the paleolatitudes at which the terrane rocks originated.

Sonomia and the Sierran Arc

We have seen how the ancient northeast-southwest-trending Paleozoic margin of western North America was truncated in the Triassic as exotic terranes were carried north by transform faults. The largest of these terranes, **Sonomia,** was sutured into its present location in Nevada by the mid-Triassic (Figs. 14.17 and 14.21). In several areas from Nevada to Alaska, there is striking evidence of this major collision. Upper Triassic strata were deposited unconformably upon a variety of Middle Permian and older rocks, and the Triassic rocks contain pebbles of the older rocks as well as granites and ultramafic rocks apparently emplaced during Permo-Triassic time. In Nevada, the Sonomia terrane was emplaced by enormous thrust faults, which today run across the western edge of the state. Known as the *Golconda thrusts* (Fig. 14.16), they took Sonomia rocks (now found in the northern Sierras of California) hundreds of kilometers eastward over much older rocks. Ironically these rocks were once part of the upper plate of the Devonian Antler thrust belt, which ran through the center of Nevada (Fig. 14.21).

Once the Sonoma orogeny concluded, most of the Cordilleran margin became a subduction zone from the Late Triassic onward (Figs. 14.10 and 14.21 bottom). By the Late Jurassic, the Cordilleran subduction zone had reached proportions comparable to those of the modern Andes Mountains of western South America. The Andes have one of the best modern examples

of a subduction zone formed between a continental plate (South America) and an oceanic plate (the Nazca plate—Fig. 14.22). Because a dense, thin oceanic plate tends to slide under buoyant continental crust, the geometry of Andean-type subduction zones is very consistent. Partial melting of the material in the upper plate, caused by heating from the downgoing slab, produces magma, which eventually reaches the surface and erupts. This eruption produces the volcanoes of the Andes and, at depth, plutonic rocks from the cooled magma chambers. On the landward side, the uplift of the arc sheds huge volumes of sediments eastward into the foreland basin. Thick clastic wedges of sediment are now being shed into the basins on the eastern edge of the Andes in Brazil, Bolivia, Peru and Argentina. Because subduction zones are formed by compression of two plates, the foreland basin region typically undergoes some compression as well. Thrust faults typically slice up and fold the growing clastic wedge in the foreland basin. The loading of the crust by these thrust blocks is probably the chief cause of the subsidence of the foreland basin (Fig. 7.36).

On the oceanic side of an Andean arc, the effects of the downgoing slab are dramatic. The boundary between the oceanic slab and the overriding continent is marked by a deep oceanic trench on the sea floor; the Peru-Chile trench lies just off the west of the Andes and reaches 8 kilometers in depth. The zone above the downgoing slab is known as the **accretionary prism** because rock is constantly being scraped off the oceanic slab (Fig. 14.28).

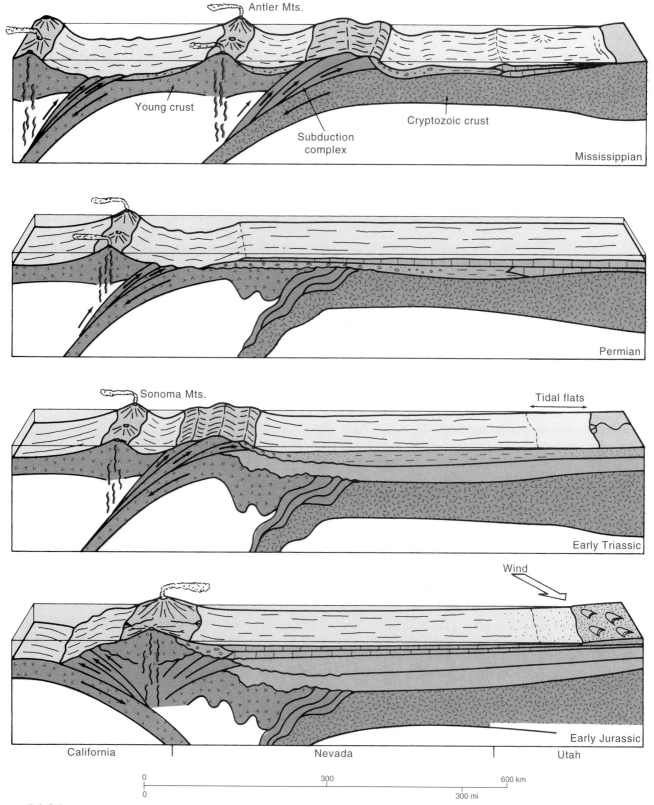

Figure 14.21   Plate-tectonic hypothesis for Mississippian to Jurassic development of western margin of North America (see Fig. 14.17 for line of cross sections). Antler and Sonoman orogenies are interpreted as results of successive arc collisions followed by reversal of subduction in Late Triassic time. Note that the Sonoma Mountains and arc are part of Sonomia of Fig. 14.17. Eastward subduction produced an Andeanlike arc on the edge of the continent, beginning in Late Triassic time. Note Navajo Sandstone deposition at lower right. *(Adapted from M. Churkin, Jr., 1974: Society of Economic Paleontologists and Mineralogist Special Publication 19; R. A. Schweickert, 1976: Nature, v. 260, pp. 586–591; F. G. Poole and C. A. Sandburg, 1977, in Paleozoic Paleogeography of western U.S., pp. 67–86; R. C. Speed, 1979, unpublished.)*

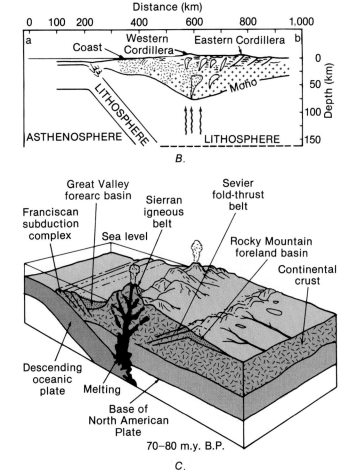

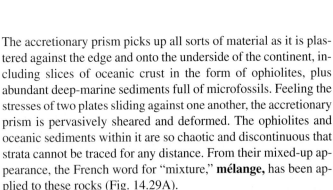

Figure 14.22   The modern Andes Mountains of South America provide an excellent analogue of the Sierran arc in the Mesozoic. *A:* Geometry of the Andean arc with respect to the subducting Nazca Plate, which originates at the East Pacific Rise and spreads eastward, going down the Peru-Chile trench. *B:* Cross section through the central Andes, showing how the subducted slab melts to produce the Andean volcanoes and magma chambers that will become batholiths. Also shown is the backarc thrusting of the eastern Cordillera. *C:* Geometry of the Sierran arc during the Mesozoic. Note the Franciscan accretionary wedge, the Great Valley forearc basin, and the backarc thrusting of the Sevier belt behind the Sierran volcanic arc. *(After D. E. James, 1971, Geological Society of America Bulletin, v. 82, p. 3325).*

The accretionary prism picks up all sorts of material as it is plastered against the edge and onto the underside of the continent, including slices of oceanic crust in the form of ophiolites, plus abundant deep-marine sediments full of microfossils. Feeling the stresses of two plates sliding against one another, the accretionary prism is pervasively sheared and deformed. The ophiolites and oceanic sediments within it are so chaotic and discontinuous that strata cannot be traced for any distance. From their mixed-up appearance, the French word for "mixture," **mélange,** has been applied to these rocks (Fig. 14.29A).

One of the characteristics of accretionary prisms is that they are continuously adding new material to the bottom of the stack, where the downgoing slab is still moving. Consequently the oldest material in the accretionary prism is at the top and the youngest is at the bottom—the reverse of Steno's law of superposition. (Of course, Steno's law applies only to normal sedimentation on the surface of the earth, not to tectonically assembled packages.) As the slab descends, it is much cooler than the surrounding mantle into which it is plunging. It is also at great depth and therefore under great pressure. Rocks in the deep part of the subduction zone experience unusual metamor-

phic conditions—high pressure yet relatively low temperature—not found anywhere else on earth. These conditions produce a peculiar suite of metamorphics known as **blueschists.** They get their name from a series of unusual minerals (such as the blue amphibole known as glaucophane) that form only under high-pressure, low-temperature conditions. As the accretionary prism builds up, blueschists from deep in the subduction zone are gradually lifted to the surface in the older slices. Whenever we find blueschists and mélange, we are looking at the remnants of an ancient subduction zone.

Finally, the region between the arc and the accretionary prism is known as the **forearc basin.** It traps volcanic sediments from the arc and reworked oceanic sediments from the uplifted accretionary prism. Due to compression, the forearc basin may subside considerably and accumulate a thick sequence of sediments. These sediments have not been transported very far or recycled, so they are typically immature sandstones with abundant rock fragments and feldspar. In many parts of the world, the forearc basin is submerged, so that the sedimentary fill is primarily marine shales and sandstones derived largely from the arc, but where sedimentation is very rapid, much of the fill may be nonmarine.

Granitic Rocks of the Sierran Arc

The Mesozoic Cordilleran margin had many similarities to the margin of the modern Andes (Fig. 14.22 bottom). Melting of the downgoing Kula and Farallon plates produced an Andean-sized volcanic arc, which must have erupted almost continuously from the Late Triassic until the Late Cretaceous. Although most of the volcanic rocks have long since been eroded away, their plutonic basement rocks are now exposed in a series of Mesozoic batholiths (Figs. 14.23 and 14.24). In fact, the late Mesozoic marks one of the greatest periods of formation of granitic rocks in Phanerozoic history. Why should this be so? We believe that these intrusions reflect rapid spreading and subduction in the Cretaceous Pacific. Recall from Chap. 8 that most plutonic granitic rocks, as well as extrusive andesite and rhyolite of similar chemical composition, form by melting of deep crustal rocks, such as old volcanic, sedimentary, or granitic ones. The necessary heat is attributed to subduction, and fast, low-angle subduction must heat and melt more crust than does high-angle subduction. Today we see many small, clearly intrusive bodies formed by magma that rose to relatively shallow levels in the crust. Recall that molten or partially molten magma is less dense than solid crust and therefore rises as gigantic, teardrop-shaped masses into the shallow crust. Where erosion has cut to greater depths, granitic rocks generally show much more complex relationships with surrounding profoundly metamorphosed rocks. Especially around the margins of large granitic batholiths, such as those of the Sierra Nevada of California and Coast Ranges of Canada, there are banded rocks with granitic and schistose layers so intimately alternating that the batholith boundaries are difficult to delineate. Clearly the temperature was great enough there to cause thorough recrystallization of all rock types. Such regions, which must be close to the sources of melting to form the granitic magmas that rose to shallower depths, seem to typify the root zones of arcs exposed by complete erosion of a volcanic edifice several kilometers thick.

The Mesozoic arc system of western North America suffered this fate. No actual Mesozoic volcanoes are left; instead, we see vast exposures of granitic batholiths and intimately associated metavolcanic and metasedimentary rocks (Fig. 14.24). Therefore, such batholiths provide our most important clue to the existence and position of countless ancient volcanic arcs that have long since been eroded to their roots. These batholiths also tend to have important ore deposits associated with them.

Foreland and Forearc Basins

East of the arc, Upper Jurassic deposits of the Rocky Mountain region form a famous nonmarine sequence called the *Morrison Formation*, best known for its treasure of dinosaur skeletons. The strata also contain invertebrate and plant fossils, which, together with the dinosaurs (Fig. 14.25A), prove that this was a nonmarine sequence. The Morrison Formation, which is made up of a varied assemblage of pastel-colored shales, sandstone, and rare conglomerate (Fig. 14.25B), constitutes a small clastic wedge similar to those of the Paleozoic in eastern North America.

Figure 14.23   Location of the major granitic batholiths formed by subduction and intrusion during the Mesozoic. They include not only the familiar example of the Sierra Nevadas but also the Southern California batholith running down the spine of Baja California, the Idaho batholith, and the Coast Range batholith of British Columbia and Alaska. At one time, these batholiths were probably lined up in a row to form a continuous volcanic arc, but later Cenozoic faulting and extension have moved the Sierran and Southern California batholith northwestward and may have rotated the Idaho batholith.

Volcanic eruptions were more important in the west than in the east, and therefore considerable ash occurs in the Morrison, a fact that provides an important clue to paleowind direction; the continent had now drifted north into the westerly wind belt. Though plant fossils and coal are rare in the Morrison, very

Figure 14.24   Granodiorites (not true granites) in the Sierra Nevada, showing the characteristic rounded outcrops typical of spheroidally weathered granitic rocks. The top of the Sierra Nevada has been ground flat by Pleistocene glaciers. This view is just north of Yosemite Valley (note Half Dome in the background) and west of Tuolumne Meadows in Yosemite National Park, California. *(Courtesy of Peter Kresan.)*

A.

B.

Figure 14.25   A: The Upper Jurassic Morrison Formation is aptly known as the "Graveyard of the Dinosaurs." Here, hundreds of dinosaur bones (mostly from large sauropods, such as *Diplodocus*) form a huge "log-jam" on what used to be a sandbar in a Late Jurassic river. After the sand buried the bones and hardened into rock, it was tilted on edge and now forms the wall of the visitor's center at Dinosaur National Monument, Utah. Originally the cliff full of bones was twice as high as shown here, but most of the early excavations removed these dinosaur specimens to the Carnegie Museum in Pittsburgh. The remaining quarry face is being slowly excavated as a public exhibit. *(D. R. Prothero.)* B: Jurassic and Cretaceous exposures in an impressive roadcut on Interstate 70 just west of Denver, Colorado. The purplish shales on the far right are floodplain deposits of the upper Jurassic Morrison Formation, overlain by tan sandstones of the lower Cretaceous Dakota Formation. This unit represents an Early Cretaceous transgression, as shown by the thin, black coal seams and the thick marine shales that overlie them out of sight on the left. *(R. H. Dott, Jr.)*

large, herbivorous dinosaur skeletons indicate that vegetation was abundant; oxidation destroyed most plants missed by the herbivores. This reasoning points to a moderately humid climate. Particularly rich concentrations of fossil dinosaur skeletons in a few localities, such as Dinosaur National Monument, Utah, suggest that unwary beasts gathered and became mired at the watering places. Then floods carried and buried carcasses downriver.

Morrison river and swamp deposits prograded to form an immense alluvial plain extending from Arizona to southern Canada. Lands were beginning to be raised in the Cordillera, especially in

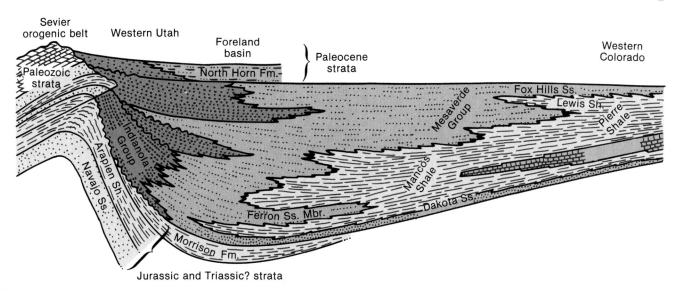

W   E

Figure 14.26   East-west cross section of the great Cretaceous foreland basin, now located in the Rocky Mountains and Great Plains (see Fig. 14.22C). To the east was the Interior Seaway and to the west was backarc thrusting and uplift along the Sevier belt. This thrusting and folding deformed Jurassic and older deposits from the basement and shed coarse conglomerates of the Indianola Group into the basin margin in western Utah. In the center of the basin, the uplift shed fluvial and deltaic sandstones, which prograded out into the marine basin and retreated when global sea level rose. This interplay between regressions from prograding deltas and transgressions from rising sea level produced a complex zigzag between tongues of shale and sandstone throughout the Rocky Mountains. (*After R. L. Armstrong, 1986, Geological Society of America Bulletin, v. 79, p. 446.*)

western Arizona, and these uplift events provided sediments to the *Morrison clastic wedge*, which was deposited in a foreland basin. Ultimately deposition crowded the epeiric sea out of much of the western craton. The sea retreated into present Canada and to the present Gulf of Mexico, but this regression was short-lived, for the sea returned to the Rocky Mountain region during Early Cretaceous time.

The eastward spread of nonmarine, varicolored Morrison sediments reflects the first major influence in the Rocky Mountain region of the Sierran orogeny, but Lower Cretaceous strata show much greater effects. All Cretaceous sediments in the eastern Cordillera coarsen westward and thicken in the same direction to form a huge clastic wedge as much as 10,000 meters thick (Figs. 14.26 and 14.27). This wedge resulted from erosion of mountains within the orogenic belt. Deposition occurred in a foreland basin, whose subsidence apparently resulted chiefly from loading of the crust by thrust faults shoved onto the craton margin from the west.

As the Sierra Nevada arc complex and its backarc sediments continued to build through the Jurassic and Cretaceous, there was also considerable activity on the Pacific side of the arc. A huge accretionary prism of mélanges and ophiolites known as the *Franciscan mélange* was scraped off the downgoing Farallon and Kula Plates. The Franciscan mélange is at least 7,000 meters thick and is highly sheared and deformed

(Figs. 14.28 and 14.29A). Many types of rocks are found in it, including marine graywackes, siltstones, black shales, and cherts derived from oceanic oozes of siliceous microfossils. There are also rocks scraped from the oceanic plate, especially ophiolitic pillow basalts and more oceanic sediments (Fig. 14.29B). Abundant blueschists are proof that parts of the prism once underwent high-pressure, low-temperature metamorphism in a subduction zone. This mélange now underlies most of the California Coast Ranges.

Behind the Franciscan accretionary prism was a great forearc basin, which was submerged during high Mesozoic sea levels. Known as the *Great Valley Group*, it is composed mostly of deep-marine shales and sandstones of Jurassic and Cretaceous age (Fig. 14.29C). Today the Great Valley Group (and younger Cenozoic sediments) underlie the Central Valley of California, one of the world's greatest agricultural regions. The accretionary prism (Coast Ranges), forearc basin (Central Valley), arc (Sierra Nevada), and backarc (foreland basin sequences in Nevada and Utah) still lie in their ancient positions, with approximately the same widths and dimensions (Fig. 14.22C).

The Sevier and Laramide Orogenies

The first major phase of Cordilleran mountain building occurred during the Late Jurassic and Early Cretaceous. Sometimes called

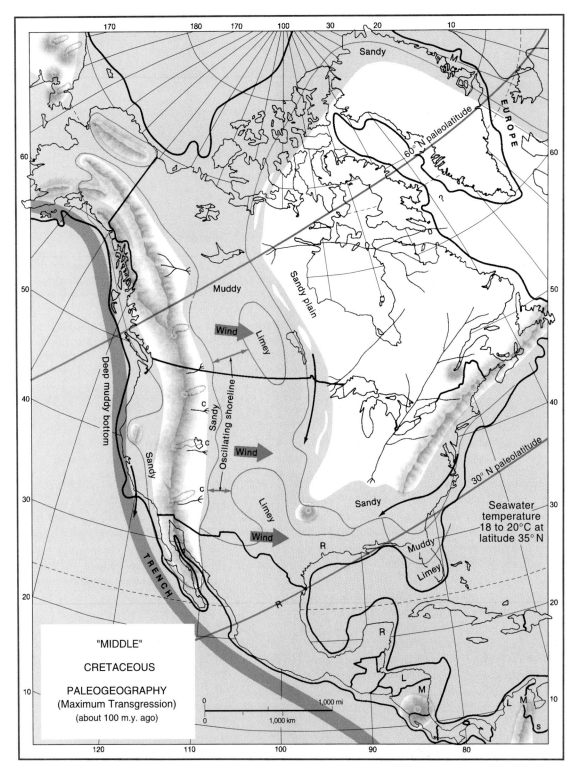

Figure 14.27 Cretaceous paleogeography at the time of maximum worldwide transgression. Westerly winds blew volcanic ash widely over the epeiric sea. Seawater temperatures are from oxygen isotope studies. Note that northward drift of the continent has placed equator and Africa off the map; Europe had not yet separated. (Compare with Fig. 14.5B.)

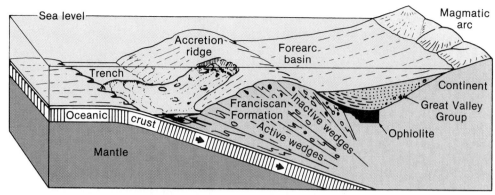

Figure 14.28 Cross section of the Mesozoic accretionary wedge, which produced the Franciscan complex, now found in the Coast Ranges of central California. As the subducting slab sank, parts of it were scraped off onto the overlying plate and became accreted to the bottom of the wedge. Ophiolites (slivers of oceanic crust) are often sliced off first, followed by wedges of oceanic sediments from the top of downgoing plate. These wedges are highly deformed by the pervasive shearing and compression, producing the chaotic rock known as mélange. Note that the youngest wedge is on the bottom of the stack, since it is accreted from the bottom. Thus, the stack gets older toward the top, in contrast to Steno's principle of superposition. In addition, the accretionary wedge bounds the Great Valley forearc basin, which filled with marine sediments and volcanics from the arc. *(After B. M. Page, 1977, Late Mesozoic and Cenozoic sedimentation and tectonics in California, San Joaquin Geological Society, p. 66.)*

the "Nevadan orogeny," this phase is inferred from the extensive clastic wedges (Fig. 14.26) in the Rocky Mountain foreland (e.g., the Morrison Formation) and from numerous Late Jurassic and Early Cretaceous radiometric dates within the batholith. Thick, coarse conglomerates and sandstones attest to vigorous erosion of Cretaceous highlands within the orogenic belt. In central Utah, coarse gravel deposits more than 3,000 meters thick accumulated on huge alluvial fans close to the mountainous highland (Fig. 14.29D).

Although Andean-style subduction continued through the entire Cretaceous, the peak of mountain building occurred in the Late Cretaceous, about 80 million years ago. Known as the **Sevier orogeny,** it is marked by a slight eastward shift of arc volcanism into Nevada and Idaho (Fig. 14.22C). The probable cause for this arc migration was a slight shallowing of the angle of subduction of the downgoing slab. This is suggested not only by the eastward migration of the arc but also by geochemistry. In modern arcs, the percentages of potassium in the magmas vary with distance from the trench; volcanic activity farthest from the trench has the highest potassium. As the Sevier arc migrated eastward, so did the potassium trends, suggesting that the depth of melting had become shallower.

The most striking effect of the Sevier orogeny was unusually high compressional forces, which led to massive backarc thrusting. Thick sequences of Paleozoic shallow-marine sandstones and limestones were sheared off their Cryptozoic basement rocks and pushed horizontally; some slid tens of kilometers eastward. The result was an extraordinary belt of stacked thrust-fault sheets, called the Sevier thrust belt, running from Nevada to the Canadian Rockies (Fig. 14.16). In many places, the mountains are built

of several stacked thrust sheets of Paleozoic sedimentary strata with remarkable folding and internal deformation (Figs. 14.30 and 14.31). The net result of this "thin-skinned" thrusting is over 100 kilometers of crustal shortening in parts of Nevada and in the Canadian Rockies.

The final phase of Cordilleran mountain building began at the very end of the Cretaceous and continued until the Eocene. Known as the **Laramide orogeny,** it did not seem to match any existing plate-tectonic analogues. Instead of Sevier-style "thin-skinned" thrusting in the foreland of western Canada, Nevada, Idaho, and Montana, a new style of mountains arose. They were formed by broad anticlinal upheavals of Cryptozoic metamorphic and plutonic basement rocks and overlying Proterozoic strata. Many of these large folds are asymmetrical, with steep thrust faults along the flank. In addition to the unusual style and source of these uplifts, they were located much farther east than the Sevier belt, running in a tract through central Colorado and Wyoming (Fig. 14.32A). Eventually these basement uplifts enclosed enormous sedimentary basins, which were filled in the Paleocene and Eocene.

Another puzzling feature of the Laramide orogeny is the fact that the eastward shift of basement deformation is accompanied by a shutoff of volcanic activity in the Sierran arc. This puzzling "magmatic null" occurs in a tract due west of the Laramide uplifts, although normal arc vulcanism persisted to the north and south of the Laramide tract (Fig. 14.32A).

William Dickinson and Walter Snyder argue that the peculiarities of the Laramide event can be explained if the angle of subduction of the downgoing Farallon Plate became so shallow that the plate scraped along horizontally beneath the continent

A.

C.

D.

B.

Figure 14.29   Characteristic rocks of the Mesozoic arc complex. *A:* Franciscan mélange south of San Simeon, California. Rounded fragments of rock are about a meter in diameter, surrounded by a pervasively sheared and deformed matrix. *(D. R. Prothero.)* *B:* Pillow lavas from a Mesozoic ophiolite, southwestern Oregon. The characteristic pillow shape indicates that these lavas erupted under water along a mid-ocean rift and were subsequently sliced off and obducted onto the continent (compare Fig. 7.16). *C:* Evenly stratified shales and turbidite sandstones *(flysch)* deposits from San Pedro Point, near the San Francisco peninsula, California. These deposits probably filled the forearc basin during the latest Cretaceous. *D:* Upper Cretaceous conglomerate (Price River Formation), Maple Canyon State Park, central Utah. Coarse fragments of lower Paleozoic and Cryptozoic quartzites were deposited in alluvial fans along the foot of mountains raised by the Sevier orogeny (see Fig. 14.26). *(Photos B–D by R. H. Dott, Jr.)*

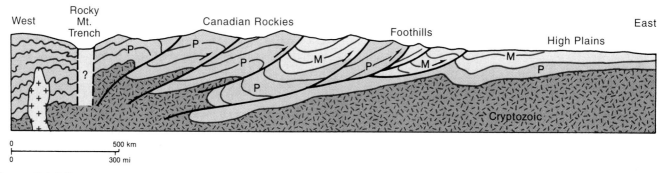

Figure 14.30   Simplified cross section through the Canadian Rocky Mountains, showing the complex stacking of thrust faults and the Rocky Mountain Trench, a possible rift or graben feature formed when the compression that caused the thrusting later relaxed. (P—Paleozoic; M—Mesozoic.)

A.

B.

Figure 14.31   Examples of the extraordinary overthrusting along the Sevier belt from Canada to Nevada. A: Sentinel Range, northern British Columbia, along the Alaskan Highway. Immense sheets of Devonian limestone have been folded and repeated by at least two thrusts, shown by the black lines; to the right of the photo, Silurian rocks have overridden Triassic deposits. (Courtesy of Lowell R. Laudon.) B: The Keystone thrust in the Spring Mountains just west of Las Vegas, Nevada. Lower Paleozoic limestones of the upper plate (dark rocks covered with vegetation on the left) have been thrust to the right over light tan Jurassic rocks of the Aztec Sandstone (equivalent to the Navajo Sandstone) in the lower plate. (© John S. Shelton.)

(Fig. 14.32C). The downgoing slab would no longer reach melting depth beneath the Sierran arc, explaining the magmatic null. In addition, the subducted plate would transfer the compressive stresses much farther eastward, effectively coupling the arc compression to the basement rocks in Colorado and Wyoming.

There is evidence to support such a model. From sea-floor magnetic anomaly patterns, it is clear that the Atlantic was spreading rapidly at this time, causing North America to ride over the Pacific Plate even faster than it had during the Triassic. Pacific sea-floor magnetic anomalies show that the convergence rate at the North American–Farallon trench was in the order of 15 centimeters/year oblique to the Cordilleran margin, an extraordinary rate of sub-

duction. According to Warren Hamilton, the continental lithosphere was "eroded" from below against a hot, young oceanic lithosphere that could not sink fast enough—due to the shallow subduction—to get out of the way of the rapidly advancing continent.

There is also some support from modern analogues. In parts of the Andes, the spreading rate of the Nazca Plate away from the East Pacific Rise is on the order of 15 to 18 centimeters/year, comparable to the rate in the North Pacific in the latest Cretaceous. Parts of the Andes are volcanically quiet, and there are, in parts of the Argentinian and Bolivian foothills, massive folds and thrusts that resemble the Laramide structures. This suggests that, in the Andes, too, there was low-angle subduction.

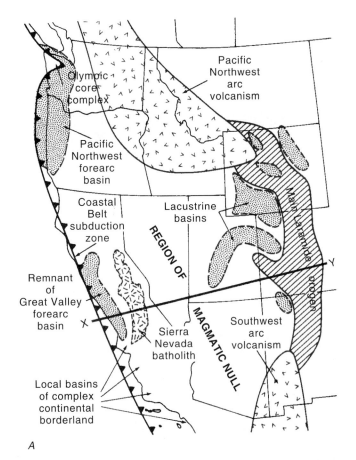

A

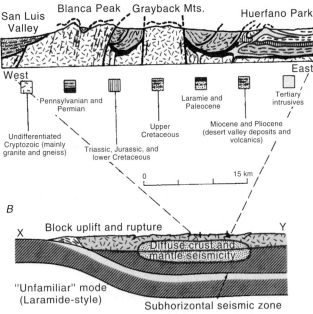

B

C

Figure 14.32   Tectonic geometry of the latest Cretaceous to Eocene Laramide orogeny. *A:* Map showing where volcanism in the Sierran arc and the entire central Rocky Mountain region ceased ("magmatic null"), although eruptions continued in the Pacific Northwest and in Mexico. At the same time, deep basement uplifts were warped and thrust upward in the formerly submerged region of Wyoming and Colorado ("main Laramide orogen"). Deep intracratonic basins subsided adjacent to the Laramide uplifts. *B:* Typical structure within the Laramide orogen near Pueblo, Colorado. *C:* One plate-tectonic explanation for peculiar Laramide geology. If the downgoing slab subducted at a nearly horizontal angle, it would not sink far enough to melt and form arc volcanoes but, instead, scrape along the base of the overlying plate and transfer its stresses far inland of the normal backarc thrust regions. *(After W. R. Dickinson and W. S. Snyder, 1978, Geological Society of America Memoir, v. 141, pp. 355–366.)*

Cretaceous Transgression and Sedimentation

Worldwide Effects

Worldwide transgression occurred during the Cretaceous Period, for marine rocks of this age unconformably overlap older rocks on practically every continent. Maximum dousing, which submerged about one-third of the present land area of the earth, occurred near the middle of the period, roughly 100 million years ago (Figs. 14.5B and 14.27). High water levels, which produced unprecedented depths of several hundred meters in several epeiric seas, occurred at slightly different times in different regions, depending upon local tectonic and topographic conditions. In broad terms, and essentially simultaneous worldwide relative rise of sea level occurred, followed by general worldwide regression down to the present time.

The Cretaceous flood affected North America profoundly, producing the last epeiric sea over the western craton and inaugurating formation of thick prisms of sediments now beneath marginal coastal plains and continental shelves. It was also during this event that the famous European and Kansas chalk was deposited (Figs. 2.6 and 14.33A).

Because there is no evidence of Cretaceous glaciation, there must have been a structural cause for this worldwide transgression. Acceleration of sea-floor spreading is known to have occurred during the Cretaceous. This caused the enlargement of ocean ridges, circum-Pacific mountain building, and wholesale continental breakup. All of those changes together displaced enough seawater to cause the transgression. Sea-floor spreading slackened at the end of the Cretaceous, and the ridges cooled and sank, causing seawater to retreat into the ocean basins.

Effects on the Craton

The craton suffered flooding both from Arctic and Gulf regions, and waters merged over the present Plains area in mid-Cretaceous time. Marine strata of this age extend from the Gulf Coast to the Arctic and from Minnesota to western Wyoming. At maximum flood, fine-grained limestones or chalks were deposited in the central part of this seaway, where terrigenous sediment was

A.

B.

C.

Figure 14.33 Sedimentary rocks of the great Cretaceous interior seaway. *A:* Cliffs of fine chalk of the Niobrara Formation from western Kansas. Thick chalks like this formed in the center of the seaway when sea level was highest, and thousands of microscopic calcareous algae rained down from the surface as they died to form a calcareous ooze. Similar chalks were common in Europe as well (see Figs. 2.5 and 2.6). *(Courtesy of J. D. Stewart.) B:* Interfingering sequence of deltaic sandstones (Ferron Sandstone) and marine shale (Mancos Shale) near Ferron in central Utah. Such complex transgressive-regressive patterns were common near the Cretaceous shoreline (see Fig. 14.26). *(D. R. Prothero.) C:* Behind the shoreline sandstones were coal swamps. Here the alternation of coals and deltaic sandstones near Helper, Utah, show that this environment fluctuated rapidly during the Cretaceous. *(Courtesy of Ken Hamblin.)*

unimportant. On the eastern side of this seaway, a widespread transgressive, sandy shoreline facies developed. Farther west, the epeiric sea lapped almost at the foot of the Sierra Nevada Mountains near the middle of the period, but marine strata became less and less widespread through the Late Cretaceous times as regression commenced. By the end of the period, the sea had retreated to the present Plains region. During early Cenozoic time, the waters parted and drained completely from the craton both northward and southward.

The great clastic wedge of Cretaceous strata on the western part of the craton (Fig. 14.26) is reminiscent of the Pennsylvanian coal-bearing wedge of the western Appalachian region. Conglomeratic facies grade cratonward into thick, massive, cross-stratified tan sandstones containing coal seams (Fig. 14.33B, C). The total volume of clastic sediments exceeds 3 million cubic kilometers. These sediments pass eastward into widespread black shales, with thin limestone layers and zones of altered volcanic ash blown from the west. Gradually sand was deposited farther and farther eastward during Late Cretaceous time, forming a classic regressive facies pattern (Fig. 14.26). The conglomerates and coal-bearing portions of sandstones are largely nonmarine, whereas the shale and some sandstone are marine. Therefore, we see that the Cretaceous coal-bearing strata have another trait in common with Pennsylvanian coal sequences—namely, repetitive sedimentation. The intertonguing of marine and nonmarine strata represents wide oscillation of the shoreline due to (1) worldwide fluctuations of sea level, (2) spasmodic uplifts in the Cordillera, and/or (3) possible cyclic climatic changes that affected weathering, erosion, and deposition of sand in a rhythmic fashion.

As the shoreline oscillated over the region, large deltas and associated swamps shifted with it, as had happened during Pennsylvanian time on the eastern craton. Sand-barrier islands paralleled the shores of marine embayments, so that the coastline must have looked much like that of Texas and Louisiana today. Abundant vegetation grew in swamps and gave rise to the widespread coal seams that constitute North American's second-largest coal reserve, which, because of a low sulfur content, may now become our most important one. "Terrible lizards" wallowed in the swamps and roamed the alluvial plain that extended back to the Sierra Nevada Mountains. Besides coal, Cretaceous strata have long been very important sources of petroleum trapped by folds and faults formed during the Laramide orogeny.

Effects in Gulf and Atlantic Coastal Plains

By Late Jurassic time, the entire Appalachian orogenic belt system from Mexico to Newfoundland had been eroded to a low-lying surface, and the present Atlantic and Gulf of Mexico basins had begun to form (Fig. 14.4). Jurassic transgression brought marine deposition over the southwestern end of the belt. Still greater transgression during Cretaceous time brought marine strata farther to cover at least half of the old eroded belt (Fig. 14.27). Cretaceous sediments must have originally extended still farther inland than now, but later erosion has stripped them from large areas of the Appalachian Mountains and southern plains. A huge barrier reef of Late Jurassic–Early Cretaceous age has recently been suggested from the results of drilling on the Atlantic and Gulf continental shelves (Fig. 14.15).

In the southern United States, quartz-bearing sandstones are common in Lower Cretaceous strata, whereas shale and limestone are more important in the Upper Cretaceous. In Mexico, limestone deposition occurred widely throughout most of the period (Fig. 14.27). Beneath the Atlantic coastal plain, Cretaceous strata are of moderate thickness (1,000 to 2,000 meters), but along the Gulf Coast, thicknesses are about 4,000 meters. The Gulf Coast region was subsiding much more rapidly than its Atlantic counterpart, but sedimentation kept pace, with the result that most of its known Cretaceous sediments are richly fossiliferous shallow-marine sands, muds, or carbonates. There is evidence of some nonmarine deposition, too. As subsidence occurred beneath the Gulf Coast region, worldwide transgression was taking place, so that younger marine Cretaceous strata

lapped farther cratonward until the latest part of the period, when worldwide regression began. Local structural warping and erosional truncations occurred, and there was considerable faulting. Small igneous intrusions (some containing diamonds) were emplaced in widely scattered areas, and a few volcanic vents in Texas, Arkansas, Louisiana, and Mexico erupted considerable ash. These intrusions and vents reflect recurring faulting along the Gulf margin and in the Mississippi aulagocen.

Paleomagnetic evidence suggests that the paleoequator was approaching its present relative position by Late Cretaceous time (Fig. 14.27). Therefore, the Gulf Coast region was still at a nearly tropical latitude. Marine invertebrates were abundant and diverse in the shallow sea there, and carbonate rocks were of great importance, with organic reefs also developing widely. These circumstances are in common with the Tethyan region to the east.

Late Mesozoic Paleoclimatology

Paleontologic and Sedimentary Evidence

Emergence and elevation of most continents in Permo-Triassic time produced a diversified "continental" climate characterized by considerable aridity (Fig. 13.20). With a much larger proportion of the earth's surface covered by water during late Mesozoic time, we would expect that more solar radiation would be absorbed by the water and heat would be efficiently distributed poleward by ocean currents, producing an overall warm, mild, "oceanic" climate with ice-free poles. Lack of evidence of late Mesozoic glaciation and widespread distribution of amphibians, reptiles, and mild-climate plants all across North America strengthen this belief.

Fossil plants, which had evolved toward a modern aspect by Late Cretaceous time, indicate mild-temperate to subtropical conditions over most continental areas. For example, conifer and ginkgo forests grew in Franz Josef Land, now at 80° N. Ancestors of breadfruit trees, laurels, *Magnolia*, and *Sequoia*, which could not stand freezing temperatures, thrived in western Greenland at 70° N. Moreover, Cretaceous dinosaur footprints were discovered on Svalbard (Spitsbergen), whose present latitude is 77° N. This frigid land today enjoys an average annual temperature range from −20 to +5°C! Although cold-blooded amphibians and reptiles can stand occasional frost and mild freezes, it is inconceivable that they could survive extended periods of subfreezing temperatures unless they hibernated in protected places. Hibernation would have been next to impossible for dinosaurs, however, because of their size. It has been suggested that some dinosaurs could regulate their body temperatures to some degree (Box 14.2). But could they handle −20°C?

Although evaporites were important in Jurassic times, suggesting high evaporative potential at many latitudes, their general absence from Cretaceous sediments suggests relatively humid conditions (or more open-marine circulation) over North America in late Mesozoic time. Cretaceous evaporites did form on the edges of Africa, however.

Oxygen-Isotope Paleothermometers

Analysis with the mass spectrometer has shown that relative proportions of the oxygen isotopes ^{18}O and ^{16}O in calcium carbonate marine shells vary according to the temperature of the seawater in which organisms grew. The ^{18}O content in shells decreases as temperature increases. Chemical relationships controlling this variation are understood sufficiently so that $^{18}O/^{16}O$ ratios in modern shells can be used for paleotemperature determination. Determinations in 1951 of $^{18}O/^{16}O$ ratios in Jurassic belemnoids from 57° N latitude (equivalent to Scotland or southern Alaska) point to an average annual seawater temperature of 14 to 20°C (54 to 68°F), or roughly 15°C warmer than is typical at that latitude today. The quantitative oxygen-isotope data, therefore, confirm qualitative fossil and sedimentary evidence pointing to a warm Jurassic climate at mid-latitudes. Cretaceous fossils from North America, western Europe, and Russia yield results (Fig. 14.34) that confirm mild ocean temperatures on the order of 20 to 25°C in present middle latitudes from 30 to 70° N. Today comparable surface ocean temperatures are characteristic of the coasts of Florida, Mexico, Central America, and northwest Africa (i.e., roughly between 5 and 32° N). Even in east-central Greenland, now at a latitude of 72° N, the sea was a balmy 17°C during mid-Cretaceous time, or about 15°C warmer than today. By extrapolation from oxygen-isotope studies of Greenlandic and Russian fossils, it is estimated that Cretaceous north polar water was no colder than 10 to 15°C.

Cretaceous Climate

If in Late Cretaceous time North America occupied approximately its present position with respect to pole and equator, and if there was little complication of ocean-current patterns, then it is possible to reconstruct a detailed paleoclimatic map from the above temperatures and what is known of the paleogeography. As just noted, both geologic and oxygen-isotope evidence indicates that total earth climate was milder and more uniform than today, and the fossil record suggests that North America was largely subtropical. Global conditions can be approximated as in Fig. 14.5B, and from this picture we can make the more detailed interpretation of North America shown in Fig. 14.35. Most of the continent would have been within the westerly wind belt, so that moist air would have blown over its western part. The Cordilleran mountains were well watered by westerly winds, and areas farther east could receive moisture from the epeiric sea east of the mountains and from the Gulf of Mexico. Widespread uniformity of Cretaceous and Eocene plants indicates a lack of sharp climatic zonation of the continent. By analogy with modern continents at similar latitudes, it is assumed that there were distinct seasonal variations of temperature and precipitation, but these cannot be fully assessed.

The Black Shale Problem

Black, organic-rich shale was the predominant Cretaceous deposit on the central part of the craton, and Cretaceous black muds are found in deep-sea cores. Less widespread black shale also occurs in Jurassic rocks, especially in Europe. Such sediment is so

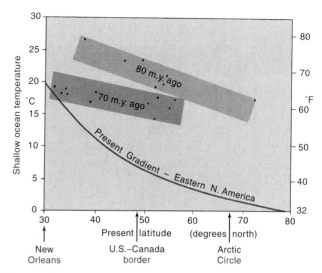

Figure 14.34 Latitudinal paleotemperature gradients for shallow-marine water at two separate times late in the Cretaceous Period based upon oxygen-isotope analysis of fossils from North America, Europe, and Russia. Note cooling trend between 80 and 70 million years ago. Present gradient (curve) is shown for comparison. *(Adapted from Lowenstam and Epstein, 1959, El Sistema Cretacico [Primer Tomo], pp. 65–76; 20th International Geological Congress.)*

widespread and characteristic—a virtual Cretaceous trademark, like the red beds of the Permo-Triassic—that it demands some special explanation.

Organic-rich mud requires both a source of organic carbon and conditions favoring preservation of that carbon from oxidation. Preservation depends upon (1) the rate of transport of carbon from surface-water plankton and/or land-plant material, (2) the relative rate of deposition of fine muds, and (3) the oxygen content of the lower water column. Two alternative extreme conditions can be envisioned: first, accelerated production of organic carbon so great that both oxidation and sedimentation are overwhelmed or, second, enhanced preservation under stagnant anaerobic conditions—even where pelagic production may be modest.

Major periods of ancient black shale deposition corresponded with times of unusually deep transgressions. Besides the Cretaceous, this was true for the Late Ordovician transgression (Maquoketa Shale of Chap. 11), for the Late Devonian transgression (Chattanooga Shale of Chap. 12), and for the Jurassic transgression in Europe (Fig. 14.4). Such floods of cratons apparently resulted in unusual production of organic carbon for reasons that are not yet clear, although some geologists believe that climatic changes caused by the earth's orbital cycle may have caused repetitive intervals of greater organic production. Enhanced preservation is easier to explain during such transgressions, and it involves important feedback between climate, paleogeography, and sedimentation. As we have seen before, great transgressions produce warm global climatic extremes, especially if the poles were simultaneously located in

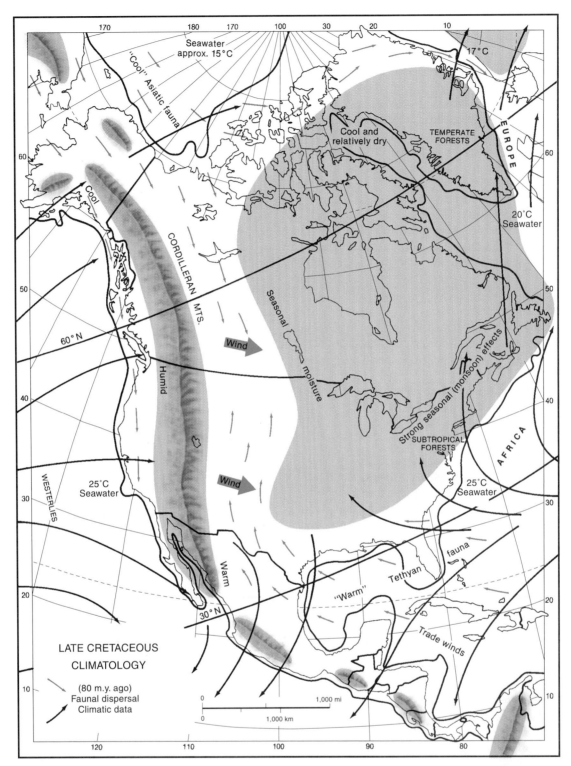

Figure 14.35 Hypothetical Late Cretaceous paleoclimatic map. Dominance of westerly winds is indicated by volcanic ash fallout; land climate from fossil plants and meteorological theory; seawater temperatures from oxygen-isotope studies; marine faunal migrations from palebiogeography; and probable latitudes from paleomagnetic evidence.

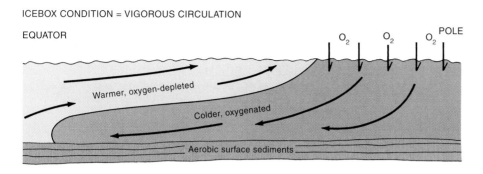

ICEBOX CONDITION = VIGOROUS CIRCULATION

Figure 14.36   Two extreme scenarios for the relationships among global climate, oceanic circulation, and oxygen content of seawater. The *icebox condition,* with sharply differing polar and equatorial water masses, characterizes the present earth. The homogeneous *greenhouse condition* fit Jurassic and Cretaceous times, when black shales were deposited widely.

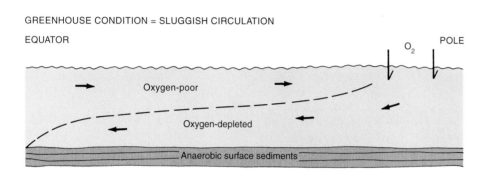

GREENHOUSE CONDITION = SLUGGISH CIRCULATION

oceanic areas, as in the Cretaceous Period (Fig. 14.5B). Because gases are less soluble in warmer water, high global temperatures would result in less oxygen uptake by seawater, and oxygen-poor water would enhance preservation of carbon in sediments. Moreover, we might expect less vigorous circulation of deep seawater on a warm earth. For example, today, with icy poles and relatively low sea level, the large-scale circulation of the deep oceans is driven primarily by the sinking of cold, dense, oxygen-rich polar water beneath warmer tropical water and the subsequent flowing of this colder water toward the equator (Fig. 14.36). During Mesozoic time, with warm poles and higher sea level, the deep circulation would have been less vigorous, which favored the preservation of organic carbon in sediments by reducing the rate of oxygen replenishment to deep waters (Fig. 14.36).

Recent modeling of global Cretaceous paleoclimate strongly supports a greenhouse world in the middle Cretaceous. A supercomputer simulated the position of Cretaceous continents, the known climatic data, and patterns of atmospheric and oceanic circulation. In such simulations, each climatic variable could be altered and the climatic consequences determined. The positions of the continents, which influence oceanic circulation and the reflectivity of the earth's surface, could account for nearly 5°C of the warming (mostly due to the absence of ice sheets). Oceanic circulation, on the other hand, was insufficient to explain the transfer of heat to the poles. Paleoclimatologists concluded that there must be some kind of greenhouse gas involved—most likely, CO_2. In these experiments, increasing the CO_2 concentration by six to eight times modern levels would be sufficient to explain Cretaceous warming. This would produce a worldwide temperature about 8°C hotter than present, or about twice what is predicted if our world goes into a greenhouse state. In other words, the Cretaceous world must have been a scorcher for millions of years!

Where would so much CO_2 come from? Unlike the modern world, there was no significant combustion of fossil fuels to produce the greenhouse effect. The only likely source is volcanic activity, which releases a variety of greenhouse gases from the mantle. Recall that the Cretaceous was a period of rapid sea-floor spreading (about three times the present rate), releasing much lava and gas along the mid-ocean ridges. In addition, there were several episodes of mantle-derived flood basalt eruptions during the Cretaceous. For example, a series of major eruptions produced the Ontong-Java Plateau in the southwest Pacific, which is more than twice the area of Alaska and has a thickness of 40 kilometers. Such massive outpourings of lava must have released enormous amounts of CO_2 into the atmosphere. By one calculation, this volcanic activity could have released enough CO_2 to raise atmospheric levels to eight to twelve times their levels just before the industrial revolution began. The calculated global temperature increase is about 10°C.

The scale of these eruptions suggests that they came from deep sources in the mantle. For example, the Hawaiian Islands are a result of a *hot spot,* a plume of hot mantle material that periodically punches up through the Pacific Plate as the plate slides over the hot spot. To erupt such prodigious amounts of lavas that could

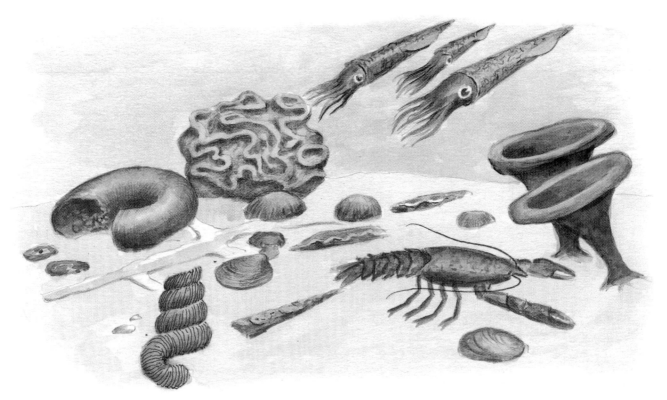

Figure 14.37   Life on the Mesozoic sea floor looked very different from the dioramas of Paleozoic life. Instead of the typical Paleozoic brachiopod-bryozoan-rugose-tabulate-crinoid communities, we see new groups in dominant roles. Corals *(center background)* were still present, but they belonged to the living group, the Scleractinia. Sponges *(far right)* were also important. But crustaceans *(the lobster on the right)* were a new element. With their shell-crushing claws, they forced most snails and clams *(center)* to swim, burrow, or become thick-shelled for protection. Swimming above them all were a variety of predatory cephalopods, especially ammonites *(buried in left foreground)* and squidlike belemnites *(swimming above). (Modified from W. S. McKerrow, 1978,* The ecology of fossils, *p. 96. MIT Press.)*

produce an oceanic plateau the size of Ontong-Java, however, more is required than a mere Hawaiian hot spot from an upper mantle source only 600 to 700 kilometers deep. Geologists have suggested that such gigantic eruptions are caused by superplumes erupting from sources deep in the inner mantle, or even from the core-mantle boundary, 2,900 kilometers below the surface of the earth. Such a massive plume would be about 1,000 kilometers wide when it reached the upper mantle and would flatten out to three times that width as it punches through the lithosphere. This enormous blister would efficiently release immense volumes of greenhouse gases (especially CO_2) into the atmosphere. The Cretaceous was a hot time in every sense!

Life in the Mesozoic

As we saw in Chap. 13, the Permian extinction was the most severe in all of earth history, wiping out perhaps 95 percent of all species in the marine realm. Characteristic elements of the typical Paleozoic fauna, particularly among the foraminiferans, corals, crinoids, ammonoids, bryozoans, and brachiopods, were severely diminished, and many groups became extinct. During the Mesozoic, a "modern" fauna arose on both the land and the sea (Fig. 11.10). Mesozoic seas were dominated by a great diversification of molluscs, sea urchins, crustaceans, and especially fish

and marine reptiles. On the land, the synapsids were replaced by a great radiation of the reptiles, including crocodiles, turtles, snakes, lizards, and dinosaurs. Indeed, the Mesozoic is commonly known as the "Age of Dinosaurs." In addition, there were both flying reptiles and birds in the skies and the earliest mammals hiding in the underbrush. By the end of the Mesozoic, flowering plants had come to dominate the land. As the name suggests, the Mesozoic was truly a time of "intermediate life," with many modern elements living side-by-side with relics of the Paleozoic.

The Mesozoic Marine Revolution

The most dramatic change took place in the oceans (Fig. 14.37). Gone were sea bottoms covered with colonies of crinoids, blastoids, bryozoans, horn corals, and clumps of spiny brachiopods. Gone were the millions of fusulinids that once built thick deposits of limestones. In their place, the molluscs began to evolve rapidly to fill the vacant ecological niches.

Both bivalves and gastropods had been present during the Paleozoic, but they were always less common than brachiopods. During the Triassic, molluscs recovered from the Permian crisis much more quickly than did the brachiopods. Orthid, strophomenid, and productid brachiopods were wiped out, and the spirifers had disappeared by the Early Jurassic. No one knows why the molluscs fared so much better than the brachiopods, but it is

Figure 14.38 During the Late Cretaceous, one group of bivalves, known as rudistids, converged on the conical body form we have already seen in rugosid corals (Figs. 12.15 and 12.26D) and richthofenid brachiopods (Fig. 13.52). These animals were remarkably asymmetrical, compared with most clams, with one shell becoming an attached cone and the other serving as a lid. They lived in huge reefs in the tropics during the Cretaceous and probably lived by lifting their lids to filter feed with their gills (and possibly fleshy mantle, shown here).

clearly not a case of ecological competition. S. J. Gould and C. B. Calloway have shown that both groups persisted throughout the Paleozoic, with no evidence of one driving the other to extinction. Besides, clams and brachiopods do not compete directly in habitat; most brachiopods live on the surface of the sea floor, and many clams were adapted for burrowing.

This last point suggests another possible reason for the disappearance of most surviving brachiopods during the Jurassic. The antipredatory adaptations of both clams and gastropods indicate that the Triassic, Jurassic, and Early Cretaceous were times of heavy predation by shell-crushing animals. The most successful groups of clams were those that developed the ability either to burrow and hide from predators or to swim as scallops do. Some gastropods became burrowers, and many of them developed spines or thickenings in the shell (especially around the lip of the opening) to prevent predators from crushing or peeling back the shell. By contrast, articulate brachiopods never developed an ability to burrow and, so, were less successful in this "arms race" of new predators. G. J. Vermeij calls this escalation of predators and antipredator defenses the **Mesozoic marine revolution.**

What predators forced this defensive response? During the Triassic, there were several groups of marine reptiles with crushing "tooth pavements," as well as a group of mollusc-eating fish. Several more groups of fish, sharks, and rays with shell-crushing teeth evolved during the Jurassic and Cretaceous. By the Late Triassic and Jurassic, crabs and lobsters with shell-crushing claws and several other groups of mollusc-eating crustaceans had appeared. The modern family of sea stars, which uses its suction-tipped tube feet to pull clams apart, appeared in the Jurassic. In addition, some molluscs developed the ability either to drill the shells of other molluscs or to insert themselves between the valves and extract their prey. Ammonites developed jaws capable of spearing or crushing prey. In short, the sea floor was no longer the relatively safe place it had been during the Paleozoic—it was a war zone of new predators. A new level of predation required new responses: great mobility; or spiny, thickened, armored shells; or the ability to burrow out of trouble. In the Mesozoic,

tiering (Fig. 11.5) reached its greatest development, with deep-burrowing bivalves exploiting the tier down to a meter below the sea floor, which had been vacant during most of the Paleozoic.

The echinoderms also showed this revolutionary change in habitat. In place of the vulnerable stalked crinoids and blastoids that had dominated the upper tier of Paleozoic sea floors, the armored and burrowing echinoids became most diverse during the Mesozoic. At first, the dominant group was round sea urchins, whose main protection was their long spines and ability to hide in crevices. By the Early Jurassic, however, the first heart-shaped urchins had appeared, a group capable of burrowing quite rapidly. This burrowing ability enabled them to escape predators and to live over a much greater area of the sea floor than spherical sea urchins. Gigantic stalked crinoids were still found in the Jurassic, but by the Early Cretaceous, they had become rare. Today the most common crinoids are stemless forms, which can move about as sea stars do. These animals are much better suited to evading mobile predators.

With the extinction of the rugosid and tabulate corals, the reef communities of the Paleozoic were wiped out. In their place arose the modern corals, or *scleractinian corals,* which have a different internal structure than Paleozoic corals. Some scientists think the scleractinians are descendants of the rugosids, but others suggest that these animals evolved independently from sea anemone–like forms that developed a hard skeleton. Either way, by the mid-Triassic, they had established small reef mounds about 3 meters above the sea floor and, by the Late Triassic, had reestablished complex reef communities (Fig. 12.21). The earliest scleractinians were found in relatively deep waters, suggesting that they had not yet developed the symbiotic relationship with algae found in tropical reefs today. Modern reef corals have algae in their tissues, a design that helps the corals secrete their calcite skeletons and produce oxygen for respiration. The appearance of large coral reefs in the Late Triassic and Early Jurassic may signal the beginning of this symbiosis.

During the Cretaceous, reef corals were displaced in some regions by a new group of reef builders, the **rudistid clams** (Figs. 14.38 and 14.39A). These oddly asymmetrical bivalves

A.

C.

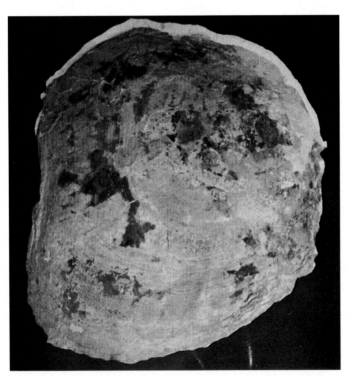

B.

D.

Figure 14.39   During the Mesozoic, there was a variety of unusual bivalves. *A:* Rudistids (Fig. 14.38) had one shell that was conical and the other that had a small, hinged cap on top. These specimens of *Coralliochama* are from the Late Cretaceous of Baja California. *B:* The largest bivalves were the inoceramids, some of which reached 2 meters in diameter. These flat, plate-shaped bivalves were so large that they served as a hard surface for many other organisms to attach to, and some had fish living symbiotically inside them for protection. This specimen of *Inoceramus* is from the Smoky Hill Chalk of western Kansas. *C: Exogyra* was a large, asymmetrically spiralled oyster, with a smaller shell *(left specimen)* serving as a flat lid on the deeper shell *(right)*. *D:* The strangest of all were the oysters with zigzag edges, known as *Alectryonia*. These specimens are from the Austin Chalk of Texas. *(D. R. Prothero.)*

grew one shell in the shape of a cone anchored to the bottom and the other shell into a small hinged lid. In their conical shape, they demonstrated evolutionary convergence on the shape of rugose corals and the bizarre conical richthofenid brachiopods of the Permian (Fig. 13.52). Rudistids attached themselves to hard objects (including other rudistids) and built huge reef complexes. Like corals and the living giant clam *Tridacna*, rudistids apparently had symbiotic algae in their tissues to allow the rapid growth necessary to compete with corals for space. By the Late Cretaceous, nearly all shallow tropical reef assemblages had been built by rudistids, not corals (Fig. 12.21), yet rudistids had become extinct by the end of the Cretaceous, and scleractinian corals and coralline algae returned to dominate Cenozoic reef communities.

In addition to housing the rudistids, the Cretaceous seas were home to a number of other bivalves. The most spectacular were huge *inoceramids*, bivalves with shells shaped like dinner plates; some individuals were over 2 meters across (Fig. 14.39B). These clams have been found with symbiotic fish and other communities that lived in their shells. Also common were a variety of oysterlike forms, including the spirally coiled *Exogyra* and weird oysters with zigzag margins (Fig. 14.39C, D). Like the rudistids, these aberrant bivalves vanished with the retreat of the shallow seaways at the end of the Cretaceous.

Swimming in the waters above the sea floor were dozens of kinds of ammonoids. Although ammonoids with zigzag goniatitic sutures (Fig. 12.3F) had been common in the Late Paleozoic, all but two ammonoid genera had become extinct at the end of the Permian. In the Early Triassic, the ammonoids underwent another great adaptive radiation from a single Permian survivor, the genus *Xenodiscus*. Within about 5 million years after the Permian extinction, a great adaptive radiation had resulted in more than 150 species and 100 genera. Triassic ammonoids typically had a U-shaped *ceratitic suture* (Fig. 14.40A). Ammonoids soon became so common and widespread around the world that they are the best index fossils for the Mesozoic.

At the end of the Triassic, ammonoids underwent another wave of extinction, and nearly all of the ceratitic forms disappeared. In the Early Jurassic, they were replaced by ammonoids that had complex, bushy, branched sutures called *ammonitic sutures* (Fig. 14.40B). Once this pattern was established, ammonitic sutures became more and more elaborate. The complexity of this suture pattern is difficult to explain because it gives only slightly more mechanical reinforcement than do the less complex sutures. Regardless of the explanation for its complexity, however, the suture pattern is diagnostic for each species of ammonite. It is the principal feature used to recognize each species and to trace its evolution.

From the few surviving Late Triassic taxa, Early Jurassic ammonites underwent another great evolutionary radiation to as many as 90 families in the Jurassic. This great radiation persisted into the Cretaceous, making the ammonites among the commonest fossils in Mesozoic marine rocks. Some ammonites, such as *Parapuzosia*, reached almost 3 meters in diameter (Fig. 14.40G). Others began to uncoil in various odd ways. One of the common-

est Cretaceous forms is the secondarily straight-shelled *Baculites* (Fig. 14.40F). The most bizarre of these forms are the heteromorph ammonites, which develop odd, J-shaped shells and knotted shells (Fig. 14.40C, D); some even spiral up an axis, like a snail shell (Fig. 14.40E).

In addition to the great success of the ammonites, another cephalopod group, the *belemnites,* was very common in the Mesozoic. Belemnites must have looked like the modern squid, and their fossilized internal supporting rods, which look like a large bullet or cigar butt, occur in great abundance in Mesozoic rocks (Fig. 14.40H). Like squids, belemnites probably swam backward with jet propulsion. Their heavy supporting rods probably served as a counterweight to keep their bodies horizontal as they swam. Similar structures are used by living squids and cuttlefish.

The Mesozoic seas had a much greater variety of life and ecological niches than the Paleozoic seas. These ranged from burrowing molluscs and echinoderms, to a variety of new shell-crushing crustaceans, to fish and reptiles, to an even greater variety of predatory swimming fish and cephalopods, and ultimately to their predators, the marine reptiles. This great expansion of life was aided by the high sea levels forming shallow epeiric seaways in the Cretaceous and by the "arms race" of predators and prey. However, the most important change was at the base of the marine food pyramid. The Mesozoic saw the expansion of several major groups of planktonic organisms, which convert nutrients and sunlight into living tissue to feed more complex organisms. Among the microscopic plants were the *diatoms,* which secreted silica-rich shells, which nested inside each other, like two petri dishes. The earliest diatoms were present in the Jurassic, but they began to radiate in the Cretaceous.

Another group of plants was the *coccolithophores,* which secreted microscopic calcareous plates known as coccoliths (Fig. 14.41). These calcareous algae were so tiny they are often called "nannoplankton" (submicroscopic plankton). In some areas of the shallow Cretaceous seas, their plates were so abundant that they made up huge volumes of fine-grained limestone known as chalk (Figs. 2.6 and 14.33A). Most of the world's great chalk deposits, including the famous White Cliffs of Dover, are made of solid microfossils. In fact, the Cretaceous was named from the Latin word *creta,* which means "chalk."

Feeding on these microscopic plants were the amoeba-like foraminiferans. In place of the fusulinids, which covered the sea bottoms in the late Paleozoic, a new group of foraminiferans emerged. These were the *globigerinids,* which built their shells out of a spiral arrangement of bubble-shaped chambers (Figs. 14.42 and 14.56). Unlike bottom-living fusulinids, however, the globigerinids were so tiny that they could float in the plankton of the sea surface, spreading around the world's oceans.

Like diatoms and calcareous nannoplankton, globigerinids first appeared in the Jurassic but underwent a huge evolutionary radiation in the Cretaceous. This great diversity of microplankton must surely have enabled the great diversification of filter-feeding molluscs and fish, a diversification that, in turn, gave a much broader food base for higher-level predators.

Figure 14.40   The Mesozoic was the heyday of ammonoids. *A:* In the Triassic, most ammonoids had a U-shaped *ceratitic suture* between the chambers. *B:* In the Jurassic, all ammonoids had a complex, florid *ammonitic* suture. Notice how the complexity of the suture is a reflection of the convolutions on the edge of the septum (exposed at the broken edge of the chamber on the left bottom of specimen). In the Late Cretaceous, ammonites began to modify their normal flat spiral in a variety of bizarre ways; these shapes are known as *heteromorphs*. *C:* The simplest heteromorphs were like *Scaphites,* which is just slightly uncoiled. These specimens are apparently male and female of the same species, with a tiny juvenile in the middle. *D: Hamites* was uncoiled into a hairpin shape. *E: Turrilites* broke the ammonoid stereotype of spiralling in a single plane; instead, it spiralled into a cone, much like a snail. Other heteromorphs were uncoiled into an upside down question mark or in a small knot. *F:* One of the commonest heteromorphs, *Baculites,* was straight-shelled, except for a curl at the tip. *Baculites* are usually preserved only as short segments of the complete, meter-long shell. Unlike the straight-shelled nautiloids (Fig. 11.2), which started out straight and later coiled up, *Baculites* clearly started out as a normally coiled ammonite (notice the complex ammonitic suture) and later straightened out. *G:* Some ammonites were truly gigantic. This specimen of the Late Jurassic ammonite *Parapuzosia* is almost 2 meters in diameter; the animal that protruded from this shell must have been nearly the size of the living giant squid. *H:* In addition to cephalopods with external shells, the squidlike belemnites (Fig. 14.37) had a rodlike internal shell for counterweighting their bodies; it typically resembles a large-caliber bullet in shape. *(Photo C courtesy of N. A. Landman; photo D courtesy of Peter D. Ward; all other photos by D. R. Prothero.)*

Figure 14.41 During the Cretaceous, the shallow surface waters of the oceans and inland seas swarmed with millions of single-celled algae known as *coccolithophorids*. These plants secrete a series of button-shaped plates *(coccoliths)* around them. These scatter on the ocean floor to form chalk when the plant dies. Coccoliths are much smaller than most other oceanic microfossils. This specimen is only a few thousandths of a millimeter in diameter; it is magnified over 4,000 times. *(Courtesy of Stanley A. Kling.)*

Marine Vertebrates

Cephalopods of various sizes were the most common swimming animals in Mesozoic seas, but there was also a great radiation of bony fish as well. Early in the Mesozoic, the seas were dominated by primitive bony fish related to the modern sturgeon (the source of caviar). These fish had asymmetrical tails resembling the tail of today's sharks, and heavy, diamond-shaped scales. By the Late Jurassic, however, the radiation of the modern *teleost* fish had begun. Virtually all living bony fish are teleosts, and there were hundreds of different species swimming in Cretaceous waters. Among their many innovations was a highly mobile jaw mechanism, which gave them great flexibility in feeding. Teleosts have a swim bladder, a gas-filled sac that allows them to regulate their buoyancy. The most spectacular of the Mesozoic teleosts was *Xiphactinus,* a gigantic (4-meter-long) predator found in the great Cretaceous inland seas (Figs. 14.43 and 14.44).

Feeding on this variety of swimming ammonoids and fish were a number of different kinds of marine reptiles. These genuine sea monsters began in the Triassic with short-legged, short-necked nothosaurs (the reptilian equivalents of seals) catching fish in near-shore waters. There were also the heavy-bodied placodonts, which used their pavementlike teeth to feed on molluscs. By the Jurassic, the placodonts had become extinct and the nothosaurs had evolved into the *plesiosaurs* (Fig. 14.44). Some plesiosaurs had short bodies with broad, paddle-shaped fins and a very long neck for catching fish. *Elasmosaurus* of the Cretaceous of Kansas had a neck 8 meters in length and was more than

A.

B.

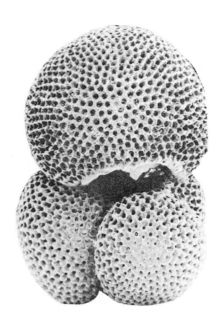

Figure 14.42 In addition to coccoliths, there were other important calcareous microplankton in the Cretaceous oceans. Although the bottom-dwelling fusulinid foraminiferans (Fig. 13.55) were extinct at the end of the Permian, another group of foraminiferans known as the *globigerinids* were tiny enough to float in the surface waters as plankton. *A:* A living globigerinid is much like an amoeba, with long, fingerlike projections of protoplasm (which it uses to capture prey) extending out from the internal shell. *(© Manfred Kage/Peter Arnold, Inc.) B:* Once the foraminiferans died, their calcareous shells, made of porous, spherical chambers arranged in a spiral, were left in huge numbers on the ocean floor. Most specimens are less than a tenth of a millimeter in diameter; this specimen of *Globigerinoides sacculifer* is magnified 100 times. *(Courtesy of Alan H. Bé.)*

15 meters long. Other plesiosaurs had short necks but long heads and a body more like that of a whale. The largest of these, *Kronosaurus*, was 17 meters (56 feet) long with a 3.5-meter jaw. It swam the Cretaceous seas of Australia.

The most completely aquatic of the reptiles were the "fish-lizards," or *ichthyosaurs* (Fig. 14.44). Their bodies were like a dolphin's, with well-developed paddles, a dorsal fin, and a tail fin like that found on sharks. Unlike sharks, however, the supporting spinal column in the tail bent downward, not upward. Ichthyosaurs had long bills with fish-catching teeth, and many had large eyes for seeing underwater. Because they were incapable of crawling out on land to lay their eggs, they must have given live birth to their young, as whales and dolphins do. Indeed, specimens have been found of ichthyosaurs preserved with a newborn emerging from the area of the birth canal. Most ichthyosaurs were less than 3 to 5 meters long, but the largest, *Shonisaurus*, reached 15 meters in length and must have weighed 36,000 kilograms. It was found in the Triassic rocks of Berlin-Ichthyosaur State Park in central Nevada.

Plesiosaurs and ichthyosaurs had become extinct by the end of the Mesozoic, but other marine predators survived and even have living relatives. There were huge marine turtles 4 meters in length. Several groups of crocodiles became fully adapted for marine life, with tail fins and feet modified into paddles. The most spectacular of these modified land reptiles were the *mosasaurs*

Figure 14.43   The enormous size of the Cretaceous teleost fish *Xiphactinus* is conveyed by the six-man crew needed to collect it from the chalk beds of western Kansas. *(Courtesy of Fort Hays State University's Sternberg Museum of Natural History, Hays Kansas.)*

Figure 14.44   The Cretaceous seas swarmed with an immense variety of swimming animals. In addition to the squidlike belemnites *(bottom)*, there were many different kinds of marine vertebrates. There was a great variety of advanced teleost fishes, including the giant *Xiphactinus (largest fish in the diorama.)* Preying upon these were many kinds of marine reptiles, including the huge sea turtle *Archelon (right)*, the long-necked paddling plesiosaurs *(three examples in the upper left)*, the fishlike ichthyosaurs *(lower left)*, and the marine lizards known as mosasaurs *(at sea surface on right)*. The giant pterodactyl *Pteranodon* flew overhead, and there was even a primitive, flightless, toothed bird, *Hesperornis*, which was adapted for a marine existence *(at sea surface on extreme right)*. *(The Age of Reptiles, a mural by Rudolph F. Zallinger. ©1966, 1975, 1985, 1989, Peabody Museum of Natural History, Yale University, New Haven, Connecticut. Courtesy of the Peabody Museum of Natural History, Yale University.)*

(Fig. 14.44). Related to the monitor lizards (such as the modern giant Komodo dragon of Indonesia), mosasaurs evolved into completely aquatic forms, with a flattened tail for underwater propulsion and feet modified into flippers. Their nostrils were located on the top of their heads, as in whales, so that they could breathe easily while swimming. Like their monitor lizard ancestors, mosasaurs had a hinged lower jaw to enable them to catch and manipulate fish, squid, and belemnites. The largest mosasaurs grew to more than 14 meters in length.

Thus, the Mesozoic marine realm was revolutionized from bottom (new microplankton) to top (marine reptiles). Although some of the animals (e.g., marine reptiles, ammonites, and some other molluscs) did not survive into the Cenozoic, most of the important elements (e.g., the plankton, corals, molluscs, crustaceans, and fish) did survive. Indeed, these are the dominant animals in present-day oceans. For all practical purposes, the modern marine realm began in the Triassic.

The Age of Dinosaurs

Although the transition from the Permian to the Triassic was dramatic in the marine realm, the Age of Dinosaurs began subtly on land. Most of the characteristic land plants, such as seed ferns, and many of the synapsids and huge amphibians that dominated Permian landscapes persisted into the Triassic. During the Triassic, however, this archaic seed fern–synapsid assemblage was replaced by the characteristic Mesozoic combination of advanced seed plants and dinosaurs.

In the Permian, we saw the beginnings of the seed plant revolution. Early gymnosperms, such as seed ferns, had become established worldwide. By the Middle Triassic, seed ferns had been replaced by more advanced gymnosperms. Two groups were particularly dominant in the Triassic and Jurassic. The *cycads,* or "sego palms" (and their close relatives, the cycadeoids), look much like palm trees, only the details of their structure show that they were gymnosperms (Fig. 14.45C, D). Today they grow only in tropical forests in Florida, Central America, and the Southern Hemisphere, but cycad stumps are the most abundant Jurassic plant fossil. Each cycad plant is either male or female, with male cones releasing pollen to be picked up by wind or beetles and deposited in female cones. The other group is represented by a living fossil, the ginkgo (Fig. 14.45E). Ginkgo trees were common in the Jurassic, but today only a single species, *Ginkgo biloba,* survives. It, too, would be extinct had it not been preserved in a monastery garden in China. Rediscovered in 1690 by a German traveler, the hardy ginkgo tree, with its distinctive leaf shaped like a duck's foot, is now common in cities all over the world and is a popular source of herbal medicine.

Early Mesozoic lowlands were shaded by a tree canopy 20 meters high. In addition to ginkgoes, there were a variety of primitive conifers. The most familiar survivor of these early conifers is *Araucaria,* familiar as the Norfolk Island pine, and the "monkey puzzle" tree (Fig. 14.45A, B). The spectacular fossilized tree trunks of Petrified Forest National Monument in Arizona (Fig. 14.11) are believed to be the wood of trees much like *Araucaria.* With its soaring canopy and its understory of cycads and ferns, this landscape was very different from our present world of flowering trees, shrubs, and grasses. Bruce Tiffney has pointed out that Mesozoic vegetation would have been very hostile to herbivorous dinosaurs. Most of the modern relatives of Mesozoic plants have fibrous, spiky leaves, and many contain mild toxins. In addition, most of these plants are slow growing, with little chance for rapid growth by vegetative reproduction. If they were munched by a large dinosaur, they would not grow back for years. Tiffney suggests that, in the absence of grasses (which did not evolve until the mid-Cenozoic), only ferns could have provided the rapid-growth, low-level fodder necessary to feed huge dinosaurs in the Jurassic.

A dramatic shift was taking place by the Middle Triassic. Advanced reptiles (including all living reptiles except turtles) began to replace the relict synapsid-amphibian fauna that had dominated late Paleozoic landscapes. Two major groups of reptiles emerged. One includes all modern lizards and snakes, as well as their extinct relatives, and possibly the ichthyosaurs and plesiosaurs. Lizards and snakes both arose during the Mesozoic but were not very common until the Cenozoic. The other, the **archosaurs,** was the dominant Mesozoic group.

The archosaurs include not only crocodiles and birds but also all the important Mesozoic animals—dinosaurs, flying reptiles, and many archaic reptiles. The Middle Triassic landscape of seed ferns and cycads, like those at Petrified Forest, was inhabited by a weird and wonderful assemblage of primitive archosaurs (formerly called "thecodonts") filling niches once occupied by synapsids. The commonest herbivorous animals worldwide were the *rhynchosaurs* (Fig. 14.46)—sprawling, pig-sized beasts with a short, beaked head. Their slicing beak and grinding tooth plates were apparently adapted for eating tough seed ferns, and they had claws well adapted for digging up roots and tubers. Their largest predators were heavy, four-legged archosaurs, which reached 6 meters long and had huge jaws. Also common were the crocodile-like *phytosaurs,* which must have lived as aquatic predators, like modern crocodiles. Unlike crocodiles, however, phytosaur nostrils were on top of the head, rather than at the tip of the snout. The earliest true crocodiles were also present in the Triassic, but they were lightly built, lizardlike animals. Only much later, after the extinction of phytosaurs, did they become large, aquatic predators.

There were even stranger beasts during this early archosaur radiation. Some were heavily armored; some had long shoulder spikes. They had leaf-eating teeth and a blunt, upturned snout, apparently for rooting up vegetation. The strangest-looking beast of all, however, was the "giraffe lizard," *Tanystropheus* (Fig. 14.46). Most of its 7-meter body consisted of a long neck, which may have been used to catch fish in the coastal waters where it lived.

Most of the archosaurs belonged to the branch represented by living crocodilians. The other branch of archosaurs became specialized for fast, bipedal running. Early Triassic forms, such as *Euparkeria,* were about a meter long, with a long tail for balancing the body on two legs (Fig. 14.46). From these bipedal forms, the first dinosaurs emerged in the early Late Triassic.

The earliest dinosaurs were small, bipedal animals less than 1 meter in length and bore little resemblance to the popular giants.

A.

B.

C.

Figure 14.45   For most of the Mesozoic, gymnosperms dominated the landscape. Typical Triassic trees (such as those of the Petrified Forest, in Fig. 14.11) included *Araucaria*, a living conifer still found in the southern continents. *Araucaria* species include *(A)* the Norfolk Island pine and *(B)* the "monkey puzzle" tree, with their distinctive long, droopy, unbranched limbs, from which their narrow leaves grow directly. Also characteristic were the cycads, or "sego palms" *(C)* , which resemble a short palm tree except that they bear male or female cones in the center of the stiff, fibrous leaves. The stumps of a close relative, the extinct cycadeoids *(D)* were very common fossils in the Jurassic. Another Mesozoic relict is the ginkgo tree *(E)* , with its distinctive, bilobed leaf shaped like a duck's foot. By the Late Cretaceous, angiosperms, or flowering plants, had begun to dominate the landscape. *Magnolia (F)* is a primitive angiosperm fossilized in Cretaceous beds. *(Photo A courtesy of C. R. Prothero; photos C, E, and F by D. R. Prothero; photos B and D courtesy of Bruce Tiffney.)*

Most people would not even guess that these animals were dinosaurs, yet they already had characteristic anatomical features. One of their major innovations was their fully upright posture. The hind limbs no longer sprawled out to the side of the body, as in typical reptiles; instead, they were straight in a line underneath the hip bones, as in mammals. Dinosaurs showed another specialization: the hinge of the ankle joint was not between the shin bone and the ankle bones (as in humans) but between the first and second row of ankle bones (Fig. 14.49).

Saurischians and Birds

From simple beginnings, the world came to be dominated by great reptiles, which reigned for 150 million years. In the Late Triassic, all but the last of the large synapsids and amphibians were extinct, and primitive archosaurs were being replaced by a great radiation of dinosaurs. Two major groups of dinosaurs emerged at this time. One group, the **saurischians,** included all the gigantic *sauropod* dinosaurs (such as "*Brontosaurus*") and the

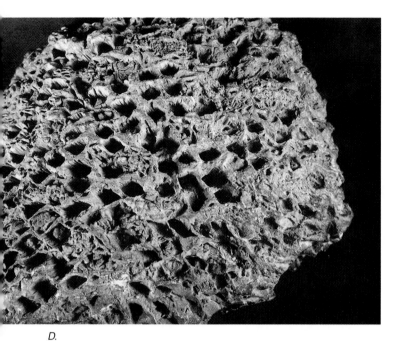

D.

F.

E.

carnivorous dinosaurs. These animals are recognized by a number of common specializations, but they still retain the primitive dinosaur hip structure. In saurischian ("lizard-hipped") dinosaurs, the pelvis consisted of three bones around the hip socket (Fig. 14.49). The pubic bone, which supported the internal organs and the muscles that pulled the thigh forward, pointed forward from the hip joint.

The other major group of dinosaurs, the **ornithischians,** or "bird-hipped" dinosaurs, includes a wide variety of herbivorous creatures, including the duckbilled dinosaurs and all armored dinosaurs. These animals can be distinguished by a number of unique specializations for herbivory, including a toothless beak with an extra bone at the front of the lower jaw and well-developed cheeks for holding vegetation while they chewed. The "bird-hipped" label comes from their specialized hip structure, which has the public bone rotated backward along the base of the ischium (Fig. 14.49). Although they are called "bird-hipped," their pelvis only superficially resembles that of advanced birds. Birds arose from carnivorous dinosaurs, and primitive birds had a saurischian hip.

Figure 14.46   When the Petrified Forest (Fig. 14.11) was alive in the Triassic, the scene might have looked like this. Amidst a flora of *Araucaria*, cycads, and other conifers lived a variety of relict synapsids and amphibians, as well as the newly diversifying reptiles. Most reptiles were primitive archosaurs, the group that included crocodiles, dinosaurs, and birds. In this landscape, the crocodile-like phytosaurs *(left center)* can be

distinguished from true crocodiles because their nostrils are over their foreheads, rather than on the tip of the snout. True crocodiles appeared in the Late Triassic, but they were small, long-legged, and without the long, flat snout that marks living crocodilians. There were also large, predatory archosaurs, such as *Erythrosuchus (right foreground),* which reached 4 meters in length. The plant-eating aetosaurs *(left foreground)* had bony armor on their backs, and spikes protruding from their shoulders. The commonest herbivore, however, was the rhynchosaur *(left center),* with its toothless beak suited for cutting fibrous vegetation. Early dinosaurs of the Late Triassic were small, bipedal animals *(left),* although *Plateosaurus,* the largest Triassic dinosaur *(right background),* was capable of browsing on treetops; it was related to the giant sauropods of the Jurassic. The most bizarre animal of all, however, was *Tanystropheus (right),* with its incredibly long neck.

Box 14.1

A Tooth, a Rib, and Some Footprints

In 1822, English physician and amateur paleontologist Gideon Mantell and his wife discovered a Cretaceous tooth in southern England. After considerable study of the tooth and of subsequently discovered bones, it was named *Iguanodon*. Originally it was misinterpreted as quadrupedal, and the thumb spike was placed as a horn on the snout. The discovery aroused much attention (Fig. 14.47), but in spite of its fame, its zoological affinity was not recognized quickly. None other than Georges Cuvier, the leading comparative anatomist, finally guessed that it was a member of an extinct group of huge reptiles later named dinosaurs ("terrible lizards"). Since then, dinosaurs have stirred the human imagination more than any other geologic phenomenon. Indeed, it is probable that their bones, together with those of Pleistocene mammoths, inspired early myths about giants. Today the stimulation powers are greater than ever, as evidenced by seeing recent movies or browsing in any toy store. Finds over the years, such as fossil dinosaur eggs polished stones thought to represent gizzard gravel, petrified dinosaur stomach contents, and nests of baby dinosaur skeletons, have prevented the extinction of interest in these ancient monsters.

In 1835, only 13 years after the discovery of the *Iguanodon* tooth but still 7 years before the term *dinosaur* was coined, huge footprints were found on upper Triassic flagstones being laid on the streets of Greenfield, Massachusetts. The tracks immediately drew attention and triggered a rash of speculation, fantasy, and even poetry. Longfellow's famous lines "Footprints in the Sands of Time" (1838) were inspired by these Triassic tracks. The discovery of bones of the barely extinct giant *Moa* bird in New Zealand lent credence to the most popular belief that the tracks had been made by some overnourished Triassic fowl (as yet, no one knew that birds appeared later). Dramatic support for the existence of such creatures was provided by famous English geologist W. E. Buckland, who "exhibited himself as a cock on the edge of a muddy pond, making impressions by lifting one leg after another." So lively was debate over footprints that it provided a great stimulus to American interest in fossils. By the end of the nineteenth century, a dozen localities in the Connecticut Valley had yielded Triassic bones that clearly were reptilian, including some of the earliest dinosaurs. The controversy was settled, with birds the losers, though the first true bird fossils (Jurassic) soon were discovered in Germany (1861).

Fossil vertebrate animals have given rise to some of the best romantic lore in the history of geology. The most colorful period was around the turn of the twentieth century, when exploration of the American West was revealing untold wealth of virgin fossil localities, many of which yielded hitherto completely unknown ancient monsters. The men who collected them were at least as colorful as their

Figure 14.47   When the first fossils of the dinosaurs were discovered in England in the 1830s, they astonished everyone. In 1853, sculptor Waterhouse Hawkins built this huge, lizardlike reconstruction of *Iguanodon* long before more complete skeletons were found to show that this dinosaur was bipedal and quite slender. To celebrate the debut of his sculptures in the Crystal Palace in Sydenham, south of London, Hawkins held a great feast inside the *Iguanodon* sculpture, with all the leading lights of British paleontology as his guests. The names of such pioneers of paleontology as William Buckland, Georges Cuvier, Richard Owen, and Gideon Mantell can be seen around the pavilion. However, only Owen was actually alive to attend this 1853 celebration. Since Owen had coined the name "Dinosauria" and was their leading expert, it was appropriate that he sat at the head of the table in the head of the sculpture. (*From* The Illustrated London News, *1854.*)

finds. It was a heroic drama on the American geologic stage and the stars were cast well for their roles.

Two of the most flamboyant participants were E. D. Cope and O. C. Marsh, who scoured arid western hills for reptile and mammal bones. These two pioneers were first harassed only by occasional unfriendly Native Americans, but later—and far more dangerously—they harassed each other in a jealous, gun-toting feud, which eclipsed even the earlier Sedgwick-Murchison altercation in Wales (Box 14.1). They falsified collecting locations to keep them secret, and each telegraphed descriptions of newly discovered skeletons back east in order to beat the other to publication. Cope even bought a scientific journal in order to publish his articles uninhibited and to have a ready medium for editorial diatribes against Marsh. As a sample of their affection for one another, consider this commentary by Marsh on Cope's extensive studies of an extinct mammal: "This mythical *Eobasileus*, under Professor Cope's domestication has changed its character more rapidly than Darwin himself ever imagined for the most protean of species. Surely such an animal belongs in the Arabian Nights and not in the records of science." And here is Cope's rejoinder: "Marsh's reputation for veracity among his colleagues is very slight in fact he has none. . . . He has never been known to tell the truth when a falsehood would serve as well. . . . He has no friends save those who do not know him well."

More recently, hordes of footprints in Cretaceous limestone exposed along the bed of the Paluxy River in Texas have created a furor because of an alleged association of human with dinosaur tracks. Many large, three-toed prints were clearly formed by dinosaurs walking across soft mud flats at the edge of a Cretaceous lagoon (Fig. 14.48A). The tracks were first publicized in the 1930s. When Creationists heard that there might be human footprints there, they seized the opportunity to discredit both evolution and the geologic time scale by arguing that humans had coexisted with dinosaurs. The Paluxy tracks were cited repeatedly during the 1960s and 1970s as a major disproof of evolution.

In addition to the large, three-toed reptilian prints, there are lines of oval-shaped depressions, which were claimed to be "man tracks." Many are large (at least 38 centimeters, or 15 inches, long) and spaced far apart, requiring that they be "giant man tracks," if humanoid at all (Fig. 14.48B). Besides their atypical size and spacing, "human" toe marks must be either imagined or enhanced. During the Great Depression of the 1930s, excavated slabs with dinosaur tracks were sold along roadsides by enterprising unemployed locals. An elderly local entrepreneur reports that "human" tracks were also carved and sold, and it is known that, as recently as the 1980s, some tracks in the river bed were enhanced.

The alleged human prints apparently had several origins including (1) bipedal dinosaurs making elongate tracks by walking on their metatarsals (soles and heels, rather than only on their toes); (2) erosional marking of the limestone; and (3) carved tracks and natural tracks altered to look more human. In 1986, Glen Kuban and Ron Hastings demonstrated conclusively that the "best" of the "man tracks" were elongate (metatarsal) dinosaur prints made indistinct by erosion, mudback-flow, or infilling. This prompted John Morris and other Creationists to backpedal reluctantly on their previous claims. (For both sides of the controversy, compare readings by Morris and Kuban.)

As Mark Twain observed in *Life on the Mississippi,*

There is something fascinating about science. One gets such wholesale returns of conjecture out of such a trifling investment of fact.

A.

B.

Figure 14.48   A: The dinosaur footprints in the Lower Cretaceous Glen Rose Limestone in the Paluxy River bed southwest of Fort Worth, Texas, have always attracted attention. This trackway shows large, circular footprints of sauropods and three-toed carnivorous theropod dinosaurs moving about. *(R. H. Dott, Jr.)* B: Close-up view of a three-toed Cretaceous therapod dinosaur track from the Paluxy River, Texas. Such elongate (metatarsal or sole-and-heel) tracks are the type often mistaken in the past by Creationists for human footprints. *(Courtesy of Glen J. Kuban.)*

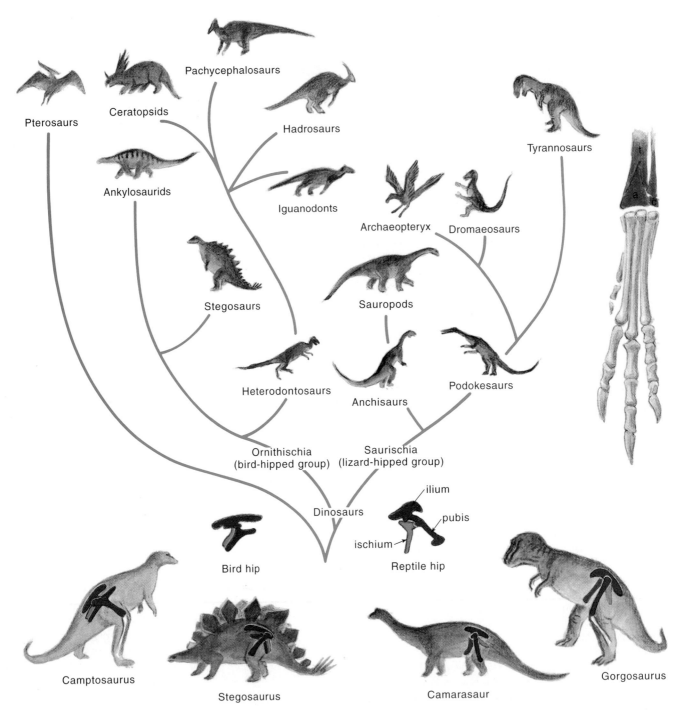

Bird hip

Reptile hip

ilium

pubis

ischium

Camptosaurus

Stegosaurus

Camarasaur

Gorgosaurus

Figure 14.49   A family tree of dinosaurs, showing some of their anatomical specializations. All dinosaurs and the flying pterosaurs share a specialized condition of the ankle *(right),* where some of the ankle bones (a and c) are fused to the end of the shin bone (t)—hence the ankle hinges in the middle of the foot. The dinosaurs are divided into two main groups. In the *saurischian* ("lizard-hipped") dinosaurs, the pelvis had a pubic bone pointing forward and down and the ischium pointing backward. Saurischians included all the carnivorous theropod dinosaurs and the huge sauropods. In *ornithischian* ("bird-hipped") dinosaurs, the pubic bone pointed backward and lies just beneath the ischium. Ornithischians were all herbivorous and included the armored stegosaurs and turtlelike ankylosaurs, the bipedal iguanodonts and duck-billed hadrosaurs, the bone-headed pachycephalosaurs, and the horned ceratopsians.

Figure 14.50 The giant sauropod dinosaurs *Brachiosaurus* and *Ultrasauros*, with a human, an elephant (the largest living land animal), and a blue whale (the largest living animal) for scale.

Primitive saurisichians and ornithischians appeared in the Late Triassic on several continents; the oldest known dinosaur was from South America. The largest dinosaurs in the Triassic were the prosauropods, which reached 9 meters in length (Fig. 14.46). These animals had the long neck and tail typical of sauropod dinosaurs, but they were not as heavy and could move on either two legs or four. By the Jurassic, sauropods had reached gigantic proportions. Familiar dinosaurs, such as *Apatosaurus* (commonly called *Brontosaurus*) were 23 meters (75 feet) in length and weighed 27,000 kilograms. In recent years, isolated bones of brachiosaurs have been found that suggest even larger sizes. Although it is known only from a few vertebrae and limb bones, *Seismosaurus* may have been over 34 meters (110 feet) long and weighed over 90,000 kilograms (Fig. 14.50).

How such huge herbivorous monsters lived is still a mystery. With their small mouths and simple, peglike teeth, it is hard to imagine how they consumed enough vegetation to stay alive. They must have been indiscriminate eaters, mowing down virtually any vegetation they could find all day long. With their long necks, they could easily reach the tops of conifers and ginkgoes, although they must have consumed far greater quantities of ferns. They probably had powerful gizzards to grind this food because many clusters of polished stones have been found inside their chests. Contrary to the popular myth, sauropods were not slow, stupid beasts hiding in swamps. Most sauropod bones come from the Upper Jurassic Morrison Formation of the western United States, which represents a well-drained upland and plains, not a swamp. In addition, we know that these dinosaurs could not hide deep in the water with just their necks exposed, as many older reconstructions portray them. If they did so, the pressure of the water on their necks and lungs at those depths would have suffocated them. Trackways indicate that sauropods moved in herds, protecting their young in their midst, as elephants do today.

Preying upon herbivorous dinosaurs were members of the other major group of saurischians, the carnivorous *theropods*. In the Triassic, theropods such as *Coelophysis* were only about 2.5 meters long and weighed only 20 kilograms (45 pounds). Even when they increased in size to become animals such as the

Box 14.2

Hot or Cold Running Dinosaurs?

When dinosaurs were discovered in 1822, they were immediately compared with familiar reptiles. They were reconstructed as oversized lizards, and most people thought that they had the physiology typical of modern reptiles. With few exceptions, reconstructions in the twentieth century also tended to portray dinosaurs as slow, stupid creatures wallowing in swamps and dragging their tails.

In the 1960s, however, John Ostrom pointed out that many of the features of dinosaurs, such as their upright posture, are inconsistent with the old view. His student, Robert Bakker, promoted the idea that dinosaurs were as warm-blooded as mammals and birds. This led to great controversy in the 1970s over dinosaur physiology. Although most of the shouting has died down, the arguments are still interesting.

First of all, the terms *hot-blooded* and *cold-blooded* are misleading. Many desert lizards, for example, thrive at body temperatures higher than those found in any mammal; and some mammals allow their body temperature to drop severely when they hibernate. It is better to contrast the *source* of the body heat and the *regulation* of that heat. Animals whose main *source* of body heat is the environment are called *ectothermic*, those whose body heat is completely obtained from metabolized food are *endothermic*. Most reptiles and amphibians are ectotherms, and most mammals and birds are endotherms. A typical lizard, for example, is sluggish when it is cold and must bask in the sun to warm its body to temperatures that allow it to hunt. An endotherm, on the other hand, must burn food continuously even when it is inactive.

Thermal *regulation* is a different matter. Animals that allow their body temperature to fluctuate with ambient temperature changes are called *poikilotherms;* those that maintain it at a constant level are *homeotherms*. Most reptiles and amphibians are poikilotherms, and most mammals and birds are homeotherms. There are exceptions, however. Pythons can shiver and generate endothermic heat to warm their nests, and some mammals behave as poikilotherms when they allow their body temperature to drop during hibernation.

The strategy of poikilothermic ectotherms is simple. They use *behavior* to regulate their body temperature. They move in and out of the sun or water to prevent becoming overheated or getting too cold. This is a very fuel-efficient way of living, but it has its drawbacks. These animals cannot function when it gets too cold, so they cannot live in extremely cold climates. Homeothermic endotherms, on the other hand, must continuously burn food to produce *metabolic* body heat. This prevents them from becoming sluggish when it is cold, but it has a great cost. Most of the food that a mammal or a bird eats goes directly to keeping it warm, rather than for other everyday activities (Fig. 14.51). This means that birds and mammals must eat much more food than a reptile of the same size and consequently must spend much more time hunting or grazing.

This distinction is apparent when one considers the ratio of predators to prey (Fig. 14.51). In carnivorous endotherms, so much food must be devoted to producing body heat that a 100-pound guard dog requires over 1,000 pounds of dog food a year to survive. By comparison, a 100-pound lizard could last a year, on only 100 pounds of food. Put another way, in the East African savannah, there is typically only one or two lions per hundred prey animals (zebra, antelopes, and so on). By contrast, there may be as many as one crocodile for every four prey animals of the same weight. A typical reptile may need to catch prey only a few times a year to survive, whereas a mammal must have hundred of prey animals available, so that it can eat continuously.

Bakker argued that the predator/prey ratios of fossil dinosaur communities are very low and so typical of endotherms. However, this argument assumes that the fossil record accurately reflects the ratios existing when the animals were alive in many cases, there are good reasons to suspect that the fossil collections are not an accurate reflection of the original living population. The most extreme examples of this are places such a the Cleveland-Lloyd Quarry in central Utah, which contains mostly predatory *Allosaurus* and few prey species. If this ratio were taken at face value, then the allosaurs must have been eating each other! Jim Farlow has shown that predator/prey ratios are highly variable and that many fossil assemblages give ratios that are not similar to those found in any living population. Also this

---

12-meter-long *Allosaurus* of the Late Jurassic, they still were able to run on their strong hind legs. Such a high level of activity suggests that many dinosaurs (especially the active predators) may have been "warm-blooded." Although there has been much controversy over this idea (see Box 14.2), if any were warm-blooded, the small predatory dinosaurs were the likeliest candidates. The last and largest of the theropods in North America was the famous *Tyrannosaurus* of the latest Cretaceous, which stood over 6 meters high and weighed up to 6,000 kilograms. Theropods probably used their enormous jaws, full of serrated teeth, and powerful hind limbs to chase down and disembowel their prey. They were highly intelligent predators and may have hunted in packs to bring down the largest prey. Trackways in Australia suggest that large theropods herded or stampeded their prey.

In addition to the familiar, larger predatory theropods, there were others, which were very ostrichlike in their proportions, although they reached 9 meters in length. Other theropods remained small and lightly built, like their Triassic ancestors. *Compsognathus* was the size of a chicken. One of the most frightening theropods was the Early Cretaceous predator *Deinonychus* (Fig. 14.52). Although it was only about 3 meters long and weighed about 60 to 75 kilograms, it was a fearsome sight. It had a long, stiffened tail for balance and agility while running and strong, grasping forelimbs. Its most impressive feature was a huge claw on each hind foot, which must have been used to slash at the underbelly of a victim or a rival. Upper Cretaceous rocks of Mongolia have yielded forelimbs over 2 meters in length of a *Deinonychus*-like animal; if scaled up proportionately, it would have been the largest predator the world has ever known.

argument is relevant only to the metabolism of the predator. It says nothing about the majority of dinosaurs, which were herbivorous.

A second argument has come from the study of the microscopic structure of bone. In the 1970s, Armand de Ricqlès showed that mammalian bones are full of open tubes called *Haversian canals* (Fig. 14.51), produced by the rapid growth rate of the bones. Most reptiles and amphibians have dense bone with no Haversian canals, but most dinosaurs have Haversian canals and other mammalian bone structures. On this evidence, Bakker has argued that they must have been endothermic.

However, on close examination, this distinction is not so simple. Some large tortoises and crocodilians have Haversian canals, and some smaller mammals and birds have none. There is a complicating factor of size, with larger size (requiring higher growth rates) promoting Haversian bone in both endotherms and ectotherms. Because most dinosaurs are so large, their bone histology may have been more influenced by their growth rate and body size than by true endothermy.

Although the upright posture of dinosaurs does resemble that of active mammals more than that of sprawling reptiles, this does not necessarily imply endothermy. Several scientists have pointed out that a sprawling lizard the size of a large dinosaur is mechanically impossible. As body size gets larger, the limbs must rotate under the body to be able to bear the load, because bone has limited resistance to bending stress. Dinosaurs have been found near the Cretaceous Arctic Circle, which some people have interpreted as evidence of endothermy. After all, no ectothermic reptiles are found that far north today. Recall, however, that the climate of the polar regions during the Mesozoic was much warmer and milder than today, as indicated by high-latitude plants and crocodilians that do not tolerate freezing.

Endothermy in large dinosaurs would also create serious problems. Heat loss in animals is related to size. In small endotherms, such as shrews and hummingbirds, heat loss is a great problem because, relative to their small volume, they have so much surface area to lose heat from. Consequently these tiny endotherms must eat almost continuously while they are awake and can die of starvation in a matter of hours. As body size gets larger, surface area increases only by a power of 2 (a length squared), but volume increases by a power of 3 (a length cubed). In very large endotherms, there is so much volume relative to surface area that these animals absorb and radiate heat very slowly. Elephants use their fan-shaped ears, and frequent mudbaths, to keep from overheating. One reason camels can survive in hot deserts is because they allow their body temperature to drop slightly on cold nights, and then it takes most of the day for them to heat up again.

For a sauropod dinosaur, which was much larger than any elephant, endothermy would be a severe problem. Sauropods apparently did not have elephant ears or any other obvious cooling mechanism, so as endotherms they would probably have overheated and died. They would have had to consume a tremendous amount of food just to compensate for their metabolic heat consumption. On the other hand, if they were ectotherms their large body size would have worked to their advantage. They would have had tremendous thermal inertia, losing heat very slowly in the night and heating up just as slowly in the day. Calculations show that sauropods could have maintained constant body heat just by virtue of their size. This is known as *inertial homeothermy,* and many scientists are convinced that this was the likely strategy of large dinosaurs.

For very small dinosaurs and for pterosaurs, however, endothermy is very likely. After all, pterosaurs could not be active fliers if they were not endotherms. In fact, one pterosaur fossil shows impressions of hair, indicating that they had furlike insulation. This would make sense only if they were small-bodied endotherms trying to prevent heat loss. Endothermy is probable for some of the smaller, more active theropods, such as *Deinonychus* (which, after all, is closely related to the endothermic birds—Fig. 14.52).

From all this debate, it is clear the dinosaurs were very different from anything living today. It is a mistake to try to fit them too closely into the rigid mold of "typical reptile" or "typical mammal." After all, there are no terrestrial animals (ectotherm or endotherm) as large as most of the dinosaurs. Dinosaurs were very diverse in ecology and behavior. It makes sense that they would also be diverse in their thermal ecology as well.

The discovery of *Deinonychus* led to another important insight: birds are descended from theropod dinosaurs. In details of the skull, wrist, ankle, and many other parts of the skeleton, advanced theropods are almost indistinguishable from birds (Fig. 14.53). Indeed, a specimen of the oldest known bird, *Archaeopteryx* (Fig. 3.1), discovered in 1861 in the Upper Jurassic Solnhofen Limestone in Germany, was once mistaken for the dinosaur *Compsognathus*. Feather impressions and a wishbone clearly show that *Archaeopteryx* was a bird, even though it still had teeth and a theropod skeleton. (An alleged bird has been reported from the Triassic of Texas, but the specimen is so poorly preserved that most scientists are not convinced it was really a bird; it may just be parts of a small theropod.) In the Cretaceous, toothed birds were present in the great shallow seaways, living as modern seabirds. One such bird, *Hesperornis*, was a flightless, fish-eating, swimming bird, such as a modern loon or grebe (Fig. 14.44). However, birds apparently never dominated the Mesozoic world as they do today. Many scientists believe that birds should be classified as a subgroup of the Dinosauria because they are descendants of dinosaurs. In that case, the dinosaurs did not die out at the end of the Cretaceous. They are alive and well and flying outside your window!

Ornithischians

Although not as gigantic as the saurischians, the ornithischian dinosaurs were very important parts of the dinosaur community (Fig. 14.49). In the Jurassic, the most familiar ornithischians were the *stegosaurs,* with armored plates on their back and a spiked tail. *Stegosaurus* was over 9 meters long and weighed 2,700 kilograms. People have long puzzled over the plates on its back, but

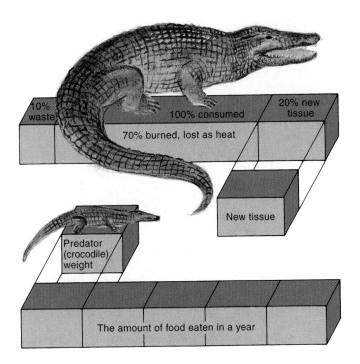

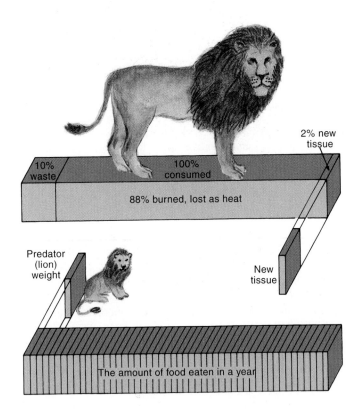

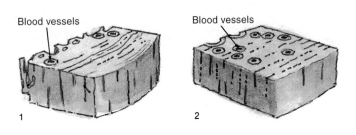

Figure 14.51   A comparison of endothermy and ectothermy. Ectothermic predators, such as a crocodile, use only 70 percent of the food they consume for metabolism, and as much as 20 percent goes to growth and new tissue. Consequently it needs to eat only about five times its body weight in a year. By contrast, an endothermic predator, such as a lion, uses 88 percent of its food for metabolism and keeping warm, and only 2 percent goes to new tissue. Consequently a lion must consume about fifty times its own body weight in a year. In other words, a much larger biomass of prey species is needed to support one lion than one crocodile. The difference between endotherms and ectotherms also shows in their bone structure. Ectothermic reptile bones (*1*) have few canals in their outer layers, but endothermic mammal bones (*2*) have abundant canals for blood vessels throughout the outer layers.

the latest research suggests that these plates were used as radiators to trap or dump body heat, rather than for defense. The plates are full of canals for blood vessels that would transport heat in the bloodstream, and they have just the right surface area, shape, and arrangement for heat radiators. In the Cretaceous, the major group of armored dinosaurs were the "dinosaur tanks," the *ankylosaurs* (Fig. 14.54). They had a heavy shield of turtlelike armor, and some had a tail club. Ankylosaurs reached 10 meters in length, and some were armored right down to their eyelids.

The most diverse ornithischians were the duckbilled dinosaurs, or *hadrosaurs* (Figs. 14.49 and 14.54). Although they are called "duckbills," they did not strain food from water, as ducks do. Instead, they had broad tooth plates made up of hundreds of prism-shaped teeth, suitable for grinding vegetation much tougher than water plants. In fact, most hadrosaurs have been found in strata from drier upland areas, not marshes. There were a great many different species during the Late Cretaceous, and many can be identified by their distinctive crests. This suggests that the elaborate crests may have been primarily for species recognition, as antelope horns are used today. In some hadrosaurs, the nasal passage looped back through the hollow crest and back over the skull, allowing it to produce deep, booming low-frequency calls.

Contrary to the older reconstructions, hadrosaurs did not drag their tails slowly through the swamps. The tail was stiffened with a complex trusswork of tendons, so that it was held straight out behind the animal. This allowed it to balance completely on its hindlegs. Most hadrosaurs were about 9 meters long, but the largest ones reached over 15 meters in length and weighed 22,000 kilograms. The family life of hadrosaurs is now well known, since Jack Horner's discovery of hadrosaur eggs and nests in the Upper Cretaceous Two Medicine Formation of Montana. The nests contained hatchlings with tooth wear, showing that they were cared for in the nest by the parents. These eggs were not abandoned immediately after laying, as are most reptilian eggs today.

A relative of the hadrosaurs was the "bone-headed" dinosaurs, or *pachycephalosaurs* (Fig. 14.49). These bipedal dinosaurs seldom reached more than 5 meters in length, but they are remarkable for their thick, bony skull roofs. Some of them have 23 centimeters of solid bone covering a fist-sized brain cavity. Because this helmet was useless against predators, it must have

Bakker '69

Figure 14.52   Early Cretaceous dinosaur *Deinonychus* ("terrible claw"). This animal was clearly a very active predator, since it had an enormous claw on each hind foot for disemboweling prey, and it kept its balance with its stiffened "balancing rod" tail while standing on one leg to slash with its feet. The obviously high energy level of this predator led John Ostrom and others to reassess whether dinosaurs were really sluggish and cold-blooded. *(Drawing by R. Bakker, Peabody Museum, Yale.)*

been used for ramming head-to-head with other pachycephalosaurs, presumably for defense of females or territory.

One of the best-known families was the horned dinosaurs, or *ceratopsians*. Beginning with the parrot-faced *Psittacosaurus* from the early Late Cretaceous of Mongolia, these animals soon radiated first into the frilled *Protoceratops* (the first dinosaur known from eggs and nests) and then into a tremendous variety of dinosaurs having different combinations of nose and eye horns and neck shields (with or without spiked edges). Ceratopsians were the reptilian rhinoceroses of the Late Cretaceous, with their horns and shields protecting them from all but the most persistent tyrannosaur. Like hadrosaurs, ceratopsians had complex tooth plates, which allowed them to chew fibrous vegetation. The last and largest of the ceratopsians, the famous three-horned *Triceratops*, was about 8 meters long and weighed up to 5,400 kilograms.

Flower Power

Why was there such a great diversification of herbivorous dinosaurs during the Cretaceous? The answer lies in a major change in the world's vegetation: the evolution of flowering plants, or **angiosperms** (Fig. 14.54). Unlike gymnosperms, which rely on wind to carry pollen to the naked seeds, angiosperms have developed flowers to attract pollinators, especially insects and birds. One of their great advantages is double fertilization: one pollen grain fertilizes the ovary, and a second triggers the growth of the nutritious kernel, nut, or fruit. These fruits attract animals, which eat and disperse the seeds. By contrast, gymnosperm nuts grow much more slowly and, so, are not as nutritious and attractive to seed dispersers.

Flowering plants have other advantages over gymnosperms. The former grow and regenerate much more quickly than the latter, so that they could recover quickly from dinosaur browsing. In fact, the oldest angiosperms originated in the mid-Cretaceous in

physically disturbed, moist habitats. At first they were "weedy" opportunistic plants, probably taking advantage of dinosaur disturbance. This ability to grow rapidly on disturbed habitat was matched by a rapid growth cycle, so the plants could sprout, grow, reproduce, and die in a single year. Most gymnosperms take 18 months or longer between reproductive cycles. Angiosperm resistance to heavy browsing and superior reproduction allowed diversification into hundreds of species by the Late Cretaceous, including primitive members of families such as the sycamore, magnolia (Fig. 14.45F), holly, palm, oak, walnut, and birch. Angiosperms filled almost every habitat, feeding hundreds of ornithischian dinosaur species.

Flowering plants could not have succeeded without the coevolution of their partners in reproduction, the insects. By developing specialized flowers and fruits, each angiosperm can attract a different pollinator and thus ensure that its pollen goes only to plants of the same species. Such selective pollination allows any new mutation to spread much more rapidly and efficiently through a restricted gene pool and speeds up angiosperm evolution. Although the insects rarely fossilize, there is good evidence that they underwent an evolutionary explosion in response to this coevolution and that more insects in turn allowed more kinds of flowering plants to develop. In fact, the most important pollinators, moths and bees, have their earliest fossil record in the Late Cretaceous.

By the Late Cretaceous, dinosaurs and angiosperms had formed a complex community, much like that of the present-day African savanna. There were the diverse hadrosaurs performing the role of antelopes and zebras, as well as ceratopsians and ankylosaurs in place of rhinoceroses and elephants. Their predators, the terrible tyrannosaurs, were much larger than lions and hyenas. There were crocodiles reaching 15 meters in length, capable of eating a small dinosaur in a single gulp. The air hummed with insects, birds, and the flying reptiles, or *pterosaurs*. Being close

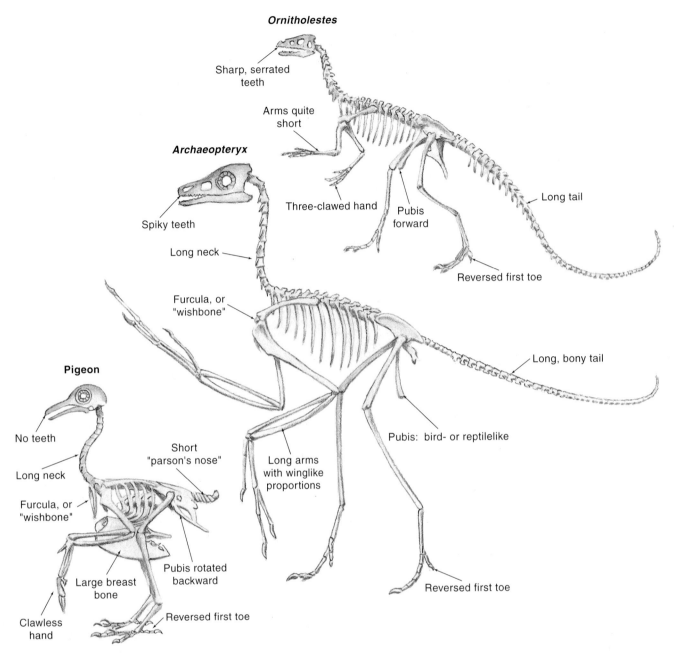

Ornitholestes

Sharp, serrated teeth

Arms quite short

Archaeopteryx

Three-clawed hand

Pubis forward

Long tail

Spiky teeth

Long neck

Reversed first toe

Furcula, or "wishbone"

Long, bony tail

Pigeon

No teeth

Long neck

Short "parson's nose"

Long arms with winglike proportions

Pubis: bird- or reptilelike

Furcula, or "wishbone"

Large breast bone

Pubis rotated backward

Reversed first toe

Clawless hand

Reversed first toe

Figure 14.53   The earliest known bird *Archaeopteryx* (see Fig. 3.1) was a clear intermediate between small predatory theropod dinosaurs and more advanced birds. In its skull, long tail, and hind limbs and details of the wrist and ankle, it is almost identical with small dinosaurs, such as *Ornitholestes*. Only in the wishbone, elongate fingers, and feathers does it show its true bird relations. Indeed, one specimen of *Archaeopteryx* was long misidentified as the small dinosaur *Compsognathus* until someone spotted the feather impressions. *(Modified from D. Norman, 1985, The illustrated encyclopaedia of dinosaurs, Crescent.)*

relatives of dinosaurs, pterosaurs showed many common evolutionary specializations, including the hinge within the ankle joint (Fig. 14.49). The earliest pterosaurs were not much larger than most birds, but by the Cretaceous, they had reached their peak size. The famous long-crested *Pteranodon* (Figs. 14.44 and 14.54) had a wingspan of 7.5 meters and could glide across the Cretaceous seaways like a giant albatross. *Pteranodon* probably plucked belemnites and fish from the surface, as do modern seabirds. Other pterosaurs had weird head crests; one species had hundreds of comblike teeth for filter feeding, like a flamingo. The most spectacular of all, however, was the great Texas pterosaur *Quetzalcoatlus*. Although it is known from only partial skeletons,

Figure 14.54   Diorama of life in the Late Cretaceous. The giant predator *Tyrannosaurus rex* dominates the scene, faced by two horned *Triceratops*. Also visible are the duckbill *Anatotitan*, the small ostrich dinosaur *Struthiomimus*, and armored *Ankylosaurus*. A huge *Pteranodon* flies overhead, and the vegetation shows a profusion of flowering plants. Although modern dinosaur artists would draw a less blimplike tyrannosaur leaning forward in a fully balanced posture, in most details this painting is accurate. Such a scene must have been common at the end of the Cretaceous, when the Lance and Hell Creek Formations of Wyoming and Montana were deposited. (*The Age of Reptiles, a mural by Rudolph F. Zallinger.* © 1966, 1975, 1985, 1989, *Peabody Museum of Natural History, Yale University, New Haven, Connecticut. Courtesy of the Peabody Museum of Natural History, Yale University.*)

its bones indicate an animal with an 11- to 12-meter wingspread, larger than that of a small airplane. The largest flying animal of all time, *Quetzalcoatlus* probably soared over the landscape like a condor, riding the updrafts and looking for prey or carrion.

Hiding in the underbrush in this dinosaurian world were frogs and salamanders (which evolved in the Jurassic), turtles (which evolved in the Triassic), and lizards (which were evolving throughout the Jurassic and Cretaceous). Snakes related to boa constrictors first appeared in the Cretaceous, although most snake diversification occurred in the Cenozoic. The most important of these smaller animals hiding from the dinosaurs, however, were their eventual successors, the mammals.

Early Mammals

Mammals were descended from the last of the synapsids in the Late Triassic. For almost 150 million years, mammals remained mouse-sized, hiding in the underbrush and coming out primarily at night (Fig. 14.55). Their teeth had high, triangular cusps for eating insects. In most features, they were clearly more advanced than their synapsid ancestors. They had larger brains relative to body size than any reptile or synapsid, and some of this brain capacity was used for more complex reproductive behavior. Most

Figure 14.55   From this origin in the Late Triassic, most Mesozoic mammals were small and shrewlike, living in the underbrush for almost 120 million years while the dinosaurs dominated the scene. Here, a Triassic morganucodont is shown preying on a lizard, another group that arose in the Triassic and stayed small. (*Painting by Zdenek Burian.*)

mammals give birth to live young and invest considerable parental care in their upbringing, feeding them with milk from the female's mammary glands. They also have a high rate of metabolism and are endothermic and homeothermic (see Box 14.2). To sustain this metabolism, they need a highly efficient mechanism of food intake. Their jaw muscles allow complex chewing motions, and their teeth are specialized for slicing, stabbing, and chewing prey. To relieve the stresses in their jaw bones caused by this chewing, the jaw is composed of a single bone, and all the extra jaw elements found in synapsids are either lost or part of the middle ear bones (Fig. 3.19). Mammals have also developed a secondary palate, so that their nasal passage does not open into the mouth cavity, preventing them from eating and breathing at the same time.

The archaic, egg-laying mammals of the Late Triassic and Jurassic are now extinct. Only the living duckbilled platypus and spiny anteater of Australia and New Guinea, which still lay soft-shelled reptilian eggs, may be descended from them. By the mid-Cretaceous, mammals had diverged into the two main groups that still survive: the marsupials and the placentals. Marsupials are the pouched mammals, such as the opossum, kangaroo, wallaby, wombat, Tasmanian devil, bandicoot, and koala. Their young are born as premature embryos and must climb up the mother's belly and attach to a nipple in her pouch to complete their development. Although this system makes the young more vulnerable to being lost, it is less dangerous for the mother. She can abandon her babies in danger or hard times without losing her own life. She also can care for three young simultaneously (one in the uterus, one in the pouch, and one in her vicinity) when times are good. **Placental mammals** range from all of those familiar to us in the zoo, to whales, to dogs, to sheep, to humans. Placental mammals carry the young in the mother's uterus until it is ready for birth. This allows the young to be protected and more completely developed at birth but also carries a greater risk to the mother. She cannot abort the baby without risking her own life, and both may perish during hard times.

By the end of the Cretaceous, both marsupials and placentals had become well established, but none was larger than a housecat. After 150 million years of waiting, they would finally get their chance when the dinosaurs died out. As we shall see in Chap. 15, they quickly exploited this new world of opportunity to evolve into hundreds of different forms. For this reason, many people call the Cenozoic the "Age of Mammals."

Late Cretaceous Extinctions

Dinosaurs fascinate so many people that the mystery of their extinction was featured on the cover of *Time* magazine not long ago. For over a century, there have been wild speculations about what killed the dinosaurs. These range from modest proposals about climatic change (either too hot or too cold) or vegetational change (dinosaurs could not digest the new angiosperms, died from plant poisoning, or died from constipation or diarrhea, or had asthma from pollen allergies), to competition (mammals ate all the dinosaur eggs), to such wild ideas as supernova explosions and

even UFOs. Most of these notions are beyond the limit of testable science and, so, do not merit further attention. Some are demonstrably false, however. Angiosperms arose in the mid-Cretaceous. Recall that, instead of causing extinction, these new plants stimulated great diversification of herbivorous hadrosaurs and horned dinosaurs with sophisticated dental grinding mills. Mammals had been around as long as the dinosaurs, and there is no reason to suspect that the former suddenly developed a craving for dinosaur eggs after 150 million years.

All of these explanations fail because they deal only with the dinosaurs. The end of the Cretaceous was the third biggest mass extinction in the Phanerozoic (after the Permo-Triassic and Late Ordovician events), killing off perhaps half of all life on earth. Organisms from microplankton to the marine reptiles were wiped out in the oceans, and some land plants were as severely decimated as the dinosaurs. Given their position at the top of the global food pyramid, a disturbance that affected all of life from the bottom up would be expected to wipe out the dinosaurs. Thus, any explanation that deals only with the dinosaurs misses the point entirely.

The discussion of the Cretaceous/Tertiary extinctions (usually abbreviated the "K/T event," after the geologic symbols for those periods) has moved out of the realm of science fiction. The much publicized discovery in 1980 of possible traces of an extraterrestrial impact created a whole new discipline of scientists seeking to understand mass extinctions. The debate has been intense over the past two decades leading to dozens of conferences, thousands of scientific papers, and even some hot tempers and insults. At the moment, two schools of thought seem to predominate, so there is no clear right answer so far. Before we analyze their arguments, however, we need to look at the raw data: the pattern of terminal Cretaceous extinctions.

The Fossil Evidence

There is no question that the Cretaceous extinctions were severe. On land, the dinosaurs, the pterosaurs, and many species of plants did not survive the Cretaceous. In the ocean, there were severe extinctions of planktonic foraminiferans and coccolithophorids, as well as the organisms that fed upon them: brachiopods, many molluscs (especially the inoceramid clams and the reef-forming rudistids), some echinoids, and some fish. The most striking extinction, however, was the total loss of ammonites. After their long history as the most common swimming predator of the seas since the Devonian, shelled cephalopods were almost entirely wiped out, and only the chambered *Nautilus* survives today. The top predators of Mesozoic seas, the marine reptiles (especially the mosasaurs and plesiosaurs) also disappeared.

However, detailed examination shows that these organisms did not all suddenly vanish at the very end of the Cretaceous. Inoceramids, rudistids, and most marine reptiles (except possibly mosasaurs) were all in decline long before the K/T event and may not have survived until the end. Although the calcareous nannoplankton (coccolithophorids) were in decline long before the end of the Cretaceous, the latest research shows that many of these

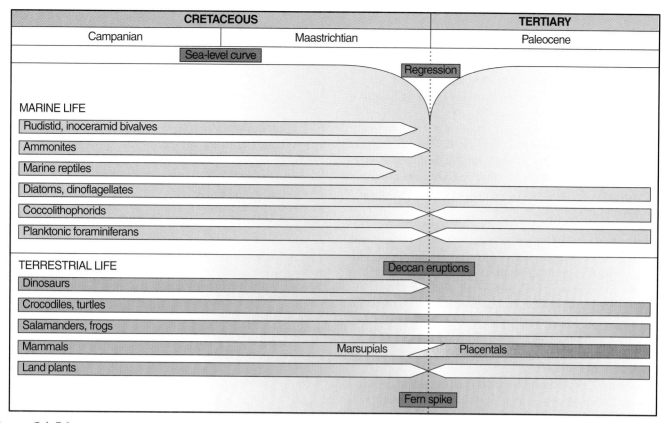

Figure 14.56   Patterns of diversity change and extinction through the latest Cretaceous and early Cenozoic. Some groups gradually declined well before the K/T event, whereas others survived with little or no effect. Only a few died out abruptly at the K/T boundary. This complex pattern of extinction suggests that the extinction mechanism is much more complex than simply the impact of an extraterrestrial body.

species survived into the Paleocene and then disappeared. Planktonic foraminiferans do not disappear abruptly but, rather, go out in several steps over a million-year interval (Fig. 14.56). Some extinctions occurred 300,000 years before the boundary, and many other foraminiferans went extinct throughout the early Paleocene. Plant microfossil groups, such as diatoms and dinoflagellates, show very little extinction, perhaps because they can survive as resting spores.

The organisms that suffered most were tropical groups, suggesting that there was a strong climatic bias in the extinction event, but the oceans yield additional evidence besides the victims of the extinction. Close examination of the oxygen and carbon isotopes incorporated into the shells of plankton shows that seawater was cooling as early as 200,000 years before the event and that the productivity of marine plankton was already shut down (Fig. 14.56). Closely tied to this cooling was a global regression event, which dried up epicontinental seas around the world during the latest Cretaceous. This regression certainly reduced the area of shallow sea bottom for marine organisms and, combined with the cooling, must have made things difficult for tropical benthic organisms, such as rudistids and inoceramids.

Ammonites were also in decline long before the boundary. Perhaps as few as twenty-two species persisted to the end. In some places, they appear to have straggled out gradually, whereas in others they disappeared abruptly at the boundary. Is the gradual pattern an artifact of poor preservation, or is the abrupt disappearance due to a big but subtle unconformity artificially chopping off the ranges of the last survivors, so that they all *appear* to have died out together? We just don't know. The pattern of brachiopod extinction also seems abrupt, but it, too, could be an artifact of an unconformity. The pattern of gradual foraminiferan decline, however, cannot be explained away as incomplete sampling because it is duplicated in at least two sections in very different parts of the world (Texas and Tunisia). Both sections are remarkably complete, showing a smooth, uninterrupted record of changes of oxygen and carbon isotopes. If there were gaps in the sequence, they would cause sudden shifts in the isotope ratios.

On land, the pattern is even more puzzling. Although dinosaurs were dying out, many scientists claim that they were in decline long before the K/T event, with as few as ten species surviving to the end. In most sections, their fossils are gone several meters below the K/T boundary. A group of scientists have argued that dinosaur fossils are too scarce to take this premature disappearance at face value. They could have died out right at the K/T boundary, and their fossils would still disappear early, simply because dinosaurs were rarely fossilized. A few scientists claim to have found Paleocene dinosaurs. However, these fossils are just isolated teeth, which could have been reworked from Cretaceous sediments into Paleocene river channels.

Besides the dinosaurs, the angiosperms were also in decline throughout the latest Cretaceous. Near the K/T boundary, there may have been an abrupt extinction event, but the plants soon recovered with some new species. There is a great abundance of fern spores right at the boundary suggesting that ferns may have responded opportunistically to unusual climatic conditions.

Except for the dinosaurs and plants, the terrestrial record is remarkably uneventful. Mammals crossed the K/T boundary with only a shift in dominance from marsupials to placentals. Although their fossil record is poor, birds survived, but the smaller dinosaurs—their closest relatives—did not. Turtles and crocodilians have an excellent fossil record, and they show almost no effect. Lizards, snakes, and amphibians were not affected, either. Any killing mechanism that wiped out the dinosaurs must also explain how turtles and crocodiles, which cannot hide or hibernate very long, did not die out.

In summary, any extinction model must answer to the fossil evidence. That evidence seems to show a gradual decline protracted over most of the Late Cretaceous, with some species persisting into the Paleocene. Most victims were extinct long before the K/T boundary. Unfortunately, some scientists have been so entranced by their own hypotheses that they ignore any fossil evidence that might contradict them. However, the "winning" model in science is not determined by a popularity poll or by the loudest publicity. Ultimately the best answer(s) will survive the scrutiny of the evidence from the fossil record.

The Impact Hypothesis

The key breakthrough in the debate occurred in 1978. A group of scientists (including geologist Walter Alvarez; his father, Nobel physicist Luis Alvarez; and chemists Helen Michel and Frank Asaro) made a surprising and accidental discovery. They were originally looking for a method to estimate the time it took to accumulate an unusual marine clay layer that represents the K/T boundary in the central Apennine town of Gubbio, Italy (Fig. 14.57). Luis Alvarez thought that, by measuring the amount of cosmic dust in the clay, this time span could be estimated. Such dust accumulates slowly, so the more dust, the longer the clay had accumulated. To measure the cosmic dust level, they looked for the element **iridium,** a platinum-group metal that is extremely rare in crustal rocks but slightly more abundant in meteorites and in the mantle.

To their surprise, the iridium concentration was far above anything that could be explained by accumulations of cosmic dust. Indeed, it was so concentrated that they could not imagine anything except an unusual extraterrestrial source. They postulated impact by an asteroid approximately 10 kilometers in di-

Figure 14.57 Close-up of the boundary layer at Gubbio, Italy, where the iridium was first discovered. Late Cretaceous limestones *(bottom)* are interrupted by a thin "boundary clay" *(below coin)* and then marine limestone deposition returned sometime later in the Paleocene *(top)*. *(Courtesy of Alessandro Montanari, Osservatorio Geologico di Coldigioco.)*

ameter. This impact supposedly caused a huge stratospheric cloud of dust to form worldwide, cooling the earth catastrophically for some period of time (like the hypothesis of nuclear winter, suggested in the early 1980s as a result of the K/T impact hypothesis). The cold and darkness would cause most plants to die and ultimately would cause the collapse of the food pyramid. Some versions of the hypothesis postulate global wildfires and huge tsunamis or massive doses of acid rain. After as few as 50 years, the world is thought to have returned to normal, albeit with far fewer organisms to populate it.

Since the original discovery, iridium has been found at K/T boundary sites all over the world; 105 examples were known as of this writing. Iridium was found in both marine and terrestrial sediments, eliminating the possibility that it was concentrated by processes peculiar to either the marine or terrestrial environment. Other rare elements, such as osmium, show the same pattern of abundance and can be explained only by a meteoritic or mantle source. In 1981, tiny spherules about a millimeter in diameter were found in the boundary layer; they were thought to be droplets of crustal basalt that were melted by the heat of the impacting asteroid and then flung into the air (Fig. 14.58A). Other localities yielded tiny grains of quartz that show shock features, apparently from the very high pressure of an impact (Fig. 14.58B). Recently a high-pressure form of quartz called stishovite was found in the boundary clay in New Mexico.

As the fossil evidence for a gradual extinction emerged, the impact group began to modify its hypotheses. Some geologists in this camp suggested a shower of comets or small meteorites spaced out over the end of the Cretaceous. Others insisted that the

A.

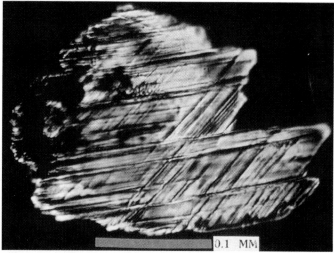

B.

Figure 14.58 Microscopic photograph of (A) spherules supposedly formed as impact droplets and (B) a quartz grain from the boundary layer, showing two sets of planar lamellae that are caused by shock metamorphism. According to meteorite advocates, this could have only resulted from an extraterrestrial impact, although the volcanism proponents suggest that high-pressure volcanic eruptions produced similar features. (A: Courtesy of Alessandro Montanari, Osservatorio Geologico di Coldigioco; B: Courtesy of Glenn Izett.)

The Volcanic Hypothesis

Although the impact hypothesis has garnered most of the press and attention, another alternative has many supporters. In 1983, studies of gases from Kilauea crater in Hawaii showed that mantle volcanoes erupt small quantities of iridium in the clouds of vapor and ash. It is well known that the end of the Cretaceous saw the eruption of the most massive volcanic event in the Phanerozoic, the Deccan traps of India and Pakistan (Figs. 14.59 and 14.60), these gigantic flood lavas erupted when India rode over the Reunion hot spot in the Indian Ocean. The eruptions covered over 10,000 cubic kilometers. Some flows are as thick as 150 meters, although most are 10 to 50 meters thick; in western India, the total thickness exceeds 2,400 meters. There were also huge volumes of Late Cretaceous flood basalts erupted in southern Brazil; some geologists say the latter exceeded the Deccan eruptions in volume. Such flows must have pumped enormous amounts of volcanic ash into the atmosphere, a condition well known to cause global cooling; they may have also pumped a lot of carbon dioxide into the atmosphere and changed oceanic chemistry. The Deccan eruptions are dated to within the last half-million years before the K/T boundary.

The Regression Hypothesis

Most of the attention has been focused on the impact hypothesis, with the volcanism scenario offering the main alternative. However, for decades it was well known that the end of the Cretaceous was marked by one of the biggest regressions in earth history. The last of the epeiric seas dried up completely on all the continents,

iridium shows only one peak at the K/T boundary and deny the possibility of additional impacts. Another problem has been finding the crater, which should be about 150 kilometers wide and easy to detect. The basaltic spherules suggest an oceanic impact, but the shocked quartz suggests an impact on continental crust. One candidate, the Manson structure in Iowa, is too small and too old to have produced the K/T extinction. Some impact supporters favor a Caribbean site, as evidenced by an impressive layer of boulder debris in Cuba, a thick layer of impact particles in Haiti, and a possible tsunami deposit in Texas.

Recently a likely impact site was found in the subsurface of the northern part of the Yucatan Peninsula of Mexico (Fig. 14.59), which might explain the debris all around the Caribbean. Known as Chicxulub (pronounced "CHIK-zu-lube"), this crater is widely regarded as the best candidate for the impact site. Dates on the melt rock from the site exactly match the K/T boundary.

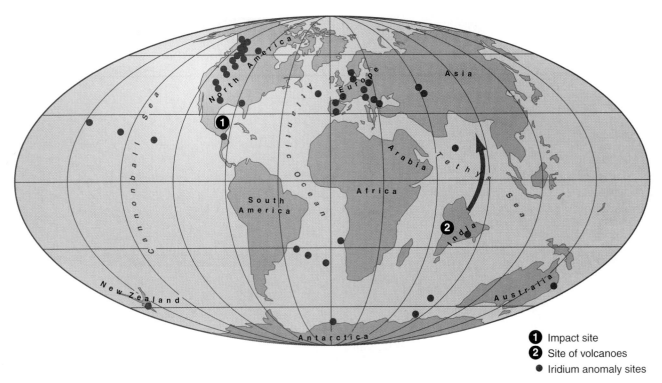

Figure 14.59 Map showing location of some of the many K/T iridium anomalies (dots), as well as the site of the Chicxulub impact (1) and the Deccan eruptions (2).

Figure 14.60 In the latest Cretaceous, there were extraordinary eruptions of mantle-derived flood basalts, known as the Deccan traps, in India. Some scientists think that these must have had a significant effect on the extinctions at the end of the Cretaceous, particularly in their effect on climate. All these cliffs near Mahabaleshwar, India, are made of hundreds of lava flows; they were erupted in a few thousand years at the end of the Cretaceous. *(Courtesy of J. Mahoney.)*

and huge areas of shallow-marine habitat disappeared around the world, especially in the tropics. Although such a cause does not seem as dramatic as a bolide impact or a gigantic volcanic eruption, major regressions do cause havoc in marine communities. Surprisingly regressions can even have an effect on land vertebrates as well. In 1996, J. David Archibald conducted a careful analysis of the environmental significance of the extinction patterns in terrestrial organisms, including the freshwater fish, amphibians, reptiles (crocodilians and turtles as well as dinosaurs), birds, mammals, and plants. Surprisingly, he found that the predictions of the global regression hypothesis were consistent with more of these paleoenvironmental indicators than was the impact scenario or the volcanic hypothesis. This scenario does not argue that volcanism or impact did not happen but that they may have been less important than originally thought and may have only exacerbated the biological stresses that were already gradually killing off many forms of life.

An Assessment

Clearly there is no simple "right" answer to the question of what killed the dinosaurs and other Late Cretaceous victims. In the years since the first iridium anomaly was discovered, the argument has gotten more complicated. New evidence appears every few months, so that any conclusion is out of date within a year. Whichever hypothesis eventually convinces the scien-

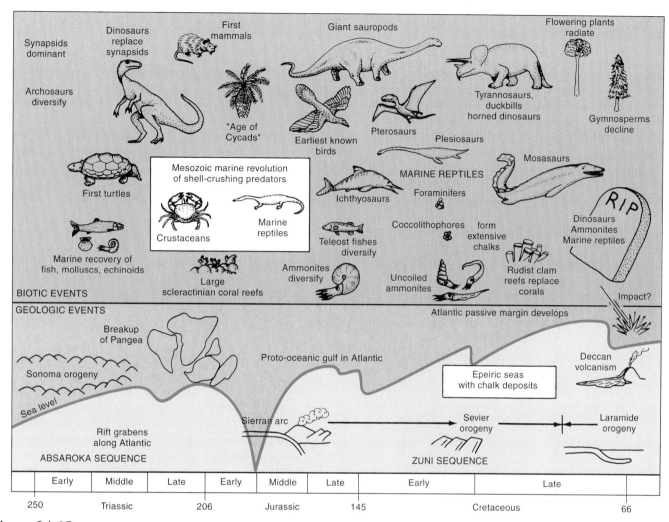

Figure 14.61 Time scale and summary of Mesozoic events.

tific community, it must account for the evidence: the protracted nature of the extinctions; the long-term change in oceanic oxygen, carbon, and strontium isotopes; the global regression; and the selective survival of many organisms that could not have lived through years of cold and darkness. It must also account for the abundance of fern spores right at the boundary, as well as the extinction of many marine and terrestrial plants. For example, the acid rain scenario fails because such highly acidic global conditions would have wiped out all amphibians. Today frogs and salamanders are declining because of the slightly increased acidity of lakes and streams. If the world had been bathed in acid (as the scenario suggests), there would be no frogs alive today.

Protracted climatic changes induced by volcanic eruptions fit most of this evidence better than a single impact. However, multiple impacts could explain the same phenomena, so the available evidence cannot distinguish between volcanism or multiple impacts. It is even possible that *both* events occurred! In any case,

there is no question that the Deccan traps are real, and they certainly must have had *some* effect. It is possible that an asteroid struck the Caribbean region at this time. Such an impact may have compounded the problems caused by a world already under severe climatic stress due to volcanism. This would fit the evidence of the abundance of fern spores right at the boundary and the decimation of both marine and other terrestrial plants that survived right up to the end of the Cretaceous.

Whatever answer finally emerges, the Cretaceous extinctions provide us with a fascinating problem, which has brought together paleontologists, geologists, oceanographers, astronomers, climatologists, and scientists from many other disciplines in a collaboration seldom seen before. The scientific results of this controversy have borne many unexpected scientific benefits, including the understanding of other extinction events and even the data for understanding a nuclear winter catastrophe. If the controversy keeps us from the brink of nuclear annihilation, then all the bitter argument has been more than worth it.

Summary

Tectonics

- Extensional tectonics is illustrated by the initial breakup of Pangea in the North Atlantic and Gulf of Mexico regions during early Mesozoic time. Aulacogens extended from the new ocean basin margins into adjacent cratons. Newly formed rift basins received first nonmarine clastic sediments and then evaporites; as the ocean basins widened, normal marine sediments were deposited in many of them. Eurasia was separated from North America at the end of the Mesozoic Era as the North Atlantic spreading ridge extended north past Greenland into the Arctic basin.

- Collage tectonics is illustrated in the Cordilleran orogenic belt of western North America. Several suspect terranes—old arcs, ocean ridges, and microcontinents—have been jumbled together and accreted to western North America tectonically at different times.

Mesozoic Geology

- The Sonoman orogeny occurred in Permo-Triassic time when a volcanic arc being subducted westward accreted to western North America. A complexly sheared subduction complex was produced by eastward overthrust faulting during collision. A reversal then produced new eastward subduction and a new Triassic-Jurassic volcanic arc on the western margin of the newly enlarged continent. This tectonic pattern persisted through Paleogene time.

- The craton remained above sea level meanwhile, and Triassic nonmarine red-bed deposition followed that of Late Permian time. Amphibians, reptiles (including early dinosaurs), freshwater fishes and molluscs, and a new flora dominated by coniferous, ginkgo, and cycad trees flourished across the continent.

- A Navajo Sandstone desert with vast dune fields covered much of the dry western craton margin during Early Jurassic time. The quartz sand was recycled by erosion of Paleozoic quartz sandstones in the central craton. Wind blew the sand southward.

- The Sundance epeiric sea flooded the western craton in Late Jurassic time. Typical cratonic fossil-rich and oolitic limestones, shale, glauconitic sandstone, and evaporites were deposited therein.

- The Morrison Formation clastic wedge prograded eastward onto the western craton at the end of the Jurassic Period in response to the erosion of embryonic mountains with volcanoes that were raised in the Cordilleran belt to the west. Volcanic ash within these sediments indicates that the continent had drifted north into the mid-latitude westerly wind belt. Morrison sediments contain some of the finest Jurassic dinosaur remains in the world.

- The Cordilleran mountain belt, which began to rise in Late Jurassic time, intensified during the Cretaceous Period and ended in Eocene time. Both uplift and volcanic activity increased within the orogenic belt, and granite batholiths formed at depth.

1. A large foreland basin subsided within the western craton due to thrust loading. The very thick clastic wedge that accumulated therein contains important coal deposits associated with repetitive progradation of sandy coastal deposits. These strata show considerable similarity to the repetitive Pennsylvanian coal-bearing strata.

2. The orogeny resulted from subduction of oceanic lithosphere beneath the continent and the collisional accretion of suspect terranes. Subduction of this type is still occurring today beneath the Andes Mountains of South America. Both systems represent Andean-type convergent plate margins, which differ from continent-continent collisional margins. Low-angle subduction such as that beneath Peru and Bolivia today can explain unusually wide zones of granitic intrusions and associated ore deposits formed in western North America during Paleogene time.

- Worldwide Cretaceous transgression was the last major flooding of continents. It was caused mainly by rapid sea-floor spreading and the resulting enlargement of ocean ridges, which displaced much seawater onto the continents. Effects of the transgression varied greatly, depending upon topography, climate, and availability of sediments in different regions. The rise of sea level was so great that pelagic sedimentation normally confined to the deep seas spread onto some cratons and produced chalk and black shale deposits.

- Global paleoclimate was profoundly influenced by feedback effects of the transgression. Enlargement of sea area coupled with location of both poles in oceanic areas and excess CO_2 from volcanic eruption explain the warm and uniform late Mesozoic climate indicated by both fossils and oxygen-isotope data. Small differences between polar and equatorial temperatures, unusual depths for epeiric seas, and retarded oxygen uptake by warm seawater resulted in sluggish oceanic circulation and oxygen depletion of bottom-ocean waters—the greenhouse condition. Resulting carbon-rich black muds became important source rocks for petroleum in many parts of the world.

Mesozoic Life

- Marine life underwent profound changes during the Triassic, following the Late Permian extinction. Fish, ammonites, and other molluscs flourished and diversified. The ammonites, worldwide in distribution, became the dominant element in most faunas. The tropical Tethyan fauna flourished in the Cretaceous Gulf of Mexico.

- Land plants underwent rapid diversification. Conifers, together with cycads and ginkgoes, dominated the forest world. Flowering plants appeared in the Late Cretaceous and rapidly spread into the temperate regions of the world.

- Reptiles underwent tremendous diversification and became the dominant animals on land. In the early Mesozoic, synapsids were the reigning group, but in the Jurassic the dinosaurs, along with swimming and flying reptiles, were most common. The first true birds appeared in the Jurassic but are rarely found as fossils. Primitive mammals began to appear commonly in some faunas by the Late Cretaceous, although they evolved in the Late Triassic.
- The end-of-Cretaceous biological crisis caused extinctions of dinosaurs and many marine groups. Theories to explain these extinctions range from the Deccan flood volcanism to the catastrophic asteroid impact hypothesis.

Readings

Alexander, R. M. 1989. *Dynamics of dinosaurs and other extinct giants.* New York: Columbia University Press.

Alvarez, L., W. Alvarez, F. Asaro, and H. V. Michel. 1980. Extraterrestrial cause for the Cretaceous-Tertiary extinction. *Science,* 208:1095–1108.

Alvarez, W., and F. Asaro. 1990. What caused the mass extinction? An asteroid impact. *Scientific American,* 78–84.

Bakker, R. T. 1986. *The dinosaur heresies.* New York: Morrow.

Barron, E. J. 1983. A warm equable Cretaceous: The nature of the problem. *Earth Science Reviews,* 19: 305–38.

———, and W. M. Washington, 1982. Cretaceous climate: A comparison of atmospheric simulations with the geologic record. *Palaeogeography, Palaeoclimatology, Palaeoecology,* 40:103–33.

Casey, R., and P. F. Rawson, eds. 1973. The boreal Lower Cretaceous. *Geological Journal,* Special Issue 5.

Courtillot, V. E. 1990. What caused the mass extinction? A volcanic eruption. *Scientific American,* 85–92.

Dickinson, W. R., et al. 1988. Paleogeographic and paleotectonic setting of Laramide sedimentary basins in the central Rocky Mountain region. *Geological Society of America Bulletin,* 100:1023–39.

———, and W. S. Snyder. 1978. Plate tectonics of the Laramide orogeny. *Geological Society of America Memoir* 141:355–66.

Fischer, A. G., and M. A. Arthur. 1977. Secular variations in the pelagic realm. *SEPM Special Publication 25*:29–50.

Glen, W. 1990. What killed the dinosaurs? *American Scientist,* 78:354–70.

Hallam, A., ed. 1973. *Atlas of palaeobiogeography.* Oxford, UK: Oxford University Press.

———. 1975. *Jurassic environments.* New York: Cambridge University Press.

———. 1984. Continental humid and arid zones during the Jurassic and Cretaceous. *Palaeogeography, Palaeoclimatology, Palaeoecology,* 47:195–223.

Hays, J. D., and W. C. Pitman III. 1973. Lithospheric plate motion, sea level changes and climatic and ecological consequences. *Nature,* 246:18–22.

Hoffman, A. 1989. *Arguments on evolution.* New York: Oxford University Press.

Howell, D. G. 1985. Terranes. *Scientific American,* 116–26.

———, and K. A. McDougall, eds. 1978. *Mesozoic paleogeography of the western United States.* SEPM Pacific Coast Paleogeography Symposium 2.

Hsü, K. J., ed. 1986. *Mesozoic and Cenozoic oceans.* American Geophysical Union Geodynamics Series 15.

Jenkyns, H. C. 1980. Cretaceous anoxic events: From continents to oceans. *Geological Society of London Journal,* 137:171–88.

Jones, D. L., A. Cox, P. Coney, and M. Beck. 1982. The growth of western North America. *Scientific American,* 70–128.

Kuban, G. J. 1986. The Taylor site man tracks. *Origins Research,* 9:1, 7–15.

Larson, R. L. 1991. Latest pulse of Earth: Evidence for a mid-Cretaceous superplume. *Geology,* 19:547–50.

Lessem, D. 1992. *Kings of creation—How a new breed of scientists is revolutionizing our understanding of dinosaurs.* New York: Simon & Schuster.

Lockley, M. 1991. *Tracking dinosaurs, a new look at an ancient world.* Cambridge, Mass.: Cambridge University Press.

Logan, A., and L. V. Hills, eds. 1974. *The Permian and Triassic systems and their mutual boundary.* Calgary: Alberta Society of Petroleum Geologists, Memoir 2.

McGowan, C. 1991. *Dinosaurs, spitfires, and sea dragons.* Cambridge, Mass. Harvard University Press.

McKee, E. D. 1954. *Stratigraphy and history of the Moenkopi Formation of Triassic age.* Geological Society of America Memoir 61.

———, et al. 1956. *Paleotectonic maps of the Jurassic system.* U.S. Geological Survey Miscellaneous Publications, Map I-175.

———, et al. 1959. *Paleotectonic maps of the Triassic system.* U.S. Geological Survey Miscellaneous Publications, Map I-300.

Morris, J. D. 1980. *Tracking those incredible dinosaurs and the people who knew them.* San Diego: Creation Life.

Moullade, M., and A. E. M. Naim. 1978. *The Phanerozoic geology of the world—II: The Mesozoic.* New York: Elsevier.

Norman, D. B. 1985. *The illustrated encyclopedia of dinosaurs.* New York: Crescent.

Ostrom, J. 1979. Bird flight: How did it begin? *American Scientist,* 67:46–56.

Padian, K., ed. 1986. *The beginning of the age of dinosaurs.* New York: Cambridge University Press.

———, and D. J. Chure, eds. 1989. *The age of dinosaurs.* Paleontological Society Short Courses in Paleontology 2.

Powell, J. L. 1998. *Night comes to the Cretaceous.* New York: Harcourt Brace.

Raup, D. M. 1986. *The Nemesis affair: The story of the death of the dinosaurs and the ways of science.* New York: Norton.

Reyment, R. P., and P. Bengston. 1981. *Aspects of mid-Cretaceous regional geology.* London: Academic Press.

Reynolds, M. W., and E. D. Dolly, eds. 1983. *Mesozoic paleogeography of the west-central United States.* Rocky Mountain Section SEPM Paleogeography Symposium 2.

Russell, D. A. 1989. *The dinosaurs of North America: An odyssey in time.* Toronto: University of Toronto Press.

Schermer, E. R., D. G. Howell, and D. L. Jones. 1984. The origin of allochthonous terranes. *Annual Reviews of Earth and Planetary Sciences,* 12:107–31.

Schlanger, S. O., and H. Jenkyns. 1976. Cretaceous oceanic anoxic events: Causes and consequences. *Geologic en Mijnbouw,* 55:179–84.

Sharpton, V. L., and P. D. Ward, eds. 1990. *Global catastrophes in earth history, an interdisciplinary conference on impacts, volcanism, and mass extinction.* Geological Society of America Special Paper 247.

Silver, L. T., and P. H. Schwartz, eds. 1982. *Geological implications of impacts of large asteroids and comets on Earth.* Geological Society of America Special Paper 190.

Spencer, A. M., ed. 1974. *Mesozoic-Cenozoic orogenic belts.* London: The Geological Society.

Thomas, B. A., and R. A. Spicer. 1987. *The evolution and palaeobiology of land plants.* London: Croon Helm.

Thomas, R. D. K., and E. C. Olson, eds. 1980. *A cold look at the warm-blooded dinosaurs.* Boulder: Westview.

Ward, P. 1983. The extinction of the ammonites. *Scientific American,* 136–47.

Weishampel, D. B., P. Dodson, and H. Osmolska, eds. 1991. *The Dinosauria.* Univ. of California Press. Berkeley, Calif.

White, M. E. 1986. *The flowering of Gondwana.* Princeton, N.J.: Princeton University Press.

(Courtesy M. Voorhies, University of Nebraska State Museum.)

Chapter 15

Cenozoic History

Threshold of the Present

The earth is not finished, but is now being, and will forevermore be remade.

C. R. Van Hise (1898)

MAJOR CONCEPTS

▶ As the Atlantic widened in Cenozoic time, compressional tectonics and mountain building affected the western margin of North America. From the latest Cretaceous to middle Eocene, unusually shallow subduction produced uplift and basin formation in the Rocky Mountains. Normal arc volcanism resumed from the late Eocene until the late Oligocene. In the Miocene, western North America was torn apart by block faulting as the Sierra Nevadas and Cascades rotated westward. The subducting margin of California was replaced by a transform margin, causing Baja California and much of coastal California to be sheared northward.

▶ The Cenozoic marked the final breakup of Pangea as India collided with Asia to form the Himalayas and Australia and South America tore away from Antarctica. The collision of Africa with Europe formed the Alps and began to close the great tropical Tethys Seaway. The Mediterranean remnant of Tethys closed and dried up for a brief time at the end of the Miocene.

▶ Following the extinction of the dinosaurs, land mammals diversified rapidly in the Paleocene and soon filled niches from bats in the air to whales in the ocean. Subtropical climates and vegetation were found almost to the Arctic Circle in the Paleocene and early Eocene; the jungles were inhabited by archaic leaf-eating and tree-dwelling mammals.

▶ In middle Eocene to Oligocene time, climate began to deteriorate as the circum-Antarctic current locked in cold water around Antarctica and triggered Antarctic glaciation. On land, the vegetation changed from jungles to grassy savannas, and the mammalian inhabitants (especially horses, camels, antelopes, cattle, and their kin) became adapted for rapid running and for eating newly evolved gritty grasses.

The magnificent mountains and canyons explored by early geologists in western North America were formed largely in Cenozoic time. In Chap. 14, we examined the Cordilleran mountain building, which produced structures typical of ancient orogenic belts. In this chapter, we shall see that there was one outstanding peculiarity—the unique reactivation of a large part of the western craton. Unusually high uplifts of relatively simple anticlinal structures were formed in the western United States during the Laramide orogeny in Paleogene time.

In this chapter, we shall see that structural disturbances of great magnitude occurred throughout the world with surprising rapidity during the Cenozoic Era. Completion of our analysis of Cordilleran history will be coupled with a brief account of the closely related tectonic evolution of the Pacific Ocean basin. Then we shall examine briefly some other global events, especially the dramatic formation of the Alpine-Himalayan mountain belt. Its upheaval was caused by continental collisions in Neogene time. Meanwhile, intense fragmentation of the crust was accompanied by widespread volcanism during Neogene time—for example, in the Basin and Range Province of the western United States. In marked contrast were the passive margins of North America and other continents, which are characterized today by coastal plains and broad continental shelves. Their tranquil facies may conceal buried structural complexity produced by rising salt domes.

Finally, we shall review the history of Cenozoic life over the past 66 million years with respect to changing geography and climate as all the continents emerged from the seas to assume their present configurations. Because Cenozoic fossils closely resemble living organisms and because both the fossil record and the rock record for the Cenozoic are more complete than those for earlier times, we are able to subdivide the Cenozoic Era more finely than older eras. Both the Paleogene (which lasted 42 m.y.) and Neogene (the last 24 m.y.) Periods were equivalent to typical pre-Cenozoic ones, but it is more common to use the smaller epoch subdivisions in discussing Cenozoic history. These latter represent much shorter time intervals than those used in previous chapters. The Eocene and Miocene Epochs were each about 20 million years long, the Paleocene and Oligocene about 10 million, and the Pliocene and Pleistocene less than 3 million years each. An apparent acceleration of geologic changes in Cenozoic time is probably only an illusion caused by a more complete record and this more refined subdivision.

Cenozoic epochs were originally defined by the relative percentage of living marine species represented among the fossils in each division. In recent years, the definition of each epoch has been greatly refined, thanks to an enormous body of data from the Deep-Sea Drilling Program (Fig. 7.1) for many fossil groups from both deep- and shallow-marine strata. The nonmarine fossil record also has been studied intensively, and its correlation with the standard marine scale is continually being improved. We also now have magnetic polarity reversal and stable-isotope, as well as radiometric, chronologies to compare with the Cenozoic paleontologic stratigraphy. All methods are more reliable and more refined for Cenozoic than for older rocks.

Figure 15.1 The "Rhino Pompeii." At a late Miocene locality called Ashfall Fossil Bed State Park, near the town of Orchard in eastern Nebraska, Michael Voorhies and his colleagues at the University of Nebraska State Museum found and excavated an extraordinary assemblage of fossil mammals, including a few three-toed horses and over a hundred complete articulated skeletons of hippolike amphibious rhinoceros called *Teleoceras*. All of the skeletons are lying as they suffocated under a volcanic ash cloud that covered their pond about 10 m.y. ago. The specimens are so complete that even the grass seeds of their last meal are preserved in their throat cavities, and some of the calves are lying in nursing position next to their mothers. Although this locality is extraordinary for its preservation, Cenozoic rocks have produced an excellent fossil record of the diversification and evolution of mammals. *(Courtesy M. Voorhies, University of Nebraska State Museum.)*

Cenozoic Cordilleran History

Phase I: Laramide Orogeny (Latest Cretaceous to Middle Eocene, 70 to 40 m.y. Ago)

As we saw in Chap. 14, the end of the Cretaceous in the Cordillera was marked by a peculiar tectonic event, the Laramide orogeny. Arc volcanism ceased in California and Nevada and deep basement folding occurred in Wyoming, Colorado, and New Mexico, far east of the normal boundary of the Cordillera, the foreland

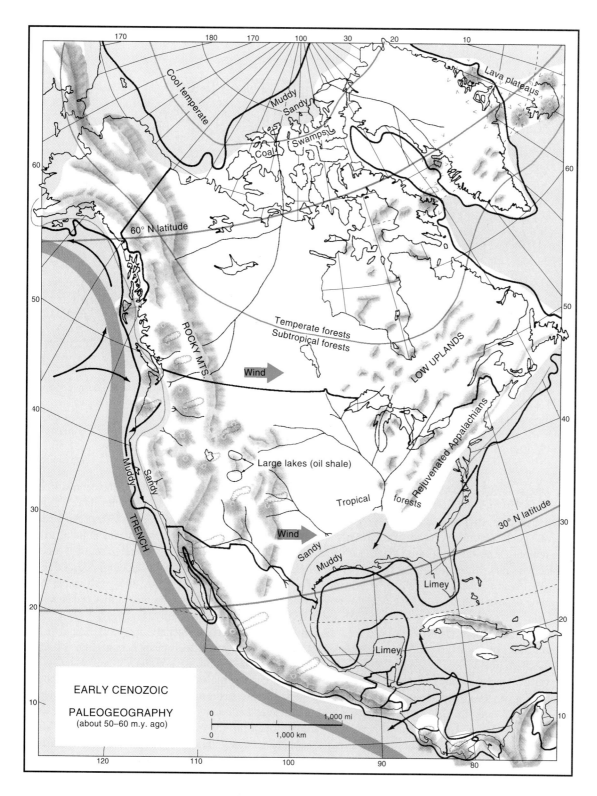

Figure 15.2
Early Cenozoic paleogeography of North America. Note that the configuration of the continent for the first time was approaching that of today, but climate still was milder. Europe had just separated, to produce extensive basalt outpourings in Greenland and Iceland. Eocene oil shales formed in the large lakes.

Map labels: Cool temperate · Muddy · Sandy · Swamps · Coal · Lava plateaus · 60° N latitude · ROCKY MTS. · Muddy · Sandy · TRENCH · Temperate forests · Subtropical forests · Wind · LOW UPLANDS · Large lakes (oil shale) · Rejuvenated Appalachians · Tropical forests · Wind · Sandy · Muddy · Limey · 30° N latitude · Limey

EARLY CENOZOIC PALEOGEOGRAPHY (about 50–60 m.y. ago)

0   1,000 mi
0   1,000 km

fold-thrust belt in Nevada and Idaho. According to the Dickinson-Snyder hypothesis, these peculiar events can be explained by a lithospheric plate that experienced shallow, almost horizontal subduction (Fig. 14.32C). Dickinson and Snyder suggest that this prevented melting above the plate to form arc volcanoes, and it transferred stresses to the base of the continental crust beneath the Rocky Mountains. This stress caused unusual folding and thrusting in the western craton.

Laramide tectonism continued throughout the Paleocene and early Eocene, from about 70 to about 40 million years ago. Between the Laramide uplifts were *intermontane basins,* which became dumping grounds for thick sequences of sediments eroded off the uplifted Cryptozoic and Paleozoic blocks. These basins filled up with up to 3,000 meters (10,000 feet) of early Cenozoic sediments (Figs. 14.32, 15.2, and 15.3). Most of these sediments were deposited either in river and floodplain settings or in gigantic freshwater lakes. During the Paleocene, some of the basins in Wyoming were filled with huge, swampy river deltas, which accumulated enormous seams of coal, some over 100 meters thick. These coal beds are close to the surface in places such as the Powder River

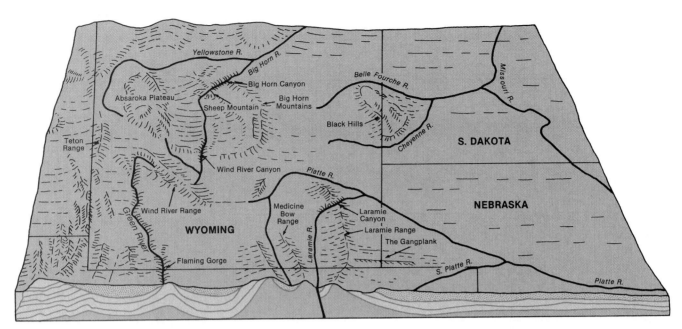

Figure 15.3   Block diagram showing some of the geometry of Laramide basins and uplifts. Uplifted during the Paleocene and early Eocene, they were filled to the top by the Miocene. While the topography was subdued, the rivers formed their drainages, oblivious to the buried mountains beneath them. Since then, renewed uplift and increased erosion has stripped away much of the basin fill, forcing the rivers to cut down through the hard cores of the mountains as they were exposed. This process is responsible for the famous water gaps of the Rocky Mountains. *(After C. O. Dunbar and K. Waage, 1969,* Historical geology, *John Wiley.)*

Basin of northeastern Wyoming and, so, can be exploited by strip mining (Fig. 15.4A). Their enormous thickness makes them very valuable, and they are low in sulfur content, so that their burning does not produce acid rain. Consequently Wyoming has become one of the foremost coal-mining regions in the world. Because the historic coal mines of the Appalachians and Midwest yield high-sulfur coal from Pennsylvanian-aged seams (many of which are thinner and deeper than the Wyoming deposits), they are gradually being displaced by Wyoming coals.

The most remarkable of the intermontane basin deposits formed in the gigantic freshwater lakes of the middle Eocene. In some places, the lakes deposited freshwater limestones, forming the spectacular pink, yellow, and red spires of Bryce Canyon National Park in Utah (Fig. 15.4B). The Uinta and Piceance (pronounced "Pee´-onts") Basins of the northern Utah-Colorado border region and the Green River Basin of southwestern Wyoming (Fig. 15.3) are filled with a thick sequence of middle Eocene lake shales, sandstones, freshwater limestones, and evaporites known as the Green River Formation (Fig. 15.4C). The most abundant rock type is hundreds of meters of finely laminated lake shales, which preserved an amazing diversity of fish, frogs, birds, plants, and other fossils with great fidelity (Fig. 15.4D). The algal productivity in the Eocene Green River lakes was so high that enormous amounts of organic matter were trapped in the shale without being oxidized. These rocks are called *oil shales* because their low-grade organic matter can be heated and distilled to produce oil. During the energy crisis of the 1970s, oil companies invested in oil shale, but depressed petroleum prices since the early 1980s have made oil-shale production too expensive to pursue.

Although Laramide tectonism dominated the central Rocky Mountains during the Paleocene and early Eocene, there was normal subduction to the north and south (Fig. 14.32A). An Eocene volcanic arc ran through Idaho and eastern Washington, to the northeast of the modern Cascades. The forearc basin was located in western Washington and Oregon, and it accumulated a thick sequence of marine graywackes and shales. In a few places, there were Eocene deltas growing from the ancestral Cascades, and these deltas produced thin coals locally (Fig. 15.4E).

In California, the Sierran volcanic arc went extinct in the Late Cretaceous as Laramide tectonism began (Fig. 14.32A). Throughout the Cenozoic, this arc continued to erode away, exposing the granitic batholiths that were once active magma chambers. The great forearc basin running along the Central Valley of California collected sediments, including marine turbidites and deltas building out from the eroding Sierras. A variety of small terranes brought in along the subduction zone formed islands along the California coast during the Paleocene and Eocene. Between these islands were small, fault-bounded basins, which also accumulated thick packages of sediment. To the south of the Laramide tract, there was extensive arc volcanism in Arizona and Mexico during the Paleocene and Eocene (Fig. 14.32A).

Phase II: Resumption of Arc Volcanism (Late Eocene–Middle Miocene, 40 to 20 m.y. Ago)

In the late Eocene, volcanic activity increased all along the Cordilleran belt (Fig. 15.5). Western Oregon had been a deep-marine basin with deltas during the middle Eocene. In the late Eocene, it was intruded by a major arc volcano complex. The volcanoes ran in a north-south belt through central Oregon, to the

A.

B.

C.

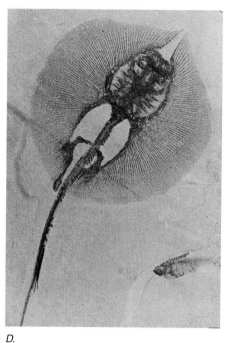

D.

Figure 15.4 Examples of Paleocene and Eocene rocks from the Cordillera. *A:* A 200-foot-thick seam of Paleocene lignite coal from a strip mine in the Powder River Basin in Wyoming. Notice the huge dump truck in the background for scale. *B:* Spectacular pinnacles of Eocene lake marls at Bryce Canyon National Park in southern Utah. *C:* Thousands of feet of finely laminated lake shales from the Green River Formation, Hells Hole Canyon, Utah. The extreme thickness and lateral continuity of these fine layers demonstrate the scale of this huge middle Eocene lake. *D:* Beautifully fossilized fish from the Green River Shale. In addition to a variety of bony fish, the Green River Shale also produces complete birds, crocodiles, turtles, frogs, rays, crayfish, and many plant fossils. *E:* Contorted strata beneath a deltaic channel in upper Eocene deposits, Coos Bay, Oregon. Sudden cutting of the channel caused cohesive muds to collapse and fold. Coal-bearing strata were deposited then along a retreating shoreline characterized by widespread swamps as early Cenozoic sediments filled embayments along the Pacific Coast. Area of view about 2 meters high. *(A–D, D. R. Prothero; E, R. H. Dott, Jr.)*

E.

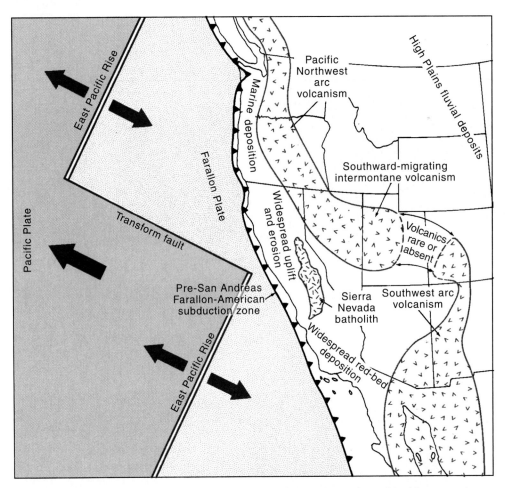

Figure 15.5   Tectonic map of the Cordillera during the Oligocene. Once the shallow-dipping Laramide slab returned to normal subduction, arc volcanism resumed in a line from Washington and Oregon to Nevada and Utah, then to the San Juan volcanics of Colorado, and on through New Mexico and Central America. Forearc basin deposition was found throughout the Pacific Coast, and sediments full of abundant volcanic debris began to bury the Laramide uplifts and spread across the High Plains to form the White River Group of the Big Badlands. Note the peculiar geometry of the ridge-transform-ridge boundary between the Farallon and Pacific Plates. As the Farallon Plate was consumed, this zigzag boundary eventually collided with the Pacific subduction zone. *(After W. R. Dickinson, 1979, Pacific Section SEPM Paleogeography Symposium, v. 3, pp. 10–11.)*

east of the modern Cascade volcanoes, forming the colorful Clarno Formation, with its abundant plant and mammal fossils (Fig. 15.6A). In the late Oligocene and earliest Miocene, the ancestral Cascades erupted a spectacular sequence of green volcanic ash deposits in central Oregon, known as the John Day Formation (Fig. 15.6B). Extensive volcanism occurred in central Idaho (the Challis volcanics) and in northwestern Wyoming (the Absaroka volcanic field) about 40 million years ago. A famous sequence of ash flows in Yellowstone National Park preserved twenty-seven separate horizons of petrified trees, each buried by a different eruption of the Absaroka volcanics (Fig. 15.6C). By the early Oligocene, volcanism had broken out across the Laramide "magmatic null" region (compare Fig. 14.32A with Fig. 15.5). Significant eruptions occurred in central Nevada and western Utah, and especially in southwestern Colorado; the latter produced the spectacular San Juan Mountains. The volcanic trend continued into New Mexico and Arizona.

Such renewed volcanism suggests that Laramide subduction, with its shallow slab that did not melt, had ceased. This is verified by the geologic relationships at the margins of the Laramide basins (Figs. 14.32B, 15.7, and 15.8B–D). In several instances, Laramide thrusts and folding deformed early Eocene rocks that filled the basins as they were subsiding. However, these deformed strata are unconformably overlapped by middle Eocene strata, demonstrating that the Laramide tectonism had ceased (Fig. 15.7).

According to Dickinson and Snyder, the downgoing plate must have resumed its normal dip and triggered crustal melting underneath North America to produce an Andean-style volcanic arc (similar to the Sevier arc—see Fig. 14.21). Renewal of the active arc spurred further forearc basin development along the Pacific Coast. In Oregon and Washington, rapidly subsiding forearc basins accumulated deepwater shales and turbidites of Oligocene age. In coastal California, the late Eocene and Oligocene were marked by a period of uplift and mostly nonmarine deposition

A.

B.

C.

Figure 15.6   Typical outcrops of the late Eocene-Oligocene arc volcanics of the western Cordillera. *A:* The middle Eocene Clarno Formation in central Oregon, which produces abundant fossil plants and mammals. The red bands are ancient soil horizons, which indicate deep tropical weathering between episodes of volcanism. *B:* The upper Oligocene volcaniclastics of the John Day Formation in central Oregon. The peculiar green color is due to alteration of the volcanic ash, and the resistant ledges are hard volcanic ash flows. *C:* Fossilized trees buried in life position in a series of middle Eocene volcanic ashes at Specimen Ridge, northeast Yellowstone National Park, Wyoming. There are twenty-seven separate forests here, showing that the region must have been buried in volcanic ash (which killed and fossilized the previous forest) and then was slowly overgrown by a new forest over two dozen times. *(A–B, D. R. Prothero; C, courtesy of Yellowstone National Park.)*

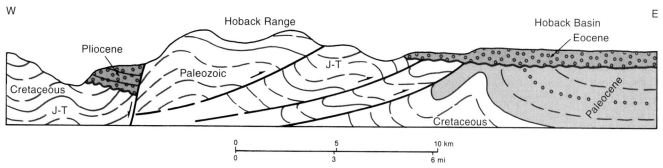

Figure 15.7   Restored cross section in western Wyoming showing structural and stratigraphic relationships that allow close dating of tectonic events during the Cenozoic. At least two thrust faults *(right)* are post-Paleocene but pre-middle Eocene. Extensional (normal) faulting *(left)* occurred during and possibly after the Pliocene. *(Adapted from Dorr, 1958; Geological Society of America Bulletin; and Eardley, 1962, Structural geology of North America.)*

RECONSTRUCTED STAGES IN THE DEVELOPMENT OF
THE MEDICINE BOW MOUNTAINS, WYOMING

By S. H. Knight
1953

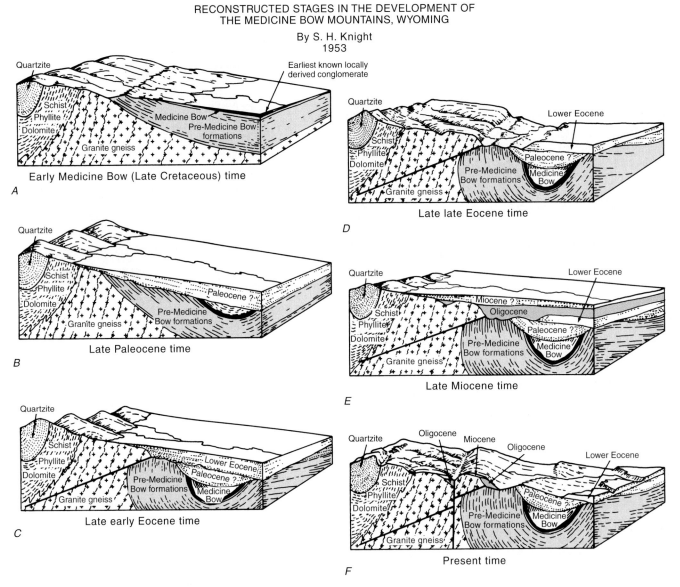

Figure 15.8   Structural and erosional development of the Rocky Mountain ranges of the Wyoming–Colorado–New Mexico region, showing the effects of the Laramide orogeny (A–D), basin filling (E), and late Cenozoic upwarping and rejuvenated erosion (F). (Adapted from S. H. Knight, 1953, Wyoming Geological Association 8th annual guidebook; by permission of S. H. Knight and Wyoming Geological Association.)

(Fig. 15.9A). Even the Central Valley, which had been mostly deep marine since the Cretaceous, was filled by nonmarine and shallow-marine sediments in the Oligocene. Since uplift was so pervasive and there was global fall of sea level, most of the Oligocene record of this region is represented by a widespread Oligocene unconformity.

At the end of the Eocene, the Laramide basins were filling rapidly. The consequence of this basin filling was unexpected. Instead of jagged peaks on top of the Rocky Mountains, there are flat surfaces with only small, eroded knobs protruding (Fig. 15.9B). The Rocky Mountain surface on top of the Colorado Front Range, the Bighorn Mountains of Wyoming, and the Beartooth Mountains of Montana are all ancient Eocene-Oligocene landscapes, now elevated above 3,000 meters (10,000 feet). During

most of the Oligocene and Miocene, the Rocky Mountains were invisible, buried beneath their own debris (Fig. 15.8E). There are still remnants of this ancient blanket on top of the Bighorns and other mountain ranges (Fig. 15.8F). Driving from Cheyenne west to Laramie, one crosses the flat-topped Laramie Range by climbing a long ramp of relict Cenozoic sediment nicknamed "The Gangplank" (Fig. 15.3). The ease of crossing the Rockies at this point made it the preferred route of pioneer trails and the first transcontinental railroad.

Although the Rockies were mostly buried, their isolated remnants still shed sediments to the east across the High Plains of Nebraska, eastern Wyoming, and the Dakotas (Fig. 15.5). In the Oligocene, the abundance of volcanic ash blowing from the west further contributed to sedimentation in the Plains region. Great

B.

A.

Figure 15.9   Typical exposures of Oligocene rocks in the western United States. *A:* Thick sequences of sandstones and red shales of the Sespe Formation were deposited by rivers, which filled basins in southern California during the middle Eocene and late Oligocene; there is an 8-m.y.-long unconformity in the middle of the sequence. *B:* Once the Laramide basins were nearly filled, the remaining peaks were deeply eroded and beveled off during the Oligocene and Miocene, producing a striking flat top on most ranges in the Rockies. This is the Beartooth Plateau in northern Yellowstone National Park, Montana and Wyoming. *C:* As the Laramide basins filled and ranges were buried, huge volumes of sediment and volcanic ash were shed eastward across the High Plains. The most familiar products of this deposition are the floodplain sediments of the upper Eocene–lower Oligocene White River Group. Rich with fossil mammals and tortoises, these beds erode into spectacular pinnacles in the Big Badlands of South Dakota. *(D. R. Prothero.)*

C.

ash-filled rivers and floodplains deposited a thick blanket of fluvial deposits over the northern High Plains. Known as the *White River Group,* these deposits are best known from the spectacular pinnacles and spires of the Big Badlands of South Dakota (Fig. 15.9C). White River sediments are found from western North Dakota all the way to central Colorado and span much of the late Eocene to the late Oligocene. They are also famous for their abundant fossils, especially of mammals and tortoises. Fossil hunters have been visiting the Big Badlands since 1846, collecting spectacular specimens of many extinct mammals. The White River Group and its fossils record the dramatic climatic changes that occurred during the late Eocene and Oligocene.

Through the late Oligocene and early Miocene, the burial of the Rockies under their blanket of sediment continued. Rivers

drained east across the High Plains, although their deposits were more restricted to sandy channels; fewer broad floodplains existed to deposit laminated badlands mudstones. In the late Oligocene, so much volcanic ash blew in from Colorado and Nevada that the *Arikaree Group* in Nebraska is composed mostly of wind-blown ash. In the early and middle Miocene, river systems continued to flow across the High Plains, entrenched in channels carved into the White River and Arikaree sediments. These deposits produce the most continuous record of fossil mammal evolution anywhere in North America during the middle Cenozoic.

The High Plains sequence is capped by fluvial sands and gravels of the upper Miocene-Pliocene *Ogallala Group,* which forms a sandstone "caprock" found today on tops of plateaus and buttes all the way from Texas to South Dakota. Ogallala caprock

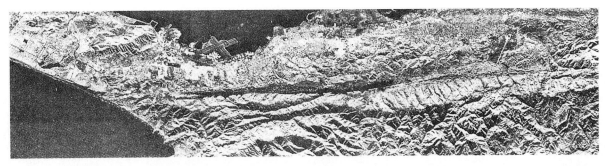

Figure 15.10   Radar image (not a photo) of the San Francisco Peninsula, showing the San Andreas and related faults delineated by straight valleys. Pacific Ocean is to lower left; San Francisco Bay at top (note airport runways bordering the bay). Expansion of greater San Francisco and the construction of huge housing tracts directly over the fault zone foretell future earthquake damage far in excess of that of 1906 or the Loma Prieta earthquake of 1989. *(NASA photo 67-6-1364; by permission of NASA.)*

is important not only as a record of sedimentation and mammal evolution during the late Miocene but also because its porous sands are the primary source of water for almost the entire Plains region. The Ogallala aquifer is now seriously threatened because excessive pumping for irrigation has depleted it of water accumulated over millions of years, and it is not being replenished rapidly enough by natural run-off. If uncontrolled pumping continues, the entire Plains agricultural belt, source of most of our wheat and cattle, could become too dry for further farming or ranching—it might revert to the infamous "dust bowl" conditions of the 1930s.

Phase III: Complex Tectonism (Early Miocene–Present, 20 to 0 m.y. Ago)

Since early in the Mesozoic, the Cordilleran margin has been a subduction zone, consuming several oceanic plates. From the Cretaceous until the early Miocene, the Farallon Plate was consumed as it moved away from a spreading ridge in the northeastern Pacific (Fig. 15.5). The Farallon Plate moved in a southeasterly direction, so it was obliquely subducted under the North American Plate Oblique subduction produced some of the unconventional features of the Cordillera, but most of the Mesozoic tectonism can be explained by an Andean-type volcanic arc and related basins, as discussed in Chap. 14.

In the early Miocene, the Cordillera experienced plate motions that have no simple modern analogue. The spreading ridge between the Farallon and Pacific Plates was not straight but offset along transform faults. The magnetic stripes on the Pacific Plate side, now found in the northern Pacific sea floor, clearly demonstrate this (Fig. 15.26). All along this spreading ridge, the sharp offset in the magnetic stripes shows that the ridge had significant "jogs" due to transform faults. When we backtrack the positions of the ridge and sea floor to their Eocene or Oligocene position, we find that one of these sharp jogs placed a corner of the Pacific Plate closer and closer to the subduction zone (Fig. 15.5). In the late Oligocene, the corner of the Pacific Plate came into contact with the subduction zone, and the Pacific Plate touched the North American Plate for the first time.

When contact was made, a remarkable new geometry resulted. Unlike the Farallon Plate, which had been steadily spread-

Figure 15.11   The San Andreas fault produces both sliding transform motion and compression and extension as irregularities in the fault move past one another. These Pliocene sediments of the Anaverde Formation just north of the fault, and south of Palmdale, California, have been intensely folded by compression along the fault (to the right and out of view). Closer to the fault zone, the rocks are more brittle and are intensely crushed and ground up, rather than folded. *(Courtesy of Kenneth O. Stanley.)*

ing eastward and then had been consumed in the subduction zone, the Pacific Plate was (and still is) moving in a northwesterly direction away from its source on the spreading ridge (now the East Pacific Rise). Relative to North America, the Pacific Plate is moving sideways, not converging down a subduction zone, as is the Farallon Plate. Such sideways motion cannot produce a subduction zone or a spreading ridge but must result in a transform fault. We know this transform as the *San Andreas fault* (Figs. 15.10 and 15.11). Today it runs from the Gulf of California to northern California, transporting the Pacific Plate northwest relative to the North American Plate. When the Pacific Plate first touched North America in the early Miocene, the San Andreas was a very short segment between the long subduction zones to the north and south (Fig. 15.12). As the Farallon Plate remnants were subducted and disappeared, the San Andreas took over a larger part of the California margin.

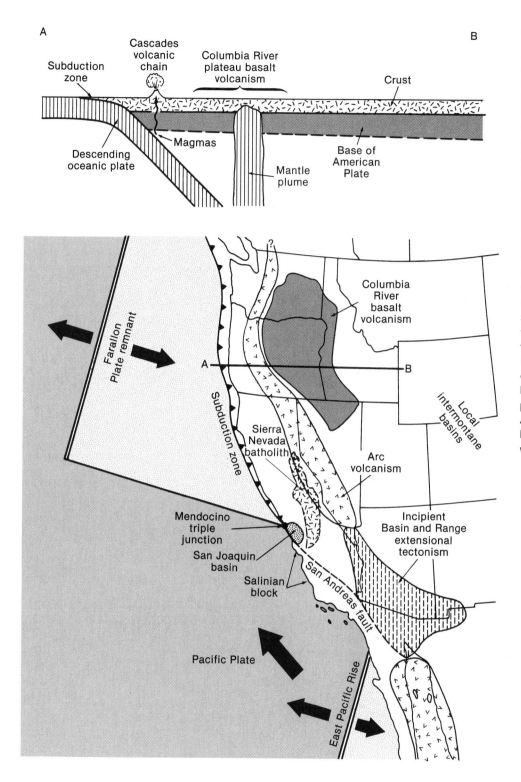

Figure 15.12   Tectonic geometry of the western United States during the middle Miocene, about 15 m.y. ago. By this time, the Farallon Plate had subducted, so that it was broken into two remnants, which continued to melt and produce a volcanic arc (compare Fig. 15.5). Between them, however, the Pacific Plate refused to go beneath the North American Plate; instead, it was sheared northwesterly along the newly formed San Andreas fault. This motion brought the Salinian block (then in Mexico) northward to its present position in the central California Coast Ranges. The region east of the San Andreas fault showed the beginning of Basin and Range extension. Note that the Sierra Nevada arc had not yet rotated southwestward to its present orientation, and Nevada was still very skinny before Basin and Range extension. The enormous eruptions of Columbia River flood basalts may have been due to a mantle plume or hot spot (see cross section A-B at top). *(After W. R. Dickinson, 1979, Pacific Section SEPM Paleogeography, v. 3, pp. 10–11.)*

The consequences of this tectonic geometry were remarkable for the Cordillera. By the middle Miocene, about 15 million years ago (Fig. 15.12), arc volcanism in southern California and Arizona had ceased. The Salinian block, today located in the central California Coast Ranges, was still mostly at the latitude of Mexico but moving north along the San Andreas transform. Arizona was stretched apart by extensional forces, initiating the **Basin and Range Province.** Basin and Range geology is characterized by a series of north-south-trending fault-block mountains with downdropped graben basins between (Fig. 15.13). On a physiographic map or satellite photo (Fig. 15.14), these mountains form a long series of parallel ranges, which C. P. Dutton long ago compared to "an army of caterpillars marching north." Most of the ranges reach 2,000 to 3,000 meters (7,000 to 10,000 feet) in elevation, and the basins are filled with as many as 3 kilometers (about 2 miles) of sediments (Fig. 15.15). Today the Basin and

Figure 15.13   Aerial view looking south at the eastern escarpment of the southern Sierra Nevada, California. Mt. Whitney (named for California's first state geologist) in the right skyline is the highest peak in the lower forty-eight states at 14,491 feet. The Alabama Hills, town of Lone Pine, and Owens River are in the foreground. More than 10,000 feet of topographic relief are shown here. The long parallel north-south trending ranges separated by narrow basins, as seen here, are typical of the Basin and Range Province. Both basins and ranges are bounded by normal faults; the faults bounding the Sierra Nevadas and the Alabama Hills have been active since the late Miocene and have moved in the last century. *(© Peter Kresan.)*

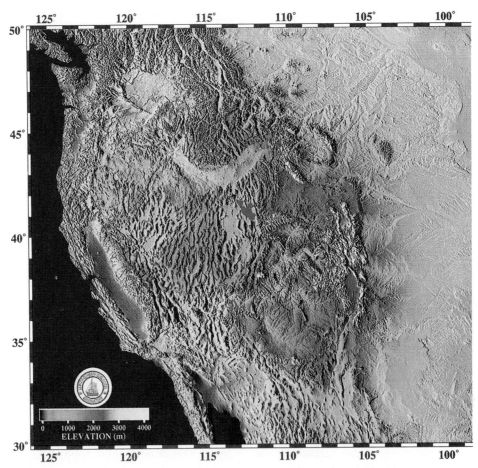

ELEVATION (m)

Figure 15.14   Computer-generated image of the western Cordillera, showing the striking topography of the region. The Sierra Nevada Mountains and the Great Valley of California stand out clearly, as does the Cascade Range. Note how the region east of the Sierras is marked by north-south-trending basins and ranges, extending from southern Arizona and California to Nevada, western Utah, and southeastern Oregon. North of the Basin and Range Province, the distinct arc of the Snake River volcanic plain in Idaho stands out. *(Courtesy of Lamont-Doherty Earth Observatory.)*

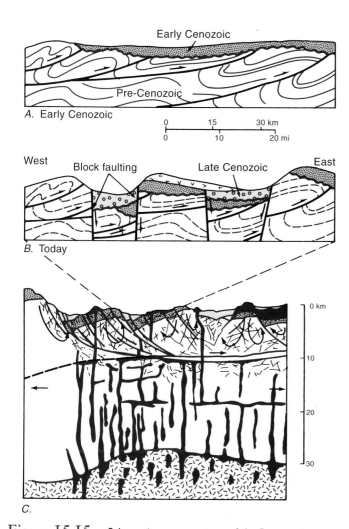

Early Cenozoic

Pre-Cenozoic

A. Early Cenozoic

West Block faulting Late Cenozoic East

B. Today

0 km

10

20

30

C.

Figure 15.15 Schematic cross sections of the Basin and Range, showing the structural complexity. *A:* Late Cretaceous thrust faults of the Sevier orogeny were later buried by early Cenozoic sediments. *B:* These thrust sheets were broken into normal fault blocks during Basin and Range extension in the late Cenozoic. *C:* The first two cross sections of the shallow crust (*A* and *B*) are placed in the context of the first 30 kilometers of crust. At greater depth, the steep normal faults become shallow as they approach a horizontal master fault, so each fault block is actually sliding and rotating as the crust stretches. Beneath the master fault, the continental crust is hot and ductile, so it is greatly stretched and thinned without faulting. It is also penetrated by intrusions from the mantle, which force their way up the shallow faults to the surface. (*C modified from G. P. Eaton, 1982,* Annual Reviews of Earth and Planetary Science, *v. 10, pp. 409–440.*)

Range Province extends through Nevada all the way to southern Oregon. The crust of Nevada is now stretched to twice its Miocene width, resulting in crust only 20 to 30 kilometers thick (as opposed to 50 to 60 kilometers in most regions). This is some of the thinnest crust on any continent, and the relatively shallow mantle supplies heat and magma to the fault zones all through the Basin and Range region. In the middle Miocene, Basin and Range extension began in the south and a much narrower Nevada was intruded by magmas that fed arc volcanoes.

Figure 15.16 Cliffs formed of multiple stacked lava flows (note the columnar jointing), Picture Gorge, near Mitchell, north-central Oregon. Each of the flows represents a separate eruption of the Columbia River flood basalts, and there are over a dozen in this area alone. Some of these flows can be traced over miles, and they covered thousands of square miles in a matter of days as they erupted out of deep fissures. *(D. R. Prothero.)*

Another remarkable phenomenon of the middle Miocene were flood basalt eruptions of eastern Washington and Oregon. Known as the *Columbia River plateau basalts,* these eruptions spewed out of crustal fissures from sources deep in the mantle (Fig. 15.16). Like the Deccan traps and many other great flood basalt eruptions, they produced enormous volumes of very fluid basaltic lava in a short period of time. Some flows covered over 40,000 square kilometers in a matter of days, moving at about 5 km/hr in flows 30 meters deep and 100 kilometers wide at temperatures of 1,100°C! In about 3.5 million years, flow after flow erupted from these fissures until the lava covered 300,000 square kilometers, filling former valleys with stacked basalt sequences as much as 4,000 meters thick. Between eruptions, the flows cooled and then weathered so deeply that forests grew on them. In central Washington, the famous Ginkgo Petrified Forest preserves hundreds of fossilized trees from these ancient woodlands.

The Columbia River eruptions cannot be explained by simple interactions of the plates and were probably the result of a large hot spot in the mantle, similar to those that produced the Hawaiian and the Deccan eruptions. In the late Miocene and Pliocene, similar eruptions occurred farther east, in the Snake River plain of southern Idaho (Fig. 15.17). Today a hot spot apparently lies under Yellowstone National Park in northwestern Wyoming. If these eruptions are all related, then they probably reflect the eastward relative migration of the hot spot as the North American Plate rode westward over it.

Through the late Miocene, the Basin and Range stretching and the shutoff of the Nevada volcanic arc continued as the boundary between the subduction zone and the San Andreas transform (known as the Mendocino triple junction) moved north. By the end of the Miocene (5 million years ago), the San Andreas reached north of San Francisco (Fig. 15.12). Basin and

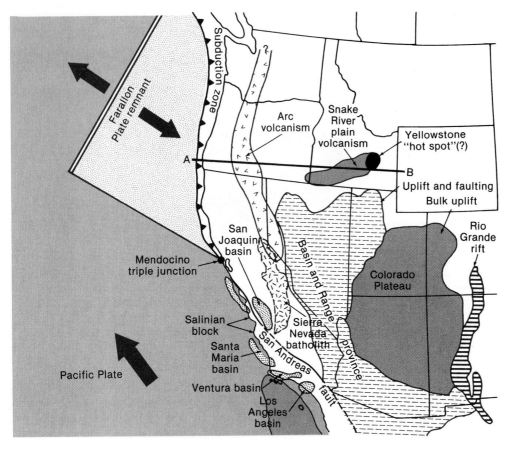

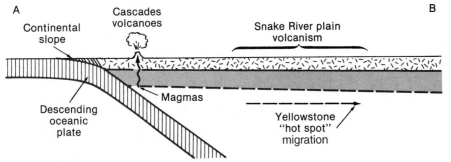

Figure 15.17 Tectonic map of the western United States in the latest Miocene-Pliocene, about 5 m.y. ago. At this time, a much smaller remnant of the Farallon Plate (called the Juan de Fuca Plate) remained west of Oregon and Washington. It continued to subduct, causing the eruption of the Cascades. South of the Mendocino triple junction, the San Andreas fault slid the Pacific Plate northwest relative to the North American Plate. All along the fault in California, deep but narrow basins were formed as shearing ripped the crust open. In the region east of the fault, the Basin and Range extension moved northward (compare Fig. 15.12), and the Sierra Nevada rotated southwestward, stretching Nevada to its modern dimensions. Between the Basin and Range and the Rio Grande rift, the Colorado Plateau began to rise as a huge, coherent crustal block, forcing the Colorado River to cut down and form the Grand Canyon. The Yellowstone hot spot, which had formed the Columbia River flood basalts in the middle Miocene (Fig. 15.12), then ripped across southern Idaho to produce the Snake River flood basalts. *(After W. R. Dickinson, 1979, Pacific Section SEPM Paleogeography Symposium, v. 3, pp. 10–11.)*

Range extension had ripped Nevada apart, breaking it into its modern geometry of north-south trending fault block mountain ranges. The stretching of Nevada meant that the Sierra Nevadas and Cascades moved west to accommodate the expansion; paleomagnetic studies of rocks from both the Sierra Nevadas and the Cascades show that they have moved westward by as much as 270 kilometers (170 miles) since the middle Miocene (Fig. 15.18). These mountains swung like a door, with the hinge located in the Olympic Peninsula of Washington; the Cascades rotated slightly clockwise around the hinge, and the southern Sierra Nevadas rotated the most. The movements are analogous to opening a paper fan, with the opening frame representing the Sierra Nevada–Cascade segment and the stretched paper folds comparable to the stretched crust of the Basin and Range in Nevada and Arizona. The earliest and most rapid stretching occurred near the tip of the fan (Arizona), followed by stretching

in the heart of the fan (Nevada); areas near the hinge (in the Olympic Peninsula) stretched the last and the least.

Late Miocene-Pliocene tectonism extended far beyond the Basin and Range. About 5 million years ago, the "four corners" region, where Arizona, Utah, Colorado, and New Mexico meet, began to be uplifted as a single coherent block known as the *Colorado Plateau* (Fig. 15.17). Unlike the Basin and Range, the Colorado Plateau did not break into hundreds of parallel normal faults. Instead, it is only slightly deformed by small, normal faults and gentle folds draped over them; most of the rocks have remained nearly horizontal, even though they have been uplifted at least 1.5 kilometers (about a mile). Some parts of the Colorado Plateau, such as the Kaibab Plateau on the north rim of the Grand Canyon, reach elevations over 2,700 meters (9,000 feet). Geophysical studies show that the crust beneath the Colorado Plateau is not thick enough to explain its uplift by isostasy

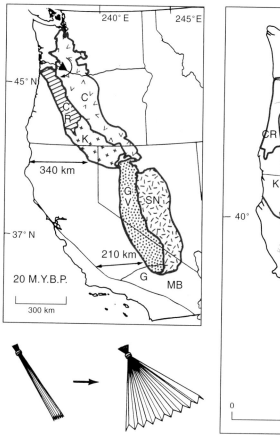

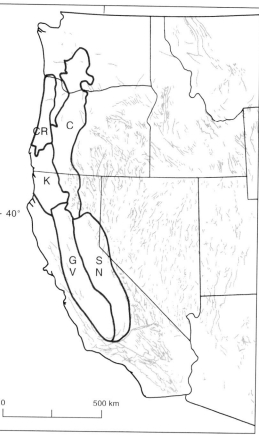

Figure 15.18 Paleomagnetic evidence shows that the Sierra Nevadas (SN) and Cascades (C) have rotated about 200 to 300 km southwestward along a hinge in western Washington (triangle) since the middle Miocene, about 20 m.y. ago. This enormous rotation stretched the crust behind it, forming the Basin and Range Province, with its many normal faults (short vertical gray lines). At the southern tip of the Sierra Nevadas, the Garlock fault (G) took up the extreme sliding at the end of the swinging "door." The Oregon-Washington Coast Ranges (CR), the Klamath Mountains (K), and the Great Valley of California (GV) also swung outward. This extension is analogous to the opening of a paper fan, with one rib of the fan (the Sierra-Cascade arc) moving outward relative to the rest of the fan; as the stretching proceeds, each fold of the fan (= Basin and Range fault blocks) separates farther. *(Modified from M. Magill and A.V. Cox, 1980, Oregon Department of Geology and Mineral Industries Special Paper, v. 10, pp. 1–67.)*

alone, so mantle upwelling must be occurring, as it is in the Basin and Range.

Today the Colorado Plateau is deeply eroded as a result of its gradual uplift, forming the spectacular "red rock" canyons and monuments of many scenic parks: Bryce, Zion, Capitol Reef, Canyonlands, Arches, Monument Valley, and especially the Grand Canyon. In the Grand Canyon and the Goosenecks of the San Juan River, the gently meandering course of the river was deeply incised into Paleozoic sandstone and limestone bedrock, producing meanders that lie in canyons almost a mile deep (Fig. 15.19). This could happen only if the meanders had been established on a low-elevation, low-relief surface during the Miocene and then cut down through hard bedrock as the Colorado Plateau rose beneath them.

A similar phenomenon occurred in the Rocky Mountains. We have seen that the great Laramide uplifts were almost completely buried during the Miocene, and only small, local basins accumulated sediments in Wyoming, Colorado, and Utah. At the end of the Miocene, the uplift of the Colorado Plateau was paired with uplift in the Rocky Mountain region. The long-buried Rockies were resurrected after nearly disappearing under their own debris. As they rose, their sedimentary cover was stripped away, and the flat Eocene Rocky Mountain surface was exposed and lifted high in the sky (Fig. 15.9B). The basins between peaks were even easier to erode because they were filled with soft Cenozoic sediments, not hard granitic and metamorphic basement. The rivers that once easily

Figure 15.19 Goosenecks of the San Juan River, southeastern Utah. Before the uplift of the Colorado Plateau, this must have been a low-lying region, since the rivers formed a meandering drainage. During Mio-Pliocene uplift, the San Juan River was forced to cut down through thick sequences of Permo-Pennsylvanian sandstones and shales. Today the river lies in a steep, V-shaped canyon, indicating relatively youthful topography, yet its meanders show that it originally developed on a low-relief, mature topographic surface. *(Courtesy of Peter Kresan.)*

Figure 15.20   Example of drainages that cut through the Rocky Mountain, forming water gaps (see Fig. 15.3). Here in Flaming Gorge in the Uinta Mountains of Utah and Wyoming, the Green River has cut through Cryptozoic and Paleozoic sedimentary rocks, which have been folded into a monocline. Major John Wesley Powell navigated this canyon en route to exploring the Colorado River in 1869, 1 year before this picture was taken by pioneer photographer William Henry Jackson. *(Courtesy of W. H. Jackson, U.S. Geological Survey #332.)*

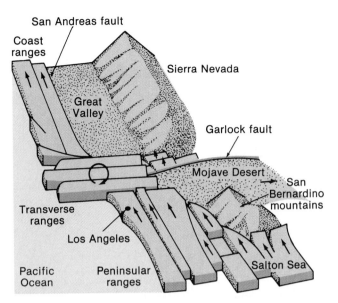

Figure 15.21   In coastal California, the consequences of the northwesterly shearing along the San Andreas fault are complex. In most places, the fault-bounded blocks slide northwestward, shearing past one another. In the Transverse Ranges north of Los Angeles, however, paleomagnetic evidence shows that the fault blocks have rotated clockwise about 90° since the mid-Miocene. This might be explained if they behaved as a "see-saw" rotating between the sheared blocks to the north and south.

crossed the buried mountain ranges were then forced to cut down rapidly as they stripped away the basin fill on either side. Such rapid downcutting produced deep canyons cut through the hard cores of mountains, so that rivers appear to take the "hard" way through mountains, rather than the "easy" way around them. The Rocky Mountains are full of examples of *water gaps* caused by drainages superimposed on a buried topography (Fig. 15.3). Spectacular canyons, such as Royal Gorge in Colorado and Split Mountain and Flaming Gorge in Utah (Fig. 15.20), are well-known examples of *superposed drainages*. In Wyoming, the Wind River moves southeast down the Wind River Basin, then jogs sharply north and disappears into Wind River Canyon in the Owl Creek Mountains. When it comes out the other side, it even has a new name: the Bighorn River. From there, it drains the Bighorn Basin into Montana and eventually reaches the Missouri River (Fig. 15.3).

A number of scientists have challenged the widely accepted idea that the Rockies have risen again in the past 5 million years. They point to fossil floras, such as Florissant in central Colorado, that indicate that the Colorado Rockies were already 2 kilometers high during the Eocene and Oligocene. This suggests that the Colorado Rockies may not have experienced renewed uplift; late Miocene-Pliocene erosion may have been due to climatic changes that caused higher erosion rates not only in the Rockies but also in the Alps and Himalayas. Whether this hypothesis about the Rockies is finally accepted or not, there seems to be little question that the Sierra Nevadas and Colorado Plateau were truly elevated in the Pliocene. In either case, the Rocky Mountains have been exhumed.

By the late Miocene, the great Columbia River basaltic eruptions had long cooled, but the same mantle hot spot was now under southern Idaho, erupting the Snake River basalt floodplain. The eastern edge of the Colorado Plateau in New Mexico was marked by another great rift valley, this one beginning to open in the middle Miocene. It is called the Rio Grande rift because the Rio Grande River now flows through it. Miocene and Pliocene sediments accumulated there and then broke into normal fault blocks, very similar to the Great Rift Valley of East Africa or the Triassic rift valleys formed when the Atlantic first opened (Fig. 14.8).

The northward expansion of the San Andreas transform meant that many pieces of what is now Mexico slid northward into California. The Salinian Block had almost reached the central Coast Ranges, where it now resides. Other blocks moved northward as well. In some places, crustal blocks could not slide smoothly northward along the faults parallel to the San Andreas. Instead, they got caught between fault zones and were sheared into a clockwise rotation (Fig. 15.21). This is common in the Transverse Ranges north of Los Angeles. As their name implies, the Transverse Ranges lie in an east-west trend that is almost perpendicular to the northwest-southeast trend of the San Andreas and the coastal ranges along it. Paleomagnetic data show that the Transverse Ranges have been rotated clockwise about 90° since the middle Miocene. This movement can be explained only if these crustal blocks had pivoted, as do ball bearings between two sliding surfaces.

As rotation proceeded, deep structural chasms opened up between the pivoting crustal blocks. These openings became very narrow, steep-walled, rapidly subsiding sedimentary basins. They were

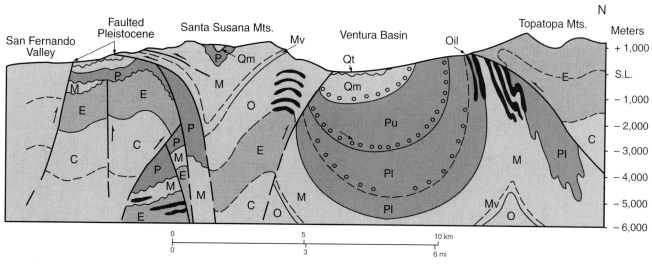

Figure 15.22   Cross section of the Ventura Basin, northwest of Los Angeles, California. Like the other fault-bounded basins in the Transverse Ranges, it is extremely deep and narrow, with close to 6 miles of upper Miocene-Pliocene sediments in a basin only 6 miles wide. Since the basin formed, compression along the many faults in the region has crumpled the beds into tight synclines and anticlines and has thrust faulted the flanks of the basin over the center. The crests of anticlines and the lower plate of the thrust faults have proven to be important traps for oil. *(After California Division of Mines Bulletin 170, 1954.)*

so deep that they quickly filled with thousands of meters of marine shales and turbidites, accumulating thicknesses of as much as 10 kilometers (30,000 feet) of sedimentary fill in the Los Angeles Basin in less than 8 million years! If you stand in downtown Los Angeles, the pre-Miocene bedrock lies more than 6 miles beneath your feet, yet all the sedimentary fill is less than 8 million years old. The Ventura Basin to the north is filled with almost 6 kilometers (over 20,000 feet) of Miocene and Pliocene sediments, and there are many smaller basins with similar histories (Figs. 15.22 and 15.23). The Ridge Basin, to the north of Los Angeles, contains 13,500 meters (44,000 feet, or over 8 miles) of fill yet is only 10 to 15 kilometers wide! These basins are remarkable not only because they are so narrow and yet very deep but also because they were deformed as they grew, so that they have steep folds and faults throughout them. Much of the oil wealth of southern California comes from these Mio-Pliocene basins. Oil migrated out of the organic-rich Miocene shales to the faults and folds along the basin edges, where it accumulated and formed great oil fields (Fig. 15.22).

An even more remarkable phenomenon is the "great rip-off" of Baja California. The East Pacific Rise makes a triple junction with the southern San Andreas transform and the Central American trench. During the Miocene, this triple junction gradually moved south along the southern California coast as the San Andreas expanded along the former subduction zone. About 5 million years ago, the triple junction reached the southern tip of Baja California, and the entire peninsula tore away from mainland Mexico (Fig. 15.23). The East Pacific Rise began to spread open in the Gulf of California, placing Baja on the Pacific Plate. Baja California continues to move north today, bringing parts of mainland Mexico with it. There are many identical desert plants and animals found in Baja California and in mainland Mexico. Originally they lived in one continuous Mexican desert. Now the in-

habitants of Baja's deserts are separated and genetically isolated from their relatives on the other side of the Gulf of California, and they are moving north as you read this.

In the last few million years, most of the events initiated in the Miocene have continued. The Yellowstone hot spot left the Snake River plain and now lies in its present location, generating geysers and hot springs in a huge caldera formed by a gigantic explosive eruption. The Basin and Range continues to rip apart, with the extension just now beginning in southeastern Oregon and Idaho (Fig. 15.23). The Cascades and Sierra Nevadas continue to rotate until they have reached their present north-south orientation. However, the Cascade arc seems to be less active that it was in the Miocene. Only a few volcanoes, such as Mount St. Helens, are still erupting in the Washington segment. The northern remnant of the Farallon Plate (known as the Juan de Fuca Plate) is still going down the subduction zone, although it is slowing down and subduction appears to be shallow. This may explain why there are fewer deep-focus earthquakes in the subduction zone beneath the Cascades. The trench off the Oregon-Washington coast is not very deep because it is very rapidly filled with sediments eroded from the humid, recently glaciated lands drained by large river systems, such as the Columbia.

Tectonic Hypotheses for the Evolution of the Cordillera

The unusual plate geometry of a subducting margin turning into a transform margin makes Cordilleran geology very complex. Before 1970, the Cordillera seemed too complicated for a plate-tectonic explanation. Then Tanya Atwater used the magnetic record of the Pacific sea floor to deduce the behavior of

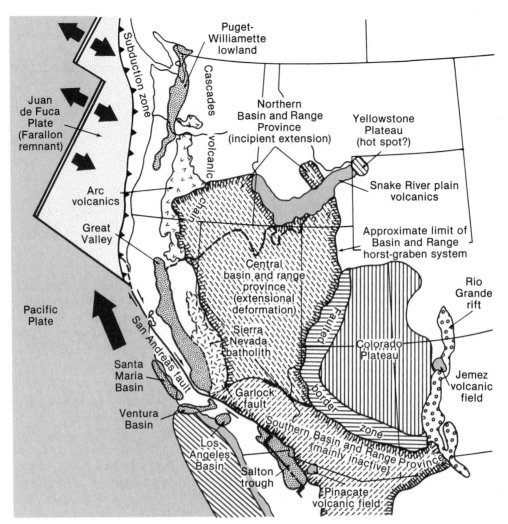

Figure 15.23   Modern tectonic geometry of the western United States. The small Juan de Fuca Plate is all that remains of the Farallon Plate, still subducting to form the Cascades. The Yellowstone hot spot has moved all the way from the Columbia River basalts, the Snake River plain, and now lies under the northwest corner of Wyoming. The Basin and Range Province continued to open northward, with the most recent activity in northern Nevada and southern Oregon. Meanwhile, the most extended part of the Basin and Range in southern Arizona, California, and New Mexico is relatively inactive. The San Andreas transform has now taken over the entire Pacific Rim from north of San Francisco to Mexico, causing Baja California to rip away from the mainland about 5 m.y. ago and move northwesterly on the Pacific Plate. Deep but narrow basins continue to form by compression in the fault zone, including the Santa Maria, Ventura, and Los Angeles Basins and the Salton Trough. (Compare with Fig. 15.14). *(After W. R. Dickinson, 1979, Pacific Section SEPM Paleogeography Symposium, v. 3, pp. 10–11.)*

Cordilleran crust. Since then, the plate-tectonic hypotheses have explained many parts of this complicated story.

Still, problems remain. We can see the coincidence in timing between the opening of the Basin and Range Province and the growth of the San Andreas, but what is the connection? A number of hypotheses have been offered. Some argue that the East Pacific Rise must lie under Nevada, producing the mantle upwelling and crustal extension. However, plate-tectonic geometries do not require that this mid-ocean ridge continue indefinitely. It is merely the expression of a two-plate boundary, and there need not be any spreading in the mantle where there is no overlying crustal spreading. Since the Pacific Plate is spreading northwestward away from it, there is no problem with the East Pacific Rise ending abruptly in a transform fault. Indeed, that geometry is characteristic of

nearly every transform fault along the mid-ocean ridges around the world. Other models postulate that the Basin and Range was produced by backarc spreading, as is now happening west of Japan. However, the Basin and Range is no longer behind an arc; today it stands behind a transform margin, which is no longer subducting. Still other models suggest that the Pacific Plate is being subducted and melting beneath Nevada, but this does not produce Basin and Range extension.

We favor the hypothesis proposed by William R. Dickinson and Walter S. Snyder in 1979 and modified by Jeff Severinghaus and Tanya Atwater in 1990. Examining plate geometries of the last 20 million years (Figs. 15.12 and 15.17), Dickinson and Snyder noticed that the beginning of the San Andreas transform means that *no plate is going down in that region*. There are remnants of

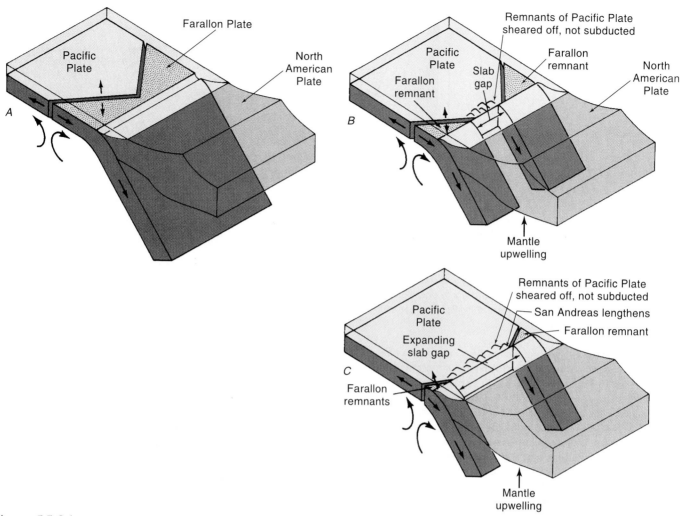

Figure 15.24   One possible explanation of the puzzles of Cordilleran tectonics in the late Cenozoic is the *slab-gap model* of Dickinson and Snyder, as modified by Severinghaus and Atwater. A: Because of the peculiarities of the plate geometry (compare Figs. 15.12 and 15.17), the subducting Farallon Plate brought a corner of the Pacific Plate closer and closer to the Pacific subduction zone. B: Once the Pacific Plate corner touched the subduction zone, it would not go down, since its motion is northwesterly relative to the North American Plate; instead, it began to shear along the San Andreas transform. Meanwhile, the two remnants of the Farallon Plate continued to subduct to the north and south. This left a gap between them unfilled by subducting plate, so mantle could rise up directly beneath the North American Plate. C: As the Farallon remnants shrank further, the San Andreas zone expanded north and south. The slab gap behind it also expanded. According to Dickinson and Snyder, this explains why the Basin and Range opened from south to north and why the Cascade volcanoes also shut down from south to north.

the Farallon Plate sinking underneath the Pacific Northwest and under Mexico, but the Pacific Plate is not subducting beneath the North American Plate. Instead, the two plates scrape sideways against each other, shearing off small fault blocks (Fig. 15.21). Consequently the North American Plate lies directly over the mantle in this region, while in the subduction zones to the north and south, the Farallon Plate lies between the deeper mantle and the North American Plate. In other words, the region east of the San Andreas transform is a **slab gap** between subducting remnants of the Farallon Plate.

The power of the slab-gap hypothesis is that it predicts so much of Cordilleran geology. Using known Pacific sea-floor-spreading geometries, the position of the gap in the Farallon Plate can be predicted through the past 20 million years (Figs. 15.24

and 15.25). As the gap expands, it successfully explains multiple events. For example, the Basin and Range opened progressively from south to north, as the expansion of the gap would generate a northward exposure of the mantle (Fig. 15.25). As the northern edge of the gap moved north, the Cascade arc was shut off in the south. The current northern edge of the gap lies east of the Mendocino triple junction across the northern part of California, just south of the southernmost active Cascade volcanoes, Mt. Lassen and Mt. Shasta. The expansion of the slab gap under the Colorado Plateau occurred between 10 and 5 million years ago, when its uplift began. Geophysical data indicate that there is mantle upwelling underneath the plateau. The Pliocene uplift of the Rockies might be indirectly caused by the general uplift of the entire region due to mantle upwelling in the west.

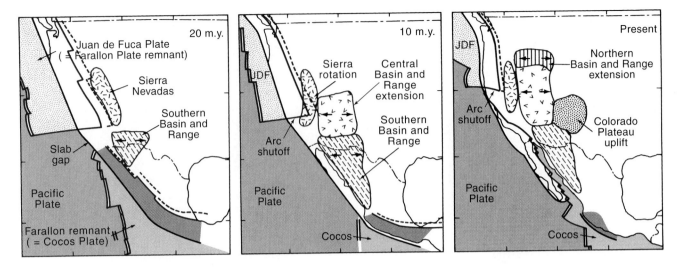

Figure 15.25   Map showing the geometry of the Farallon Plate and its remnants, as well as the slab gap underneath the North American Plate in the late Cenozoic; note the south-to-north opening of the Basin and Range and shutoff of the Cascade arc. Compare with Figs. 15.12, 15.17 and 15.24. JDF—Juan de Fuca Plate. *(Modified from J. Severinghaus and T. Atwater, 1990, Geological Society of America Memoir, v. 176, pp. 1–22.)*

Figure 15.26   Evolution of the northern Pacific Ocean basin showing progressive collision of North America with ocean ridges (compare Figs. 15.12 and 15.17). Bend in Emperor-Hawaiian seamount chain reflects change of motion of Pacific Plate about 40 million years ago at the same time as initial collision of North America with the East Pacific ridge (SA—San Andreas fault; GC—future site of Gulf of California; SF—San Francisco; H.I.—present position of Hawaiian Islands; numbers—dates m.y. ago). *(Adapted from Atwater, 1970: Geological Society of America Bulletin; Grow and Atwater, 1970: ibid.; Dalrymple, Silver, and Jackson, 1973: American Scientist; Larson and Chase, 1972: Geological Society of America Bulletin; Winterer, 1973: Bulletin American Association of Petroleum Geologists; Pitman, Larson, and Herron, 1974, The age of the ocean basins, special map by Geological Society of America.)*

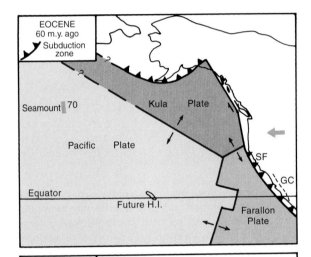

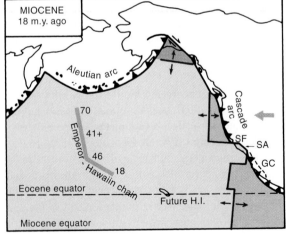

Pacific Tectonics

The Hawaiian Mantle Plume

Chains of volcanic islands and submerged volcanic seamounts, such as the Emperor-Hawaiian chain, also provide clues for Pacific motions. The volcanic centers along this chain become progressively younger toward the southeast (Fig. 15.26). It has been proposed that a plume of hot material rises from the base of the mantle and produces volcanic eruptions through the Pacific lithosphere plate (Fig. 15.27; see also Fig. 7.18). It is argued that the plume remains fixed in position for a long time and, because the plate is moving continuously over the plume, the line of extinct volcanic centers traces the history of plate motion.

Fossils dredged from the northernmost of the Emperor Seamounts are Late Cretaceous. Isotopic dates for volcanic rocks farther southeast indicate ages from 46 million years just north of the prominent bend in the chain to 18 million years at Midway Island, as well as from 6 million to 0 years for the Hawaiian Islands themselves (again with younger dates southeastward). If the

Emperor-Hawaiian chain indeed was formed by passage of the Pacific Plate over a hot mantle plume, then the sharp bend in the chain must record a major change of direction of plate motion

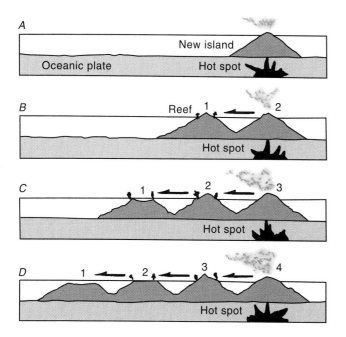

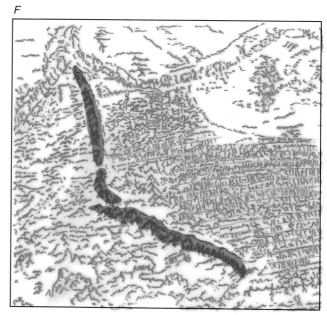

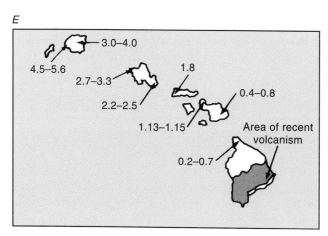

Figure 15.27 Evolution of the Hawaiian hot spot. *A–D:* Cartoon showing how each Hawaiian island was formed as the hot spot punched through the moving Pacific Plate. Each Island then moved off the hot spot, sank down into the ocean, eroded, and may have also become home for coral reefs. Eventually they formed the flat-topped seamounts of the Emperor-Hawaiian seamount chain *(F). E:* Note that the radiometric ages of each island in the Hawaiian chain get progressively older as they move northwest from the island of Hawaii, which is currently over the hot spot and still an active volcano. *(Page 497, Figure 20.18 from EARTH'S DYNAMIC SYSTEMS, 6/e by W. Kenneth Hamblin. Upper Saddle River, NJ: Prentice Hall.)*

around 35 to 40 million years ago (Oligocene). Each volcanic center was built from the abyssal sea floor 6,000 meters below sea level in only a few million years. The older, northern islands later subsided beneath sea level to form a submarine ridge.

Western and Southern Pacific Tectonics

A mid-Cenozoic date for a major change in the direction of motion of the Pacific Plate, from south to north to southeast to northwest, is very close to the time of initial collision between the American Plate and the East Pacific ridge. The mid-Cenozoic date also coincides with a renewal of activity in the Aleutian and all the western Pacific island arcs (Fig. 15.26). This matching of dates seems to be one more bit of evidence of major mid-Cenozoic tectonic changes that were of global importance. There is considerable evidence, for example, that Japan and several of the other western Pacific island arcs were formed from slivers of continental crust torn from mainland Asia and that New Zealand was torn from Australia. The result of both these events was the formation of small marginal ocean basins

behind them. The separation of New Zealand occurred in late Mesozoic time, but the separation of Japan and formation of the Marianas and other arcs began in mid-Cenozoic time, when other major tectonic changes were occurring in the Pacific. Local sea-floor spreading behind certain arcs seems to produce the marginal basins, but the ultimate cause of that spreading is not yet proven.

In the days when continents were seen as fixed and as always growing larger by in-place accretion, the western Pacific arcs were imagined to be embryonic orogenic belts that would eventually become additions to Asia or Australia. It is an indication of how profoundly our views of global evolution have changed that now many of the same arcs are seen as slivers *removed* from continents, rather than as potential new additions.

Southeastern Pacific Tectonics

On the southeastern Pacific margin, the last Gondwana separation was completed in mid-Cenozoic time. Until the Miocene, South America and Antarctica were structurally connected via

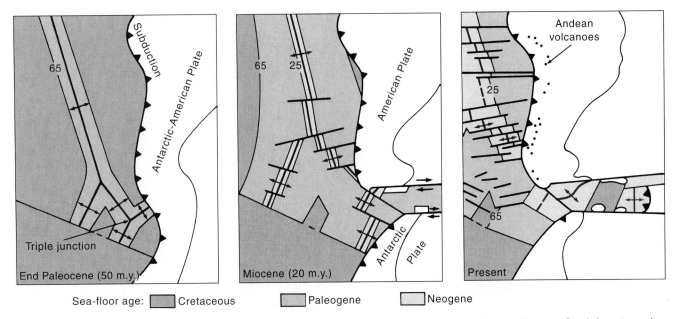

Sea-floor age: ▮ Cretaceous   ▮ Paleogene   ▯ Neogene

Figure 15.28   Cenozoic history of the southeastern Pacific region showing opening of the Drake Passage between South America and Antarctica, the last stage in the breakup of Gondwanaland. Apparently Gondwanaland collided with a spreading-ridge triple junction at the beginning of Miocene time (23 m.y. ago), somewhat like the continent-ridge collision of California. This collision broke the narrow continental connection and created several microcontinents, which then drifted eastward. (All numbers indicate m.y. ago.) *(After R. H. Dott, Jr., 1976: American Geophysical Union Monograph 19, pp. 299–310; W. C. Pitman III et al., 1974, Geological Society of America sea floor age chart.)*

the Antarctic Peninsula. Sea-floor magnetic anomaly patterns, deep-sea drilling, and comparisons of land geology show that the separation occurred in early Miocene time (23 m.y. ago; Fig. 15.28). Microcontinents torn from both continents and set adrift may someday collide with one or more other continents and become suspect terranes.

The final separation of the Gondwanan continent allowed the full development of a vigorous circum-Antarctic ocean current through the Drake Sea south of Cape Horn. This cold, encircling flow blocks any close approach of warmer waters to the polar continent. The very strong Circum-Antarctic "Roaring Forties" westerly winds isolate it meteorologically as well. Without occasional warming air masses, the climate of Antarctica turned glacial, as confirmed by Oligocene and Miocene tillites in Antarctica and ice-rafted pebbles in Oligocene marine sediments near the continent. Again we see an example of a feedback between tectonics and climate.

Meanwhile, the northern end of South America continued separating from North America as the Caribbean region widened. A branch of the mid-Atlantic ridge (Figs. 14.5 and 14.6) had separated North America from South America in the Jurassic and Cretaceous. Not until the Pliocene did volcanism build a continuous Isthmus of Panama between the two continents. Major faults fragmented the Caribbean region during the late Cenozoic, moving Cuba, Hispaniola, and Jamaica eastward. Simultaneously the eastern Caribbean, or Lesser Antilles, were formed as an island arc complex.

Cenozoic Tectonics of Eurasia

Early Cenozoic Events

Major Cenozoic plate reorganizations also occurred outside the Pacific region (Fig. 15.29). Recall from Chap. 14 that Greenland began separating from Europe in Paleocene time as the northern mid-Atlantic ridge formed. Widely scattered Eocene basalt flows and swarms of dikes mark this event in Ireland, Scotland, Iceland, and Greenland. On the other side of the world, Antarctica and Australia had separated and gone their opposite ways, and India had completed its separation from Africa with outpourings of huge volumes of basalts in northwestern India. India, Africa, and Australia were soon to collide with Eurasia, but internally each southern continent was very stable, as evidenced by vast, well-preserved erosion surfaces with very old lateritic soils.

The Alpine-Himalayan Belt of Eurasia

While the Pacific margin was experiencing intense deformation and volcanism associated with late Mesozoic and Cenozoic mountain building, the Tethyan region between Gondwanaland and Eurasia continued to suffer deformation, and this crushing culminated in late Cenozoic time with the upheaval of the Alps, Himalayas, and many other less famous ranges of Asia Minor (Fig. 15.29). This was the **Alpine-Himalayan orogeny.** Old continental crust on the southern margin of Eurasia, which had been disturbed previously by the late Paleozoic Hercynian orogeny,

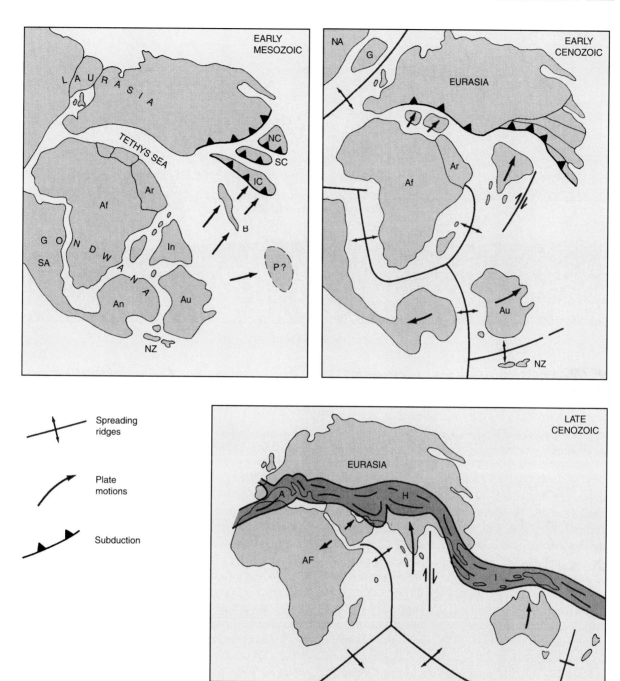

Figure 15.29   Formation of the Alpine (A), Himalayan (H), and Indonesian (I) orogenic belts and destruction of the Tethys seaway by sea-floor spreading and successive collisions of continents and microcontinents with Eurasia (AF—Africa; Ar—Arabia; An—Antarctica; Au— Australia; B—Burma-Thailand; G—Greenland; In— India; NC—North China; SC—South China; IC—Indochina; NA—North America; SA—South America; NZ—New Zealand; P—hypothetical microcontinent Pacifica). *(Adapted from Carey, 1959; Klimetz, 1983; Mitchell, 1981; Schermer et al., 1984; Stauffer, 1985; Zhang et al., 1984.)*

was involved in this Cenozoic upheaval. In addition, Mesozoic oceanic (Tethyan) rocks were thrust northward onto the old continental margin and crushed together with the Hercynian basement. As tectonic activity intensified in Cretaceous time, islands were pushed up near the northern side of the shrinking Tethys seaway. Their rapid erosion produced clastic debris, which periodi-

cally was swept by turbidity currents into the deep-water basins between the islands, much as sediments are now being deposited in basins among the many islands of the Aegean Sea between Greece and Turkey. The result was rhythmic alternations of Cretaceous to Eocene sandstone and shale originally named *flysch* in Switzerland (see Fig. 14.29C).

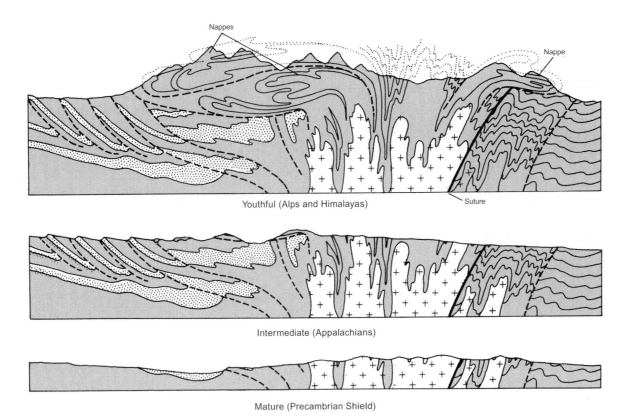

Figure 15.30   Diagrammatic comparison of effects of different levels of erosion on exposed characteristics of orogenic belts. Nappe folds shown at top seem unique to the Alpine-Himalayan belt only because it is so youthful and shallowly eroded. (Compare Fig. 13.26.)

As compression increased, the area and height of uplifted terrain increased rapidly. The culmination of compressive deformation occurred in Oligocene and Miocene time, when sedimentation changed to very thick nonmarine gravel and sands as well as some deltaic, coal swamp, and shallow-marine deposits, all collectively named *molasse* long ago in Switzerland (see Fig. 11.28). Some of the world's most complex thrust faulting, accompanied by a plastic folding of strata into immense recumbent folds called *nappes*, characterizes the Alps. Similar structures are inferred for older mountain belts as well, but erosion has cut more deeply into those belts and has removed much of the upper parts of the great folds (Fig. 15.30). Alpine granites formed around 30 million years ago, but metamorphism continued locally until 11 million years ago. The great vertical uplift totaling 7 to 8 kilometers in the earth's youngest mountains continues today.

In Chap. 14, we noted that Mesozoic fragmentation of the southern margin of the Tethys seaway accompanied the formation of a series of spreading sites. One result was the formation of a number of microcontinents separated by new, small-ocean basins. As these microcontinents moved northward toward Eurasia, there was subduction of considerable oceanic material; andesitic volcanism occurred along the continental margin. Finally, just as F. B. Taylor suggested long ago (see Fig. 7.5), India collided with Asia to produce the Himalayas (Fig. 15.29). Arabia and Iran also collided with Asia to produce the Taurus, Zagros, and other mountains of Asia Minor (Fig. 15.31), and parts of

north Africa collided with Europe to give birth to the Alpine mountain systems (Fig. 15.29). Corsica and Sardinia were torn from Spain and France and rotated counterclockwise to their present positions near Italy. Spain and Italy apparently had previously been torn from northern Africa.

Where two continents collided along the old Tethyan zone, the resulting orogenic belt is more or less structurally symmetrical. Old oceanic ophiolite rocks commonly mark a suture zone between two formerly separate continents, and outward thrust faulting occurred toward each craton (Fig. 15.30). Although most of the former Tethyan oceanic material was consumed by subduction, some was thrust up onto the sedimentary rocks. Gravity measurements in the Himalayan region indicate that the thickest known continental crust in the world underlies these highest of all mountains—about 70 kilometers, or twice the normal thickness. Collision of the continents there drove Indian crust beneath Asiatic. A repeat performance may occur within the next few million years as north-traveling Australia collides with the Indonesia arc and crushes Indonesia against Indochina (Fig. 15.29).

Effects of India's Collision with Southeastern Asia

Eastern Asia consists of small cratons representing ancient microcontinents accreted together during Phanerozoic time. Much of that accretion occurred in early Mesozoic time, when at least two microcontinents rifted from Gondwanaland collided to

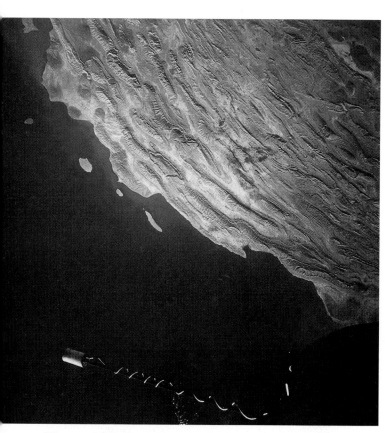

Figure 15.31 Eroded folds of the Zagros Mountains on the northeastern shore of the Persian Gulf in southern Iran. These mountains are part of the central Alpine-Himalayan orogenic belt. Thick Mesozoic and Paleogene strata deposited on the northern side between the Arabian craton (south, out of view) and the Asian craton (north, shown here). (See Fig. 13.26.) *(Taken from* Gemini *manned space capsule; NASA photo 66-63483.)*

form Indochina and Burma-Thailand (Fig. 15.29). Then in mid-Cenozoic time, the collision of India with southern Asia and the subduction of oceanic lithosphere beneath the arcs of the western Pacific and Indonesia produced the complex tectonics seen today (Fig. 15.32).

The Indian collision was so violent that several blocks of continental crust were squeezed toward the Pacific. Among the many effects of this squeeze play was the elevation of the Tibetan plateau in southern China, a plateau that is being deeply dissected to produce one of the world's most spectacular landscapes, made famous by centuries of Chinese painters.

When the Tethys Dried Up

Subduction is continuing today in the northern Mediterranean region beneath Italy and the Greek Islands, as well as in Indonesia, where oceanic trenches, deep-focus earthquakes, and andesitic volcanism are still active. In both cases, oceanic material is being underthrust northward (Fig. 15.29). It seems a safe bet that the Mediterranean will eventually be closed by the ultimate collision of northern Africa with Europe. In other words, the

Mediterranean–Black Sea region represents the final stage in the destruction of the Tethyan realm, a process that has already been completed farther east—from Turkey to Thailand.

In early Miocene time, the northward drift of Africa toward Europe was restricting the western Tethys Sea (Fig. 15.29). Mountain building had closed most of the eastern seaway and split the western Tethys into two parts, the Mediterranean Tethys and a Paratethys ("beside Tethys") inland sea across the present region of the Black, Caspian, and Aral seas (Fig. 15.33A). The early Miocene marine fauna was similar in both parts, but as the basins became more isolated, their faunas became dissimilar. About 6 to 6.5 million years ago, collision temporarily closed the main connection(s) between the Mediterranean and the Atlantic Ocean, and the Mediterranean dried up. Fossil plant pollen indicates that the climate of the region was becoming cooler and drier, a picture that fits with late Miocene evaporites long known around the Mediterranean basin. Recent deep drilling has shown that these evaporites underlie virtually the entire basin. But a single drying event would produce only about 100 meters of salt and gypsum (see Table 12.1), whereas the maximum thickness drilled is a staggering 1,000 to 2,000 meters. Apparently the sea had an intermittent connection to the Atlantic Ocean, which resupplied water repeatedly; possibly as many as forty fillings and dryings occurred within a period of only 1 or 2 million years.

The lowering of the Mediterranean Sea level caused the Nile River to cut a very deep gorge, which later filled up with alluvium as the sea eventually returned in Pliocene time (Fig. 15.33C).

About 5 to 5.5 million years ago, fresh- to brackish-water faunas appeared in several parts of the Mediterranean basin, indicating that fresh water from somewhere caused lakes to form. It may have come from the landlocked Paratethys via channels in the Yugoslavia-Greece region (Fig. 15.33B). About 5 million years ago, at the beginning of Pliocene time, the Strait of Gibraltar opened. Atlantic water gushed in to refill the Mediterranean and eventually to reconnect it with the Black Sea.

During Pleistocene glacial episodes, the worldwide drop of sea level temporarily isolated the Black Sea from the Mediterranean repeatedly, but the Caspian and Aral Seas have never been reconnected since Pliocene time.

Cratonic Rifting and Mantle Plumes

While Pacific arcs were being rejuvenated and the Alpine-Himalayan orogeny was in progress, extensive Neogene rift faulting accompanied by outpourings of lavas affected large parts of the cratons of Africa, southern Siberia (Baikal rifts), central Europe (Rhine graben), and Antarctica. This volcanism was unusual in that it occurred within, rather than at the edges of, lithosphere plates. Large-scale upwarping (perhaps caused by mantle plumes) preceded rifting. In Africa, some plateau areas were raised 2,000 meters, and then their crests collapsed along normal faults to form deep, narrow valleys (see Fig. 7.18). The resulting rift valleys in East Africa have become famous for their remains of early humans.

Most of the African rifting began in Oligocene or Miocene time and continues today. Evaporite deposits were the first

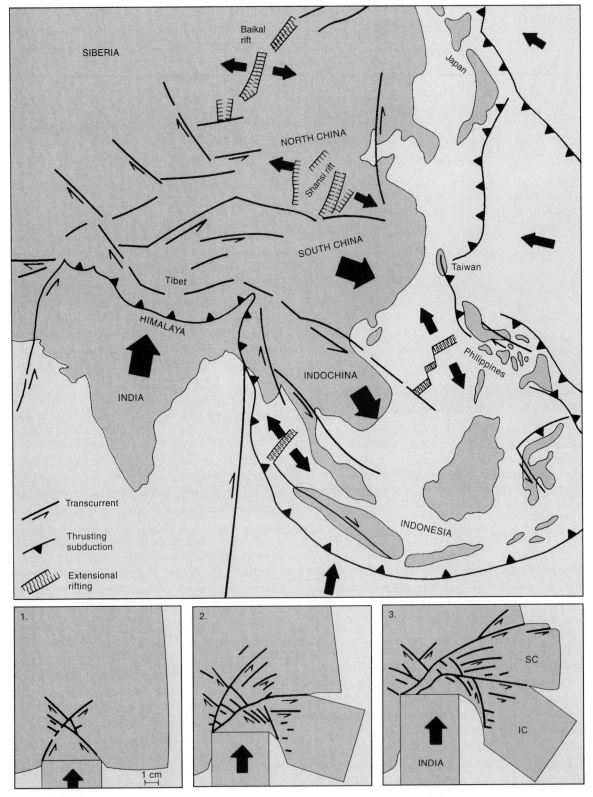

Figure 15.32   The profound effects of the collision of India with Asia beginning about 25 to 30 million years ago. *Upper:* present tectonics of southeast Asia, showing major crustal blocks and types of faulting among them. *Lower:* Three stages in a scale-model indentation experiment. A rigid block representing India was intruded into the edge of deformable plasticine "modeling clay" representing Asia, which was confined on the left and top. Note the striking similarity of deformation to the major crustal blocks and transcurrent faults of southeast Asia today *(upper).* This suggests that the collision of India has controlled most of the complex Cenozoic structure of eastern Asia, although subduction of oceanic lithosphere has also affected the continental margin. *(After P. Tapponnier, G. Peitzer, A. Y. Le Deain, and R. Armijo, 1982: Geology, v. 10, pp. 611–616. Reprinted by permission of Geological Society of America via Copyright Clearance Center.)*

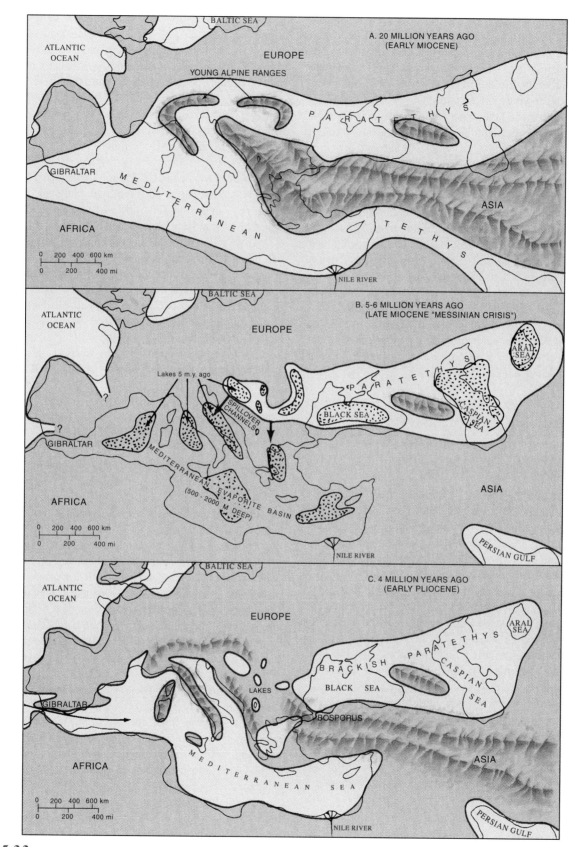

Figure 15.33   Paleogeography of southern Europe during Neogene time, showing destruction of the western Tethys Sea at the peak of the Alpine orogeny due to collision of Africa with Europe. *(Adapted from R. Brinkmann, 1960, Geologic evolution of Europe: Verlag-Hafner; L. J. Wills, 1951, A Palaeogeographical atlas: Blackie; K. J. Hsü, 1978: Scientific American, May, pp. 52–63.)*

prominent sediments to form (Fig. 14.3). Associated volcanoes, including Africa's tallest, Kilimanjaro (5,995 meters), produced basalts like those typical of oceanic regions, together with peculiar alkaline (calcium- and sodium-rich) lavas. The rifts average 50 kilometers in width, but the Red Sea and Gulf of Aden rifts are twice as wide, and they extend to oceanic depths (2,000 meters below sea level). The latter have oceanic crust beneath their axes and are zones of unusually high rates of heat flow through the crust; hot solutions beneath the central Red Sea are now creating ore deposits there. The East African rifts to the south may be classed as aulacogens, or "failed arms" of a triple junction located at the Red Sea–Aden junction (Figs. 7.18 and 7.20).

Thirty years ago, the suggestion by Australian geologist S. W. Carey that the Red Sea and Aden rifts resulted from Africa's being pulled 200 kilometers away from Arabia, tearing open a long scar in the continental crust, was regarded as outrageous. Now it is universally accepted, and paleomagnetic evidence has confirmed that Africa was rotated 11° counterclockwise away from Arabia (Fig. 15.29). The Jordan rift, discussed in Chap. 1, represents a dominantly lateral, or strike-slip, offshoot from the Red Sea rift. It is most like the much larger Gulf of California. Offset streams and lavas indicate a total of 100 kilometers of lateral movement since Miocene time (Fig. 1.8).

Coastal Plains and Continental Shelves

Passive Successors of and Predecessors to Orogenic Belts

We have now traced the history of the earth from Cryptozoic through Cenozoic time. This study has shown that orogenic belts arise either by subduction or by continental collisions. They typically evolved through long histories of mountain building, generally accompanied by formation of granitic batholiths, to become stabilized finally as parts of cratons. Our analysis of continental evolution would be incomplete, however, without a look at the young coastal plain strata and their extensions beneath the continental shelves; these strata formed partly over the sites of old orogenic belts (Fig. 15.2).

Coastal plains are especially prominent on passive continental margins, as in the eastern and southern United States and in the Arctic. Although passive today, such margins are always vulnerable to compressive tectonics when lithosphere plate patterns change, as we know they have many times in the past. Recall that all margins of North America and those of the Tethys Sea began as passive ones. Thus, coastal plain regions represent tranquility between episodes of great tectonic activity. Remember what Heraclitus said long ago: "There is nothing permanent except change."

Gulf of Mexico Coastal Province

General Patterns After Appalachian-Ouachita mountain building ceased in Triassic time, marine sedimentation commenced again in the Gulf Coast region during the Jurassic Period (Fig. 14.5). As discussed in Chap. 14, the Gulf of Mexico was formed in Jurassic time when Gondwanaland split from North America. Local structural warping and erosional truncations occurred on the margins of the young gulf, and there was considerable faulting. Small igneous intrusions (some containing diamonds) were emplaced in widely scattered areas, and a few volcanic vents related to faults in Texas, Arkansas, Louisiana, and Mexico erupted considerable ash.

The present lower Mississippi River valley is underlain by an aulacogen (Fig. 14.5), which developed perpendicular to the larger Gulf of Mexico rift. Occasional earthquakes reflect continuing movement along the marginal faults of the Mississippi aulacogen, including the most severe earthquakes in United States history, at New Madrid, Missouri, in 1811–1812.

As we found in discussing the breakup of Gondwanaland, the first deposits formed in the young Gulf of Mexico rift were evaporites. By the end of Jurassic time, normal marine deposition had ensued and continued through the Cretaceous Period, with considerable carbonate sedimentation. Cenozoic deposits are largely terrigenous sands and shales with only minor carbonates. Enlargement of the continent, coupled with a gradual change of climate, which accelerated erosion, caused this change. The Mississippi and other modern river systems had formed at least by Miocene time. Having large drainage basins, these streams brought immense volumes of sediment to the northern Gulf and began building the great deltas we know today.

Following the great Cretaceous transgression, a Cenozoic regression commenced, and the axis of maximum sedimentation continuously shifted southward. The continental shelf prograded seaward, for it is but a sedimentary embankment produced by the spreading of material in shallow, agitated water. At the same time, subsidence of the crust occurred as a result of broad, regional, isostatic downbending accompanied by some faulting under the load of such a large volume of material (see Fig. 7.35).

The northern margin of the Gulf of Mexico contains beneath it by far the thickest Cenozoic sequence of any of the coastal plains. The strata are almost 12,000 meters thick near the Mississippi delta, whereas in Mexico and along the Atlantic Coast, they are only one-fifth as thick. Under the Arctic coastal plain, Cenozoic sediments are thinner still. In all cases, the strata dip gently seaward and are thickest near present coastlines (Fig. 15.34).

Evaporite Intrusions The most interesting features of the Gulf Coast province are hundreds of intrusions of evaporite material through Cretaceous and Cenozoic strata (Fig. 15.34). Such intrusions, sometimes called *salt domes,* occur over an immense area, even well beyond the continental shelf margin. Pollen contained in the salt in Louisiana indicates a Jurassic and Triassic(?) age for the parent evaporites.

The Gulf Coast region has the largest North American petroleum reserves, most of which were trapped by evaporite intrusions, but it also is important for salt and sulfur. Salt domes have a peculiar caprock composed of carbonate, evaporite, and pure sulfur. Pressure from the upward rise of the intrusions plays a role

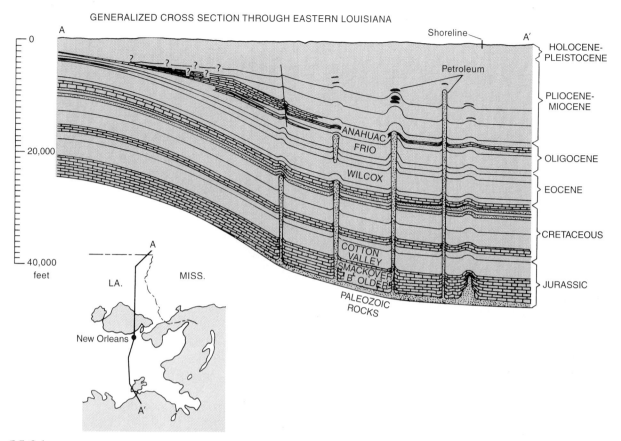

GENERALIZED CROSS SECTION THROUGH EASTERN LOUISIANA

Figure 15.34 Diagrammatic cross section of the thick Gulf Coast stratigraphic sequence punctured by evaporite intrusions (diapirs), or "salt domes," due to density disequilibrium of the deeply buried, less dense Jurassic salt beneath denser, younger strata. Salt rises in ridges, cylinders, and "teardrops" until isostatic equilibrium is achieved. Black crescents indicate oil accumulations. (See also Fig. 7.34.) *(From J. B. Carsey, 1950 in Bulletin of AAPG, v. 34, pp. 361–385. Tulsa, OK: AAPG.)*

in the formation of caprock through a series of chemical reactions. Bacterial attack of petroleum also may be important. Calcium sulfate ($CaSO_4$) from the evaporites probably reacts with carbon dioxide (CO_2) to form carbonate rock ($CaCO_3$), releasing the sulfate ion, which in turn reacts with water trapped in the sediments to produce pure sulfur.

The Atlantic Coastal Province and Appalachian Rejuvenation

The Atlantic coastal plain is the simplest of our three examples (Fig. 7.34). Cenozoic strata are only about 1,000 meters thick, and clastic sediments predominate; carbonate facies are confined to southernmost Georgia and Florida (Fig. 15.2). During early Cenozoic time, a tropical marine fauna thrived, implying a clear, warm sea. Texture and composition of sands suggest that the adjacent Appalachian region was low (Fig. 15.2) and undergoing mature weathering. Fossil floras of the southeastern states and the development of bauxite in Arkansas also attest to very warm, humid conditions.

After an Oligocene regression, the sea returned, bringing with it a rich, cooler-water Miocene fauna (Fig. 15.35). A mid-Cenozoic stratigraphic unconformity is also recognized in the Gulf Coast, Europe, and many other regions. It probably reflects the growth of large mid-Oligocene ice sheets in Antarctica.

In the present Appalachian Mountains, there is conspicuous topographic evidence that, after Mesozoic beveling, regional upwarping caused rivers to reexpose the buried old landscape (Fig. 15.36). As the area rose, gradients of major rivers were rejuvenated, and soft early Cenozoic and Cretaceous strata were stripped off, exposing the harder, deformed rocks of the old orogenic belt. Rivers managed to maintain their flow directions originally established on younger deposits, and their drainage patterns became superimposed upon old rocks as the region slowly rose. Entrenched meanders winding through anomalously narrow valleys and water gaps in resistant ridges are the results. Tributaries took advantage of relatively softer strata to carve out long, narrow valleys between parallel mountain ridges to produce the ridge and valley topography so characteristic of the Appalachians today.

A.

B.

Figure 15.35 Middle Miocene (about 13 m.y. old) richly fossiliferous, shallow-marine strata in the Calvert Cliffs of Chesapeake Bay, Maryland. A: Outcrop view of the cliffs. The Drumcliff Member of the Choptank Formation begins at the student's feet, unconformably overlying older Miocene rocks. The lower third of the Drumcliff is made of silty material and finely crushed shells. It is overlain by unbedded masses of complete, articulated shells, followed by distinctly bedded sands with smaller shells. B: Close-up of the massive shellbed, showing the extraordinary density and preservation of the shallow-marine molluscs, including scallops, venus clams, and moon snails. *(Courtesy of Susan Kidwell.)*

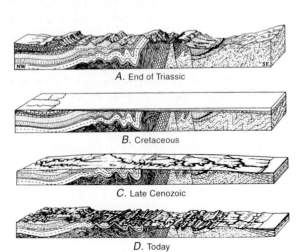

A. End of Triassic

B. Cretaceous

C. Late Cenozoic

D. Today

Figure 15.36 Diagrammatic evolution of modern Appalachian Mountains by epeirogenic rejuvenation and entrenchment of rivers. *(After Douglas Johnson, 1931, Stream sculpture on the Atlantic slope; by permission of Columbia University Press.)*

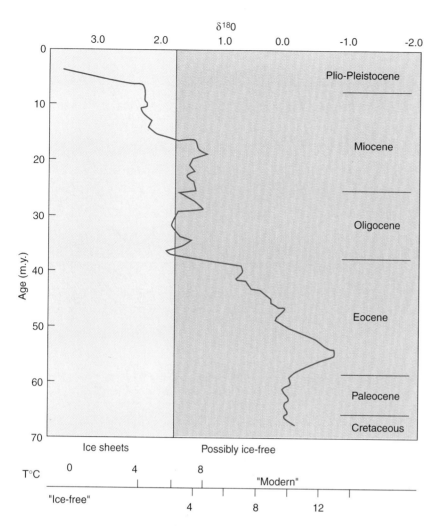

Figure 15.37 Oxygen isotope evidence for Cenozoic cooling and climatic deterioration. The warmest climates since the Mesozoic occurred in the early Eocene, and global average temperature has been steadily declining since then. A particularly sharp drop occurred in the earliest Oligocene, when major ice sheets first developed on Antarctica. Temperatures then moderated in the late Oligocene and early Miocene, but about 15 m.y. ago the ice cap returned to Antarctica for good. In the mid-Pliocene (about 3 m.y. ago), temperatures dropped to the bottom of the scale, since the Arctic ice cap appeared about then and the modern ice ages commenced. Note the threshold line between ice-free *(right)* and glaciated *(left)* conditions. The temperature scales are different for the two zones, since the "ice-free" world temperature is calculated in one manner from the oxygen isotopes, but the addition of ice changes the calculation. *(Adapted from K. G. Miller et al., 1987, Paleoceanography, v. 2, pp. 1–19.)*

Cenozoic Climate

As discussed in Chap. 14, Cretaceous climates were much warmer than those of today. Greenland and Siberia, which were about 15°C warmer, were covered by temperate vegetation instead of polar ice caps and were inhabited by dinosaurs, crocodiles, and turtles. Some of this warmth can be explained by the positions of the continents and their effect on oceanic circulation, but most of the warming must have been caused by a significant greenhouse effect from abundant CO_2 in the atmosphere.

Although the oceans and atmosphere were perturbed by the K/T extinction event, climate soon returned to its greenhouse state in the Paleocene. Eocene rocks above the Arctic Circle yield temperate-zone plants, and palms and cycads occurred as far as 60° N latitude. Alligators and tortoises also lived above 77° N latitude, in a region that today experiences extreme cold during 6 months of darkness and has no trees. To explain this puzzle, some scientists have argued that the Arctic Circle must have been smaller; in other words, the earth's tilt was much less than its present inclination of 23°. This much change of the earth's tilt creates enormous geophysical problems, however, and some climatic models of an earth with a shallower tilt produces *colder* po-

lar regions. Most of the plants, reptiles, and mammals found above the Eocene Arctic Circle probably could have survived months of darkness if the winters were sufficiently warm (especially under greenhouse conditions).

The end of the Paleocene was marked by a major warming event, which made it the warmest event since the Mesozoic thermal maximum. Oxygen isotope ratios in deep-sea microfossil shells suggest that the earth warmed by as much as 5°C at the Paleocene-Eocene boundary (Fig. 15.37). Some of the strongest evidence of Cenozoic temperatures comes from the shapes and sizes of leaves (Fig. 15.38). Paleobotanist Jack Wolfe has shown that warm, humid climates propagate plants with large, thick, smooth-edged leaves having a "drip tip" on the point for shedding abundant rainfall. Leaves from cool-climate plants are thinner and smaller, with jagged edges, and are shed every winter; plants in strongly seasonal climates do not have time to grow large, thick leaves that are not shed. By analyzing ancient leaf fossils, Wolfe produced temperature estimates for the Cenozoic in North America (Fig. 15.38). The late Paleocene warming is dramatic in the leaf data, just as it is in the marine record, and it is apparent that the early and middle Eocene was the warmest interval of the Cenozoic.

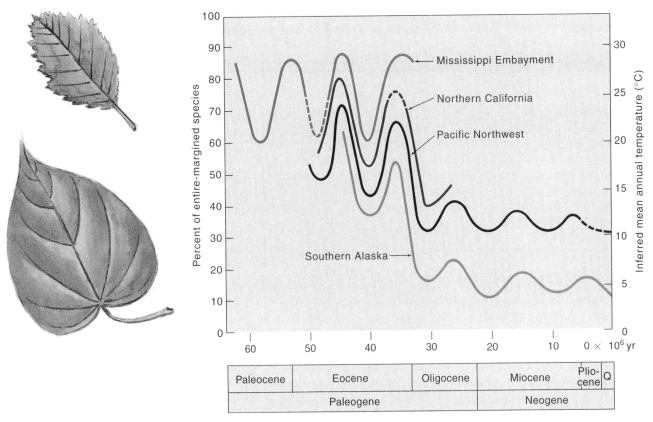

Figure 15.38   North American land plants were particularly sensitive to Cenozoic cooling. Jack Wolfe has shown that large, thick leaves with drip tips and smooth, "entire" margins (bottom leaf) are highly correlated with warm tropical temperatures, and smaller, thinner deciduous leaves with jagged margins (upper leaf) are typical of cooler climates. By plotting the percentage of species with entire-margined leaves, Wolfe obtained paleotemperature estimates for floras from Alaska to the Mississippi Embayment. Note that the curves for all four regions show the same overall shape, but each is slightly different in average temperature, depending upon its latitude. Two distinct cooling events occurred in the middle Eocene after the early Eocene peak of warmth. The most dramatic cooling occurred in the earliest Oligocene about 33 m.y. ago, when mean annual temperature dropped 8°C in about a million years, and the annual range of temperature doubled. This appears to correspond to the first pulse of glaciation in Antarctica (see Fig. 15.37). Temperatures fluctuated slightly from the late Oligocene onward, but no event was as severe as the early Oligocene cooling. *(Adapted from J. A. Wolfe, 1978,* American Scientist, *v. 66, pp. 694–703.)*

The events of the late Paleocene warming are just now being deciphered. Evidence from planktonic microfossils, and from their oxygen and carbon isotopes, suggests that the Paleocene ocean had vigorous circulation between surface and deep waters. This circulation was produced by an Antarctic source, similar to conditions in our modern ocean. In the latest Paleocene, circulation became sluggish, so that the ocean became uniformly warm, with no significant cold, deep-water circulation. Most deep-water, cold-loving foraminiferans went extinct at the end of the Paleocene, and only warm-water microplankton thrived in the early Eocene. The tropical Tethyan belt seems to be the source of this warm water.

The cause of this late Paleocene warming is still puzzling. The major tectonic event at the Paleocene-Eocene boundary was the collision of India with the Asian continent (Figs. 15.29 and 15.32). As a consequence, spreading ridges in the Indian Ocean and between Australia and New Zealand (related to India's northward motion) became inactive. In their place, Australia began to move rapidly away from Antarctica. The late Paleocene also

marked the beginning of the closure of the Mediterranean (Fig. 15.29). All these changes were accompanied by a worldwide acceleration in sea-floor-spreading rates and a rise in sea level as plate motions were reorganized in response to India's grinding to a halt and the closing of the Mediterranean. Some scientists attribute the late Paleocene warming to an increase in greenhouse gases released from the mantle by the rapidly erupting mid-ocean ridges. Others prefer a paleoceanographic explanation. According to Jim Kennett and Lowell Stott, the deep waters formed in the Tethys were warm and saline. If the tectonic changes along the old Tethyan belt of the Mediterranean and India suddenly enhanced warm Tethyan source waters at the expense of cold southern ocean waters, this could explain the late Paleocene warming (Fig. 15.29).

The extraordinary rapidity (only 20,000 years in duration) and the peculiarity of the warming of deep-ocean waters have led to another explanation of the late Paleocene thermal maximum. The most striking evidence comes from the ratio of carbon-12 (the common isotope of carbon) to carbon-13 (a rare isotope of carbon). In two short pulses, each less than a thousand years in duration, the oceans became extremely rich in carbon-12, suggesting that some source had released enormous amounts of carbon in a geologic instant. This rules out the slow oceanographic or plate-tectonic changes. Instead, many scientists now think that methane (CH_4, or natural gas) was trapped on a chemical form known as methane hydrates (methane combined with water in complex compounds). These compounds can form huge volumes of trapped carbon, nearly frozen on the ocean floor. According to the late Paleocene scenario, as much as 1.12×10^{18} grams of CH_4 was locked up on the sea bottom. When the methane hydrates broke down and their methane was released, it flooded the deep ocean with excess methane and carbon dioxide (hypercapnia, as in the events implicated in the end-Permian extinction; see Chap. 13). This carbon-rich water nearly wiped out many of the bottom-dwelling organisms (especially the benthic foraminifera). Eventually this methane escaped to the atmosphere to produce a warm "super-greenhouse." Although some of the carbon was eventually returned to the crust in the form of carbonates and coals, much remained throughout the early Eocene and was responsible for the extraordinary global warming we have just discussed.

As dramatic as the late Paleocene warming was, the events that followed were even more important. The late Eocene and Oligocene witnessed one of the most dramatic climatic deteriorations in earth history. In a little over 15 million years, mean global temperature declined by more than 10°C, more than during any of the Ice Ages. By the end of this episode, the earth had changed from a warm greenhouse world to our modern "icehouse" world of polar glaciers and large temperature differences between poles and equator.

The details of the Eocene-Oligocene transition are just now coming into focus. The first event occurred about 37 million years ago, at the end of the middle Eocene. Dramatic cooling occurred both in the oceans (Fig. 15.37) and on land (Fig. 15.38). This triggered many extinctions in the warm-water microfossils and in many land animals and plants (discussed further in the next section). Some scientists have argued that this cooling was caused by the beginning of Antarctic glaciation. Sediment from melted icebergs has been found in the late Eocene deposits of the southern Indian Ocean.

Whatever the cause of the events at the middle–late Eocene boundary, the events around the Eocene-Oligocene boundary are better understood. Oxygen isotope evidence (Fig. 15.37) and abundant glacial deposits show that Antarctica was partially glaciated in the early Oligocene. The most dramatic cooling in North American floras (Fig. 15.38) also occurs in the early Oligocene. According to Wolfe, mean annual temperature dropped by almost 8°C in a little over a million years, and the mean annual range of temperature went from less than 10°C in the Eocene to more than 20°C in the Oligocene. The cooling and extreme temperature fluctuation were accompanied by drying. Dense forest vegetation turned into a mixture of forests and grasslands. Water-loving crocodiles and aquatic turtles disappeared and were replaced by land tortoises. Large, subtropical land snails were replaced by smaller, drought-tolerant species.

By the middle Oligocene, the climatic deterioration had reached its low point. Major glaciations in Antarctica locked

enormous amounts of seawater in the ice caps and caused the largest drop in sea level during the entire Cenozoic. Late Oligocene microplankton were almost entirely cold-adapted species; the warm-water plankton so characteristic of the Eocene had gone extinct.

What could have caused such dramatic climatic changes? As with the K/T boundary event, impacts have been suggested as a cause. Unfortunately, the only evidence of iridium or impact particles is very sparse and occurred during the middle of the late Eocene, when there were no extinctions of any significance. The strongest characteristic of the Eocene-Oligocene transition is cooling and glaciation from about 40 to 30 million years ago, so the likely cause of the climatic change must be related to cooling. The obvious culprit is the opening of the passage between Australia and Antarctic as they rifted apart. Today this passage allows cold waters to circulate around Antarctica, locking in the cold and turning the South Pole into a giant icebox. The Circum-Antarctic current prevents cold southern ocean waters from mingling with the warm equatorial currents, exaggerating the temperature differences between pole and equator (Fig. 15.39). In the Eocene, the lack of a significant passage around Antarctica meant that South Polar waters could circulate northward and mingle with the equatorial warm waters. Consequently the temperature differences between pole and equator were much less in the Eocene, and the world was much warmer and milder at nearly every latitude.

The Circum-Antarctic current developed in steps. There may have been some kind of cooling associated with the widening of the passage between Australia and Antarctica at the end of the middle Eocene to produce the possible evidence of late Eocene Antarctic glaciation. Paleoclimatic modeling suggests that the Antarctic was already cold enough for ice caps in the Eocene, but the opening of the shallow gulf between Australia and Antarctica provided the necessary moisture for late Eocene ice deposits. In the early Oligocene, cold water circulated through a restricted passage south of Tasmania (Fig. 15.39). By the middle Oligocene, the passage had become so wide that both cold and warm currents could flow through, producing an oceanic front that further accentuated the Antarctic icebox. By the early Oligocene, the passage between South America and Antarctica had also opened, completing the circulation of cold waters around Antarctica.

The world began to recover from the Oligocene refrigeration in the early Miocene. Temperatures warmed slightly (Fig. 15.37), and the Antarctic glaciers temporarily disappeared until their final return in the middle Miocene, about 15 million years ago. The middle Miocene cooling appears to be related to the appearance of deep, cold bottom waters from the North Atlantic. This was caused by the sinking of the Iceland-Faeroe ridge, which had bottled up cold Arctic waters prior to the middle Miocene. When the ridge sank, it released these cold waters into the Atlantic and triggered the growth of the present Antarctic ice cap.

The next climatic event marked the Miocene-Pliocene boundary. A major expansion of the Antarctic ice sheet caused another dramatic cooling (Fig. 15.37). Carbon isotope data indicate that this cooling intensified oceanic circulation, bringing up nutrients from the deep bottom waters. By locking up more ice in

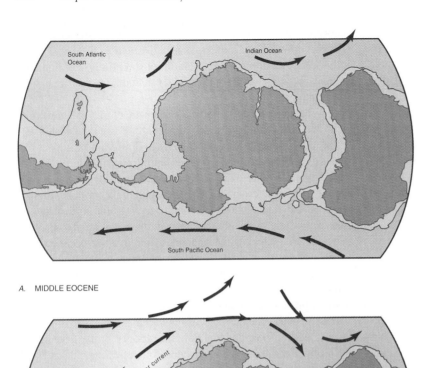

A.   MIDDLE EOCENE

B.   EARLY OLIGOCENE

Figure 15.39   The onset of cooling and glaciation in Antarctica may have had several causes, but the most important was plate tectonics. In the early and middle Eocene, Antarctica started rifting from Australia and South America as a remnant of Gondwanaland. These connections separated the South Pacific, South Atlantic, and Indian Oceans, so that polar waters in each were forced to mix with warmer waters from the subtropics. This caused mixing of warmer and cooler oceanic water masses and kept the difference in temperatures between pole and equator at a minimum. In the early Oligocene, the separation between Australia and Antarctica allowed a Circum-Antarctic current to flow eastward between the continents. This new current trapped cold water around the Antarctic, preventing it from mixing with the equatorial waters and allowing the Antarctic to become colder. When the Drake Passage between Antarctica and South America opened in the early Miocene (see Fig. 15.28), the Circum-Antarctic circulation was complete. Today this current flows in a clockwise fashion around Antarctica, trapping its cold water around it and thermally isolating it from other oceans, so that the continent is covered in the earth's largest ice cap. (*Modified from L. Frakes, 1979,* Climate through geologic time, *Elsevier.*)

the Southern Hemisphere, this cooling event triggered another significant drop in sea level. Recall that the Gibraltar gateway to the Mediterranean was closing due to the collision of Africa with Spain. When sea level abruptly dropped about 6.3 million years ago, the rising Gibraltar Mountains became a dam, which prevented Atlantic water from entering the Mediterranean. This led to the complete drying of the Mediterranean, followed by the catastrophic refilling in the early Pliocene, about 5 million years ago (Fig. 15.33). When the Gibraltar dam was breached, it must have been an awesome sight—a waterfall with a flow thousands of times as voluminous and powerful as Niagara Falls!

The last step in Cenozoic climate leads to the Ice Ages. During the late Pliocene, about 3 million years ago, we find the first evidence of Arctic ice caps. At the same time, glaciers began to advance in the northern continents. This glacial onset coincided with the closure of the Panamanian land bridge between the Americas. Although the two events seem only coincidental, in fact, the closure of the Panama gap prevented the mixing of Atlantic and Pacific waters (Fig. 15.40). This allowed the Gulf

Stream, a warm current moving north off the Atlantic Coast, to intensify its flow and drop more warm Caribbean moisture across eastern North America and in northwestern Europe. Eventually the moisture built up as snow across Newfoundland and Scandinavia, producing the first of the Pleistocene ice sheets. We shall discuss these changes further in Chap. 16.

Life in the Marine Realm

During the Cenozoic, the modernization of marine life was essentially completed. Most of the common invertebrates we find on the seashore today—gastropods, bivalves, and echinoids—were dominant from the beginning of the Cenozoic. In fact, many of the living genera were present in the Paleocene. After the Cretaceous extinctions, the microplankton recovered and soon swarmed in the Paleocene oceans. Globigerinid foraminiferans and calcareous nannofossils (Figs. 14.41 and 14.42) continued to be the dominant planktonic groups. Diatoms and radiolarians were also abundant.

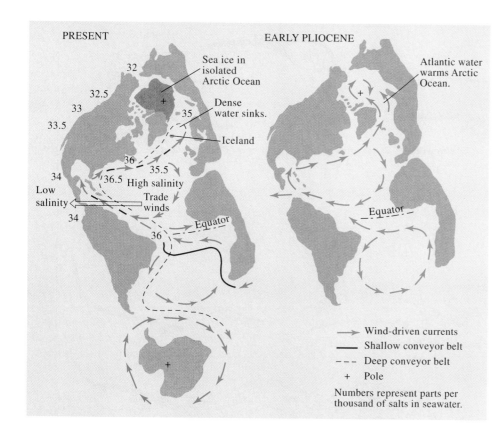

PRESENT   EARLY PLIOCENE

Sea ice in isolated Arctic Ocean
Dense water sinks.
Iceland
High salinity
Trade winds
Low salinity
Equator

Atlantic water warms Arctic Ocean.
Equator

→ Wind-driven currents
— Shallow conveyor belt
- - - Deep conveyor belt
+ Pole

Numbers represent parts per thousand of salts in seawater.

Figure 15.40   The conveyor belt of the present world *(left)* and the oceanographic pattern that may have existed in the North Atlantic during the early Pliocene *(right)*, before the modern Ice Age of the Northern Hemisphere began. Today the trade winds carry to the Pacific water that they evaporate from the surface of the Atlantic. (Numbers show salinities for the two oceans.) The dense, saline Atlantic water sinks north of Iceland, driving the conveyor belt. During early Pliocene time, before the Isthmus of Panama was in place, mixing with Pacific waters should have lowered the salinity of the Atlantic. The more buoyant Atlantic waters may then have flowed northward into the Arctic, keeping the polar region warm. The uplift of the Isthmus of Panama may have triggered the Ice Age by elevating the salinity of Atlantic waters and causing them to sink north of Iceland, as they do today; this change would have deprived the Arctic Ocean of heat from the Atlantic. *(After S. M. Stanley, Journal of Paleontology 69:999–1007, 1995. Reprinted by permission of The Paleontological Society.)*

A.

B.

C.

Figure 15.41   During the early and middle Eocene, the commonest animals in the warm Tethys seaway from the Mediterranean region to Indonesia (see Fig. 15.29) were the nummulitid foraminiferans. A: Slab of nummulitids showing their typical abundance. Each is about the size and shape of a large coin—remarkably big, considering that they were secreted by single-celled, amoeba-like protozoans. *(Courtesy of T. Aigner.)* B: Close-up view of a sectioned nummulitid, showing the spiral arrangement of the internal chambers. *(D. R. Prothero.)* C: Middle Eocene nummulitic limestones are so abundant in the Tethys region that, in some places, such as the Gizeh Plateau of Egypt, they occur by the millions in limestones. When the Egyptians quarried local rocks *(foreground)* to build the Great Pyramids *(background)*, they used large quantities of nummulitic limestone. *(Courtesy of Robert M. Schoch.)*

The rudistid reef community of the Cretaceous was replaced by modern scleractinian corals in the Paleocene (Fig. 12.21). The ammonites, however, were never completely replaced by other shelled cephalopods. Instead, new teleost fishes took over their role as the middle-level marine predators.

There were also new additions to the Cenozoic sea floor. Sand dollars evolved from biscuit-shaped urchins during the early Tertiary and quickly became adapted to very shallow burrowing on sandy beaches. Several new groups of clams exploited the shifting beach sands as well. Some of the most peculiar animals were the coin-sized *nummulitid* foraminiferans (Fig. 15.41). They secreted shells built of many tiny chambers in a flat spiral reaching several centimeters in diameter. Like the late Paleozoic fusulinids, this is an amazing adaptation for a single-celled, amoeba-like animal. Apparently their incredible growth was supported by incorporating unicellular algae into their tissues. Nummulitids were so

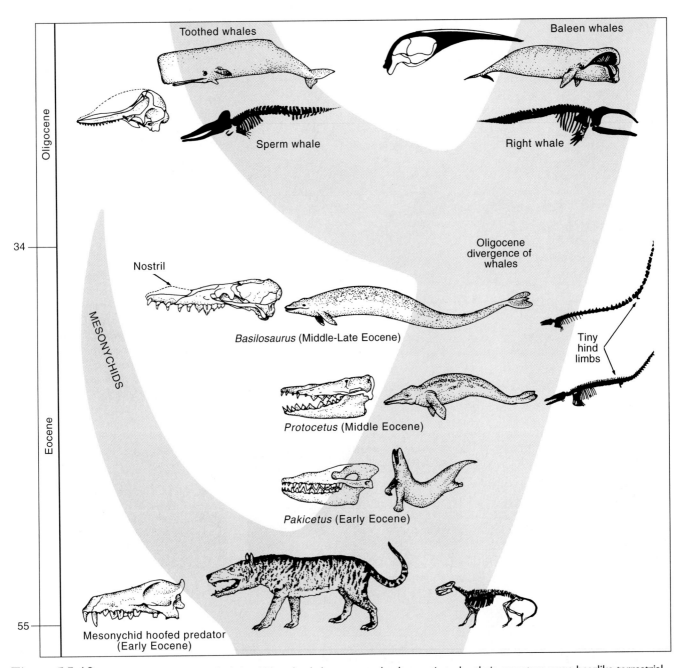

Figure 15.42 Evolutionary history of whales. Although whales are completely aquatic today, their ancestors were bearlike terrestrial hoofed mammals known as mesonychids. This can be demonstrated from mesonychid teeth, which are identical to those of early whales, and other details of the mesonychid skeleton. A series of discoveries from lower Eocene rocks of Pakistan has provided several intermediate evolutionary steps, which show progressively more whalelike skulls and ears. The best-known middle Eocene whales were the zeuglodonts, which still had a primitive skull but were fully aquatic with front flippers, tail fin, and hind limbs reduced to tiny vestiges. Other middle Eocene whales from West Africa had walking hind limbs but front feet modified into flippers. In the Oligocene, these archaic whales were replaced by the two modern groups, the toothed whales (including dolphins, porpoises, killer whales, and sperm whales) and the baleen whales (which have a screen of baleen, or "whale-bone," in their mouths to filter feed krill and plankton).

common in the Tethys seaway during the middle Eocene that it has been called the "Nummulitic Epoch." Nummulitic limestones were used to build the Egyptian pyramids, and Greek geographer Strabo thought that nummulitids were the petrified remains of lentils from the meals of the slaves who built the pyramids. Like many other warm-water species, most nummulitids became extinct in the late Eocene as a result of worldwide cooling.

Although the marine reptiles of the Cretaceous were gone, the niche of top marine predator was eventually filled by marine mammals. In the late Oligocene and early Miocene, seals, sea lions, walruses, and their relatives evolved from the bear family. Even more amazing is the evolution of whales. The earliest whales are known from the Eocene, and they already had streamlined bodies with front flippers (Fig. 15.42). They did not yet have blowholes on the top of their heads or many other typical whale specializations. Recently even more primitive whales have been found that still had tiny hind limbs. Whale ancestry can be traced to a group of bearlike hoofed mammals, the *mesonychids,* which had skulls and teeth like primitive whales, although they were not yet aquatic. By the Oligocene, whales had become even more specialized into the two main groups found today. One group, the baleen whales, include the gigantic beasts such as the blue whale, specialized for filter feeding to extract huge volumes of tiny plankton and crustaceans from the seawater. The other group, the toothed whales, eat fish and squid. They range from the giant sperm whale to the dolphins, killer whales, and porpoises.

Whales were not the only big marine predator in the ocean. During the Miocene, there was a gigantic species of the great white shark that may have reached 17 meters (55 feet) in length (Fig. 15.43). This shark was so huge that only whales and other sharks would have been suitable prey!

The Age of Mammals

The Early Cenozoic Adaptive Explosion

The Cenozoic is best know as "The Age of Mammals." Although birds, crocodiles, turtles, snakes, lizards, and amphibians were abundant, the dominant roles on land were occupied by mammals. After their origin in the Late Triassic, mammals spent nearly 150 million years as shrew-sized beasts, hiding from the dinosaurs. As we saw in Chap. 14, the two major groups of mammals are the marsupial (pouched) mammals and placental mammals (which carry their young in the uterus until birth). Marsupials and placentals diverged in the Early Cretaceous, and the two are found side-by-side in Late Cretaceous deposits that also include *Tyrannosaurus* and *Triceratops.* Late Cretaceous marsupials were very much like the modern opossum. Most early placentals also looked like opossums or shrews, but they were already beginning to diversify. Although few Cretaceous placentals can be placed in modern orders of mammals, some showed telltale signs of ancestry of modern groups. The earliest known hoofed mammal, *Protungulatum,* and the earliest relative of the primates, *Purgatorius,* were among these.

When the K/T event wiped out the dinosaurs, a whole new world was opened to mammals. In the early Paleocene, they began an explosive evolutionary radiation, diversifying and adapting to many different ecological niches and ranges of body sizes. The spectacular radiation of placental mammals took place during the Paleocene and early Eocene, a period of less than 15 million years (Fig. 15.44). It is justly considered one of the most dramatic examples of evolutionary diversification in life history. From shrewlike beasts of the Late Cretaceous (including only a few modern orders), mammals had diversified into more than twenty-two orders by the Eocene, including forms as diverse as bats, whales, and a variety of hoofed mammals, carnivores, rodents, and rabbits. In the Late Cretaceous,

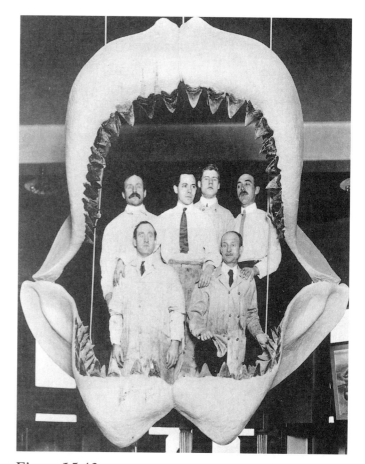

Figure 15.43   Restoration of the jaws of *Carcharocles megalodon,* the giant Miocene relative of the great white shark. Since sharks have skeletons made only of cartilage, not bone, the jaws had to be reconstructed using the fossil teeth. Unfortunately, these scientists did not realize that shark teeth get progressively smaller near the hinge of the jaw. They used only the largest teeth throughout the jaws, so they reconstructed them about a third too large. Even so, the living animal was probably about 17 meters (55 feet) in length, an impressive fish by any standards! *(Image #: 336000 American Museum of Natural History Library.)*

no mammal was larger than a house cat. By the Eocene, nearly all modern orders of mammals had evolved, and some were the size of a rhinoceros or elephant.

Although this was a very rapid diversification, it was not instantaneous. The mammals of the Paleocene and Eocene looked very different from those living today, and few could be recognized by the average zoo visitor (Fig. 15.45). The primary reason was that the mammals were all either members of archaic groups now extinct or very primitive members of groups still living. Another reason was that paleoclimate and vegetation were very different from what they are today. The world was still in its Cretaceous greenhouse state, with warm, almost subtropical climates from pole to equator. Even during the Eocene, there were alligators, lemurlike primates, and temperate plants living north of the Arctic Circle. Montana and Wyoming had climate and vegetation resembling Panama, and Europe was an archipelago of tropical islands resembling Indonesia.

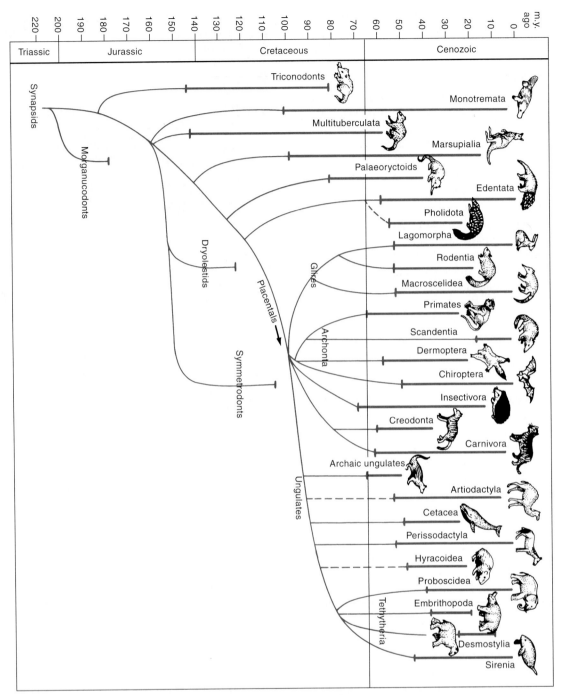

Figure 15.44   Evolutionary radiation of the mammals. Although mammals originated in the Late Triassic, they remained small, shrewlike beasts for the entire Mesozoic, with several different experiments in evolution that did not survive. By the Late Cretaceous, three lineages were established, which continued into the Cenozoic. One lineage, the squirrel-like multituberculates, died out at the end of the Eocene. The second group, the pouched marsupials, carry their young in the uterus for a while, then give birth while it is premature; the embryo must then crawl up the mother's belly, find the pouch, and clamp onto a nipple inside. Living marsupials are represented by animals such as the opossum, kangaroo, and koala. The third group, the placentals, carry their young in the mother's uterus for the full term, so they are less vulnerable when they are born. In the Paleocene, the placentals underwent a huge evolutionary diversification, splitting into almost two dozen orders. These include the glires (rodents, rabbits, and elephant shrews); archontans (primates, tree shrews, colugos, and bats); insectivorans (shrews, moles, and hedgehogs); carnivorans (cats, dogs, bears, raccoons, weasels, and seals) and their extinct flesh-eating relatives, the creodonts; and the tremendous diversification of hoofed mammals, including the even-toed artiodactyls (pigs, hippos, camels, deer, antelope, cattle and giraffes), the whales (see Fig. 15.41), the odd-toed perissodactyls (horses, rhinos, and tapirs), and the tethytheres (elephants, manatees, and several extinct relatives). The relationships of living aardvarks and hyraxes are still controversial. *(Modified from M. J. Novacek, 1992,* Nature, *356, p. 122. Reprinted by permission.)*

Figure 15.45 This is how Wyoming might have looked in the middle Eocene, about 47 m.y. ago. Amidst the lush, tropical vegetation was a fauna that included mostly archaic mammals that are now extinct. A few were ancestors of later groups, although they are so primitive that you would not easily connect them to their descendants. Among the survivors shown in this diorama are early rodents, tapirs, horses, rhinos, and carnivorans. Crocodiles and pond turtles flourished in the wet jungles, and *Mesonyx (lower left)* was related to whales (see Fig. 15.41). Most of the others had no living descendants. *(Painting by J. H. Matternes, courtesy of National Museum of Natural History.)*

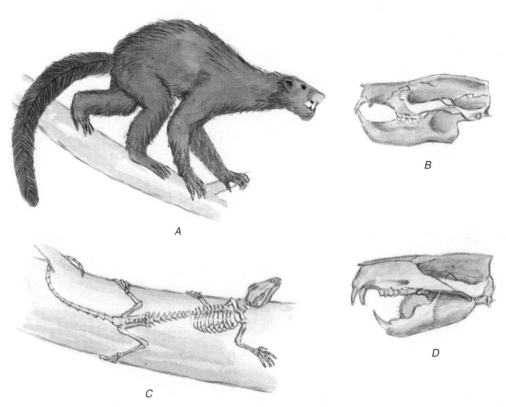

A

B

C

D

Figure 15.46   Paleocene and Eocene faunas included many archaic groups of mammals that were adapted to the abundant tropical forests on all the continents. Two groups were particularly squirrel-like in their ecology, even though neither was related to the rodents. Similarities included not only their body shape and feet adapted to climbing but also even the chisel-like front teeth for gnawing and specialized grinding teeth in the back of the jaw for eating seeds and fruits. Both groups declined after the early Eocene and were extinct by the late Eocene as climates changed and the forests disappeared. The appearance of rodents in the early Eocene may have also hastened their disappearance. *A–B:* One group, the plesiadapids, was one of the commonest fossils in the Paleocene. Once thought to be related to primates, more recent research shows that plesiadapids are extinct relatives of the living colugo, or "flying lemur" of the Philippines (see Fig. 15.45). *C–D:* The other group, the multituberculates, first arose in the Jurassic, so they lasted almost 100 m.y., until their extinction in the late Eocene. This was the longest time span of any mammalian order. Multituberculates had a specialized, blade-shaped lower tooth, which apparently helped them slice seeds and fruits.

The Paleocene-Eocene subtropical world supported a great diversity of mammals, but most were adapted to living in dense forest. The most common mammals were small tree dwellers, such as lemurlike primates and rodentlike multituberculates (Figs. 15.45, 15.46, and 16.29). Both of these groups must have fed on the abundant leaves and seeds in the forest canopy or on the floor beneath. In the Eocene, both rodents and rabbits migrated to North America from Asia, where they originated. By the end of the Eocene, they had taken over the ecological niche of fruit and nut eaters in the trees and on the ground, driving the multituberculates and most lemurlike primates to extinction. Forests support a great variety of insects, so insectivorous mammals were very common (as they had been since the Jurassic).

On the ground, the dominant mammals were all leaf-eating (browsing) mammals, most of them small to medium size. The most typical were the archaic hoofed mammals, once known as "condylarths." Although most of these archaic forms are now extinct, some were related to the living hoofed mammals. There were also a number of extinct archaic mammals that have no living relatives. For example, the *pantodonts* were the largest browsing ground dwellers of the Paleocene and Eocene. The last of their lineage, *Coryphodon,* was a cow-sized animal from the early Eocene. Another bizarre group, the uintatheres (Fig. 15.45), had six knoblike horns on their heads and huge tusks. These rhino-sized beasts were the largest animals of the middle Eocene.

The predators were also unusual. There were no dogs, cats, bears, hyenas, or weasels. Instead, one of the largest predators was not a mammal at all but, rather, a gigantic carnivorous bird almost 2.4 meters (8 feet) tall! There were also the mesonychids (Figs. 15.41 and 15.45), which had a bearlike build but were hoofed mammals; their only living descendants are whales. The predatory *creodonts,* an archaic group of carnivorous mammals, were very bearlike or wolflike (*Patriofelis* in Fig. 15.45). The modern order Carnivora (including all the living carnivorous mammals, such as dogs, cats, bears, weasels, and hyenas) appeared in the late Paleocene, but the first members of this order were small, weasel-like animals and did not flourish until the Oligocene.

If we took a time machine back to the Eocene of Wyoming, we would be struck by several things. The subtropical climate and vegetation would remind us of Central America or the Amazon basin. The jungles would be filled with small, tree-dwelling mammals; sheep-sized browsing mammals would shuffle through the underbrush. Because of the dense cover, most mammals would depend on stealth, camoflage, and a good sense of smell and hearing for safety, and their predators would have to hunt by scent and ambush. Neither predators nor prey needed to be good runners in such dense forests. This tropical paradise extended from equator to high latitudes in the Northern Hemisphere, with many of the same genera of mammals found in North America and Europe during the early Eocene. Animals crossed the narrow North Atlantic via a corridor through Iceland and Greenland while it was still green!

Paradise Lost

Today this warm, lush paradise has vanished except in the tropics. What happened to it? The most important event was a worldwide cooling, which began in the late Eocene and Oligocene. Recall that global temperatures dropped as much as 10°C during this time. By the late Oligocene, the grasses we know so well today—including most cereal crops as well as lawns—had become common. The subtropical forests were gradually replaced by more open forests mixed with grasslands, and in the Miocene there were savannas like those of today's East Africa (Fig. 15.47).

These climatic and vegetational changes had several important effects on land mammals. Clearly the tree dwellers, such as the lemurlike primates and multituberculates, no longer had a dense forest canopy to shelter and feed them. Similarly the archaic hoofed mammals that browsed on leaves were without their food source or cover. In their place, several groups of modern hoofed mammals (including horses, rhinos, camels, deer, and early elephants) became the dominant herbivores. The earliest members of each of these groups were also browsers, but during the Oligocene and Miocene, they became more and more specialized for living and feeding in open grasslands.

Two important adaptations appeared in all the hoofed herbivores that moved into the grassy savannas. As they left the protection of thick forest cover, they had to depend on good eyesight and fast running to protect them from predators. Under the survival pressure of such environmental changes, each of these groups developed longer and longer limbs, a change that increased their running efficiency. In addition, a diet of grass is much more abrasive to the teeth than a diet of leaves because grass contains siliceous grit in its tissues, as well as grit on the surface. This abrasive material makes grasses more difficult to eat, initially discouraging any animal from eating too much. Any mammal with primitive, low-crowned teeth for eating soft, leafy vegetation eventually wears its teeth to the gums and starves to death. Thus, horses (Fig. 3.15A), rhinos, camels, and many others developed teeth with long crowns and roots that grow continuously through life; as the top layer wears off, more of the tooth grows in to replace it. By the Miocene, almost all herbivorous mammals had developed high-crowned teeth.

Finally, the change from subtropical forest to temperate savanna affected not only the herbivores but their predators as well. Archaic, small-brained ambush predators, such as mesonychids and creodonts, became extinct. During the Oligocene and Miocene, their place was taken by the earliest dogs, cats, bearlike animals, and weasels. These predators were much better runners, or they had better brains and eyesight for stalking and ambushing their prey in the open savanna. In the Miocene, the dog family provided not only predatory wolflike forms but also scavenging, bone-crushing "hyena-dogs." There were also archaic sabertooth cats, primitive weasels and raccoons, and bearlike carnivores in the Miocene.

Indeed, a time machine trip to Nebraska in the middle Miocene would remind us of the modern East African savanna (Fig. 15.47). The largest herbivores were mastodonts, distantly related to modern elephants found in Africa. North American rhinos were both browsers (analogous to the living black rhinoceros) and hippolike grazers that lived in ponds and rivers. Camels evolved into forms that were the ecological equivalents of giraffes, gazelles, and many African antelopes. Horses, too, were very diverse (Fig. 3.15B); as many as twelve species of grazers and browsers lived side-by-side (as African antelopes do today). Because true antelopes never made it to North America, there was a great diversity of pronghorn "antelopes." Today only one species of pronghorn still survives, but there were many different kinds, with a variety of horn shapes, in the Miocene.

Family Trees

The individual histories of these groups demonstrate the trend toward grassland specialization. Horses, for example, began in the early Eocene as beagle-sized forest browsers with four toes on their front feet and three on their hind feet (Fig. 3.15). Recent discoveries have shown that these horses (along with all other odd-toed hoofed mammals, or *perissodactyls*) and elephants were descended from a common ancestor, which lived in the late Paleocene of Asia. Primitive Eocene horses became progressively more specialized for grassland life in the Oligocene and Miocene. Their limbs became longer and better adapted for running. Their side toes were used less and less while running on the hard plains and, so, became reduced. In Oligocene horses such as *Mesohippus* and *Miohippus*, there were only three toes on both front and hind feet. Three-toed horses dominated in the Miocene, although with further reduction of the side toes. The living horse, *Equus*, has all but lost its side toes, as this animal has become an extreme specialist for running (Fig. 3.7A).

In addition to the limbs, horse heads also changed. The eyes and brain both became larger for spotting and recognizing ambushing predators. The teeth became more high-crowned for eating grasses. As this change in tooth structure developed, the face and jaw became longer to support the enlarged battery of grinding teeth. However, these changes did not occur throughout the horse family in a single, unbranched evolutionary lineage, as old diagrams suggest (Fig. 3.15A). Instead, there were many different kinds of horses, including side branches that became specialized for forest browsing (Fig. 3.15B). These browsing horses

Figure 15.47   By the Oligocene and Miocene, the jungles of the Eocene had disappeared, and open grasslands and savanna began to replace them. This is how Nebraska or Texas might have looked in the late Miocene, about 7 m.y. ago. Almost all of the mammals are members of living groups, including rabbits, rodents, dogs, cats, peccaries, camels, horses, rhinos, and mastodonts (related to elephants). A few extinct groups are shown, such as the "slingshot-beast" *Synthetoceras* (distantly related to camels) and the deerlike *Cranioceras*. In many ways, this scene resembles the modern East African savanna, except that horses, camels, and pronghorns performed the roles that antelopes and giraffes do in Africa today. In the Miocene of North America, dogs were the main predators and scavengers (like African hyenas and lions), and the short-legged grazing rhino *Teleoceras* (see Fig. 15.1) occupied the niche of the hippopotamus. *(Painting by J. H. Matternes, courtesy of American Museum of Natural History.)*

lived alongside the fast grazers but retained low-crowned teeth and three toes. By the end of the Miocene, however, changing conditions had wiped out most of this great diversity of horses, so that only one-toed *Equus* survives today as living horses, donkeys, asses, and zebras.

Other perissodactyls, such as the rhinoceroses and tapirs, had a similarly long and distinguished history. The earliest ancestors of rhinos and tapirs lived in the early Eocene and were virtually indistinguishable from early horses. By the late Eocene, tapirs had become the dominant leaf-eating browser, with a well-developed proboscis, and they have lived in that ecological niche without much change since then. Rhinos, on the other hand, diversified into many niches. One group became an aquatic hippolike grazer in the late Eocene and early Oligocene before dying out. Another group became specialized for running; in North America, there were long-limbed forms about the size of a Great Dane (Fig. 15.48). One of their Asian descendants grew to gigantic proportions in the Oligocene. Known as *Paraceratherium* (formerly called *Indricotherium* or *Baluchitherium*), this beast reached 6 meters (18 feet) at the shoulder, weighed 20,000 kilograms (22 tons), and could browse on the tops of trees (Fig. 15.48). The largest land mammal that ever lived, it was hornless, as rhinoceroses have been throughout most of their history.

Whereas the aquatic and running rhinos became extinct in the early Miocene, true rhinoceroses became the dominant large herbivores in the Oligocene and Miocene. The first true rhinos appeared in the late Eocene, and by the late Oligocene two groups of rhinos had independently developed small paired horns on their noses. In the Miocene, some rhinos became specialized for browsing leaves and bushes and even developed a prehensile lip or proboscis for this purpose. Others became fat, amphibious grazers, like the modern hippopotamus, living in the water in the day and going out at night to graze (Fig. 15.1). Rhinos vanished from North America in the early Pliocene, presumably in response to the same climatic events that nearly wiped out horses and many other typically Miocene groups.

Today perissodactyls are very rare in the wild. The dominant group of modern hoofed mammals consists of the even-toed ("cloven-hoofed") *artiodactyls,* which have two or four toes on each foot. Some groups, such as the piglike mammals (including pigs, peccaries, hippos, and several extinct groups), have changed little since the Eocene. Others, such as the camels, deer, pronghorns, and cattle, have changed dramatically. Camels, for example, immigrated to North America in the middle Eocene and were already longer-limbed and had higher-crowned teeth than any of their contemporaries. They became the North American analogues of antelopes and giraffes during the Oligocene and Miocene. Some also developed into short-legged, browsing forms that lived in thick underbrush. Camels, too, nearly vanished at the end of the Miocene but survived in their native North America until the end of the last Ice Age. Late in their history, they migrated to Asia and Africa, where they survive today as the dromedary and Bactrian camels, and to South America, where they flourish as llamas, alpacas, and their kin.

The spread of Miocene savannas favored many other artiodactyls besides camels. Advanced artiodactyls, such as deer, pronghorns, giraffes, cattle, antelopes, and sheep, are cud chewers, or *ruminants.* They have a complex, four-chambered stomach to feed more efficiently on the indigestible cellulose in grasses and other woody plants. After the vegetation is swallowed, it is fermented in the first stomach chamber, then regurgitated to the mouth to be chewed as cud. Then it is swallowed a second time and digested completely. Ruminant artiodactyls gradually replaced perissodactyls in the Miocene, with camels, deer, and pronghorns dominating in North America and antelopes and cattle diversifying in the Old World. Some of these Old World ruminants, such as bison and sheep, migrated to North America during

Figure 15.48   Life-sized restoration of the hornless rhinoceros *Paraceratherium* (once called *Indricotherium* or *Baluchitherium*), the largest land mammal that ever lived. Found in the Oligocene beds of Mongolia and Pakistan, it was so large that it could browse on the tops of trees, and it reached 6 meters (18 feet) at the shoulder. It evolved from a dog-sized ancestral running rhino, shown between it and the elephant. Note how it dwarfs a modern African elephant. *(Courtesy of University of Nebraska State Museum.)*

the Ice Ages and became very successful until their slaughter by white hunters in the past two centuries. Today domesticated ruminants (especially cattle, sheep, and goats) have taken over the world, thanks to human intervention, and provide us with most of our meat, milk, and leather.

Along with perissodactyls and artiodactyls, elephants have a long and distinguished history (Fig. 15.49). Originating in Africa in the late Eocene while it was still an island continent, the earliest elephants looked more like tapirs. From these beasts evolved a variety of mastodonts with four tusks and a short trunk. In the early Miocene, mastodonts escaped the isolation of Africa and soon spread to Eurasia and North America. There was a great diversity of early mastodonts, including beasts with shovel-like jaws and others with tusks that curved downward from the lower jaw. By the Pleistocene, only the forest-browsing American mastodonts and the great grazing mammoths

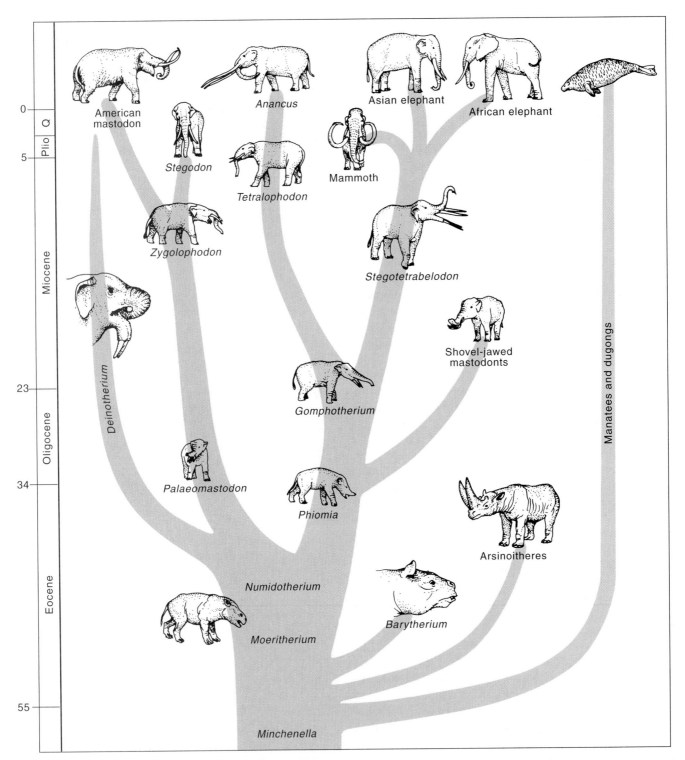

Figure 15.49   Family tree of the order Proboscidea, or the elephants and their relatives. The earliest known proboscideans are found in the late Paleocene of China and early Eocene of Pakistan. Piglike *Moeritherium* is known from late Eocene rocks of North Africa. By the early Oligocene, several different lineages of proboscideans had evolved, including the deinotheres (with the downward-pointing lower tusks) and the first mastodonts with short trunks. In the early Miocene, proboscideans escaped their isolation in Africa and spread all over Eurasia, reaching North America by the middle Miocene. Mastodonts continued to diversify in the Miocene, producing the American mastodont that survived to the end of the Pleistocene and the shovel-tusked mastodonts of the late Miocene. The late Miocene and Pliocene also witnessed the radiation of the elephant family, including the long-tusked anancines, the mighty mamoths, and the two living species, the African and Asian elephants.

remained. Mammoths repeatedly migrated back and forth between the northern continents and Africa. The woolly mammoth was one of the most widely traveled of all mammals, but it and mastodonts became extinct at the end of the Pleistocene. Today elephants are being decimated by poachers in their last strongholds in Africa and Southeast Asia because of the demand for their ivory tusks.

The Southern Continents

Australia, Africa, Antarctica, and South America are remnants of the ancient Gondwana continent, which broke up mostly in the Jurassic and Cretaceous. All these continents, along with islands such as Madagascar and New Zealand, have been relatively isolated from the Northern Hemisphere during most of their history. Consequently their land faunas evolved in isolation and developed many unique forms. Although most of these bizarre animals are now extinct, some survive today in their island domains.

A good example of this phenomenon is flightless birds (Fig. 15.50). Diverging during the Cretaceous, before the evolution of most modern bird families, today these birds are found only on the Gondwana remnants. Each Gondwana continent has its own flightless bird: Africa, the ostrich; South America, the rhea; Australia, the emu and cassowary; and New Zealand, the kiwi. In addition to the living flightless birds, there were two kinds of gigantic "elephant birds" (up to 4 meters tall and weighing about 200 kilograms), which once lived on Madagascar and New Zealand. As Gondwanaland broke up and its parts drifted away from each other, each of these birds evolved on its own continent from a primitive flightless ancestor. A similar remnant Gondwana pattern is shown by many families of reptiles and amphibians.

The most spectacular examples, however, come from the mammals. By the Late Cretaceous, marsupials and placentals had begun their separate evolutionary histories. The two were found together in the Americas, but for some reason marsupials were far more abundant than placentals in Australia. Consequently Australia's marsupials evolved to fill most of the ecological niches occupied by placentals in the Northern Hemisphere, including marsupial "wolves," "wolverines" (the Tasmanian devil), "cats," "anteaters," "woodchucks" (the wombats), "moles," "flying squirrels," "rabbits" (the bandicoots), "mice," and many other astonishing examples of ecological convergence (Fig. 3.9). Although all these marsupials look superficially like their placental counterparts on the northern continents, they are all pouched and therefore unrelated to placentals. Other marsupials, such as the kangaroos, have no counterparts among either placental mammals or marsupials on other continents, but their ecological niche is similar to the one filled by the placental hoofed mammals elsewhere.

Although we cannot rule out an Australian or Antarctic origin for marsupials, the fossil record suggests that they probably either originated in North America and migrated into South America in Cretaceous time or vice versa. From South America, apparently they migrated via Antarctica to Australia. Exactly

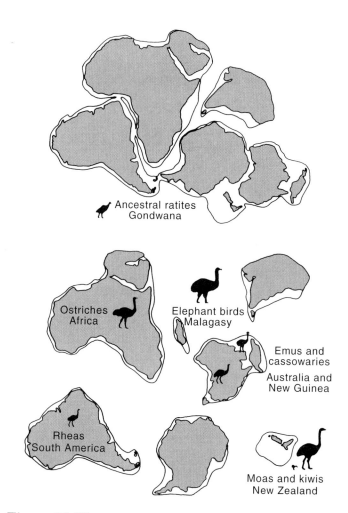

Figure 15.50   Evolution of the flightless birds, or ratites. Ratites are the most archaic group of living birds, and apparently they evolved in the Cretaceous before the Gondwana continents had completely separated. As a result, each of the southern continents has its own species of flightless birds (the ostrich in Africa, the rhea in South America, the emu and cassowary in Australia, and the kiwi in New Zealand) or had a ratite in the Pleistocene (the elephant bird of Madagascar). Only Antarctica (which is now glaciated) and India (which docked with Asia in the Eocene, so it was less isolated from the Northern Hemisphere than the other southern continents) lack ratites. (From B. McCulloch, 1982, No Moa, Canterbury Museum, Christchurch.)

when they arrived "down under" is not clear, since there is a gap in the Australian fossil record from the Triassic to the Eocene. Presumably their arrival occurred before the end of Cretaceous time, when these two continents separated from one another.

The absence of any marsupials from New Zealand suggests that that island had already separated from Australia before their arrival. The lack of snakes in New Zealand may also indicate isolation of the island before snakes evolved in the latest Cretaceous. New Zealand, together with New Caledonia and Norfolk Island (between New Zealand and Australia), illustrates a commonly

Figure 15.51   The Great American Interchange. For most of the Cenozoic, South America was an Island continent, with its own distinctive assemblage of mammals that had evolved in Isolation from the rest of the world. These included a number of carnivorous marsupials (including one remarkably convergent on saber-toothed cats—see Fig. 3.16), edentates (anteaters, sloths, and armadillos), and an amazing spectrum of native hoofed mammals that converged on horses, elephants, camels, hippos, and other northern hoofed mammals. In the Oligocene, the native fauna was invaded by rodents and monkeys, presumably rafted from Africa. During the mid-Pliocene, about 3.5 m.y. ago, the Panamanian land

observed situation whereby the most (or longest) isolated biotas tend to be the most primitive and least diverse.

South America was an island continent throughout much of the Cenozoic, and the same kind of ecological convergence with Northern Hemisphere mammals for Australia occurred in South America. This continent was colonized not only by marsupials (which evolved into the dominant predators, including equivalents of wolves, hyenas, and even sabertooths—Fig. 3.16) but also by placental mammals. These placentals probably were isolated in South America during the Cretaceous and evolved in complete isolation from Northern Hemisphere placentals (Fig. 15.51). One group of South American placental natives was the *edentates,* which evolved into the anteaters, sloths, and armadillos. One of the first placental mammal orders to diverge (Fig. 15.44), edentates evolved into a variety of huge ground sloths and armored glyptodonts. The other orders were uniquely South American hoofed mammals, distantly related to archaic hoofed mammals of the Paleocene of North America. These animals evolved some amazing parallels with Northern Hemisphere hoofed mammals, including South American equivalents of mastodonts, hippos, giraffes, camels, antelopes, and even a horse-like form that was more completely one-toed than the modern horse.

During the Oligocene, two immigrants reached South America from the outside world: the caviomorph rodents (porcupines, chinchillas, guinea pigs, and their relatives) and the New World monkeys. The source of these immigrants is controversial. Some researchers argue that both groups island-hopped from North America, and others think both groups were rafted from Africa; the latest research suggest that they have their nearest relatives in

Africa. Except for these chance immigrants, South America remained isolated until the middle Pliocene. Then the *Panamanian land bridge* rose up and reconnected South and North America (Fig. 15.51).

The consequences of this new land bridge were dramatic. Because North America had few tropical habitats in the Pliocene, few South American natives migrated much farther north than Mexico. The ground sloths and armadillos were about the only successful immigrants to go north of the U.S./ Mexican border. On the other hand, North American mammals (including mammoths and mastodonts, cats and dogs, horses and camels, tapirs and peccaries, and many others) migrated south with great success. This **Great American Interchange** led to competition between the two different faunas, and most South American natives lost out. Some South American opossums and edentates survived, but the native hoofed mammals were driven to extinction by horses and camels (which became llamas and alpacas) and by such North American predators as dogs, sabertooths, and other cats (such as the jaguar and ocelot).

The early isolation of Australia and South America provides two examples of natural experiments in evolution, with extraordinary ecological parallelism between natives of the two continents and Northern Hemisphere mammals. In the case of South America, nature ended the experiment when the Panamanian land bridge arose in the Pliocene. In Australia, however, the native marsupials were unthreatened until humans (and their dogs, cats, sheep, rabbits, and rats) invaded and changed the habitat forever. Tragically, these native mammals, so successful for tens of millions of years, may become extinct after only a few centuries of human intervention.

bridge rose and broke the isolation. Although some South American natives (ground sloths, armadillos, and porcupines) successfully invaded the north, many more North American mammals (mastodonts, llamas, horses, tapirs, peccaries, dogs, cats, bears, and deer, among others) went south and displaced the South American natives. By the Pleistocene, the last of the unique South American hoofed mammals and large carnivorous marsupials were gone, so only edentates and small marsupials still remain of their original native fauna. *(Modified from L. G. Marshall, 1988, American Scientist, v. 76, pp. 380–381.)*

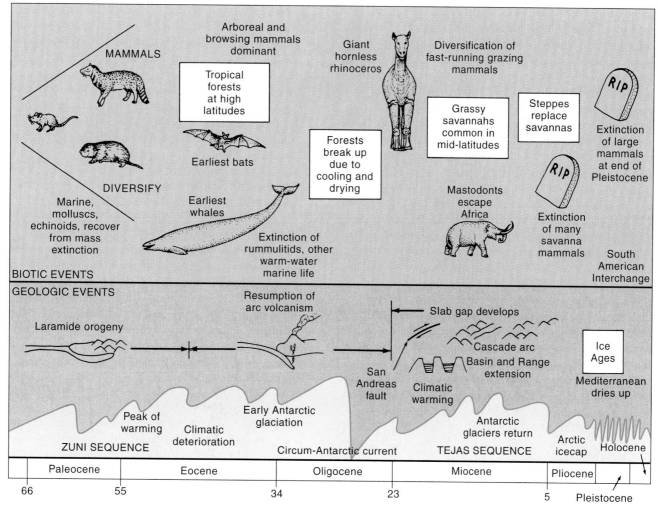

Figure 15.52   Time scale and summary of Cenozoic events.

Summary

Cenozoic History

- The Cordilleran mountain belt in western North America went through continual tectonism during the Cenozoic.

  1. From the latest Cretaceous to middle Eocene, the Rocky Mountains were formed by flexure and faulting of Cryptozoic, Paleozoic, and Mesozoic basement rocks during the Laramide orogeny. Intermontane basins were filled with river and lake deposits, including the famous Eocene Green River oil shale. The Sierran arc ceased erupting. Such peculiar tectonics might be explained if the plate subducting under the Pacific margin of North America underwent very shallow subduction, so that it no longer melted under the Sierran arc and, instead, transferred its stresses inland to the crustal basement of the Rocky Mountains.

  2. From the middle Eocene to early Miocene, normal arc volcanism returned to a region running from central Oregon through Nevada, Utah, Colorado, and New Mexico, east of the old Sierran arc. This suggests that the shallow-dipping subducted plate of the Laramide orogeny had returned to a steeper dip and produced a normal subduction zone. The Laramide basins in the Rockies were filled to the brim, and the isolated tops of the Rocky Mountain peaks were beveled off, forming a Rocky Mountain summit surface. Volcanic-rich sediments spilled out of the basins and eastward across the High Plains, depositing the famous fossiliferous sediments of the White River Group of the Big Badlands.

  3. From the early Miocene until the present, the Cordilleran region has been subjected to complex tectonic forces, resulting in a variety of unusual geologic phenomena. The Basin and Range Province of Utah-Nevada-Arizona–eastern California was formed when the crust was stretched to twice its original width, resulting in numerous north-south-trending block-faulted basins and ranges. At the same time, the Sierra Nevadas and Cascades swung west and clockwise, as a door on a hinge centered in the Olympic Peninsula of Washington. About 15 million years ago, massive flood basalts erupted from fissures in eastern Washington and Oregon, forming the Columbia River Plateau. Coastal California switched from a subduction zone to a transform margin as the San Andreas fault formed between the northwest-spreading Pacific Plate and the North American Plate. Most rocks of the California coast were transported north from Mexico, and some crustal blocks in the Transverse Ranges rotated clockwise, like ball bearings trapped between two sliding surfaces. About 5 million years ago, Baja California ripped away from mainland Mexico and began moving north, creating the Gulf of California. At the same time, the Rocky Mountains and Colorado Plateau began to rise up thousands of meters, causing deep canyon cutting and superposed drainages. The cause of all these complex events in the Miocene and Pliocene is hotly debated. It is almost certainly related to the fact that no Pacific Plate was being subducted, allowing mantle in the region east of the San Andreas transform fault to rise up directly beneath the thin crust of the Basin and Range.

- The Pacific Ocean basin also experienced major changes from Paleogene to Neogene time. The distribution of equatorial sediments and the track of the Hawaiian mantle plume indicate a change of Pacific Plate motion about the same time that California collided with the ridge.

  1. Most modern western Pacific island arcs either were born or were modified in mid-Cenozoic time, as was the Cascade volcanic arc of western North America.

  2. The modern configuration of the Central America–Caribbean region was achieved during the Neogene, and the Drake Sea was formed by the separation of South America from Antarctica by another continent-ridge collision. The full development of the Circum-Antarctic current resulted.

- Eurasian tectonics also is highlighted by major late Cenozoic events.

  1. The Alpine-Himalayan orogeny was caused by collisions of Africa and Turkey with Europe as well as Arabia, India, and Thailand with Asia. Very complex folded structures, including recumbent nappes, resulted, and southeastern Asia was squeezed eastward toward the Pacific along several large transcurrent fault zones by the force of India's collision.

  2. Australia is currently on a collision course with the Indonesian arc and, ultimately, with southeastern Asia.

  3. The Tethyan seaway disappeared as these collisions occurred. The Mediterranean Sea, a partial remnant of Tethys, is still being closed. Blocking of the Strait of Gibraltar in late Miocene time led to repeated evaporation of the Mediterranean, followed by Pliocene refilling.

  4. Cratonic Eurasia, Africa, and Antarctica all experienced important rift faulting and volcanism as a by-product of Neogene global lithosphere plate reorganization. Most important was the separation of Africa from Arabia along the Red Sea–Aden rifts.

- Passive continental margin coastal-plain and continental-shelf strata succeeded many Paleozoic active-margin structures. The Atlantic Coast Province of North America also experienced a slight rejuvenation of the Appalachian Mountains by upwarping and river downcutting. The Gulf of Mexico Province possesses the thickest sequence of any passive margin and is highlighted by abundant salt domes, whose upward intrusion produced important petroleum traps.

Cenozoic Climate and Life

- During the Paleocene and Eocene, the warm greenhouse climates of the Cretaceous persisted. The early Eocene was marked by unusually warm conditions, so that crocodiles and temperate-zone plants lived near the Arctic Circle.
- In the marine realm, most of the dominant Cretaceous groups—clams, snails, crustaceans, sea urchins, bony

fish—returned to populate the Cenozoic seas; many families and genera of these groups are still alive today. Only ammonites and marine reptiles were missing. By the Eocene, whales had evolved from land-dwelling hoofed mammals to become the top predator in the oceans.

- After the mass extinction at the end of the Cretaceous, the surviving fauna of mammals, birds, and insects underwent an adaptive radiation and came to dominate the terrestrial realm. By the early Eocene, most of the major groups of mammals living today, from bats to whales to carnivores to horses, had evolved.

- From the middle Eocene to the middle Oligocene, the world underwent climatic deterioration, triggered by cooling and the first Antarctic glaciation. This refrigeration was apparently caused by rifting of Australia and South America from Antarctica, which allowed Circum-Antarctic circulation to trap cold air over the South Pole. Global cooling destroyed the warm tropical Eocene forests and caused the extinction of many mammals adapted to these forests, especially tree dwellers and primitive leaf eaters.

- The early Miocene was marked by slight warming, but by the middle Miocene, glaciers had returned to Antarctica permanently. In North America and Eurasia, cooling and drying led to widespread grassy savannas, populated by mammals adapted for eating grasses and running from predators. The end of the Miocene was marked by the Messinian event, when the Mediterranean dried up multiple times and then catastrophically refilled again and again. This further perturbed the climate and triggered the extinction of many savanna mammals. In the Pliocene, cold steppe climates prevailed in the Northern Hemisphere. By the mid-Pliocene, the Arctic ice cap had formed and the Ice Ages had begun.

- Australia, South America, and Antarctica were relatively isolated from the northern continents during the Cenozoic, and their unique mammals evolved to fill the ecological niches of Northern Hemisphere placentals—an outstanding example of convergent evolution. Most of South America's native mammals were driven to extinction in the mid-Pliocene when the Panamanian land bridge rose and allowed North American placentals to migrate south and compete with the native mammals. Australia's native marsupials are now becoming extinct as humans and domesticated placental mammals destroy their habitats.

Readings

Armentrout, J., et al., eds. 1979. Cenozoic paleogeography of the western United States. *SEPM Pacific Section Symposium Volume 3.*

Aubry, M.-P., S. G. Lucas and W. A. Berggren, eds. 1998. *Late Paleocene–Early Eocene Climatic and Biotic Events.* New York: Columbia University Press.

Berggren, W. A., and J. A. Van Couvering. 1974. *The Late Neogene.* Amsterdam: Elsevier.

Christiansen, R. L., and R. S. Yeats. 1992. Post-Laramide geology of the U.S. Cordilleran region. In *The Cordilleran Orogen: Conterminous U.S.* Edited by B. C. Burchfiel, P. W. Lipman, and M. L. Zoback. Boulder: Geological Society of America.

Curtis, B. F., ed. 1975. Cenozoic history of the southern Rocky Mountains. *Geological Society of America Memoir 144.*

Dickens, G. R., M. M. Castillo, and J. C. G. Walker. 1997. A blast of gas in the latest Paleocene. *Geology,* 25:259–62.

Dickinson, W. R., and W. S. Snyder. 1979. Geometry of subducted slabs related to the San Andreas transform. *Journal of Geology,* 887:609–27.

———, et al. 1988. Paleogeographic and paleotectonic setting of Laramide sedimentary basins in the central Rocky Mountain region. *Geological Society of America Bulletin,* 100:1023–39.

Eaton, G. P. 1982. The Basin and Range Province: Origin and tectonic significance. *Annual Reviews of Earth and Planetary Science,* 10:409–40.

Flores, R. M., and S. Kaplan, eds. 1985. *Cenozoic Paleogeography of the west central United States.* Special Publication of the Rocky Mountain Section, SEPM.

Frazier, W. J., and D. R. Schwimmer. 1987. *Regional Stratigraphy of North America.* New York: Plenum Press.

Galloway, W. E., D. G. Bebout, W. L. Fisher, J. B. Dunlap, Jr., R. Cabrera-Castro, J. E. Lugo-Rivera, and T. M. Scott. 1991. Cenozoic. In *The Gulf of Mexico Basin.* Edited by A. Salvador. Boulder: Geological Society of America.

Hamilton, W., and W. B. Myers, 1966. Cenozoic tectonics of the western United States. *Reviews of Geophysics,* 4:509–47.

Hsü, K. J. 1983. *The Mediterranean was a Desert.* Princeton, Princeton University Press.

———, ed. 1986. *Mesozoic and Cenozoic oceans.* American Geophysical Union Geodynamics Series 15.

Kennett, J. P. 1977. Cenozoic evolution of Antarctic glaciation, the Circum-Antarctic ocean, and their impact on global paleoceanography. *Journal of Geophysical Research,* 82:3843–60.

———. 1982. *Marine Geology.* Englewood Cliffs, N.J.: Prentice-Hall.

———, ed. 1985. *The Miocene Ocean.* Geological Society of America Memoir 153.

———, and L. D. Stott. 1991. Abrupt deep-sea warming, paleoceanographic changes and benthic extinctions at the end of the Palaeocene. *Nature,* 353:225–29.

Lemoine, M. 1978. *Geological atlas of Alpine Europe* New York: Elsevier.

Lipman, P. W., and A. F. Glazner, eds. 1991. Mid-Tertiary Cordilleran magmatism. *Journal of Geophysical Research,* 96:13193–735.

Luyendyk, B. P., M. J. Kamerling, and R. R. Terres. 1980. Geometric model for Neogene crustal rotations in southern California. *Geological Society of America Bulletin,* 91:211–17.

Magill, J., and A. V. Cox. 1980. Tectonic rotation of the Oregon western Cascades. *Oregon Department of Geology and Mineral Industries Special Paper,* 10:1–67.

McGowran, B. 1990. Fifty million years ago. *American Scientist,* 78:30–39.

McKenna, M. C. 1975. Toward a phylogenetic classification of the Mammalia. In *Phylogeny of the primates: a multidisciplinary approach.* Edited by W. P. Luckett and F. S. Szalay. New York: Plenum Press.

———. 1983. Holarctic landmass rearrangement, cosmic events, and Cenozoic terrestrial organisms. *Annals of the Missouri Botanical Garden,* 70:459–89.

Miller, K. G., R. G. Fairbanks, and G. S. Mountain. 1987. Tertiary oxygen isotope synthesis, sea level history, and continental margin erosion. *Paleoceanography,* 2:1–19.

———, J. D. Wright, and R. G. Fairbanks. 1991. Unlocking the Ice House: Oligocene-Miocene oxygen isotopes, eustasy, and margin erosion. *Journal of Geophysical Research,* 96:6829–48.

Molnar, P., and P. England. 1990. Late Cenozoic uplift of mountain ranges and global climate change: Chicken or egg? *Nature,* 346:29–34.

Novacek, M. J. 1992. Mammalian phylogeny: Shaking the tree. *Nature,* 356:121–25.

Pomerol, C. 1982. *The Cenozoic Era.* New York: Wiley.

Prothero, D. R. 1994. *The Eocene-Oligocene Transition: Paradise lost.* New York: Columbia University Press.

———. 1994. The chronological, paleogeographic, and paleoclimatic background to North American mammalian evolution. In *Tertiary Mammals of North America.* Edited by K. M. Scott, C. Janis, and L. Jacobs. Cambridge, UK: Cambridge University Press.

———, and W. A. Berggren, eds. 1992. *Eocene-Oligocene Climatic and Biotic Evolution.* Princeton, N.J.: Princeton University Press.

———, L. C. Ivany, and E. A. Nesbitt, eds. 2003. *From Greenhouse to Icehouse: The Marine Eocene-Oligocene Transition.* New York: Columbia University Press.

Savage, D. E., and D. E. Russell. 1983. *Mammalian paleofaunas of the world.* Reading, Mass. Addison-Wesley.

Savage, R. J. G., and M. R. Long. 1986. *Mammal evolution, an illustrated guide.* New York: Facts-on-File.

Scott, K. M., C. Janis, and L. Jacobs, eds. 1994. *Tertiary mammals of North America.* Cambridge, U.K.: Cambridge University Press.

Severinghaus, J., and T. Atwater. 1990. Cenozoic geometry and thermal state of the subducting slabs beneath western North America. *Geological Society of America Memoir,* 176:1–22.

Smith, R. B., and G. P. Eaton, eds. 1978. *Cenozoic tectonics and regional geophysics of the western Cordillera.* Geological Society of America Memoir 152.

Snyder, W. S., W. R. Dickinson, and M. L. Silberman. 1976. Tectonic implications of space-time patterns of Cenozoic magmatism in the western United States. *Earth and Planetary Sciences Letters,* 32:91–106.

Spencer, A. M., ed. 1974. *Mesozoic-Cenozoic orogenic belts.* The Geological Society Special Publication 4.

Stehli, F. G., and S. D. Webb, eds. 1985. *The great American biotic interchange.* New York: Plenum Press.

Van Andel, T. J., et al. 1975. *Cenozoic history and paleoceanography of the central equatorial Pacific.* Geological Society of America Memoir 143.

Voorhies, M. R. 1981. Ancient ashfall creates Pompeii of prehistoric animals. *National Geographic,* 66–75.

Webb, S. D. 1977. A history of savanna vertebrates in the New World. Part I: North America. *Annual Reviews of Ecology and Systematics,* 8:355–80.

———. 1978. A history of savanna vertebrates in the New World. Part II: South America and the Great Interchange. *Annual Reviews of Ecology and Systematics,* 9:393–426.

———. 1983. The rise and fall of the Late Miocene ungulate fauna in North America. In *Coevolution.* Edited by M. D. Nitecki. Chicago: Chicago University Press.

Wolfe, J. A. 1978. A paleobotanical interpretation of Tertiary climates in the Northern Hemisphere. *American Scientist,* 66:694–703.

Woodburne, M. O., ed. 1987. *Cenozoic mammals of North America: Geochronology and biostratigraphy.* Berkeley, University of California Press, Berkeley, Calif.

(Courtesy of Peter Kresan.)

Chapter 16

Pleistocene Glaciation and the Advent of Humanity

The more it snows

(Tiddely pom),

The more it goes

(Tiddely pom),

The more it goes

(Tiddely pom),

On snowing.

A. A. Milne, *The House at Pooh Corner* (1928)
(By permission of E. P. Dutton and Co. Inc., and Methuen and Co., Ltd., 1928)

▶ The Pleistocene was characterized by multiple episodes of glaciation, which caused ice to advance and retreat approximately every 100,000 years beginning about 2 million years ago. These rapid glacial-interglacial cycles caused dramatic climatic changes, worldwide sea-level changes, and fluctuations of lakes and vegetation zones on land.

▶ The Pleistocene was the culmination of a long-term Cenozoic cooling trend, and eventually the earth cooled down below a critical threshold. Once this limit was exceeded, subtle changes in the earth's orbital motions, which ever so slightly vary the amount of solar radiation received on the earth, caused glacial-interglacial cycles.

▶ The climatic fluctuations produced a distinctive, cold-adapted vegetation, including steppes and tundra, in regions that are far south of today's glaciers. Many distinctive types of mammals, including woolly mammoths, mastodons, woolly rhinos, giant bison, ground sloths, and saber-toothed cats, were typical of the Pleistocene. Most of these large mammals died out completely at the end of the last Ice Age due to climatic changes and/or human hunting.

▶ The first members of our hominid family, *Sahelanthropus tchadensis* appeared in East Africa about 7 million years ago. For about 6–7 million years after this, there were multiple lineages of hominids, including the oldest species of our genus, *Homo,* which first appeared about 1.8 million years ago. The first "modern" humans, *Homo sapiens,* appeared about 90,000 years ago in Africa but did not arrive in Europe until they displaced the Neanderthals about 40,000 years ago.

The Pleistocene Epoch of the Neogene Period represents a truly unique interval in geologic history. Not only did it encompass the Great Ice Age, but also during its brief 2-million-year span, humans evolved both biologically and culturally to take their preeminent position on the ladder of life. Because of its recency, the record of the Pleistocene can be read in fascinating detail. The presence of humans makes Pleistocene study the most interdisciplinary facet of geology. Anthropologists, climatologists, botanists, zoologists, and even biblical scholars are very much concerned with latest Cenozoic history.

We shall begin this chapter by summarizing briefly the initial recognition, around 1830, that glaciers once had covered a large part of northern Europe. Then we shall review some of the very diverse effects of continental-scale glaciation, from warping of the crust to changes in both lake and sea levels. This review emphasizes the truly profound impact of Pleistocene climate upon practically every corner of the earth. Next we shall review the methods geologists use to date Pleistocene events and to make correlations between glaciated and nonglaciated land areas and between both types of land areas and the marine realm.

Then we shall review several hypotheses proposed to explain occasional continental-scale glaciation during earth history. This discussion will build upon the fundamentals of paleoclimate discussed earlier. We shall conclude the chapter with a discussion of the profound effects of Pleistocene climate upon life; special attention will be given to the evolution of hominids, which led to the development of our human species during the Pleistocene Epoch.

What Is the Pleistocene?

Long ago, Lyell suggested a biologic definition of the Pliocene-Pleistocene boundary in Italy: Pleistocene marine strata contain only species still living, whereas Pliocene strata contain a few extinct forms. Problems arise, however, in trying to extrapolate this definition from shallow-marine deposits (the type used to formulate the definition) to deep-marine and nonmarine successions. Today, therefore, we rely on isotopic methods. Isotopic dating places this boundary about 1.6 to 1.8 million years ago. At the other end of the interval, the Pleistocene-Holocene boundary is best defined as the end of the last rapid rise of sea level, which affected all parts of the earth between 6,000 and 8,000 years ago.

In most minds, the Pleistocene is synonymous with glaciation, but that turns out to be an oversimplification. It is obvious that, solely on the basis of glaciation, polar and alpine parts of the earth would be classed as still in the Pleistocene today (Fig. 16.1), whereas mid-latitudes emerged from the ice age between 10,000 and 11,000 years ago. Similarly ice caps began forming in polar areas at least as early as Oligocene time. Thus, neither beginning nor end of glaciation provides a wholly satisfactory defining time limit. This is why we have chosen to define the limits as specified above.

Figure 16.1 A landscape such as this dominated much of the world during the peaks of glaciation in the Pleistocene. Huge ice caps covered much of North America, Eurasia, the high Himalayas, the Rocky Mountains, the Andes, and the tip of South America. This scene, from Kenai Fjord in Alaska, gives the proper impression of the thick ice sheets that once influenced the whole world. *(Courtesy of Peter Kresan.)*

Recognition of Continental-Scale Glaciation—A Triumph of Reasoning

Erratic boulders—large boulders lying helter-skelter on the landscape of northern Europe (Fig. 16.2) and composed of rock types foreign to their surroundings—puzzled people for generations. As noted in Chap. 4, the most popular explanation until the mid-nineteenth century was to attribute them to the biblical Flood. Mariners had seen boulders frozen in icebergs drifting from the Arctic across the North Atlantic shipping lanes. It was ingenious reasoning to connect these two things and explain the erratics as having been carried in icebergs that drifted across Europe during the Flood. This *diluvial hypothesis* led to the still-common general term *drift* for the unsorted, boulder-bearing clay deposits that mantle much of northern Europe in association with the large erratics. Today these deposits are called *till*. In northern Scotland, a famous set of conspicuous parallel benches along a valley, the Parallel Roads of Glen Roy, also was interpreted as having been formed by the Deluge pouring down a narrow defile; in fact, they are glacial lake beaches.

It is interesting to question whether widespread continental glaciation would ever have been thought of if no glaciers existed today for comparison. James Hutton in 1795 was among the first to publish a suggestion that Alpine glaciers in the past had been much more extensive than today. He made this outlandish suggestion after reading of erratic boulders on hilltops in the Swiss Plain near Geneva and on the slopes of the Jura Mountains to the north of that city, both localities more than 50 kilometers from the nearest modern glacier. This was a lucky bit of

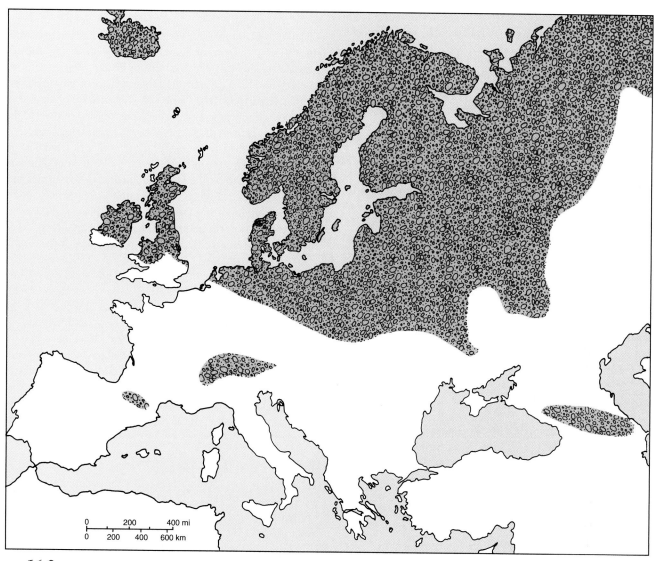

Figure 16.2   Distribution of the drift or Diluvium containing erratic boulders associated with unsorted "boulder clay" over Europe and long considered to have been drifted in by icebergs during the biblical Flood. After 1840, it was attributed to glacial transport. (*Adapted from L.J.Wills, 1951,* A Palaeogeographical atlas; *by permission of Blackie and Son Ltd.*)

Huttonian intuition, but it is a curious irony that he never recognized the abundant evidence of glaciation in his own Scottish homeland.

J. Esmark, working in Norway in 1824, was the first to recognize that a huge, now vanished ice sheet had once covered most of the European continent, although great German writer Goethe apparently anticipated Esmark somewhat earlier. Esmark's suggestion was largely overlooked, and it was a Swiss engineer, Ignace Venetz-Sitten, who, beginning in 1821 and joined about 1830 by Jean de Charpentier, confirmed that Alpine glaciers once extended out of the mountains onto the Swiss Plain, leaving moraines and erratic boulders far north of present glaciers. A prominent Swiss paleontologist, Louis

Agassiz, felt compelled to see what manner of heresy these men were promulgating. To his own surprise, he became convinced in 1836 that they were correct. In 1840, after gathering evidence of a widespread ice cover over nearly all of northern Europe, including Britain, Agassiz was the leading champion of continental glaciation.

The composition of the erratic boulders of northern Europe proved that most of the ice had not come from the Alps but, rather, had flowed south from Scandinavia, the only place where such rocks were known to occur. In 1846, Agassiz visited the United States and was elated to find evidence of similar continental glaciation in New England. He later joined the faculty of Harvard University.

By mid-century, other geologists were becoming interested in continental glaciation. Patterns of ice flow were being recognized:

> The rock scorings are the trails left by the invader. Their character should reveal the nature of the icy visitant as tracks reveal the track maker. (T. C. Chamberlin, 1886)

In the midwestern United States, as well as in Scotland, several levels of glacial till superimposed upon one another and separated by obvious buried soils containing plant and animal remains were differentiated. This layering indicated that there had been not one but multiple advances and retreats of the ice.

It had already been suggested in 1842 that a major, worldwide lowering of sea level must have accompanied Agassiz's past expansion of glaciers. With the recognition of multiple advances, it became apparent that sea level must have fluctuated several times. Soon geologists also recognized that the huge masses of ice must have depressed the crust isostatically. The final triumph in the recognition of continental glaciation was the demonstration in the 1860s that a much more ancient glacial episode had occurred in India. Soon this late Paleozoic Gondwana glaciation was also recognized in South Africa and Australia.

Diverse Effects of Glaciation

General

With nearly one-third of the present land area of the globe covered by roughly 43 million cubic kilometers of ice during maximum Pleistocene glaciation (Figs. 16.3 and 16.4), it is not difficult to imagine that profound effects resulted. No place—not even the deep seas—escaped at least some influence of Pleistocene climate (Figs. 16.4 and 16.5); however, paleobotanical studies show that the tropics were least affected.

Direct influences of glacial ice—its great erosive power, deposition of erratic boulders and moraines, and the like—are well known, and so we shall not dwell upon them. Other phenomena deserve special mention, however.

Isostasy

Depression of the crust by ice caps 3,000 meters thick and the subsequent crustal rebound upon their melting (Fig. 16.6) provide convincing proof of the general validity of the principle of isostasy and of the ability of mantle material to flow plastically as the crust is warped epeirogenically (see Chap. 7). Evidence of postglacial rebound consists primarily of raised ocean and lakeshore features such as beaches. Around the Great Lakes and the northern Baltic Sea, such raised features are not level but, rather, are tilted southward. This tilt reflects thicker ice at the northern end of these areas, thus greater crustal depression there followed by greater rebound.

Rebound continues today, indicating that it is a slow process by human standards, spanning at least 10,000 years. If Greenland and Antarctica were to be deglaciated, they, too, would rise. Figure 1.9 shows the differential rise of the Baltic region.

Lakes

Glaciation had a marked effect upon drainage patterns and produced many temporary lakes. Preglacial drainage valleys were gouged by the ice, and the resulting large depressions subsequently filled with water; the Finger Lakes of New York are examples (Fig. 16.6). The Great Lakes basins are of similar origin and had a complex history during advances and retreats of the ice. Huge postglacial Lake Agassiz—four times as large as Lake Superior—formed temporarily in the southern Canadian Plains when the ice began to retreat (Fig. 16.6). With natural drainage to the north blocked by the ice, it drained at various times into the Mississippi River (Fig. 16.6), later into Lake Superior, and finally into Hudson Bay. Similarly the Great Lakes spilled over and drained in various directions as the ice retreated (Fig. 16.6). As lower outlets emerged from beneath the ice, former drainage routes were abandoned.

Immense terraces of coarse gravel and sand over 30 meters high line the upper Columbia River valley in western Canada, attesting to staggering volumes of meltwater flowing through river valleys to the Pacific. Large lakes formed at least five times in the northern Rocky Mountains. In western Montana, a large basin filled several times to form Lake Missoula, which was as much as 300 meters deep (Fig. 16.7). This lake drained several times as glaciers to the northeast enlarged and then shrank. The most spectacular draining occurred about 18,000 years ago, when the basin's ice-moraine dam in northern Idaho broke (Box 2.4). Water rushed across eastern Washington with incredible velocity. This gigantic flood scoured channels and deposited immense gravel ripples 10 meters high over a large part of the Columbia Plateau, now called channeled scablands because of the peculiar topography left by the flood. Evidence suggests that Lake Missoula rose and spilled again and again, causing a great series of floods. So much water rushed down the Columbia River valley that a large temporary lake was backed up into the tributary Willamette River valley above Portland in western Oregon. River-borne icebergs carried erratic boulders, which were deposited helter-skelter along the shores of this lake as well as along the Columbia valley.

In Utah, Lake Bonneville and a smaller lake on the Snake River overspilled their lava dams about 30,000 years ago to create an enormous flood down the Snake River (Fig. 16.7). In all these floods, flow velocities probably reached 80 kilometers per hour (50 miles per hour); huge boulders were tossed about as if they were tiny sand grains. Such floods lasted only days or a few weeks at most. Wherever great volumes of water were released, canyons were deepened rapidly, and some were abandoned after the ice disappeared (Fig. 16.8).

During much of Pleistocene time, accumulated water in low depressions also formed large lakes far from continental glaciers. Most notable were those of the Great Basin region of the western United States (Fig. 16.7). Today, under a warmer and more arid climate, only small, saline bodies, such as Great Salt Lake, Pyramid Lake, and playa or alkali flats, remain as remnants of once much larger and far more numerous lakes (Fig. 16.9). These early lakes were formed in closed basins between the youthful block-faulted ranges of the Utah-Nevada-California region. The largest were glacial Lake Bonneville in Utah and Lake Lahontan in western Nevada, each of which covered up to 50,000 square kilometers and was as much as 300 meters deep. Lake levels fluctuated enormously several times as climate oscillated (Fig. 16.9). Such playas today are important sources of rare elements, such as boron and lithium.

Wind Effects

With much meltwater pouring from wasting glaciers near the end of glacial episodes, tremendous quantities of sediment were dumped in front of receding ice fronts. Rivers became so sediment-choked that they formed mosaics of braided channels. Wind then worked over the sediment bars along such channels, winnowing out fine sand and silt. Huge dune fields were formed, as in the Sand Hills of Nebraska, and large areas of silt called loess accumulated near major river valleys (Fig. 16.6). As the ice front retreated farther, vegetation became reestablished and wind effects gradually decreased.

Worldwide Changes in Sea Level

Worldwide fluctuations in sea level in response to glacial oscillations was one of the most profound phenomena of the Pleistocene. Submerged beach ridges far out on continental shelves indicate a maximum reduction of sea level on the order of 100 to 140 meters. Elephant teeth 25,000 years old have been dredged by fishermen from more than forty sites as far as 130 kilometers off the Atlantic Coast and in water as much as 120 meters deep. On the other hand, dead coral reefs on several oceanic islands indicate sea levels much higher than now. Melting of all existing glaciers, especially those of Antarctica, which today contain 90 percent of the globe's ice, would raise sea level about 65 meters above present level; this would flood nearly 15 percent of the existing land area!

Major drops in sea level occurred *at least* four times, with the greatest about 40,000 years ago. Short-term oscillations were superimposed upon more subtle, long-term worldwide Cenozoic regression of the sea. It is interesting to note that the last rise (Fig. 16.10) produced universal transgression, in opposition to the regressive tendency caused by isostatic rebound along many high-latitude coasts. Hudson Bay, most of which lies less than 600 meters below sea level today, was drowned by rapid rise, but it appears destined to become largely land as crustal rebound continues (see Fig. 16.6). This is an important example of the interaction of local rebound versus worldwide (sea-level) effects operating on different time scales.

Sea-level changes have widespread effects, from well inland to the deep seas. Reduction of sea level lowers the effective base level of major rivers; in other words, it steepens their gradient. This steepening causes downcutting, which produces river terraces that correlate with former marine beaches and deltas (Fig. 16.11). In coastal regions, vast expanses of former shallow seas were laid bare, in several cases forming important land bridges for migration of organisms between adjacent con-

tinents or islands. Finally, lowering of sea level made possible a greater than normal delivery of relatively coarse clastic material to the deep seas. The total rate of sedimentation in large portions of the ocean basins must have accelerated greatly. The greatest drops in sea level carried the shoreline out to continental-shelf margins (Figs. 16.3 and 16.4), so that rivers swollen many times during early melting of the ice could carry vast quantities of sand to the very edge of the shelf. Ordinarily most such sediment would be trapped on the inner shelves, but with little or no submerged shelf, much sand could escape to the deep seas. Gushing, muddy meltwater torrents stimulated the cutting of the submarine canyons that nick most continental shelves, though the deeper ends of the canyons extend far below even the lowest glacial sea level. Below that level, they must have been cut by submarine currents and mudflows. Once cut, the canyons acted as funnels through. which turbidity currents could continuously export sediment from the shelves onto abyssal plains.

Pleistocene Chronology

As glacial features in North America and Europe were mapped during the early twentieth century, the various glacial and interglacial deposits were differentiated and named. In central North America, four major glacial advances were recognized, and in Europe, from three to five advances, depending upon location. Apparently each major glacial episode was made up of lesser oscillations of the ice. Details are, of course, much clearer for the last, or Wisconsinan, glaciation, than for earlier episodes. Full understanding of Pleistocene history requires correlations of deposits, first among glaciated areas and then with unglaciated ones. The same principles and problems encountered in correlation and interpretation of more ancient strata apply.

Perfection of carbon-14 isotopic dating methods around 1950 provided a breakthrough for late Pleistocene chronology. Deposits from widely separated localities containing wood, charcoal, bone, or calcareous shells could be dated, and correlation was thus enhanced. Even carbon-bearing deep-sea sediments could be correlated with the glacial standard time scale established on the continents. It became possible to compare glacial conditions on the continents with conditions in the oceans.

At the same time, ^{14}C dating became very important to archaeologists studying the past 80,000 years of human history. Efforts are continuously underway to extend isotopic dating back further in time from 80,000 to at least 100,000 years, at which age the K-Ar method becomes applicable. Protactinium 231–thorium 230 dating is being used for deep-sea clays back to 300,000 years ago. Fission-track dating is also used (see Chap. 5), as is dating by the degree of alteration of the volcanic glass obsidian, which is not an isotopic method. Obsidian alteration dating is new but offers promise for glass fragments in volcanic ash and in certain human artifacts. Dating by comparing magnetic polarity reversals in Pleistocene deposits is also used.

Figure 16.3   Map of global climate during the last glacial maximum 20,000 years ago. Note how far each of the ice sheets extended toward the equator, both on land and in the ocean. This changed the oceanic circulation patterns and caused all the other temperate climatic belts to shift to lower latitudes. The tropical belt did not cool, but it did shrink in size. The effect on climate was surprising. Note how the Sahara region and the western United States, which are now deserts, were once wet and fertile. Expanded glaciers also locked up so much seawater that much of the continental shelf was exposed, enlarging the coastal plains of many regions. Important migration of humans and cultural breakthroughs are also shown. *(Modified from N. Calder, 1983,* Timescale, *Viking, New York.)*

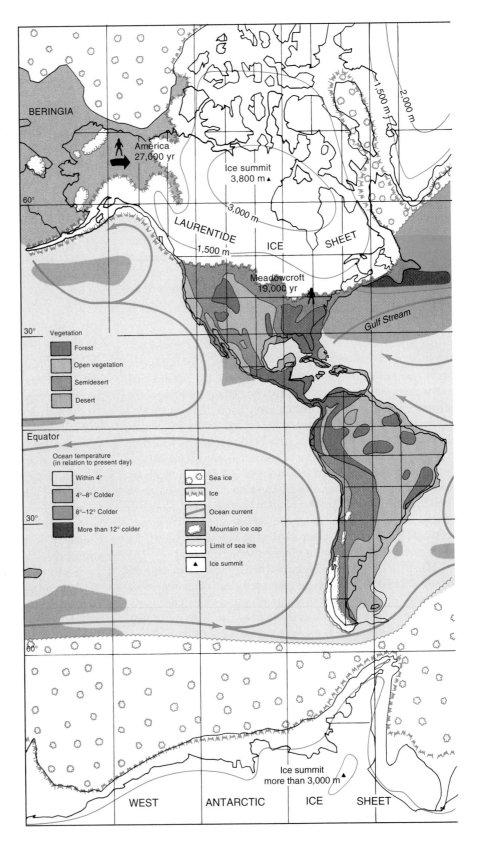

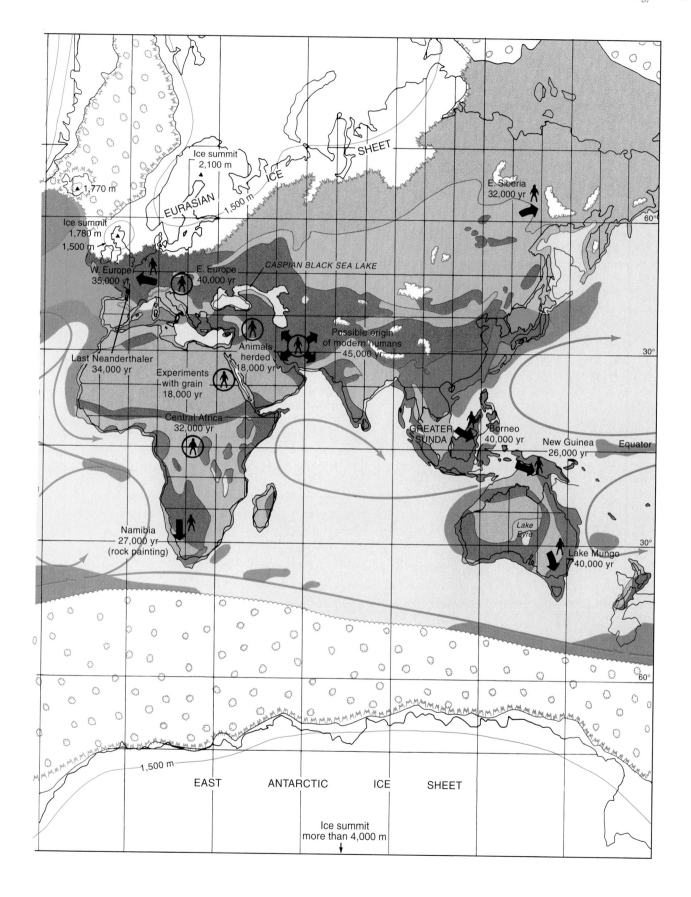

Ice summit
2,100 m

EURASIAN   ICE   SHEET

1,500 m

▲ 1,770 m

E. Siberia
32,000 yr

Ice summit
1,780 m
1,500 m

W. Europe
35,000 yr

E. Europe
40,000 yr

CASPIAN BLACK SEA LAKE

Last Neanderthaler
34,000 yr

Experiments
with grain
18,000 yr

Animals
herded
18,000 yr

Possible origin
of modern humans
45,000 yr

Central Africa
32,000 yr

GREATER
SUNDA

Borneo
40,000 yr

New Guinea
26,000 yr

Equator

Namibia
27,000 yr
(rock painting)

Lake
Eyre

Lake Mungo
40,000 yr

60°

30°

30°

60°

EAST   ANTARCTIC   ICE   SHEET

1,500 m

Ice summit
more than 4,000 m
↓

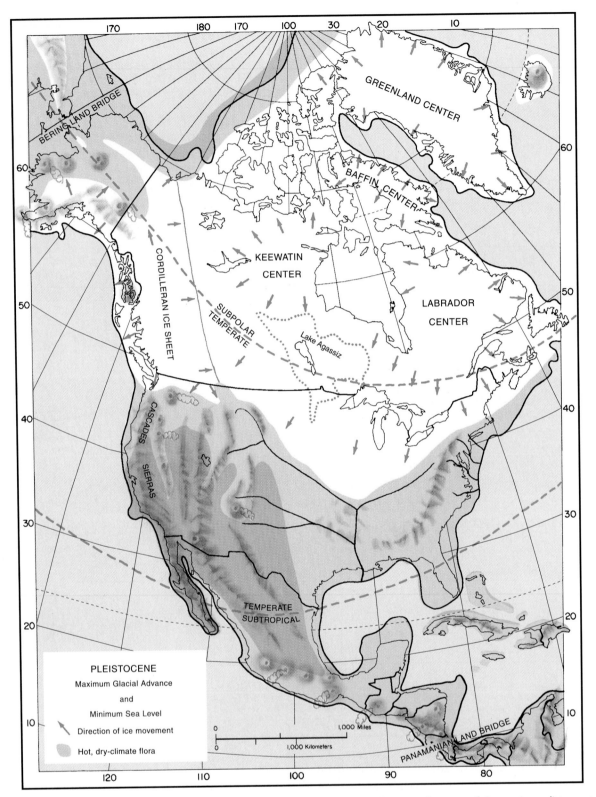

Figure 16.4   Pleistocene paleogeography of North America showing maximum ice advance and retreat of the sea, immediate postglacial Lake Agassiz, Cascade and other volcanoes, and late Pleistocene desert floras in the southwest. Joining of Cordilleran ice caps with main Canadian Shield cap in western Canadian plains blocked migrations of organisms to and from Asia during glacier advances. Note Bering and Panamanian land bridges.

Figure 16.5 Iceberg-rafted cobbles as much as 10 cm in diameter dropped onto the sea floor 4,000 m below sea level between Antarctica and South America; such cobbles have been carried as far as 3,000 km from Antarctica. Similar ancient rafted pebbles also are known (e.g., Figs. 8.32 and 13.37). *(Official NSF photo, USNS Eltanin, Cruise 10; courtesy Smithsonian Oceanographic Sorting Center.)*

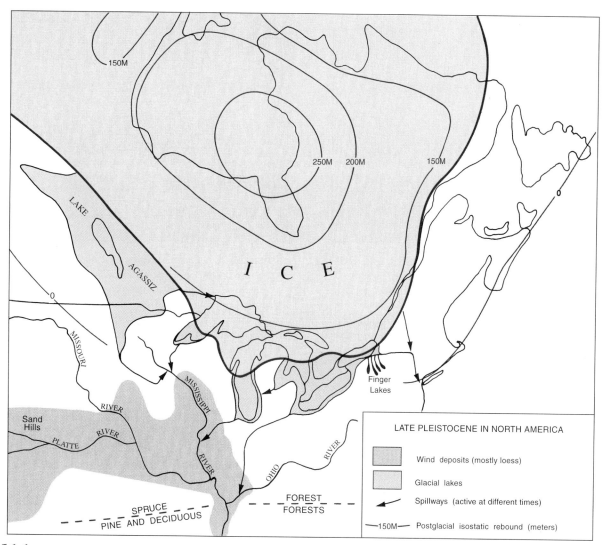

Figure 16.6 Some late Pleistocene phenomena associated with retreat of the Wisconsinan ice sheet (approximately 10,000 to 15,000 years ago). Ice-front lakes overflowed via temporary spillways, but these were not all active simultaneously. Forest zones are shown at the bottom; these shifted northward as ice retreated. Main areas of wind-blown silt (called *loess*) surround major glacial meltwater drainage systems. *(After Hough, 1958, Geology of the Great Lakes; King, in Wright and Frey, eds., 1965, The Quaternary of the United States; Whitehead, 1973: Quaternary Research, v. 3.)*

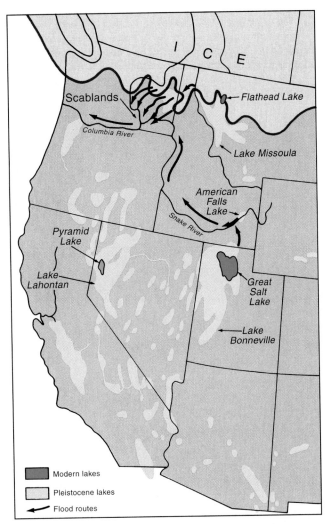

Figure 16.7   Pleistocene lakes and giant floods in western United States. Lake Missoula was dammed by ice (see Fig. 2.19), Lakes Bonneville and American Falls by lava flows, and the two westernmost ones by debris temporarily choking tributary valleys. All others resulted from drainage into closed valley in the Basin and Range region; they rose during colder episodes and fell during warmer ones, due mainly to changing evaporation rates. *(After Feth, 1961;* U.S. Geological Survey Professional Paper 424-B; *Weiss and Newman, 1973,* The channeled scablands of eastern Washington; *U.S. Geological Survey pamphlet.)*

Figure 16.8   Ouimet Canyon near Nipigon, Ontario, on the north shore of Lake Superior. Nearly 200 meters deep, this gorge was cut by meltwater and then completely abandoned after the deglaciation of southern Canada. It was one temporary outlet of glacial Lake Agassiz (see Fig. 16.4). (Person standing in upper left corner provides an idea of scale.) *(R. H. Dott, Jr.)*

Figure 16.9   Antelope Island, in the middle of the Great Salt Lake, Utah. The lake is a briny relict of much larger Pleistocene Lake Bonneville (Fig. 16.7). Notice how the island is notched by a series of abandoned shorelines, which mark higher levels of the lake during wet periods in the Pleistocene. White salt deposits cover the ground along the shoreline. Evaporation due to warmer, drier postglacial climate has produced many smaller saline lakes and playas (alkali flats) in the Basin and Range region. Some contain economically important salt and other alkali deposits. Bonneville salt flats is famous for another reason—the miles and miles of straight, flat desert are perfect for racing rocket cars and breaking the land speed record. *(© John S. Shelton.)*

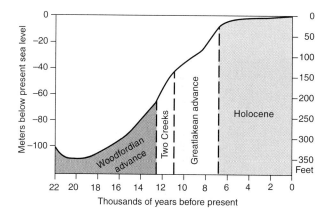

Figure 16.10   Curve representing the last rise of sea level to its present position in late Pleistocene time (compare Table 16.1).

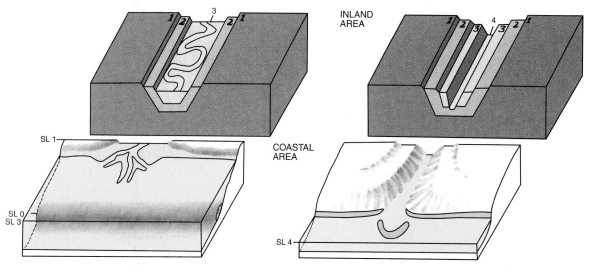

Figure 16.11   Effects of sea-level changes on river valleys and coastlines. Sea levels and river terraces are numbered from oldest to youngest *Left:* Initial drop of sea level (SL 0) caused cutting of main valley and formation of now submerged shore features; a rise, then a second fall, and a second rise also are recorded. The final rise (3) produced a branching delta, present beaches, and valley alluvium. *Right:* Recent drop in sea level (SL 4) caused downcutting into all previous valley alluvium and seaward migration of crescentic delta bar and left several high terrace levels reflecting earlier events. *(Adapted from* Scientific American, The Bering Strait land bridge, W. G. Haag, January, 1962; Copyright © 1962 by Scientific American, Inc.; used by permission.)

Even with numerical dating becoming increasingly available and more reliable, it is still necessary to correlate Pleistocene strata by other, more conventional means as well. Figure 16.12 shows some ways that correlations are accomplished between different types of Pleistocene sequences. In deep-marine sediments, temperature-induced reversal of coiling direction of foraminiferal shells (Fig. 16.13), together with **oxygen-isotope analyses,** provides a basis for constructing temperature fluctuation curves, which can be used for correlation together with ^{14}C dating of the same shells (Fig. 16.12). Changes in polarity of the earth's magnetic field leave an imprint on magnetically susceptible minerals in sediments. Using the standard time scale of magnetic-polarity-reversal events, these polarity changes can now be correlated from place to place (see Fig. 4.24).

Several persistent and obvious Pleistocene volcanic ash layers occur over western North America, and they provide invaluable time lines for correlation because they were deposited instantaneously. Some of the best known extend through the Great

Plains from Texas to North Dakota. Most of these widespread ash layers were derived either from Yellowstone or from the Cascade volcanoes. Careful geologic detective work has shown that the ashes are mineralogically distinctive enough so that their sources can be pinpointed accurately (Fig. 16.14). The final eruption of Mt. Mazama in Oregon, which created Crater Lake (Fig. 16.15), was witnessed by Native Americans, for its ash buried some sandals that have been dated isotopically as 6,700 years old.

Lavas interstratified with glacial tills may provide indirect means for dating glaciations. For example, K-Ar dates on lavas in the Sierra Nevada date an interstratified till as 3 million years old, probably the oldest known Cenozoic glacial deposit in middle latitudes.

Table 16.1 compares the classic stratigraphic classifications of Pleistocene records for central North America and Europe, as determined through the application of the dating procedures just discussed. The chronologies apply strictly only to the glaciated areas where they were first established. Not

Figure 16.12   Sequences of strata in four different types of regions showing different records of glacial and interglacial episodes in each as well as the criteria for correlating among them. Vertical time scale is the same for all four (A—glacial advance; R—glacial retreat; C—cold; W—warm; numbers represent ^{14}C dates in years before present).

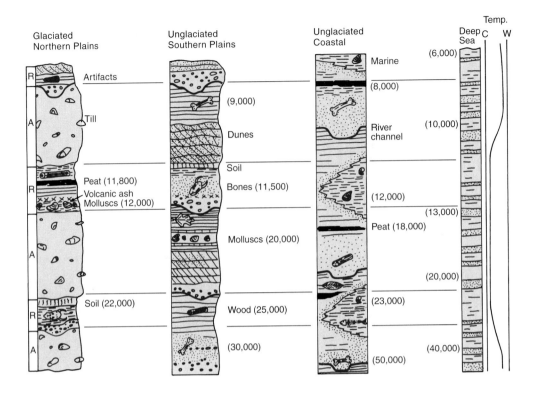

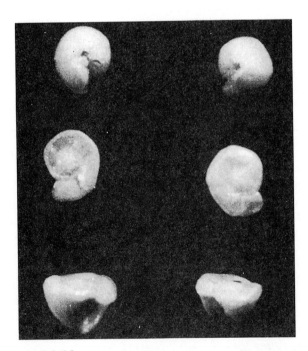

Figure 16.13   Contrast of coiling directions of microscopic shells of planktonic Foraminifera *(Globorotalia truncatulinoides)* from deep-sea Pleistocene sediments of the North Atlantic; because of some unknown physiological mechanism, right-hand coiling *(left)* occurs in waters warmer than 8 to 10°C, whereas left-hand coiling *(right)* characterizes cooler temperatures. *(After D. B. Ericson et al., 1954, Deep sea research; by permission of Pergamon Press.)*

many years ago, the Pleistocene Epoch was assumed to have lasted a bare 1 million years, but deep-sea cores and K-Ar dating now suggest that it was perhaps twice that long—that is, 1.5 to 2.0 million years. This revelation has required some drastic revision in subdivisions.

As the use of ^{14}C dating increased in the 1960s, it became apparent that the record for the Wisconsinan glacial episode was much more complete than for earlier ones. Perhaps this should not have been surprising, for repeated erosion both by ice and by glacial rivers—together with oxidation—would inevitably degrade the older deposits. But the degree of disparity between the two records was unexpected. In the 1970s, doubts increased about the earlier episodes (Illinoian, Kansan, Nebraskan). Did the deposits for these represent a complete record of glacial-interglacial events, as long assumed? Were those events of equal duration? In short, what are the numerical ages of those events? Widespread volcanic ash called Pearlette in the Plains region was found to include four or five different ash layers rather than a single one, as long believed (Fig. 16.14). This new information threw many correlations into doubt, and recent dating using fission tracks, obsidian alteration, and magnetic polarity reversals suggest that most of the earlier half of the Pleistocene Epoch may have no record whatsoever in the glaciated region of North America (Table 16.1). Therefore, the classic glacial sequence now must be questioned, for neither the numbers nor the exact numerical dates of pre-Wisconsinan events are clear.

The deep-marine sequences should provide a more complete record of alternations between colder and warmer intervals, and from that record it appears that there were many more than

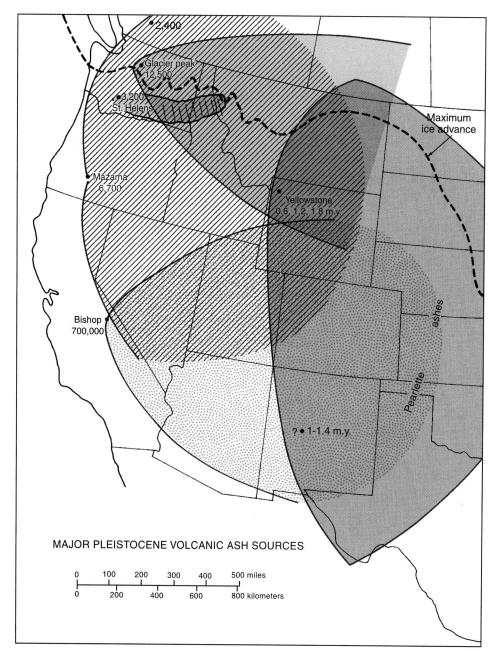

Figure 16.14   Major sources and fallout patterns of Pleistocene volcanic ash in western North America. Ash layers provide "markers" for correlation of Pleistocene deposits between widely scattered areas, especially from unglaciated into glaciated country. Glass and mineral fragments in the ash are datable isotopically. *(Adapted from Wilcox, in Wright and Frey, eds., 1965, The Quaternary of the United States.)*

the classic four or five Pleistocene mid-latitude glacial episodes (Fig. 16.16). For polar regions, we expect more, and the record verifies glaciation back to Oligocene time. But what about lower latitudes? What is now needed are reliable correlations between well-documented deep-sea sequences and the classic glacial sequences on land (compare Table 16.1 and Fig. 16.16).

Although progress has been made, it will be a while before the correlations can be fully accepted, for the marine record has its problems, too. It tends to be compressed because of the relatively slow rate of sedimentation in the deep seas, and this compression makes fine discrimination difficult. Also, the oxygen-isotope data are not straightforward. For example, the $^{18}O/^{16}O$ ratio is sensitive

to ocean salinity as well as to temperature. Therefore, some of the observed variation reflects changes in the oxygen-isotope composition caused by episodic influxes of fresh glacial meltwater and not just temperature. With care, it is possible to adjust for this effect.

Another problem is the selective dissolving of shells deposited on the deep-sea floor, where calcium carbonate is unstable. The shells most affected are those of animals that lived at shallower—and warmer—depths; therefore, loss by solution results in an error that makes the apparent temperature appear too cold. Finally, even if we can accept each marine oxygen-isotope cold interval as valid, we cannot yet be certain which ones resulted in significant glaciation (Fig. 16.16).

Table 16.1	Classic Pleistocene Stratigraphic Classification Standards for Glaciated North America* and Europe (Numbers Indicate Years before Present, Where Known)	

North America (Upper Mississippi Valley)	*(0–7,000 or 8,000 years ago)*	**Europe (Alps and Southern Baltic Regions)**
Holocene Epoch		*Holocene Epoch*

Pleistocene Epoch

North America		Europe
Wisconsin glacial	Greatlakean advance (8,000–11,000)	Younger Dryas advance (10,000–11,000)
	Twocreekan retreat (11,000–12,500)	*Allerod retreat (11,000–12,000)*
	Woodfordian advance (12,500–22,000)	Older Dryas advance (13,000–20,000)
	Farmdalian retreat (22,000–28,000)	*Riss-Würm retreat (20,000–30,000+)*
	Altonian advance (28,000–70,000+)	Riss advance (30,000–60,000+)

(Würm glacial, spanning the Wisconsin/Würm interval in the central column)

Sangamonian Interglacial		*Mindel-Riss Interglacial*
Illinoian glacial (ended ca. 130,000 years ago)		Mindel glacial
Yarmouthian interglacial		*Gunz-Mindel interglacial*
Kansan glacial		Gunz glacial
Altonian interglacial (0.6–0.7 m.y.)		*Donau-Gunz interglacial*
Nebraskan glacial		Donau glacial
Volcanic ash (2.2 m.y.)		

—————————— (1.6–1.8 m.y.) ——————————

Pliocene Epoch		*Pliocene Epoch*

Source: *After Blact et al., 1973; Boellstorff, 1978.*

*Common present-day usage in America: Wisconsinan, Sangamonian, and Illinoian are as shown, but the pre-Illinoian record is now considered too uncertain to subdivide formally.

Figure 16.15   Looking north across Crater Lake, Oregon, showing north-south alignment of Cascade volcanoes in the distance. Crater Lake formed when former Mt. Mazama suffered a cataclysmic eruption. Immense volumes of ash were blown hundreds of miles eastward by prevailing westerly winds (see Fig. 16.14). The top of the mountain then collapsed, forming a caldera, which filled with water. Ash from Mt. Mazama and other Cascade volcanoes provides important marker layers for correlation of late Pleistocene deposits over the western states. *(R. H. Dott, Jr.)*

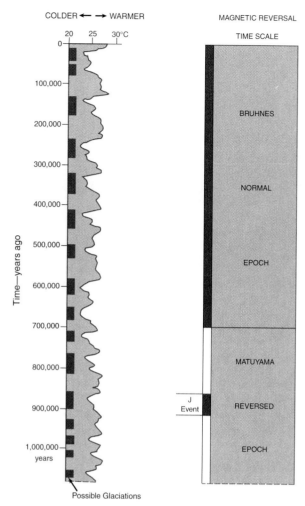

COLDER ← → WARMER

MAGNETIC REVERSAL
TIME SCALE

BRUHNES

NORMAL

EPOCH

MATUYAMA

J
Event

REVERSED

EPOCH

Possible Glaciations

Figure 16.16   Late Pleistocene standard marine paleo-temperature curve *(left)* based upon oxygen-isotope analyses of calcium carbonate in microfossil shells from deep-sea cores of three oceans. Magnetic polarity measurements on the same cores *(right)* and limited isotopic dating of cores provide a time scale. Note that, for the last 600,000 years, cold intervals had a periodicity of about 100,000 years; from then back to about 1.4 million years, the period was about 40,000 years (J—Jaramillo brief normal polarity event). *(Adapted from Emiliani and Shackleton, 1974: Science, v. 183, pp. 511–514; and Shackleton and Opdyke, 1976: Geological Society of America Memoir 145, pp. 449–464.)*

What Caused Glaciations?

Climatologic Background

By 1850 or so, the principal phenomena associated with Pleistocene glaciation had been recognized. Naturally speculation about the cause began, but even more than a century and a half later, there is still much doubt. Discussion of several hypotheses provides an excellent illustration of scientific analysis by the method of multiple working hypotheses. Before discussing specific hypotheses, however, it is important to state some important climatologic background.

First, *continental glaciations have been relatively rare events in earth history.* Besides the well-known late Cenozoic ice ages

(0 to 15 m.y. ago), there were widespread glaciations in the southern continents in late Paleozoic time (230 to 350 m.y. ago), in South America in Siluro-Devonian time (around 410 m.y. ago), and in North Africa during the Ordovician (about 440 m.y. ago), as well as the very widespread pre-Vendian ones (circa 700 m.y. and 2 b.y. ago) (see Figs. 8.33 and 8.34). The time period between the recognized great glacial events is irregular, varying from 50 to 1,000 million years, so it is difficult to claim that glaciation is a regularly cyclic phenomenon of the earth. Because of its rarity, glaciation must have required a very special combination of conditions.

A second important observation is that all available evidence, incomplete as it is, indicates that *the "normal" climate during the past 1 billion years was milder than that of the past 20 million years.* Present lower-latitude regions offer the closest modern analogies with typical climate of the past, but even they were affected by late Cenozoic cooling.

The third point, really a corollary to the second, is that *past climatic conditions typically were not only milder but also more uniform over the earth.* Although the overall mean temperature of the earth cannot have changed greatly throughout earth history, temperature and moisture gradients were less abrupt during much of the past, and climatic zones tended to be broader. Nonetheless, because of the geometry of the earth-sun system, polar areas must *always have been cooler* than lower latitudes, although it does not follow that there always was ice at the poles. Polar areas would be coldest when continents lay at or near them because land retains less heat than does water. Conversely poles would be relatively milder whenever open, well-mixed seas lay there. This important paleoclimatic principle was appreciated and discussed at length by Lyell over 100 years ago.

The fourth point to note is that the *magnitude of overall average yearly temperature differential between a normal and a glacial climate is less than 10°C.* That is to say, the earth enjoys a very critical temperature position in the solar system and has a sensitive thermal budget. It is estimated that a drop on the order of only 4 to 5°C from the present mean annual temperature could cause renewal of continental glaciation (the present must be considered as interglacial rather than truly normal, however). This prediction is confirmed by $^{18}O/^{16}O$ ratios and from coiling directions of Foraminifera from deep-sea cores (Figs. 16.13 and 16.16), which indicate that the average sea-surface temperature 18,000 years ago—near the peak of the last glacial advance—was only 2.3°C lower than now; maximum differentials were 6 to 10°C lower. Such modest differences suggest that the earth's heat budget is very sensitive, which in turn might lead one to wonder why glaciations have not been more common. The answer must be that a rather stable thermal equilibrium has characterized most of geologic time, and because the total heat budget has varied only within a narrow range, it follows that glaciation requires not wholesale change in kind but only in degree.

Late Cenozoic Climatic Deterioration

Long-known land-fossil evidence of cooling, alluded to in Chap. 15, now is supplemented with morphologic and oxygen-isotope paleotemperature data from calcareous shells in marine sediments. Figure 15.37 shows the general trend of marine temperatures for the past 70 million years. The striking shifts of plant communities in arid southwestern United States and Mexico dis-

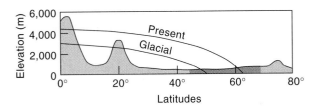

Figure 16.17 Elevation versus latitude of terminal moraines (lines) during maximum glaciation and at present, superimposed on an equator-to-pole topographic profile; gray band shows chief latitudinal span of major Pleistocene continental ice sheets.

Figure 16.18 Grand Teton Range and Snake River viewed across Jackson Hole valley. The Tetons were carved by glaciation out of Cryptozoic metamorphic rocks, which had been raised along a late Cenozoic normal fault, which forms the flat face of the range. Large glaciers once flowed onto the edge of the valley; only one puny relict glacier remains since the warming of climate began about 10,000 years ago. Prominent river terraces (the green benches in the valley) along the Snake River reflect the climatic oscillations (compare with Fig 16.9). *(R. H. Dott, Jr.)*

cussed in Chap. 15 continued into Pleistocene time. Whatever the cause, it is clear that a steady temperature decline occurred beginning 35 to 40 million years ago, until a *critical threshold* level was reached when ice caps began to form and grow.

Glaciation began at both poles in mid-Cenozoic time. Oligocene and Miocene fossil plants suggest cooling temperatures, and local Miocene tills are known near both poles; some ice-rafted pebbles appear in Miocene deep-sea sediments near Antarctica. D. L. Clark of Wisconsin has shown that the Arctic Ocean has been largely ice-covered for approximately the past 15 million years. It is inescapable that, in Pliocene time, polar and alpine ice caps were sizable, as evidenced by the prominent appearance of ice-rafted pebbles in Pliocene deep-sea sediments surrounding Antarctica (Fig. 16.5) and by the 3-million-year-old till in the Sierra Nevada.

The critical threshold necessary for glaciation was reached when, for a number of consecutive years, annual snow precipitation slightly exceeded summer wastage by melting and evaporation. This amounts to saying that, if the permanent snow line (above which not all snow melts in the summer) descends sufficiently in either elevation or latitude, ice can begin to form. The difference between present levels of glacier descent and levels during past glacial episodes shows how fluctuations of overall temperature change the snow line (Figs. 16.17 and 16.18). Apparently the snow line was lowered from 600 to 1,000 meters during glacial episodes. Although the snow-line level is critical in the original area of ice-cap formation, once a glacier becomes large, it may flow to either elevations or latitudes well below the snow line. If rate of flow is sufficient, the ice front will advance in spite of relatively warm temperatures. This accounts for the seeming anomaly of finding mild-climate fossils very near obvious glacial moraines. A notable example exists on the equator in East Africa, where glaciers from the 5,113-meter-high Ruwenzori Mountains left terminal moraines on the edge of eastern Congo jungles.

Fossil crocodiles and giant land turtles occur in Pliocene and Pleistocene sediments of the Plains and the southern United States. Distribution of these ectotherms suggests relatively warm, moist conditions as far north as South Dakota in early Pliocene time. In late Pliocene, these animals still ranged as far north as Kansas; the Plains region was more humid than now and had scattered forests. Even during the first two glacial episodes, turtles and crocodiles still lived in the southwestern Plains region, and during interglacial times they again ranged north to Kansas and Nebraska. It is apparent that a really harsh climate did not grip most of the United States

until the last (Wisconsinan) glacial advance. That episode involved three well-dated fluctuations of the ice (Table 16.1).

The many fluctuations of climate in historic times have shown widely differing time periods. At least the shorter ones seem to be controlled by subtle meteorological changes in atmospheric circulation patterns. Historically, alpine glaciers have been extremely sensitive to such shorter-term changes; therefore, large continental ice sheets must have responded similarly (but on a much larger scale) to longer-period changes.

Types of Glacial Hypotheses

Two principal phenomena associated with Pleistocene glaciation require explanation. The first is a general cooling of climate down to a threshold for ice-cap accumulation; the second is the oscillatory, or periodic, nature of glacial expansion and retreat. Note that the major colder and warmer episodes that have occurred in the past 600,000 years appear to have been rhythmic, with a period of roughly 100,000 years (Fig. 16.16). It is natural to assume that both aspects of Pleistocene glaciation were produced by one mechanism, but the rarity of continental glaciation suggests an *unusual chance combination of several factors.*

Many working hypotheses and speculations have been offered to explain glaciation, but they can be conveniently grouped into a relatively few types:

1. Changes in solar radiation
2. Astronomical or orbital effects involving changes in earth-sun geometry

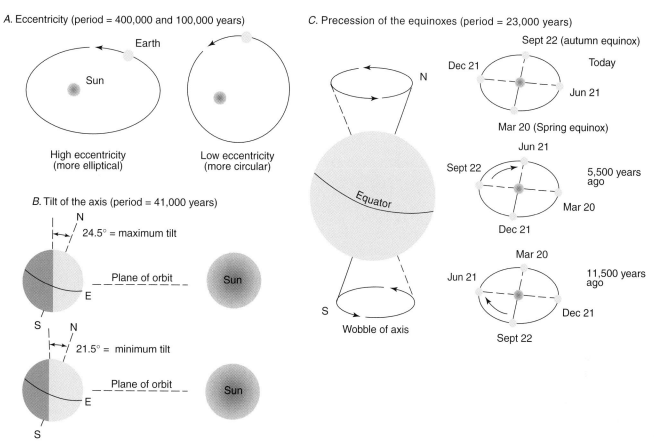

Figure 16.19   Variations in the earth's orbit, which affect global climate. The slowest cycle is caused by the changes in the shape of the earth's orbit around the sun. Known as the *eccentricity* cycle, it takes about 100,000 years for the earth's orbit to change from almost perfectly circular to slightly more elliptical. Note how a more elliptical orbit puts the earth farther away from the sun's heat during the summer, so not all the previous winter's ice can melt. The second cycle is the *tilt* of the earth's axis relative to its plane of motion around the sun. This tilt varies by only a few degrees from the present 23.5° in a cycle lasting about 41,000 years, but the change is enough so that, when the tilt is shallow, more snow can accumulate, since the sun's rays strike less of the polar regions in the winter; conversely when the tilt is steeper, more of the polar snow can melt in the summer. The fastest cycle is known as *precession*, or wobble. Like a spinning top, the earth's axis wobbles, so that it moves through a conical path every 21,000 to 23,000 years. This means that, if the axis points to Polaris, the North Star, today, it pointed to another star (Vega) about 11,000 years ago. Depending upon how the wobble moves the polar regions during winter and summer, the precession cycle can cause the earth to receive either more or less sunlight on its ice caps.

3. Terrestrial changes affecting net heat budget:
   a. Changes in atmospheric transparency
   b. Changes in reflectivity of the earth's surface
4. Variations in heat-exchange rates from equatorial to polar regions due to paleogeographic changes

Hypotheses of the first type, although plausible, are difficult to test because no adequate absolute measure of solar radiation has been made. Moreover, such measurements would be needed for a very long time span. Apparent fluctuations of ^{14}C generation suggest solar fluctuations, but appeal to sunspot cycles as the cause of the fluctuations seems invalid because no corresponding patterns have been verified by analysis of tree rings or annual layers, called *varves*, in glacial lake sediments (Fig. 5.2). It is possible that solar variations are the major cause of glaciation and other climatic changes, but it is fruitful to see if other factors can be discovered that are more readily testable by geologic means.

Astronomical, or Orbital, Hypotheses

The second category of hypotheses has received a great deal of attention for more than a century. In the 1860s, a humble Scot named James Croll, who was then working as a janitor in a Glasgow university, proposed that variations in the earth's orbit caused the alternations of glacial and interglacial climate. His ingenious hypothesis was largely ignored, however, until in 1920 a Yugoslavian meteorologist, Milutin Milankovitch, made further calculations of the net solar radiation variations due to three known cyclic properties of the earth's orbit (Fig. 16.19). These are variations of *eccentricity* of the orbit, *tilt* of the axis, and *wobble* of the axis due to changing gravitational interactions of the moon and sun on the earth's equatorial bulge. The periods of these perturbations in the earth's movements are shown in Table 16.2. Because the three periods of these phenomena are different from each other, there can be only a few times when their effects

Table 16.2	Milankovitch Orbital Factors	
Parameter	**Relative Variation**	**Approximate Periods**
Eccentricity of the orbit (ellipticity)	0.017–0.053	100,000 years
Tilt of the axis (obliquity)	21½–24½°	41,000 years
Precession of the axis (wobble)	0–360°	23,000 years

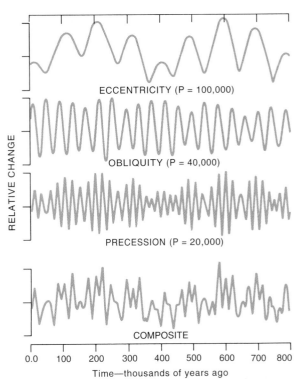

Figure 16.20   Variations in the Milankovitch orbital factors, eccentricity of the earth's orbit, obliquity of the axis, and precession of the equinoxes. The different approximate periods (P) for each of these factors are indicated (see Table 16.2), and a composite curve shows the result of adding all three curves together. *(Adapted from Berger, 1976: Celestial Mechanics, v. 15, pp. 53–74.)*

reinforce one another to cause unusually great or small net solar radiation. For example, if the earth simultaneously had its maximum orbital eccentricity, had its smallest angular tilt, and was at its greatest annual distance from the sun during the Northern Hemisphere summer, then a small but significant lowering of net radiation in the north would occur (Fig. 16.20).

The reality of the **Milankovitch cycles** as the control of Ice Age alternations is strongly supported by isotope studies of late Cenozoic ocean temperature variations. Those studies show a period of about 100,000 years between times of coldest temperatures (Figs. 16.16 and 16.20). Presumably the effect has operated over all of geologic time, yet glaciation has been relatively rare in earth history. Because the total effect is small, causing a change of annual temperature of only 10°C or less, it could leave a conspicuous mark in the historical record *only when some other factors* were also operating to reduce overall average temperature to a critical threshold. In other words, only when the earth was made susceptible to glaciation by other factors could the Milankovitch effect cause expansion and contraction of glaciers.

The Milankovitch cycles have been refined since they were first proposed in 1920. In more recent years, there has been a great deal of interest in this possible relationship, and a number of hypotheses have been invoked to help explain glaciation. In the 1960s, Cesare Emiliani pioneered oxygen-isotope paleotemperature determinations for deep-sea sediments (Fig. 16.16). He and others compared the resulting marine cold-warm alternations with the Milankovitch curve and noted a close correspondence between the periods of variation. During the 1970s, there was a massive cooperative research program called CLIMAP, which involved dozens of scientists in the investigation of Pleistocene climate, primarily using oceanographic data. One result of CLIMAP was a sophisticated mathematical analysis of the frequencies of several parameters observed in deep-sea cores. In 1976, it was shown that the principal period of variation for the past 600,000 years has had a frequency of about 100,000 years (see Fig. 16.16). This dominant variation seems to be due primarily to the variations of orbital eccentricity, but it is doubtless also influenced by other variations. Less pronounced periodic features of the marine sequences have periods of 23,000 and 42,000 years

each, durations that are very close to those of the other two Milankovitch factors (Table 16.2 and Fig. 16.20).

J. Kutzbach of the University of Wisconsin has developed computer simulations of global climate for different times during the past 18,000 years (that is, since the middle of the Woodfordian ice advance, noted in Table 16.1). His work uses the Milankovitch theory to determine changes of seasonality in a climate model. Given the earth's eccentric orbit, the precession of the earth's rotational axis causes the date of nearest approach to the sun (perihelion) to change by about 1 day per 60 years. Today it occurs in January, making northern winters warmer than they were 10,000 years ago, when nearest approach occurred in July, which made northern summers warmer and winters colder than now (Fig. 16.19). Kutzbach's models show that this variation of seasonal timing is very important and can produce approximately a 7 percent difference in annual radiation reaching the earth, which is just the order-of-magnitude difference postulated for glacial versus nonglacial conditions. Kutzbach's simulations also show the wind patterns expected for the glaciated Northern Hemisphere 18,000 years ago. The results compare favorably with such wind-pattern indicators as Pleistocene dunes and patterns of fallout of dust (loess), and they show that, because of the presence of large ice masses, the wind patterns of 18,000 years ago were markedly different from modern patterns.

Hypotheses of Changing Heat Budget

Any change that shifts the balance between incoming and outgoing radiation—the heat budget—would affect the overall temperature of the earth. If such a change is of sufficient magnitude, it might contribute to forming (or terminating) a glacial climate. The changes envisioned include variations in the transparency of the atmosphere to either incoming or outgoing radiation and variations in the reflectivity of the earth's surface—albedo. For example, when continents are unusually large, as they are today, global albedo is high, and much solar radiation is reflected back to space, causing cooling. Conversely when great transgressions produced larger total sea surface than now, albedo was low and climate was warmer.

Long ago, it was suggested that unusual volcanic activity might have brought on glaciation by filling the atmosphere with much fine volcanic dust, which would inhibit incoming solar radiation. There are many examples of measured reductions of average temperature for a few years following very large historical eruptions. Although the Cenozoic thus far has been characterized by much volcanism, it is not clear that it has been dramatically more volcanic than either the warm Mesozoic or the warm intervals of the Paleozoic, when no glaciation occurred.

Several gases—especially carbon dioxide, oxygen, ozone, and water vapor—inhibit the escape of long-wavelength radiation from the earth's surface. The result is the greenhouse effect, which causes a warming of the lower atmosphere. A suggestion that a changing carbon dioxide content in the atmosphere might be a major factor in climatic change dates from 1861, when it was proposed by British physicist John Tyndall. The carbon dioxide budget is complex. Additions come through igneous and hot spring activity, animal respiration, decay, and combustion. Carbon dioxide is consumed by plant photosynthesis and deposition of carbonate rocks. Although the modern increase of CO_2 levels due to combustion seems to be causing a warming trend, we have only a few measurements from deep cores taken from glaciers and the Antarctic ice cap that may give some hints of CO_2 content for the past several thousand years. These data suggest that there was 35 to 40 percent less CO_2 in the atmosphere during the last episode of glaciation, which seems to support Tyndall's hypothesis, but much more data, extending back 40 or 50 million years, are needed to test the idea fully.

The second factor able to alter the earth's heat budget is albedo, the measure of surface reflectivity. Let us now turn to an interesting glacial hypothesis that invokes changes in albedo to explain the long-term temperature change that triggered glaciation. The *Emiliani-Geiss hypothesis* couples albedo changes with the Milankovitch cycles to explain smaller-magnitude temperature fluctuations indicated by the glacial and interglacial episodes. As continents have become larger and higher during the Cenozoic Era, Emiliani and Geiss argue, total albedo has increased. This albedo increase could produce general temperature decline until the threshold for glaciation was reached about 20 to 30 million years ago in polar areas. Once a critical heat-budget level was reached, the Milankovitch effect then could become significant in varying average temperature slightly, but critically,

so that ice caps have expanded and contracted over the millennia. As was shown earlier, the Milankovitch effect has roughly the same time period as do at least the last five or six cold-warm episodes (Fig. 16.16).

Emiliani and Geiss invoked a feedback effect such that, once an ice cap formed, it caused further cooling and assured its own further growth. An equilibrium would be reached, however, when oceanic and atmospheric temperatures became too cold for appreciable further evaporation of moisture to nourish the cap. Recently it has been shown that ice caps have an inherent tendency toward instability, which has also affected their regimes. As caps expand, the wastage area increases as the ice flows outward and downward into warmer areas. Meanwhile, the accumulation area has remained unchanged. A small ice cap may grow slowly to a critical size and then expand suddenly, but it may then shrink suddenly if it becomes unstable. Paleo-temperature curves suggest that, indeed, glacial episodes did end suddenly.

Hypotheses of Paleogeographic Causes

It was suggested long ago that the great amount of mountain building and continent enlargement during late Mesozoic and early Cenozoic time may have altered atmospheric circulation enough to bring on glaciation. The average elevation of continents is now 800 meters above sea level, whereas at the end of the Mesozoic Era, it probably was closer to 200 meters.

American Pleistocene geologist R. F. Flint emphasized that an essentially worldwide uplift of continents in Miocene and Pliocene time may have contributed to the onset of glaciation simply by elevation alone. Mountains and plateaus deflect moist air upward to colder levels and induce considerable atmospheric turbulence. The net result is cloud formation and local precipitation. Interestingly this large-scale uplift also was accompanied by retreat of seas and therefore an increase of worldwide albedo. The result was a double cooling effect. As discussed in Chap. 15, mountain building dramatically affected local climate in western North America, but, because most large continental ice sheets accumulated on lowland areas far from high mountains, Flint's hypothesis is of secondary importance.

A number of scientists have proposed that geographic factors causing variations in equator-to-pole heat-exchange rates controlled glacial and nonglacial climates. In all probability, such changes also were only of secondary importance. Of the several variations of these hypotheses, we shall examine only one, that of geophysicists M. Ewing and W. L. Donn, who in the 1950s used a combination of geographic factors to account for Pleistocene glaciation. To explain the general Cenozoic temperature decline, they pointed to the paleomagnetic evidence of continental drift. While the North Pole apparently lay in or near the North Pacific until Mesozoic time, a warm, mild climate persisted. As the pole then became thermally isolated in the nearly enclosed Arctic Ocean, temperatures declined. Meanwhile, Antarctica had migrated over the South Pole, an even more thermally isolated position.

Ewing and Donn believed continental glaciation began in Antarctica during Miocene or Pliocene time, with these first glaciers chilling the entire globe. Continental glaciation then

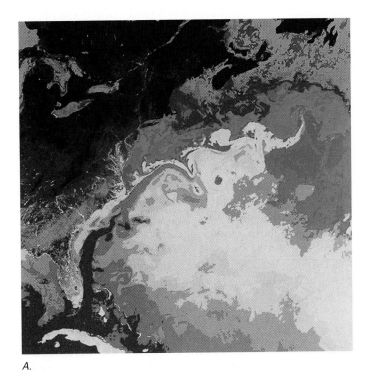

A.

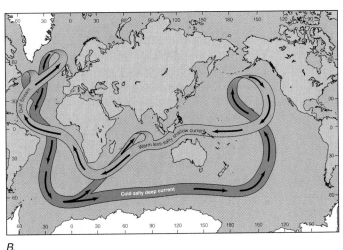

B.

Figure 16.21   *A:* Satellite image of the Gulf Stream, a current of warm surface water flowing north along the East Coast. The colors represent different temperatures (red is warmer). Note the large eddies that form along the margin of the Gulf Stream. *B:* Exchange between upwelling deep water and downwelling surface water creates a global conveyor belt, which circulates water throughout the entire ocean; this takes hundreds of years to millennia. *(A: From: Physical Geography 2e, by Strahler. John Wlley & Sons, Inc., p. 194 Fig 7.33b: From: The Dynamic Earth: An Introduction to Physical Geology 4e, by Skinner and Porter. Wiley, 1995 p. 377 Fig. 14.6b.)*

occurred on the chilled northern continents through evaporation of moisture from a presumably ice-free Arctic Ocean. Finally, the Arctic Ocean froze over. Ice caps ceased to advance for lack of nourishment, and melting along their southern margins soon caused retreat. As ice melted, total climate warmed, more ice melted, and so on. However, if the Arctic Ocean were reopened by climatic warming and rise of sea level, then it again might provide moisture for new northern ice caps. Presumably, glacial and interglacial fluctuations should continue indefinitely until some major change either of pole positions or of polar ocean water exchange occurs.

From archaeological records, we know that, between A.D. 800 and 1200, whales could move eastward along the Arctic Coast from Alaska to northern Greenland, where they were hunted by Eskimos. From historic records, we also know that at the same time Scandinavia enjoyed a very favorable climate, and the Vikings settled in Iceland, Greenland, and Newfoundland; they met (and were raided by) Eskimos in Greenland. Viking sailors reported nearly ice-free conditions as far north as Svalbard (Spitsbergen)—79° N latitude. Although this suggests a semi-open Arctic coastal condition, it hardly proves a completely ice-free ocean.

Many meteorologists doubt that significant evaporation of moisture could occur *even if* the Arctic Ocean were ice-free, and studies of oxygen isotopes and foraminifera in Arctic Ocean sediments, as well as land plant fossils, all suggest a continuous ma-

jor ice cover throughout the Pleistocene. Moreover, centers of the ice caps lay well south of the Arctic Ocean (Figs. 16.3 and 16.4), and practically all moisture today goes to these areas from elsewhere—the Gulf of Mexico for North America and the North Atlantic Gulf Stream for Europe.

The Thermohaline Conveyor Belt and Dansgaard-Oeschger Cycles

So far, we have seen how differences in solar radiation cause the 100,000-year glacial-interglacial cycles, once the earth has cooled past certain climatic thresholds due to the effects of ocean currents, modified by tectonic conditions. In the past decade, however, another factor has come to be better appreciated: oceanic circulation. In the 1980s, scientists, including Wallace Broecker of Columbia University and the Lamont-Doherty Earth Observatory in Palisades, New York, discovered the great "conveyor belt" of thermohaline circulation. Waters that are warmed in the tropical Pacific and Indian Oceans flow around South Africa, across the tropical Atlantic, and then up the East Coast of North America as the Gulf Stream (Fig. 16.21). As this huge, warm surface current reaches the North Atlantic, it begins to cool, and its dense, salty waters begin to sink. Today the sinking occurs north of Iceland. From there, the dense, salty, cold water flows south again as the North Atlantic Deep Water and flows all the way to the South Atlantic, eastward across the deep

Antarctic-Indian Ocean, and eventually across the South Pacific bottom to the North Pacific. There it is finally warmed enough that it rises to the surface and rejoins the conveyor belt of surface waters on their path back through the Indian Ocean to the Atlantic (Fig. 16.21).

As we saw in Chap. 15, the system has not always worked this way. Today the warm tropical Gulf Stream helps bring moisture up to the already cold Arctic region, providing the snow and ice that form the Arctic ice cap. But in the Pliocene, the gap through the Isthmus of Panama (Fig. 15.40) caused mixing with the low-salinity surface waters of the Caribbean and Pacific. This made the surface waters fresher and more buoyant, so they flowed farther north into the Arctic without sinking. The warm current kept the Arctic warmer and prevented the development of the full Arctic ice cap. The closure of the Panamanian gap shut off this freshwater flow and forced the much saltier Gulf Stream to sink as it does today, initiating the modern thermohaline conveyor belt among the Atlantic, Indian, and Pacific Oceans.

Such a mechanism works well at explaining the long-term cooling of the Pliocene and Pleistocene, but it yielded even more surprising answers. Scientists had long been puzzled by the tiny "wiggles" of climatic change on a cycle of thousand-year durations or less, too short for Milankovitch orbital variations. These short-duration cycles came to be known as the **Dansgaard-Oeschger** (DO) cycles, after the scientists who first described them. For example, as the present interglacial began about 11,000 years ago, there was a short, final burst of cold called the Younger Dryas (Table 16.1) before the final interglacial warming began in full. Could there be another way of changing the thermohaline conveyor belt and causing changes in Arctic ice volume? Careful measurements of chemical isotopes in the North Atlantic showed there were several times when massive amounts of freshwater poured into the North Atlantic. One of the most important occurred from the draining of glacial lakes on North America, such as the Great Lakes and glacial Lake Agassiz, which flooded down the St. Lawrence River. Formerly they had drained down the Mississippi to the Gulf of Mexico when ice covered the St. Lawrence valley (Fig. 16.6). This torrent of freshwater and icebergs rushed out of the St. Lawrence and formed another freshwater lid, shutting off the conveyor belt for almost a thousand years. With the conveyor belt shut off, there was no source of salty water to cool and sink into the deep ocean. The deep saline waters no longer flowed back out of the North Atlantic, so there was very little flow of warm tropical waters back to the surface of the North Atlantic, and thus there was very little flow of warm tropical waters back to the surface of the North Atlantic as well, because the Gulf Stream was blocked by a freshwater "dam." Without this warmth reaching the North Atlantic, Canada, and Europe, the region quickly froze over, and the 1,000-year Younger Dryas cooling interval began. It ended when the freshwater dam finally stopped flowing, and the thermohaline conveyor belt resumed, allowing warm, salty waters to reach the North Atlantic and the present interglacial warming to resume.

The frightening thing about this discovery is that it confirmed what many ice cores had shown: glacial and interglacial events can start rapidly, in just a few years or decades, rather than slowly over thousands of years. This thermohaline conveyor belt is so sensitive to small changes that it takes only a local event, such as a freshwater flood in a critical place, to shut it off and start a new ice age. Think of how humans have already changed global climate with the greenhouse gases, which have caused global warming and a rise in sea level. Instead of slowly changing climate, our meddling with climate might trigger a sudden change, such as a new ice age, in just a few years. Considering how hard it is for us to find the political will to cope with a slowly developing crisis such as global warming, is it likely that we will be able to react quickly enough if an ice age develops in just a few years?

Where Do We Stand?

We still do not have a fully acceptable theory of glaciation. Several hypotheses have not been tested yet and, so, can be neither accepted nor rejected. Circumstantial evidence from the history of the earth, however, conforms well enough with several ideas to make a geologically attractive working hypothesis as follows.

Movement of continents to thermally isolate the poles coupled with the enlargement of total land area to produce high albedo could explain the long-term Neogene chilling of global climate down to the critical threshold for glaciation. Then the Milankovitch cycles, which produce a relatively small temperature effect by themselves, could have become critical in causing oscillation between glacial and interglacial conditions. Changes in atmospheric CO_2 levels provided a secondary feedback effect, which contributed to those oscillations. This working hypothesis predicts a continuation of oscillations indefinitely into the future until a major, long-term change of pole positions and/or shrinkage of total land area occurs to tip the overall heat budget the other way. It is much as the famous American banker J. P. Morgan said when asked to forecast the behavior of the stock market: "In all probability, it will fluctuate."

Pleistocene Climatic Effects upon Life

Ranges of Organisms

Continental glaciation in the Northern Hemisphere is reflected in Pleistocene marine faunas as well as among land fossils. For example, species living today along the northern California coast ranged only as far north as San Diego during glacial episodes.

Worldwide changes in sea level during Pleistocene time had a great effect on coral reef growth. For example, in the West Indies, fossil coral reefs occur from 1.5 to 5 meters above present low-tide level. These reefs have been dated as interglacial, when water may have stood slightly higher than at present (alternatively the islands may have risen). Most nearby living reefs are postglacial, being no older than 5,000 years, and form veneers growing on fossil reefs. Furthermore, the biota of the living reefs is unique to the West Indies. During earlier Cenozoic time, the biota of the reefs was closely related to Indo-Pacific and Mediterranean (Tethyan) faunas. Alternate heating and cooling, plus many other factors, had caused the recent reef fauna to become more restricted geographically and faunally. Other reefs of the world also show the ravages of sea-level fluctuations, but not as clearly or to the same extent as in the West Indies.

On land, there were even more dramatic effects upon the distribution of plants and animals. For example, large logs, cones, and tree leaves dated as 30,000 years old have been dredged off western Mexico. Because no such trees grow along this very dry coast today, it is inferred that the climate was much wetter there 30,000 years ago.

Throughout the hot, arid southwest, outposts of cool-climate floras are found today on widely scattered mountain ranges surrounded by desert lowlands. Although we cannot rule out seeding of these vegetation "islands" by wind or birds, it is far more likely that a continuous, cool-adapted flora covered much of the intervening lowlands during glacial episodes.

Faunal and floral changes due to glaciation were most extreme near the southern limit of the ice. There, alternately glacial and nonglacial conditions prevailed. The Great Lakes region provides a convenient example. The last major ice advance (Woodfordian substage) culminated there about 17,000 years ago. Ice had retreated from the southern Great Lakes by 12,000 years ago, and cold-climate spruce forests, much like those of central Canada today, reoccupied the region. A famous buried forest at Two Creeks, Wisconsin, about 11,850 years old, gave the name to this warmer interval (Table 16.1). Then, about 10,500 years ago, the ice again advanced south a short distance to cover part of the northern Great Lakes region. The Two Creeks Forest was overridden, with the spruce trees toppled and buried beneath till (Fig. 16.22). Spruce forests persisted farther south, however. About 10,000 years ago, the glacial climate ended and ice began its last retreat. The ice front paused briefly in the western Hudson Bay region about 8,000 years ago and finally completed its retreat from that region by 6,000 years ago. As the ice left the Great Lakes region, northern spruce forests (cold-climate trees) migrated northward. By about 8,000 years ago, spruce forests were replaced by pine (warmer-climate trees) in the Great Lakes region. From 8,000 to about 1,500 years ago, a relatively dry period set in, which is reflected in a change from pine to widespread oak-hemlock forests (Fig. 16.23).

The postglacial climate culminated about 5,000 years ago, but in the past 1,500 years there have been several small fluctuations in climate, which are reflected in soils, vegetation, and human and animal activities. Today the northern hardwood forest covers much of the Great Lakes region, with boreal spruce forests just touching its northern limits; prairie-oak forests characterize the southwestern corner of the region. Boundaries between these entities have fluctuated, as is indicated by plant types found in vertical sequences of late Pleistocene and Holocene sediments in lakes and bogs. Wood, leaves, fruits, pollen, animal remains, and soil profiles all are employed in their study (Fig. 16.24). The magnitude of the changes suggested by such data is indicated in Fig. 16.23 for North America and in Fig. 16.25 for Europe. By far the harshest climatic stress occurred during the last (Wisconsinan) glacial episode. Cold-blooded reptiles retreated farther south during (and reinvaded less far north after) each advance. Only the Wisconsinan episode was able to relegate them completely to the Gulf Coast region during glaciation.

Figure 16.22   Excavation of Two Creeks Forest on Lake Michigan shore, northeastern Wisconsin. Dozens of spruce logs were buried in lake sediments, which in turn were covered by late Wisconsinan glacial till. Some stumps are *in situ*; prostrate logs were driftwood. Carbon-14 dating indicates that the logs are about 11,850 years old. *(Courtesy of D. M. Mickelson.)*

Land Bridges

Formation of land bridges by glacial lowering of sea level was of paramount importance to the present biogeography of the earth. The most important examples include the Bering bridge between Siberia and Alaska, the North Sea bridge between Europe and Britain, the Sunda bridge between western Indonesia and Asia, and the New Guinea–Australia bridge. The Isthmus of Panama, second only to the Bering bridge in biological importance, was formed by structural rather than sea-level changes.

The Bering bridge deserves special mention. Fossil plant and invertebrate evidence indicates that Alaska and Siberia were connected during most of Cenozoic time. Two-way traffic occurred through most of the Cenozoic, but Asia-to-America flow was twice as heavy as the opposite. In late Miocene time, the bridge was flooded. Then, during Pleistocene glaciation, Alaska was reconnected to Asia, and it was biologically more like Siberia because of its isolation from the remainder of North America by a wide ice barrier in Canada (Fig. 16.4). During interglacial high sea levels, its fauna showed more affinities with that in the interior of North America because an ice-free corridor was opened in the western Canadian Plains. The Bering bridge was drowned 13,000 years ago and remains so today. The 90-kilometer expanse of water that separates Siberia from Alaska hardly constitutes a great barrier, but the harsh climate now effectively blocks temperate-dwelling organisms from migrating.

The End of an Era

As with the close of the other eras, there was (and is) a period of extinction in late Cenozoic time. More than 200 genera of mammals became extinct between late Pliocene and recent time. One

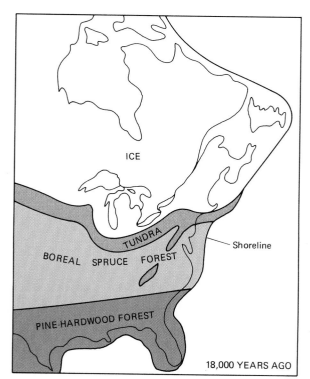
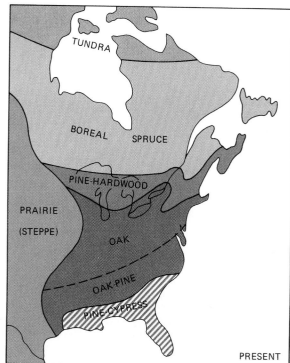

Figure 16.23 Effects of climate change on plant communities of eastern North America during the last maximum ice advance (Woodfordian) *(left)*, compared with modern native vegetation *(right)*. Past distribution at left is based largely upon studies of fossil pollen distributions. *(Adapted from Mayewski et al., 1981, in* The last ice sheets: *Wiley, pp. 67–178; Delcourt and Delcourt, 1984:* Natural History, v. 93, no. 9, *p. 24; Goode's world atlas:* Rand McNally.)

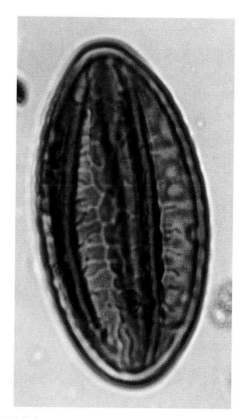

Figure 16.24 Microscopic photo of pollen grain from *Ephedra*, or "Mormon tea," a plant found in semiarid regions. Climate must be interpreted carefully from pollen, because some grains are known to be blown hundreds of miles. This grain is 0.05 to 0.07 mm long. *(Courtesy of Estella Leopold.)*

might expect that the extreme climatic variations brought about by glaciation would cause these widespread extinctions. The picture, however, is not that clear, for the climate almost to the ice edge in some areas was relatively mild, especially before the most recent glacial episode.

An alternative theory has gained some support. It suggests that humans were responsible for the extinction of the larger mammals. Earliest Pleistocene mammalian extinctions in Africa also seem to be related to early Stone Age hunters.

However, the record from fossil sites is ambiguous. In Eurasia, artifacts are found along with large mammal remains, yet only four genera of mammals became extinct during human existence there. In view of the small populations of early humans, it seems unlikely that they alone could have erased so many mammalian species.

Scientists who reject the human hunter "overkill" hypothesis point out that the end of the last glaciation was very different and much more extreme than any other deglaciation. Humans may have killed a few large mammals, but the rest that were not human prey were probably done in by climate.

Mammalian extinction began in the late Miocene with the loss in North America of some artiodactyl families and the rhinos. It is among the larger types that we find the most striking extinction patterns. Although some artiodactyls, such as the bison, survived, we lost giant beavers, the mammoth and mastodon, saber-toothed cats, camels, horses, and many others. A magnificent cross section of a late Pleistocene terrestrial fauna of about 40,000 years ago was recovered from the classic Rancho La Brea tar pits of Los Angeles, which are preserved in a park (Fig. 16.26).

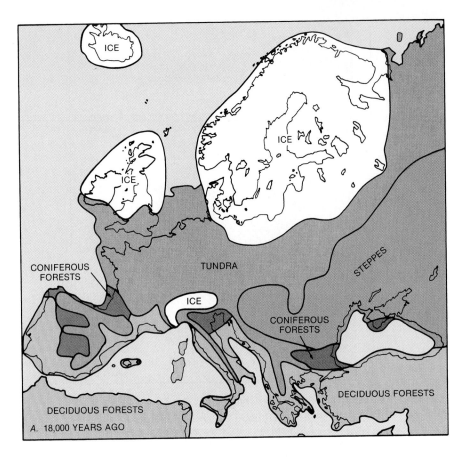

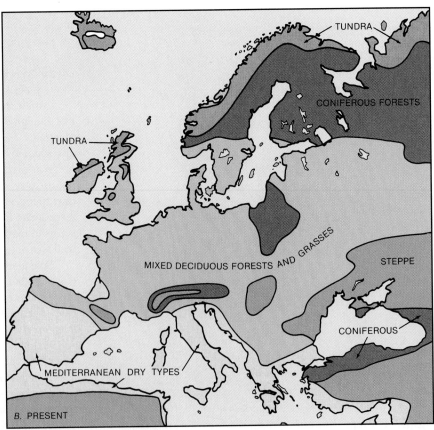

Figure 16.25 Effects of climatic changes on plant communities of Europe during the last maximum ice advance about 18,000 years ago (A), compared with native vegetation of modern Europe (B). (A adapted from Brinkmann, 1960, Geological evolution of Europe: *Ferdinand Enke; Wills, 1951,* A Palaeogeographical atlas: *Blackie & Son Ltd.; B after Goode's world atlas: Rand McNally Corp.*)

Figure 16.26   Reconstruction of life in Los Angeles about 40,000 years ago, as preserved in the famous tar pits of Rancho La Brea. The tar came from oil seeping up from the Miocene rocks in the Los Angeles Basin and forms natural ponds, which are typically covered by rainwater. Thirsty animals waded into the water and became trapped by the sticky tar. Typically the panicked cries of a single trapped animal would attract many predators and scavengers, so the tar pits have many more fossils of saber-toothed cats, dire wolves, lions, bears, and vultures than they have bison, camels, horses, ground sloths, or mammoths and mastodonts. However, in nature there is much more biomass of prey species than endothermic predators (see Fig. 14.51), so the tar pits were a selective death trap that preserves a biased sample of life. In this famous reconstruction by Charles R. Knight, prey species (such as the ground sloth, camels, horses, and mammoths) are in the background relative to the predators and scavengers (saber-toothed cats, lions, dire wolves, and giant vultures). *(Mural by Charles Knight. Courtesy of George C. Page Museum.)*

The Evolution of Primates and Humans

As we shall see in Chap. 17, the most destructive and influential mammal of the Holocene fauna is *Homo sapiens.* The subject of human origins has been of intense interest ever since the first fragmentary fossil skull was unearthed in the Neander Tal (Valley) near Düsseldorf, Germany, in the 1840s.

The story of human evolution is fraught with high emotions and irrational responses, and whole libraries have been written about it; much can be classified as rank speculation. To this day, the pathway leading from a primitive ancestor to modern humans and their relatives remains incomplete. There is good reason for this mystery, for the fossil record of the human family—the

Hominidae—is small and fragmentary; most of it consists of broken jaws, broken skulls, and teeth. No classification or family tree, no matter how sophisticated, has remained undebated for long. Teams of paleoanthropologists have scoured the earth intensely in the past 50 years, and the best fossils older than 1.6 m.y. old recovered to date are the *Homo erectus* find by Leakey and "Lucy," an immature female skeleton 40 percent complete, found at Hadar in the Afar Triangle of Ethiopia in 1972 (Fig. 16.27). This exquisite skeleton, along with many other fossils at this site, represents the earliest occurrence of *Australopithecus afarensis* (Fig. 16.28), an early species of the line of *Australopithecus* species leading to *Homo.*

To approach the problem of the descent of humans, we need a set of characteristics to define humanness. Two features accepted

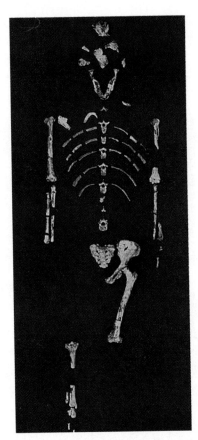

Figure 16.27   The nearly complete skeleton of "Lucy," a young female *Australopithecus afarensis* from 3.0–3.4-m.y.-old rocks of the Afar Basin of Ethiopia, discovered in 1973. Along with other specimens of this species, *A. afarensis* is one of the oldest members of the human lineage. *(Courtesy of Institute of Human Origins.)*

Figure 16.28   Restoration of *Australopithecus afarensis* on display in the British Museum of Natural History in London. Adult females were typically just over a meter (3.5 feet) in height, and adult males about 30 percent taller (4.5 feet). The hip and knee bones of several specimens clearly show that this species was fully erect and bipedal. Bipedal gait can also be established from the 3.5-m.y.-old footprints at Laetoli in Tanzania (see Fig. 1.1). *(D. R. Prothero.)*

by most workers are brain size and bipedalism, which may be interrelated. Most morphological changes in the hominids can be traced to these two features and to the basic primate character. The human brain is huge in proportion to body size; for example, it is almost three times the size of a gorilla's. An elephant weighing several tons might have a brain weighing 10 pounds, whereas the brain of a 150-pound person might weigh as much as 5 pounds.

Brain size has been linked to bipedalism on the basis that, sometime after their hands were freed from locomotion, stone toolmaking occurred. This, along with the later development of active hunting, required reasoning and complex neurological hookups that gave a survival advantage to those with the largest brains (greatest reasoning ability). Bonding of a society followed cooperation in hunting, and from that point languages developed, or so one scenario goes. As we shall see, bipedalsim appeared at the beginning of human evolution, but large brains were a late development.

The exact origin of the primates is unknown, but they were derived from early placental mammals that probably resembled shrews. The earliest primates were squirrel-like nut eaters repre-

sented in the Paleocene of Europe and North America. By Eocene time, these groups had been mainly replaced by the ancestors of the modern lemurs and tarsiers (Fig. 16.29), which were widespread in Asia, Africa, and North America. The primates possess many primitive mammalian characteristics; for example, they retain the primitive five-digit feet, which are well suited for arboreal (tree-dwelling) life. One important early characteristic was the development of stereoscopic vision, which is an advantage for judging distances and is especially important for arboreal creatures.

Fossil primates are a very rare part of most Cenozoic faunas, mainly because they probably inhabited forested uplands, which were very subject to erosion. After the prosimian primates (lemurs and tarsiers) experienced a rapid adaptive radiation in Paleocene and Eocene times, they became somewhat restricted and fairly conservative throughout the later Cenozoic. It is in Africa, where Eocene prosimians became well established, that major events of anthropoid evolution evidently occurred.

The earliest true apes (Fig. 16.30A) have been found in the Oligocene Fayum beds of Egypt, a remarkably rich deposit of

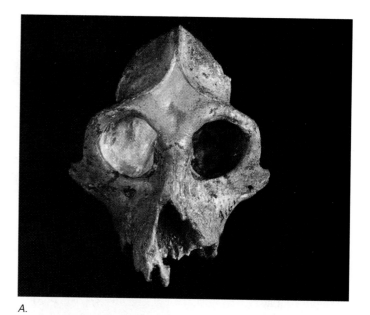

A.

B.

Figure 16.29   The tarsierlike primate *Tetonius*, known from early Eocene (about 50 m.y. ago) deposits in the Rocky Mountains. Living lemurs of Madagascar are remarkably similar to the great radiation of Paleocene and Eocene primates that flourished when the earth was warm and subtropical even at high latitudes (see Figs. 15.44 and 15.45). By the Oligocene, the cooling and drying that destroyed these forests had greatly reduced primate diversity in the Northern Hemisphere. Instead, Africa became their refuge, and from this source most apes, Old World monkeys, and possibly the New World monkeys evolved and later dispersed outward.

Figure 16.30   A: Skull of *Aegyptopithecus*, one of the earliest known fossil apes, from the Oligocene of Egypt. *(Photograph by W. Secco, courtesy of David Pilbeam.)* B: Partial skull of *Sivapithecus sivalensis* from late Miocene deposits about 8 m.y. old in Pakistan. Along with *Ramapithecus* (now thought to be the same creature), these late Miocene apes were long placed in the ancestry of hominids. But more complete specimens such as this skull now show that *Sivapithecus* was more closely related to orangutans and predates the split between great apes and humans. According to molecular evidence, that split occurred about 5 m.y. ago. *(Courtesy of D. Tab Rasmussen.)*

fossil early primates, 100 kilometers southwest of Cairo, in an area that is desert today and is located on the edge of the fertile and historically important Nile valley. During the Oligocene, this region was the site of a rich, tropical forest with broad rivers and vast swamps inhabited by crocodiles, primitive elephants, and ever present carnivores.

The apes of the Oligocene and Miocene were arboreal fruit eaters. At the end of the Miocene, two lineages of apes spent at least part of their lives in open country. A group of great apes (dryopithecines) appeared during the Miocene in widely scattered areas; they are known mostly from skull fragments. These apes had rows of evenly spaced teeth and other features suggesting that they were related to the hominids; they probably were early forms that might have given rise to later great apes, such as the gorilla. The Miocene saw the main radiation of ho-

minid primates culminating in *Sivapithecus* (and related, or perhaps synonymous, *Ramapithecus*), which has many characteristics of living apes and humans (Fig. 16.30B). These primates apparently lived in forested areas with stretches of open grasslands near rivers and lakes. They probably ate seeds and roots. It is possible that they also ate meat they scavenged, but they were not hunters.

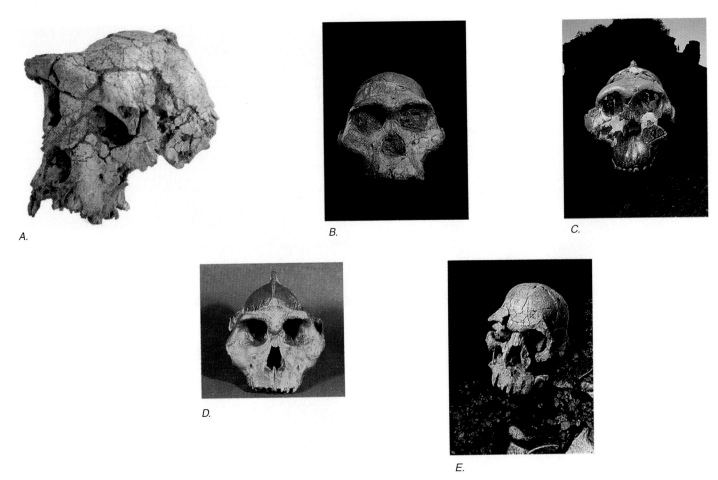

Figure 16.31   Several species of fossil hominids evolved in Africa. A: *Sahelanthropus,* from rocks 6–7 million years old in Chad. B: *Australopithecus africanus,* known from 3.0–2.3 m.y. ago, had a lower forehead and a more projecting face than later humans. However, it was less robust than most of the hominids that followed it. C: The "Nutcracker Man," discovered by the Leakeys in Olduvai Gorge in beds dating between 1.2 and 2.2 m.y. ago. This fellow had robust jaws, large molars, and a strong crest along the top of the skull, indicating a diet that required jaw strength to crush seeds or nuts. Originally called *Zinjanthropus boisei* by Louis Leakey, many scientists now refer it to *Paranthropus,* the robust lineage of australopithecines. Another robust species, *Paranthropus robustus,* is known from beds 1.6–1.9 m.y. in age. (It should be noted that other paleoanthropologists place all of these fossils in a very broadly drawn concept of *Australopithecus.*) D: The controversial "Black Skull," discovered by Alan Walker in 1975 on the shores of West Lake Turkana. Dated at about 2.5 m.y., it seems to be a very early robust form; some paleoanthropologists consider it the primitive relative of the robust *Paranthropus* lineage. If it is a member of this group, its proper name is *Paranthropus aethiopicus.* E: In the midst of all these robust species of australopithecines was the first member of our genus, *Homo habilis* ("handy man"). This specimen was found by Richard Leakey in East Turkana in 1972 and is best known by its museum catalogue number, KNM-ER 1470. Known from beds dated from 2.1–1.8 m.y. in age, *Homo habilis* was clearly much less robust in proportions, with a less projecting face, larger brain, and more rounded forehead, all hallmarks of our genus. (*A: M.P. F.T/Corbis Sygma; B.,C: ©John Reader/Science Photo Library/Photo Researchers, Inc.; D: Courtesy of Alan Walker ©National Museum of Kenya; E: ©John Reader/Science Photo Library/Photo Researchers, Inc.*)

Some species of *Sivapithecus* have such hominid features as a broad jawbone, low-crowned molars with thick enamel, and reduced canine tooth size. They also share many features with the orangutan. In fact, molecular evidence comparing *Homo* and the orangutan indicates that the two diverged 10 to 11 million years ago. The same evidence suggests that the African apes and the hominids diverged sometime near the end of the Miocene, 6 million years ago. The earliest known hominid fossil, however, is just under 7 million years old.

The Early Hominid Record

The earliest record of the Hominidae is at several sites in Africa. In 2002, an extraordinary skull was reported from rocks between 6 and 7 million years old in Chad, western Africa. Called *Sahelanthropus tchadensis,* the skull (Fig. 16.31A) is very chimplike in its small size, small brain, and large brow ridges (so large that it suggests the skull belonged to a male). However, it shows some remarkably humanlike features, such as a short, flattened face; reduced canine teeth; and enlarged cheek teeth with heavy crown

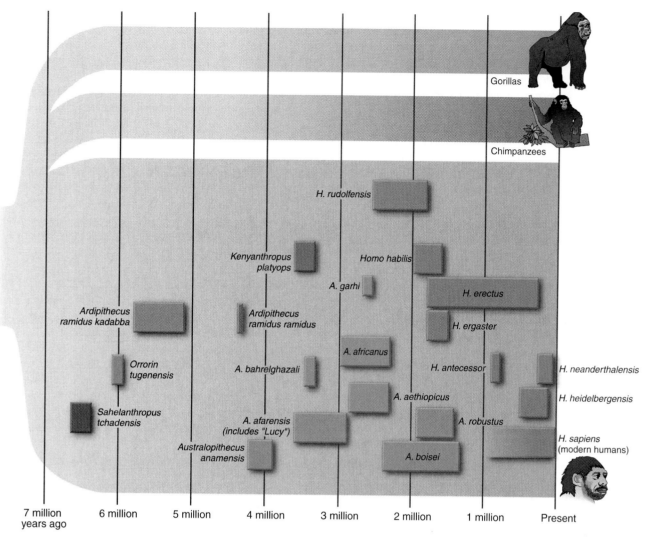

Figure 16.32 Geologic time ranges of the many species of hominids in the past 7 million years. Contrary to popular myth, there was no single lineage of hominids leading to modern humans. Instead, we went through *Australopithecus* to a great diversity of australopithecines, robust australopithecines, and the earliest *Homo habilis* and then back to a single lineage of *Homo erectus*, which led to Neatherthals and modern humans. (see Fig. 16.31D).

wear like those of early hominids. Most important, the hole through which the spinal cord exits the braincase is directly below the skull, showing that *Sahelanthropus* had upright posture at the very beginning of hominid evolution.

Sahelanthropus is significant for another reason—its great age. At 6 to 7 million years old, it is almost half a million years older than the previous record holder, *Ororrin tugenensis* (Fig. 16.32). For years, molecular biologists had been saying that the split between the human lineage, and the lineages of the gorilla and chimpanzee, had occurred between 5 and 7 million years ago. Now we have fossils representing the earliest member of the hominid lineage, dated at the time when the ape-human split took place. A better "missing link" between apes and humans could not be imagined!

The next youngest specimens are the fossils of *Ororrin tugenensis,* described in 2000. These fossils came from the upper

Miocene Lukeino Formation in the Tugen Hills of Kenya and are now dated between 5.72 and 5.88 million years old. Although known from the fragmentary remains of six individuals in four different sites, the teeth have thick enamel, like many other primitive hominids (and unlike apes), and the thighbones and shinbones clearly showed that *Ororrin tugenensis* walked upright (no longer surprising, considering that *Sahelanthropus* was also upright).

After *Sahelanthropus* and *Ororin,* the next youngest fossils are those of *Ardipithicus ramidus,* discovered in the Afar Depression of Ethiopia in 1992. Dated at 4.4 million years old, the jaws and teeth had a mixture of apelike features, as well as humanlike reduced canine teeth and a U-shaped lower jaw (rather than a V-shaped, apelike jaw). Eventually almost half the skeleton was found, showing clearly that it, too, was bipedal. Shortly after this discovery, an even older subspecies, *Ardipithicus ramidus kadabba,* was found in Ethiopia in deposits dated between 5.2 and 5.8

million years old. The most important feature of these specimens are the foot bones, which show that humans used the "toe off" manner of upright walking as early as 5.2 million years ago.

Rocks of 3.9 to 4.2 million years in age near Lake Turkana in Kenya yield *Australopithecus anamensis,* the first member of our ancestral genus, *Australopithecus.* These specimens are only slightly older than *Australopithecus afarensis* ("Lucy"), yet they were fully bipedal, with a narrow, apelike lower jaw. At about the same time, there are nonanatomical signs of bipedalism. The most fascinating find is two sets of human footprints, along with the tracks of a great number of mammals still seen today on the plains of Laetoli, Tanzania (Fig. 1.1). The prints show that the hominids were bipedal with a shuffling gait and a short stride. They are dated at 3.6 to 3.8 million years.

At Hadar, in Tanzania, which has been dated at 3.0 to 3.4 million years, slightly younger than Laetoli, a very rich treasure trove of hominid specimens *(Australopithecus afarensis)* has been found, containing fragments of at least 40 individuals, including Lucy (Figs. 16.27 and 16.28). Nearly every type of bone in the body has been found. The nature of the hand and foot bones together with the pelvic bones and knees clearly shows that *Australopithecus afarensis* was bipedal, but they were not as erect as *Homo erectus.* They had large canine and incisor teeth and a large, overhung jaw. The opening for the spine is much more anteriorly located than in the apes. The brain size is unknown at present, but it apparently was similar to or slightly smaller than that of later species of *Australopithecus* (500 to 600 cm^3). The individual known as Lucy was about 20 years old and stood at just under 1 meter. Sexual differences are pronounced in these as well as all other species of hominids.

These early forms, *A. afarensis,* have primitive features like the two lineages that are found in the next younger group of fossil sites dated at 3.0 to 1.5 million years, at or near the Pliocene-Pleistocene boundary. At a number of sites in southern Africa and East Africa, two groups of hominids have been recognized. The first of these was described in 1925 by Dart from an immature skull as *Australopithecus africanus* (Fig. 16.31B). This group is called the gracile lineage; *gracile* refers to a relatively dainty jaw, with rather small cheek teeth, no skull crest, and a centrally located spinal opening, indicating that the spinal column was in a vertical position and hence that gracile individuals were better walkers than their predecessors. The brain was about 450 cm^3 in size.

In addition to gracile *A. africanus,* two robust australopithecine species have been found in beds dated between 1.6 and 1.9 million years. *Australopithecus robustus* (originally *"Paranthropus" robustus*) has a massive jaw with very large molars and premolars, indicating a tough diet with more nutcracking or bone cracking. The skull has a crest across the top and a brain capacity of about 530 cm^3, roughly the same size as in *A. africanus.* The skeleton is much larger, and some individuals weighed about 75 to 150 pounds.

The most robust australopithecine of all was *A. boisei* (formerly *"Zinjanthropus" boisei*) (Fig. 16.31C). Nicknamed "Nutcracker Man" by Louis Leakey, who found it at Olduvai Gorge in 1959, *A. boisei* has massive robust jaws and widely flaring cheekbones, a strong crest on the top of the head, and massive molars. Most specimens of *A. boisei* occur in beds between 1.2 and 2.2 million years in age.

Further complicating the picture is yet a third robust australopithecine, discovered by Alan Walker on the western shore of Lake Turkana in 1975. Known as the "Black Skull" (or by its catalogue number, WT 17000) (Fig. 16.31D), it is very controversial. Although the skull is robust, like that of *A. boisei,* it is much older (about 2.5 million years) and has many primitive features that resemble those of the gracile *A. afarensis.* It also has a dish-shaped face, a feature that was thought to have originated much later. Some anthropologists argue that WT 17000 is a primitive link between *A. afarensis* and *A. boisei;* others argue that it is the primitive relative of both *A. robustus* and *A. boisei.* Those who relate it to *A. boisei* and *A. robustus* argue that all three form a distinct genus of robust hominids that should be separated from *Australopithecus.* These workers resurrect the old genus *Paranthropus* for the robust hominids, altering their names to *P. robustus* and *P. boisei.* Some anthropologists think WT 17000 should be assigned to *boisei,* but others think it resembles a long-forgotten specimen, *Australopithecus* (or *Paranthropus*) *aethiopicus.*

Although there is no shortage of opinions about the relationships of early hominids, there is little question that, in the late Pliocene of Africa, about 2 million years ago, there were at least four species of hominids living side-by-side: the robust *Australopithecus* (or *Paranthropus*) *boisei* and *A.* (or *P.*) *robustus,* the gracile *A. africanus,* and the earliest species of our own genus, *Homo.*

The Genus *Homo*

Among the fossils found by Louis Leakey at Olduvai and dated at 1.75 million years is specimen number OH 7, which has a dramatically large brain size of 750 cm^3. The skull has no crest, and the jaws and cheek teeth are moderately robust. Leaky named it *Homo habilis* (the "handy man") because tools were found with it. Up to that time (1960s) it was believed that *Australopithecus* was the direct ancestor of *Homo.* The problem is that both are found together at Olduvai, so then who is the ancestor and who is the descendant?

Another *H. habilis* skull, along with stone implements, was found by Richard Leakey (the son of Louis) in the Koobi Fora Formation at Lake Turkana in East Africa (Fig. 16.31E). These rocks were dated at just under 2 million years. This skull is even more humanlike, with a brain size of nearly 800 cm^3, more than one-half that of *H. sapiens.*

In Java and China during the 1920s and 1930s, a number of specimens of a tall, large-brained (over 900 cm^3), and small-faced human form were found in mid-Pleistocene deposits dated at about 1.0 to 0.25 million years. These fossils were given a variety of names but are now placed in the species *H. erectus.* Legs and associated bones show that *H. erectus* walked fully erect. In 1984, Richard Leakey and Alan Walker found, on the western shore of Lake Turkana, a remarkably complete skeleton of a boy, missing only the hands and feet. At 1.6 million years, it is one of the oldest known specimens of the species. It has provided valuable clues to the growth patterns of *H. erectus.*

Additional fossils found in Africa are somewhat younger (1.3 million years), with smaller brains, but in any event it is likely that *H. erectus* evolved into modern *Homo sapiens*. The late-appearing *H. erectus* seems to overlap features sufficiently with *H. sapiens* so that some workers are inclined to consider them as a single species (Fig. 16.33). During the 1.3 million years of *H. erectus*'s existence, the brain size increased to over 1,000 cm³, and this was accompanied by an evolution of complex behavior and the development of social patterns. Evidence for these conclusions is based on the presence of tools, use of fire for cooking and warmth, construction of stone and wood shelters, and arrangement of these shelters into communities.

The earliest members of *H. sapiens* we know about so far were found in Africa and Europe in middle Pleistocene sites dated at about 500,000 to 300,000 years. Certainly one of the first dispersals of *Homo* was that of *H. erectus* migrating as far as China and Java to the east and, later, *H. erectus* and *H. sapiens* migrating together to England and Germany to the north. The European *H. sapiens* were called Neanderthals (Fig. 16.33), and their brain size averaged around 1,300 cm³. Thus, in less than 1 million years, the brain size doubled in *Homo* compared with essentially no brain size increase in *Australopithecus* during the entire span of about 2 million years. Surely a large brain with all the attendant features is the most important difference between humans and all other species of animals. Neanderthals had heavy eyebrow ridges, as did all earlier hominids, and a pronounced chin and were rather stocky. Their general build, in fact, resembles that of the Lapps and Eskimos; perhaps Neanderthals were adapted for colder climates. And despite characterizations as brutish, Neanderthals in reality had a well-developed society, as evidenced by complex burial sites and a tool culture.

Modern humans (*Homo sapiens sapiens;* Fig. 16.33) first appeared over 90,000 years ago in southern Africa and in the Middle East and 45,000 years ago in Europe; like the Neanderthals, they were hunters and gatherers. Unlike other species, though, modern humans have such features as a high, flattened brow; a higher, shorter skull; and a more rounded face. Neanderthals and *H. sapiens sapiens* coexisted for perhaps 10,000 years, but modern humans, perhaps because of more highly developed technical skills, arts, and language, dominated and rapidly spread throughout the world.

Because of their enormous adaptive abilities, humans have been able to move into more diverse environments than any other organism by the simple device of constructing an environment if adverse conditions exist. As a result of this geographic ubiquity, distinct races have appeared. Many racial characteristics clearly were adaptive for particular environments. One example is skin pigmentation. No one is sure why different skin types evolved, but the strong suggestion is that it was in response to ultraviolet light intensity at different latitudes. Pigmented skin is an effective ultraviolet light filter, which not only protects delicate tissues but also prevents overproduction of vitamin D. It is postulated that, as humans moved into higher latitudes, where ultraviolet radiation is less intense, there was insufficient light to promote the synthesis of vitamin D that occurs in the skin. Selective pressures operated toward light skin, hence the evolution of white skin.

Most anthropologists believe that one ancestral stock spread from Africa or Asia Minor to other, more or less isolated regions

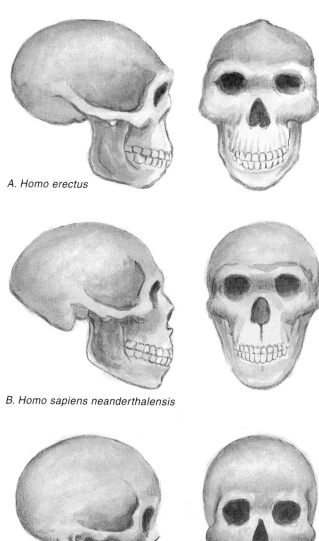

A. *Homo erectus*

B. *Homo sapiens neanderthalensis*

C. *Homo sapiens sapiens*

Figure 16.33   A: About 1.6 m.y. ago, most of the diversity of African hominids were replaced by the most long-lived (surviving over 1.3 m.y.) species, *Homo erectus*. These tall, large-brained (almost 1,000 cm³ brain capacity) people dispersed from their African homeland about 1.0 m.y. ago and spread over most of Eurasia. Indeed, they were first discovered in Southeast Asia ("Java man") and China ("Peking man"). B: Even more advanced were the Neanderthals, here considered a subspecies of *Homo sapiens*. First appearing about 170,000 years ago, they were widespread throughout the glaciated regions of Europe and western Asia until their mysterious disappearance about 40,000 years ago. Although Neanderthals had the same or larger brain size as we do, they had much stronger, more robust skeletons suited to their existence in cold climates. Their distinctive skull has a protruding face and brow ridges, and the rear of the skull stretches backwards to a point. C: About 100,000 years ago, the first fossils of truly modern-looking *Homo sapiens* appeared in Africa. *Homo sapiens* migrated into Eurasia about 45,000 years ago; they are called "Cro-Magnon men," although they are anatomically identical to us.

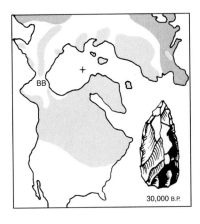
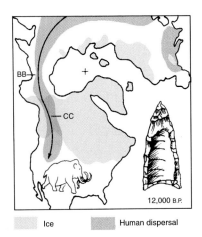
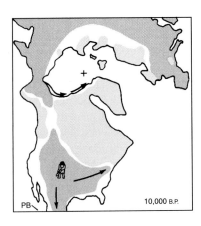

Figure 16.34   Probable history of entry and dispersal of humans into the New World, via the Bering land bridge (BB) and Canadian Plains corridor (CC) and then into South America via the Panamanian bridge (PB). Characteristic early projectile points are shown for Asia 30,000 years B.P. and for North America 12,000 years B.P. *(Adapted from P. S. Martin, 1967,* Natural History, *December, pp. 32–38, and personal communication from J. Stoltman, 1987, University of Wisconsin.)*

and then became racially diversified in response to local environmental selective pressures. By no later than 10,000 years ago, all major races of modern humans had appeared and had occupied their primary distribution areas (i.e., their distributions as of the beginning of historic times).

Final agreement on a precise phylogenetic tree for the hominids is a long way off (Fig. 16.32). Each summer produces a new find that alters the lineage of one part or another of the tree. Intensive collecting by a number of anthropological teams in Africa and elsewhere will continue to fill in gaps and make the picture of the evolution of *Homo sapiens* more complex and controversial.

Human Entry into America

Although *Homo sapiens* is known to have appeared in the Old World about 500,000 to 300,000 years ago, our species was a late-comer to the New World. Evidence of first entry is scant, but it appears that humans did not cross the Bering land bridge from Siberia into Alaska until 20,000 to 30,000 years ago during the lowered sea level of the Wisconsinan glaciation (although some say as old as 50,000 years ago). This was the same route used earlier by such other mammals as the woolly mammoth and the reindeer, mammals the early humans in America hunted very effectively.

The first comers would have been isolated in Alaska by the ice sheet that covered practically all of Canada (Fig. 16.34). Only when the ice retreated sufficiently to open the *Canadian Plains corridor* several thousand years later could the Stone-Age hunters reach the United States in pursuit of big game.

The earliest confident dates, 12,000 years ago, are for the Clovis and Folsom cultures of the southwestern states, which were characterized by the distinctive, fluted projectile points that evolved in North America from unfluted types found in Asia (Fig. 16.34). Mammal-hunting peoples with similar projectiles also migrated via the Isthmus of Panama bridge into South America

and reached the southern end of that continent (6,500 km away) only 2,000 years later, apparently causing extinctions of several mammal species as they migrated. Ten thousand years ago, humans occupied most parts of both Americas, and new projectile types as well as other tools were beginning to appear. This dispersal seems remarkably rapid, and there are recent claims of earlier human occupation of South America going back at least 13,000, and perhaps to 30,000 to 33,000, years ago. A few anthropologists speculate that people may have boated to South America before humans crossed the Bering bridge, but most still feel that entry was only via Alaska. The date of human arrival continues to be hotly debated.

Eskimo people dispersed across the Arctic during the post-Wisconsinan warming. They migrated as far eastward as southern Greenland, where they had come in contact with Vikings from Europe by A.D. 1000. In northern isolation, Eskimos adapted both biologically and culturally to one of nature's harshest environments. Farther south, various groups also were developing distinctive cultures in semi-isolation from one another. The most sophisticated of these were the agricultural civilizations of the Mississippi valley, Central America, and Peru, which were discovered and plundered by the sixteenth-century European conquistadors.

Human migrations into the Americas under the influence of two land bridges, the fluctuating Canadian Plains corridor, and diverse new ecological niches provide further vivid illustrations of the basic principles of biogeography and evolutionary adaptation that we discussed for earlier life forms.

Climate in Human History

The Climatic Optimum

We have seen how climate has influenced prehistoric humans and affected the development of human cultures, especially during the Pleistocene. Climatic cycles also had a profound effect in

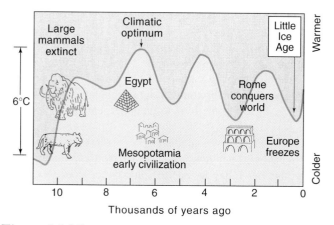

Figure 16.35 The effects of climatic cycles on the past 10,000 years of human history.

Holocene time. From the peak of glaciation about 18,000 years ago, the earth warmed rapidly over the next 7,000 to 8,000 years, so that the present interglacial conditions began about 12,000 to 10,000 years ago. This was when humans apparently migrated to the Americas in large numbers, and the large Pleistocene mammals also disappeared. The climate continued to warm until it reached a peak known as the *Climatic Optimum* about 7,000 to 6,000 years ago (Fig. 16.35). It is no coincidence that the first great civilizations of Egypt, Mesopotamia, and the Indus valley flourished about this time. Today the Middle East is a harsh, forbidding place, and it is difficult to imagine that it was once a "Promised Land" flowing with milk and honey, as described in the Bible. But during the Climatic Optimum, these regions were much wetter and more hospitable—truly the "Fertile Crescent" described in history books.

About 3,000 years ago (between 1200 and 800 B.C.), the climate began to change again, and the Middle East experienced long droughts punctuated by annual flooding of the rivers. Under these conditions, the great river valley civilizations needed to contain and distribute floodwaters for irrigation, store the annual harvests to survive the drought, and distribute these food supplies. This apparently triggered the rise of highly complex, centralized governments (such as those of the Egyptian pharaohs and Babylonian kings) to coordinate all these activities. Civilization may not have arisen in the same way (or possibly at all) without the appropriate climatic conditions for large, agriculturally based cultures with major cities, complex religions and governments, and all the other features we associate with advanced cultures.

Other civilizations did not respond as effectively. The great Mycenaean culture in Greece collapsed, shortly after it had conquered and destroyed Troy. Archeological research at the site of Troy shows that the surviving city was so besieged by famine that it never fully recovered (as Herodotus had written). Elsewhere in Turkey, the great Hittite empire, which once beat back the Egyptians, also collapsed. During this period of drought and famine in the Middle East, many other weakened civilizations fell prey to foreign invaders. The bloodthirsty Assyrian empire reached its peak about 800 B.C. after subduing nearly every

other Mesopotamian culture. It even overran Egypt and captured the capital at Memphis; not until the Assyrians fell was Egyptian unity restored, but the Pharaonic dynasties were never again as strong.

Another important climatic shift, known as the *Subatlantic Deterioration,* began around 2,500 years ago (500 to 400 B.C.). Apparently changes in wind and moisture patterns in northern Europe ended the Bronze Age there and spurred Germanic invasions of southern Europe and Scandinavia. The Mediterranean world of the classic Greek civilization was considerably colder and wetter than it had been during the time of the Minoan and Mycenean civilizations. Even the art shows this climatic change: Minoan paintings show sparsely clothed people living in buildings with flat roofs suitable for a warm, dry climate, but the classical Greeks wore heavier clothing and had pitched gabled roofs for the wet winters. In early Rome, the climate was much colder, and the Tiber River froze frequently.

A century later, climate began to warm in southern Europe, and the frozen Alpine passes began to thaw. Hannibal's Carthaginian armies and their elephants took advantage of this thaw to cross the Alps and invade the plains of Rome; a few decades later, the Romans invaded what had recently been the cold regions of northern Europe. Some historians believe that the end of this warming period, from about A.D. 450 to 500, produced a prolonged freeze and drought in central Europe, which stimulated the barbarian migrations that ended the Roman Empire.

Medieval Climatic Fluctuations

After an extended cool period, climate began to warm again during the *Medieval Warm Period,* about A.D. 950. With this thaw, the icy North Atlantic began to open up, and Vikings such as Erik the Red and Leif Eriksson were able to settle in Greenland and Labrador. Contrary to school books, they, not Columbus, were the first Europeans to reach North America. Simultaneously the great Mayan civilizations of Yucatan, which were at the peak of their power in 950, collapsed within a few decades. Several people have suggested that this collapse was due to the global climatic shifts, which apparently affected tropical rainfall patterns.

The Medieval Warm Period ended in the 1300s, and the Viking outposts soon disappeared due to the cold and famine. The Greenland colonies show this decline particularly clearly. In A.D. 1200, Greenland had a population of about 4,000 with a successful agricultural economy, but soon the settlements declined. The growing season became shorter and shorter, so the more northerly outposts had to be abandoned by 1350. The sea ice became increasingly impassable, cutting off the colonists for months. Hard times can be clearly seen in the archeological remains. Early burials were in deep graves, with coffins made of imported wood. As the colonists became poorer, they were buried in shrouds in shallow graves. Eventually most of the burials were of young people, suggesting a short life expectancy. The adult survivors were stunted and misshapen in growth, not tall and strapping like their Viking ancestors, and had extraordinarily worn teeth, suggesting a coarse, gritty vegetable diet. By 1410, the last colonists had abandoned Greenland.

During this cooling, Europe experienced unusually wet years, which caused flooding and rotted harvests. These triggered the famines of 1315 to 1317 and caused an unprecedented number of deaths during what historian Barbara Tuchman calls the "calamitous fourteenth century." In fact, the famine indirectly stimulated even a greater calamity, the infamous Black Death, or bubonic plague, which decimated Europe about 30 years later. To avert another great famine, great quantities of grain were imported from the Middle East; plague-carrying rats were inadvertantly imported, too. The rats and their plague originated in China, where plague deaths were reported as early as 1333. In the same year, China experienced severe floods of the Yellow River, including the largest flood of the Middle Ages, which killed 7 million Chinese.

The wet weather of this period triggered another equally grim disease. During the ninth to fourteenth centuries, a strange madness accompanied by a devastating illness decimated whole villages. Hundreds of people suffered convulsions, hallucinations, gangrenous rotting of the limbs, and often death; pregnant women miscarried. In chronic cases, the victims developed an icy chill; then their limbs darkened, shriveled, and fell from the body. Because of the blackened limbs that appeared to be burned, the disease was known as "St. Anthony's fire."

By 1596, the cause of the disease had been discovered. It was due to a fungus (*Claviceps purpurea*) responsible for ergot blight, which affected the rye used in making bread. A few blackened diseased kernels could infect a whole sack of flour and sicken everyone who ate it. Ergot blight was most severe during the early Middle Ages but disappeared almost completely in the 1500s. Later research showed that it grew only during the cold, damp conditions that prevailed until the fifteenth century; today it is virtually unknown.

The effects of this climatic change were not confined to the Old World. In the American Southwest, the Native American Anasazi culture had built large pueblos into the cliffsides and had developed an agricultural economy based on maize. The years 1271 to 1285 were extremely dry, and the crops failed. The problems of the Anasazi were accentuated by their clear-cutting of the trees on the plateau above them. This intensified the effects of flash floods, which washed away their crops in the dry creekbeds below. By 1300, most of the cliff houses had been abandoned, and their people had moved on to more reliable sources of water along the Rio Grande and on the Hopi Mesas. Two hundred years later, Coronado explored the region, seeking gold from the "Seven Cities of Cibola," but the great pueblos that may have spawned this myth had been long abandoned.

The Little Ice Age

The cooler and wetter conditions of the Middle Ages culminated in a period known as the *Little Ice Age,* which lasted from about 1550 to about 1850. Old lithographs of the Alps from this period showed that the glaciers had advanced far beyond their present extent. The paintings by the Dutch and English masters of this period frequently show people skating on frozen canals, lakes, and rivers that do not freeze today. The famous painting of Washing-

ton crossing the Delaware also shows large ice floes; today the Delaware River rarely freezes.

Once again, the main consequence of such climatic changes was famine. The great famine of 1594 to 1597 hit Europe particularly severely, causing cannibalism and food riots; in some countries, people ate everything in sight, including cats, dogs, and even snakes. In 1693, one-third of Finland's population died from another period of famine. In Scotland, cod fisheries failed, and the Scots experienced repeated famine. By 1691, 100,000 Scots, or 10 percent of the population, had fled to Northern Ireland to escape the famine. Today we see the bloody consequences of the conflict between their descendants and the Catholic Irish.

The effects of the Little Ice Age were felt all over the world. Northern Africa had been particularly wet and fertile before the 1500s and had developed a number of civilizations on the fringe of the Sahara Desert in Mali and Ethiopia. But climatic cooling and weather extremes decimated these civilizations. During the late 1500s and early 1600s a series of floods and famines caused a complete collapse of their cultures. In 1628, snow was reported far below modern levels in the equatorial mountains of Ethiopia.

In India, the great Moghul city of Fatepur Sikri was abandoned in 1588, only 16 years after it was finished. Historical records show that the monsoonal circulation was disrupted, and the region experienced a severe drought. China suffered through a series of severe winters between 1654 and 1676, causing severe famines and floods and millions of deaths.

As the Little Ice Age came to a close, it culminated in 1816 with one more extremely wet and unpleasant year in Europe. That summer, 19-year-old Mary Shelley wrote *Frankenstein,* and John William Polidori wrote *The Vampire.* Both authors spent their summer with Lord Byron and Percy Shelley near Lake Geneva, Switzerland. Cooped up in their house, they spent hours writing ghost stories and produced two of the most famous characters in horror fiction. The following year, the wine grapes were ruined, the harvests were very poor, and hunger trigered yet more riots and revolution in a world just recovering from the Napoleonic wars.

There were many other direct effects of this refrigeration on history, but some are less obvious. In the late 1700s, a number of factors caused the population of Ireland to nearly double, yet most of this population was poor and landless. They came to depend for survival on potatoes, which thrived in the colder conditions of the late Little Ice Age. But as world temperatures rose at the end of the Little Ice Age, conditions became wet and warm enough for the fungus that causes potato blight to thrive. In modern laboratory studies, this fungus (*Phytophthora infestans*) requires 12 hours at relative humidities of 90 percent or more, temperatures higher than 10°C, and free water on the potato leaves for at least 4 hours. The summer and winter of 1845 experienced record-breaking moisture and warmth, and the blight spread so quickly in 1846 that more than a million Irish died of famine within a few years.

The Present and Future

Since the end of the Little Ice Age, the world has experienced almost two centuries of relatively warm, mild climate. The warmest period in recent history occurred in the middle of the twentieth century; not surprisingly, these years correspond to some of the most productive periods in modern agriculture. In fact, this recent period was the mildest in the past 1,000 years since the Middle Ages, and after an optimum in the 1940s, climate has been declining ever since. The most severe weather in recent history occurred in the early 1970s, when droughts occurred in Russia, India, and Africa. In 1972, the Soviets were forced to buy American grain to replace their poor harvests, a landmark event in the history of the Cold War era.

Over the long run, one of the most obvious effects of the past century of climatic change has been the continual drying of northern Africa. In the Sahel region south of the Sahara, continual desertification has produced year after year of famine, and our television screens are often filled with scenes of starving African children in Biafra, Sudan, Ethiopia, or Somalia. Although the desertification is accentuated by overgrazing and bad agricultural practices (with civil wars amplifying the famine), the beginning of the drying was clearly triggered by climatic shifts.

What of the future? If the Milankovitch cycles were allowed to run their course without human interference, then we should be near the end of the present interglacial. During the last peak of warming about 125,000 years ago, the actual interglacial period lasted only about 7,000 to 10,000 years and then declined into glaciation. We have been experiencing interglacial conditions for over 10,000 years now, so we should be reaching full glacial conditions in about 23,000 years. If so, then the great ice sheets could return to the northern continents, and New York City would again be covered in ice (Fig. 16.36).

We used to think that the change to a glacial climate happened over a few thousand years, but it now appears that this can occur so rapidly that human societies would have little time to adjust. In 1993, the completion of two holes drilled through the thickest part of the Greenland ice cap provided a jolting surprise. Analyses of two 3,000-meter-long Greenland ice cores indicate that, during the past 250,000 years, the average global temperature changed many times by as much as 10°C, which is enough to trigger either a glacial or an interglacial episode! By contrast,

Figure 16.36 Which way will future climate go? Will Increasing atmospheric dust cause readvance of glaciers, or will increasing carbon dioxide cause melting of existing ice and a substantial rise in sea level? Either way, we lose the Big Apple. *(R. H. Dott, Jr.)*

the data show that the climate of the Holocene has been unusually stable.

However, humans are now interfering with global climate. As we discuss in Chap. 17, the greenhouse warming triggered by our burning of fossil fuels may warm the earth far beyond the effects of the normal glacial-interglacial cycles. If so, we might experience a "superinterglacial," with abnormally warm conditions and extensive melting of ice caps, triggering a global rise in sea level. Either way, the coastal cities of the Northern Hemisphere will be flooded, and the Statue of Liberty (Fig. 16.36) has two possible fates, both unpleasant: immersion by the rising sea or burial in glacial ice!

Summary

Glaciation

- Recognition of continental-scale glaciation in the 1830s replaced the diluvial hypothesis for erratic boulders and boulder clay (now called till), which were postulated to have been drifted over northern Europe by the biblical Flood.
- Effects of glaciation (in addition to erratic boulders, scratched bedrock surfaces, and moraines) include
  1. Isostatic rebound of crust formerly warped down by the load of large ice sheets thousands of meters thick
  2. Lakes formed far from glaciated regions in closed valleys of arid western states as well as along the ice margins. The latter contained annual varve laminations that can be counted like tree rings. All the lakes rose and

fell with glacial-interglacial changes. Some of them drained very suddenly and violently.
  3. Wind blew tremendous volumes of loess (silt) away from valleys that drained the melting glaciers.
  4. Worldwide sea-level changes responding to glacial-interglacial oscillations resulted in the following:
     a. River terraces along valleys
     b. Dead coral reefs now high and dry
     c. Marine beaches and deltas now high and dry
     d. Temporary land bridges drowned by high sea levels
     e. Submerged beach ridges on continental shelves
     f. Submarine canyons extending from continental shelves to the deep sea

- Pleistocene correlation and chronology are established by
  1. Numerical dating techniques (e.g., radiometric, magnetic polarity, oxygen-isotope curves), reversal of coiling of Foraminifera, and volcanic ash layers.
  2. The classic four glacial episodes are now suspect; oxygen-isotope curves suggest many more cold episodes extending back more than 1 million years. On land, younger glacial episodes obliterated most of the evidence left by older ones.
- Explanation of continental glaciation must account for these geologic facts:
  1. Glaciations have been relatively rare.
  2. Normal past climate was warmer and more uniform than Pleistocene climate.
  3. A drop of average annual global temperature of the order of only 10°C below the present level could cause a renewal of glaciation.
  4. Glaciation would have begun when global temperature declined to a critical threshold for long periods; more snow would then accumulate above the permanent snow line in the winter than would melt in the summer. This would have occurred earliest in always colder polar regions.
- Hypotheses for glaciation have appealed to such things as
  1. Solar changes (geologically untestable)
  2. Changes in the earth's orbit (Milankovitch cycles)
  3. Terrestrial changes in the global heat budget
     a. Changes of atmospheric transparency
     b. Albedo changes
  4. Changes in heat exchange rates between equator and poles

A plausible working hypothesis for glaciation that is consistent with geologic evidence invokes two unrelated elements:
  1. Global cooling beginning 30 to 40 million years ago triggered by continental drift, which thermally isolated the poles, and by enlargement of land area, which increased total earth albedo.
  2. Always acting Milankovitch orbital effects could then cause the small oscillations between glacial and interglacial climates.

Pleistocene Life

- Pleistocene life was profoundly affected by cold climate; at no other period in the fossil record is the effect of cold climate more clearly shown. Organisms migrated or became extinct in direct response to the glacial and interglacial cycle.

Human Evolution

- Human evolution began with the first hominid, *Sahelanthropus,* which appeared about 6 to 7 million years ago in eastern Africa. The first species of the genus *Homo* was *Homo habilis,* dated at 1.75 million years. It was first found at Olduvai Gorge by Louis Leakey. The first modern human, *Homo sapiens sapiens,* did not appear until 90,000 years ago in south Africa, and humans arrived in Europe only 35,000 years ago.

Historical Climate Changes

- Since the beginning of the present interglacial 10,000 years ago, human history has been profoundly influenced by global climate fluctuations. During the Climatic Optimum about 7000 years ago, great civilizations arose in the Middle East. In the Little Ice Age between 1550 and 1850, the world experienced unusually cold conditions, and Alpine glaciers advanced. Since then, world climate has been unusually warm, reaching a peak in the 1940s.
- If natural climatic cycles are allowed to continue, then we should be entering the next glacial period. However, the warming of the earth by greenhouse gases may cause a "superinterglacial" instead.

Readings

Black R. F., R. P. Goldthwait, and H. B. Willman, eds. 1973. *The Wisconsin Stage.* Geological Society of America Memoir 136.

Bolles, E. B. 1999. *The ice finders: How a poet, a professor, and a politician discovered the Ice Age.* New York: Counterpoint.

Bowen, D. Q. 1978. *The Quaternary.* Oxford, UK: Pergamon Press.

Broecker, W., and G. Denton. 1990. What drives glacial cycles? *Scientific American,* January, pp. 49–56.

Brunet, M., et al. 2002. A new hominid from the upper Miocene of Chad, central Africa. *Nature,* 418:145–51.

Bryson, R. A., and T. J. Murray. 1977. *Climates of hunger, mankind and the world's changing weather.* Madison: University of Wisconsin Press.

Charlesworth, J. K. 1957. *The Quaternary Era.* 2 vols. London: E. Arnold.

Cline, R. M., and J. D. Hays. 1976. *Investigation of Later Quaternary paleoceanography and paleoclimatology (CLIMAP).* Geological Society of America Memoir 145.

Delson, E. 1985. *Ancestors: The hard evidence.* New York: Liss.

Diamond, J. 1992. *The third chimpanzee.* New York: HarperCollins.

Dort, W., Jr., and J. R. Jones, Jr. 1970. *Pleistocene and recent environments of the Central Great Plains.* Lawrence: University of Kansas Press.

Eldredge, N., and I. Tattersall. 1982. *The myths of human evolution.* New York: Columbia University Press.

Emiliani, C., and J. Geiss. 1957. On glaciations and their causes. *Geologische Rundschau,* 46:576–601.

Fagan, B. 1990. *The journey from Eden.* New York: Thames and Hudson.

Fleagle, J. G. 1988. *Primate adaptation and evolution.* San Diego: Academic Press.

Flint, R. F. 1971. *Glacial and Pleistocene geology.* New York: Wiley.

Gibbons, A. 2002. In search of the first hominids. *Science,* 295:1214–19.

Gribbin, J. 1972. *The climatic threat.* New York: Walker.

———. 1982. *Future weather.* New York: Penguin.

Guthrie, R. D. 1990. *Frozen fauna of the mammoth steppe.* Chicago: Chicago University Press.

Hays, J. D., J. Imbrie, and N. J. Shackleton. 1976. Variations in the earth's orbit: Pacemaker of the Ice Ages. *Science,* 194:1121–32.

Imbrie, J., and K. P. Imbrie. 1979. *Ice Ages: Solving the mystery.* Short Hills, N.J., Enslow.

Johanson, D., and M. Edey. 1981. *Lucy, the beginnings of humankind.* New York: Simon and Schuster.

Klein, R. 1989. *The human career: Human biological and cultural origins.* Chicago: Chicago University Press.

Kurtén, B. 1968. *Pleistocene mammals of Europe.* New York: Columbia University Press.

———. 1988. *Before the Indians.* New York: Columbia University Press.

———, and E. Anderson. 1980. *Pleistocene mammals of North America.* New York: Columbia University Press.

Ladurie, E. L. R. 1971. *Times of feast, times of famine.* New York: Doubleday.

Lamb, H. H. 1972. *Climate: Past, present, and future.* London: Methuen.

———. 1982. *Climate, history, and the modern world.* London: Methuen.

Leakey, R., and R. Lewin. 1992. *Origins reconsidered, in search of what makes us human.* New York: Doubleday.

Levenson, T. 1989. *Ice time: Climate science and life on earth.* New York: Harper & Row.

Lewin, R. 1988. *The age of mankind.* Washington, D.C.: Smithsonian.

———. 1997. *Bones of contention.* Chicago: Chicago University Press.

———. 1997. *Principles of human evolution.* Boston: Blackwell.

———. 1998. *Human evolution: An illustrated introduction.* Boston: Blackwell.

———. 1998. *The origin of modern humans.* New York: Freeman.

———, C. C. Swisher, and G. H. Curtis. 2000. *Java man.* New York: Simon and Schuster.

Loomis, W. G. 1967. Skin pigment regulation of vitamin-D biosynthesis in man. *Science,* 157: 501–6.

Martin, P. S., and R. G. Klein, eds. 1984. *Quaternary extinctions, a prehistoric revolution.* Tucson, AZ: University Press.

Matsch, C. L. 1976. *North America and the great Ice Age.* New York: McGraw-Hill.

Mellars, P., and C. Stringer. 1989. *The human revolution: Behavioural and biological perspectives on the origins of modern humans.* Edinburgh: Cambridge University Press.

Pielou, E. C. 1991 *After the Ice Age, the return of life to glaciated North America.* Chicago: Chicago University Press.

Rankama, K., ed. 1965. *The Quaternary.* New York: Wiley.

Schneider, S. H., and R. Londy. 1984. *The coevolution of climate and life.* San Francisco: Sierra Club.

Smith, F. H., and F. Spencer, eds. 1984. *The origins of modern humans.* New York: Liss.

Sutcliffe, A. J. 1985. *On the track of Ice Age mammals.* Cambridge, UK: Harvard University Press.

Tattersall, I. 1996. *The fossil trail: How we know what we think we know about human evolution.* Oxford, UK: Oxford University Press.

———. 1999. *Becoming human.* New York: Harcourt.

———. 2001. *The human odyssey, four million years of human evolution.* Englewood Cliffs, N.J.: Prentice Hall.

———, and J. Schwartz. 2001. *Extinct humans.* New York: Harcourt.

———, and N. Eldredge. 1977. Fact, theory, and fantasy in human paleontology. *American Scientist,* 65(2):204–11.

Willis, D. 1989. *The hominid gang.* New York: Viking.

Wright, H. E., and D. G. Frey, eds. 1965. *The Quaternary of the United States.* Princeton, N.J.: Princeton University Press.

(Charlie Ott/Photo Researchers.)

Chapter 17

The Best of All Possible Worlds?

Pangloss proved admirably that there is no effect without a cause, and that, in this best of all possible worlds, My Lord the Baron's castle was the finest of castles. "Things cannot possibly be otherwise, for everything being made for an end, everything is necessarily for the best end."

Voltaire, *Candide* (1759)

Life can only be understood backward, but must be lived forward. Sören Kierkegaard

▶ The explosive growth of human population on this planet threatens not only ourselves and our environment but also many other species, which have been around millions of years longer than we have.

▶ Humans are fouling our environment in many ways. Some of the most serious changes include the increase in greenhouse gases, which threatens to warm this planet past normal interglacial limits, and the loss of stratospheric ozone, which protects us from dangerous ultraviolet radiation.

▶ Humans are exploiting many renewable resources much faster than they can be replaced and exhausting nonrenewable resources so rapidly that many are nearly completely gone; others will be depleted within a century or two.

▶ The Western industrial world (particularly the United States) is extraordinarily wasteful of its resources, spending not only our generation's allotment but also much of the share of future generations and of the less developed world.

▶ Our high standard of living in the industrialized world has been bought at an incredible price in environmental damage and depleted resources, which clearly cannot be sustained for much longer. The rate of these changes is unprecedented in geologic terms. We need to ask ourselves now (when it is still possible to make a difference) whether our standard of living and economic growth are worth the heavy price that future generations must pay.

At the end of our long journey through the maze of geologic time, it is appropriate to ask, What are the greatest implications of earth history? We believe that there is a message of lasting value for all, and in this closing chapter we enumerate what seem the few most important and transcendent implications.

From a scientific standpoint, we have attempted to plot a course through the historical maze based upon an objective relating of evidence and the formulation and testing of working hypotheses in a search for general explanations of the development of the physical earth and its life. However, though we have succeeded in formulating several comprehensive explanations, it should be obvious that the scientific process is never completed, especially in a young science such as geology. Hypotheses are transitory; we should nurture them only until we can construct better ones. It is not in the least surprising that some of the speculations presented in our first six editions are now outdated. By the time you read this page, perhaps some of the ideas incorporated in this edition will already be out of date. Bernard Shaw observed that "all progress is initiated by challenging current conceptions."

It is common to all intellectual pursuits that each question answered tends to raise new questions. This situation is so universally true that what we may question if *anything*, strictly speaking, is fact, for so-called facts have an annoying tendency to be reinterpreted without notice. Skeptical minds question the most basic premise of all—namely, that there is an inherent order in nature to be discovered, described, and explained. Voltaire once suggested that "if life has no meaning, man will invent one," and existentialist writer Albert Camus wrestled with the dilemma presented by the seeming chaos of constant change and lack of fixed absolutes. Is there a hopeless futility in the ceaseless quest for understanding, or does the search itself give meaning to human activity? Robert Louis Stevenson already had anticipated Camus with the affirmation that "to travel hopefully is a better thing than to arrive."

Three Important Maxims

Faced with the reality of uncertainty, it is important to ask if there are not at least a few relatively durable geologic tenets. We believe that the study of earth history carries three maxims of far greater significance and timelessness for humanity as a whole than any alleged facts of history or the hypotheses we have developed to explain them.

First, a study of geologic history necessarily develops *wholly new concepts of time and rates of change.* It clarifies the human position on the scale of life history. Instead of having appeared together with other life forms soon after the origin of the earth, as was assumed until the eighteenth century, humanity has resided on the earth for but a brief 2 to 2.5 million years—only 0.04 percent of geologic time!

Second, the history of the earth and life shows forcefully that, indeed, *"there is nothing permanent except change."* More important, however, is that the change is of a special kind. It is not

the perfectly cyclic, or steady-state, change envisioned a century ago but, rather, an *irreversible, cumulative, evolutionary change.* "Nature created ever new forms; what exists has never existed before; what has existed returns not again," wrote Goethe as early as 1781.

The third major implication follows from a study of evolution through geologic time, which is what this book is all about. This, most important maxim is the great *ecologic interaction, or feedback, between the living and nonliving realms of the earth throughout history.* Because an evolutionary sequence represents a series of unique events, history cannot, after all, repeat itself exactly. Although organic evolution has been possible because of random mutations in genes, the *results of evolution were not random.* The way that life developed on the earth is the result of natural selection operating on mutants through a 3.5- or 4-billion-year series of changes. Involved were changes in both the living and the nonliving realms *in a particular order and at particular rates.* Had some of these occurred in a different fashion, the course of evolution would have been different, yet no one can say exactly how it would have differed.

One of the chief goals of science is *prediction of future consequences,* and if a study of earth history truly has relevance to the lives of humans, then we feel obliged to make some extrapolations from history and to suggest some implications for the future of humans and their environment (Fig. 17.1). We shall do this by reviewing briefly the evolutionary highlights in earth history as they affected life, and we shall conclude by applying our third maxim to the contemporary human world.

Evolution of the Earth

The Idea of Evolution

The concept of evolutionary change is one of the greatest of human ideas, and it applies as much to the nonliving as to the biologic realm. Evolution, we feel, provides a powerful unifying basis for the study of earth history. By now, it should be clear that nature cannot be understood fully without special attention to evolution, which is most completely revealed in the historical sciences of geology, astronomy, and evolutionary biology.

There was little interest in history, however, until the 1700s, when Pompeii, Mayan ruins, and other ancient cities were discovered and excavated and fossils were finally understood. It was also in the 1700s that people really began to think of natural change in an evolutionary fashion (see Chap. 3). Previously, practically everything was assumed to be fixed in form and in relation to everything else. All life was assumed to have been created nearly simultaneously around 4000 B.C. No extinctions or new creations of species had yet been demonstrated, though Robert Hooke and G. L. L. de Buffon suspected them.

Seventeenth-century cosmogonist Leibniz first preached a principle of grand continuity in nature, a principle that led to a notion called the Great Chain of Being, from atoms to God. According to this view, nature abhors discontinuities; rather, everything is

Figure 17.1   "When you have seen one redwood, you have seen them all," said the then governor of California (and later president) during a 1960s controversy over establishment of Redwood National Park. *(Charlie Ott/Photo Researchers.)*

arranged in gradational sequences. Development of the calculus by Leibniz and Newton doubtless lent impetus to embryonic evolutionary thought by focusing on the integration of small changes in both space and time. In the nineteenth century, Lord Kelvin, using newly discovered principles of thermodynamics, anticipated the modern conception of the physical earth as a dynamic, irreversibly changing sphere when he challenged Lyell's ultrauniformitarian, steady-state earth. Then in 1896, 37 years after Darwin's momentous *On the Origin of Species* appeared, the discovery of radioactivity provided an important example of *inorganic evolution* through transformation of one element to another, as described in Chap. 5. It also provided an important clue to an overall chemical evolution of the earth in which radioactive decay has played a major role, as emphasized in Chap. 6.

Chemical Evolution of the Earth

Terrestrial evolution likely began with the aggregation and densification of the protoearth. Internal heating, especially through radioactivity, facilitated the density differentiation of core and mantle between 4.5 and 4.7 billion years ago, and later the crust. The primitive atmosphere and seawater began accumulating and underwent changes in composition as a result of igneous activity, weathering, and photochemical reactions.

As one result of chemical changes on the earth's surface, life developed—apparently as an inevitable consequence of the evolution of complex carbon-bearing molecules (see Chap. 9). Three or four billion years ago, *this certainly was the best of all possible worlds for life* in our solar system, for it contained abundant carbon, hydrogen, nitrogen, oxygen, and phosphorus; had a critical surface temperature range such that much water was present in the liquid state; and possessed a strong magnetic field to shield the surface from most cosmic radiation. It did not yet have an ozone shield from ultraviolet radiation, however. To the famous Russian biochemist, A. I. Oparin, matter is always changing from one form of motion to another, each more complex and harmonious than before. Life, to him, appeared as a very complex form of matter in motion that arose at a particular stage in the general evolution of matter on Earth. As we have seen, the origin of life required very special conditions, especially moderately warm temperatures, high-energy radiation, and lack of abundant free oxygen in the environment.

Evolutionary Feedback

Life developed in an irreversibly evolving physical environment. Once photosynthesis became possible, free oxygen, and eventually ozone, began to accumulate in the atmosphere. With the addition of free oxygen, life brought about a change in the physical realm that irreversibly precluded life being created anew.

Organisms are also important in the cycling of both carbon and nitrogen through the atmosphere, water, and soil. Atmospheric argon evolved continuously through decay of radioactive potassium 40. It is the only atmospheric gas wholly independent of life processes.

As noted in Chaps. 6 and 8, an oxygen-bearing atmosphere is indicated no later than about 2.0 to 2.5 billion years ago; by at least 0.7 billion years, marine animals also had appeared, indicating that oxygen respiration had become possible. But neither plants nor animals had invaded the land as yet, though large land areas had existed for long times, especially at the end of Cryptozoic time. Only after much free atmospheric oxygen accumulated could the important ultraviolet-filtering ozone layer develop high above the land surface. Such development may have taken until middle Paleozoic time, when land organisms first appeared (see Chaps. 12 and 13). Significantly mountain building at that same time formed a variety of potential ecological niches, and many of these were quickly filled by a sudden invasion of the land by both plants and animals. Here we see a clear example of physical changes influencing the path of organic evolution. Conversely once organisms were established on land, they in turn markedly influenced the land by greatly modifying processes of weathering, erosion, and sedimentation. Furthermore, we see that plate tectonics has also affected topography, climate, and thus biogeography and organic evolution. These are examples of the grand *evolutionary feedback* that has characterized the overall history of the earth.

Increased mountain building on all continents in late Paleozoic time created more new habitats for expanding land life. As land emerged from the sea in Carboniferous time, great coastal coal swamps formed on a colossal scale. Relative land area and elevation are major factors in the earth's heat budget, as are atmospheric and oceanic circulation. As the continents achieved their largest total area and greatest elevation since Cryptozoic time, climate changed in response. Permo-Triassic red-bed deposits became nearly universal on northern lands, which were then located in low latitudes, while glaciation set in on southern, high-latitude ones.

Late Paleozoic and early Mesozoic land animal life was homogeneous on most continents, reflecting the connections among most land areas. Plants were somewhat less uniform, apparently due to climatic zonation. By Cretaceous time, all land organisms had begun to be more differentiated and diverse, on present continental units. The fossil record indicates that, as the Mesozoic separation of continents occurred, all land life continued to become more diverse and endemic. The presence of several tenuous land bridges has maintained some interchange of land organisms between the different continents up to the present, but Cenozoic land life was far more differentiated than earlier life. In the south, isolation has been more complete and of longer duration than in the north, with the result that more disjunctive distributions of organisms are found in the south (as discussed in Chaps. 15 and 16). Thus, there is a global feedback between plate tectonics and the history of life.

The late Cenozoic tectonic episode has been more extreme and rapid than geologists appreciated until recently. A global Miocene revolution caused many profound changes. These included inception (or rejuvenation?) of all island-arc systems as we know them today, the Alpine-Himalayan mountain building, changes in the ocean-ridge-spreading regime leading to changes in plate motions, and the beginning of large-scale rifting on five continents. As all this occurred, continents enlarged and rose.

During Cenozoic time, mammals and plants have become greatly diversified, and some major extinctions have occurred, doubtless influenced by physical changes in another great feedback process. Finally, the atmospheric heat budget tipped in favor of glaciation, which began in the Oligocene in polar regions and the Pleistocene in mid-latitudes. It was during this geologically recent climatic and structural turmoil that *Homo* evolved in Africa or Asia Minor and within the brief span of 2 million years dispersed throughout all land areas. In only the past two centuries, humans have modified the earth's surface, atmosphere, and water to such a phenomenal degree as to represent the most rapid and potentially catastrophic organic event in all of geologic history. Anthropologist Loren Eiseley mused that, whereas evolution of the human body from that of an aggressive ape is essentially completed, humans seem to possess an unfinished mind that "could lead humanity down the road to oblivion." In the remainder of this chapter, we shall explore the implications of Eiseley's warning.

Our Place in the Global Ecosystem

Effects on Other Life

Some astronomers estimate that there are about 10^{20} planets in the universe capable of supporting intelligent life. Such life has one chance per planet, however, for the resources of each planet are finite and evolution is irreversible. The human, though not nearly as strong as many other animals, is the cleverest and most adaptable organism that has yet evolved on Earth. Through the development of tools and the utilization of stored energy (i.e., fuels), humans have more than compensated for lack of brawn. Tools, however, have also given us an enormous capability for manipulating and contaminating the environment, leading to a common belief that a complete control of the environment will soon be possible. Indeed, Western humanity long ago appointed itself the predestined master over all of nature. Thus, the natural environment came to be thought of as a *commodity* possessed by the human race.

Humans have brought about the extinction of hundreds of species, with an extinction rate averaging about one species per year over the past century. We now threaten an additional 500 species. By the end of this century, humans may have caused extinctions of tens of thousands of species. At about 14 extinctions per week, by the twenty-second century, humans may have caused the greatest mass extinction in earth history.

Fur coats, for example, take such a merciless toll on the wild cat family (e.g., six leopard skins per coat) that several of the cats are all but extinct because of human vanity. Besides overtly killing and inadvertently poisoning organisms, humans also threaten many forms by destroying their habitats—often quite unwittingly. In some cases, we have introduced new predators and, in other cases, competitors for food. It is estimated, for example, that in the past the average life span for isolated island bird species was about 200,000 years. Since aboriginal humans have populated islands, the bird species have averaged only 30,000 years, and since Europeans have arrived, the span has dropped to only 12,000 years.

Many of the results of human tampering with natural communities have been quite unexpected. Some species doubtless become extinct without ever being named—or even being discovered! The effects of upsetting ecological balance are difficult to predict with present limited knowledge. For example, oak forests have *increased* in area since Anglo-Saxon settlement of the northern Mississippi valley region, just the reverse of the general pattern of conquest of the frontiers. In this case, cessation of the Native Americans' habit of regularly burning prairie grass has allowed oak saplings to thrive and mature as never before. Here the settlers had to cut down forests *after* settlement, rather than before, as a consequence of their own activities.

After many years of fire control in western American forests, it is apparent that former periodic burning greatly enhanced the propagation of certain desirable forest trees that do not reproduce well in shady, thickly forested areas. Examples include the Ponderosa, or western yellow pine, one of our major lumber trees. This serves as an example of unforeseen results of well-meaning actions. Although no one is advocating wanton burning, foresters now let some natural fires burn themselves out with no interference, and they also do controlled burning.

Some deliberate manipulations of animal communities were motivated only by sentimentality, such as the introduction of the English sparrow into North America and the rabbit into Australia.

Serious devastation of rangelands resulted from the latter within only two decades and prompted the construction of thousands of kilometers of rabbit fences. All this stresses how little we really can predict of the long-term consequences of ecological manipulation and suggests some parallels with the results of natural competition among species in past ecosystems.

Many changes induced by humans have been beneficial to a few organisms. The skunk, opossum, coyote, raccoon, deer, rabbit, cow, robin, dog, and sparrow, to name but a few, have thrived in company with *Homo*. And no doubt urban slums and dump grounds have been great for rats. Roadsides also have produced a whole new ecological niche successfully exploited by rabbits, foxes, pheasants, skunks, ground squirrels, toads, and a host of other birds, mammals, insects, and, of course, so-called weeds.

Most of the changes cited so far seem relatively innocuous, but many others have been more detrimental and have adversely affected humanity itself. Degradation of the Great Lakes provides an outstanding example. Those lakes are said to hold nearly one-third of the world's freshwater supply, but most are hardly fresh anymore. Insecticides and herbicides, among other things, find their way into the lakes and then into the food chain—including that of humans (DDT has even been found in Antarctic penguins).

The entry of the parasitic sea lamprey into the upper Great Lakes produced catastrophic effects upon fish populations. Early in the twentieth century, the Welland Canal provided lampreys from Lake Ontario with a route around the Niagara Falls barrier and into the upper Great Lakes. By 1921, they had been found in Lake Erie; by 1937, in Lakes Michigan and Huron; and, in 1946, in Superior. They multiplied rapidly in the three upper lakes and soon decimated the fish populations, upon which they are parasitic. Of most direct concern were the quick inroads in lake trout fisheries. In Lakes Huron and Michigan, large trout production dropped to nothing; Superior's annual production shrank from 4.5 to 0.3 million pounds in only 15 years. Since 1953, a joint Canadian–United States control program has been in effect and has successfully reduced the lamprey population in Lake Superior so that trout have increased. Remedial steps have begun in some of the other lakes.

It can be argued that human-caused extinctions of other life forms, as well as acceleration of modifications of the natural environment, simply are contemporary examples of selection in action—"survival of the fittest." According to such a view, history is merely repeating itself as the human species evolves and expands its domain at the expense of others, producing a colossal *ecological replacement* analogous to many examples illustrated for past geologic periods.

However, although people live more and more in artificially controlled subenvironments of steel and concrete, our species is still the product of billions of years of evolution in the natural environment. Moreover, *Homo sapiens* always will be dependent upon the mineral and biological resources of the earth, as well as upon its water and atmosphere. Just as plants help to shape their neighbors, humanity is influenced by other animals, plants, and microbes as surely as individual people are affected by their cultural and intellectual atmospheres. People are part of a fantastically complex ecosystem—unable to extract more from the earth than it can produce and ultimately subject to selection processes acting upon their genetic makeup, as are other organisms. It behooves us humans to attempt to maintain an ecological balance that will ensure the well-being of future generations of people, as well as of the rest of the system upon which we are dependent. Rather than regarding nature as a commodity to be exploited, we should think of nature as a complex community, of which our species is both a member and the custodian. It is good, therefore, to remind ourselves of some of the consequences to us of membership in the community of nature.

The Impending Ecological Crisis

Fouling of the Environment

Environmental awareness began in 1961 with the publication of Rachel Carson's book *Silent Spring,* in which the author sounded the alarm of the insidious effects of DDT. It was rekindled in 1970 when then Senator Gaylord Nelson of Wisconsin suggested the annual observation of Earth Day. Since that first Earth Day, pollution of the environment has received much publicity, although the problem has been with people a very long time. Ancient Rome was perpetually encased in a smelly dust cloud, thanks to heavy horse-chariot traffic. And it is estimated that New York City had 150,000 horses at the turn of the twentieth century, each producing about 9 kilograms (20 pounds) of manure a day. Apparently in those good old days, horses far outdid today's dogs in littering urban streets. Pets continue to contribute significantly today. The 500,000 dogs in New York City deposit on the streets about 70,000 kilograms (150,000 pounds) of feces and 320,000 liters (90,000 gallons) of urine *each day.*

Broadly defined, pollution includes everything from dumping of raw sewage, DDT, and fertilizers into streams to the "uglification" of the landscape with billboards, junked cars, and other solid wastes (Figs. 17.2 and 17.3) and electric utility poles and wires. Even noise from industry, aircraft, vehicles, and television constitutes environmental pollution. The sonic boom generated by jet aircraft, for example, not only annoys urbanites by rattling their windows but also has triggered massive rock slides in western national parks.

Unfortunately, indifference toward degradation of the environment still is the rule. Seemingly only a garbage collectors' strike can arouse very many people. For example, as reported by *The Times* of London some years ago, the city council of an English coastal resort voted to continue pumping raw sewage into their bay because "that is the cheapest, and the proper method for the resort. We have been doing this for ages past and it can safely be done for ages in the future."

In recent years, chemical insecticides, fertilizers, and detergents have caused serious damage to many aquatic ecosystems. The 1986 spill of mercury and other toxic chemicals into the Rhine River served as a grim reminder of the problem. Lakes rapidly are becoming fertile with nutrients from sewage plant effluent and dissolved agricultural fertilizers carried by runoff. What once were clear fishing and swimming havens are being converted rapidly to smelly slime ponds as algae thrive. Nitrates from

Figure 17.2   Roadside art. A road cut in Cretaceous sandstones on California Highway 128, 35 miles west of Sacramento. Ironically, very old autographs on rocks (as along the Oregon Trail) are prized and protected, but surely wanton autographing by millions of people is intolerable. *(R. H. Dott, Jr.)*

Figure 17.3   Roadside art. A "sculpture" made from litter collected by Boy Scouts along only four blocks of a typical street in Madison, Wisconsin. *(R. H. Dott, Jr.)*

fertilizers ultimately can have an ironic deleterious effect upon humans because, even from drinking water, some nitrate inevitably is taken into the body. Intestinal bacteria convert it to nitrite, which hinders hemoglobin's ability to transport oxygen in the bloodstream and can cause the suffocation of infants.

No objective conservationist arbitrarily advocates complete outlawing of all fertilizers and pesticides, for clearly they are of great benefit, but the consequences of their application are being studied intensively, as well as alternative control measures. For example, there are clever biological ways to cope with insect pests. Release of sterile males into populations impedes propagation, and introduction of natural predators can control them. In Australia, after a century of rangeland devastation by European rabbits, introduction of a virus quickly reduced the rabbit population by about 90 percent. Methods such as these are preferable for several reasons. Not only do they pose no threat of poisoning, but they also do a more efficient job of extermination *over the long term.* Insects have such short life cycles and propagate so rapidly that chemical insecticides produce unnaturally rapid selection in their populations. Only the fittest survive, so evolution of hardier strains is *accelerated* by the insecticides. Soon the chemicals become less effective, even in dosages that begin to be toxic to humans. Thus, an initially spectacular treatment becomes self-defeating. Unfortunately, chemicals may not be very selective, either: when roadsides are sprayed, wild flowers as well as ragweed are killed—and an ugly brown fire hazard is created as a by-product.

Atmospheric pollution is well known, especially as a result of concern over radioactive fallout and killing smogs (Fig. 17.4). Fallout across Europe from the 1986 Chernobyl nuclear reactor disaster dramatized the problem. It is a classic irony that the tech-

nology developed by the western civilization first spawned in Greece now threatens the famous Acropolis and other priceless relics in Athens. All of the ancient buildings and statues there were made of marble, which is devastated by acid fog and rain resulting from smog. Effects of acid rain long familiar in industrialized urban areas are now showing up in wilderness regions, too, where they threaten stream and lake ecology. For example, Scandinavian freshwaters are almost devoid of fish today. And the damage to trees in Germany's Black Forest as well as in other European and eastern American forests is a recent cause for alarm.

There is much ignorance about long-term climatic and biological effects of the pollutants pumped into the air. (Some possible effects were mentioned in Chaps. 6 and 16). Until recent years, the atmosphere, like the oceans, seemed an infinite reservoir into which wastes could be dumped forever without harm. In reality, both of these reservoirs can take on and disperse only finite amounts of pollutants before lives become adversely affected.

The atmosphere has acquired untold quantities of pollutants from human activities—both agricultural and industrial—much of it as smoke. In addition, combustion has released carbon dioxide and toxic gases in ever increasing quantities since the Indus-

Figure 17.4 Effects of atmospheric pollution in Los Angeles, California. *Upper:* City Hall on a clear day in 1956. *Lower:* same view on a smoggy day in the same year. A temperature inversion—warm air above a cool layer near the ground—prohibits diffusion of contaminated air into the higher atmosphere. The inversion, seen here at an altitude of about 300 feet, is present approximately 320 days per year. Recent control measures have cut Los Angeles smog back to approximately the level at the time these photos were taken. *(Courtesy of Los Angeles County Pollution Control District.)*

trial Revolution. The most serious culprit is exhaust from internal combustion engines. Tetraethyl lead released from burning of gasoline affects our nervous systems and can be fatal. Over the past few decades, the average person's lead content has risen a hundredfold, bringing it near present estimates of the human tolerance level. In more recent years, however, unleaded gasolines have begun to decrease atmospheric lead levels. Lead also is accumulating in the shallow seas, which are our major fishing

grounds; apparently it diffuses into the deeper ocean much more slowly than it is being added.

Carbon monoxide is a more familiar product of exhaust pipes, and nitrous oxide gives rise through photochemical reactions to the chemicals that cause the ugly color of smog (Fig. 17.4). Both of these are also prominent in tobacco smoke, along with the carcinogenic components.

Besides known toxic effects of gaseous wastes in the atmosphere, gross climatic changes might be triggered by atmospheric pollution. Local meteorological effects are well known around large, industrialized cities, where rainfall is increased by smoke and dust particles (cities received about 10 percent more rain and fog than do their surroundings), yet, in cities, most rain water runs off because more than half the ground surface is covered with pavement and buildings, so infiltration to the water table is correspondingly reduced.

Cities are not alone in causing atmospheric pollution. Rural areas also contribute, especially where low-technology agriculture is prevalent. Widely practiced burning of brush and overgrazing contribute so much smoke and dust that the ground commonly is invisible from a plane in some parts of the world. Worldwide increase of turbidity of the lower atmosphere results in diminished incoming solar radiation, thus a climatic cooling tendency. A 10 percent increase in turbidity causes an estimated reduction of average world temperature of 1°C. From Chap. 16, recall that reduction of annual temperature of a mere 4 to 5°C over many years would be enough to cause the readvance of ice caps.

Global Warming Of all the atmospheric pollutants produced by our civilization, greenhouse gases, such as carbon dioxide, have the most serious long-term effects. As we saw in Chap. 6 and repeatedly throughout geologic history, these greenhouse gases can form a thermal "lid" over the planet. The earth receives nearly half of its solar radiation in the medium wavelengths of visible light, which penetrate the atmosphere easily (Fig. 6.13). But the earth reradiates most of this energy back out into space in the infrared and other long wavelengths. Our atmosphere is not transparent to these wavelengths and becomes less so when carbon dioxide, methane, and ozone become more abundant in the atmosphere. The result is increased absorption of this heat. The same principle causes the solar heating in greenhouses and closed automobiles. Glass transmits shortwave solar radiation, but surfaces inside absorb and reradiate this energy in longer wavelengths, to which glass is opaque. Thus, the air inside a greenhouse or car is heated.

Although carbon dioxide is only about 0.03 percent of the molecules in the air (about 345 parts per million—ppm), even slight increases in this percentage have large effects in the transparency of the atmosphere. A doubling to about 600 ppm could result in a global mean atmospheric temperature increase of as much as 9°C, which is greater than the extremes between glacials and interglacials (only about 5 to 6°C between maximum glacial and interglacial states). This was understood over a century ago by Nobel-prize–winning chemist Svante Arrhenius. Since the 1800s, the atmospheric greenhouse levels have gone from about 280 ppm to over 345 ppm today, and they have increased 9 percent (from 315 ppm) just since 1958!

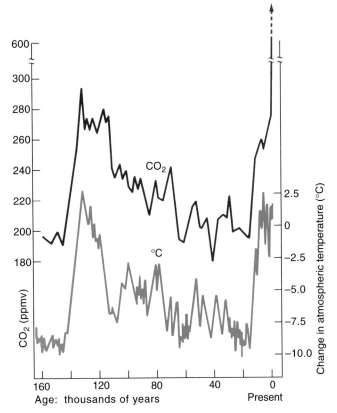

Figure 17.5   The last 160,000 years of global temperature and atmospheric CO_2, as recorded by gas bubbles frozen into Antarctic ice cores. Drilling almost 2 miles into the Antarctic ice cap, scientists were able to retrieve ancient samples of the atmosphere trapped in the ice and measure the gas content. The peak of the last interglacial (about 130,000 year ago) shows up clearly, as does the decline to the coldest period of the last glacial maximum about 18,000 years ago. Global temperatures then warmed up steadily, reaching a stable level over the past few thousand years, but marching upward in this century. Even more dramatic is the increase in CO_2 due to human burning of fossil fuels. Prior to the past two centuries, the concentration was only about 280 ppm (parts per million), but as of 1992, it had already reached 355 ppm, higher than any previous interglacial peak of warmth. Many scientists project that the level will reach 600 ppm in less than 40 years, even with major efforts to cut back on our burning of fossil fuels. *(After J. Barnola et al., 1991, in* Trends' '91: A compendium of data on global change, *ORNL/CDIAC-46, Carbon Dioxide Information Analysis Center, Oak Ridge National Laboratory, Tennessee.)*

Even more alarming is the projection of this rate of increase into the future. Unless measures are taken to slow down the pumping of carbon dioxide into the atmosphere, we shall reach levels of 600 ppm by the middle of the twenty-first century. The magnitude of this change is particularly striking, compared with the levels of carbon dioxide trapped in air bubbles from Antarctic ice cores, which span the past 160,000 years of climatic fluctuations (Fig. 17.5). Even during the warmest period of the last interglacial about 125,000 years ago, the carbon dioxide level did not exceed 300 ppm. We exceeded that level in the middle of the nineteenth century, and we are likely to double it in the next few decades.

The evidence for increasing greenhouse gases was once controversial, but today almost no dispute remains among scientists. Neither is there any doubt that global annual temperature has increased over 1°C in just the past century. Although there is still some possibility that this warming could be due in part to the earth's natural glacial-interglacial cycles, most scientists agree that it is primarily due to greenhouse gases—especially because the Milankovitch cycles predict that we should be cooling into the next glacial by now. There are reservoirs for carbon dioxide, such as oceans, ice sheets, and plants, but most scientists agree that we are pumping far too much carbon dioxide into the system for the natural "sinks" to absorb. Our rush to deforest the world has decreased its capacity for carbon storage, and as the rain forest trees are burned, they release much of their organic carbon into the atmosphere. Indeed, the slow absorption of greenhouse gases by the oceans has a perverse effect. Because the measurable changes are delayed, sometimes by decades, we have been misled into thinking that there is no problem yet. We may be seeing less warming now because of absorption by carbon sinks; when they are overwhelmed, we could see dramatic increases, and by then it would be too late to do anything about it.

The exact effects of global warming on the planet are still unclear, but some trends are clear. Many areas that are now the critical "breadbasket" regions of the world (especially the farm belts of the Northern Hemisphere) will be too hot and dry for this kind of agriculture. Most projections suggest that the northern areas of Canada and Siberia might become warmer and wetter, but these areas have thin soils, not the rich loams of the Midwest or the Ukraine that now support world agriculture. Other areas might become wetter, but the balance of agricultural power will certainly change.

There will be many other significant changes in local climate, but the most serious effect will be the melting of the ice caps. Recall that greenhouse conditions in the mid-Paleozoic and Cretaceous completely melted the polar ice caps on this planet, resulting in great epeiric seas, which covered almost all the continents. Even a 1-meter rise in sea level would inundate most of the coastal wetlands of the world, especially in places such as the Atlantic or Gulf Coast or the Netherlands. A 6-meter rise would completely swamp coastal cities such as New York and London, so that only the tall buildings would be visible. If the ice caps melted completely, sea level would rise by about 70 meters, and virtually all coastal plains (all of Florida, most of the Gulf and Atlantic regions, and many other low-lying parts of the world) would be gone. Cities such as St. Louis would become the mouth of the Mississippi, even as New Orleans was drowned.

Unfortunately, the increase in greenhouse gases is due to factors that are very difficult to curb or control. These include burning of fossil fuels (coal, oil, gas), deforestation, and intensive agriculture, which releases carbon dioxide and methane from the humus in the soil. As recent experience has shown, getting people to conserve even slightly in their energy use, let alone abandon oil, gas, and coal altogether, is extremely difficult—yet these forms of energy have been the primary cause of global warming ever since we began burning them at the start of the Industrial Revolution. Similarly, reversing the deforestation trend and replanting the trees would help the situation greatly, but the pressures to clear-cut the tropical rain forests are so intense that all the ecological alarms we have raised over years have not slowed down the rate of deforestation.

The Ozone Hole   Another long-term atmospheric change has rapidly become almost as serious as the problem of global warming. Our planet is protected from most of the sun's ultraviolet (UV) radiation by a thin layer of ozone (O_3) in the stratosphere (Fig. 6.13). In 1974, Sherwood Rowland and Mario Molina first showed that chlorofluorocarbons (CFCs), complex molecules that have chlorine or fluorine bonded to an organic carbon backbone, are capable of breaking down ozone at a remarkable rate. When a CFC molecule reaches the stratosphere, sunlight splits off the chlorine, which then reacts with ozone to form O_2 and ClO (chlorine monoxide). Each ClO molecule repeatedly splits apart, releasing the chlorine to break up additional ozone molecules again and again. Scientists estimate that 1 chlorine atom can destroy 100,000 molecules of ozone. Unfortunately, CFCs released today may take 15 years to reach the stratosphere. The damage we are now measuring is from gases released decades ago, and it will take decades before the supply of CFCs stops completely. However, some scientists calculate that what we have already sent up there is more than sufficient to destroy the ozone layer.

Although this frightening chemistry was recognized in the 1970s and confirmed again and again, the problem has taken on a new urgency in the past few years. As satellite, airplane, and ground-based measurements of the stratosphere over the Antarctic have become available, the loss of ozone from this region has far in excess of what anyone expected. Every year since it was first discovered, the Antarctic ozone hole has become wider and deeper, and it now covers an area three times as big as the forty-eight contiguous United States. At the end of the southern winter in September and October, the depletion of ozone reaches 40 to 60 percent, so that it is truly a "hole in the stratosphere."

Several factors conspire to make ozone depletion so severe at the end of the Antarctic winter. The cold Antarctic stratosphere is full of ice particles of nitric acid and water, which speed up the chemical reactions that destroy ozone. Second, these gases circulate around the Antarctic in a tight vortex, so they become concentrated and do not mix or diffuse with adjacent layers of the stratosphere. As the sunlight returns after the long Antarctic winter, it stirs up strong circulation of gases and ice crystals in the stratosphere, which increases the mixing and rate of chemical reaction. The sunlight also causes the breakdown of ozone directly, so that within weeks after the Southern Hemisphere spring has begun, the ozone is vanishing rapidly (Fig. 17.6). Eventually the stratosphere warms up enough so that the ozone-depleted region breaks up, mixing with higher latitudes and spreading the ozone depletion around the globe. Sometimes, however, "bubbles" of the ozone hole break off and float northward, causing serious problems for people in the Southern Hemisphere.

How serious is the ozone hole? The excess bombardment by UV has several effects, including increased skin cancer, as well as damage to our vision and to our immune systems. For every 1 percent decrease in ozone, there is a 2 percent increase in UV radiation and a 4 percent increase in skin cancer. Residents of Ushuaia, Argentina, on the southern tip of South America, are advised to stay indoors during September and October, and blind rabbits and salmon are frequently found by hunters and fishermen in Patagonia. (Ironically, the second largest employer in Ushuaia is a company that makes CFCs!) Skin cancer and cataracts are now increasingly common in Australia, New Zealand, South Africa, and Patagonia. Even in Queensland, in northeastern Australia, more than 75 percent of citizens over age 65 have some form of skin cancer, and children are required by law to wear large hats and scarves to and from school. The Australian health service estimates that four out of five of its citizens will develop skin cancer.

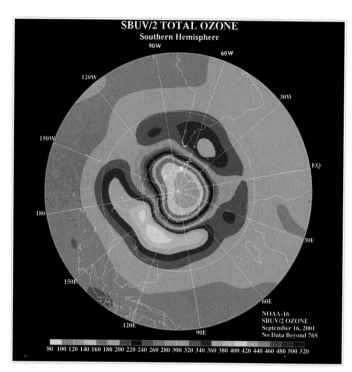

Figure 17.6   Computer-generated image of the ozone hole over Antarctica as measured from the Nimbus-7/Total Ozone Mapping Spectrometer (TOMS) satellite and probes. Dark purple and blue colors denote the greatest areas of ozone depletion. Chlorofluorocarbons (CFCs) released by our hairsprays, refrigerators, air conditioners, and styrofoam rise to the stratosphere, where they break up to form highly reactive molecules. These combine with the ozone in the stratosphere and break it up; the CFCs are then released to carry out the reaction again and again. Even if no further CFCs were released again, this destruction would continue for decades with what we have already sent up there. The Antarctic ozone hole is particularly intense in September, the beginning of the Southern Hemisphere spring. When the first rays of sunlight end the Antarctic winter, they trigger a rapid warming and strong cyclonic atmospheric circulation around the pole, which stirs up the CFCs and eats up enormous amounts of ozone in a short period of time. As the season progresses, the cyclonic winds weaken, and air from other parts of the world begins to mix in and decrease the contrast in ozone depletion. This image was taken on November 30, 1992, long after the ozone hole normally breaks up. It covered over 1.7 million square miles. *(Courtesy of Goddard Space Flight Center/NASA.)*

Relatively few people live in the southernmost Southern Hemisphere, but the depletion in the Northern Hemisphere has begun to alarm scientists as well. Although the Arctic does not form as strong a vortex or as distinct an ozone hole, already the Northern Hemisphere ozone layer has been depleted by about 10 percent in just four decades. In 1993, the World Meteorological Organisation reported that February ozone levels in the Arctic were 20 percent below normal. Because much of the world's population lives in the Northern Hemisphere, further depletion of the Arctic ozone layer would have very serious effects.

Recent studies have proved conclusively that humans are the source of these treacherous CFCs. Most come from the propellants in hair sprays, fire extinguishers, and deodorants, the freon in air conditioners and refrigerators, plastic foams, and many industrial processes that use CFCs. People were so fond of their styrofoam containers and air conditioners (and the chemical industry found CFCs so profitable) that there was almost a decade of resistance to the evidence of the ozone hole. Eventually, however, the depletion became so alarming that some actions were taken. Although CFCs in hairsprays were banned in the United States in 1979, most European countries and Japan still permit them. In the mid-1980s, even the United States stalled and resisted tougher measures. Du Pont, the largest manufacturer of CFCs, which had been investing millions in finding substitutes for CFCs, halted its research shortly after the Reagan election, anticipating a probusiness climate with few regulations that might force them to find substitutes. Production of CFCs increased through the 1980s until 1986, when Du Pont suddenly reversed its position and admitted the need to control production and phase out CFCs. But humanity may pay dearly for that delay: between 1978 and 1988, nearly 19 billion pounds of CFCs were produced worldwide.

In 1985, the Vienna Convention for the Protection of the Ozone Layer began to take steps in the right direction. The breakthrough came in March 1987, when a very strongly worded plan was proposed, and most countries agreed to it. In September 1987, twenty-four nations and the European Community signed the Montreal Protocol, which stipulated CFC reductions of about 40 percent over the next decade. For the first time, a major international agreement was signed that addressed an important global ecological problem. Unfortunately, existing CFCs will still continue to float up to the stratosphere for the next 15 years before the effects of the CFC curbs begin to show, and CFCs may stay up there for 50 to 100 years. Many scientists fear that the Montreal Protocol was too little and too late. If the political will had come a decade earlier, it might have made much more of a difference.

As seen in the CFC problem, changes in atmospheric (or oceanic) composition are sluggish processes that have a certain momentum; thus, they cannot be changed quickly even if methods of cleaning were developed tomorrow. Damage produced over more than a century probably cannot be corrected in less time. Pollution, as serious as it is, is really a symptom—rather than a cause—of our environmental crisis. It reflects the deeper problems of rapid *growth of population* and still more rapid *increase in resource consumption*. Only by balancing these factors in the total-demand equation with the earth's finite resources can we avoid calamity.

Limits of the Earth's Resources

The Population Explosion In the spring of the year 2000, world population passed 6 billion people. This number is even more staggering when one considers the accelerating rate of increase (Table 17.1). For humankind to reach a population of 1 billion took at least 500,000 years. The next billion was added between 1850 and 1930, only 80 years. The third billion arrived between 1931 and 1960, only 30 years. The fourth billion came along almost instantaneously between 1961 and 1975, just 15 years, and the next billion took only 11 years (in 1986). In 1990 alone, 92 million people were born, more than the population of Mexico, or ten New York Cities. In the 45 seconds you took to read this paragraph, *135 people were born!* Each hour there are 11,000 more mouths to feed; each day, about a quarter of a million; each year, more than 100 million.

Clearly the fundamental problem behind all the other dilemmas discussed in this chapter is human population growth. As long as population continues to explode, no miraculous breakthroughs in energy production or agricultural yields will be able to keep pace with it. Exploding population is the fundamental cause behind the grinding poverty of the less developed countries, which in turn leads to greater overpopulation as people have more kids to help support them and to compensate for high infant mortality. Exploding population is the main factor behind the rapid deforestation of the tropical rain forests, the exponential growth in our consumption of resources and production of wastes and pollutants. If any of the problems we have discussed are to be solved, the population problem must be solved first.

There have been some encouraging signs. Much of the industrialized world has slowed down growth rates, and some countries have reached zero population growth. Countries such as Thailand have mounted aggressive campaigns for birth control, and their population growth rate was cut in half in just 15 years. China has gone to extreme measures, limiting women to one child, with severe penalties for disobedience, in order to slow down its growth rate and contain its population at 1.4 billion by the year 2010. Countries that make all-out efforts to promote population planning and birth control have had success. With proper education and access to birth control methods, most people are willing to limit the size of their families. However, most of the countries of the world are still too poor to educate their populations, or to reduce the infant mortality that drives up the birthrate, let alone make birth control available to all of their people.

Unfortunately, the success stories have not made a dent in the fact that most countries of the world are still growing too fast. Kenya, for example, is expected to grow from 23 million to 79 million in just the next 30 years. Nigeria will soar from 112 million to 274 million in the same time span. Although growth is slower in population giants such as India, Brazil, Indonesia, and China, even their slower rates mean billions more people simply because of the sheer size of existing populations. In 1991, India added 17 million, China 16 million, and Brazil 3 million people.

Ultimately, the question boils down to the limits of growth on this planet. How many people can it sustain? Considering how impoverished most of the world is at present with only 6 billion

Table 17.1	The Growth of Human Population.* Note Especially the Rapidly Increasing Rate of Population Growth			
Year	**World Population**	**Doubling Time**	**Year**	**U.S.A. Population**
A.D. 1	1/4 billion	—		
1650	1/2 billion	1,650 years	1790	4 million
1850	1 billion	200 years		
			1915	100 million
1930	2 billion	80 years		
1960	3 billion	44 years		
1975	4 billion	40 years		
2000	6 billion	<35 years	2000	300 million
2100	10 billion (?)	(?)		

*After S. Cain, 1967: *Geoscience News,* v. 1, and United Nations data, 1992.

people, most environmentalists doubt that we can support 10 billion people on Earth, let alone the 100 billion that the extreme optimists contemplate. However, unless current trends are reversed soon, world population will reach 10 billion by 2020, and UN estimates now project 14 billion by the end of the twenty-first century. Even if we never reach numbers as alarming as these, we must support the people now living on the planet. The basis for our survival can be thought of in terms of renewable and nonrenewable resources.

Renewable Resources Some of the earth's resources, such as lumber and food, are renewable up to a point. Seemingly any plant-based resource can be regrown, and water can be cleaned up (as it has been in Lake Erie and elsewhere) if we are willing to pay the costs. In reality, however, even so-called renewable resources are limited by the earth's ultimate productive capacity, which is the limit of how much and how quickly energy can be supplied to life in usable forms. The total earth energy reservoirs are limited, like a bank savings account; how long the savings will last depends upon how rapidly we make withdrawals. And the picture is even more complicated because the difficulty of extracting the resources increases continually.

The human food chain is fantastically complex. Primary producers of food for the chain are, of course, plants, the only organisms capable of manufacturing food. In the total energy budget of ecosystems, however, some of the energy stored by plants must be released for recycling in the system. Perpetual recycling affects every single creature. Your body contains carbon, which perhaps was used for awhile by a dinosaur, later loaned to some mammal, and possibly borrowed again by a cockroach or dandelion before reincarnation in you.

Recycling is accomplished largely by decomposition to maintain an energy balance. Plants would quickly deplete all carbon dioxide for the atmosphere if fire and animal respiration did not recycle it. Certain bacteria could deplete atmospheric nitrogen in about a million years, but other organisms release it back

to the atmosphere as ammonia. Much of the decomposition is accomplished by lowly "pests," which, like it or not, are essential to the whole system. Primary consumers of plant production are herbivorous animals and decomposing microbes. Secondary consumers, in turn, eat either the primary consumers or their products, such as milk and honey, and so on. Organisms are notoriously inefficient, however, and the last link in the chain, such as humans, may consume the equivalent of hundreds of millions of pounds of primary plant food annually.

There is no immediate prospect that we humans can short-circuit photosynthesis by coming up with an artificial substitute for plant foods and lumber. If the human population grows unchecked, however, we may be forced to short-circuit the food chain by eliminating all intermediate animal and plant consumers that are not absolutely essential in the human food chain so as to gain exclusive use of earth resources. Shade trees, garden flowers, and pets would not be permissible. Already, China has severely limited the number of pets people are allowed to own. Moreover, diet eventually would be entirely vegetarian, for even now, the total existing human population cannot be supported on a largely meat diet. Even an intensive, rigidly controlled agricultural system still would be limited in total productivity by soil fertility, which is extended only by a limited amount and for a limited time through chemical treatment. And we face a serious fertilizer shortage because much of it is manufactured from petroleum!

Well over half the arable land of the earth is now in use, and most of what remains is marginal. The cost of producing food on marginal land is enormous, and production tends to decrease yearly as such land is used—even with fertilizers. This is especially true of soils in the humid tropics, where a large portion of the human population resides. When rain forest is cleared, the soil produces only one or two good crop-years before its fertility rapidly declines. In 1950, each additional ton of fertilizer produced 10 tons more of grain, but by 1960 only 8 tons, and in 1980, only about 6 tons. Already, conflicts over the use of land for housing and industry versus agriculture are upon us; they surely will intensify in the future. Some visionaries believe that we could ease both dilemmas by depending upon algae grown in glass water tanks on the roofs of huge apartment houses or upon bacteria, fungi, and yeast grown under conditions requiring little land and little energy.

The tropical rain forests are singled out by biologists as especially crucial both because they represent the earth's most diverse nonmarine ecosystem and because they affect the climate of the entire globe. Although they cover only 7 percent of the present land surface, tropical forests contain more than half of all species of plants and animals. For example, a mere 25 acres of rain forest may contain nearly 1,000 species of trees. In the tropics of Peru, Harvard biologist Edward O. Wilson has counted no less than 43 species of ants on a *single tree,* which is "roughly equivalent to the entire ant fauna of the British Isles." Although about 1.7 million species of organisms have been named and described so far, it is estimated that there are 40 to 50 million more as yet undescribed! Half the known species live in the tropical forests, but the forests are now being cut down to feed both the indigenous societies, which make up more than half the earth's human population, and the greedy temperate-zone ones that comprise only one-fourth the population—an example of *the tragedy of the commons* (see Hardin, 1968, in the Readings list). *Each day,* 80 square miles of rain forest are lost, mostly in Africa, Southeast Asia, and South America. *Each year* an area of rain forest the size of Maine or Indiana is cut down. At this rate, virtually all will be gone before the middle of the twenty-first century!

Estimates of how fast tropical plant and animal species are becoming extinct as a result of deforestation approach in magnitude the great end-of-Mesozoic and end-of-Paleozoic mass extinctions. Such a devastating reduction of biodiversity not only would deprive us of food, fuel, medicinal plants, and plant pollinators but also would upset the entire global chemical and climatic equilibrium. The rain forest not only captures significant solar energy through photosynthesis but also contributes importantly to the regulation of atmospheric oxygen, carbon dioxide, and water vapor. As population biologist Paul Ehrlich of Stanford University has said, likening the earth to a bank savings account, we are living off the earth's biological principal rather than just the interest. As any student of the most elementary economics knows (unlike many politicians), such practice cannot continue long without disaster.

The tropics are rightly receiving special attention from ecologists, but overdrafts on resource capital are the rule outside the tropics as well. For example, in the 1980s, the government proposed to double the United States harvest of timber over the next 50 years. This drastic move was thought necessary to accommodate the increase of human population and a growing appetite for woodland resources. Conservationists are skeptical of the projected timber need and point out the destructive by-products of such greatly expanded lumbering. These include damage from road building, destruction of wildlife habitats, and damage to watersheds.

These are all tough issues without any simple solutions, but surely Ehrlich is correct in arguing that ecological maintenance of the earth must be valued as much as economic development if we are to sustain an acceptable quality of life for the future—even on a time scale of only the next century or two, to say nothing of a more geologically significant time span.

Nonrenewable Resources

Water   Water is vital for all organisms. The supply of suitable water is already an important limiting factor for population growth and economic development in many parts of the world, and it will become so in additional areas in the not too distant future. For thousands of years, people have managed their water supplies with elaborate engineering projects that are marvels to behold. Reservoirs help balance irregular precipitation rates over time, and irrigation systems help distribute the water geographically. More recently, desalination of seawater has provided fresh water to some desert regions, but the distillation process requires much energy.

As human population grows and develops more sophisticated agricultural and industrial processes, demands for water increase greatly. Quality as well as abundance is crucial for most uses, a fact that further limits the total available usable resource. Some polluted water can be purified at considerable cost, but some cannot be purified at any cost. Surely it is best to avoid pollution! Geologists are increasingly occupied both in the search for new reserves of high-quality water and in the research necessary to avoid water pollution from human and agricultural sewage and toxic and radioactive wastes.

Mineral Resources   Ultimate limits of nonrenewable mineral resources, such as ores of the metals, building materials, and the natural hydrocarbon fuels (with all of which geologists have had intimate concern) should be even more obvious than the limits of renewable resources. The greatest service anyone interested in geology can perform for society in the next quarter century is to participate in the development of new water, fuel, and mineral resources.

For many nonrenewable raw materials, supply exceeds current demand by so wide a margin that there may seem to be no immediate problem. But all such resources are finite, and most people have no inkling of the magnitude of our conspicuous consumption. If we had to depend only upon presently known mineral reserves, our industrial society would collapse in a mere two or three decades. On the other hand, estimates of this sort are elusive because definitions of reserves are based so much upon economics. Ore that is too low-grade to be considered worthwhile today may become workable in the future as demand forces economic redefinitions and recovery techniques are improved (Fig. 17.7). Ore deposits represent unusual natural concentrations of rare elements. The "ore" curve of Fig. 17.7 cannot be extrapolated indefinitely to ordinary rocks because concentrations of valuable elements in them is so small that commonly they must be measured in parts per million or even parts per billion. Moreover, lower-grade raw materials require more energy for processing. (New York sewer sludge might become an "ore deposit" one day, for it contains a greater percentage of silver, chromium, copper, tin, lead, and zinc than do ordinary sedimentary rocks.) Are not new

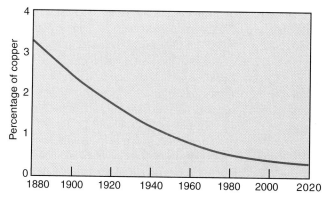

Figure 17.7   Decreasing grade of profitable copper ore over the past 120 years. Development of more efficient mining techniques, economies of scale, and market factors combine to determine the definition of "ore," which changes as these factors change. *(Data from U.S. Bureau of Mines.)*

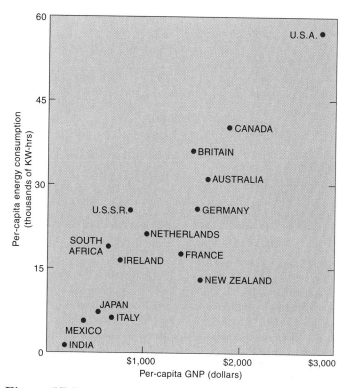

Figure 17.8   Relation between per-capita wealth expressed as gross national product and per-capita energy consumption expressed in thousands of kilowatt hours. Obviously climate also plays a role in these figures; for example, India's energy consumption is low partly because of its warm climate. *(Adapted from data in Cook, 1971, Scientific American and The Plain Truth, July–August, 1973.)*

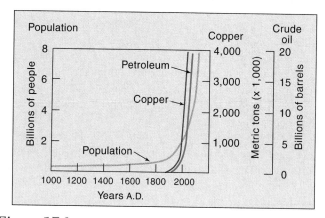

Figure 17.9   Graph comparing world population increase with a typical example of the increasing demand for metals (copper) and the increasing demand for petroleum. The latter two rates of increase are even more rapid than that for population. Such exponential increases of rate are the most alarming facets of the human ecological dilemma. *(Adapted from C. F. Park, Jr., 1968, Affluence in jeopardy, and other sources.)*

mineral deposits still being generated in the earth? Yes, of course, they are but at rates far slower than demand. Here we are once again confronted with the human versus the geologic time scale.

When we compare present demands upon nonrenewable resources with population increase, then the picture becomes more sobering. On the average, every member of the human race is increasing his or her usage of energy by 3 percent per year. From 1935 to 1967, world petroleum production grew from 6.5 to 30 million barrels per day; within the next decade, requirements exceeded 60 million barrels per day (one barrel is 42 gallons). It is estimated that, since 1955, North Americans have burned more mineral fuels than did the *entire world in all of previous history.* Today North America alone requires five or six times as much fuel on a per-capita basis as the rest of the world (Fig. 17.8). Consumption of metals in the United States is increasing at about double the rate of population growth (Fig. 17.9). Industrially developed societies with roughly 20 percent of the world's population consume at least half of all the earth's annual total mineral raw material production, and a great deal of this is imported from the other countries. North America alone, with only about 7 percent of the world's population, consumes nearly 30 percent of the mineral raw materials produced by the entire world! From such figures, it is clear that each person in a very industrial nation puts a much greater stress on world resources than does a person in a nonindustrial country.

The United States has a material wealth one-third greater than the next wealthiest nations (Fig. 17.8). Will such disparities be tolerated? Can we sustain an ever increasing gross national product, or will we be forced to adjust to a lower material standard of living? In the 1970s, the latter began to happen. If nonindustrial countries should bring their populations under control, their demands for metals and fuels might increase exponentially if they, too, aspire to our style of living. With such further increasing rates of demand, we would be much closer to ultimate resource limits, especially of certain rare metals, than we now think.

Salvage and reuse of discarded materials will become increasingly necessary, which is as it should be, for we are frightfully wasteful (Fig. 17.3). The average American, who is the world's worst litterbug, disposes of 3 to 4 kilograms (6 to 8 pounds) of solid wastes per day, or nearly 0.9 metric ton (1 ton) per year (and we already have littered the moon, too). As is illustrated by the composition of New York sewer sludge, a fundamental redefinition of "waste" as "resources out of place" is being

Figure 17.10   Bingham open-pit copper mine in the Oquirrh Mountains near Salt Lake, Utah. This is one of the largest open-pit mines in the world and a major producer of copper from low-grade ore disseminated in a late Mesozoic granitic pluton. Low-grade ore of this sort already is the major world source of copper. Still lower-grade ores will be mined in the future for the many metals essential to complex, industrialized societies. *(Bureau of Mines, U.S. Department of Interior.)*

called for by many authorities because we can no longer afford to waste anything. Development of substitutes already has alleviated some resource pressures and no doubt will become increasingly necessary. Substitutes may have to supersede hydrocarbon fuels before ultimate supplies are threatened for at least two reasons: first, burning of fossil fuels may become intolerable because of the atmospheric pollution it produces; second, hydrocarbons may become too precious for lubricants and as raw materials for the chemical industry (e.g., for plastics, textiles, and fertilizers). As use of fossil fuels declines, nuclear, solar, geothermal, and perhaps even wind and tidal energy will become more important. Of more importance, ultimately, may be the earth's great internal thermal energy. Besides building mountains and displacing continents, it might be tapped by deep heat wells to drive our society.

Other by-products of the extraction of nonrenewable resources are atmospheric, water, and sound pollution, as well as modification of the landscape through extensive excavations (Fig. 17.10). These ills can be alleviated, however, if society is willing to pay for corrective measures through higher costs. The producer cannot be expected to carry the entire burden of reclamation. The magnitude of excavations and water required to mine and process coal and oil shale in the amounts needed to supplant declining petroleum reserves staggers the imagination, however, and makes one wonder if other energy sources might be preferable.

Nonfuel Minerals   Impending shortages of nonfuel mineral resources have been less publicized than have energy shortages, but they are no less real. Figure 17.11 shows just how dependent you as a citizen of an industrial society are upon such prosaic materials as stone and gravel, salt and clay, to say nothing of the more obvious metals. Figures 17.9 and 17.12 show how dramatically

our appetite for such resources has increased. Most significant is that the increase of demand now exceeds the rate of population growth! Yet another facet of the mineral resource problem is revealed in Fig. 17.12, which shows that, over recent decades, consumption has exceeded domestic production of eighteen major minerals in the United States. In short, industrial nations, of which the United States is but one example, are importing more and more of the minerals they need (Table 17.2). Use of substitute materials and recycling of scrap can help alleviate this undesirable situation, and recycling efforts have been increasingly significant. For examples, roughly half of the iron, lead, and antimony, and about 25 percent of the copper, zinc, and tin used in manufacturing in the United States is now obtained from scrap. Recycling has the added dividend, of course, of alleviating the environmental problems of waste disposal. Many minerals, however, are dissipated during use or are converted to products such as concrete, which cannot be recycled.

The importing of some minerals is inescapable because we simply have little or no domestic supply. For example, Table 17.2 reveals why there is so much interest in mining the sea floor for manganese nodules. In many cases, however, importation occurs only because the domestic mineral industry does not compete favorably with foreign suppliers. Indeed, some of the domestic companies have themselves gone abroad to obtain ores more cheaply. Most deplorable, however, is the importing of fabricated products, such as construction steel, which we are perfectly capable of manufacturing at home. University of Wisconsin expert E. N. Cameron emphasizes the little appreciated fact that, when we turn from manufacturing things from mineral raw materials to importing finished goods, in effect we are exporting part of our industrial economy and employment opportunities for our own people. There are many reasons that the situation has arisen, but not the least are rising domestic labor costs and declines in grade of ore.

Obviously these factors are complex and are not intrinsically bad, but it is important for an enlightened citizenry to recognize the facts. First, industrial societies are utterly dependent upon mineral resources, and, second, it behooves them, therefore, to have healthy mineral industries. Only one of several reasons for doing so is that the reliability of some foreign sources is doubtful (consider cobalt, chromium, and tin; Table 17.2). Moreover, it is irresponsible to export all the socially unpopular aspects of mineral production to other, especially less developed countries. We should accept our dependence upon such resources and take full responsibility, which means paying the entire social cost for developing the resources and then rehabilitating the environment as necessary.

Energy—The Ultimate Limit

Energy is required for all human activities, including utilization of mineral resources and the growing, harvesting, processing, and transportation of the food from which our bodies derive energy. Indeed, today there is ten times as much energy input to the total food systems as is gained by humans from their food! More alarming, however, is the fact that our increasingly technological food system now requires four times as much energy for approximately

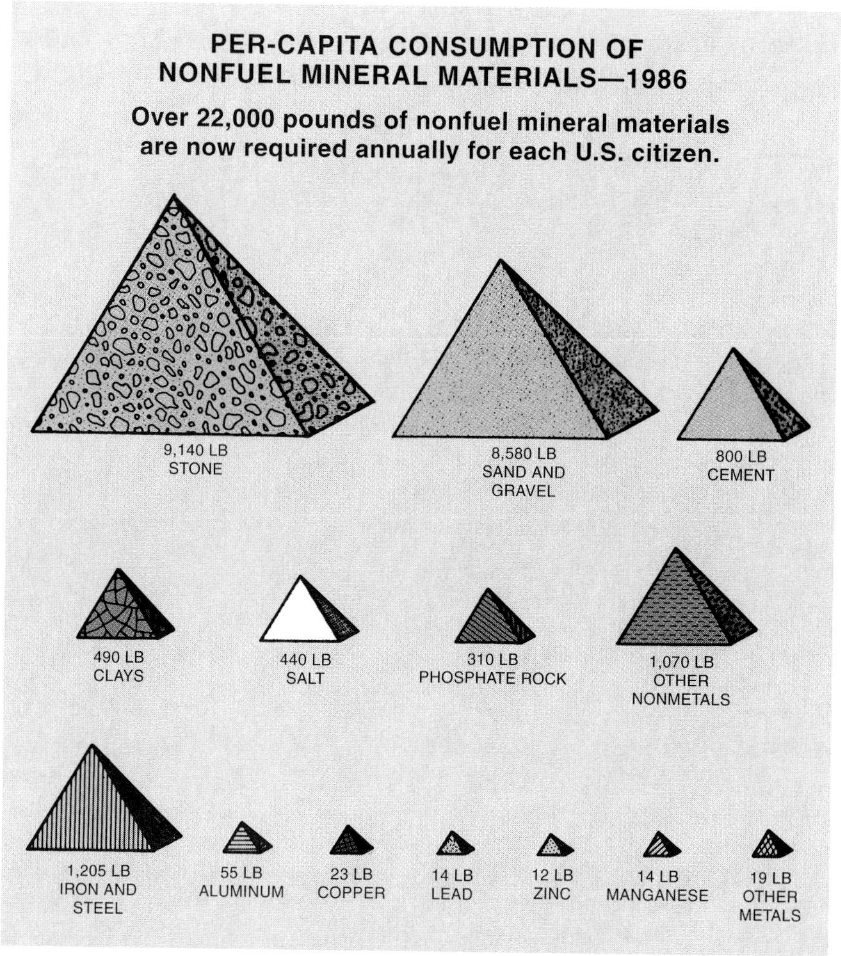

PER-CAPITA CONSUMPTION OF NONFUEL MINERAL MATERIALS—1986

Over 22,000 pounds of nonfuel mineral materials are now required annually for each U.S. citizen.

9,140 LB
STONE

8,580 LB
SAND AND
GRAVEL

800 LB
CEMENT

490 LB
CLAYS

440 LB
SALT

310 LB
PHOSPHATE ROCK

1,070 LB
OTHER
NONMETALS

1,205 LB
IRON AND
STEEL

55 LB
ALUMINUM

23 LB
COPPER

14 LB
LEAD

12 LB
ZINC

14 LB
MANGANESE

19 LB
OTHER
METALS

Figure 17.11   Every man, woman, and child has a stake in mineral resources. Over 22,000 pounds of these materials are required annually for each person. The total of such minerals used in the United States is about 2 billion tons. *(From U.S. Bureau of Mines.)*

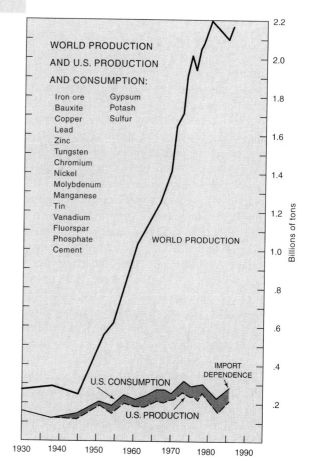

WORLD PRODUCTION
AND U.S. PRODUCTION
AND CONSUMPTION:

Iron ore	Gypsum
Bauxite	Potash
Copper	Sulfur
Lead	
Zinc	
Tungsten	
Chromium	
Nickel	
Molybdenum	
Manganese	
Tin	
Vanadium	
Fluorspar	
Phosphate	
Cement	

WORLD PRODUCTION

IMPORT DEPENDENCE

U.S. CONSUMPTION

U.S. PRODUCTION

Billions of tons

2.2 2.0 1.8 1.6 1.4 1.2 1.0 .8 .6 .4 .2

1930 1940 1950 1960 1970 1980 1990

Figure 17.12   World production compared with United States production and consumption of eighteen important minerals. Note the widening gap between domestic production and consumption, making the United States increasingly dependent upon foreign sources. *(Data from U.S. Bureau of Mines; courtesy E. N. Cameron.)*

Table 17.2	Net Import Reliance for the United States for Some Important Minerals*	
Minerals[†]	**Percent Imported**	**Major Foreign Sources**
Columbium	100	Brazil, Thailand, Canada
Manganese	98	Gabon, Brazil, South Africa
Cobalt	97	Zaire, Zambia, Finland
Aluminum	93	Jamaica, Australia, Surinam
Chromium	92	South Africa, U.S.S.R., Rhodesia, Turkey
Platinum	91	South Africa, U.S.S.R.
Tin	81	Malaysia, Bolivia, Thailand, Indonesia
Nickel	77	Canada, New Caledonia, Dominican Republic
Zinc	62	Canada, Mexico, Australia
Gold	54	Canada, South Africa, U.S.S.R.
Tungsten	50	Canada, Bolivia, Peru, Thailand
Silver	44	Canada, Mexico, Peru
Iron ore	29	Canada, Venezuela, Brazil, Liberia
Copper	19	Canada, Chile, Peru, Zambia
Iron and steel products	13	Japan, Europe, Canada

*Data from Status of the Mineral Industries, 1979, United States Bureau of Mines.

[†]Includes refined metals as well as raw ore in some cases.

the same level of per-capita caloric food energy value as in 1940. Increasing mechanization apparently is a dubious blessing. The U.S. Department of Agriculture has shown that the most energy-efficient agriculture is the small, single-family farm.

Because every aspect of life is ultimately dependent upon energy, much attention has been focused upon it in recent years. Every activity must be examined in terms of its energy budget. For example, wise as it may seem to recycle materials, in some cases it can be more costly in total energy expended to recycle than to process the raw material. Glass is a case in point. There is no shortage of the raw material silica, and when all hidden energy costs (e.g., transportation) are added to the energy required for reprocessing glass, the result is discouraging. It simply does not make sense to recycle materials that are not rare when to do so costs more in energy than to start from scratch. (However, reuse of existing bottles saves both energy and money.)

Figure 17.13 shows the prediction for the history of fossil fuel resources on Earth. First developed by geologist M. King Hubbert nearly 50 years ago, it reflects the history of the most important nonrenewable mineral resources. Hubbert knew that production of some minerals that had been heavily exploited for years, such as copper and silver, continued to decline despite increasing demand and prices. Lower and lower grades of ore were extracted, but the amount produced did not keep up with demand. This illustrates a basic principle of limited resources, such as minerals: *no matter how much the demand, the supply is limited and eventually will be exhausted.* Unlike manufactured goods or renewable resources, which can be produced as demand dictates, limited resources do not obey the laws of supply and demand over the long run. *No more oil or coal is being produced in the short*

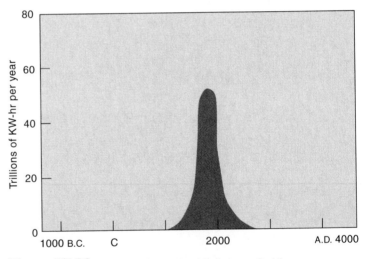

Figure 17.13   Exploitation of fossil fuel energy (in kilowatt hours). According to this estimate, world consumption will peak only about 1,200 years after humans lit their first coal fire. *(From Hubbert, 1962, National Research Council Publication/Washington, DC. National Academy of Sciences 1000-D.)*

time span of human existence, so when it is gone, there is no more, no matter how high we price it!

The history of scarce resources such as copper and silver displays a distinctive "bell-shaped curve" similar to that in Fig. 17.13. In the early stages of production, the easily obtained surface deposits are mined cheaply and rapidly, and the demand increases as people find uses for the cheap resource. This happened during the

1860s, when the abundant deposits of silver from the rich Comstock Lode in Nevada were found near the surface and in very shallow mines. The production curve climbs exponentially in this early stage, and soon the easily extracted deposits are exhausted. But demand has now been whetted by the early boom in supply, so miners must look harder and harder for an increasingly valuable resource. In the late 1800s, silver miners resorted to deeper and deeper shaft mines to find sufficient silver. The increased cost of production was more than compensated by the rise in prices.

Eventually even these deposits were exhausted, and then miners were driven to more and more desperate measures, and must extract larger and larger quantities of low-grade ore to meet the demand. The gigantic open-pit mine at Bingham Canyon, Utah (Fig. 17.10), is literally mining away an entire mountain of extremely low-grade copper ore, with less than 1 percent of copper in every ton of rock extracted. In the case of silver mining, miners had to process old mine dumps through baths of cyanide to extract the little remaining silver that had been left by an earlier generation of miners. Cyanidization peaked early in the twentieth century, and since that time silver production has been in decline *despite increased demand.* Currently, at this point in time, the production bell curve goes into a steep decline, and continues to decrease despite higher demand and prices. Setting aside the silver demands of jewelers, photographic films and papers use huge amounts of silver, and these have become increasingly expensive. Eventually we shall be forced to abandon all but the most essential uses, and computerized digital images will have to replace film emulsions and their demand for silver.

When Hubbert examined the history of other mineral resources that had been exploited to near total depletion, he realized the same history must be true of oil. In the late 1950s, he predicted that oil production would show the characteristic bell-shaped curve of all limited resources and predicted that the peak would occur between 1966 and 1971. His prediction was remarkably accurate; the peak for crude oil production in the United States did happen in 1971, as the first edition of this book was published. The peak for natural gas production in the United States occurred 5 years later, about the time that the second edition of this book appeared. For the world as a whole, the same peaks were reached around the turn of the century. In other words, *despite increased demand for oil during the two energy crises of the 1970s, our domestic oil production has been declining!* In 2001, Princeton geologist Deffeyes reevaluated Hubbert's analysis by incorporating petroleum reserve data for the previous 40 years. He concluded that world production will peak by about 2010.

It is sobering, indeed, to realize that 80 percent of the total fossil fuel reserves that took 700 million years to accumulate on this planet will have been exhausted in just a few generations of human history. Considered on a geologic time scale, this is truly a catastrophic event.

What about the apparent oil glut, you might ask? Indeed, over the very short term, it is possible to pump too much oil, and prices and supplies will respond to supply and demand. Both the twin "energy crises" of the 1970s and the oil glut of the 1980s were artifacts of short-term political events (the OPEC oil embargo, the Iranian crisis, and then the overproduction that they

stimulated). But oil is being formed so slowly on this planet that early in this century the real limits will be felt. We will have another oil shortage, and this time higher prices will not be sufficient to give us enough supply, let alone a glut. Depending upon how big the estimates of reserves and the estimates of demand, most projections indicate that we will be permanently running out of oil sometime in the twenty-first century, *within the lifetime of most students reading this book.*

Indeed, the effects of oil gluts are even more insidious. By giving people the mistaken impression that oil is abundant, we revert to our wasteful ways when oil was abundant and cheap. In the United States, recent oil gluts led to the decline of our domestic petroleum industry. Aside from the human cost of this decline, the most serious consequence of the glut is that the large oil companies have severely curtailed exploration in this country, making us even more dependent on foreign oil than we were before the first OPEC embargo. Another serious consequence is that much of our domestic production gets shut down when the low price of crude makes it too expensive to pump marginal wells. This is especially true of the low-grade "stripper" wells. When these are shut down, many are lost permanently because no rise in prices will make it economical to revive them. In short, all the efforts of the 1970s to reduce our energy dependency have gone for nothing, and we are now more vulnerable than ever.

Clearly, alternative energy sources must be developed. Nuclear energy does not now seem as promising as in the flush of post–World War II atomic optimism. Not only were the Atomic Energy Commission and the power industry naïve in overestimating available nuclear ore reserves, but waste-disposal and safety problems were underestimated. Perfection of breeder reactors would help greatly, but this is still uncertain.

Even if solar, geothermal, and other new energy sources can ultimately provide our major needs, we shall still be dependent on oil for decades. We may be able to substitute solar-powered cars and nuclear plants for much of our use of coal and oil, but there are many other uses for oil besides energy. Anyone who lives in the farm belt is acutely aware of how dependent we have become on agricultural chemicals. The commercials during the dinnertime news (when farmers are tuned in to hear about tomorrow's weather) are nonstop ads for pesticides and fertilizers. Without these chemicals, our agricultural abundance would collapse, and there are no easy ways to obtain these chemicals from anything except oil. Alternative approaches to agriculture, which are less chemically dependent, are under investigation.

Look around and note the hundreds of examples of synthetic materials, from the ubiquitous plastics to synthetic fabrics such as nylon, rayon, and dacron. All these are made from crude oil. Modern society has become so dependent on these oil-based chemicals that we cannot imagine living without them. Visit a historical museum and see how life was lived early in the twentieth century with only natural substances, such as wood, iron, and cotton. You quickly realize how your every activity is now dependent on plastics and synthetic fabrics. None of these things would be possible without petroleum.

Because the demand for fossil fuels is increasing at phenomenal rates, we face serious environmental and economic problems in addition to the fuel pinch itself. We are resorting to more and more elaborate measures to extract oil, such as gigantic offshore drilling platforms (Fig. 17.14). To obtain coal, we must strip-mine huge areas of land, and much of this coal produces acid rain, as well as greenhouse gases. Consider these figures. World petroleum consumption is doubling twice as fast as is population (Fig. 17.9). In recent years, the number of automobiles per capita has increased 25 percent in the United States, 50 percent in Britain, and 90 percent in Japan, yet the private automobile, which consumes half the total energy expended for transportation, is notoriously inefficient. A municipal bus is about five times more efficient for carrying people than the car, and urban railways are still more efficient. Similarly the train is five times more efficient than a jumbo jet and ten times more so than a supersonic transport. Ironically the extinct electric street car, which was more efficient and less polluting than the bus, was killed deliberately by industry when the diesel bus came into production after World War II. Equally ironic has been the subsequent construction of costly freeways alongside existing railroad tracks, which are falling into disrepair and being abandoned. The freeways have been self-fulfilling prophecies. As more were built, feedback produced more cars and suburban sprawl, thus requiring still more freeways. If fuel costs make commuting by auto prohibitive, then we shall rue the day that we let our railroads deteriorate.

Canada and Mexico are well fixed with domestic petroleum reserves adequate for their demands for the foreseeable future, but the energy-hungry United States now has to import 60 percent of its petroleum needs. When the famous 15-million-barrel (630-billion-gallon) Prudhoe Bay oil field in northern Alaska, came into production in the 1970s, it could supply the total United States demand for only 3 years and the total world demand for only 6 months inspite of being the largest field in North America. Only 30 years later, Prudhoe Bay production was declining rapidly. A few more large oil fields like this one may yet be discovered, especially beneath continental shelves (Fig. 17.14), but with any finite, nonrenewable resource, the race between discovery and consumption cannot be won in the long run. Also, we have to determine if it would cost more energy to extract these remote reserves than we would gain from them. In 1973, the Arab embargo on petroleum exports gave us both a warning and an opportunity, but due to political wrangling we muffed our chance to begin getting our energy house in order. Meanwhile, time runs on, and it is critical that we develop soon some alternative ways of satisfying our energy needs. The remaining fossil fuel reserves, especially coal, can buy time, but new sources cannot be developed overnight.

Conservation of energy in the United States has been impressive. Industry, always sensitive to a cost pinch, has done very well by reducing its energy-consumption growth rate to 0.4 percent, while still increasing output by about 3 percent, since 1973. Recycling of scrap, for example, not only stretches raw ore supplies but also saves enormous amounts of energy. Savings through manufacturing products from recycled metal rather than raw ore is about 50 percent for steel, 90 percent for copper, and

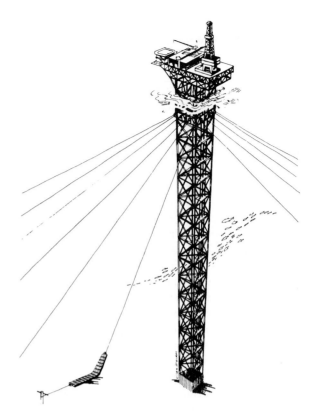

Figure 17.14   Deep-water oil-drilling platform about as tall as the tallest skyscraper, the Sears Tower (440 meters). It is supported with guy wires for use in outer shelf areas of the Gulf of Mexico in water depths of the order of 400 meters. The cost of such a structure is more than $500 million. *(Courtesy Exxon Corp., by permission.)*

95 percent for aluminum. Conservation also involves more efficient energy use. Different sources of energy should be tailored to their best uses. For example, electricity should not be used directly for heating but, rather, for lighting and running machinery to do other tasks.

Authorities such as Amory Lovins advocate both conservation and emphasis upon so-called soft technology. The argument is that hard technology is less energy efficient and requires huge capital resources at the expense of labor resources. Thus, it aggravates not only the energy problem but also the social problem of unemployment, which is a growing concern. Lovins argues that electricity can be produced more cheaply by smaller plants; they require less investment of capital and can be built faster, which reduces costs due to interest and inflation; shorter distribution distances also reduce costs and energy loss. Also, the hot water exhausted from them can be more readily used for local heating than can that from huge, remote plants. Lovins, J. S. Steinhart, and many others also advocate more emphasis upon solar, wind, and other alternative energy sources. Better insulation of buildings also is urged. Lovins talks of houses so well insulated that body heat, electric lights, and the sun could satisfy all heat requirements. "If one feels chilly in the evening, just have your pet chase a ball every so often to give off a little more body heat," he says.

Alternative Futures

What sort of life will our children and grandchildren lead? Will it be a steak-and-convenience-blessed paradise or mere existence at a bland algae-and-tenement subsistence level? Is concern for lower life forms and for untrammeled open spaces free of billboards and beer bottles mere sentimentality irrelevant to preeminent humans of the twenty-first century? We have seen that the only real certainty in nature is change. Among laypersons, there are two prevalent and opposing views of social and technological change. One, the blindly optimistic view, assumes uncritically that *all* change is good and represents "progress." This has been aptly dubbed the Gospel of Growth, and part of its creed reads "That which can be built must be." On the other hand, the pessimistic, or "good-old-days," view holds that most change is bad because it undermines a glorified status quo.

All of this reflects in a curious way our culture's definition of progress, which dates back a couple of centuries to those medieval times dubbed the "Dark Ages." Today we see that those ages were not so dark after all, in either the technological or the cultural sense. In a similar way, we should be thinking of the future in other than just the two extreme, simplistic options. There should be an acceptable middle ground of limited growth. It is as absurd to assume that all technological and environmental changes that humans produce, even if masquerading as "advances," are beneficial in the long run as it is to weep, ostrichlike, for a sentimentalized past. Therefore, if change *is* as inevitable as we claim, then surely it behooves us to study, plan, and adapt or try to control biological and other environmental changes for the best long-run effects on the total ecosystem.

Certainly humans are the most adaptable creatures ever, and perhaps someday people can control the environment absolutely and manipulate it purely for self-convenience, but we doubt it. One inescapable truth, however, is that the amount and location of mineral resources and of agriculturally favorable climatic and soil conditions have already been fixed by geologic history. Faced with the complex ecological interdependence both among the life forms and between life forms and their physical environment, and knowing what we do about the intricacies of ecosystems, it is unwise to plot a course for the future on blind faith.

History teaches that evolution is irreversible, and once any creature becomes extinct, it can never return, no matter how beneficial it may have been to humans. This law applies even to the lowliest microbes, some of the most beneficial of which probably have not yet even been discovered (much less named) by us. Examples cited of our unwitting degradation of the environment provide ample evidence of the folly of any policy of blind faith that "everything will work out for the best because science will find a way."

Famous pioneer conservationist Aldo Leopold said that the height of ignorance is the person who asks what good is an animal or a plant. "If the land mechanism as a whole is good, then every part is good, whether we understand it or not. If the biota, in the course of aeons, has built something we like but do not understand, then who but a fool would discard seemingly useless parts?" Here we have a hard-headed rather than a sentimental reason for the preservation of land and organisms. For example, many nondescript plants and bacteria manufacture substances of great value. To name just one example, a compound that looks hopeful for ovarian cancer, taxol, has been found in the Pacific yew tree. Nearly half our drugs were first discovered in nature, yet only a small fraction of plants and microbes have been assayed for useful chemicals.

It is a basic principle of biology that *diversity* is the key both to species survival and to ecological stability. The more genetic diversity a species has, the less susceptible it is to complete extinction in the face of environmental change. The highly touted Green Revolution created very productive hybrid crop plants through careful breeding, but those plants require more care and are far less resistant to disease, pests, and climatic adversity than are the natural strains. Because of their narrow range of genetic diversity (a monoculture), the hybrids could be completely decimated overnight. Diversity is also a boon to ecological communities, for it ensures a greater degree of stability for the entire community, which suggests why we should take Leopold's advice very seriously. Increasing pressure for food to feed the world's ever increasing population will drive us to seek new plant food sources with desirable genetic traits. At present, we use only twenty crop plants as major food suppliers. Wild, native plants are the reservoir from which new genetic strains of crop foods can be developed. Thus, there are sober reasons for us to beware of disrupting natural communities and triggering further extinctions. We need those wild weeds!

Let us suppose that complete ecological understanding could be achieved and with it the ability to manipulate safely everything in the environment for the benefit of humanity. Would the quality of life be acceptable in what would likely be some kind of Orwellian world? Should we presume to judge what is best for future generations?

The amount of space available on our planet is rarely discussed as a resource. In North America today, natural landscapes are being lost to development at the rate of 1 acre per minute! Because of land filling, the area of San Francisco Bay is only two-thirds of what it was 200 years ago. In 1970, there was an average of about 60 persons per square mile of land on Earth, though distribution was very irregular (70 percent of Americans live on 10 percent of the available land). At the present rate of population growth, by about the year 2025 there would be 10,000 persons per square mile. Many, if not most, people require contact with the natural environment for their emotional well-being, so space may turn out to be one of the most precious commodities as population increases. For example, it took more than 40 years for national parks in the United States to tally their first billion visitors, but only 12 years for the second billion (Fig. 17.15). Already reservations are required long in advance for camping in many of the parks.

Figure 17.15   The pressure of a growing population on space. *(Reprinted from* Burdened acres—The people problem, *by R. Wendolin with drawing by Monroe Bush, in* The Living Wilderness, *Spring-Summer 1967, ed. by M. Nadel. By permission of the Wilderness Society, Washington, D.C.)*

Summary—What Quality of Life?

Any way that we examine the future of humans, even for the geologically minuscule span of the next century, the quality of life will be dependent first and foremost upon food, water, and energy resources. But the scenario really depends upon population growth and the resource demand resulting from it. Problems of food, raw materials, pollution, and space are all secondary to population growth. Figs. 17.9 and 17.15 and Table 17.1 illustrate the magnitude of this problem. It is the phenomenal *acceleration of the growth rate* that is so alarming, together with the resulting *acceleration of demands placed upon the environment.*

World population now doubles in a mere 20 to 30 years. It is now approaching 8 billion. India, with more than 1,000 million now, will add at least 200 million more people—equivalent to the entire 1970 population of the United States before 2020. Unless nutritional standards were dropped, even North America would have difficulty feeding 200 million more mouths if added in such a short period. Today nearly half of the world's population is undernourished, and famine is occurring in some countries.

Studies of exponential rates of change in past geologic times provide a hint of the dramatic impact of humans upon this earth in the past two centuries. Consider the phenomenal acceleration of technological changes alone. For at least 2 million years, *Homo* was a hunter and gatherer; the agricultural revolution occurred no more than 10,000 years ago; the Industrial Revolution is but 200 years old; and only within the past century have the electric light, automobile, airplane, television, nuclear reactor, antibiotics, electric toothbrushes, electric matches, computers, pocket calculators, and genetic engineering appeared. But is the human prepared either emotionally or socially to cope with change at still faster rates? Should we not, however, take a more positive view toward change and try to plan to influence change for the better? We might as well, for change will occur, like it or not.

The population problem is not just the result of increase in absolute numbers of births, but also improvement of birth-survival expectancy and longevity. A little appreciated consequence of disease control is the perpetuation in the breeding portion of the population, or gene pool, of all types, including the mentally and physically deficient, which by natural selection would have been reduced. Together with chemicals and radiation, such perpetuation presents a genetic (thus evolutionary) problem for the race and, at the same time, complex sociological problems. It cannot be ignored, for humans, like so many predecessors on Earth, could become overspecialized in our complex, artificial civilization and perhaps become genetically weakened. On the optimistic side, however, there are predictions of genetic engineering just around the corner that may correct defects and even provide prenatal controls.

As a product of a 3.5-billion-year history of genetic mutations and natural selection by countless changes of environment, apparently *Homo sapiens* must reform if a tolerable ecological equilibrium is to be achieved. And there will be but one chance. At least for the foreseeable future, we humans will not be omnipotent but, rather, will remain very much dependent upon, and a part of, the total environment. Above all, we need to achieve a balance of human population in order to conserve that environment and the present way of life, for unlimited growth disrupts stability of ecosystems. Remember Darwin's early observation that most natural populations tend to be stable in size, but as a comic strip philosopher named Pogo put it in the 1960s, "we have met the enemy, and he is us."

Conservation, simply defined, is applied ecology, and it does not mean only to preserve or lock up resources. About 40 years ago, Wisconsin geologist C. K. Leith aptly described conservation as the "balancing of natural resources against human resources and the rights of the present generation against the rights of future ones." Conservation should ensure maximum present and future benefit from resources. Both public and private machinery are required to make it succeed, but inevitably serious conflicts arise between public and private interests. The conflicts involve long-cherished property rights and freedom to exploit private property as one sees fit, yet who is to say what interest is the more important? Should the public interest take precedence over individual self-interest and private initiative? It seems self-evident that, as societies increase in size and complexity, some individual freedoms must be restricted, and it has been popular to cry for governmental control of every socioeconomic problem that arises. But, as George Orwell warned in his prophetic novel *Nineteen Eighty-Four,* one of the greatest threats to the quality of individual lives is the gradual, insidious restriction of personal freedoms by large organizations of all kinds—governmental, military, business, and even educational. Bigness itself may, after all, be the greatest threat to individualism; therefore, the development of responsible environmental and of equitable socioeconomic policies without making a brainwashed slave out of the individual is the greatest social challenge facing all nations. Diversity of people is as important socially as it is genetically.

It has been observed that ours is the first generation that could foresee a major cultural change within its span. The decade of the 1970s will prove to have been a major watershed in the history of industrial societies when such a change began. *Postindustrial society* and *stagflation* were added to our vocabulary, and the United States standard of living dropped for the first time since the Great Depression of the 1930s. Classical economic theories do not seem to be working very well, and dubious experiments are being tried, such as attempting to balance the books by selling sophisticated armaments abroad. Ironically, seemingly good policies designed to conserve energy mean less tax revenue to run the government. Our economic ailments are very complex, but much of the problem can be traced to the imbalance of trade, which is aggravated chiefly by the importation of much petroleum, construction materials, and consumer goods that we used to supply for ourselves. During the 1970s, the United States increased petroleum imports from 33 to 50 percent, and now 60 per-

cent must be imported. The magnitude of costs for obtaining energy are now so large that they are as mind boggling as geologic time. For example, a single new deep-water offshore drilling platform to tap new petroleum reserves costs between $120 and $500 million (Fig. 17.14), and one supertanker to carry the oil costs over $80 million. In the face of such staggering expenses, capital must now be regarded as a critical resource, and *capital formation* could become the most severe limit on the acquisition of needed natural resources. This challenge of capital is reflected in the many mergers of oil companies in recent years. It may also force hard choices among competing large projects that can be supported by society, all of which should provide added incentive to diminish our external dependence for resources.

Respect for the natural environment is required on several grounds. Apart from the esthetic and sentimental arguments for conservation, the steepness of the population growth curve is somber warning that already it is later than we think. Our present course may be like Russian roulette with a bullet in all but one chamber. But any really meaningful program for maintenance of an acceptable ecological balance with the environment seemingly will require first a major change in centuries-old attitudes toward nature. Leopold's eloquent plea for a new *land ethic* characterized by an ecological conscience with human beings behaving more as stewards than exploiters seems even more relevant today than when it was published in 1949.

In the past century, industry's increasing ability to produce goods has led to an *ethic of consumption*—"more is better." We need to replace this with an *ethic of conservation* so as to achieve a manageable equilibrium, or *steady state,* between population demand and resource supply, especially of food and energy. Chancellor and Goss have analyzed computer models and shown that if near-zero population growth, reduced resource consumption, and near-zero energy growth can be realized by early in the twenty-first century, such a food and energy equilibrium could be sustained even with an acceptable standard of living. Their model presupposes an increasing reliance upon alternative energy sources such as the sun, however.

Fundamental overhaul on a social level of some of the most cherished tenets of Western industrial society seems to be the prerequisite to future welfare of the earth's surface. Inasmuch as history reveals, above all else, the inevitability of change, it also suggests that the thinking human has the choice of how to influence future changes in the environment. Can we accept more difficult and uncertain times? Can we innovate creatively to deal with them? Most organisms are at their best when challenged. Let us hope that we are. Surely an appreciation of geologic time, of the inevitability of change, and of the operation of natural selection in the past should help to prepare anyone to cope better with the future. Ultimately all our problems converge upon the same great question: *Will this continue to be the best of all possible worlds?*

Readings

Adams, D., and M. Carwardine. 1990. *Last chance to see.* New York: Harmony Books.

Bartlett, A. 1980. The forgotten fundamentals of the energy crisis. *Journal of Geological Education,* 28:4–35.

Brown, L. R. 2003. *State of the world, a Worldwatch Institute report on progress toward a sustainable society.* New York: Norton (issued annually).

———. 2003, C. Flavin, and H. Kane, *Vital signs 2003, the trends that are shaping our future.* New York: Norton (issued annually).

Cain, S. 1967. The problem of people. *Geoscience News,* 1:6.

Cameron, E. N. 1986. *At the crossroads: The mineral problems of the United States.* New York: Wiley.

Carson, R. 1961. *Silent spring.* Boston: Houghton Mifflin.

Chancellor, W. J., and J. R. Goss. 1976. Balancing energy and food production, 1975–2000. *Science,* 192:213–18.

Cook, E. 1975. The depletion of geologic resources. *Technology Review,* 22:15–27.

Deffeyes, K. F. 2001. *Hubbert's curve: The impending world oil shortage.* Princeton, N.J.: Princeton University Press.

Ehrlich, P., and A. Ehrlich, 1990. *The population explosion.* New York: Simon and Schuster.

Gore, A. 1992. *Earth in the balance: Ecology and the human spirit.* New York: Penguin.

Gribben, J. 1982. *Future weather and the greenhouse effect.* New York: Delacorte.

Hardin, G. 1968. The tragedy of the commons. *Science,* 162:1243–48.

Hirsch, R. L. 1987. Impending United States energy crisis. *Science,* 235:1467–72.

Hubbert, M. K. 1969. Energy resources. In *Resources and man.* San Francisco: Freeman.

Leopold, A. 1949. *A Sand County almanac.* Oxford, UK: Oxford University Press.

Lovins, A. B. 1977. *Soft energy paths—Toward a durable peace.* San Francisco: Friends of the Earth International.

McKibben, B. 1989. *The end of nature.* New York: Random House.

Meadows, D. H. 1991. *The global citizen.* Washington, D.C: Island Press.

Oppenheimer, J., and R. Boyle. 1990. *Dead heat: The race against the greenhouse effect.* New York: Basic Books.

Roan, S. L. 1989. *The ozone crisis: The fifteen-year evolution of a sudden global emergency.* New York: Wiley.

Schneider, S. H. 1976. *The genesis strategy: Climate and global survival.* New York: Delta.

———. 1990. *Global warming: Are we entering the greenhouse century?* San Francisco: Sierra Club.

Schumacher, E. F. 1973. *Small is beautiful, economics as if people mattered.* New York: Harper & Row.

Weiner, J. 1990. *The next one hundred years: Shaping the future of our living earth.* New York: Bantam Books.

Wilson, E. O., ed. 1988. *Biodiversity.* Washington, D.C: National Academy Press.

World Resources Institute. 2003. *World resources 2002–2003.* New York: Oxford (revised annually).

Appendix I

The Classification and Relationships of Living Organisms

Any classification of organisms is really a progress report on the status of our understanding of their relationships. Several things must be kept in mind when using any classification. First of all, it is a partly subjective concoction. Even though there is general agreement among specialists on many details, other points remain controversial even after decades of study. Second, the "rank" of many taxa is also a subjective proposition. Many biologists try to compare different orders or classes or families of animals, whereas others insist that these ranks have no true biological reality outside the convenience of classification, and so such comparisons are meaningless. You will find that nearly every classification contains the same groupings of organisms, but there is much disagreement as to where to rank them.

In recent years, a new school of thought has challenged many of the conventions of classification. According to the systematic philosophy known as phylogenetic systematics, or **cladistics,** *classification should be a direct reflection of the evolutionary history of organisms.* For the most part, older, conventional classifications are largely consistent with the cladistic philosophy and show the branching pattern of evolutionary history quite well. But in some areas, they differ greatly from a cladistic classification.

Cladistics argues that the branching pattern of evolution ought to be read directly from the arrangement of names in the classification. For example, birds are descended from dinosaurs, which are in turn descended from more primitive reptiles, then from amphibians and lobe-finned fish, and back to their ultimate ancestors (Fig. A1). Expressed cladistically, the birds are a subgroup within the Dinosauria, which are in turn a subgroup of the Reptilia, which are a subgroup of the Amphibia, and so on. Traditional classifications, on the other hand, are not so rigidly phy-

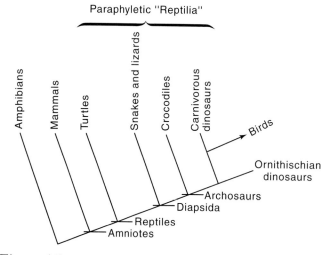

Figure AI

logenetic. They mix in a bit of ecology or attempt to reflect how much evolutionary divergence has occurred within a group. For example, because birds have diversified more than any other descendant of the reptiles, conventional classifications place the birds in their own class, Class Aves, equal in rank with Class Reptilia. Although this may satisfy some scientists' sense of balance, this classification no longer shows that birds are descended from a group of reptiles.

Another convention in cladistics is that *natural evolutionary groups must include all descendants of common ancestors.* For example, if we split off the birds from their reptilian ancestors, then we create two kinds of taxa: a natural monophyletic group (in this case, the birds, or Class Aves), which includes the earliest birds and all their descendants, and another, unnatural paraphyletic group, (Class Reptilia), which is made up of all other vertebrates that lay land eggs and have other reptilian specializations (such as lizards, snakes, crocodiles, turtles, and all their extinct relatives)—*but not birds*. Most biologists agree that we should use monophyletic taxa wherever possible, but traditional taxonomists still prefer to retain some paraphyletic groups for convenience. For example, the "invertebrates" are not a natural group—they are just animals that do not happen to have a backbone. If reptiles are not considered a subgroup of amphibians, then the amphibians become paraphyletic, because they gave rise to reptiles. Cladists reject these "nongroups," because they are not defined by unique evolutionary specializations. Such nongroups are considered "wastebaskets," used as a last resort in the absence of careful analysis or when there are insufficient data to resolve relationships better.

The dispute here is not over the *pattern of life's history*. Both schools of classification agree upon the basic events in the evolution of life. Instead, the disagreement is over the *goals of classification*. To a cladist, a classification should reflect evolutionary history and nothing else. Any other components confuse the reader, who will not know which groups are natural and which ones are wastebaskets. To a traditional taxonomist, the classification should reflect some evolutionary history, but it should also include the degree of ecological specialization or evolutionary divergence. Under these premises, birds have been such a success in terms of number of species, or anatomical specializations, that they deserve equal rank with amphibians, reptiles, and mammals as separate classes within the vertebrates.

Let us consider one more example to make the point clear. All taxonomists agree that humans are descended from a common ancestor shared with great apes, such as the gorilla and chimpanzee (Fig. A2). Cladistically this would be expressed by putting humans in the ape family Pongidae or by placing the human family, Hominidae, as a subgroup of the Pongidae. But traditional classifications have always placed the Hominidae and Pongidae as families of equal rank, largely because we humans are so impressed with the success of our own family.

To cladists, this makes the apes into a wastebasket group and might suggest to someone reading the classification that humans were not related to chimps and gorillas. In some ways, traditional taxonomists are inconsistent in their aversion to including hominids with the apes. They do not want to place hominids within the Pongidae, but they classify the hominids within a higher group (the catarrhine primates), which includes great apes and our distant relatives, the Old World Monkeys (Family Cercopithecidae). Indeed, Linneaus, the father of classification, put humans with apes and monkeys in the Order Primates when he first set up the modern scheme of classification in 1758.

The classification that follows reflects this uncomfortable tension between two schools of thought. A large number of practicing taxonomists reject traditional classification schemes, but no standardized cladistic classification of all living organisms has emerged to replace it. However, many people who were trained some years ago would find elements of a cladistic classification unfamiliar, because many traditional groups are unnatural wastebaskets or worse. The poor student, torn between the warring ideologies, does not know whom to believe and finds the welter of names and ranks utterly confusing (especially when they change radically from one year to the next).

To avoid confusion, we have adopted a compromise. Instead of attempting to place every taxon in either a traditional or cladistic scheme, we have shown the modern conception of their evolutionary relationships with a series of branching diagrams (Fig. A3). The named branching points are useful to remember, but we have not labeled them with taxonomic rank in many cases. We have done our best to avoid unnatural paraphyletic groups, although in a few cases this is impossible. This system may be a bit more cumbersome than the symmetrical traditional classifications seen in other textbooks, but at least it contains fewer misleading or erroneous implications. Besides, the branching diagrams are much easier to understand than pages and pages of phyla, classes, and other groups whose validity is doubtful.

For simplicity, groups mentioned in the text are emphasized, and many living groups with no fossil record are left out.

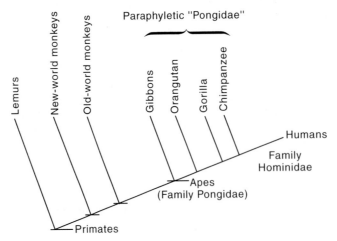

Figure A2

KINGDOM ARCHAEBACTERIA

Prokaryotic organisms that have the most primitive genetic codes of all life. These include sulfur-reducing and methane-dependent bacteria, as well as bacteria that tolerate environments with extreme temperatures or salinities, such as hot springs. Many live in the absence of oxygen (Fig. 9.6).

KINGDOM EUBACTERIA

Advanced prokaryotes, including the more familiar types of digestive and disease-causing bacteria. The *cyanobacteria* (or cyanophytes, or "blue-green bacteria") are among the earliest fossils known.

KINGDOM PROTISTA

Single-celled or simple multicellular eukaryotes, including most algae.
Phylum Haptophyta *(calcareous nannoplankton)*
   Photosynthetic single-celled planktonic algae (coccolithophorids) that are covered by small, calcareous, button-shaped plates (coccoliths) (Fig. 14.41).
Phylum Sarcodina *(amoebas and their relatives)*
   Class Foraminifera—Single-celled, amoeba-like organisms with a calcareous skeleton surrounded by protoplasm (Fig. 14.42).
   Class Radiolaria—Single-celled, amoeba-like, planktonic organisms that secrete an intricate internal skeleton of silica.

KINGDOM PLANTAE

Multicellular, sexually reproducing eukaryotes that obtain their energy by photosynthesis with the aid of chloroplasts. Except for mosses, most plants have vascular tissues for conducting fluids. Most of these groups are paraphyletic, since they are ancestral to later groups. However, botanical classification has not yet fully come to terms with cladistic thinking.
Phylum Bryophyta (mosses and liverworts)
   Small plants that lack well-developed vascular tissues but have complex reproductive cycles that allow them to live on land.
Phylum Psilophyta
   Small vascular plants with simple stems but no true roots or leaves. Common in the mid-Paleozoic, but only a few survive today (Fig. 12.10A, B).
Phylum Lycopodophyta (club mosses)
   Spore-bearing plants with true roots and leaves, which are arranged in a spiral pattern around the stem. Most living lycopsids are small and low-growing, but they were the dominant trees of the Carboniferous, as represented by *Lepidodendron* and *Sigillaria* (Fig. 12.10C, D).
Phylum Sphenophyta (horsetails)
   Spore-bearing plants whose stems are divided into nodes that bear whorls of branches. The living species are mostly small water-loving plants known as "scouring rushes," but some Carboniferous species were tree-sized (Fig. 12.10E).
Phylum Filicinophyta (true ferns)
   Spore-bearing plants with branching leaves, with their spores on the undersides of the leaves (Fig. 12.10H).
Phylum Gymnospermae (seed ferns, cycads, conifers and their relatives)
   The most primitive seed-bearing plants. They include a number of fernlike forms (such as the extinct seed fern *Glossopteris*); the *Cordaites* trees of the late Paleozoic; the palmlike cycads (or "sego palms") and their extinct relatives, the cycadeoids; the ginkgo (or "maidenhair tree"); and many kinds of conifers (*Araucaria,* pines, spruces, firs, and sequoias) (Fig. 13.60 and 14.45).
Phylum Angiospermae (flowering plants)
   Plants with internal covered seeds, which are fertilized with the aid of flowers. They arose in the Early Cretaceous and underwent an explosive radiation, so that today they make up the vast majority of plants. They include all hardwood trees and other flowering shrubs, as well as all grasses.

KINGDOM MYCOTA (OR FUNGI)

Single-celled or multicellular eukaryotes that break down decaying organic matter for their nutrition. The Mycota include molds, slime molds, yeasts, true fungi, and mushrooms. Fungi help break down dead tissues and return their nutrients to the food chain. Recent studies of their molecular sequences suggest that they are more closely related to the animal kingdom than they are to the plants.

KINGDOM ANIMALIA

Multicellular eukaryotes that obtain their nutrition by consuming other organisms. The phylogeny of multicellular animals is becoming much better known, allowing us to present the family tree shown in Fig. A3 (from Morris, S. Conway. 1993. *Nature,* 361:219, and based largely on the 18S ribosomal RNA sequence of Lake, J. A. 1990. *Proceedings of the National Academy of Sciences,* 87:763). Because this family tree makes many of the old ranks obsolete, we shall not place all the animals in formal ranked taxa. Instead, the clustering of taxa is indicated by the indentation of the descriptions but is better shown by their position on the branching diagram (Fig. A3).

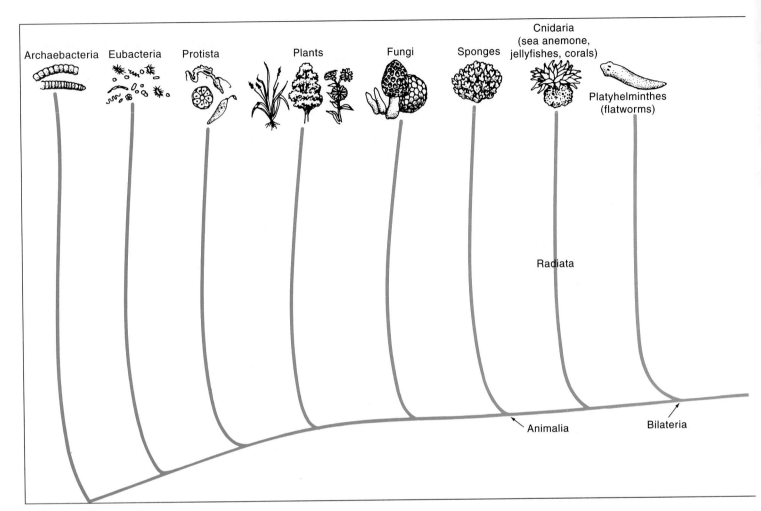

Figure A3

Superphylum Porifera (sponges)
  Simple colonial marine organisms that secrete a common skeleton and then filter food through the pores in the walls. The individual cells carry out most of their own digestive, respiratory, and reproductive functions; they have no true internal organs.

Superphylum unknown
  Archaeocyatha *(archaeocyathids)* (Fig. 9.17A–I)
  Vase-shaped, double-walled colonial filter feeders that formed Early Cambrian reeflike mounds but became extinct by the Middle Cambrian. Some paleontologists place them in the sponges, but their true relationships are enigmatic.

Superphylum Radiata
  Animals with radial symmetry and no true head, tail, or one-way digestive tract.

Phylum Cnidaria (sometimes called **Coelenterata** in older classifications)
  Simple, saclike animals with a three-layered body wall surrounding their body cavity and tentacles around the opening. They capture food with the stinging cells in the tentacles.
        Class Scyphozoa—Jellyfishes
        Class Anthozoa—Corals and sea anemones. They include two groups of extinct corals, the tabulates and rugosids, and the living scleractinian corals (Fig. 11.6F, G).

Superphylum Bilateria
  Animals with bilateral symmetry, a definite front and back end, and a one-way digestive tract. Most have other specialized organs for reproduction, respiration, and other functions.

Infraphylum Platymorpha (flatworms)
  Bilaterally symmetrical animals with no internal body cavity, or coelom. Flatworms have no known fossil record.

Infraphylum Protostoma
  Bilaterally symmetrical animals with or without a coelomic cavity. They are recognized by a number of embryonic features, in addition to their molecular similarities. For example, when the embryo develops into a hollow ball of cells, the opening (blastopore) eventually develops into the mouth. As the first cells cleave, they do so in a spiral fashion, and if broken apart, they form an incomplete animal.

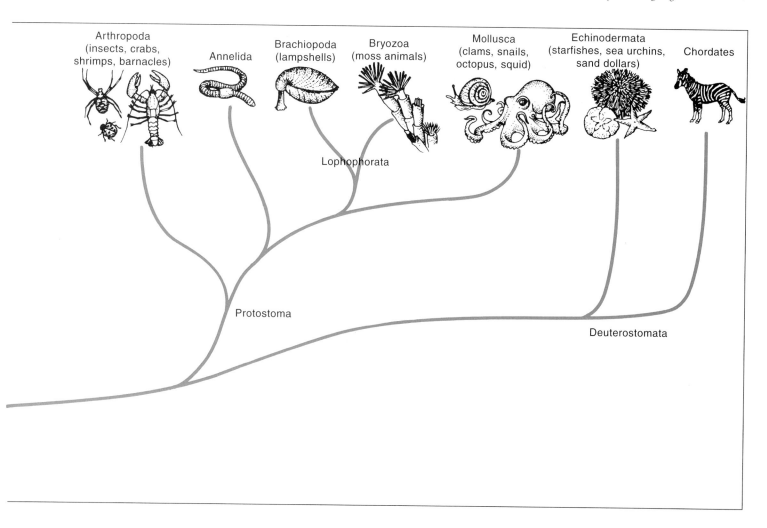

Phylum Arthropoda

The largest phylum of animals, including millions of species. They all have a segmented external skeleton made of the protein chitin and jointed appendages specialized for locomotion, predation, respiration, and reproduction. The subphyla are sometimes listed as superclasses in other classifications. Recent molecular data suggest that the crustaceans and chelicerates form one group, and the insects are closer to the millipedes. However, we list them as taxa of equal rank.

Subphylum Trilobitomorpha (trilobites)

Marine arthropods with a three-lobed body (a central lobe and two side lobes) and hard skeletons made of both chitin and calcite. Most had a well-developed head segment, called a cephalon, and a fused tail plate, known as a pygidium (Figs. 9.15, 11.6I–L, 12.3D–E, and 13.66).

Subphylum Chelicerata (spiders, scorpions, ticks, mites, horseshoe crabs, and many extinct arthropods, such as the eurypterids) (Fig. 12.4)

Arthropods with appendages developed into pincers and six frontal segments fused into a head section.

Subphylum Crustacea (crabs, lobsters, shrimps, barnacles, sowbugs, ostracodes, and many important types of zooplankton)

Mostly aquatic arthropods with specialized antennae and mouthparts. Next to insects, they are the most abundant and successful group (Fig. 11.8).

Subphylum Myriapoda (millipedes and centipedes)

Multisegmented arthropods with many legs and relatively unspecialized appendages. Millipedes first crawled onto land in the Late Ordovician, making them the earliest land animals. With their relatively thin chitinous exoskeleton, they are rarely fossilized (Fig. 12.12C).

Subphylum Hexapoda or Insecta (insects)

Mostly terrestrial arthropods with six legs and a well-defined head, thorax, and abdomen. The most abundant and diverse group of animals on earth, consisting of millions of species that make up about 75 percent of the animal kingdom. They are the most common organisms in every terrestrial habitat from the tropics to the deserts to the tundra, and a few have become marine as well (Fig. 2.1).

Phylum Annelida (segmented worms)

Soft-bodied worms whose body is divided into a number of segments with separation of the internal body cavity into sealed chambers, which give them hydraulic rigidity for burrowing. They live in marine and fresh waters (polychaetes) and in the soil (earthworms). Although they rarely fossilize, their burrows are common from the late Vendian onward.

Phylum Lophophorata
Animals that feed by means of a featherlike filtering device called a lophophore.
  Subphylum Brachiopoda ("lamp shells") (Fig. 9.18)
  Marine filter feeders with two shells enclosing them; the lophophore is supported on a loop-shaped structure.
    Class Inarticulata
      Primitive brachiopods that have no tooth-and-socket joint between their shells; they hold their shells together with muscles instead. Shell composed of chitinophosphate (Fig. 9.18A).
    Class Articulata
      The most common brachiopods since the Cambrian, with calcareous shells articulated with a tooth-and-socket mechanical hinge (Fig. 9.18B).
        Order Orthida—With wide hinge line or oval-shaped biconvex shells. Most common during lower Paleozoic (Figs. 11.3 and 11.6A).
        Order Pentamerida—Short hinge line, large internal platform for muscles and other structures. Important index fossils in the Silurian (Fig. 12.3A).
        Order Strophomenida—One of the very common lower and middle Paleozoic groups. Has a wide hinge line. One valve is planar or convex. Pedicle opening very small or absent (Figs. 11.3 and 11.6B).
        Order Productoidea—A large, varied, and sometimes bizarre group most common and important in the upper Paleozoic. Interarea (between valves) small or absent, hinge line is moderately long. The shell is generally covered by spines. Attachment to hard substrate with spines, or spines used as stilts to raise shell off a muddy bottom (Fig. 13.52).
        Order Rhynchonellida—Convex shells with sharp ribbing. The beaks may be large; hinge line is short. A relatively conservative group.
        Order Spiriferida—Typically with a long hinge line and with the interarea on pedicle valve only. Internally there is a spiral lophophore. An important group particularly in the Devonian (Fig. 12.3B, C).
        Order Terebratulida—Biconvex, punctate shells with a short hinge line and an interarea as in the spiriferids. Internally there is a "looped" gill support (Fig. 9.18B).
  Subphylum Bryozoa ("moss animals")
  Minute colonial filter feeders that live in tiny chambers on a calcareous colony, which may be massive, branching, or lacy in structure; most post-Paleozoic types have less heavily calcified colonies and may encrust on hard surfaces (Figs. 11.6C, D and 13.51).

Phylum Mollusca
Soft-bodied animals with a muscular foot for locomotion, feathery gills in an internal chamber, and a fleshy flap covering the body that can secrete a calcareous shell in many groups. They have a complex digestive system, a two- or three-chambered heart, and a complex nervous system with well-developed eyes, which approach those of vertebrates in complexity.
    Class Monoplacophora—A segmented, bilaterally symmetrical form with a single, cap-shaped shell. Fairly common in certain lower Paleozoic strata, not known after the Devonian until Recent. One of the best examples of a "living fossil" is *Neopilinia,* which was dredged up from deep water near Central America some years ago.
    Class Polyplacophora—This group also is called *Amphineura* or chitons. Its forms have bilateral symmetry; most possess a shell consisting of eight separate calcareous plates. They are adapted for living on rocks in the surf zone or hard substrate. They are rare in the fossil record, being represented only by isolated plates.
    Class Gastropoda—Single-shelled, generally coiled forms, which may have an operculum for protection. Distinct from all other molluscs by having the gastrointestinal tract contorted into a figure-eight pattern so that the anus and mouth are close together. One of the most successful of all invertebrate classes (Fig. 11.6M).
    Class Cephalopoda—Molluscs that have evolved into highly efficient swimmers. Because of their mobility, they have developed excellent eyes and a jet-propulsion system of locomotion in addition to having tentacles for feeding and locomotion. Includes the living octopus, squid, and chambered nautilus. Fossil cephalopods include the subclass Nautiloidea, dominant in the lower Paleozoic, and the subclass Ammonoidea, dominant in Mesozoic seas (Fig. 14.40).
    Class Bivalvia (*Pelecypoda*)—Molluscs have two shells that are mirror images in contrast to the brachiopods in which the left and right sides of each shell reflect bilateral symmetry. Clams share with gastropods the honor of being among the most diverse of the invertebrates in today's seas (Figs. 14.38 and 14.39).

Infraphylum Deuterostoma
Bilaterally symmetrical animals with a true coelomic cavity, which originally develops from the gut. They are recognized by a number of embryonic specializations, in addition to their molecular similarities. For example, when the embryo develops into a hollow ball of cells, the mouth originates on the opposite side of the original opening (blastopore). As the first cells cleave, they do so in a radial fashion, and if broken apart in the four-cell stage, they can still form a complete animal out of any one of the four cells.

Phylum Echinodermata
  Marine animals that have a fivefold radial symmetry superimposed on their embryonic bilateral symmetry. Their skeletons are made of monocrystalline plates of calcite. Locomotion is by an internal hydraulic system, which powers their movements and their tube feet by means of fluid pressure. Most have a series of arms radiating out from a central disk, which includes the mouth.
  Subphylum Echinozoa—Globose echinoderms that do not develop arms; most forms are unattached.
    Class Helicoplacophora—A unique Lower Cambrian group that is the earliest echinozoan known. They are free living, top-shaped with their plates arranged as helical spirals (Fig. 9.19).
    Class Edrioasteroidea—A Paleozoic group, may be attached or free living. From three to five feeding areas are twisted into sigmoid shapes resembling starfish.

Class Holothuroidea—The sea cucumbers, sediment feeders having a reduced skeleton consisting of calcareous spicules imbedded in a tough skin. Radial symmetry is observable internally. Known mainly from curiously shaped disarticulated spicules in the fossil record.

Class Echinoidea—The largest group of echinoderms. Dominantly with a bun-shaped exoskeleton composed of fused or articulated polygonal plates. The anus is on the upper surface, the mouth on the lower. Typically with protective spines on the intraambulacral plates. The echinoids tend to be sessile or move very slowly by moving their spines. One group is more mobile and tends to develop a secondary bilateral symmetry. When this occurs, the anus and mouth migrate to posterior-anterior positions.

Subphylum Homalozoa—A small group of lower Paleozoic forms with flattened, asymmetrical bodies. Another name for the group is Carpoidea; representatives are rare.

Subphylum Crinozoa—Forms attached by means of a calcified jointed stem.

Class Cystoidea—Globular to pear-shaped forms attached directly to the substrate by means of a stem. The porous head (calyx) plates are not stabilized in numbers or shape.

Class Blastoidea—A relatively small class of Paleozoic attached echinoderms having rather short stems. The calyx has prominent ambulacra and the thirteen interambulacral plates are arranged in three circles. The top of the calyx has five openings leading to complex, calcified internal circulatory structures (Fig. 13.50B).

Class Crinoidea—The most important class of Paleozoic echinoderms, which has gradually diminished in numbers and kinds until today. The majority of the crinoids were attached by a well-developed and jointed stem and a root system. The viscera are contained in a calyx, which is variable in structure but with a regular system of plates and a system of food-gathering arms (Figs. 13.49 and 13.50A).

Subphylum Asterozoa—A group of stemless echinoderms that are vagrant benthonic. Mobility is achieved by arms and tube feet.

Class Somasteroidea—Starfish with flattened, petal-shaped arms and lacking an anus. The tube feet are nonextendable.

Class Asteroidea—The common starfish with a water vascular system carrying water by radial tubes from near the mouth to the tips of the arms. The mouth is on the underside and anus on the central dorsal surface; water for the circulatory system enters a sieve plate near the anus.

Phylum Hemichordata

Deuterostomes with segmentally organized muscles, a pharynx with round gill slits, and a hollow dorsal nerve chord. Living examples include the acorn worms and filter-feeding colonial pterobranchs. The extinct graptolites were extremely similar in wall structure to the pterobranchs and, so, are usually assigned to this phylum (Fig. 11.7).

Phylum Chordata

Deuterostomes with a stiffened rod of cartilage (the notochord) along the back, just above a hollow nerve cord. In some groups, this embryonic nerve chord is replaced by a bony vertebral column. Hard parts are made of cartilage, replaced by phosphatic bone in most groups. Most chordates have gill slits and a pharynx for feeding.

(**NOTE:** *Within the chordates, it is easier to show the branching pattern of relationships than to put them all in standard paraphyletic ranks, with many confusing infraphyla, subclasses, and superorders. In the following listing, all ranks have been dropped, and their relative branching order is shown in Fig. A4.*)

Urochordata (tunicates, or "sea squirts")

Soft-bodied colonial organisms that use their large pharynx for filter feeding. As larvae, they have a tail with segmented muscles, a dorsal hollow nerve chord, and a notochord. These are lost when they attach and metamorphose into sessile adults (Fig. 3.23).

Cephalochordata (amphioxus, or the lancelet)

Elongate, fishlike animals with a long notochord, segmented muscles, and dorsal hollow nerve cord. They have tentacles around their mouths for filter feeding when they lie half-buried with their heads protruding from the sand.

Craniata

Chordates with a highly specialized head region, including a distinct brain, specialized hearing, seeing, and smelling organs, and semicircular canals in the ear for balance. They also have paired kidneys and excretory system, a well-developed heart, and many other unique specializations.

Myxinoidea (hagfishes)

Slimy, eel-like marine chordates that burrow in the sea floor to eat polychaete worms or feed on the internal organs of dead fish.

Vertebrata (Fig. 12.7)

Chordates that can develop phosphatic bone from the embryonic skin cells, or dermis. They have many other unique evolutionary specializations.

Jawless vertebrates (Figs. 11.9 and 12.5)

At this level, there are a number of extinct jawless "fish" that used to be placed in the wastebasket taxon "Agnatha." As can be seen from Fig. A4, some are more closely related to higher vertebrates than others, so they cannot all be placed in a single group. Jawless vertebrates include not only the many extinct armored forms but also the living lamprey. Recent discoveries suggest that the eel-like conodont animals (Fig. 4.19), with their microscopic phosphatic toothlike conodont fossils in their throat region, were probably vertebrates.

Gnathostomata

Vertebrates with true jaws formed from their cartilaginous gill arches. They have many other unique features, include teeth containing dentin, a specialized gill apparatus, three semicircular canals for balance, pectoral and pelvic girdles in their bodies to support their fins, and dozens more specializations.

Placodermi (Fig. 12.1)

Extinct armored fish with bone in their armor plating and a specialized joint in their neck region. They are among the earliest jawed vertebrates and were very successful before their final extinction at the end of the Devonian.

Chondrichthyes (sharks, skates, rays, ratfish, and their relatives) (Figs. 12.1 and 13.53)

Cartilaginous fish with bony spines in their skins, distinctive teeth, and a complex fin support system.

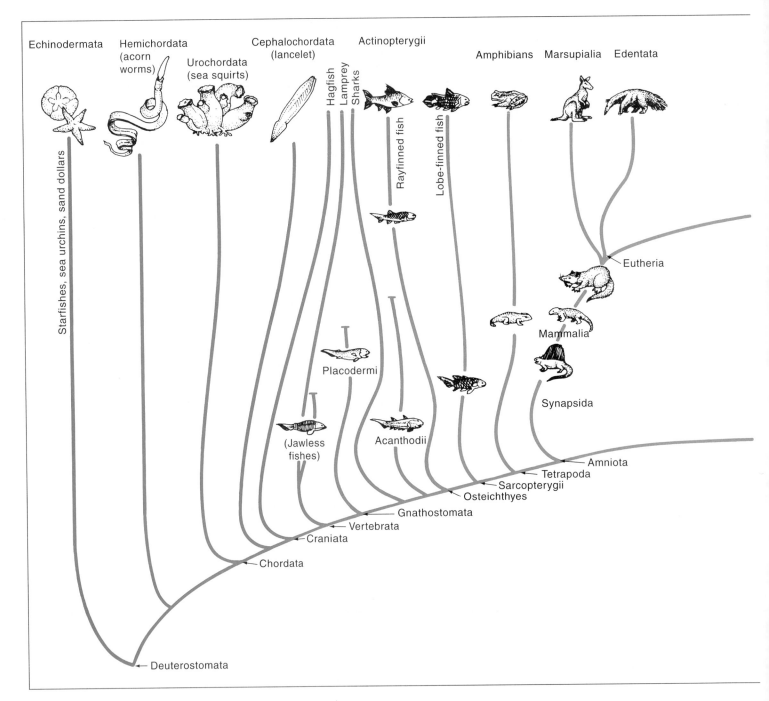

Figure A4

Acanthodii (Fig. 12.6)

The "spiny sharks," an extinct group of large-eyed fish with many fins supported by bony spines. They are the earliest group of jawed vertebrates, with fossils known from the Silurian.

Osteichthyes (bony fish and their descendants)

Vertebrates with bone replacing their internal cartilaginous skeletons and many other unique specializations.

Actinopterygii (ray-finned fish) (Fig. 13.53)

Bony fish with a series of bony rays that support their fins. They include a number of archaic forms, such as the sturgeon, garfish, and bowfin, plus the great diversity of modern bony fish known as *teleosts,* which make up 90 percent of living fish.

Sarcopterygii (lobe-finned fish and their descendants) (Figs. 12.5 and 12.13)

Vertebrates with their muscular fins or limbs supported by simple bony structures, complex enamel in the teeth, and many unique features. Most have the capability to breath air with fully developed lungs. The lobe-fins include the coelacanth, the lungfishes, and the advanced groups that were related to tetrapods.

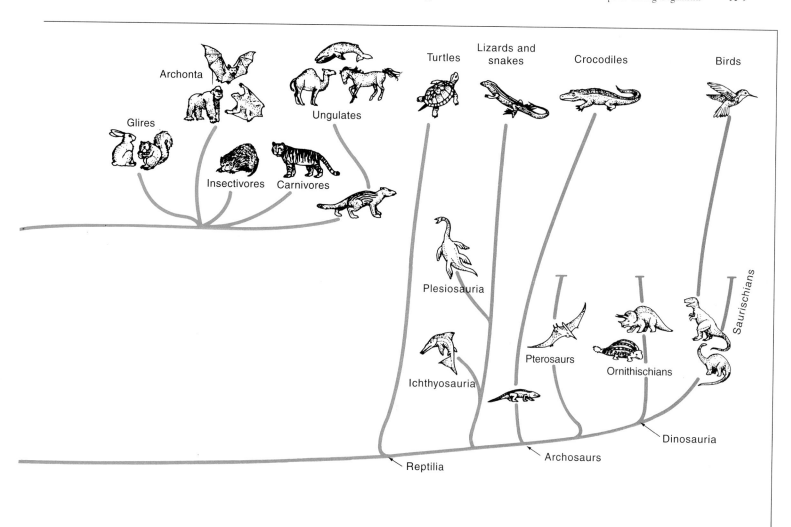

Tetrapoda (four-legged vertebrates)

Vertebrates with four limbs built of a single bone near the body and a pair of bones in the outer segment. Their shoulder and hip joints are attached to the spinal column for support, and they have many other advanced features related to life on land.

Amphibia (Figs. 12.13 and 13.61)

The oldest terrestrial vertebrates. They include not only the living frogs and salamanders but also the various extinct amphibians known as *temnospondyls,* large, flat-skulled creatures common in the late Paleozoic and Triassic. The relationships of other extinct groups, including the lizardlike microsaurs, the eel-like aistopods, and the boomerang-headed nectrideans, are less certain. The more advanced anthracosaurs, however, are much closer to amniotes than they are to amphibians.

Amniota

Vertebrates that lay a land egg, with membranes that allow it to develop out of water (Fig. 13.62).

Reptilia

Amniotes (excluding synapsids and their mammalian descendants) characterized by a number of unique features of the skull bones.

Testudomorpha (turtles and their extinct relatives)

Archaic group of reptiles with no temporal openings on the side of their skull. Turtles have many unique evolutionary features related to their life inside a protective shell.

Diapsida

Reptiles with two openings on the side of their skull for expansion and attachment of jaw muscles.

Lepidosauria (Figs. 14.44 and 14.55)

The living tuatara, lizards, and snakes and their extinct relatives. Some scientists argue that the fishlike ichthyosaurs, the paddling plesiosaurs, and several other Mesozoic marine reptiles are distantly related to lepidosaurs.

Archosauria (Fig. 14.46)

Advanced reptiles with a number of shared specializations, especially in the skull openings on the snout. There are many archaic archosaurs in the Triassic, which used to be lumped into the taxonomic wastebasket known as "thecodonts." Of these, three major groups became important in the Jurassic and Cretaceous.

Crocodylia—Crocodiles and their relatives.

Pterosauria—Flying reptiles.

Dinosauria (Fig. 14.49)

Ornithischia—"Bird-hipped" dinosaurs with the pubic bone rotated backward parallel to the ischium. They also have a unique bone in front of the snout, the predentary, and teeth recessed in the skull so they had fully developed cheeks. All were herbivorous. Familiar examples include the duckbilled hadrosaurs, the iguanodonts, the armored stegosaurs and ankylosaurs, the bone-headed pachycephalosaurs, and the horned ceratopsians.

Saurischia—"Lizard-hipped" dinosaurs, including the giant sauropods, the carnivorous theropods, and their descendants, the birds. All have the primitive pelvic structure, with the pubic bone pointing forward, but other specializations of the skull.

Aves—Birds are descended from carnivorous theropod dinosaurs, such as *Deinonychus* (Fig. 14.52); cladistically speaking, they are a group within the Saurischia (Fig. 14.53).

Synapsida ("mammal-like reptiles" and mammals)

Amniotes with a single opening on the side of the skull and many other specializations of the skeleton. In older books, they are called "mammal-like reptiles," but they have nothing to do with true reptiles, which originated as a separate lineage at the same time in the Late Carboniferous. A number of different synapsid groups, such as the Early Permian finbacks, or the various advanced Late Permian carnivorous and herbivorous beasts, dominated the late Paleozoic and Early Triassic. In the Late Triassic, they were replaced by archosaurs (especially dinosaurs), leaving only tiny mammals as their descendants.

Mammalia (Fig. 15.43)

Synapsids with hair, mammary glands, and dentary-squamosal jaw joint.

Multituberculata—Extinct squirrel-like mammals with distinctive complex upper molars and lower bladed premolars (Fig. 15.45).

Monotremata—The platypus and echidna, the only living mammals that lay eggs.

Marsupialia (= Metatheria)—Pouched mammals, including the opossums, kangaroos, koalas, and their relatives (Fig. 3.9).

Eutheria—Placental mammals, which give birth to fully developed young after a long gestation in the uterus.

Edentata—Anteaters, armadillos, sloths, and pangolins (Fig. 15.50).

Glires—Rodents, rabbits, and elephant shrews.

Archonta—Primates (lemurs, monkeys, apes, and humans; Fig. 16.29), plus tree shrews, colugos ("flying lemurs"), and bats.

Insectivora—Moles, shrews, and hedgehogs.

Carnivora—Carnivorous mammals, including dogs, bears, cats, hyenas, raccoons, weasels, and their relatives (Fig. 3.16).

Ungulata—Hoofed mammals, including the artiodactyls (even-toed beasts, such as pigs, hippos, camels, deer, antelopes, and cattle), whales (descended from hoofed ancestors), tethytheres (Fig. 15.48) (elephants, sea cows, and their extinct relatives), and perissodactyls (odd-toed beasts, such as horses, rhinos, tapirs, and their extinct relatives) (Figs. 3.15 and 15.47).

Appendix II

English Equivalents of Metric Measures

Units of Length

1 millimeter (mm) = 0.1 centimeter (cm) = 0.039 inch (in.)

1,000 mm = 100 cm = 1 meter (m) = 39.37 in. = 3.28 feet (ft) = 1.09 yard (yd)

1,000 m = 1 kilometer (km) = 0.62 mile (mi)

1 in. = 2.54 cm

1 mi = 1.61 km

Units of Area

1 m^2 = 10.76 ft^2

1 km^2 = 0.39 mi^2

1 ft^2 = 0.09 m^2

1 mi^2 = 2.58 km^2

Units of Volume

1 m^3 = 35.32 ft^3

1 km^3 = 0.24 mi^3

1 ft^3 = 0.03 m^3

1 mi^3 = 4.17 km^3

Units of Weight

1 gram (g) = 0.001 kilogram (kg) = 0.0022 pound (lb)

1 kg = 1,000 g = 2.2 lb

Temperature Scales

1°F = 0.56°C

1°C = 1.8°F

0°C = 32°F (freezing point of water)

100°C = 212°F (boiling point of water)

Glossary

"When I use a word, it means just what I choose it to mean—nothing more or less."

Humpty Dumpty *in Lewis Carroll's Through the Looking Glass*

A

Acadian orogeny. A major mountain-building event affecting eastern North America in Late Devonian and Early Mississippian time. Apparently it was caused by collision with the Avalonia microcontinent.

Accretionary prism. The thick sequence of rocks sliced off the downgoing plate in a subduction zone and stuck onto the overlying plate.

Acritarchs. Organic-walled microfossils thought to be the cysts of early eukaryotes. They were common in the late Proterozoic and early Paleozoic but became extinct shortly thereafter.

Activity. The process of emission of particles and/or radiation from the nuclei of unstable atoms during radioactive decay (see also **Emissions**). Rate of activity or decay provides a basis for calculating isotopic dates.

Actualism. The assumption that present laws of science would apply at all times; thus, presently known processes are presumed to have always acted in the same way but with greatly varying intensities.

Adaptation. A feature of an organism that makes it better fit for its environment.

Adaptive radiation. Diversification by means of speciation into a new environment and subsequent specialization by new species in the new environment.

Aerobic. An environment rich in oxygen; also an organism that requires free oxygen in its environment.

Albedo. Reflectivity of surfaces (e.g., water, forest, snow), measured by percent of incoming solar radiation that is reflected; an important factor in global climate.

Allele. One of the two or more different states of a given gene.

Allopatric speciation. Small populations isolated from the main population tend to become genetically different from their ancestors and eventually can become new species.

Alpine-Himalayan orogeny. A major mountain-building event extending from Spain through the Alps, southeastern Europe, southwestern Asia, and Himalayan Ranges to Indonesia during Oligocene and Miocene time. It resulted from several continent continent collisions.

Amino acids. Relatively simple organic compounds that are the building blocks of proteins.

Anaerobic An environment poor in oxygen; also organisms that live without oxygen.

Andean-type plate margin. A margin like that of western South America with oceanic crust being subducted beneath continental crust to produce a volcanic arc (Andes volcanoes) on the edge of a continent.

Andesite. Volcanic rock of intermediate composition like that of coarser diorite; characteristic of volcanic arcs above subduction zones.

Angiosperms. Vascular plants that produce flowers containing ovaries in which capsuled seeds are formed.

Antler orogeny. A mountain-building event in western North America during Late Devonian to Early Mississippian time. It may have resulted from the collision of a volcanic arc or microcontinent. It is unusual in that neither significant granitic plutons nor metamorphism occurred.

Appalachian orogeny. The last major mountain-building event that affected eastern North America (Permian and possibly Early Triassic time). It produced most of the folding and thrust faulting so prominent in the present Appalachian Mountains; however, the present topography there resulted from Cenozoic upwarping, which rejuvenated rivers to cause incising of their valleys.

Arches. Large-scale cratonic structures that have a broad anticlinal form; total deposition was less and was interrupted by more unconformities than in their adjacent basins.

Archaebacteria. The most primitive group of single-celled organisms; typically found in extreme environments, such as hot springs.

Archosaurs. The group of reptiles that includes birds, dinosaurs, crocodiles, and their extinct relatives.

Asteroids. Extraterrestrial stony bodies, meters to 1,000 kilometers in diameter, that are larger than meteorites but smaller than planets. They are distributed throughout the solar system, but the best known ones orbit the sun between Mars and Jupiter. Most asteroids are thought to be either leftover protoplanetary bodies or pieces of a disrupted planet; however, some may be the burned-out cores of comets. Asteroids may collide with one another or with planets and their satellites.

Atolls. Ring-shaped groups of islands composed of coral reefs and coral sands built on submerged volcanoes.

Atomic number. Number of protons in an atomic nucleus. (See also **Mass number.**)

Aulacogen. Fault-bounded trough (graben) generally extending into a craton margin and filled with thick strata; thought to form when a supercontinent begins to break apart.

Autotrophs. Organisms that manufacture their own food.

Axis of spreading. A hypothetical axis that defines the rotation of one or a pair of lithosphere plates; not the same as geographic spin axis of the earth.

B

Basalt. Fine-grained, dark (mafic) igneous rock composed chiefly of calcium-rich feldspar and pyroxene.

Basins. Large-scale structures with synclinal form and especially (though not exclusively) important in cratons; deposition thicker and with fewer unconformities than in adjacent areas, such as arches.

Basin and Range Province. Area of Nevada and western Utah (and parts of adjacent states) characterized by north-south-trending fault-block ranges and grabens and by extreme crustal thinning.

Benthonic. Living on the bottom of lakes, streams, or oceans. They are either sessile (stationary) or vagrant (can move around) and may live on the bottom (epifaunal) or beneath the surface (infaunal).

Bentonite. Volcanic ash that settled on the sea floor and became altered to clay.

Biotic. Pertaining to plants *and* animals.

Block faulting. Fragmentation or rifting of the crust due to tension, which results in parallel fault blocks dropped down relative to one another.

Blocking temperature. As cooling occurs, it is the temperature below which a given mineral becomes a closed chemical system for a particular radioactive decay series. The parent-daughter isotope ratio dates the time of this closure.

Blueschists. Unusual metamorphic rocks composed of blue-colored minerals that are produced in regions of high pressure but relatively low temperature, primarily in subduction zones.

C

Caledonian orogeny. A middle Paleozoic mountain-building event (culminated in Siluro-Devonian time) that affected northwestern Europe and East Greenland. It resulted from a collision between these two continental masses.

Carnivore. An animal (rarely a plant) that eats animals for its basic food.

Catastrophism. An explanation of phenomena such as great unconformities and extinction of fossil species by appeal to supernatural causes acting rapidly and violently.

Chemical sediments. Sediments accumulated by precipitation of grains from chemically saturated water (e.g., the evaporite deposits, salt, and gypsum).

Chemostat. The chemical regulatory system among seawater, atmosphere, solid earth, and life, which helps to maintain a chemical balance, or equilibrium, among all four.

Chromosome. A structure in a cell nucleus carrying genes. During cell division, the chromosome splits into two sections, so that identical genetic information is passed on to the daughter cells.

Cladistics. The study of the relationship between organisms by analyzing derived and primitive features.

Classification. Any system of grouping species into a more inclusive category—the genera into families and families into orders, classes, and phyla, each more inclusive than the preceding category.

Clastic sediments. Fragmental sediments formed of broken grains of older minerals, rocks, shells, or other skeletal materials. *Terrigenous clastic* sediments were derived by erosion of older rocks exposed in land areas; *nonterrigenous clastic* materials were derived from skeletons formed in the sea but then washed about on the sea floor.

Clastic wedge. A prism of clastic sediments with wedge-shaped cross section, which was deposited adjacent to rising mountains. As the mountains rose, their debris was deposited on either adjacent lowlands or in the sea, depending upon relative subsidence and sedimentation. (See also **Foreland sedimentary basin.**)

Collisions. The crashing together of various large divisions of the earth's crust or lithosphere to form orogenic belts. They may involve continent-continent, microcontinent-continent, arc-continent, ridge-continent, arc-arc, etc., collisions. (See also **Suture zone** and **Suspect terrane.**)

Conformable. Lacking any discontinuities (unconformities), thus representing an apparently complete record of continuous deposition.

Conodonts. Toothlike phosphatic microfossils, now thought to be part of the feeding apparatus of a primitive, eel-like relative of the vertebrates; common throughout the Paleozoic, then vanished in the Triassic.

Continental accretion. An increase through time of the volume and area of continental crust by the formation of new granitic and andesitic rocks within orogenic belts as well as by the collision of volcanic arcs and microcontinents with continents.

Continental drift. The theory of Taylor, Wegener, and others that continents may be broken and displaced large distances; where they split apart, new ocean basins are formed and where they collide, mountains are formed.

Convergence. Two or more organisms, living in a similar environment but not closely related yet having similar characteristics (e.g., streamlining of the body and fins in sharks and dolphins).

Convergent plate margin. The margin along which two lithosphere plates move toward each other. In some cases, an oceanic plate is subducted (or consumed) beneath the other, causing andesitic volcanism and granite in-

trusion, whereas in other cases, two continents or arcs may converge and collide.

Core. The central zone of the earth, with a radius of 3,471 km and comprising 16 percent by volume; average density is 10.7 g/cm³. Seismic waves passing through the core indicate inner solid and outer liquid portions. Composition is inferred to be largely iron and nickel with minor cobalt, sulfur, and possibly other elements.

Cordillera. The mountainous regions of western North America, including the Rocky Mountains, the Cascades, and the Sierra Nevadas.

Correlation. A comparison of strata at two or more localities to establish a similarity of age. Most commonly, correlation is established by comparing index fossils, but it may also be based upon similar isotopic dates, magnetic polarity reversals, isotope variations, etc.

Cosmic particles. High-energy particles from extraterrestrial space, which may shatter nuclei of oxygen and nitrogen atoms in the upper atmosphere, releasing neutrons that then produce ¹⁴C by collision with other nitrogen atoms.

Cosmogonists. Those who developed all-encompassing theories for the origin of the earth and solar system, such as Buffon and Descartes.

Cosmopolitan. Found over vast stretches of the world (such as the Norwegian rat).

Craton. Large, tectonically stable nucleus of a continent that has relatively subdued topography. Most existing continental cratons have been relatively stable throughout most of Phanerozoic time (past 700 m.y.).

Cross-cutting age relationships. Any igneous intrusion or fault must be younger than all rocks penetrated by it but older than any that cut it. Inference of relative age from such relationships is an extension of the *superposition principle.*

Cross-stratification. Smaller-scale lamination at an angle (typically 20° to 30°) to that of the major stratification and produced by the migration of ripples or dunes.

Crust. The outer 1.4 percent of the solid earth, whose base is defined by the Mohorovicic seismic discontinuity. The crust's average density is 2.7 to 2.8. *Oceanic crust* averages 5 km in thickness and is composed of basalt; *continental crust* averages 35 km in thickness and is a complex mixture of granitic, metamorphic, and sedimentary rocks.

Cyanobacteria. (Also called blue-green algae.) The most primitive organisms known. They lack an organized nucleus (as do procaryotes) and are closely related to bacteria.

Cyclothem. See **Repetitive sedimentation.**

D

Daughter-isotope dating. The comparison of ratios among different radiogenic daughter isotopes to determine rock ages. Because different daughter isotopes are generated at different rates, their ratios to one another change progressively through time, providing a basis for dating.

Decay series. A radiogenic parent isotope and all daughter isotopes derived therefrom by decay, such as ^{40}K to ^{40}Ar, ^{87}Rb to ^{87}Sr, and ^{238}U to ^{206}Pb series.

Deccan traps. Voluminous masses of basalts erupted from deep fissures in western India during the latest Cretaceous to Paleocene.

Derived characters. Specialized characters that evolved subsequent to the ancestral (or primitive) characters; used in the cladistic classification method.

Differentiation of the earth. The chemical and physical separation of the earth's components from a homogeneous protoplanet to form core, mantle, crust, oceans, and atmosphere. Most of this separation occurred very early, but some differentiation of crust continues today through volcanism and orogenesis.

Diluvialists. People who hold that most strata and fossils were deposited by the biblical Flood.

Diorite. Coarse-grained igneous rock of intermediate composition; contains feldspar and amphibole, but little or no quartz.

Discordant isotopic dates. Two or more differing isotopic dates obtained by different methods for the same rock, generally due to selective resetting of one decay series that is more susceptible to heating (e.g., K-Ar); a valuable clue to complex metamorphic histories.

Disjunct faunas and floras. Portions of the same fauna or flora now isolated but originally spread over all the area between present separated outposts.

Divergence. The process of the evolution of organisms away from the general adaptive mode of the ancestral group, such as the amphibians becoming progressively land-oriented (in contrast to fish) by developing strengthened backbones and limbs.

Divergent plate margin. The margin of lithospheric plate that is moving away from a spreading zone; characterized initially by tensional faulting and basaltic volcanism followed by stable tectonic behavior. New oceanic lithosphere is created here.

DNA (deoxyribonucleic acid). A complex molecule arranged in a double-stranded helix. The backbone strands are made of sugars and phosphates, and they are held together by combinations of four bases. The sequence of these bases gives the information for the genetic code.

Dolomite. Sedimentary rock composed of calcium-magnesium carbonate [$CaMg(CO_3)_2$]; most commonly formed by replacement of limestone ($CaCO_3$) through introduction of magnesium ions carried in solution in pore water.

Dominant genes. Genes that are expressed at the expense of recessive genes.

Dynamo theory. An explanation of the earth's magnetic field by analogy with an electric dynamo. An electrical conductor moving in a magnetic field has an electrical current generated within it; the earth's metallic core is a conductor, and current generated therein as it spins maintains the magnetic field (as in an electric dynamo with some of the current it generates being fed back into its electromagnets to maintain the field required for further generation of current in the moving conductor).

E

Ecological niche. A habitat occupied by a species or group of organisms especially adapted for existing in that habitat.

Ecological replacement. The filling of a vacated ecological niche by the evolution of a new species or the migration of a different species.

Ecosystem. All living and nonliving things in a given area that have a self-renewing relationship.

Ectothermy. The ability to obtain body heat from the surrounding environment.

Ediacaran fauna. The late Proterozoic (about 600 m.y. old) soft-bodied fauna of multicellular organisms found in many parts of the world.

Ellesmerian orogeny. A major mountain-building event in Arctic Canada and northern Greenland (Late Devonian to Early Mississippian age). Its cause is unclear but may have involved a microcontinent collision.

Emissions. Spontaneous expulsions from an atomic nucleus of one or more of the following caused by radioactive decay: *alpha particle* (He nucleus), *beta particle* (electron), or *gamma ray* (similar to X ray). Emission rate reflects decay rate. (See also **Activity.**)

Endemic species. A species known only from a restricted geographic region.

Endothermy. The ability to produce body heat from the metabolic burning of food.

Epeiric sea. Shallow sea that flooded continental cratons frequently through geologic time; also called *epicontinental seas* (*epi* = above, thus "seas above or over continents").

Epifauna. Animals living on the bottom of aqueous bodies, such as oceans, lakes, and streams, but above the bottom sediments.

Episodic events. Phenomena that occur irregularly in time (i.e., nonperiodically) and involve large deviations from the average intensity of processes for a given environment. They include so-called rare events, such as the "500-year flood" or "200-year typhoon."

Equilibrium. A balance between two or more processes, such as uplift and erosion, new snowfall and glacier melting, or rise of sea level and upward growth of reefs.

Eubacteria. The more familiar advanced bacteria, including those that are responsible for digestion and many diseases.

Eukaryotes. All cells that have a nucleus and certain organelles. (See also **Prokaryotes.**)

Evaporite intrusions (salt domes). Upward intrusions by plastic flow of salt or gypsum from deeply buried evaporite layers, which become isostatically unstable because they are less dense than overlying strata.

Evaporites. Sedimentary rocks, such as salt (NaCl) and anhydrite ($CaSO_4$), formed by the evaporation of large volumes of seawater.

Evolution. The anatomical changes in populations of organisms through time, largely due to changes in their genes.

Exponential change. Change at an increasing (or decreasing) rather than a constant rate (e.g., population growth).

F

Facies. See **Sedimentary facies.**

Facies fossils. Types of fossils that tend to be restricted to a single lithology or facies (such as graptolites in black shales), so they are not useful for correlation from that facies into other facies.

Facies map. A map that shows the distribution of different sedimentary facies (i.e., lithologies) over a geographic area for a specified moment or interval of geologic time; important for interpreting ancient depositional environments and paleogeography.

Fatty acids. The basic molecular building block of fats and oils; they combine to form lipids.

Fauna. All the animals in a given region or time period or all the species of a phylum in a given region or time period.

Feedback. The result of a process may "feed back into" the system and modify the further development of the same process. There may be either amplification or suppression of the process, so feedback changes the conditions of equilibrium in the system. Feedback occurs in both living and nonliving systems as well as between the two; thus, the evolution of the one realm has influenced the evolution of the other.

Fermentation. Bacterial action causing the breakdown of large organic molecules, usually under anaerobic (oxygen-depleted) conditions.

Filter feeders. Organisms that can strain water to remove food particles.

Fission tracks. Imperfections in minerals and volcanic glass caused by spontaneous fission of an unstable atomic nucleus, which propels energy particles through surrounding material. Density of tracks is a function of numbers of atoms that have undergone fission, thus also of age.

500-year flood. Concept of violent natural phenomena, such as floods, that are rare events on the human time scale, but not on the geologic scale.

Flora. All the plants in a given region or time period.

Flysch. Collective term for very evenly layered, alternating thin sandstones and shales; characteristic of sequences in which the sandstones were deposited by turbidity currents before an orogenic collision.

Focus of earthquakes. The actual location of an earthquake source beneath the surface; *shallow-focus* events characterize ocean ridges, and *deep-focus* events characterize only volcanic arcs.

Forearc basin. A basin formed between the volcanic arc and the accretionary prism in a subduction zone.

Foreland sedimentary basin. A sedimentary basin formed by subsidence of the margin of a craton in front of ("fore")—and apparently because of loading by—the overthrusting of an orogenic belt onto the craton. Generally such basins are filled by clastic wedges whose sediments have been derived from erosion of the orogenic belt. Depending upon rate of initial subsidence, the basin may begin with relatively deep- or shallow-marine water over it; rapid sedimentation generally overtakes subsidence, causing a progression to nonmarine conditions.

Formation. The most fundamental local rock division of stratigraphic classification, which has some distinctive homogeneity of color, texture, fossil content, or the like; generally named formally for a geographic locality (e.g., Morrison Formation for Morrison, Colorado).

Fossil. Any record of past life, such as actual bones, shells, and teeth but also including impressions, footprints, burrows, etc.

Fossil zone. A restricted thickness of strata characterized by a distinctive *index fossil* (q.v.), which constitutes a biological datum for correlation.

Founder effect. When a small population becomes genetically isolated from its ancestral population, it can have unusual gene combinations. These come to dominate the genes of all the descendants of the founder population, so that they could become a new species.

G

Gene. The fundamental inheritance unit that carries a characteristic from parent to offspring; composed of a linear segment of the DNA molecule.

Gene pool. The actively breeding portion of a population in which genes are exchanged as a result of reproduction.

Genetics. The study of heredity and the causes of variation of organisms.

Genotype. The total genetic information that codes for an individual.

Genus. The category name for a group of closely related species. *Genera* is the plural.

Geosyncline. A relatively thick sequence of sedimentary and/or volcanic rocks that was deposited within a subsiding linear zone of the crust (generally an orogenic belt) and that ultimately was upheaved to form mountains.

Glauconite. A green, micalike silicate mineral that forms within a marine sedimentary environment wherein sedimentation is very slow. This allows slow chemical reactions between seawater and clay or mica minerals on the sea floor. Because it contains some potassium, glauconite can provide direct K-Ar isotopic dates of sedimentation.

Glossopteris **flora.** The dominant gymosperm tree found in the southern temperate forests in Gondwanaland during the late Paleozoic and early Mesozoic. (See also **Lycopsid flora.**)

Gondwana rock sequence. A late Paleozoic–early Mesozoic nonmarine stratigraphic sequence unique to the five southern or Gondwana continents but very different from contemporaneous sequences and fossils on other continents. Tillites occur in the lower part in close association with coal and the *Glossopteris* flora; nonmarine vertebrates occur above these, and basaltic lavas and dikes complete the sequence.

Gondwanaland. The Paleozoic southern supercontinent composed of South America, Africa, India, Australia, Antarctica, New Zealand, and Southeast Asia.

Gondwanan orogeny. A major Permo-Triassic mountain-building event recognized in South America, South Africa, Antarctica, and Australia.

Graded bedding. A regular gradation from coarser to finer grain size upward through a clastic stratum; implies settling of grains at rates proportional to their size or mass.

Gradualism. A view championed by Lyell and Darwin that changes of species and of the earth in general have been slow and steady through geologic time. (See **Punctuated equilibrium.**)

Granitic rocks. Coarse-grained, light-colored (silicic) igneous rock with quartz and potassium feldspar predominating over plagioclase feldspar, therefore relatively rich in K, Al, and SiO_2.

Graywacke. A heterogeneous, poorly sorted (i.e., texturally immature) dark sandstone.

Great American Interchange. The exchange of mammals between South America and North America when the Panamanian land bridge rose in the mid-Pliocene.

Greenhouse effect. The heating of air because some solar radiation reaching the earth's surface is radiated back as long-wavelength infrared radiation, most of which is trapped in the atmosphere by H_2O and CO_2 molecules (or by glass in a greenhouse).

Greenstone belts. Zones of the earth's crust (chiefly Archean) characterized by mildly metamorphosed volcanic rocks (greenstones) and associated immature clastic sedimentary rocks.

Gymnosperm. A class of plants that originated in the middle Paleozoic and that has seeds borne on leaves or in special cones.

H

Half-life. The time required for decay of one-half the parent isotope present in a radioactive sample. After one half-life, 50 percent of the parent is left; after two half-lives, only 25 percent; and so on.

Haversian canals. Abundant canals in the bone of endotherms that are filled with blood vessels.

Herbivore. An animal that eats plants for its primary source of food.

Hercynian orogeny. A major late Paleozoic mountain-building event that affected most of Europe and southern Asia. It is approximately equivalent to the Appalachian orogeny of North America as well as to the upheaval of the Ural Mountain belt between Europe and Siberia in western Russia.

Heterotrophs. Organisms that must obtain their food by consuming nutrients from their environment (as in animals).

Homeothermy. The ability of an animal to maintain a constant body temperature.

Hominid. A member of the zoological family Hominidae, including modern humans and their ancestors.

Homology. Organs in two or more different organisms that have different functions but the same evolutionary origin.

Hummocky stratification. A distinctive undulatory, lamination in fine sandstones formed in marine shelf environments (e.g., epeiric seas) by deposition influenced by large storm waves.

Hypothetico-deductive method. A method of scientific inquiry wherein hypotheses are proposed to explain a phenomenon and are then eliminated by testing (deduction).

I

Inclination angle. Vertical angle of inclination of rock remanent magnetization (*fossil magnets*) relative to the surface of the earth; indicative of paleolatitude of a specimen because steep inclination of the earth's magnetic field is characteristic of high latitudes and low inclination is characteristic of low latitudes.

Included-fragment age relationship. Any fragment included within a clastic sediment or an igneous rock must be older than the enclosing rock; an extension of the *superposition principle* (q.v.) for relative age.

Index fossils. Fossils unique to a limited thickness of strata, widespread geographically, and easily recognized. As a result, they are important for correlation of strata. (See also **Fossil zone.**)

Infauna. Organisms living within the sediment of the bottoms of aqueous bodies, such as oceans, lakes, or streams.

Inheritance of acquired characteristics. The origin of new species by passing on to offspring new habits (hence, ultimately new characteristics) acquired during the lifetime of an adult because of changing (new) needs; the theory usually associated with Lamarck, although held by many eighteenth and nineteenth century naturalists, including Darwin.

Inorganic evolution. Irreversible evolutionary change in the nonliving realm, such as change from an oxygen-free early atmosphere to an oxygen-rich later one and from no initial continental crust to the present configuration.

Intertonguing. Two sedimentary facies, such as sandstone and shale, have a complex lateral gradation from one to the other, so that long "tongues" (or fingers) of one penetrate into the other. Such tongues reflect repeated lateral shifting of adjacent sedimentary environments during deposition of both facies.

Iridium. A rare-earth element in the platinum group of metals, depleted in the earth's crust but more abundant in the mantle and in meteorites. Its abundance at the Cretaceous/Tertiary boundary has been used as evidence for an asteroid impact.

Iron formation. Evenly laminated or "banded" iron-rich layers alternating with either chert or carbonate layers; unique to Cryptozoic sequences.

Isolation. The separation of one unit of population of a species from other populations, so that there no longer is gene flow between them. Isolated subpopulations tend to evolve and become more distinct with time.

Isostasy. Gravitational equilibrium such that large parts of the brittle crust "float" on the plastic mantle; relative elevations are functions of thickness and density of the crustal blocks.

Isostatic rebound. Rise of the crust in response to removal of a load that had previously depressed the crust, as in Scandinavia and around Hudson Bay after the melting of huge glaciers.

Isotope. A nuclear species of a chemical element having a different number of neutrons in the atomic nucleus than other isotopes of the same element. Unstable or radioactive isotopes decay to stable ones, producing a spontaneous change in the number of neutrons. Stable isotopes (such as ^{16}O and ^{18}O) do not decay.

Isotopic or (**radiometric**) **dating.** Mineral or rock dating using radioactive isotopes. In the usual case, the ratio of radiogenic daughter isotope to its unstable parent isotope contained in a mineral is multiplied by the rate of decay of the parent to determine how long the parent isotope has been incorporated in that mineral. (See also **Total-rock dating** and **Daughter-isotope dating.**)

Iterative evolution. The repeated evolution of a certain body form after the first species of that shape became extinct.

K

Kimberlite. A species of ultramafic peridotite with high-pressure minerals, such as diamond. Kimberlite is derived by rapid explosive intrusions into the crust from the mantle and occurs in vertical cylindrical, pipelike bodies.

Komatiite. A fine-grained ultramafic volcanic rock with the same composition as coarser-grained, deep-seated peridotite; found only among early Cryptozoic (Archean) volcanic rocks.

L

Land bridge. A narrow strip of land connecting two continents and permitting land organisms previously restricted to one continent to migrate to the other (such as the Bering and Panamanian bridges).

Laramide orogeny. The latest Cretaceous–middle Eocene episode of mountain building in the Cordillera, which uplifted deep basement rocks in the Rocky Mountain region, forming large mountains with deep basins in between.

Laurasia. The northern supercontinent of late Paleozoic time, composed of North America, Europe, and Asia.

Limestone. Sedimentary rock composed of calcium carbonate ($CaCO_3$); made up largely of invertebrate fossil skeletal debris.

Lipids. The major biochemical component of all fats and oils; made up of many fatty acids.

Lithology. The total characteristics of a rock, including texture, composition, fossil constituents, color, etc.

Lithosphere. The rigid outer portion of the earth (100 to 250 km thick) above a low-seismic-velocity zone; includes continental and oceanic crust in its top part.

Lithosphere plates. Large, irregular, plate-like units of lithosphere whose margins are delineated today by the major zones of active earthquakes. (See also **Plate tectonics.**)

Low-seismic-velocity zone. The zone in upper mantle (below 100 to 250 km) that defines the base of the lithosphere. This zone has a seismic velocity lower than those above or below, indicating less rigidity; seems to be the zone in which most lithosphere plate motion is concentrated.

Lycopsid flora. The dominant tree found in the northern tropical forests in Laurasia during the late Paleozoic and early Mesozoic. (See also *Glossopteris* **flora.**)

M

Macroevolution. The evolution of large-scale differences between organisms, such as the origin of jaws or of flight.

Magnetic anomalies. Any local deviations from normal (either greater or lesser) intensity of the earth's magnetic field, such as the narrow positive and negative sea-floor-anomaly stripes parallel to ocean ridges.

Mantle. Eighty-two percent of the volume of the earth, which lies between the crust and the core with a radius of almost 2,900 km. It is composed of magnesium-iron- and calcium-rich ("mafic") silicate minerals. Average density is 4.5 g/cm^3. The upper 700 km is moderately rigid, but the lower 2,000 km apparently can flow plastically.

Mantle plume. A hypothetical plume of partially molten material rising up through the mantle and doming the crust above; may cause initiation of rifts, such as the Red Sea, and probably causes mid-plate volcanism in the Hawaiian Islands.

Marsupials. Pouched mammals, which give birth to their young prematurely (unlike placental mammals). The embryo then climbs up to the mother's pouch and attaches to a nipple inside, where it continues to mature. Familiar marsupials include opossum, kangaroos, koalas, and wombats.

Mass extinction. The relatively sudden loss of large numbers of organisms in relation to the number of new species being added to the fauna; usually occurs near the end of a geologic system.

Mass number. The total sum of the number of protons and neutrons in any atomic nucleus; generally written at top left of chemical symbol for an element; thus, ^{14}C is the isotope of carbon with a mass number of 14.

Mélange. Chaotically mixed-up and sheared mass of rocks formed by the shearing due to subduction; usually found in the accretionary wedge.

Mesozoic marine revolution. The Triassic-Jurassic decline of vulnerable invertebrates (especially brachiopods) that lived on the sea bottom and their replacement by armored, mobile, or burrowing invertebrates (especially molluscs). It is presumed to be the result of the appearance of a number of shell-crushing predators.

Metazoans. Multicellar organisms with differentiated tissues, which are used for reproduction, respiration, food processing, etc.

Meteorites. Small, solid objects believed to be fragments of asteroids moving in space; occasionally they collide with planets or their moons. Two major types include *metallic* (Fe-Ni-Co-rich) and *stony* (Ca-Mg-Fe-rich). Metallic meteorites are thought to resemble the earth's core and stony ones the earth's mantle.

Microbiotic crusts (or **cryptogamic soils**). A community of fungi, bacteria, and algae that form a soft, spongy mat on otherwise barren clays and silts.

Microcontinents. Crustal blocks smaller than a continent with a continental-type crust (e.g., Japan).

Microevolution. Small-scale evolutionary changes, usually within the normal range of variation of a species.

Microspheres. Minute globular masses of complex organic molecules found in the Precambrian and believed to be precursors of cells.

Milankovitch cycles. The cyclic change of temperature of the earth's surface due to combined effects of variations of tilt of axis, wobble of axis (precession), and ellipticity of orbit.

Mineral. A naturally formed chemical element or compound with a definite composition, a characteristic crystal form, and other distinctive physical properties.

Mohorovicic discontinuity (or **Moho**). Discontinuity in seismic velocity long used to define the base of the crust. Seismic velocity increases abruptly below the Moho from about 6 or 7 km/sec to 8 km/sec (P-wave velocities).

Molasse. A collective term for coarse, clastic sediments with large-scale cross-stratification, channel structures, coal, and in some cases red beds; generally nonmarine deposits accompanying the uplift of a mountain belt.

Mutations. Genetic changes resulting in heritable differences of offspring from a parent.

N

Nappe structure. Very large, recumbent (flat-lying) folds associated with the thrust-fault zones where orogenic belts impinge upon craton margins. In young belts such as the Alps and Himalaya, nappes are obvious, but in older belts, erosion has removed all but their roots.

Natural selection. The process postulated by Darwin that results in the differential reproductive success of those individuals best adapted for particular changes in the environment.

Nebula. See **Solar nebula.**

Nektonic. Swimming.

Neptunism. An archaic theory, championed by Werner, that held that all rocks (including granite and basalt) were deposited from an early, universal ocean.

Neutralism. The idea that some genetic changes are unaffected by natural selection, thus are selectively neutral.

Normal fault. A fault along which the upper block has moved down relative to the lower block along a steeply inclined surface; characteristic wherever crust has been subjected to tension (e.g., Red Sea rift, mid-ocean ridge crests, Basin and Range region).

O

Obduction. The overthrusting of a thick slab of one plate over another at a convergent lithosphere plate margin. Commonly it has involved sea-floor (ophiolite) material obducted upon continental or arc crust.

Ontogeny. The historical changes in an individual's development from fertilized egg to death.

Oolites. Sedimentary rock commonly of limestone composition consisting of spheres approximately 1 mm in diameter called oolites with microscopic concentric layers, which were precipitated from shallow, supersaturated waters as the grains were agitated vigorously by waves or currents.

Ophiolite suite. The European term for sequences found in many mountain belts that consist of peridotite overlain successively by basalt, chert, and black shale. Ophiolite rocks represent uplifted deep sea floor.

Organelles. Structures within a cell that process food, eliminate waste, trigger reproduction, and perform many other functions.

Original horizontality. Strata are originally deposited in horizontal layers because they accumulate grain by grain under the influence of gravity. Inclined strata imply post-depositional deformation.

Original lateral continuity. Inference of original extent of strata now partially eroded. Such extent must be estimated in order to infer past expanse of land and sea or to reconstruct deeply eroded structures, such as cratonic arches.

Ornithischians. "Bird-hipped" dinosaurs, including the duckbills, horned dinosaurs, armored dinosaurs, stegosaurs, and their relatives.

Orogenic belt. Also called *mobile belt.* Linear or arcuate zones of the crust that experienced unusually severe compressional tectonics for long spans of time. They are sites of convergence and subduction of lithosphere plates, of andesitic volcanism, granitic batholiths, and mountain ranges.

Orogeny. Mountain building, especially a mountain-building event in history, such as the *Taconian orogeny.*

Outgassing. The migration via volcanoes and thermal springs of volatile gases and vapors from the interior to the surface of the earth to form the atmosphere and seawater.

Overprinted. Superimposing of later deformation and/or metamorphism upon earlier ones; commonly results in discordant isotopic dates by the resetting (reequilibration) of "isotopic clocks."

Oxygen-isotope analysis. Ratio of ^{18}O to ^{16}O in shells composed of $CaCO_3$ provides indication of paleotemperature of seawater when shell was formed.

Ozone layer. The zone in the lower atmosphere between 15 and 30 km above the earth's surface that is enriched in ozone (O_3). It is formed by splitting of O_2 molecules by ultraviolet radiation followed by the combining of the stray oxygen atoms with other O_2 molecules. Because O_3 is unstable, however, it soon breaks up again. The process is a steady-state one because ozone is forming at the same rate as it breaks down. There is also an important feedback effect, for the ozone layer filters most of the lethal high-energy ultraviolet radiation so that it does not reach the earth's surface; this makes land areas habitable for life.

P

Paleoecology. The study of fossil organisms in relation to past environments.

Paleogeography. The restoration of geography of any past time, including land and sea distributions, and such things as mountains and lowlands, paleocurrents and paleowinds, hot tropics and cold areas, etc.

Paleolatitude. The latitude of a continent or any other part of the crust at a past time. Due to continental drift, a continent may have changed latitude significantly over time.

Paleontology. The study of past life.

Pan-African orogeny. Widespread metamorphism and mountain building between 400 and 600 m.y. ago in all southern continents except India.

Pangea. The single supercontinent of Permian and Triassic time formed by collision of Gondwanaland with Laurasia.

Panselectionism. The belief that natural selection operates on every feature of an organism and its genes.

Paratethys Sea. The relict of northern part of Tethys seaway now represented by Black Sea, Caspian Sea, and Aral Sea.

Partial melting. Only a portion of the mantle or crust is melted to form magma; for example, andesitic magma seems to represent selective, partial melting of sodium, aluminum, and silica from an iron-magnesium-rich source.

Passive continental margin. A continental margin within a single lithosphere plate and fused to adjacent oceanic crust—that is, *not* a tectonically active convergent or divergent margin; also known as a trailing edge.

Pelagic. Living in open seawater; also may refer to very fine sediment particles that settle slowly from seawater.

Peridotite. Dark, ultramafic rock composed of pyroxene and olivine; characteristic of mantle.

Phenotype. The anatomical results of growth and development from a given genotype.

Photic zone. The upper 100 meters or so of water in which sufficient sunlight can penetrate to allow photosynthesis.

Photochemical reactions. Chemical reactions that are triggered by solar radiation, such as the formation of some of the gases in smog by the impingement of sunlight on atmospheric hydrocarbon pollutants; postulated

also as one mechanism for the generation of CO_2 and N_2 in the early atmosphere.

Photosynthesis. The method by which plants chemically transform carbon dioxide and water into simple carbohydrates (basic food) utilizing the sun's radiation. Free oxygen (O_2) is a by-product.

Phyletic gradualism. The gradual transformation of fossil species through time.

Phylogeny. The evolutionary history of a group of related organisms.

Pillow (or ellipsoidal) lava. Lava flows with distinctive internal ellipsoidal structures resembling a pile of pillows; forms when lava is erupted in or flows into water.

Placental mammals. Mammals that allow their young to develop in the uterus of the mother, then give birth when they are nearly fully developed (unlike marsupials).

Planetesimals. Meteorite-like objects thought to have condensed from the solar nebula and to have aggregated together with gases and ices to form protoplanets between 4.6 and 5.0 b.y. ago.

Planets. *Terrestrial,* or *inner, planets* are relatively small, dense, stony bodies orbiting around the sun and thought to have lost most of their original light gases early in solar system history (Mercury, Venus, Earth, and Mars). *Gaseous,* or *outer, planets* are larger, low-density bodies that have retained most of the original gases derived from the solar nebula (Jupiter, Saturn, Uranus, Neptune, and Pluto).

Plankton. Microscopic organisms that float in water.

Plate tectonics. The arrangement of the outer earth in ten or twelve large, rigid plates of lithosphere all in motion and "floating" upon plastic mantle material. Plate margins are defined by zones of seismicity. Movement of the plates is away from one another along *divergent spreading ridges* and toward one another at *convergent subduction zones.*

Plateau basalts. Basalts formed by eruptions of lava flows from many fissures, resulting in great thicknesses of flow-upon-flow to produce large plateaus.

Plutonism. Hutton's theory of the origin of all granitic rocks (and mountains) from molten magmas forcibly intruded upward into the crust due to subterranean heat.

Poikilothermy. The ability to let body temperature fluctuate to conserve energy.

Polarity reversals. Nonperiodic reversals of polarity of the earth's magnetic field (the last of which occurred about 780,000 years ago); provide the basis for a magnetic reversal time scale.

Polymers. Molecules made up of simpler components linked into complex chains and other three-dimensional structures.

Population density. The number of individuals occupying a given area or stratum.

Preadaptation. Evolutionary features that apparently arose for one purpose and then were taken over for another purpose.

Precambrian borderlands. Hypothetical high land composed of Precambrian metamorphic and igneous rocks postulated by Americans before 1940 to be the source of all Paleozoic and younger clastic sediments deposited in geosynclines. (See also **Tectonic land.**)

Primordial lead ratios. The ratios among four lead isotopes found in meteorites. Because neither parent Th nor U are present in the meteorites, these ratios are assumed to represent the primary or primordial ratios in the solar system 4.5 to 4.7 b.y. ago.

Progradation. Lateral expansion of any sedimentary environment as deposition occurs—for example, the regression of the sea along a shoreline (e.g., a delta front) where sedimentation is rapid enough to push the sea back; the shore progrades, or "migrates forward."

Prokaryotes. Primitive, single-celled organisms lacking a cell nucleus. (See also **Eukaryotes.**)

Protein. Large organic molecules constructed of amino acids; the basic molecule in all cells.

Proteinoids. Droplets of proteins surrounded by a lipid membrane that have many of the properties of simple cells.

Punctuated equilibrium. The geologically abrupt appearance of new species followed by very long periods of no visible change.

R

Radioactive decay. Spontaneous change of an unstable atomic nucleus to a stable one. By emission of particles or radiation, an unstable parent isotope changes to a stable daughter isotope. (See also **Emissions.**)

Recessive genes. Genes that are not expressed unless no dominant genes are present in the genotype.

Recombination. The mixing of genetic material from both parents when egg and sperm meet. This mixing results in new genetic variability among the offspring.

Red beds. Any sedimentary rocks that are red-colored due to disseminated, fully oxidized iron coatings on grains.

Reefs. Most densely populated environment of the sea floor where calcareous skeletons of many animals and algae build mounds. Optimal growth conditions occur in the warm, shallow, lighted (photic) zone with oxygenation of the water and abundant nutrients. *Fringing reefs* grow directly adjacent to shore; *barrier reefs* are separated from shore by a shallow shelf; *atolls* are rings of reefs with a central shallow lagoon (see also **Atolls**). Today most reefs grow only in shallow seas between 30° N and 30° S latitude.

Regressive facies pattern. The retreat of the sea away from a land area results in deposition of facies that characteristically become coarser upward at any one locality and generally have an unconformity at the top.

Remanent magnetism. A component of a rock's magnetism induced by the earth's magnetic field when that rock formed. Measurement of its orientation provides a clue to paleolatitude of the sample (thus of continents) at time of formation. (See also **Inclination angle.**)

Repetitive sedimentation. Changing conditions, such as fluctuations of sea level or climatic changes, implied by a distinct succession of sediment types; this succession is repeated many times in a vertical sequence; especially conspicuous in upper Paleozoic strata; may or may not be regularly (periodically) recurring.

Resetting of isotopic clocks. Loss of daughter isotopes from minerals generally due to heating by metamorphism (e.g., loss of ^{40}Ar gas), which results in apparent ages younger than true age; valuable in dating metamorphic events. (See also **Discordant isotopic dates.**)

Respiration. The release of energy by the controlled oxidation of food molecules.

Restored cross section. A cross section that shows thickness and facies of strata with any later structural deformation and erosion ignored (i.e., restored); valuable for portraying inferred conditions at a past time.

Ribosomes. Knots of RNA and protein in the cell that help decode the DNA.

Ribozymes. RNAs from the ribosomes that not only contain a genetic code but also have the ability to catalyze the reactions to reproduce themselves.

Rifts. Long, narrow zones of normal faulting due to tension of the crust, as in **aulacogens,** ocean-ridge crests, etc.

RNA (ribonucleic acid). Single-stranded, helically coiled molecule with sugars and phosphates making up the strand and a variety of bases lined up along it. In the double-stranded form, it is DNA. Some organisms (especially viruses) have no DNA and use RNA for their genetic code, and all organisms use a variety of RNAs for transferring and copying genetic information.

Rock sequences. Very widespread, relatively thick, formally designated intervals of strata definable chiefly on cratons. They are bounded by profound, widespread unconformities and include many formations.

Rudistid clams. Large, irregular, and conical organisms that formed coral-like reefs in the Cretaceous.

S

Sabkha. An arid supratidal salt flat, typically found along the margins of a shallow tropical sea, such as the Persian Gulf.

Samfrau belt. Permo-Triassic geosyncline or orogenic belt extending across Gondwanaland from South America along southern Africa and Antarctica to eastern Australia.

Sandstone maturity. The relative purity of a sandstone in terms of *compositional maturity* or of relative uniformity of texture in terms of *textural maturity.*

Sauk Sequence. The first Phanerozoic unconformity-bounded stratigraphic sequence of the North American craton. The base varies in age from Vendian beyond the craton margin to Late Cambrian in the center. The top is a cratonwide unconformity eroded on Early Ordovician (and older) strata.

Saurischians. The "lizard-hipped" dinosaurs, including the giant sauropods and the carnivorous theropods.

Sea-level-fluctuation (or **Vail**) **curve.** A graph of apparent worldwide changes of sea level through Phanerozoic time of 200 to 300 m above or below present level, which resulted in half a dozen profound unconformities that seem to be present on several continents.

Sedimentary basins. See **Basins.**

Sedimentary facies. Overall lithology of strata reflecting environment of deposition; facies characteristic of one environment, such as beach sand, grade laterally into facies of another environment, such as offshore mud.

Sevier orogeny. A Late Cretaceous phase of mountain building in the Cordillera due to subduction that produced the Sierran arc. It is responsible for extensive backarc thrusting from southern Nevada to Alberta.

Slab gap. The hypothesis that the tectonics of the western Cordillera is due to a gap between the two remnants of the Farallon Plate, where no Pacific Plate was subducted to fill the gap. Consequently in this region the North American Plate lies directly over hot mantle.

Solar nebula. The spinning, disc-shaped cloud of gas, dust, and ices from which the entire solar system condensed 4.6 to 5.0 b.y. ago.

Sonoman orogeny. A Permo-Triassic mountain-building event in the western United States thought to have resulted from the collision of a volcanic arc with the continental margin.

Sonomia. The terrane that makes up the northwestern part of Nevada and the northern Sierras, sutured onto North America in the Triassic.

Sorting. The range of grain sizes in clastic sediments; a restricted range of sizes, as in beach sands and dune sands, indicates good sorting.

Speciation. The formation of a new species, thought to be the result of mutation and reproductive isolation.

Spontaneous generation. A discredited belief that new life originates from dead material in dark, damp places.

Steady state. A dynamic system (such as a beach) is said to be in a steady state when energy is being dissipated, but the system appears the same from time to time; also called *dynamic equilibrium.*

Stratification. The layering characteristic of sedimentary deposits, which reflects differences of color, texture, or composition.

Strike-slip fault. See **Transcurrent fault.**

Stromatolites. Complex microbial ecosystem in the form of mounds and columns of layered anaerobic and aerobic cyanobacteria, algae, and fine sediments.

Subduction. The process whereby one lithosphere plate is thrust beneath another along a convergent plate margin—generally a deep-ocean trench and an adjacent volcanic arc result.

Subduction complex. Intensely sheared and fractured assemblages of rocks with many overthrust faults, characteristic of lithosphere plate margins above subduction zones.

Subsidence. The sinking of a large part of the earth's crust due to such causes as loading by ice or thick sediments, gradual increase of density as by cooling of young oceanic crust, downward convective flow in the mantle beneath, or loading of the crust by a thick pile of thrust faults.

Supercontinent. The result of joining two or more continents by collision. Gondwanaland was an example before it broke up to spawn modern India, Africa, South America, Antarctica, and Australia.

Superposition. A relationship among strata such that the lowest layer in a sequence is the oldest; the basis of relative age of strata. For deformed strata, original superposition must be determined with fossils, isotopic dating, or features such as graded bedding that indicate the original up direction.

Suspect terrane. A tectonic region (terrane) whose original location is uncertain (suspect);

some examples have moved thousands of kilometers and have collided with a continent or another suspect terrane.

Suture zone. The zone of collision between continental plates, which is generally characterized by intense deformation, metamorphism, and (especially) strips of mafic and ultramafic oceanic rocks.

SWEAT hypothesis. The idea that southwestern North America and East Antarctica were once connected in the late Proterozoic, then rifted apart.

Sweepstakes routes. Dispersal routes across hostile regions so that migration is rare and difficult. Chance plays a major role in determining which organisms will make it across (e.g., an insect blown across a mountain range).

Symbiosis. A cooperative relationship between two different organisms; each provides an important benefit for the other.

Sympatric speciation. Species formed within populations that are not physically isolated from one another.

Synapsids. The lineage of land vertebrates that eventually led to mammals (formerly called "mammal-like reptiles").

Synthetic theory of evolution. (or **Neo-Darwinism**). The modernized version of Darwin's theory incorporating genetics, systematic zoology, ecology, and paleontology.

T

Taconian orogeny. A major Late Ordovician mountain-building event on the southeastern margin of North America. The previously passive continental margin was converted to an active convergent margin in Middle Ordovician time, and the orogeny then resulted from collision of a volcanic arc.

Taxonomy. The grouping together of related organisms into categories arranged in a classification.

Tectonic. The structural behavior of large parts of crust through significant geologic time (e.g., orogenic tectonic belts, rifted tectonic zones, stable tectonic areas).

Tectonic land. Large, mountainous land raised within an orogenic belt by an orogeny and composed of only slightly older sedimentary, volcanic, and plutonic rocks; principal source of so-called geosynclinal sediments.

Template. A framework or scaffold that helps support and organize the assembly of larger, more complex structures, as in clays, pyrite, or zeolites aligning organic molecules to form complex polymers.

Terrane. Any large portion of the earth's crust, such as a mountain belt, a volcanic

terrane, an oceanic island terrane, etc. (See also **Suspect terrane.**)

Tethys Sea. The Permian through Eocene seaway between Gondwanaland and Laurasia with a distinctive, very diverse tropical fauna.

Thermal convection. The slow circulation or turnover of mantle material by plastic flow due to differential heating as air rises from a radiator, cools, and then sinks on the opposite side of a room.

Thrust fault. Surface or zone inclined less than 45° along which the upper block moved over the lower one perhaps for tens of kilometers; characteristic of portions of the crust subjected to intense compression.

Thrust-loading subsidence. See **Foreland sedimentary basin.**

Tiering. The subdivision of the regions above and below the sea floor into different feeding levels, or tiers, which allows more different kinds of organisms to live in the same area of sea floor.

Tillite. Lithified glacial till characterized by heterogeneous composition, unsorted texture, and a lack of stratification.

Tippecanoe Sequence. The second Phanerozoic unconformity-bounded stratigraphic sequence of the North American craton. The basal unconformity is overlain in the central craton by the Middle Ordovician St. Peter Sandstone; the top is a cratonwide unconformity developed on Early Devonian and older rocks.

Tommotian. One of the earliest stages of the Cambrian Period; marked by abundant small shelly fossils but no trilobites yet.

Total-rock dating. Isotopic dating using an entire rock specimen rather than a single mineral. Thus, parent-daughter isotope ratios are determined for a specimen containing several different minerals.

Trace fossils. Indirect evidence of life, such as footprints, trails, and burrows.

Transcurrent fault. A lateral or strike-slip fault with horizontal displacement, such that one side moves laterally past the other (e.g., San Andreas fault of California).

Transform faults. A type of transcurrent or strike-slip fault characteristic of oceanic spreading ridges, which offsets the ridge axes as spreading progresses.

Transgressive facies pattern. Advance of the sea over a land area results in deposition of *transgressive facies,* which overlie an unconformity and are characterized by deposits that become finer upward at any one locality.

Triple junction. A junction of three linear tectonic elements, such as three spreading ridges or three subduction zones.

Tsunamis. The Japanese name for giant breaking waves generated by submarine earthquakes; very damaging to coastal structures.

Turbidity currents. Currents whose movement is driven by the action of gravity pulling denser, sediment-laden, turbulent water masses downslope beneath less dense, clear water.

U

Ultramafic rocks. Rocks "ultra-rich" in magnesium, iron, and calcium, which make up the earth's mantle. Examples are *peridotite* (including *kimberlite*) and fine-grained *komatiite.*

Unconformity. A break or discontinuity in a stratigraphic sequence, which implies an interval of erosion.

Uniformitarianism. In the strict nineteenth-century sense of Lyell, this was an assumption that present processes acted the same way through all time and *with about the same intensity.* (See also **Actualism.**)

Uplift. The raising of large regions by structural forces, such as upward convective flow in the mantle rise of masses of magma through the crust, or compression of the crust by subduction or continental collision.

V

Vendian. The period just before the Cambrian Period when the Ediacaran soft-bodied fauna was dominant.

Vestigial organs. Organs that are no longer of any use to an organism, but their presence indicates that organisms once had these organs in their evolutionary past.

Volcanic arc. An arcuate chain of volcanic islands or volcanoes along a continental margin, which erupt mainly andesitic lavas and ash (but including also basalts and rhyolites); generally associated with topographic trenches and much seismicity. All these phenomena are the results of subduction of lithosphere beneath the arc.

W

Walther's Law. The observation that, when environments change position, adjacent sedimentary facies will succeed each other in vertical depositional sequences. Vertical changes of lithology due to transgression or regression are mirrored by similar lateral changes (e.g., coarsening laterally as well as upward).

Index